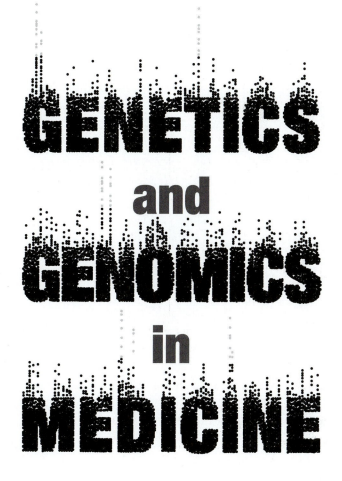

GENETICS and GENOMICS in MEDICINE

Division of Clinical
and Metabolic Genetics

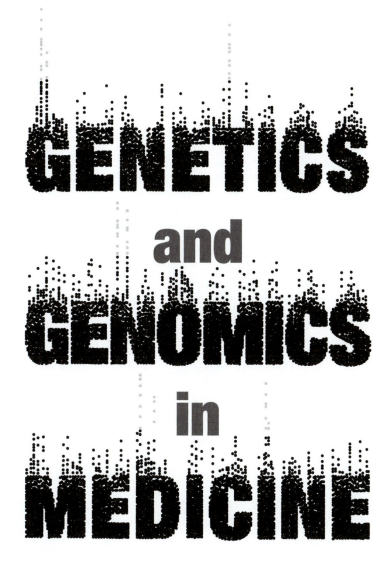

GENETICS and GENOMICS in MEDICINE

Tom Strachan

Judith Goodship • Patrick Chinnery

Garland Science

Taylor & Francis Group

NEW YORK AND LONDON

Garland Science
Vice President: Denise Schanck
Senior Editor: Elizabeth Owen
Assistant Editor: David Borrowdale
Production Editor: Ioana Moldovan
Typesetting: EJ Publishing Services
Illustrator, Cover, and Text Designer: Matthew McClements, Blink Studio Ltd.
Copyeditor: Bruce Goatly
Proofreader: Chris Purdon
Indexer: Bill Johncocks

ISBN 978-0-8153-4480-3

Library of Congress Cataloging-in-Publication Data
Strachan, T., author.
  Genetics and genomics in medicine / Tom Strachan, Judith Goodship, Patrick Chinnery.
     p. ; cm.
  Includes bibliographical references.
   ISBN 978-0-8153-4480-3 (paperback : alk. paper)
  I. Goodship, J. (Judith), author. II. Chinnery, Patrick F., author.
III. Title.
  [DNLM: 1. Genetic Phenomena. 2. Genetic Diseases, Inborn--therapy. 3. Genomics. 4. Individualized Medicine. QU 450]
  RB155.5
  616'.042--dc23
                                        2014010769

Published by Garland Science, Taylor & Francis Group, LLC,
an informa business,
711 Third Avenue, New York, NY, 10017, USA,
and 3 Park Square, Milton Park, Abingdon, OX14 4RN, UK.

Printed in the United States of America

15 14 13 12 11 10 9 8 7 6 5 4 3 2 1

Tom Strachan is Emeritus Professor of Human Molecular Genetics at Newcastle University, UK. He was the founding Head of Institute at Newcastle University's Institute of Human Genetics (now the Institute of Genetic Medicine) and is a Fellow of the Royal Society of Edinburgh and a Fellow of the Academy of Medical Sciences. Tom has made many important research contributions in medical, evolutionary, and developmental genetics. After publishing a short textbook on the Human Genome in 1992 he conceived the idea of a larger, follow-up textbook, Human Molecular Genetics, currently in its fourth edition and co-written with a former colleague, Andrew Read. Their achievement was recognized by the award of the European Society of Human Genetics' Education Prize in 2007.

Judith Goodship is Professor of Medical Genetics at Newcastle University and a Clinical Geneticist. She has identified the molecular basis of a number of Mendelian disorders including Seckel syndrome and Ellis-van Creveld syndrome. Her main research is on development of the heart and congenital heart disease. As a practicing clinician her aim is to improve management of inherited disorders; she particularly wants to improve support and information for those with genetic disorders around the transition from childhood to adulthood.

Patrick Chinnery is Director of the Institute of Genetic Medicine at Newcastle University, where he is Professor of Neurogenetics and a Clinical Neurologist. He has been a Wellcome Trust Senior Fellow in Clinical Science for over ten years, and is a Fellow of the Academy of Medical Sciences. His laboratory and clinical research has focussed on understanding the molecular basis of inherited neurological diseases, with a particular interest in mitochondrial disorders. In addition to defining several new mitochondrial disease genes, he has defined the molecular mechanisms underpinning the mtDNA genetic bottleneck, and carried out clinical trials developing new treatments for mitochondrial disorders.

Garland Science
Taylor & Francis Group

Visit our website at http://www.garlandscience.com

PREFACE

Enduring high hopes that genetics might transform medicine have been tempered by technical challenges that initially restricted the pace of developments. Even until quite recently, the geneticist's view of our genome was typically limited to small regions of interest, a gene-centered vista.

The Human Genome Project and subsequent technological developments—notably, genomewide microarray technologies and massively parallel DNA sequencing—changed all that. Now, genome-centered perspectives are readily attained and are transforming the research landscape and clinical applications. In January 2014, the long-sought $1000 genome became a reality at last, genomewide screening and diagnosis are becoming routine, and ambitious sequencing projects are underway to decode the genomes of very large numbers of affected individuals. Might we soon live in societies in which genome sequencing of citizens becomes the norm?

It's a time of transition: the era of medical genetics—with a focus on chromosomal abnormalities, monogenic disorders, and genes—is giving way to the era of clinical and public health genomics; genomewide analyses of genetic variation are beginning to link genome to phenome in a comprehensive way. Genetic and genomic technologies are being deployed in a wide range of medical disciplines, and debate has begun as to what conditions might make it appropriate to begin routine genome sequencing of newborns. That, inevitably, raises many ethical questions.

In this book we try to summarize pertinent knowledge, and to structure it in the form of principles, rather than seek to compartmentalize information into chapters on topics such as genetic variation, epigenetics, population genetics, evolutionary genetics, immunogenetics, and pharmacogenetics. To help readers find broad topics that might be dealt with in two or more chapters, we provide a road map on the inside front cover that charts how some of the broad themes are distributed between different chapters.

We start with three introductory chapters that provide basic background details. Chapters 1 and 2 cover the fundamentals of DNA, chromosomes, the cell cycle, human genome organization, and gene expression. Chapter 3 introduces the basics of three core molecular genetic approaches used to manipulate DNA: DNA amplification (by DNA cloning or PCR), nucleic acid hybridization, and DNA sequencing, but we delay bringing in applications of these fundamental methods until later chapters, setting them against appropriate contexts that directly explain their relevance.

The next three chapters provide some background principles at a higher level. In Chapter 4 we take a broad look at general principles of genetic variation, including DNA repair mechanisms and some detail on functional variation (but we consider how genetic variation contributes to disease in later chapters, notably chapters 7, 8, and 10). Chapter 5 takes a look at how genes are transmitted in families and at allele frequencies in populations. Chapter 6 moves from the basic principles of gene expression covered in Chapter 2 to explaining how genes are regulated by a wide range of protein and noncoding RNA regulators, and the central role of regulatory sequences in both DNA and RNA. In this chapter, too, we outline the principles of chromatin modification and epigenetic regulation, and explain how aberrant chromatin structure underlies many single-gene disorders.

The remainder of the book is devoted to clinical applications. We explain in Chapter 7 how chromosome abnormalities arise and their consequences, and how mutations and large-scale DNA changes can directly cause disease. In Chapter 8 we look at how genes underlying single-gene disorders are identified, and how genetic variants that confer susceptibility to complex diseases are identified. Then we consider the ways in which genetic variants, epigenetic dysregulation, and environmental factors all make important contributions to complex diseases. Chapter 9 briefly covers the wide range of approaches for treating genetic disorders, before examining in detail how genetic approaches are used directly and indirectly in treating disease. In this chapter, too, we examine how genetic variation affects how we respond to drug treatment. Chapter 10 deals with cancer genetics and genomics, and explains how cancers arise from a combination of abnormal genetic variants and epigenetic dysregulation. Finally, Chapter 11 takes a broad look at diagnostic applications (and the exciting applications offered by new genomewide technologies), plus ethical considerations in diagnosis and gene therapies.

As genetic and genomic technologies have an increasing impact on mainstream medicine, and huge numbers of people have their genomes decoded for medical reasons, we really are moving into a new era. How far will we move from the commonplace one-size-fits-all approach to disease treatment toward an era of personalized or precision medicine? At the very least, we can expect an era of stratified medicine in which, according to the genetic variants exhibited by patients with specific diseases, different medical actions are taken.

We have tried to convey the excitement of fast-moving research in genetics and genomics and their clinical applications, while explaining how the progress has been achieved. There is a long way to go, notably in fully understanding complex disease and in developing effective treatments for many disorders, despite some impressive recent advances in gene therapy. But the new technological developments have engendered an undeniable sense of excitement and optimism.

We would like to thank the staff at Garland Science, Elizabeth Owen, David Borrowdale, and Ioana Moldovan, who have undertaken the job of converting our drafts into the finished product. We are also grateful to the support of family members: to Meryl Lusher and James Strachan for help on proofreading and choice of questions, and to Alex Strachan for assistance on various text issues.

Tom Strachan, Judith Goodship, and Patrick Chinnery

Literature access

We live in a digital age and, accordingly, we have sought to provide electronic access to information. To help readers find references cited under Further Reading we provide the relevant PubMed identification (PMID) numbers for the individual articles—see also the PMID glossary item. We would like to take this opportunity to thank the US National Center for Biotechnology Information (NCBI) for their invaluable PubMed database that is freely available at: http://www.ncbi.nlm.nih.gov/pubmed/. Readers who are interested in new research articles that have emerged since publication of this book, or who might want to study certain areas in depth, may wish to take advantage of literature citation databases such as the freely available Google Scholar (scholar.google.com).

For background information on single gene disorders we often provide reference numbers to access OMIM, the Online Mendelian Inheritance in Man database (http://www.omim.org). For the more well-studied of these disorders, individual chapters in the University of Washington's GeneReviews series are highly recommended. They are electronically available at the NCBI's Bookshelf at (http://www.ncbi.nlm.nih.gov/books/NBK138602/) and within its PubMed database. For convenience we have given the PubMed Identifier (PMID) for individual articles in the GeneReviews series (which are also collected as an alphabetic listing of all disorders at PMID 20301295).

Online resources

Accessible from www.garlandscience.com, the Student and Instructor Resource Websites provide learning and teaching tools created for Genetics and Genomics in Medicine. The Student Resource site is open to everyone, and users have the option to register in order to use bookmarking and note-taking tools. The Instructor Resource site requires registration, and access is available only to qualified instructors. To access the Instructor Resource site, please contact your local sales representative or email science@garland.com. Below is an overview of the resources available for this book. On the Websites, the resources may be browsed by individual chapters and there is a search engine. You can also access the resources available for other Garland Science titles.

For students

(available directly at www.garlandscience.com/ggm-students)

Quiz A multiple-choice quiz is given with answers and guidance for self-testing.

Answers and Explanations The answers and explanations to the end-of-chapter questions are provided for further analysis of student knowledge and understanding.

Flashcards Each chapter contains a set of flashcards that allow students to review key terms from the text.

Glossary The complete glossary from the book can be searched and browsed as a whole or sorted by chapter.

For instructors

Figures The images from the book are available in two convenient formats: PowerPoint® and JPEG. They have been optimized for display on a computer.

Question Bank A further set of questions and answers are provided for instructors to use as homework, tests, and examinations.

ACKNOWLEDGMENTS

In writing this book we have benefited greatly from the advice of many geneticists, biologists, and clinicians. We are grateful to many colleagues at Newcastle University and the NHS Northern Genetic Service at Newcastle upon Tyne who advised on the contents of the chapters and/or commented on some aspects of the text, notably the following: Lyle Armstrong, David Bourn, Nick Bown, Gareth Breese, Oonagh Claber, Steven Clifford, Heather Cordell, Ann Daly, David Elliott, Jerome Evans, Fiona Harding, Michael Jackson, Majlinda Lako, Herbie Newell, Caroline Relton, Miranda Splitt, Louise Stanley, Josef Vormoor, and Simon Zwolinski.

We would also like to thank external advisers and reviewers for their suggestions and advice in preparing the text and figures.

Sayeda Abu-Amero (University College London Institute of Child Health, UK); S.S. Agarwal (Sanjay Gandhi Postgraduate Institute of Medical Sciences, India); Robin Allshire (Edinburgh University, UK); Barbara Birshtein (Albert Einstein College of Medicine, USA); Daniel Brazeau (University at New England, USA); Hsiao Chang Chan (Chinese University of Hong Kong, Hong Kong); Frederic Chedin (University of California, USA); Ken-Shiung Chen (Nanyang Technological University, Singapore); David N. Cooper (Cardiff University, UK); Ashwin B. Dalal (Centre for DNA Fingerprinting and Diagnostics, India); Caroline Dalton (Sheffield Hallam University, UK); Shoumita Dasgupta (Boston University School of Medicine, USA); Josh Deignan (University of California, Los Angeles, USA); Donna Dixon (New York Institute of Technology, USA); Diane Dorsett (Georgia Gwinnett College, USA); George Edick (Rensselaer Polytechnic Institute, USA); Mark S. Elliot (George Washington University, USA); David Elliott (Newcastle University, UK); Robert Fowler (San Jose State University, USA); Mary Fujiwara (McGill University, Canada); K.M. Girisha (Kasturba Medical College, India); Jack R. Girton (Iowa State University, USA); Neerja Gupta (All India Institute of Medical Sciences, India); Adrian Hall (Sheffield Hallam University, UK); Lise Lotte Hansen (Aarhus University, Denmark); Sankar V. Hariharan (Government Medical College, India); Graham Heap (Queen Mary, University of London, UK); Chew-Kiat Heng (National University of Singapore, Singapore); Simon Hettle (University of the West of Scotland, UK); Matthew Hurles (Wellcome Trust Sanger Institute, Hinxton, UK); Mary O. Huff (Bellarmine University, USA); Howard N. Hughes (Manchester Metropolitan University, UK); Daniela Iacoboni (Michigan State University, USA); David Iles (University of Leeds, UK); Miho Ishida (University College London Institute of Child Health, UK); Leigh Jackson (Plymouth University, UK); Maria Jackson (University of Glasgow, UK); Suman Kapur (Birla Institute of Medical Sciences, India); Susan Karcher (Purdue University, USA); Robert Koeleman (University Medical Center Utrecht, the Netherlands); Michael Ladomery (University of the West of England, UK); Zhi-Chun Lai (Pennsylvania State University, USA); Janine Lamb (University of Manchester, UK); Alan Lehmann (University of Sussex, UK); Cathy W. Levenson (Florida State University College of Medicine, USA); Qintong Li (Sichuan University, China); Dick Lindhout (University Medical Center Utrecht, the Netherlands); Anneke Lucassen (Southampton University, UK); Alasdair MacKenzie (University of Aberdeen, UK); Khadijah Makky (Marquette University, USA); Elvira Mambetisaeva (University College London Genetics Institute, UK); Elaine Mardis (Washington University School of Medicine, USA); Sarabjt Mastana (Loughborough University, UK); Cynthia J. Moore (Illinois State

University, USA); Gudrun Moore (University College London Institute of Child Health, UK); Tom Moore (University College Cork, Ireland); Claire Morgan (Swansea University, UK); Kenneth Morgan (McGill University, Canada); Yuguang Mu (Nanyang Technological University, Singapore); William Newman (University of Manchester, UK); Alvaro Cantini Nunes (Federal University of Minas Gerais, Brazil); Neil Osheroff (Vanderbilt University School of Medicine, USA); Anthony Otsuka (University of Hawaii at Hilo, USA); Siddaramappa Jagdish Patil (Narayana Hrudayalaya Hospitals, India); Shubha R. Phadke (Sanjay Gandhi Postgraduate Institute of Medical Sciences, India); André Ramos (Federal University of Santa Catarina, Brazil); Prajnya Ranganath (Nizam's Institute of Medical Sciences, India); Michael Reagan (College of Saint Benedict/St John's University, USA); Charles Sackerson (California State University, Channel Islands, USA); R.C. Sample (Mississippi College, USA); Malcolm von Schantz (University of Surrey, UK); Stephanie C. Schroeder (Webster University, USA); Ge Shan (University of Science and Technology of China, China); Alan Shanske (Albert Einstein College of Medicine, USA); Andrew Sharp (Mount Sinai School of Medicine, USA); Andrew Shelling (University of Auckland, New Zealand); Rita Shiang (Virginia Commonwealth University, USA); Heather Skirton (Plymouth University, UK); Vincent E. Sollars (Marshall University, USA); Howard M. Steinman (Albert Einstein College of Medicine, USA); Mike Stratton (Wellcome Trust Sanger Institute, Hinxton, UK); Stefan Surzycki (Indiana University, USA); Chris Talbot (University of Leicester, UK); John Taylor (Newcastle University, UK); Anna Thomas (University College London Institute of Child Health, UK); Patricia N. Tonin (McGill University, Canada); Helga Toriello (Michigan State University, USA); Robert Trumbly (University of Toledo, USA); Andrew Walley (Imperial College London, UK); Tracey Weiler (Florida International University, USA); Feng Zhang (Fudan University, China).

CONTENTS

CHAPTER 6   Principles of Gene Regulation and Epigenetics   149

CHAPTER 9   Genetic Approaches to Treating Disease   309

CHAPTER 10   Cancer Genetics and Genomics   373

Fundamentals of DNA, Chromosomes, and Cells

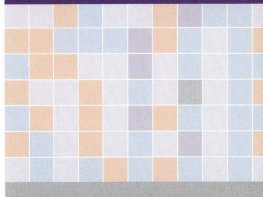

We start this book by describing key aspects of three structures that are the essence of life: cells, chromosomes, and nucleic acids. All living organisms are composed of cells. Cells receive basic sets of instructions from DNA molecules that must also be transmitted to successive generations. And DNA molecules work in the context of larger structures: chromosomes.

Many organisms consist of single cells that can multiply quickly. They are genetically relatively stable, but through changes in their DNA they can adapt rapidly to changes in environmental conditions. Others, including ourselves, animals, plants, and some types of fungi, are multicellular.

Multicellularity offers the prospect of functional specialization and complexity: individual cells can be assigned different functions, becoming muscle cells, neurons, or lymphocytes, for example. All the different cells in an individual arise originally from a single cell and so all nucleated cells carry the same DNA sequences. However, during development the chromosome architecture can be changed in particular ways to determine a cell's identity.

Growth during development and tissue maintenance requires cell division. When a cell divides to produce daughter cells, our chromosomes and the underlying DNA sequences must undergo coordinated duplication and then careful segregation of these structures to the daughter cells.

Some of our cells can carry our DNA to the next generation. When that happens, chromosomes swap segments and DNA molecules undergo significant changes that make us different from our parents and from other individuals.

1.1 The Structure and Function of Nucleic Acids

General concepts: the genetic material, genomes, and genes

Nucleic acids provide the *genetic material* of cells and viruses. They carry the instructions that enable cells to function in the way that they do and to divide, allowing the growth and reproduction of living organisms. Nucleic acids also control how viruses function and replicate. As we describe later, viruses are highly efficient at inserting genes into human cells, and modified viruses are widely used in gene therapy.

Nucleic acids are susceptible to small changes in their structure (**mutations**). Occasionally, that can change the instructions that a nucleic acid gives out. The resulting genetic variation plus mechanisms for shuffling the genetic material from one generation to the next explains why individual organisms of the same species are nevertheless different from each other. And genetic variation is the substrate that evolutionary forces work on to produce different species. (But note that the different types of cell in a single multicellular organism cannot be explained by genetic variation—the cells each contain the same DNA and the differences in cell types must arise instead by **epigenetic** mechanisms.)

In all cells the genetic material consists of double-stranded DNA in the form of a double helix. (Viruses are different. Depending on the type of virus, the genetic material may be double-stranded DNA, single-stranded DNA, double-stranded RNA, or single-stranded RNA.) As we describe below, DNA and RNA are highly related nucleic acids. RNA is functionally more versatile than DNA (it is capable of self-replication and can also direct the synthesis of proteins). RNA is widely believed to have developed at a very early stage in evolution; subsequently, DNA evolved and because it was much more stable than RNA it was more suited to being the store of genetic information in cells.

Genome is the collective term for all the different DNA molecules within a cell or organism. In prokaryotes—simple unicellular organisms, such as bacteria, that lack organelles—the genome usually consists of just one type of circular double-stranded DNA molecule that can be quite large and has a small amount of protein attached to it. A very large DNA–protein complex such as this is traditionally described as a **chromosome**.

Eukaryotic cells are more complex and more compartmentalized (containing multiple organelles that serve different functions), and they have multiple different DNA molecules. As we will see below, for example, the cells of a man have 25 different DNA molecules but a woman's cells have a genome that is composed of 24 DNA molecules.

In our cells—and in those of all animals and fungi—the genome is distributed between the nucleus and the mitochondria. Many different, extremely long DNA molecules are typically found in the nucleus; they are linear DNA molecules and are complexed with a variety of different proteins and some types of RNA to form highly organized chromosomes. However, in mitochondria there is just one type of small circular DNA molecule that is largely devoid of protein. In plant cells, chloroplasts also have their own type of small circular DNA molecule.

Genes are the DNA segments that carry the genetic information to make proteins or functional RNA molecules within cells. The great bulk of the genes in a eukaryotic cell are found in the chromosomes of the nucleus; a few genes are found in the small mitochondrial or chloroplast DNA molecules.

The underlying chemistry of nucleic acids

Each nucleic acid strand is a polymer, a long chain containing many sequential copies of a simple repeating unit, a **nucleotide**. Each nucleotide in turn consists of a sugar molecule, to which is attached a nitrogenous base and a phosphate group.

In DNA the sugar is deoxyribose, which has five carbon atoms that are labeled 1′ (one *prime*) to 5′. It is very closely related to ribose, the sugar molecule found in RNA—the only difference is that a hydroxyl (–OH) group at carbon 2′ of ribose is replaced by a hydrogen atom in deoxyribose (**Figure 1.1**).

Individual nucleotides are joined to their neighbors by the phosphate group that links the sugar components of the neighboring nucleotides. As a result, a nucleic acid has a *sugar–phosphate backbone*. The presence of the negatively charged phosphate groups means that nucleic acids are polyanions.

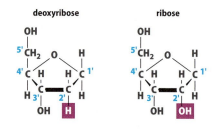

Figure 1.1 Structure of deoxyribose (left) and ribose (right). The five carbon atoms are numbered 1′ (one *prime*) to 5′. The magenta shading is meant to signify the only structural difference between deoxyribose (the sugar found in DNA) and ribose (the sugar found in RNA): ribose has a hydroxyl (–OH) group in place of the highlighted hydrogen atom attached to carbon 2′. The more precise name for deoxyribose is therefore 2′- deoxyribose.

The sugar–phosphate backbone is asymmetric: each phosphate group links a carbon 3′ from the sugar on one nucleotide to a carbon 5′ on the sugar of a neighboring nucleotide. Internal nucleotides will be linked through both carbon 5′ and carbon 3′ of the sugar to the neighboring nucleotides on each side. However, in linear nucleic acids the nucleotides at the extreme ends of a DNA or RNA strand will have different functional groups. At one end, the **5′ end**, the nucleotide has a terminal sugar with a carbon 5′ that is not linked to another nucleotide and is capped by a phosphate group; at the other end, the **3′ end**, the terminal nucleotide has a sugar with a carbon 3′ that is capped by a hydroxyl group (**Figure 1.2**).

Unlike the sugar molecules, the nitrogenous bases come in four different types, and it is the sequence of different bases that identifies the nucleic acid and its function. Two of the bases have a single ring based on carbon and nitrogen atoms (a **pyrimidine**) and two have a double ring structure (a **purine**). In DNA the two purines are adenine (A) and guanine (G), and the two pyrimidines are cytosine (C) and thymine (T). The bases of RNA are very similar; the only difference is that in place of thymine there is a very closely related base, uracil (U) (**Figure 1.3**).

Base pairing and the double helix

Cellular DNA exists in a double-stranded (duplex) form, a double helix where the two very long single DNA strands are wrapped round each other. In the double helix each base on one DNA strand is noncovalently linked (by hydrogen bonding) to an opposing base on the opposite DNA strand, forming a **base pair**. However, the two DNA strands fit together correctly only if opposite every A on one strand is a T on the other strand, and opposite every G is a C. Only two types of base pairs are tolerated in double-stranded DNA: A–T and G–C base pairs. G–C base pairs, which are held together by three hydrogen bonds, are stronger than A–T base pairs, which are held together by two base pairs; see **Figure 1.4**.

There is one additional restriction on how two single-stranded nucleic acids form a double-stranded nucleic acid. In addition to a sufficient degree of base pairing, for a duplex to form, the two single strands must be anti-parallel; that is, the 5′→3′ direction of one strand is the opposite of the 5′→3′ direction of the other strand.

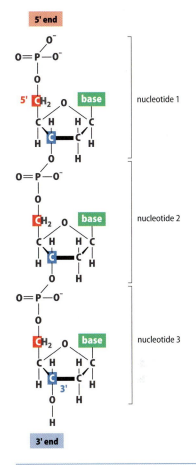

Figure 1.2 Repeating structure and asymmetric 5′ and 3′ ends in nucleic acids. All nucleic acid strands are polymers of a repeating unit, a nucleotide that consists of a sugar with an attached base and phosphate. The sugar–phosphate backbone is asymmetric because phosphate groups connect the carbon 5′ (red shading) of a sugar with the carbon 3′ (blue shading) of the neighboring sugar. This results in asymmetric ends: a 5′ end where the carbon 5′ is attached to a phosphate group only, and a 3′ end where the carbon 3′ is attached to a hydroxyl group only.

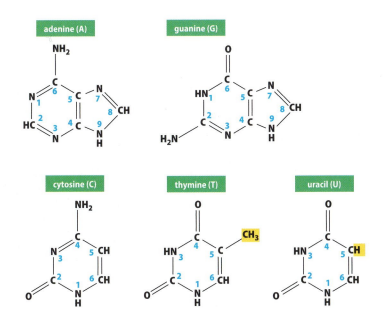

Figure 1.3 Structure of the bases found in nucleic acids. Adenine and guanine are purines with two interlocking rings based on nitrogen and carbon atoms (numbered 1 to 9 as shown). Cytosine and thymine are pyrimidines with a single ring. Adenine, cytosine, and guanine are found in both DNA and RNA, but the fourth base is thymine in DNA and uracil in RNA (they are closely related bases—carbon atom 5 in thymine has an attached methyl group, but in uracil the methyl group is replaced by a hydrogen atom).

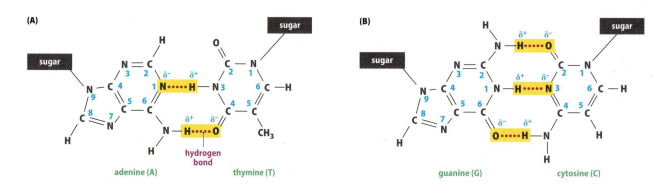

Figure 1.4 Structure of base pairs. In the A–T base pair shown in (A), the adenine is connected to the thymine by two hydrogen bonds. In the G–C base pair shown in (B), three hydrogen bonds link the guanine to the cytosine; a G–C base pair is therefore stronger than an A–T base pair. δ^+ and δ^- indicate fractional positive charges and fractional negative charges.

Two single nucleic acid strands that can form a double helix with perfect base matching (according to the base pairing rules given above) are said to have **complementary sequences**. As a result of base pairing rules, the sequence of one DNA strand in a double helix can immediately be used to predict the base sequence of the complementary strand (see **Box 1.1**). Note that base pairing can also occur in RNA; when an RNA strand participates in base pairing, the base pairing rules are more relaxed (see Box 1.1).

DNA replication and DNA polymerases

Base pairing rules also explain the mechanism of DNA replication. In preparation for new DNA synthesis before cell division, each DNA double helix must be unwound using a helicase. During the unwinding process the two individual single DNA strands become available as templates for making complementary DNA strands that are synthesized in the 5′→3′ direction (**Figure 1.5**).

DNA replication therefore uses one double helix to make two double helices, each containing one strand from the parental double helix and one newly synthesized one (semi-conservative DNA replication). Because DNA synthesis occurs only in the 5′→3′ direction, one new strand (the leading strand) can be synthesized continuously; the other strand (the lagging strand) needs to be synthesized in pieces, known as Okazaki fragments (**Figure 1.6**).

Mammalian cells have very many kinds of DNA-directed DNA polymerases that serve a variety of different roles, including DNA replication initiation, synthesis of the leading and lagging strands, and also, as

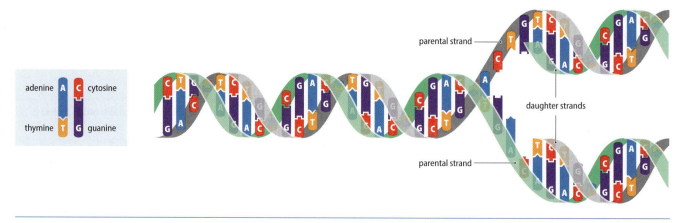

Figure 1.5 DNA replication. The parental DNA duplex consists of two complementary DNA strands that unwind to serve as templates for the synthesis of new complementary DNA strands. Each completed daughter DNA duplex contains one of the two parental DNA strands plus one newly synthesized DNA strand, and is structurally identical to the original parental DNA duplex.

BOX 1.1 Base Pairing Prevalence, Sequence Complementarity, and Sequence Notation for Nucleic Acids.

The prevalence of base pairing

The DNA of cells—and of viruses that have a double-stranded DNA genome—occurs naturally as double helices in which base pairing is restricted to A–T and C–G base pairs.

Double-stranded RNA also occurs naturally in the genomes of some kinds of RNA viruses. Although cellular RNA is often single-stranded, it can also participate in base pairing in different ways. Many single-stranded RNAs have sequences that allow intramolecular base pairing—the RNA bends back upon itself to form local double-stranded regions for structural stability and/or for functional reasons. Different RNA molecules can also transiently base pair with each other over short to moderately long regions, allowing functionally important interactions (such as base pairing between messenger RNA and transfer RNA during translation, for example; see Section 2.1). G–U base pairs are allowed in RNA–RNA base pairing, in addition to the standard A–U and C–G base pairs.

RNA–DNA hybrids also form transiently in different circumstances. They occur when a DNA strand is transcribed to give an RNA copy, for example, and when an RNA is reverse transcribed to give a DNA copy.

Sequence complementarity

Double-helical DNA within cells shows perfect base matching over extremely long distances, and the two DNA strands within a double helix are said to exhibit **base complementarity** and to have *complementary sequences*. Because of the strict base pairing rules, knowing the base sequence of just one DNA strand is sufficient to immediately predict the sequence of the complementary strand, as illustrated below.

Sequence notation

Because the sequences of bases govern the biological properties, it is customary to define nucleic acids by their base sequences, which are always written in the 5′→3′ direction. The sequence of a single-stranded oligonucleotide might be written accurately as 5′ p-C-p-G-p-A-p-C-p-C-p-A-T-OH 3′, where p = phosphate, but it is simpler to write it just as CGACCAT.

For a double-stranded DNA it is sufficient to write the sequence of just one of the two strands; the sequence of the complementary strand can immediately be predicted by the base pairing rules given above. For example, if a given DNA strand has the sequence CGACCAT, the sequence of the complementary strand can easily be predicted to be ATGGTCG (in the 5′→3′ direction as shown below, where A–T base pairs are shown in green and C–G base pairs in blue).

Given DNA strand: 5′ CGACCAT 3′
                      | | | | | | |
→ Complementary strand: 3′ GCTGGTA 5′

described in Section 4.2, multiple roles in DNA repair. Our cells also contain specialized DNA polymerases that use RNA as a template to synthesize a complementary DNA; see **Table 1.1**.

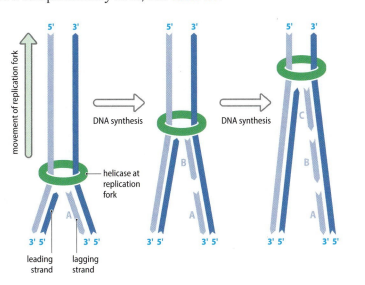

Figure 1.6 Semi-discontinuous DNA replication. The enzyme DNA helicase opens up a **replication fork**, where synthesis of new daughter DNA strands can begin. The overall direction of movement of the replication fork matches that of the continuous 5′→3′ synthesis of one daughter DNA strand, the *leading strand*. Replication is semi-discontinuous because the *lagging strand*, which is synthesized in the opposite direction, is built up in pieces (Okazaki fragments, shown here as fragments A, B, and C) that will later be stitched together by a DNA ligase.

DNA POLYMERASES	ROLES
Classical DNA-dependent DNA polymerases	**Standard DNA replication and/or DNA repair**
α (alpha)	initiates DNA synthesis (at replication origins, and also when priming the synthesis of Okazaki fragments on the lagging strand)
δ (delta) and ε (epsilon)	major nuclear DNA polymerases and multiple roles in DNA repair
β (beta)	base excision repair (repair of deleted bases and simply modified bases)
γ (gamma)	dedicated to mitochondrial DNA synthesis and mitochondrial DNA repair
RNA-dependent DNA polymerases	**Genome evolution and telomere function**
Retrosposon reverse transcriptase	occasionally converts mRNA and other RNA into complementary DNA, which can integrate elsewhere into the genome. Can give rise to new genes and new exons, and so on.
TERT (telomerase reverse transcriptase)	replicates DNA at ends of linear chromosomes, using an RNA template

Table 1.1 Classical DNA-dependent and RNA-dependent mammalian DNA polymerases. The classical DNA-dependent DNA polymerases are high-fidelity polymerases—they insert the correct base with high accuracy; however, we also have many non-classical DNA-dependent DNA polymerases that exhibit comparatively low fidelity of DNA replication. We will consider the non-classical DNA polymerases in Chapter 4, because of their roles in certain types of DNA repair and in maximizing the variability of immunoglobulins and T-cell receptors.

Genes, transcription, and the central dogma of molecular biology

As a repository of genetic information, DNA must be stably *transmitted* from mother cell to daughter cells, and from individuals to their progeny; DNA replication provides the necessary mechanism. But within the context of individual cells, the genetic information must also be *interpreted* to dictate how cells work. **Genes** are discrete segments of the DNA whose sequences are selected for this purpose, and gene expression is the mechanism whereby genes are used to direct the synthesis of two kinds of product: RNA and proteins.

The first step of gene expression is to use one of the two DNA strands as a template for synthesizing an RNA copy whose sequence is complementary to the selected template DNA strand. This process is called *transcription*, and the initial RNA copy is known as the primary transcript (**Figure 1.7**). Subsequently, the primary transcript undergoes different processing steps, eventually giving a mature RNA that belongs to one of two broad RNA classes:

- *Coding RNA.* RNAs in this class contain a coding sequence that is used to direct the synthesis of polypeptides (the major component of proteins) in a process called translation. This type of RNA has traditionally been called a messenger RNA (mRNA) because it must transport genetic instructions to be decoded by the protein synthesis machinery.

- *Noncoding RNA.* All other mature functional RNAs fall into this class, and here the RNAs, not proteins, are the functional endpoint of gene expression. Noncoding RNAs have a variety of different roles in cells, as described in Chapter 2.

In all forms of life, genetic information is interpreted in what initially seemed to be one direction only: DNA→RNA→protein, a principle that became known as the central dogma of molecular biology. However,

Figure 1.7 Transcription. Transcription results in the synthesis of an RNA transcript in the 5′→3′ direction. The nucleotide sequence of the primary RNA transcript is complementary to that of the *template strand* and so is identical to that of the *sense strand*, except that U replaces T.

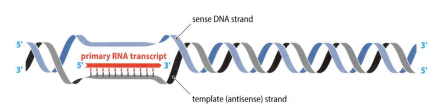

certain DNA polymerases, known as reverse transcriptases, were found initially in certain types of viruses, and as their name suggests they reverse the flow of genetic information by making a DNA copy of an RNA molecule. Cells, too, have their reverse transcriptases (see Table 1.1). In addition, RNA can sometimes also be used as a template to make a complementary RNA copy. So, although genetic information in cells mostly flows from DNA to RNA to protein, the central dogma is no longer strictly valid.

We will explore gene expression (including protein synthesis) in greater detail in Chapter 2. And in Chapter 6 we will focus on both genetic and epigenetic regulation of gene expression.

1.2   The Structure and Function of Chromosomes

In this section we consider general aspects of the structure and function of our chromosomes that are largely shared by the chromosomes of other complex multicellular organisms. We will touch on human chromosomes when we consider aspects of the human genome in Chapter 2, when we first introduce the banded pattern of human chromosomes. In Chapter 7 we consider how disease-causing chromosome abnormalities arise. We describe the methodology and the terminology of human chromosome banding in Box 7.4, and diagnostic chromosome analyses in Chapter 11.

Why we need highly structured chromosomes, and how they are organized

Before replication, each chromosome in the cells of complex multicellular organisms normally contains a single, immensely long DNA double helix. For example, an average-sized human chromosome contains a single DNA double helix that is about 4.8 cm long with 140 million nucleotides on each strand; that is, 140 million base pairs (140 megabases (Mb)) of DNA.

To appreciate the difficulty in dealing with molecules this long in a cell that is only about 10 μm across, imagine a model of a human cell 1 meter across (a 10^5-fold increase in diameter). Now imagine the problem of fitting into this 1-meter-wide cell 46 DNA double helices that when scaled up by the same factor would each be just 0.2 mm thick but on average 4.8 km (about 3 miles) long. Then there is the challenge of replicating each of the DNA molecules and arranging for the cell to divide in such a way that the replicated DNA molecules are segregated equally into the two daughter cells. All this must be done in a way that avoids any tangling of the long DNA molecules.

To manage nuclear DNA molecules efficiently and avoid any tangling, they are complexed with various proteins and sometimes RNAs to form **chromatin** that undergoes different levels of coiling and compaction, forming chromosomes. In interphase—the stages of the cell cycle other than mitosis (see Section 1.3)—the nuclear DNA molecules are still in a very highly extended form and normally the very long slender interphase chromosomes remain invisible under the light microscope. But even in interphase cells, the 2-nm-thick double helix is subject to at least two levels of coiling. First, the double helix is periodically wound round a specialized complex of positively charged histone proteins to form a 10 nm nucleosome filament. The nucleosome filament is then coiled into a 30 nm chromatin fiber that undergoes looping and is supported by a scaffold of non-histone proteins (**Figure 1.8**).

During interphase most chromatin exists in an extended state (**euchromatin**) that is dispersed through the nucleus. Euchromatin is not uniform,

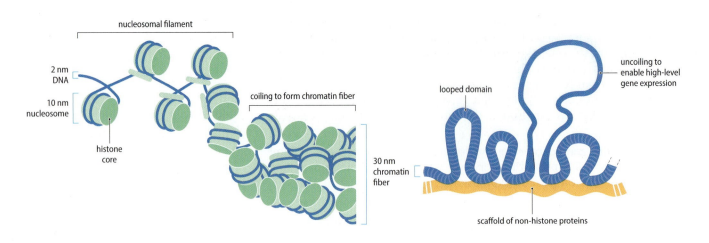

Figure 1.8 From DNA double helix to interphase chromatin. Binding of basic histone proteins causes the 2 nm DNA double helix to undergo coiling, forming first a 10 nm filament studded with nucleosomes that is further coiled to give a 30 nm chromatin fiber. In interphase, the chromatin fiber is organized in looped domains, each containing about 50–200 kilobases of DNA, that are attached to a central scaffold of nonhistone proteins. High levels of gene expression require local uncoiling of the chromatin fiber to give the 10 nm nucleosomal filaments. The diagram does not show structural RNAs that can be important in chromatin. (Adapted from Grunstein M [1992] *Sci Am* 267:68–74; PMID 1411455. With permission from Macmillan Publishers Ltd; and Alberts B, Johnson A, Lewis J et al. [2008] Molecular Biology of the Cell, 5th ed. Garland Science.)

however—some euchromatic regions are more condensed than others, and genes may or may not be expressed, depending on the cell type and its functional requirements. Some chromatin, however, remains highly condensed throughout the cell cycle and is generally genetically inactive (**heterochromatin**).

As cells prepare to divide, the chromosomes need to be compacted much further to maximize the chances of correct pairing and segregation of chromosomes into daughter cells. Packaging of DNA into **nucleosomes** and then the 30 nm chromatin fiber results in a linear condensation of about fiftyfold. During the M (mitosis) phase, higher-order coiling occurs (see Figure 1.8), so that DNA in a human metaphase chromosome is compacted to about 1/10,000 of its stretched-out length. As a result, the short, stubby metaphase chromosomes are readily visible under light microscopes.

Chromosome function: replication origins, centromeres, and telomeres

The DNA within a chromosome contains genes that are expressed according to the needs of a cell. But it also contains specialized sequences that are needed for chromosome function. Three major classes are described below.

Centromeres

Chromosomes must be correctly segregated during cell division. This requires a **centromere**, a region to which a pair of large protein complexes called kinetochores will bind just before the preparation for cell division (**Figure 1.9**). Centromeres can be seen at metaphase as the primary constriction that separates the short and long arms. Microtubules attached to each kinetochore are responsible for positioning the chromosomes correctly at metaphase and then pulling the separated chromosomes to opposite poles of the mitotic spindle.

The DNA sequences at centromeres are very different in different organisms. In a mammalian chromosome, the centromeric DNA is a heterochromatic region dominated by highly repetitive DNA sequences that often extend over megabases of DNA.

Replication origins

For a chromosome to be replicated, it needs one or more replication origins—DNA sequence components to which protein factors bind in preparation for initiating DNA replication. The chromosomes of budding yeast can be replicated using a single very short highly defined DNA

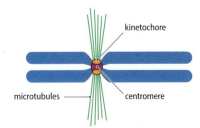

Figure 1.9 Centromere function relies on the assembly of kinetochores and attached microtubules.

sequence, but in the cells of complex organisms, such as mammals, DNA is replicated at multiple initiation sites along each chromosome; the replication origins are quite long and do not have a common base sequence.

Telomeres

Telomeres are specialized structures at the ends of chromosomes that are necessary for the maintenance of chromosome integrity (if a telomere is lost after chromosome breakage, the resulting chromosome end is unstable; it tends to fuse with the ends of other broken chromosomes, or to be involved in recombination events, or to be degraded).

Unlike centromeric DNA, telomeric DNA has been well conserved during evolution. In vertebrates, the DNA of telomeres consists of many tandem (sequential) copies of the sequence TTAGGG to which certain telomeric proteins bind. The G-rich strand (TTAGGG) has a single-stranded overhang at its 3' end that folds back and base pairs with the C-rich strand (CCCTAA repeats). The resulting T-loop is thought to protect the telomere DNA from natural cellular exonucleases that repair double-stranded DNA breaks (**Figure 1.10**).

1.3   DNA and Chromosomes in Cell Division and the Cell Cycle

Differences in DNA copy number between cells

Like other multicellular organisms, we have cells that are structurally and functionally diverse. In each individual the different cell types have the same genetic information, but only a subset of genes is expressed in each cell. What determines the identity of a cell—whether a cell is a B lymphocyte or a hepatocyte, for example—is the pattern of expression of the different genes across the genome.

As well as differences in gene expression, different cells can vary in the number of copies of each DNA molecule. The term *ploidy* describes the number of copies (*n*) of the basic chromosome set (the collective term for the different chromosomes in a cell) and also describes the copy number of each of the different nuclear DNA molecules.

The DNA content of a single chromosome set is represented as *C*. Human cells—and the cells of other mammals—are mostly **diploid** (2*C*), with nuclei containing two copies of each type of chromosome, one paternally inherited and one maternally inherited. Sperm and egg cells are **haploid** cells that contain only one of each kind of chromosome (1*C*). Human sperm and eggs each have 23 different types of chromosomes and so *n* = 23 in humans.

Some specialized human cells are nulliploid (0*C*) because they lack a nucleus—examples include erythrocytes, platelets, and terminally differentiated keratinocytes. Others are naturally polyploid (more than 2*C*). Polyploidy can occur by two mechanisms. The DNA might undergo multiple rounds of replication without cell division, as when the large megakaryocytes in blood are formed (they have from 16 to 64 copies of each chromosome, and the nucleus is large and multilobed). Alternatively, polyploid cells originate by cell fusion to give cells with multiple nuclei, as in the case of muscle fiber cells.

Mitochondrial DNA copy number

Whereas all diploid cells contain two copies of each nuclear DNA molecule, the number of mtDNA molecules can vary from hundreds to many

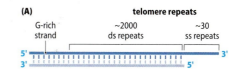
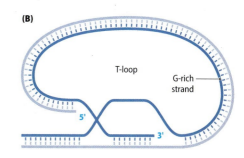

Figure 1.10 Telomere structure and T-loop formation. (A) Human telomere structure. A tandem array of roughly 2000 copies of the hexanucleotide TTAGGG is followed by a single-stranded protrusion (by the G-rich strand). Abbreviations: ss, single-strand; ds, double-strand. (B) T-loop formation. The single-stranded terminus can loop back and invade the double-stranded region by base pairing with the complementary C-rich strand. (C) Electron micrograph showing formation of a roughly 15-kilobase T-loop at the end of an interphase human chromosome. (From Griffith JD et al. [1999] *Cell* 97:503–514; PMID 10338214. With permission from Elsevier.)

thousands according to the cell type, and can even vary over time in some cells. The two types of haploid cells show very large differences in mtDNA copy number: a human sperm typically has about 100 mtDNA copies, but a human egg cell usually has about 250,000 mtDNA molecules.

The cell cycle and segregation of replicated chromosomes and DNA molecules

Cells also differ according to whether they actively participate in the cell cycle and undergo successive rounds of cell division. Each time a cell divides, it gives rise to two daughter cells. To keep the number of chromosomes constant there needs to be a tight regulation of chromosome replication and chromosome segregation. Each chromosome needs to be replicated just once to give rise to two daughter chromosomes, which must then segregate equally so that one passes to each daughter cell.

During normal periods of growth there is a need to expand cell number. In the fully grown adult, the majority of cells are terminally differentiated and do not divide, but stem cells and progenitor cells continue to divide to replace cells that have a high turnover, notably blood, skin, sperm, and intestinal epithelial cells.

Each round of the cell cycle involves a phase in which the DNA replicates—S phase (synthesis of DNA)—and a phase where the cell divides, the M phase. Note that M phase involves both nuclear division (mitosis) and cell division (cytokinesis). In the intervals between these two phases are two gap phases: G_1 phase (gap between M phase and S phase) and G_2 phase (gap between S phase and M phase)—see **Figure 1.11**.

Cell division takes up only a brief part of the cell cycle. For actively dividing human cells, a single turn of the cell cycle might take about 24 hours; M phase often occupies about 1 hour. During the short M phase, the chromosomes become extremely highly condensed in preparation for nuclear and cell division. After M phase, cells enter a long growth period called **interphase** ($= G_1 + S + G_2$ phases), during which chromosomes are enormously extended, allowing genes to be expressed.

G_1 is the the long-term end state of terminally differentiated nondividing cells. For dividing cells, the cells will enter S phase only if they are committed to mitosis; if not, they are induced to leave the cell cycle to enter a resting phase, the G_0 phase (a modified G_1 stage). When conditions become suitable, cells may subsequently move from G_0 to re-enter the cell cycle.

Changes in cell chromosome number and DNA content

During the cell cycle, the amount of DNA in a cell and the number of chromosomes change. In the box panels in Figure 1.11 we follow the fate of a single chromosome through M phase and then through S phase. If we were to consider a diploid human cell this would be one chromosome out of the 46 (2*n*) chromosomes present after daughter cells are first formed. We also show in the box panel in S phase how a single chromosome (top) relates to its DNA double helix content at different stages in S phase. The progressive changes in the number of chromosomes and the DNA content of cells at different stages of the cell cycle are listed below.

- From the end of the M phase right through until DNA duplication in S phase, each chromosome of a diploid (2*n*) cell contains a single DNA double helix; the total DNA content is therefore 2*C*.

- After DNA duplication, the total DNA content per cell is 4*C*, but specialized binding proteins called cohesins hold the duplicated double helices together as **sister chromatids** within a single chromosome.

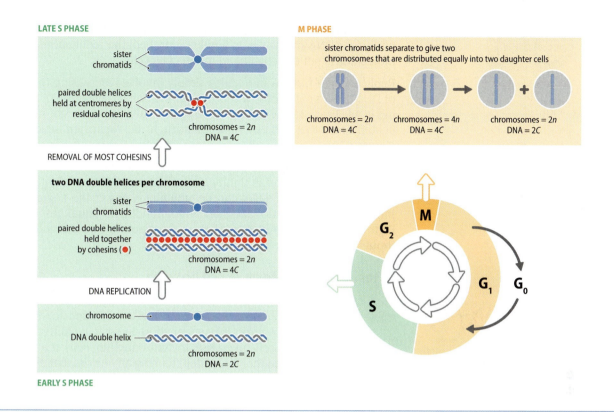

Figure 1.11 Changes in chromosomes and DNA content during the cell cycle. The cell cycle consists of four major phases as shown at the bottom right (in the additional G_0 phase a cell exits from the cell cycle and remains suspended in a stationary phase that resembles G_1 but can subsequently rejoin the cell cycle under certain conditions). In the expanded panels for M and S phases we show for convenience just a single chromosome, and we illustrate in the S-phase panel how a single chromosome (top) relates to its DNA molecule (bottom) at different stages. Chromosomes contain one DNA double helix from the end of M phase right through until just before the DNA duplicates in S phase. After duplication, the two double helices are held tightly together along their lengths by binding proteins called cohesins (red circles), and the chromosome now consists of two sister chromatids each having a DNA double helix. The sister chromatids becomes more obvious in late S phase when most of the cohesins are removed except for some at the centromere, which continue to hold the two sister chromatids together. The sister chromatids finally separate in M phase to form two independent chromosomes that are then segregated into the daughter cells. Note that the S-phase chromosomes in the left panel are shown, purely for convenience, in a compact form, but in reality they are enormously extended.

The chromosome number remains the same ($2n$), but each chromosome now has double the DNA content of a chromosome in early S phase. In late S phase, most of the cohesins are removed but cohesins at the centromere are retained to keep the sister chromatids together.

• During M phase, the residual cohesins are removed and the duplicated double helices finally separate. That allows sister chromatids to separate to form two daughter chromosomes, giving $4n$ chromosomes. The duplicated chromosomes segregate equally to the two daughter cells so that each will have $2n$ chromosomes and a DNA content of $2C$.

Figure 1.11 can give the misleading impression that all the interesting action happens in S and M phases. That is quite wrong—a cell spends most of its life in the G_0 or G_1 phase, and that is where the genome does most of its work, issuing the required instructions to make the diverse protein and RNA products needed for cells to function.

Mitochondrial DNA replication and segregation

In advance of cell division, mitochondria increase in mass and mtDNA molecules replicate before being segregated into daughter mitochondria

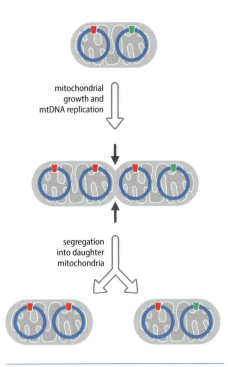

mitochondrial
growth and
mtDNA replication

segregation
into daughter
mitochondria

Figure 1.12 Unequal replication of individual mitochondrial DNAs. Unlike in the nucleus, where replication of a chromosomal DNA molecule normally produces two copies, mitochondrial DNA (mtDNA) replication is stochastic. When a mitochondrion increases in mass in preparation for cell division, the overall amount of mitochondrial DNA increases in proportion, but individual mtDNAs replicate unequally. In this example, the mtDNA with the green tag fails to replicate and the one with the red tag replicates to give three copies. Variants of mtDNA can arise through mutation so that a person can inherit a mixed population of mtDNAs (heteroplasmy). Unequal replication of pathogenic and nonpathogenic mtDNA variants can have important consequences, as described in Chapter 5.

that then need to segregate into daughter cells. Whereas the replication of nuclear DNA molecules is tightly controlled, the replication of mtDNA molecules is not directly linked to the cell cycle.

Replication of mtDNA molecules simply involves increasing the number of DNA copies in the cell, without requiring equal replication of individual mtDNAs. That can mean that some individual mtDNAs might not be replicated and other mtDNA molecules might be replicated several times (**Figure 1.12**).

Whereas the segregation of nuclear DNA molecules into daughter cells needs to be equal and is tightly controlled, segregation of mtDNA molecules into daughter cells can be unequal. Even if the segregation of mtDNA molecules into daughter mitochondria is equal (as shown in Figure 1.12), the segregation of the mitochondria into daughter cells is thought to be stochastic.

Mitosis: the usual form of cell division

Most cells divide by a process known as mitosis. In the human life cycle, mitosis is used to generate extra cells that are needed for periods of growth and to replace various types of short-lived cells. Mitosis ensures that a single parent cell gives rise to two daughter cells that are both genetically identical to the parent cell (barring any errors that might have occurred during DNA replication). During a human lifetime, there may be something like 10^{17} mitotic divisions.

The M phase of the cell cycle includes both nuclear division (mitosis, which is divided into the stages of prophase, prometaphase, metaphase, anaphase, and telophase), and also cell division (cytokinesis), which overlaps the final stages of mitosis (**Figure 1.13**). In preparation for cell division, the previously highly extended duplicated chromosomes contract and condense so that, by the metaphase stage of mitosis, they are readily visible when viewed under the microscope.

The chromosomes of early S phase have one DNA double helix; however, after DNA replication, two identical DNA double helices are produced and held together by cohesins. Later, when the chromosomes undergo compaction in preparation for cell division, the cohesins are removed from all parts of the chromosomes apart from the centromeres. As a result, as early as prometaphase (when the chromosomes are now visible under the light microscope), individual chromosomes can be seen to comprise two **sister chromatids** that remain attached at the centromere (bound by some residual cohesins).

Later, at the start of anaphase, the remaining cohesins are removed and the two sister chromatids can now disengage to become independent chromosomes that will be pulled to opposite poles of the cell and then distributed equally to the daughter cells (see Figure 1.13).

Meiosis: a specialized reductive cell division giving rise to sperm and egg cells

The **germ line** is the collective term for all cells that pass genetic material to the next generation. It includes haploid sperm and egg cells (the **gametes**) and all the diploid precursor cells from which they arise by cell division, going all the way back to the zygote. The non-germ-line cells are known as **somatic cells**.

In humans, where $n = 23$, each gamete contains one sex chromosome plus 22 nonsex chromosomes (**autosomes**). In eggs the sex chromosome is always an X; in sperm it may be either an X or a Y. After a haploid

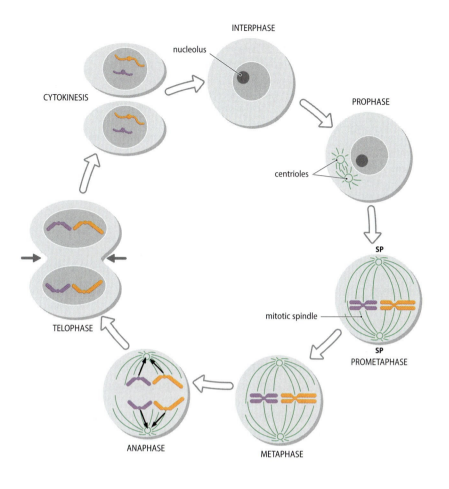

INTERPHASE

nucleolus

CYTOKINESIS

PROPHASE

centrioles

SP

SP

mitotic spindle

PROMETAPHASE

TELOPHASE

ANAPHASE

METAPHASE

Figure 1.13 Mitosis (nuclear division) and cytokinesis (cell division). Early in prophase, centrioles (short cylindrical structures composed of microtubules and associated proteins) begin to separate and migrate to opposite poles of the cell to form the spindle poles (SP). In prometaphase, the nuclear envelope breaks down, and the now highly condensed chromosomes become attached at their centromeres to the array of microtubules extending from the mitotic spindle. At metaphase, the chromosomes all lie along the middle of the mitotic spindle, still with the sister chromatids bound together (because of residual cohesins at the centromere that hold the duplicated DNA helices together). Removal of the residual cohesins allows the onset of anaphase: the sister chromatids separate and begin to migrate toward opposite poles of the cell. The nuclear envelope forms again around the daughter nuclei during telophase, and the chromosomes decondense, completing mitosis. Before the final stages of mitosis, and most obviously at telophase, cytokinesis begins with constriction of the cell that will increase progressively to produce two daughter cells.

sperm fertilizes a haploid egg, the resulting diploid **zygote** and almost all of its descendant cells have the chromosome constitution 46,XX (female) or 46,XY (male).

Diploid primordial germ cells migrate into the embryonic gonad and engage in repeated rounds of mitosis, to generate spermatogonia in males and oogonia in females. Further growth and differentiation produce primary spermatocytes in the testis and primary oocytes in the ovary. The diploid spermatocytes and oocytes can then undergo **meiosis**, the cell division process that produces haploid gametes.

Meiosis is a *reductive* division because it involves two successive cell divisions (known as meiosis I and meiosis II) but only one round of DNA replication (**Figures 1.14** and **1.15**). As a result, it gives rise to four haploid cells. In males, the two meiotic cell divisions are each symmetric, producing four functionally equivalent spermatozoa. Huge numbers of sperm are produced, and spermatogenesis is a continuous process from puberty onward.

Female meiosis is different: cell division is asymmetric, resulting in unequal division of the cytoplasm. The products of female meiosis I (the first meiotic cell division) are a large secondary oocyte and a small cell, the *polar body*, which is discarded. During meiosis II the secondary oocyte then gives rise to the large mature egg cell and a second polar body (which again is discarded).

In humans, primary oocytes enter meiosis I during fetal development but are then all arrested at prophase until after the onset of puberty. After puberty in females, one primary oocyte completes meiosis with each

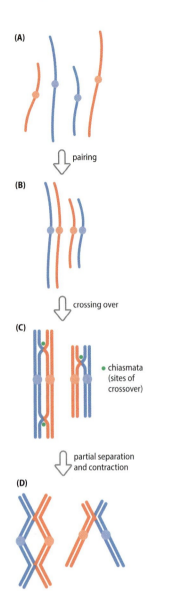

Figure 1.14 Prophase stages in meiosis I. (A) In leptotene, the duplicated homologous chromosomes begin to condense but remain unpaired. (B) In zygotene, duplicated maternal and paternal homologs pair, to form bivalents, comprising four chromatids. (C) In pachytene, recombination (crossing over) occurs through the physical breakage and subsequent rejoining of maternal and paternal chromosome fragments. There are two chiasmata (crossovers) in the bivalent on the left, and one in the bivalent on the right. For simplicity, both chiasmata on the left involve the same two chromatids. In reality, more chiasmata may occur, involving three or even all four chromatids in a bivalent. (D) During diplotene, the homologous chromosomes may separate slightly, except at the chiasmata. A further stage, diakinesis, is marked by contraction of the bivalents and is the transition to metaphase I. In this figure, only 2 of 23 possible pairs of homologs are illustrated (with the maternal homolog colored pink, and the paternal homolog blue).

menstrual cycle. Because ovulation can continue up to the fifth and sometimes sixth decades, this means that meiosis can be arrested for many decades in those primary oocytes that are not used in ovulation until late in life.

Pairing of paternal and maternal homologs (synapsis)

Each of our diploid cells contains two copies (**homologs**) of each type of chromosome (except in the special case of the X and Y chromosomes in males). One homolog is inherited from each parent, so we have a maternal chromosome 1 and a paternal chromosome 1, and so on.

A special feature of meiosis I—that distinguishes it from mitosis and meiosis II—is the pairing (*synapsis*) of paternal and maternal homologs. After DNA replication (when the chromosomes comprise two sister chromatids), the maternal and paternal homologs of each type of chromosome align along their lengths and become bound together. The resulting *bivalent* has four strands: two paternally inherited sister chromatids and two maternally inherited sister chromatids (see Figures 1.14C, D and 1.15).

The pairing of homologs is required for recombination to occur (as described in the next subsection). It must ultimately be dictated by high levels of DNA sequence identity between the homologs. The high sequence matching between homologs that is required for pairing does not need to be complete, however: when there is some kind of chromosome abnormality so that the homologs do not completely match, the matching segments usually manage to pair up.

Pairing of maternal and paternal sex chromosomes is straightforward in female meiosis, but in male meiosis there is the challenge of pairing a maternally inherited X chromosome with a paternally inherited Y. The human X chromosome is very much larger than the Y, and their DNA sequences are very different. However, they do have some sequences in common, notably a major *pseudoautosomal region* located close to the short-arm telomeres. The X and Y chromosomes cannot pair up along their lengths, but because they have some sequences in common they can always pair up along these regions. We will explore this in greater detail in Chapter 5 when we consider pseudoautosomal inheritance.

Recombination

The prophase of meiosis I begins during fetal life and, in human females, can last for decades. During this extended process, paternal and maternal chromatids within each bivalent normally exchange segments of DNA at randomly positioned but matching locations. This process—called

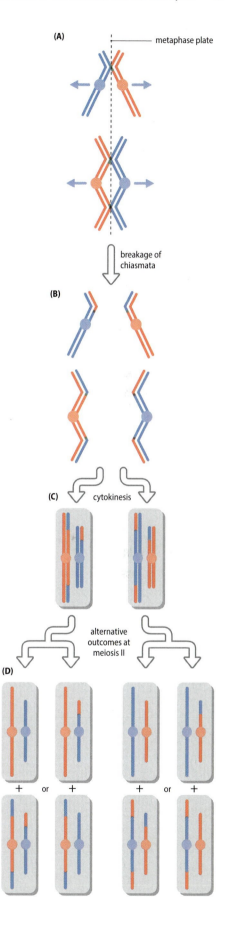

Figure 1.15 Metaphase I to production of gametes. (A) At metaphase I, the bivalents align on the metaphase plate, at the centre of the spindle apparatus. Contraction of spindle fibers draws the chromosomes in the direction of the spindle poles (arrows). (B) The transition to anaphase I occurs at the consequent rupture of the chiasmata. (C) Cytokinesis segregates the two chromosome sets, each to a different primary spermatocyte. Note that, after recombination during prophase I (see Figure 1.14C), the chromatids share a single centromere but are no longer identical. (D) Meiosis II in each primary spermatocyte, which does not include DNA replication, generates unique genetic combinations in the haploid secondary spermatocytes. Only 2 of the possible 23 different human chromosomes are depicted, for clarity, so only 2^2 (that is, 4) of the possible 2^{23} (8,388,608) possible combinations are illustrated. Although oogenesis can produce only one functional haploid gamete per meiotic division, the processes by which genetic diversity arises are the same as in spermatogenesis.

recombination (or crossover)—involves physical breakage of the DNA in one paternal and one maternal chromatid, and the subsequent joining of maternal and paternal fragments.

Recombined homologs seem to be physically connected at specific points. Each such connection marks the point of crossover and is known as a chiasma (plural chiasmata—see Figure 1.14C). The distribution of chiasmata is non-random and the number of chiasmata per meiosis shows significant sex differences, and also very significant differences between individuals of the same sex, and even between individual meioses from a single individual. In a large recent study of human meiosis, an average of 38 recombinations were detected per female meiosis and an average of 24 meioses occurred on male meiosis but with very significant variation (as shown in Figure 8.3 on page 252). In addition to their role in recombination, chiasmata are thought to be essential for correct chromosome segregation during meiosis I.

There are hotspot regions where recombination is more likely to occur. For example, recombination is more common in subtelomeric regions. In the case of X–Y crossover there is an obligate crossover within the short 2.6 Mb pseudoautosomal region. This region is so called because it is regularly swapped between the X and Y chromosomes and so the inheritance pattern for any DNA variant here is not X-linked or Y-linked but instead resembles autosomal inheritance.

Why each of our gametes is unique

The sole purpose of sex in biology is to produce novel combinations of gene variants, and the instrument for achieving this aim is meiosis. The whole point of meiosis is to produce *unique* gametes by selecting different combinations of DNA sequences on maternal and paternal homologs.

Although a single ejaculate may contain hundreds of millions of sperm, meiosis ensures that no two sperm will be genetically identical. Equally, no two eggs are genetically identical. Each zygote must also be unique because at fertilization a unique sperm combines with a unique egg. However, a unique fertilization event can occasionally give rise to two genetically identical (**monozygotic**) twins if the embryo divides into two at a very early stage in development (monozygotic twins are nevertheless unique individuals—genetics is not everything in life!).

The second division of meiosis is identical in form to mitosis; meiosis I is where the genetic diversity originates, and that involves two mechanisms. First, there is independent assortment of paternal and maternal

Figure 1.16 Independent assortment of maternal and paternal homologs during meiosis. The figure shows a random selection of just 5 of the 8,388,608 (2^{23}) theoretically possible combinations of homologs that might occur in haploid human spermatozoa after meiosis in a diploid primary spermatocyte. Maternally derived homologs are represented by pink boxes, and paternally derived homologs by blue boxes. For simplicity, the diagram ignores recombination—but see Figure 1.17.

diploid primary spermatocytes

| 1 | 2 | 3 | 4 | 5 | 6 | 7 | 8 | 9 | 10 | 11 | 12 | 13 | 14 | 15 | 16 | 17 | 18 | 19 | 20 | 21 | 22 | X | maternal |
| 1 | 2 | 3 | 4 | 5 | 6 | 7 | 8 | 9 | 10 | 11 | 12 | 13 | 14 | 15 | 16 | 17 | 18 | 19 | 20 | 21 | 22 | Y | paternal |

⬇ meiosis

haploid sperm cells

1	2	3	4	5	6	7	8	9	10	11	12	13	14	15	16	17	18	19	20	21	22	Y	sperm 1
1	2	3	4	5	6	7	8	9	10	11	12	13	14	15	16	17	18	19	20	21	22	X	sperm 2
1	2	3	4	5	6	7	8	9	10	11	12	13	14	15	16	17	18	19	20	21	22	Y	sperm 3
1	2	3	4	5	6	7	8	9	10	11	12	13	14	15	16	17	18	19	20	21	22	X	sperm 4
1	2	3	4	5	6	7	8	9	10	11	12	13	14	15	16	17	18	19	20	21	22	X	sperm 5

homologs. After DNA replication, the homologous chromosomes each comprise two sister chromatids, so each bivalent is a four-stranded structure at the metaphase plate. Spindle fibers then pull one complete chromosome (two chromatids) to either pole. In humans, for each of the 23 homologous pairs, the choice of which daughter cell each homolog will enter is independent. This allows 2^{23} or about 8.4×10^6 different possible combinations of parental chromosomes in the gametes that might arise from a single meiotic division (**Figure 1.16**).

The second mechanism that contributes to genetic diversity is recombination. Whereas sister chromatids within a bivalent are genetically identical, the paternal and maternal chromatids are not. On average their DNA will differ at roughly 1 in every 1000 nucleotides. Swapping maternal and paternal sequences by recombination will therefore produce an extra level of genetic diversity (**Figure 1.17**). It raises the number of permutations from the 8.4 million that are possible just from the independent assortment of maternal and paternal homologs alone, to a virtually infinite number.

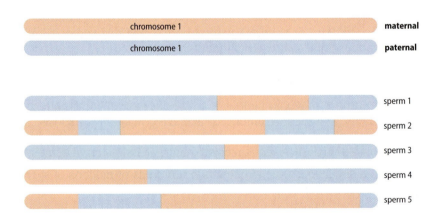

Figure 1.17 Recombination superimposes additional genetic variation at meiosis I. Figure 1.16 illustrates the contribution to genetic variation at meiosis I made by independent assortment of homologs, but for simplicity it ignores the contribution made by recombination. In reality each transmitted chromosome is a mosaic of paternal and maternal DNA sequences, as shown here. See Figure 8.2 on page 252 for a real-life example.

Summary

- Nucleic acids are negatively charged long polymers composed of sequential nucleotides that each consist of a sugar, a nitrogenous base, and a phosphate group. They have a sugar–phosphate backbone with bases projecting from the sugars.

- Nucleic acids have four types of bases: adenine (A), cytosine (C), guanine (G), and either thymine (T) in DNA or uracil (U) in RNA. The sequence of bases determines the identity of a nucleic acid and its function.

- RNA normally consists of a single nucleic acid chain, but in cells a DNA molecule has two chains (strands) that form a stable duplex (in the form of a double helix). Duplex formation requires hydrogen bonding between matched bases (base pairs) on the two strands.

- In DNA two types of base pairing exist: A pairs with T, and C pairs with G. According to these rules, the two strands of a DNA double helix are said to have complementary base sequences.

- Base pairing also occurs in RNA and includes G–U base pairs, as well as G–C and A–U base pairs. Two different RNA molecules with partly complementary sequences can associate by forming hydrogen bonds. Intramolecular hydrogen bonding also allows a single RNA chain to form a complex three-dimensional structure.

- DNA carries primary instructions that determine how cells work and how an individual is formed. Defined segments of DNA called genes are used to make a single-stranded RNA copy that is complementary in sequence to one of the DNA strands (transcription).

- DNA is propagated from one cell to daughter cells by replicating itself. The two strands of the double helix are unwound, and each strand is used to make a new complementary DNA copy. The two new DNA double helices (each with one parental DNA strand and one new DNA strand) are segregated so that each daughter cell receives one DNA double helix.

- RNA molecules function in cells either as a mature noncoding RNA, or as a messenger RNA with a coding sequence used to make the polypeptide chain of a protein (translation).

- Each nuclear DNA molecule is complexed with different proteins and some noncoding RNAs to form a chromosome that condenses the DNA and protects it.

- Packaging DNA into chromosomes stops the long DNA chains from getting entangled within cells, and by greatly condensing the DNA in preparation for cell division it allows the DNA to be segregated correctly to daughter cells and to offspring.

- Our sperm and egg cells are haploid cells with a set of 23 different chromosomes (each with a single distinctive DNA molecule). There is one sex chromosome (an X chromosome in eggs; either an X or Y in sperm) and 22 different autosomes (non-sex chromosomes).

- Most of our cells are diploid with two copies of the haploid chromosome set, one set inherited from the mother and one from the father. Maternal and paternal copies of the same chromosome are known as homologs.

- There is one type of mitochondrial DNA (mtDNA); it is present in large numbers in cells. Both the replication of mtDNA and its segregation to daughter cells occur stochastically.

- Cells need to divide as we grow. In fully formed adults, most of our cells are specialized, nondividing cells, but some cells are required to keep on dividing to replace short-lived cells, such as blood, skin, and intestinal epithelial cells.

- Mitosis is the normal form of cell division. Each chromosome (and chromosomal DNA) replicates once and the duplicated chromosomes are segregated equally into the two daughter cells.

- Meiosis is a specialized form of cell division that is required to produce haploid sperm and egg cells. The chromosomes in a diploid spermatogonium or oogonium replicate once, but there are two successive cell divisions to reduce the number of chromosomes in each cell.

- Each sperm cell produced by a man is unique, as is each egg cell that a woman produces. During the first cell division in meiosis, maternal and paternal homologs associate and exchange sequences by recombination. Largely random recombination results in unpredictable new DNA sequence combinations in each sperm and in each egg.

Questions

Help on answering these questions and a multiple-choice quiz can be found at www.garlandscience.com/ggm-students

1.   The sequence GATCCAGGACCATGTTATCCAGGATAA is part of a protein-coding gene. Write out the equivalent sequence on the template strand and on the mRNA that is produced.

2.   The nuclear DNA molecules in our cells need to be organized as complex structures called chromosomes. Why?

Further Reading

More detailed treatment of the subject matter in this chapter can be found in more comprehensive genetics and cell biology textbooks such as:

Alberts B, Johnson A, Lewis J & Raff M (2008) Molecular Biology of the Cell, 5th ed. Garland Science.

Strachan T & Read AP (2010) Human Molecular Genetics, 4th ed. Garland Science.

Fundamentals of Gene Structure, Gene Expression, and Human Genome Organization

Our genome is complex, comprising about 3.2 Gb (3.2×10^9 base pairs; Gb = gigabase) of DNA. One of its main tasks is to produce a huge variety of different proteins that dictate how our cells work. Surprisingly, however, **coding DNA**—DNA sequences that specify the polypeptides of our proteins—account for just over 1.2% of our DNA.

The remainder of our genome is noncoding DNA that does not make any protein, but a significant fraction of the noncoding DNA is functionally important. Functional noncoding DNA sequences include many different classes of DNA regulatory sequences that control how our genes work, such as promoters and enhancers, and DNA sequences that specify regulatory elements that work at the RNA level.

We now know that in addition to our protein-coding genes (which make a messenger RNA that makes a polypeptide) we have many thousands of genes that do not make polypeptides; instead they make different classes of functional noncoding RNA. Some of these **RNA genes**—such as genes encoding ribosomal RNA and transfer RNA—have been known for decades, but one of the big surprises in recent years has been the sheer number and variety of noncoding RNAs in our cells. In addition to the RNA genes, our protein-coding genes frequently make noncoding RNA transcripts as well as messenger RNAs (mRNAs). They include both linear and circular RNAs, and many of these RNAs have important regulatory roles.

Like other complex genomes, our genome has a large proportion of moderately to highly repetitive DNA sequences. Some of these are important in centromere and telomere function; others are important in shaping how our genome evolves.

The Human Genome Project delivered the first comprehensive insights into our genome by delivering an essentially complete nucleotide sequence by 2003. Follow-up studies have compared our genome with other genomes, helping us to understand how our genome evolved. The comparative genomics studies, together with genomewide functional and bioinformatic analyses, are providing major insights into how our genome works.

2.1 Protein-Coding Genes: Structure and Expression

Proteins are the main functional endpoints of gene expression. They perform a huge diversity of roles that govern how cells work, acting as structural components, enzymes, carrier proteins, ion channels, signaling molecules, gene regulators, and so on.

Protein-coding genes come in a startling variety of organizations and encode polypeptide chains—long sequences of amino acids, as described below. A newly synthesized polypeptide undergoes different maturation steps, usually involving chemical modification and cleavage events, and often then associates with other polypeptides to form a working protein.

Gene organization: exons and introns

The protein-coding genes of bacteria are small (on average about 1000 bp long) and simple. The gene is transcribed to give an mRNA with a continuous coding sequence that is then translated to give a linear sequence of about 300 amino acids on average. Unexpectedly, the genes of eukaryotes turned out to be much bigger and much more complex than anticipated. And, as we will see, our protein-coding genes often contain a rather small amount of coding DNA.

A major surprise was that the coding DNA of most eukaryotic protein-coding genes is split into segments called **exons** that are separated by noncoding DNA sequences called **introns**. The number of exons and introns varies considerably between genes and there seems to be little logic about precisely where introns insert within genes.

Excluding single-exon genes (that is, genes that lack introns), average exon lengths show moderate variation from gene to gene. However, another great surprise was to find that in complex genomes, such as the human genome, introns show extraordinary size differences; our genes are therefore often large, sometimes extending over more than a megabase of DNA (**Table 2.1**).

RNA splicing: stitching together the genetic information in exons

Like all genes, genes that are split into exons are initially transcribed by an RNA polymerase to give a long RNA transcript. This primary transcript is identical in base sequence to the transcribed region of the sense DNA strand, except that U replaces T (the transcribed region of DNA is called a **transcription unit**). Thereafter, the primary RNA transcript undergoes a form of processing called **RNA splicing**.

RNA splicing involves first cleaving the RNA transcript at the junctions between transcribed exons and introns. The individual transcribed intron sequences do not seem to have any useful function and are degraded, but the transcribed exon sequences are then covalently linked (spliced) in turn to make a mature RNA (**Figure 2.1**). RNA splicing is performed within the nucleus by spliceosomes, complex assemblies of protein factors and small nuclear RNA (snRNA) molecules.

We do not fully understand how the spliceosome is able to recognize and cut the primary transcript at precise positions marking the start and end of introns. However, we do know that certain sequences are important in signaling the splice sites that define exon–intron boundaries. For

Table 2.1 Examples of differential gene organization for human protein-coding genes. Items in brackets show the protein name, where this is different from the gene symbol. kb, kilobases (= 1000 bp).

HUMAN GENE	SIZE IN GENOME (kb)	NO. OF EXONS	AVERAGE SIZE OF EXON (bp)	AVERAGE SIZE OF INTRON (bp)
SRY	0.9	1	850	–
HBB (β-globin)	1.6	3	150	490
TP53 (p53)	39	10	236	3076
F8 (factor VIII)	186	26	375	7100
CFTR (cystic fibrosis transmembrane regulator)	250	27	227	9100
DMD (dystrophin)	2400	79	180	30,770

example, almost all introns begin with a GT dinucleotide on the sense DNA strand and end with an AG so that the transcribed intron sequence begins with a GU (that marks the **splice donor site**) and ends in an AG (marking the **splice acceptor site**). The GT (GU) and AG end sequences need to be embedded in broader splice site consensus sequences that we will describe in Section 6.1 when we consider how gene expression is regulated. As we will see in Chapter 7, mutations at splice sites are important causes of disease.

Figure 2.1 might give the erroneous impression that all protein-coding genes undergo a specific, single type of RNA splicing. However, close to 10% of our protein-coding genes do not undergo RNA splicing at all (and so have a single, uninterrupted exon)—notable examples include histone genes. And most of those genes that do undergo RNA splicing undergo alternative RNA splicing patterns; a single gene can therefore produce different gene products that may be functionally different. We consider the concept of alternative splicing in greater detail in Chapter 6, in the context of gene regulation.

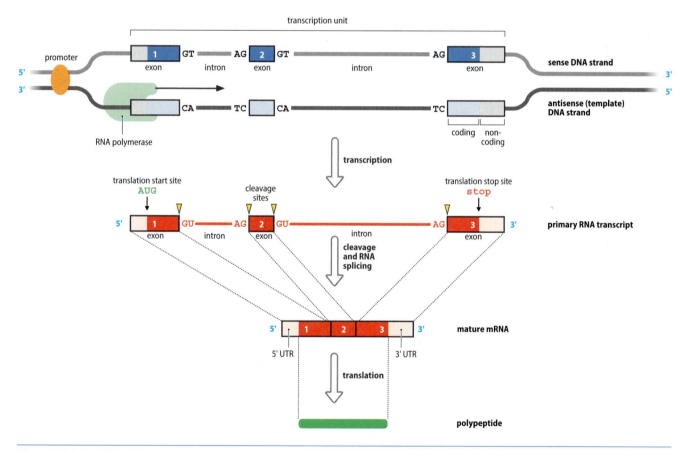

Figure 2.1 RNA splicing brings transcribed exon sequences together. Most of our protein-coding genes (and many RNA genes) undergo RNA splicing. In this generalized example a protein-coding gene is illustrated with an upstream promoter and three exons separated by two introns that each begin with the dinucleotide GT and end in the dinucleotide AG. The central exon (exon 2) is composed entirely of coding DNA, but exons 1 and 3 have noncoding DNA sequences (that will eventually be used to make untranslated sequences in the mRNA). The three exons and the two separating introns are transcribed together to give a large primary RNA transcript. The RNA transcript is cleaved at positions corresponding to exon–intron boundaries. The two transcribed intron sequences that are excised are each degraded, but the transcribed exon sequences are joined (*spliced*) together to form a contiguous mature RNA that has noncoding sequences at both the 5′ and 3′ ends. In the mature mRNA these terminal sequences will not be translated and so are known as *untranslated* regions (UTRs). The central coding sequence of the mRNA is defined by a translation start site (which is almost always the trinucleotide AUG) and a translation stop site, and is read (*translated*) to produce a polypeptide. Figures 2.3 and 2.4 provide a brief outline of translation.

The evolutionary value of RNA splicing

As we will see in Section 2.2, many RNA genes also undergo RNA splicing. At this stage, one might reasonably wonder why RNA splicing is so important in eukaryotic cells, and so especially prevalent in complex multicellular organisms. Why do we need to split the genetic information in genes into sometimes so many different little exons? The answer is to stimulate the formation of novel genes and novel gene products that can permit greater functional complexity during evolution.

The huge complexity of humans and other multicellular organisms has been driven by genome evolution. In addition to periodic gene duplication, various genetic mechanisms allow individual exons to be duplicated or swapped from one gene to another on an evolutionary timescale. That allows different ways of combining exons to produce novel hybrid genes. An additional source of complexity comes from using different combinations of exons to make alternative transcripts from the same gene (alternative splicing).

Translation: decoding messenger RNA to make a polypeptide

Messenger RNA (mRNA) molecules produced by RNA splicing in the nucleus are exported to the cytoplasm. Here they are bound by ribosomes, very large complexes consisting of four types of ribosomal RNA (rRNA) and many different proteins.

Although an mRNA is formed from exons only, it has sequences at its 5′ and 3′ ends that are noncoding. Having bound to mRNA, the job of the ribosomes is to scan the mRNA sequence to find and interpret a central coding sequence that will be translated to make a polypeptide. The noncoding sequences at the ends will not be interpreted; they are known as untranslated regions (UTRs; see Figure 2.1).

A polypeptide is a polymer made up of a linear sequence of **amino acids** (**Figure 2.2A**). Amino acids have the general formula NH_2-CH(R)-COOH, where R is a variable side chain that defines the chemical identity of the amino acid. There are 20 common amino acids (Figure 2.2C). Polypeptides are made by a condensation reaction between the carboxyl (COOH) group of one amino acid and the amino (NH_2) group of another amino acid, forming a peptide bond (see Figure 2.2B).

To make a polypeptide, the coding sequence within an mRNA is translated in groups of three nucleotides at a time, called **codons**. There are 64 possible codons (four possible bases at each of three nucleotide positions makes $4 \times 4 \times 4$ permutations). Of these, 61 are used to specify an amino acid; three others signal an end to protein synthesis. The universal **genetic code**, the set of rules that dictate how codons are interpreted, therefore has some redundancy built into it. For example, the amino acid serine can be specified by any of six codons (UCA, UCC, UCG, UCU, AGU, and AGG), and, on average, an amino acid is specified by any of three codons. As a result, nucleotide substitutions within coding DNA quite often do not cause a change of amino acid. We discuss the genetic code in some detail in Section 7.1 when we consider the effects of single nucleotide substitutions.

The process of translation

Translation begins when ribosomes bind to the 5′ end of an mRNA and then move along the RNA to find a translational start site, the initiation codon—an AUG trinucleotide embedded within the broader, less well defined Kozak consensus sequence (GCC**Pu**CC**AUGG**; the most conserved bases are shown in bold, and Pu represents purine).

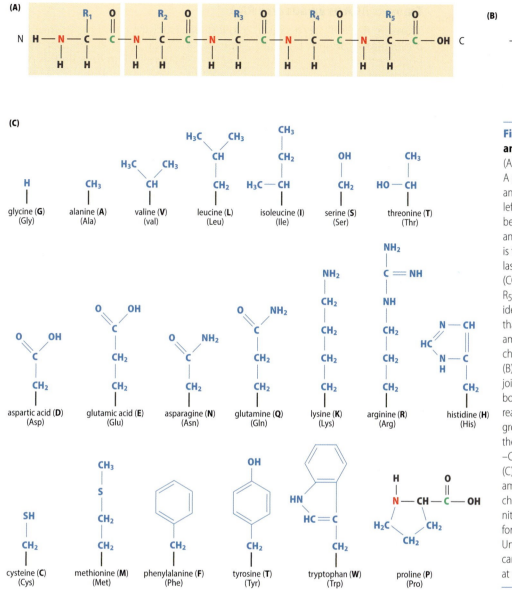

Figure 2.2 Polypeptide and amino acid structure.
(A) Polypeptide primary structure. A pentapeptide is shown with its five amino acids highlighted. Here, the left end is called the N-terminal end because the amino acid has a free amino (NH_2) group; the right end is the C-terminal end because the last amino acid has a free carboxyl (COOH) group. The side chains (R_1 to R_5) are variable and determine the identity of the amino acid. Note that at physiological pH the free amino and carboxyl groups will be charged: NH_3^+ and COO^- respectively. (B) Neighboring amino acids are joined by a peptide bond. A peptide bond is formed by a condensation reaction between the end carboxyl group of one amino acid and the end amino group of another: $-COOH + NH_2- \rightarrow -CONH- + H_2O$. (C) Side chains of the 20 principal amino acids. Note that the side chain of proline connects to the nitrogen atom of the amino group, forming a five-membered ring. Unlike as shown here, free amino and carboxyl groups would be charged at physiological pH.

The initiation codon is the start of an **open reading frame** of codons that specify successive amino acids in the polypeptide chain (see **Box 2.1** for the concept of translational reading frames). As described below, a family of transfer RNAs (tRNAs) is responsible for transporting the correct amino acids in sequence. Individual types of tRNA carry a specific amino acid; they can recognize and bind to a specific codon, and when they do so they unload their amino acid cargo.

As each new amino acid is unloaded it is bonded to the previous amino acid so that a polypeptide chain is formed (**Figure 2.3**). The first amino acid has a free NH_2 (amino) group and marks the N-terminal end (N) of the polypeptide. The polypeptide chain terminates after the ribosome encounters a **stop codon** (which signifies that the ribosome should disengage from the mRNA, releasing the polypeptide; for translation on cytoplasmic ribosomes, there are three choices of stop codon: UAA, UAG, or UGA). The last amino acid that was incorporated in the polypeptide chain has a free COOH (carboxyl group) and marks the C-terminal end (C) of the polypeptide.

Transfer RNA as an adaptor RNA

Transfer RNAs have a classic cloverleaf structure resulting from intra-molecular hydrogen bonding (**Figure 2.4A**). They serve as adaptor RNAs because their job is to base pair with mRNAs and help decode the coding sequence messages carried by mRNAs. The base pairing is confined to a three-nucleotide sequence in the tRNA called an anticodon, which is complementary in sequence to a codon. According to the identity of their anticodons, different tRNAs carry different amino acids covalently linked to their 3′ ends. Through base pairing between codon and anti-codon, individual amino acids can be sequentially ordered according to the sequence of codons in an mRNA, and sequentially linked together to form a polypeptide chain (see Figure 2.4B).

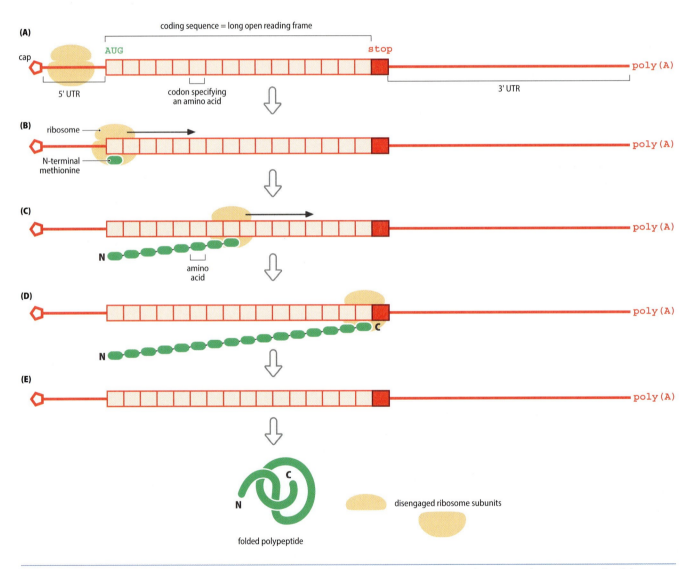

Figure 2.3 The basics of translation. (A) A ribosome attaches to the 5′ untranslated region (5′ UTR) of the mRNA and then slides along until (B) it encounters the initiation codon AUG, at which point a methionine-bearing transfer RNA (not shown) engages with the AUG codon and deposits its methionine cargo (green bar). (C) The ribosome continues to moves along the mRNA and as it encounters each codon in turn a specific amino-acid-bearing tRNA is recruited to recognize the codon and to deposit its amino acid, according to the *genetic code*. The ribosome catalyses the formation of a *peptide bond* (Figure 2.2B) between each new amino acid and the last amino acid, forming a polypeptide chain (shown here, for convenience, as a series of joined green bars). (D) Finally, the ribosome encounters a stop codon, at which point (E) the ribosome falls off the mRNA and dissociates into its two subunits, releasing the completed polypeptide. The polypeptide undergoes *post-translational modification* as described in the text.

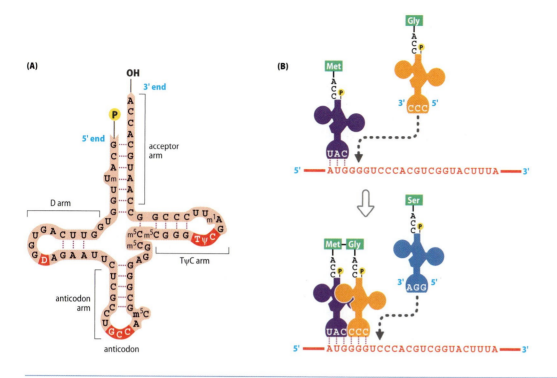

Figure 2.4 Transfer RNA structure, and its role as an adaptor RNA in translation. (A) *Transfer RNA structure*. The tRNA^{Gly} shown here illustrates the classical cloverleaf tRNA structure. Intramolecular base pairing produces three arms terminating in a loop plus the acceptor arm that binds amino acids by covalent bonding to the 3′ end. The three nucleotides at the centre of the middle loop form the *anticodon*, which identifies the tRNA according to the amino acid it will bear. Minor nucleotides are: D, 5,6-dihydrouridine; ψ, pseudouridine (5-ribosyluracil); m5C, 5-methylcytidine; m1A, 1-methyladenosine; Um, 2′-O-methyluridine. (B) *Role of adaptor RNA*. Different tRNAs carry different amino acids, according to the type of anticodon they bear. As a ribosome traverses an mRNA it identifies the AUG initation codon and a tRNA bearing methionine; the complementary anticodon sequence 5′ CAU 3′ engages with the ribosome so that the CAU anticodon base pairs with the AUG codon—for ease of illustration we show the tRNAs in the opposite left–right orientation to the standard form shown in (A). Thereafter, a tRNA bearing glycine engages the second codon, GGG, by base pairing with its CCC anticodon. The ribosome's peptidyltransferase then forms a peptide bond between the N-terminal methionine and glycine. The ribosome moves along by one codon and the tRNA^{Met} is cleaved so that it can be reused and the process continues with an incoming tRNA carrying a serine and an anticodon GGA to bind to the third codon UCC, after which the incoming serine will be covalently bonded to the glycine by the ribosome's peptidyltransferase.

Untranslated regions and 5′ cap and 3′ poly(A) termini

As illustrated in Figure 2.3, each mature mRNA has a large central coding DNA sequence flanked by two **untranslated regions**, a short 5′ untranslated region (5′ UTR) and a rather longer 3′ untranslated region (3′ UTR). The untranslated regions regulate mRNA stability and, as described in Chapter 6, they also contain regulatory sequences that are important in determining how genes are expressed.

As well as having sequences copied from the gene sequence, mRNA molecules usually also have end sequences that are added post-transcriptionally to the pre-mRNA. At the 5′ end a specialized cap is added: 7-methylguanosine is linked by a distinctive 5′–5′ phosphodiester bond (instead of the normal 5′–3′ phosphodiester bond) to the first nucleotide. The cap protects the transcripts against 5′→3′ exonuclease attack, and facilitates transport to the cytoplasm and ribosome attachment. At the 3′ end a dedicated poly(A) polymerase sequentially adds adenylate (AMP) residues to give a poly(A) tail, about 150–200 nucleotides long. The poly(A) helps in transporting mRNA to the cytoplasm, facilitates binding to ribosomes, and is also important in stabilizing mRNAs.

BOX 2.1 Translational Reading Frames and Splitting of Coding Sequences by Introns.

Translational reading frames

In the examples of different translational **reading frames** below, we use sequences of words containing three letters to represent the triplet nature of the genetic code. Here, the reading frames (RF) are designated 1, 2, or 3 depending on whether the reading frame starts before the first, second, or third nucleotide in the sequence.

As can be seen from **Figure 1**, reading frame 1 (RF1) makes sense, but making a shift to reading frames 2 or 3 produces nonsense. The same principle generally applies to coding sequences. So, for example, if one or two nucleotides are deleted from a coding sequence the effect is to produce a **frameshift** (a change of reading frame) that will result in nonsense.

Splitting of coding sequences by introns

At the DNA level, introns may interrupt a coding sequence at one of three types of position: at a point precisely between two codons, between the first and second nucleotides of a codon, or between the second and third nucleotides of a codon.

An internal exon may be flanked by introns that insert at the same type of position; in an exon like this (sometimes called a 'symmetric' exon) the number of nucleotides is always exactly divisible by three. An internal 'asymmetric' exon has a total number of nucleotides that is not exactly divisible by three and so is flanked by introns that insert at different types of positions.

In **Figure 2A** we imagine that the same sequence is split by two introns, 1 and 2. Two alternatives are

sequence: THEOLDMANGOTOFFTHEBUSANDSAWTHEBIGREDDOGANDHERPUP

RF1: THE OLD MAN GOT OFF THE BUS AND SAW THE BIG RED DOG AND HER PUP

RF2: T HEO LDM ANG OTO FFT HEB USA NDS AWT HEB IGR EDD OGA NDH ERP UP

RF3: TH EOL DMA NGO TOF FTH EBU SAN DSA WTH EBI GRE DDO GAN DHE RPU P

Figure 1 The importance of using the correct translational reading frame. We use three-letter words to illustrate how meaning can be altered by changing the translational reading frame. The sequence of letters at the top can be grouped into sets of three (codons) that make sense in reading frame 1 (RF1) but make no sense when using reading frame 2 (RF2) or reading frame 3 (RF3).

From newly synthesized polypeptide to mature protein

The journey from newly synthesized polypeptide released from the ribosome to fully mature protein requires several steps. The polypeptide typically undergoes post-translational cleavage and chemical modification. Polypeptides also need to fold properly, and they often bind to other polypeptides as part of a multisubunit protein. And then there is a need to be transported to the correct intracellular or extracellular location.

Chemical modification

We describe below one type of chemical modification that involves crosslinking between two cysteine residues within the same polypeptide or on different polypeptides. Usually, however, chemical modification involves the simple covalent addition of chemical groups to polypeptides or proteins. Sometimes small chemical groups are attached to the side chains of specific amino acids (**Table 2.2**). These groups can sometimes be particularly important in the structure of a protein (as in the case of collagens, which have high levels of hydroxyproline and hydroxylysine).

In other cases, dedicated enzymes add or remove small chemical groups to act as switches that convert a protein from one functional state to another. Thus specific kinases can add a phosphate group that can be

BOX 2.1 (continued)

shown in green and orange. The green introns both interrupt the coding sequence between codons and so the second exon in this case is symmetric (Figure 2B). However, the two orange introns interrupt codons at different positions, between nucleotides 1 and 2 (intron 1) and between nucleotides 2 and 3 (intron 2). In this case the second exon is asymmetric (Figure 2C).

Mutations that cause an internal exon to be deleted or not included in splicing have less effect in the case of symmetric exons. Loss of the symmetric exon 2 in Figure 2B leads to the loss of five words (codons), but the sense is maintained in exon 3. Loss of asymmetric exon 2 in Figure 2C leads to the loss of the words in exon 2 but also produces a frameshift that causes scrambling of the words in exon 3.

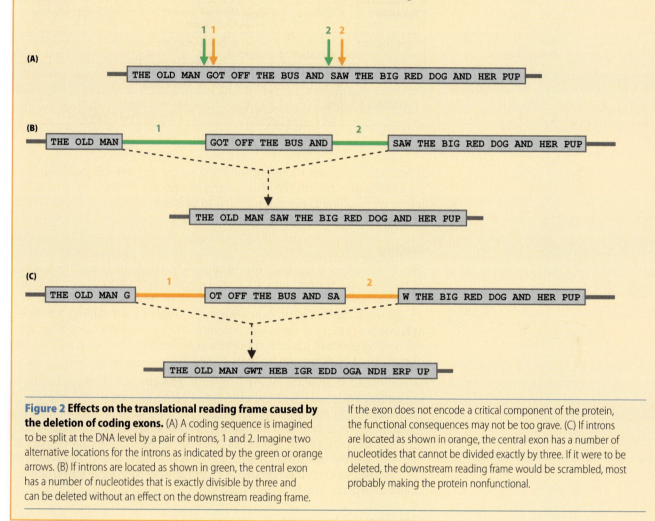

Figure 2 Effects on the translational reading frame caused by the deletion of coding exons. (A) A coding sequence is imagined to be split at the DNA level by a pair of introns, 1 and 2. Imagine two alternative locations for the introns as indicated by the green or orange arrows. (B) If introns are located as shown in green, the central exon has a number of nucleotides that is exactly divisible by three and can be deleted without an effect on the downstream reading frame. If the exon does not encode a critical component of the protein, the functional consequences may not be too grave. (C) If introns are located as shown in orange, the central exon has a number of nucleotides that cannot be divided exactly by three. If it were to be deleted, the downstream reading frame would be scrambled, most probably making the protein nonfunctional.

removed by a dedicated phosphatase. The change between phosphorylated and dephosphorylated states can result in a major conformational change that affects how the protein functions. Similarly, methyltransferases and acetyltransferases add methyl or acetyl groups that can be removed by the respective demethylases and deacetylases. As we will see in Chapter 6, they are particularly important in modifying histone proteins to change the conformation of chromatin and thereby alter gene expression.

In yet other cases, proteins can be modified by attaching complex carbohydrates or lipids to a polypeptide backbone. Thus, for example, proteins that are secreted from cells or that are destined to be part of the excretory

Table 2.2 Common types of chemical modification of proteins by covalent addition of chemical groups to a side chain.

TYPE OF CHEMICAL MODIFICATION	TARGET AMINO ACIDS	COMMENTS
Addition of small chemical group		
Hydroxylation	Pro; Lys; Asp	can play important structural roles
Carboxylation	Glu	especially in some blood clotting factors
Methylation	Lys	specialized enzymes can add or remove the methyl, acetyl, or phosphate group, causing the protein to switch between two states, with functional consequences
Acetylation	Lys	
Phosphorylation	Tyr; Ser; Thr	
Addition of complex carbohydrate or lipid group		
N-glycosylation	Asn	added to the amino group of Asn in endoplasmic reticulum and Golgi apparatus
O-glycosylation	Ser; Thr; Hydroxylysine	added to the side-chain hydroxyl group; takes place in Golgi apparatus
N-lipidation	Gly	added to the amino group of an N-terminal glycine; promotes protein–membrane interactions
S-lipidation	Cys	a palmitoyl or prenyl group is added to the thiol of the cysteine. Often helps anchor proteins in a membrane

process of cells are routinely modified so as to have oligosaccharides covalently attached to the side chains of specific amino acids. And different types of lipids are often added to proteins that function within the lipid bilayer of a membrane (see Table 2.2).

Folding

The amino acid sequence, the primary structure, dictates the pattern of folding, and specific regions of polypeptides can adopt specific types of secondary structure that are important in how a protein folds (**Box 2.2** gives an outline of protein structure). Until correct folding has been achieved, a protein is unstable; different chaperone molecules help with the folding process (careful supervision is needed because partly folded or misfolded proteins can be toxic to cells).

When placed in an aqueous environment, proteins are stabilized by having amino acids with hydrophobic side chains located in the interior of the protein, whereas hydrophilic amino acids tend to be located toward the surface. For many proteins, notably globular proteins, the folding pattern is also stabilized by a form of covalent cross-linking that can occur between certain distantly located cysteine residues—the thiol groups of the cysteine side chains interact to form a disulfide bond (alternatively called a disulfide bridge—see **Figure 2.5**).

Cleavage and transport

The initial polypeptide normally undergoes some type of N-terminal cleavage (additional cleavage can occur but is rare). The N-terminal cleavage can sometimes mean removing just the N-terminal methionine (the AUG initiation codon specifies methionine, but the amino acid sequence of mature polypeptides rarely begins with methionine).

In other cases N-terminal cleavage can be more extensive. For proteins that are secreted from cells, the polypeptide precursor carries an N-terminal leader sequence (signal peptide) that is required to assist the protein to cross the plasma membrane, after which the signal peptide is cleaved at the membrane, releasing the mature protein. (The signal peptide, which is often 10–30 amino acids in length, carries multiple hydrophobic amino acids.)

BOX 2.2 A Brief Outline of Protein Structure.

Four different levels of structural organization have been defined for proteins, as listed below.

- primary structure—the linear sequence of amino acids in constituent polypeptides

- secondary structure—the path that a polypeptide backbone follows within local regions of the primary structure; common elements of secondary structure include the α-helix, the β-sheet, and the β-turn (see below)

- tertiary structure—the overall three-dimensional structure of a polypeptide, arising from the combination of all of the secondary structures

- quaternary structure—the aggregate structure of a multimeric protein (composed of two or more polypeptide subunits that may be of more than one type)

Elements of secondary structure

The *α-helix* is a rigid cylinder that is stabilized by hydrogen bonding between the carbonyl oxygen of a peptide bond and the hydrogen atom of the amino nitrogen of a peptide bond located four amino acids away (**Figure 1A**). α-Helices often occur in proteins that perform key cellular functions (for example in transcription factors, where they are usually represented in the DNA-binding domains).

Like the α-helix, the *β-sheet* (also called the *β-pleated sheet*) is stabilized by hydrogen bonding. In this case, the hydrogen bonds occur between opposed peptide bonds in parallel or antiparallel segments of the same polypeptide chain (Figure 1B). β-Sheets occur—often together with α-helices—at the core of most globular proteins.

The *β-turn* involves hydrogen bonding between the peptide-bond carbonyl (C=O) of one amino acid and the peptide-bond NH group of an amino acid located only three places farther along. The result is a hairpin turn that allows an abrupt change in the direction of a polypeptide, enabling compact globular shapes to be achieved. β-Turns can connect neigboring segments in a β-sheet, when the polypeptide strand has to undergo a sharp turn.

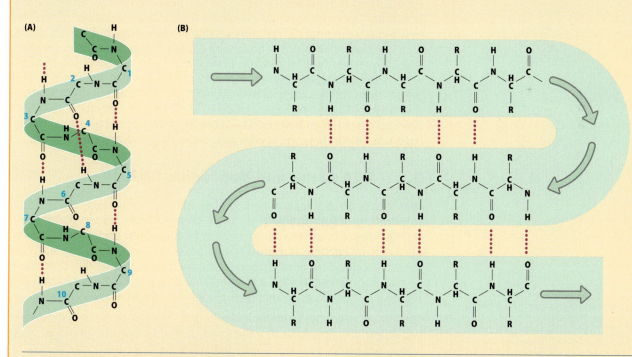

(A) (B)

Figure 1 Elements of protein secondary structure. (A) An α-helix. The solid rod structure is stabilized by hydrogen bonding between the oxygen of the carbonyl group (C=O) of each peptide bond and the hydrogen on the peptide bond amide group (NH) of the fourth amino acid away, yielding 3.6 amino acids per turn of the helix. The side chains of each amino acid are located on the outside of the helix; there is almost no free space within the helix. Note: only the backbone of the polypeptide is shown, and some bonds have been omitted for clarity. (B) A β-sheet (also called a β-pleated sheet). Here, hydrogen bonding occurs between the carbonyl oxygens and amide hydrogens on adjacent segments of a sheet that may be composed either of parallel segments of the polypeptide chain or, as shown here, of antiparallel segments (arrows mark the direction of travel from N-terminus to C-terminus).

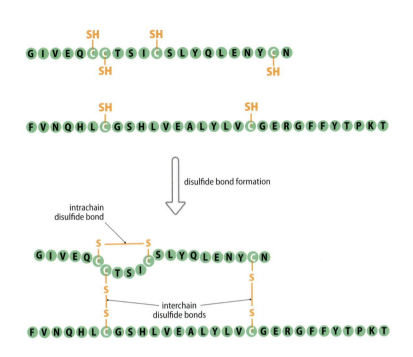

Figure 2.5 Intrachain and interchain disulfide bridges in human insulin. Human insulin is composed of two peptide chains, an A chain with 21 amino acids, and a B chain with 30 amino acids. Disulfide bridges (–S–S–) form by a condensation reaction between the sulfhydryl (–SH) groups on the side chains of cysteine residues. They form between the side chains of cysteines at positions 6 and 11 within the insulin A chain, and also between cysteine side chains on the insulin A and B chains. Note that here all the cysteines participate in disulfide bonding, which is unusual. When disulfide bonding occurs in large proteins, only certain cysteine residues are involved.

Other short internal peptide sequences can act simply as address labels for transporting proteins to the nucleus, mitochondria, plasma membrane, and so on. They are retained in the mature protein.

Binding of multiple polypeptide chains

Proteins are often made of two or more polypeptide subunits. Occasionally, constituent polypeptides are covalently linked with disulfide bridges (as in the case of joining the different chains of immunoglobulins; see Figure 4.13 on page 105). Often, however, constituent polypeptides are held together mainly by noncovalent bonds, including nonpolar interactions and hydrogen bonds. For example, hemoglobins are tetramers, composed of two copies each of two different globin chains that associate in this way. Collagens provide a good example of very intimate structural association between polypeptides, consisting of three chains (two of one type; one of another) wrapped round each other to form a triple helix.

2.2 RNA Genes and Noncoding RNA

RNA genes do not make polypeptides; instead, they make a functional **noncoding RNA (ncRNA)** as their end product. Like proteins (and mRNA), ncRNAs are made as precursors that often undergo enzymatic cleavage to become mature gene expression products. They are also subject to chemical modification: minority bases such as dihydrouridine or pseudouridine and various methylated bases are quite common—see Figure 2.4A for some examples of modified bases in a tRNA.

Until quite recently, ncRNAs were largely viewed as having important but rather dull functions. For the most part, they seemed to act as ubiquitous accessory molecules that worked directly or indirectly in protein production. After ribosomal and transfer RNAs, we came to know about various other ubiquitous ncRNAs that mostly work in RNA maturation: spliceosomal small nuclear RNAs (snRNAs); small nucleolar RNAs (snoRNAs) that chemically modify specific bases in rRNA; small Cajal-body RNAs (scaRNAs) that chemically modify spliceosomal snRNA; and certain RNA enzymes (ribozymes) that cleave tRNA and rRNA precursors. All of these types of RNA can be viewed as general accessory molecules that, like rRNA and tRNA, are ultimately engaged in supporting protein synthesis. In stark contrast to RNA, proteins were viewed as the functionally important endpoints of genetic information, the exciting pace-setters that performed myriad roles in cells.

The view that RNAs are mostly ubiquitous accessory molecules that assist protein synthesis in a general way is no longer tenable. Over the past decade we have become progressively more aware of the functional diversity of ncRNA and of the many thousands of ncRNA genes in our genome. Multiple new classes of regulatory RNAs have very recently been discovered that are expressed only in certain cell types or at certain stages of development. Working out what they do has become an exciting area of research.

With hindsight, perhaps we should not be so surprised at the functional diversity of RNA. DNA is simply a self-replicating repository of genetic information, but RNA can serve this function (in the case of RNA viruses) and can also have catalytic functions. The 'RNA world' hypothesis views RNA as the original genetic material and that RNA performed executive functions before DNA and proteins developed. That is possible because, unlike naked double-stranded DNA (which has a comparatively rigid structure), single-stranded RNA has a very flexible structure and can form complex shapes by intramolcular hydrogen bonding, as described below. As will be described in later chapters, the relatively recent understanding of just what RNA does in cells and how it can be manipulated is driving some important advances in medicine. Mutations in ncRNA genes are now known to underlie some genetic disorders and cancers, and RNA therapeutics offers important new approaches to treating disease.

The extraordinary secondary structure and versatility of RNA

The primary structure of nucleic acids and proteins is the sequence of nucleotides or amino acids, which defines their identity; however, higher levels of structure determine how they work in cells. Single-stranded RNA molecules are much more flexible than naked double-stranded DNA, and like proteins they have a very high degree of secondary structure where intramolecular hydrogen bonding causes local alterations in structure.

The secondary structure of single-stranded RNA depends on base pairing between complementary sequences on the same RNA strand. Intervening sequences that do not engage in base pairing will loop out, producing stem-loop structures (called hairpins when the loop is short). Extraordinarily intricate patterns can develop—see **Figure 2.6** for an example. Higher-level structures can form when, for example, a sequence within the stem of one loop base pairs with another sequence. Note that base pairing in RNA includes G–U base pairs as well as more stable A–U and G–C base pairs.

Stem-loop structures in RNA have different functions. As described in Chapter 6, they can serve as recognition elements for binding regulatory proteins, and they are crucially important in determining the overall structure of an RNA that can be important for function.

In general, because of the flexible structure of single-stranded RNA, different RNAs can adopt different shapes according to the base sequence; this enables them to do different jobs, such as working as enzymes. Many different classes of RNA enzyme (ribozyme) are known in nature, and some originated very early in evolution. For example, the catalytic activity of the ribosome (the peptidytransferase responsible for adding amino acids to the growing polypeptide chain) is due solely to the large RNA of the large subunit (28S rRNA). In recent years RNAs have been found to work in a large variety of roles (**Figure 2.7**).

Figure 2.6 Highly developed secondary structure in RNA. Unlike DNA, which has a stiff rod-like structure, cellular RNAs are very flexible. They have a very high degree of secondary structure caused by intrachain hydrogen bonding that can produce a very highly folded structure, evident in the ribosomal RNA shown in (A); the 5′ and 3′ ends of the RNA are highlighted for clarity. The basic motifs are *stem-loop structures* (B), which are formed when the RNA folds back on itself so that two short regions can base pair to form the stem while intervening sequences loop out. Note that G–U base pairs form in RNA, in addition to G–C and A–U base pairs.

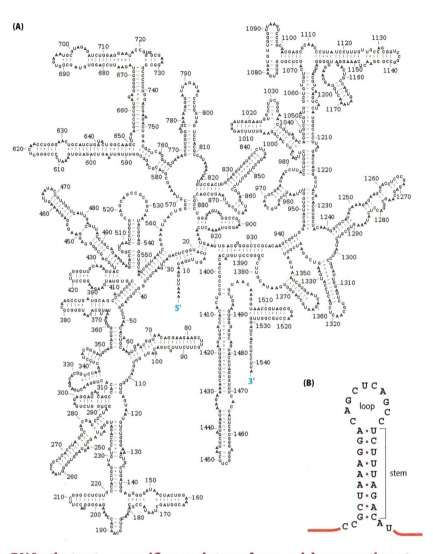

RNAs that act as specific regulators: from quirky exceptions to the mainstream

The first examples of more specific regulatory RNAs were discovered more than 20 years ago, and for a long time they were considered to be interesting but *exceptional* cases. They included RNAs that work in epigenetic regulation to produce monoallelic gene expression, where it is *normal* that only one of the two parental alleles is expressed. This unusual situation applies in the case of certain genes that are expressed or silenced according to whether they are paternally or maternally inherited, and to genes on one of the two X chromosomes in the cells of female mammals (X-chromosome inactivation). We describe the underlying mechanisms in Chapter 6.

The Human Genome Project had been crucially important in identifying many novel protein-coding genes, but not so effective at identifying RNA genes. However, subsequent studies of genomewide transcription plus cross-organism genome comparisons and bioinformatics analyses revealed many thousands of novel RNA genes. Many of these genes make regulatory ncRNAs that are expressed in certain cell types only, including large families of long ncRNAs, tiny ncRNAs, and circular RNAs.

Long ncRNAs

Long regulatory ncRNAs come in two broad classes. **Antisense RNAs** are transcribed using the *sense* strand of a gene as a template and are

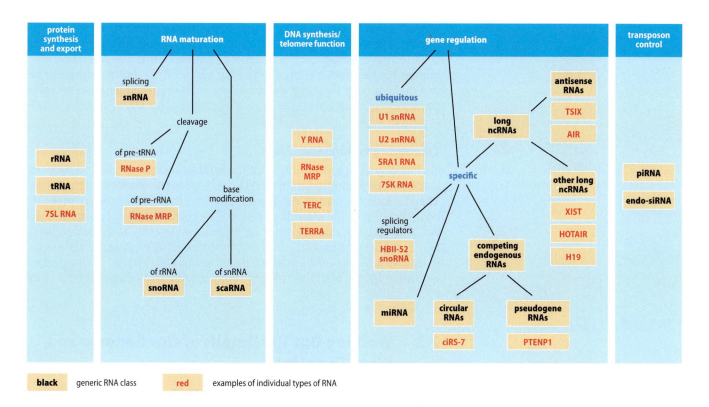

| **black** | generic RNA class | **red** | examples of individual types of RNA |

Figure 2.7 The versatility of ncRNA. The two panels on the left show ubiquitously expressed RNAs that are important in generally assisting protein production and export, including RNA families that supervise the maturation of other RNAs: snRNA, small nuclear RNA; snoRNA, small nucleolar RNA; scaRNA, small Cajal body RNA. The central panel includes RNAs involved in DNA replication (the ribozyme RNase MRP has a crucial role in initiating mtDNA replication, as well as in cleaving pre-rRNA), and developmentally regulated telomere regulators (TERC is the RNA component of telomerase; TERRA is telomere RNA). Diverse classes of RNA regulate gene expression. In addition to ubiquitous RNAs, such as the listed RNAs that have general roles in transcription, many classes of RNA regulate *specific* target genes and are typically restricted in expression. They work at different levels: transcription (such as antisense RNAs), splicing, and translation (notably miRNAs, which bind to certain regulatory sequences in the untranslated regions of target mRNAs). Some RNAs, notably the highly prevalent class of circular RNAs, regulate the interaction between miRNAs and their targets. piRNAs, and to a smaller extent endogenous short interfering RNAs (endo-siRNA), are responsible for silencing transposable elements in germ line cells. We describe how RNAs regulate gene expression in detail in Chapter 6.

not subject to cleavage and RNA splicing. As a result they can be quite large, often many thousands of nucleotides long. They work by binding to the complementary sense RNA produced from the gene, down-regulating gene expression.

A second class of long regulatory RNAs are formed from primary transcripts that are typically processed like the primary transcripts of protein-coding genes (and so usually undergo RNA splicing). Many of these RNAs regulate neighboring genes, but some control the expression of genes on other chromosomes. We consider the details of how they work in Chapter 6.

Tiny ncRNAs

Thousands of tiny ncRNAs (less than 35 nucleotides long) also work in human cells. They include many cytoplasmic microRNAs (miRNAs) that are usually 20–22 nucleotides long and are expressed in defined cell types or at specific stages of early development. As described in Chapter 6, miRNAs bind to defined regulatory sequences in mRNAs, and each miRNA specifically regulates the expression of a collection of defined target genes. They are important in a wide variety of different cellular processes.

Human germ cells also make many thousands of different 26–32-nucleo-tide Piwi protein-interacting RNAs (piRNAs). The piRNAs work in germ cells to damp down excess activity of **transposons** (mobile DNA elements; see Section 2.4). Active mobile elements in the human genome can make a copy that migrates to a new location in our genome and can be harmful (by disrupting genes or inappropriately activating some types of cancer gene).

Competing endogenous RNAs

Certain regulatory RNAs that contain multiple miRNA-binding sequences work by competing with an mRNA for access to its miRNA regulators. They include RNA transcripts made by some defective copies of protein-coding genes, called **pseudogenes**, such as the *PTENP1* pseudogene—we describe pseudogenes in more detail below. Recently identified circular RNAs, a highly abundant class of conserved RNA regulators, also work in this way. Most of the circular RNAs overlap with protein-coding genes and arise by circularized splicing. We describe the details of how they regulate genes in Chapter 6.

2.3 Working Out the Details of Our Genome and What They Mean

The human genome consists of 25 different DNA molecules partitioned between what are effectively two genomes, one in the nucleus and one in the mitochondria. In the nucleus there are either 23 or 24 different linear DNA molecules (one each for the different types of chromosome: 23 in female cells or 24 in male cells). These DNA molecules are immensely long (ranging in size from 48 Mb to 249 Mb; see **Table 2.3**). In the mitochondria there is just one type of DNA molecule: a circular DNA molecule that is comparatively tiny (only 16.6 kilobases (kb) long, or just a bit more than about 1/10,000 of the size of an average nuclear DNA molecule). Unlike the nuclear DNA molecules (which are each present in only two copies in diploid cells), there are many mitochondrial DNA copies in a cell.

In what was a heroic effort at the time, the mitochondrial DNA (often called the mitochondrial genome) was sequenced by a single research team in Cambridge, UK, as far back as 1981. Despite its small size, it

Table 2.3 DNA content of human chromosomes. Chromosome sizes are taken from the ENSEMBL Human Map View at http://www.ensembl.org/Homo_sapiens/Location/Genome (click on individual chromosome ideograms to obtain information on DNA content, for example). Although numbering of the autosomes was historically done on the basis of perceived size, there are some anomalies. For example, chromosome 21 is the smallest chromosome by DNA content, not chromosome 22.

CHROMOSOME	TOTAL DNA (Mb)	CHROMOSOME	TOTAL DNA (Mb)
1	249	13	115
2	243	14	107
3	198	15	103
4	191	16	90
5	181	17	81
6	171	18	78
7	159	19	59
8	146	20	63
9	141	21	48
10	136	22	51
11	135	X	155
12	134	Y	59

is packed with genes. The complexity of the nuclear genome—roughly 200,000 times the size of the mitochondrial genome—posed a much more difficult challenge. That would require an international collaboration between many research teams, as described in the next section.

Working out the nucleotide sequence was only the first step. The next challenge, which is still continuing, is to work out the details of how our genome functions and what all the component sequences do. That has begun with a combination of comparative genome analyses, functional assays and bioinformatics analyses.

The Human Genome Project: working out the details of the nuclear genome

For decades, the only available map of the nuclear genome was a low-resolution physical map that was based on chromosome banding. Chromosomes can be stained with certain dyes, such as Giemsa, to reveal an alternating pattern of dark and light bands for each chromosome, as represented by the image shown in **Figure 2.8**.

We describe the methodology and terminology of human chromosome banding in Box 7.4. For now, there are two salient points to note. First, the alternating pattern of bands reflects different staining intensities. That in turn reflects differences in chromatin organization along chromosomes (as a result of differences in base composition), and differences in gene and exon density (see the legend to Figure 2.8). Secondly, the resolution of the map is low—in even a high-resolution chromosome map the average size of a band is several megabases of DNA. What was needed was a map with a 1 bp resolution, a DNA sequence map.

Obtaining the complete DNA sequence of the human nuclear genome was a long-drawn-out process that was driven by an international collaborative effort between many research teams, the Human Genome Project. After it was realized that many regions of our DNA have highly variable DNA sequences it was possible to define multiple polymorphic DNA markers and map them to individual chromosomes. Panels of mapped polymorphic markers were then used to build a genetic map by looking at their segregation in defined reference families. The genetic map was then used as a scaffold to map short DNA sequences to chromosomal regions; they were in turn used to fish out clones with much longer DNA sequences that could be allocated to defined positions on each chromosome.

Ultimately, for each chromosome, a map was constructed based on many DNA clones with long inserts that could be ordered as a series of clones with partly overlapping DNA sequences. Finally, the DNAs from selected clones were sequenced and used to build chromosome-wide DNA sequences.

Not all regions of DNA were sequenced: a low priority was given to the DNA of **heterochromatin**, the regions of the genome where the chromatin is highly condensed throughout the cell cycle, which were believed to be transcriptionally inactive (these regions account for close to 7% of our genome; see Figure 2.8 for the distribution of heterochromatin across the human chromosomes). Heterochromatin was believed to have very few genes and the underlying DNA is technically difficult to sequence (it largely consists of highly repetitive noncoding tandem repeats).

Instead, the focus was on sequencing the **euchromatin** component of our genome, which was known to be transcriptionally active and suspected to contain virtually all of our genes. A draft sequence for the human nuclear euchromatin genome was published in 2001 (after collation of all the different chromosome DNA sequences). Thereafter, essentially complete sequences of the euchromatin region of all 24 nuclear DNA molecules were obtained and published by 2003–2004. The DNA of the euchromatin component was found to represent about 93% of the nuclear genome. Subsequently, significant components of heterochromatin DNA have been sequenced.

Interrogating our genome

To manage the voluminous and ever-increasing information about our genome, many different databases and interactive genome browsers have been developed. The US National Center for Biotechnology Information (NCBI) provides a useful web section on Human Genome Resources. More detailed information can be found using the ENSEMBL browser, a collaboration between the European Bioinformatics Institute and the Wellcome Trust Sanger Institute, and at the University of California at Santa Cruz (UCSC) Genome Bioinformatics site (Box 2.3).

What the sequence didn't tell us

With hindsight, we can appreciate that obtaining the human genome sequence was the easy part. The hard part was to work out what the

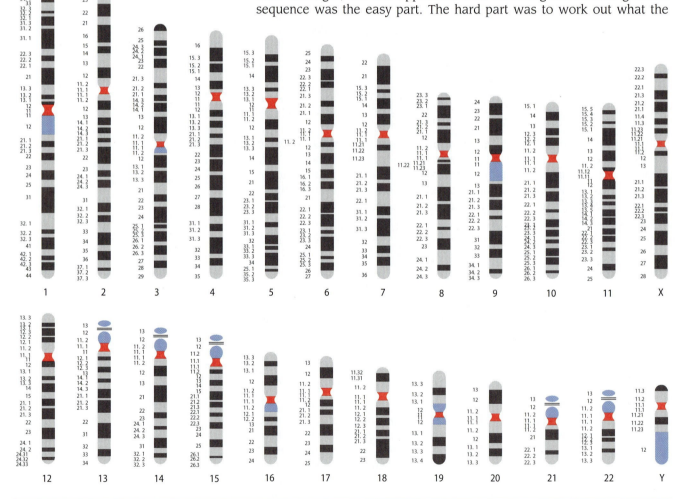

Figure 2.8 Ideogram showing a 550-band Giemsa banding pattern and constitutive heterochromatin within human metaphase chromosomes. Dark bands represent DNA regions where there is a low density of G–C base pairs and a generally low density of exons and genes. Pale bands represent DNA regions where there is a high density of G–C base pairs and a generally high density of exons and genes. Centromeric heterochromatin is illustrated by red blocks, and other constitutive heterochromatin is shown as blue blocks. Note the large amounts of non-centromeric heterochromatin on the Y chromosome, the short arms of the acrocentric chromosomes (13, 14, 15, 21, and 22), and on chromosomes 1, 9, and 16. Numbers to the left are the numbers of individual chromosome bands; for the nomenclature of chromosome banding, see Box 7.4.

sequence means. When our genome sequence was first obtained, it did deliver many previously unstudied genes, mostly protein-coding genes. Most of our RNA genes were initially not identified, however, and the functional significance of most of our genome remained a mystery. To address that puzzle, post-genome studies focused on two major approaches—evolutionary sequence comparisons and functional assays—underpinned by comprehensive bioinformatics analyses. We describe these in the next two sections. Other post-genome studies looked at variation in the genome sequence between individuals; we examine that aspect in greater detail in later chapters.

Identifying genes and other functionally important DNA elements through evolutionary conservation

Comparisons of human DNA sequences and inferred protein sequences with all other sequences in the databases allow us to identify the extent to which DNA sequences have been highly or poorly conserved during evolution. Starting from a single query sequence, computer programs such as BLAST and BLAT readily allow comparison across sequence and genome databases (Table 2.4). Over the past decade or so, the genomes of large numbers of species have been sequenced, and powerful programs allow the alignment of whole genome sequences from different organisms.

Identifying genes and their orthologs

If a human DNA sequence has an important function, it will often have been highly conserved during evolution—that is, it will have a recognizably similar counterpart (**homolog**) in other species. For genes, the homologs in other species will quite often be examples of an **ortholog**, a direct equivalent that shares a common evolutionary origin with the human gene (that is, they both originated from the same ancestral gene that was present in some common evolutionary ancestor).

When the human genome was first sequenced, thousands of novel putative protein-coding genes were identified (on the basis of detected open reading frames). They needed to be confirmed, and sequence comparisons showing strong evolutionary conservation provided strong evidence. When the genomes of large numbers of other species were obtained, detailed information could be obtained on orthologs of human genes. That information has been important in constructing animal models of disease that provide insights into the molecular basis of disease, and that allow testing of novel disease therapies, including gene therapies.

The HomoloGene resource (see Table 2.4) maintains lists of genes in other organisms that are highly related to individual human genes and inferred proteins, and provides sequence alignments with sequence similarity scores. If, for example, we use the gene *TP53* (encoding the p53

PROGRAM/ DATABASE	INTERNET ADDRESS	DESCRIPTION
BLAST programs	http://blast.ncbi.nlm.nih.gov	suite of programs for comparing query nucleotide sequences or amino acid sequences against each other or against databases of recorded sequences
BLAT	http://genome.ucsc.edu	for rapid matching of a query sequence to a genome; available through the University of Santa Cruz genome server
HomoloGene	http://www.ncbi.nlm.nih.gov/homologene	for a gene of interest, lists homologs found in other species

Table 2.4 Examples of programs and databases used to find homologs for a DNA or protein sequence query.

BOX 2.3 Access to Public Genome Databases and Browsers.

Various resource centres provide genome information; three of the more popular ones are listed below. A user's guide to the human genome was published by *Nature Genetics* in 2002 (volume 32, supplement 1, pp. 1–79), providing the reader with an elementary step-by-step guide for browsing and analyzing human genome data. Genome browsers have graphical interfaces (usually with zoom-in and zoom-out facilities) that show nucleotide coordinates for each gene or selected chromosome region, with aligned exons and introns, transcripts, and so on. A variety of click-over tools allow the user to jump to other databases listing reference sequences, polymorphisms, and so on.

NCBI Human Genome Resources (http://www.ncbi.nlm.nih.gov/genome/guide/human)

This major human genome resource includes a human gene database at http://www.ncbi.nlm.nih.gov/gene. Information is provided on specific genes, including a gene summary; the genomic context (with a graphical interface called MapViewer); genomic regions, transcripts, and proteins (with links to reference sequences at genome, RNA, and protein levels); a standard bibliography; and a curated gene reference into function (GeneRIF) bibliography.

ENSEMBL (http://www.ensembl.org)

The human genome can be browsed at http://www.ensembl.org/Homo_sapiens/Info/index. In the search box you can search by gene name, genome nucleotide coordinates, or disease. Useful video tutorials are available at http://www.ensembl.org/info/website/tutorials/index.html.

UCSC Genome Bioinformatics (http://www.genome.ucsc.edu)

The human genome browser can be accessed by clicking on 'genome browser' at the top left; a multitude of different queries are possible, as shown in an examples list. Clicking on 'BLAT' in the tool bar at the top brings you into the useful BLAT program, which allows very fast searching of the genome for sequences that are very closely related in sequence to a query sequence of interest.

A worked example: using ENSEMBL to investigate hemophilia genes

After accessing the human genome browser page at ENSEMBL (see above), select Gene under Search categories (at the top left) and type 'hemophilia' into the query box. Different gene entries are returned. If you then click on 'genes,' they are revealed to be *F8* (coagulation factor VIII) and *F9* (coagulation factor 9), which can be followed up.

To pursue the *F9* gene for example, click on its genome location (given as X:138612917–138645617—that is, the region from nucleotide position 138,612,917 to nucleotide position 138,645,617 on the X chromosome). Now you will see graphical displays including those shown in **Figure 1**, where the open red box in the

---

tumor suppressor) as a query, the top two HomoloGene results are for *TP53* itself (HomoloGene entry #460) and for *TP53RK* (HomoloGene entry #6042). The former entry shows that *TP53* has been quite strongly conserved, having orthologs in bony vertebrates (eutelostomes), but *TP53RK* is extraordinarily highly conserved. It encodes a serine kinase that regulates target proteins, notably including p53, and has clear counterparts in all eukaryotes, including fungi and plants. By contrast, some human genes, such as *TCP10L*, have orthologs in other primates only.

If we click on individual HomoloGene entries we can access detailed information, including protein alignments, both multiple alignments and pairwise alignments. In the case of *TP53* (entry #460), the pairwise protein alignments show that the sequence of human p53 is identical to the ortholog in chimpanzee, and displays 78.6% sequence identity to both mouse and rat orthologs and 59.9% sequence identity to a zebrafish ortholog (see **Figure 2.9** for the comparison between human and mouse).

Sequence conservation due to selection

Although mutation can potentially change any nucleotide in DNA, there must be constraints on changing a functionally important sequence

BOX 2.3 (continued)

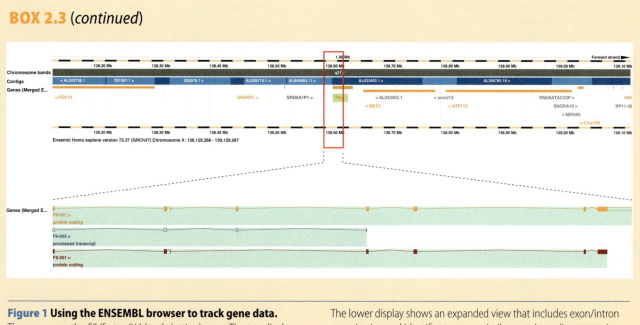

Figure 1 Using the ENSEMBL browser to track gene data.
The query was the *F9* (factor IX blood clotting) gene. The top display shows a 1 Mb chromosome environment on the X chromosome that encompasses the *F9* gene (shown in the middle within the tall red box).

The lower display shows an expanded view that includes exon/intron organization and identifies two very similar protein-coding transcripts (F9-001 and F9-201) and one noncoding transcript (F9-002) from the *F9* gene.

---

centre top of the top panel image shows the position of the *F9* gene and the bottom image shows different *F9* transcripts plus exon–intron organization. Two transcripts, F9-001 and F9-201, are predicted to produce a protein, but the smaller one, F9-201, lacks an exon; the third transcript, F9-002 is noncoding.

To obtain the sequences of these transcripts and their exon-intron organization, click on the transcript.

Say we click on the major full-length transcript F9-001 that takes us to a table with the transcript ID ENST00000218099. Now click on this symbol; you will get a transcript-based display panel at the left, and under 'Sequence' you can click on 'Exons,' 'cDNA,' or 'Proteins' to obtain relevant nucleotide/ amino acid sequences.

---

during evolution. If we take a protein-coding sequence, for example, even a single amino acid change might quite often result in loss of the protein's function or produce an aberrant protein that might contribute to disease.

Mutations that result in adverse changes to the phenotype are effectively removed from populations over generations. That happens by a type of Darwinian natural selection called **purifying** (or **negative) selection**: compared with normal alleles, the mutant allele is not efficiently transmitted to subsequent generations because some people that carry it will not reproduce as well as people with normal alleles. Over long periods of evolutionary time, therefore, protein-coding sequences are constrained because of the need to maintain function, and they change slowly in comparison with most other DNA sequences.

One of the big surprises from sequencing the human genome was the discovery that only just over 1.2% of our DNA is coding DNA. Comparative genome studies also produced a surprise. When our genome was aligned with multiple other mammalian genomes, the DNA sequences that showed strong conservation were found to make up only about 5% of the

Figure 2.9 Pairwise alignment of the human p53 protein and the corresponding mouse p53 protein. The alignment was generated by the BLASTP program as a selectable function within the HomoloGene resource. The output lists the human p53 protein sequence as the query sequence, and the mouse p53 sequence as the subject (Sbjct) sequence. The alignment is slightly skewed because extra amino acids can be found in one sequence that are lacking in the other. For example, the mouse p53 protein has an extra three amino acids at its N-terminus (the sequence from amino acid number 4 onward matches the start of the human sequence) but lacks six amino acids in the region encompassing amino acids 33–58 (shown as dashes in the mouse sequence). The middle lines between the query and subject lines show identical residues (at 304 out of the 393 matched positions). The + symbol in the middle lines denote functionally similar amino acids; they occur here at 22 positions, which when added to 304 identities gives a total of 326 positive matches out of 393 (the term *protein similarity* is sometimes applied to this type of ratio).

score	expect	method	identities	positives	gaps
574 bits(1480)	0.0	Compositional matrix adjust.	304/393(77%)	326/393(82%)	6/393(1%)

```
Query    1    MEEPQSDPSVEPPLSQETFSDLWKLLPENNVLSPLPSQAMDDLMLSPDDIEQWFTEDPGP    60
              MEE QSD S+E PLSQETFS LWKLLP  ++L P P   MDDL+L P D+E++F   GP
Sbjct    4    MEESQSDISLEPLPLSQETFSGLWKLLPPEDIL-PSP-HCMDDLLL-PQDVEEFFE---GP    57

Query   61    DEaprmpeaappvapapaaptpaapapapSWPLSSSVPSQKTYQGSYGFRLGFLHSGTAK   120
              EA R+  A      P    P P APAPA  WPLSS VPSQKTYQG+YGF LGFL SGTAK
Sbjct   58    SEALRVSGAPAAQDPVTETPGPVAPAPATPWPLSSFVPSQKTYQGNYGFHLGFLQSGTAK   117

Query  121    SVTCTYSPALNKMFCQLAKTCPVQLWVDSTPPPGTRVRAMAIYKQSQHMTEVVRRCPHHE   180
              SV CTYSP LNK+FCQLAKTCPVQLWV +TPP G+RVRAMAIYK+SQHMTEVVRRCPHHE
Sbjct  118    SVMCTYSPPLNKLFCQLAKTCPVQLWVSATPPAGSRVRAMAIYKKSQHMTEVVRRCPHHE   177

Query  181    RCSDSDGLAPPQHLIRVEGNLRVEYLDDRNTFRHSVVVPYEPPEVGSDCTTIHYNYMCNS   240
              RCSD DGLAPPQHLIRVEGNL  EYL+DR TFRHSVVVPYEPPE GS+ TTIHY YMCNS
Sbjct  178    RCSDGDGLAPPQHLIRVEGNLYPEYLEDRQTFRHSVVVPYEPPEAGSEYTTIHYKYMCNS   237

Query  241    SCMGGMNRRPILTIITLEDSSGNLLGRNSFEVRVCACPGRDRRTEEENLRKKGEPHHELP   300
              SCMGGMNRRPILTIITLEDSSGNLLGR+SFEVRVCACPGRDRRTEEEN RKK    ELP
Sbjct  238    SCMGGMNRRPILTIITLEDSSGNLLGRDSFEVRVCACPGRDRRTEEENFRKKEVLCPELP   297

Query  301    PGSTKRALPNNTSSSPQPKKKPLDGEYFTLQIRGRERFEMFRELNEALELKDAQAGKEPG   360
              PGS KRALP  TS+SP KKKPLDGEYFTL+IRGR+RFEMFRELNEALELKDA A +E G
Sbjct  298    PGSAKRALPTCTSASPPQKKKPLDGEYFTLKIRGRKRFEMFRELNEALELKDAHATEESG   357

Query  361    GSRAHSSHLKSKKGQSTSRHKKLMFKTEGPDSD   393
               SRAHSS+LK+KKGQSTSRHKK M K  GPDSD
Sbjct  358    DSRAHSSYLKTKKGQSTSRHKKTMVKKVGPDSD   390
```

genome. The remaining 95% or so of our genome is poorly conserved, with large amounts of highly repetitive noncoding DNA (**Figure 2.10**).

The ENCODE Project: functional assays to determine what our genome does

To define the functional elements in our genome and to assess how much of our genome is functionally significant, a second large international project was needed. The ENCODE (Encyclopedia of DNA Elements) Project took 10 years, reporting late in 2012. It consisted of a series of assays to identify functional DNA elements in the human genome. Operationally, a functional DNA element was defined as a discrete genome segment that makes a defined product (protein or ncRNA) or displays a reproducible biochemical signature (such as a protein-binding capacity or a specific chromatin structure).

Transcript analysis

Previous studies—notably the Human Genome Project and comparative genomic studies—had identified and confirmed protein-coding genes, and comparative genome studies plus bioinformatic analyses were enormously helpful in identifying novel human genes that made tiny RNAs, such as microRNAs. But there was a need to perform comprehensive analyses to identify RNA genes, and to conduct genomewide mapping of RNA transcripts.

After surveying 15 different human cell lines, the ENCODE consortium concluded that about 75% of the human genome is transcribed in at least one of the cell types studied (but the percentage is significantly lower in any one cell type). RNA transcripts from both strands were found to be common (both within genes and in intergenic regions close to genes), and a protein-coding gene was found to produce, on average, about six different RNA transcripts, of which four contain protein-coding sequences and two are noncoding transcripts.

Biochemical signatures

The ENCODE Project devoted major effort to the use of different analyses to define genomewide binding sites for transcription factors in various human cell lines and genomewide patterns of chromatin structure. As we detail in Chapter 6, the different patterns of gene expression in different cell types depend largely on different patterns of DNA methylation and the use of different histone variants and chemically modified histones. ENCODE sampled only 119 of 1800 known transcription factors, and just over 20% of the known DNA and histone modifications; 95% of the genome was found to lie within 8 kb of a DNA–protein interaction and, as expected, the patterns of DNA methylation, histone variants, and histone modification were found to vary extensively between different cell types.

The outcome

The ENCODE Project provided a guide to the functional significance of our genome, but much remains to be done and the continuing work to analyze how our genome works will take a very long time. A principal conclusion of ENCODE—that a large part of our genome is functionally significant—has been strongly resisted by evolutionary biologists and has been the subject of very heated debate and controversy. We consider that aspect in the next section.

2.4   The Organization and Evolution of the Human Genome

A brief overview of the evolutionary mechanisms that shaped our genome

The widely accepted endosymbiont hypothesis proposes that our two genomes originated when a type of aerobic prokaryotic cell was endocytosed by an anaerobic eukaryotic precursor cell, at a time when oxygen started to accumulate in significant quantities in the Earth's atmosphere. Over a long period, much of the original prokaryote genome was excised, causing a large decrease in its size, and the excised DNA was transferred to the genome of the engulfing cell. The latter genome increased in size and went on to undergo further changes in both size and form during evolution, developing into our nuclear genome; the much reduced prokaryotic genome gave rise to the mitochondrial genome.

The theory explains why mitochondria have their own ribosomes and their own protein-synthesizing machinery and why our mitochondrial DNA closely resembles in form a reduced (stripped-down) bacterial genome. But how did the genome of the engulfing cell become so large and complex? That largely happened by a series of different mechanisms that copied existing DNA sequences and added them to the genome. After some considerable time, the copies can acquire mutations that make them different from the parent sequences; ultimately new genes, new exons, and so on can be formed in this way.

Whole-genome duplication is a quick way of increasing genome size, and comparative genomics has provided very strong evidence that this mechanism has occurred from time to time in different evolutionary lineages. There is compelling evidence, for example, that whole-genome duplication occurred in the early evolution of our chordate ancestors just before the appearance of vertebrates.

Additional duplications of moderately large to small regions of DNA occur comparatively frequently on an evolutionary timescale, and they also give rise ultimately to novel genes, novel exons, and so on. They occur by copying mechanisms that work at the level of genomic DNA, or by using reverse transcriptases to make DNA copies of RNA transcripts that then insert into the genome.

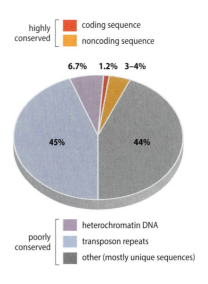

Figure 2.10 Human genome organization: extent of evolutionary conservation and repetitive sequences. Only just over 1.2% of our genome is coding DNA that specifies protein sequences, and another roughly 3–4% or so of our genome is made up of noncoding DNA sequences that have been highly or moderately conserved during evolution (as determined by looking at nucleotide substitutions in mammalian sequence alignments). Some of this conserved sequence is present in multiple copies and includes different types of repeated genes (gene families). The 6.7% of our genome that is located in constitutive heterochromatin is very largely made up of poorly conserved repetitive DNA sequences that include sequences responsible for centromere function. Transposon repeats include highly repetitive interspersed repeats such as Alu and LINE-1 repeats; it is thought that during evolution some of these repeats contributed to the formation of new exons and regulatory elements, including some long ncRNAs.

Comparative genomics can reveal when new genes were formed in evolution by screening multiple species to identify those that possess versions of that gene. In the examples given in Section 2.3, the gene encoding the p53 tumor suppressor first appeared at the time of bony vertebrates, but the *TCP10L* gene seems to have originated in catarrhine primates.

The duplication mechanisms that led to a progressive increase in genome size and to the formation of novel genes and other novel functional sequences are, to a limited extent, offset by occasional loss of functional DNA sequences, including genes. After whole-genome duplication, for example, many of the new gene copies pick up mutations that cause them to be silenced, and they are eventually lost. And the Y chromosome is believed to have shed many genes over hundreds of million years. Gene loss can happen on a smaller scale, too.

Gene birth and gene loss are comparatively infrequent events; even though humans and mice diverged from a common evolutionary ancestor about 80 million years ago, our gene repertoire is extremely similar to that of the mouse. However, *cis*-acting regulatory sequences such as enhancers often evolve rapidly, and although we have much the same set of genes as a mouse, they are often expressed in different ways. Differential gene regulation is a primary explanation for the differences between species—such as mammals—that are evolutionarily closely related.

How much of our genome is functionally significant?

We will end this chapter by looking in some detail at different facets of our genome. But first let us step back and take a broad look at its design. Here is one perspective attributed to the evolutionary biologist David Penny: "I would be quite proud to have served on the committee that designed the [*Escherichia*] *coli* genome. There is, however, no way that I would admit to serving on a committee that designed the human genome. Not even a university committee could botch something that badly."

The *E. coli* genome is a sleek genome that is tightly packed with gene sequences. By contrast, like the genome of many complex organisms, our genome seems rather flabby, with many DNA sequences of questionable functional value (see Figure 2.10). For decades much of our genome had seemed to be 'junk DNA,' an idea supported by the *C*-value paradox, the lack of correlation between genome size and organism complexity (for example, the genome of the diploid onion is more than five times the size of our genome, and genome size varies enormously in different fish).

The ENCODE Project seemed to offer a different perspective. There seemed to be pervasive transcription of the genome, and 80.4% of the human genome was claimed to participate in at least one RNA-associated or chromatin-structure-associated event in at least one cell type. However, the possible conclusion that much of our genome might be functionally significant has been strongly resisted by many evolutionary biologists. Part of the difficulty in interpreting the ENCODE data is that much of the 80.4% figure comes from the observed representation of RNA transcripts, but many RNAs are produced at such low levels that they might alternatively represent transcriptional background 'noise.'

Estimating functional constraint

The amount of the human genome that is highly conserved (under functional constraint as a result of purifying selection) was initially estimated to be about 5% (see Figure 2.10). However, that figure came from comparisons with many different mammals. Additional functional constraint became apparent in the 1000 Genomes Project, in which multiple human genomes were compared.

Functionally important DNA sequences that are rapidly evolving might not be seen to be conserved (and therefore constrained) when comparisons are made between a broad range of mammalian species. Although some noncoding DNA sequences are very strongly conserved—sometimes more than coding DNA—it is clear that many regulatory DNA sequences and RNA genes are rapidly evolving. They do, however, make important contributions to functional constraint in narrower evolutionary lineages, including primate and then human lineages in our case.

Taking that into account, the proportion of our genome that is subject to purifying selection is now thought to be of the order of 10%; on that basis, most of our genome does not seem to have a valuable function.

The mitochondrial genome: economical usage but limited autonomy

David Penny's comment about the human genome (see above) certainly does not apply to the mitochondrial genome. Our mitochondrial DNA closely resembles in form a reduced (stripped-down) bacterial genome. Mitochondria, like chloroplasts, have their own ribosomes and their own protein-synthesizing machinery and almost certainly originated when a prokaryotic cell was engulfed by an anaerobic eukaryote precursor cell, allowing aerobic eukaryotes to develop.

The human mitochondrial genome has a total of 37 genes. Of these, 24 are RNA genes that make all the RNA required for protein synthesis in the mitochondrial ribosomes: the two rRNAs and 22 tRNAs (**Figure 2.11**). The remaining genes make 13 out of the 89 polypeptide subunits of the oxidative phosphorylation system (OXPHOS). (The other 76 OXPHOS subunits, like all other mitochondrial proteins, are encoded by nuclear genes and

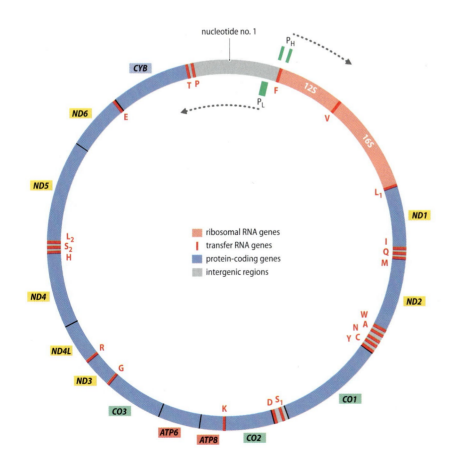

Figure 2.11 Organization of the human mitochondrial genome. The circular 16.6 kb genome has 24 RNA genes, two making the 12S and 16S rRNAs and 22 making tRNAs (shown as thin red bars with a letter corresponding to the amino acid specified; there are two tRNA^{Leu} and two tRNA^{Ser} genes). The 13 protein-coding genes make a few components of the oxidative phosphorylation system: seven make NADH dehydrogenase subunits (*ND1–ND6* and *ND4L*), two make ATP synthase subunits (*ATP6* and *ATP8*), three make cytochrome *c* oxidase subunits (*CO1–CO3*), and one makes cytochrome *b* (*CYB*). The pale colored large intergenic region at the 12 o'clock position has important regulatory sequences; the reference nucleotide sequence begins in the middle of this region and increases in nucleotide number in a clockwise direction. Two promoters, P$_H$ and P$_L$ (green boxes), which transcribe respectively the heavy and light strands in opposite directions (clockwise and counterclockwise), generate large multigenic transcripts from each strand that are subsequently cleaved. For further information, see the MITOMAP database at http:// www.mitomap.org. For the revised Cambridge reference sequence, see http://mitomap.org/MITOMAP/HumanMitoSeq

synthesized on cytoplasmic ribosomes before being imported into the mitochondria.)

Because of the need to make just 13 different proteins, the genetic code used by mitochondrial DNA has been allowed to drift a little from the 'universal' genetic code that is used for nuclear DNA. It uses four stop codons, for example. We consider the genetic code in some detail in Section 7.1; for the differences between the nuclear and mitochondrial genetic codes see Figure 7.2 on page 192.

None of the mitochondrial genes is interrupted by introns, and the genome is a model of economical DNA usage: close to 95% of the genome (all except 1 kb out of the 16.6 kb of DNA) makes functional gene products. Note that transcription of the two DNA strands occurs using one promoter each to generate large multigenic transcripts that are subsequently cleaved to generate individual mRNAs and ncRNAs.

Gene distribution in the human genome

More than 90% of the mitochondrial DNA sequence directly specifies a protein or functional ncRNA, and there is one intronless gene every 450 bp on average. The nuclear genome is very different: the gene density is much lower, genes are frequently interrupted by introns (see Table 2.1), and a sizable fraction of the genome is made up of repetitive DNA, notably highly repetitive noncoding DNA.

Close to 7% of the nuclear genome is located in constitutive **heterochromatin** that remains highly condensed throughout the cell cycle (see Figure 2.8 for a map of the chromosomal locations of human heterochromatin). Constitutive heterochromatin is essentially devoid of genes (but does contain a tiny number of genes that are inactive in somatic cells but can be expressed in germ cells). The remaining 93% of our genome is accommodated in **euchromatin**, which is less condensed and contains the vast majority of our genes.

Protein-coding genes are comparatively easy to identify (the DNA sequence can be scanned for evidence of convincing open reading frames, and predicted protein products are often evolutionarily well conserved). The most recent GENCODE data estimates that there are close to 20,400 human protein-coding genes (the number can never be exact because of variation between individuals in the copy number of some repeated genes).

Identifying RNA genes is, however, much more problematic. Three characteristics made it difficult: there is no associated open reading frame, RNA gene sequences are often not as well conserved, and they can also be tiny and easily overlooked. That makes it sometimes quite difficult to establish the functional significance of many DNA sequences that are used to make transcripts. The most recent GENCODE data might suggest that there are close to 22,000 RNA genes (**Table 2.5**), but the functional status of many putative RNA genes remains unproven. For example, the status of many long ncRNAs that are expressed at very low levels in cells is uncertain—they might possibly represent background 'transcriptional noise.'

Genome sequencing showed that gene and exon density in the euchromatic regions can vary enormously. Some chromosomes are gene-rich, such as chromosomes 19 and 22; others are gene-poor, notably the Y chromosome (which makes only 31 different proteins that mostly function in male determination). Within a chromosome, the pattern of alternating dark and light bands reflects different base compositions, and also differences in gene and exon density (see the legend to Figure 2.8).

CLASS	NUMBER
Protein-coding genes	20,345
RNA genes	22,883
Making long ncRNA	13,870
Making short ncRNA	9013
Pseudogenes	14,206
Processed	10,535
Unprocessed	2942
Other	729

Table 2.5 A snapshot of the numbers of human genes and pseudogenes listed by GENCODE in January 2014. The data are from version 19 (July 2013 freeze) and were obtained at http://www.gencodegenes.org/stats.html. Note that whereas the protein-coding gene number is likely to remain stable, the status of many RNA genes, notably many genes encoding long ncRNA, is provisional.

The extent of repetitive DNA in the human genome

Our large nuclear genome is the outcome of periodic changes that have occurred over very long timescales during evolution, including rare whole-genome duplication and intermittent chromosome re-arrangements, localized DNA duplications, DNA duplication followed by dispersal to other genome locations, and loss of DNA sequences. The net result has been a gradual increase in DNA content and gene number through evolution.

Previous whole-genome duplications were followed by a gradual loss of most of the duplicated sequences; accordingly, unique sequences make up quite a sizable fraction of our genome. Nevertheless, highly repetitive DNA sequences that are derived from transposons (mobile DNA elements; see below) plus the repetitive DNA families found in hetero-chromatin account for more than 50% of our genome (see Figure 2.10).

In addition to transposon-derived repeats, our euchromatin contains clear evidence of localized DNA duplications. In some cases, the repeats have diverged considerably in sequence—the duplication occurred many tens or hundreds of millions of years ago in evolution, and subsequent muta-tions have led to divergence in sequence between the repeats. But other localized duplications are quite striking because they have occurred very recently in evolution. For example, about 5% of our euchromatin DNA consists of neighboring duplicated segments that are more than 1 kb long and show more than 90% sequence identity. Many of these **segmental duplications** are primate-specific, and they are particularly common close to telomeres and centromeres (about 40% occur in subtelomeric regions; about 33% occur at pericentromeric regions).

There is a significant amount of repetitive coding DNA within genes and also repeated genes. Within a gene, repetitive coding DNA may be found in an individual exon (usually as a tandem duplication of one or more nucleotides), or one or more exons has been repeated (**Figure 2.12A, B**).

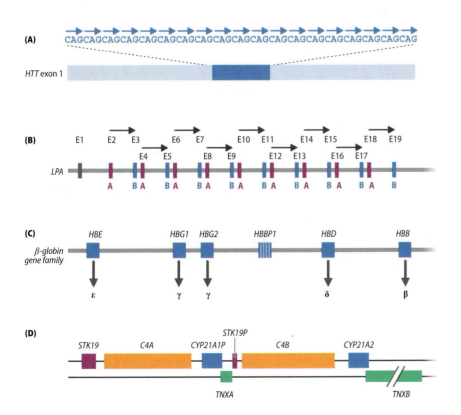

Figure 2.12 Examples of tandemly repetitive coding DNA and clustered gene families.
(A) Normal alleles of the *HTT* huntingtin gene have an array of tandemly repeated CAG codons in exon 1 that varies in number up to 35 repeats (having more than 36 repeats results in Huntington disease). (B) The *LPA* gene encodes lipoprotein Lp(a), a protein with multiple kringle domains that are each 114 amino acids long and extremely similar in sequence. Each kringle repeat is encoded by a tandemly repeated pair of exons (here labeled A and B) that can be present in different copy numbers. The example shown here has nine pairs of A and B exons, starting with exons 2 and 3 (E2 and E3) and continuing through to exons 18 and 19 (E18 and E19). (C) The β-globin gene family has six highly related genes. Four genes make alternative globins used in hemoglobin, but the status of *HBD* is uncertain (δ-globin is never incorporated into a hemoglobin protein) and *HBBP1* is a pseudogene. (D) The HLA (human leukocyte antigen) region of normal individuals has a tandemly duplicated unit containing four gene sequences, encoding serine threonine kinase 19, complement C4, cytochrome P450 21-hydroxylase, and tenascin-X (transcribed from the opposite strand). Subsequently, three of the genes became pseudogenes, having acquired inactivating mutations (*CYP21A1P*) or having also lost significant amounts of sequence (*STK19P* and *TNXA*).

GENE FAMILY	COPY NUMBER	GENOME ORGANIZATION
β-Globin	6 (includes one pseudogene)	clustered within 50 kb at chromosome 11p15 (see Figure 2.12C)
Class I human leukocyte antigen (HLA)	17 (includes many pseudogenes and gene fragments)	clustered over 2 Mb at 6p21.3
Neurofibromatosis type I	1 functional gene; 8 unprocessed pseudogenes	functional gene, *NF1*, at 17q11.2; pseudogenes dispersed to pericentromeric regions on several other chromosomes
Ferritin heavy chain	1 functional gene; 27 processed pseudogenes[a]	functional gene, *FTH1*, at 11q13; pseudogenes dispersed over multiple chromosome locations
U6 snRNA	49 genes; 800 processed pseudogenes[a]	scattered on many chromosomes

Table 2.6 Examples of multi-gene families in the human genome. [a]A processed pseudogene is a copy of a gene transcript and so has counterparts of exon sequences only, unlike an unprocessed pseudogene (which is a copy of the genomic sequence and so also has sequences corresponding to introns and upstream promoters); see Box 2.4.

On a larger scale, the repeated unit can consist of a whole gene or occasionally two or more unrelated genes (Figure 2.12C, D). The resulting multi-gene families contain two or more genes that produce related or even identical gene products. The more recently duplicated genes are readily apparent because they make very closely related or identical products. Genes originating from more evolutionarily ancient duplications make more distantly related products.

The organization of gene families

Different classes of multigene families exist in the human genome, and the number of genes in a gene family can range from two to many hundreds (**Table 2.6**). Some are clustered genes confined to one subchromosomal region. They typically arise by tandem gene duplication events in which chromatids first pair up unequally so that they become aligned out of register over short regions. The mispaired chromatids then exchange segments at a common breakpoint (**Figure 2.13**).

The α- and β-globin gene clusters on chromosomes 16 and 11, respectively, arose by a series of tandem gene duplications. Some of the duplicated genes (such as the *HBA1* and *HBA2* genes, which make identical α-globins, or the *HBG1* and *HBG2* genes, which make γ-globins that differ at a single amino acid) are the outcome of very recent gene duplication (see Figure 2.12C). Other globin genes are clearly related to each other but have more divergent sequences. The different globin classes have slightly different properties, an advantage conferred by gene duplication (see below).

Other gene families are distributed over two or more different chromosomal regions. In some cases they originally arose from duplicated genes in gene clusters that were then separated by chromosome rearrangements. In other cases, cellular reverse transcriptases were used to make natural complementary DNA (cDNA) copies of the mRNA produced by a gene, and the cDNA copies were able to insert successfully elsewhere in the genome of germ-line cells. Because the cDNA copies of mRNA lacked promoters, as well as intron sequences, they very frequently degenerated into nonfunctional **pseudogenes**. For some genes, however, cDNA

Figure 2.13 Tandem gene duplication. Gene duplication can occur after sister chromatids or non-sister chromatids of homologous chromosomes mispair so that they are slightly out of alignment. In this example, the result is that gene A on one chromatid is out of register with gene A on the opposing chromatid. Subsequent breakage of both chromatids at the position marked by the cross (X) and swapping of fragments between chromatids can result in a chromatid that has two copies of gene A (the left fragment of the top chromatid joins to the right fragment of the bottom chromatid). The exchange may be facilitated by base pairing between highly related noncoding repetitive DNA sequences (orange boxes).

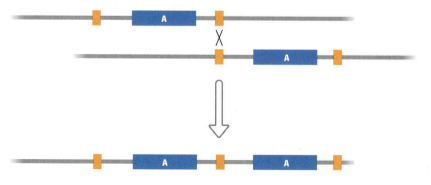

copies have integrated during evolution at other chromosomal locations to produce functional genes (Box 2.4).

Many RNA genes are also members of large gene families. For example, the short arms of chromosomes 13, 14, 15, 21, and 22 each have 30–40 tandem repeats of a 45 kb DNA sequence that specifies 28S, 18S, and 5.8S ribosomal RNA. During mitosis, the megabase-sized clusters of ribosomal DNA on the different chromosomes can pair up and exchange segments, a type of interchromosomal recombination.

Noncoding RNAs can also be reverse transcribed to give cDNA copies that can integrate elsewhere in the genome, like the mRNA-derived cDNAs in the figure in Box 2.4. During evolution, cDNA copies of certain classes of RNA genes were particularly successful in integrating into the genome. RNA genes that are transcribed by RNA polymerase III (such as tRNA genes and some others such as the 7SL RNA gene) have internal promoters (the promoter sequence is located within the transcription unit rather than upstream of it as in protein-coding genes). The cDNA copies of these genes therefore carry with them a promoter that will help the integrated cDNA to be expressed. In some cases this resulted in a huge increase in copy number for some gene families, and as explained below it gave rise to the most commonly occurring repetitive DNA sequence in the human genome, the Alu repeat.

The significance of gene duplication and repetitive coding DNA

Over long periods of evolutionary time there seems to have been a relentless drive to duplicate DNA in complex genomes. That has meant that whole genes have been duplicated to give gene families as described above. Tandem duplication of exons (which occurs by the same mechanisms that produce tandem gene duplication) is also evident in about 10% of human protein-coding genes. So what are the advantages of DNA sequence duplication?

Dosage

Duplication of genes can be advantageous simply because it allows more gene product to be made. Increased gene dosage is an advantage for genes that make products needed in large amounts in cells—we have hundreds of virtually identical copies of genes that make individual ribosomal RNAs and individual histone proteins, for example. Exon duplication might also be an advantage when an exon (or group of exons) encodes a structural motif that can be repeated, allowing proteins such as collagens to extend the size of structural domains during evolution.

Novel genetic variants

Once a gene or exon has duplicated, there are initially two copies with identical sequences. When that happens, the constraints on changing the sequence imposed by Darwinian natural selection may be applied to one of the two sequences only. The other sequence is free from normal constraints to maintain the original function; it can diverge in sequence over many millions of years to produce a different but related genetic variant. Divergent exons allowed the formation of different but related protein domains and the possibility of alternative splicing to produce transcripts with different exon combinations. Additionally, as described below, certain types of mobile element allow the copying of exons from one gene to another (exon shuffling) to produce novel combinations of exons.

Divergent genes produced by tandem gene duplication allow the production of variant but related proteins. The vertebrate globin superfamily

BOX 2.4 Pseudogenes.

One common consequence of gene duplication is that a duplicated copy diverges in sequence but instead of producing a variant gene product it gradually accumulates deleterious mutations to become a **pseudogene**. A pseudogene copy of a protein-coding gene can usually be detected by identifying deleterious mutations in the sequence that corresponds to the coding DNA sequence; RNA pseudogenes are less easy to identify as pseudogenes.

GENCODE lists more than 14,000 pseudogenes in the human genome (see Table 2.5). According to their origin, pseudogenes can be divided into two major classes: unprocessed pseudogenes and processed pseudogenes. Despite their name, some pseudogenes are known to have functionally important roles, as described below.

Unprocessed pseudogenes

An unprocessed pseudogene arises from a gene copy made at the level of genomic DNA, for example after tandem gene duplication (see Figure 2.13). Initially, the copied gene would have copies of all exons and introns of the parental gene plus neighboring regulatory sequences including any upstream promoter. Acquisition of deleterious mutations could lead to gene inactivation ('silencing') and subsequent decay, and sometimes instability (substantial amounts of the DNA sequence can be lost, leaving just a fragment of the parental gene).

Unprocessed pseudogenes are typically found in the immediate chromosomal vicinity of the parental functional gene (see the example of *HBBP1* in Figure 2.12C). Sometimes, however, they are transposed to other locations because of instability of pericentromeric or subtelomeric regions. For example, the *NF1* neurofibromatosis type I gene is located at 17q11.2, and eight highly related unprocessed *NF1* pseudogenes are found (with one exception) in pericentromeric regions of other chromosomes as a result of interchromosomal exchanges at pericentromeric regions.

Processed pseudogenes

A processed pseudogene (also called a retropseudogene) arises by reverse transcription of an RNA from a parental gene followed by random integration of the resulting cDNA copy elsewhere in the genome (**Figure 1**). The cDNA copy lacks any sequences corresponding to introns and regulatory sequences occurring outside exons, such as upstream promoters. Integration of a cDNA copy of a protein-coding gene will usually mean that the cDNA is not expressed and it will acquire deleterious mutations to become a retropseudogene. If, however, the cDNA integrates at a position adjacent to an existing promoter it may be expressed and acquire some useful function to become a **retrogene** (see Figure 1).

Several protein-coding genes have associated retropseudogenes (see Table 2.6), but the highest numbers of retropseudogenes derive from RNA genes. RNA genes that are transcribed by RNA polymerase III have *internal promoters*; that is, the sequences needed to attract the transcriptional activation complexes (which then go on to recruit an RNA polymerase) are located within the transcription unit itself, instead of being located upstream like the promoters of protein-coding genes. When transcripts of these genes are copied into cDNA, the DNA copies of their transcripts also have promoter sequences, giving the potential to reach very high copy number. Human Alu repeats originated as cDNA copies of 7SL RNA and the mouse B1 and B2 repeats sequences are diverged cDNA copies of tRNAs.

Functional pseudogenes

Comparative genomic studies indicate that some pseudogene sequences have evolved under purifying selection (they are more evolutionarily conserved than would have been expected for a functionless DNA sequence). Many pseudogenes are known to be transcribed, and there is good evidence that the transcripts of some pseudogenes have important regulatory roles. For example, the *PTEN* gene (which is mutated in multiple advanced cancers) is located on chromosome 10 and is regulated by an RNA transcript from a closely related processed pseudogene *PTENP1* located on chromosome 9. As described in Chapter 6, the *PTEN* mRNA and *PTENP1* RNA compete for binding by some regulatory miRNAs (they have the same miRNA-binding sites), and by regulating cellular levels of *PTEN*, *PTENP1* acts as a tumor suppressor (see Figure 6.9 and the cross-referencing text at the end of Section 6.1).

BOX 2.4 (continued)

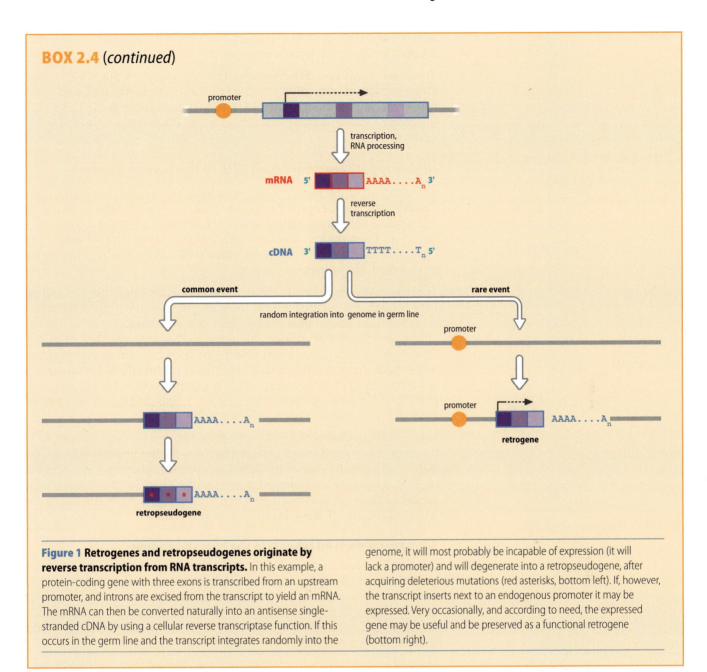

Figure 1 Retrogenes and retropseudogenes originate by reverse transcription from RNA transcripts. In this example, a protein-coding gene with three exons is transcribed from an upstream promoter, and introns are excised from the transcript to yield an mRNA. The mRNA can then be converted naturally into an antisense single-stranded cDNA by using a cellular reverse transcriptase function. If this occurs in the germ line and the transcript integrates randomly into the genome, it will most probably be incapable of expression (it will lack a promoter) and will degenerate into a retropseudogene, after acquiring deleterious mutations (red asterisks, bottom left). If, however, the transcript inserts next to an endogenous promoter it may be expressed. Very occasionally, and according to need, the expressed gene may be useful and be preserved as a functional retrogene (bottom right).

provides illustrative examples. Over a period of 800 million years or so, a single ancestral globin gene gave rise to all existing globin genes by a series of periodic gene duplications. Early duplications led to diverged gene copies that ultimately came to be expressed in different cell types, producing globins that were adapted to work in blood (hemoglobins), in muscle (myoglobin), in the nervous system (neuroglobin), or in multiple cell types (cytoglobin).

More recent duplications in the α- and β-globin gene clusters (see Figure 2.12C for the latter) led to different varieties of hemoglobin being produced at different stages of development. Thus in early development, zeta (ζ)-globin is used in place of α-globin, while epsilon (ε)-globin is used instead of β-globin in the embryonic period, and γ-globin is used instead of β-globin in the fetal period. The globins incorporated into hemoglobin in the embryonic and fetal periods have been considered to be better adapted to the more hypoxic environment at these stages.

There are, however, disadvantages to DNA sequence duplication. One consequence of repetitive coding DNA and tandemly repeated gene sequences is that the repeated DNA sequences can be prone to genetic instability, causing disease in different ways. We examine this in detail in Chapter 7.

Highly repetitive noncoding DNA in the human genome

Just over half of the human genome is made up of highly repetitive noncoding DNA sequences, of which a minority (about 14%) is found in constitutive heterochromatin (which accounts for a total of about 7% of our DNA). Euchromatin accounts for about 93% of our DNA, of which just under half is made up of highly repetitive noncoding DNA (accounting for 45% of the total genome).

The repetitive noncoding DNA in heterochromatin is a mixture of repetitive DNA sequences that are found in both heterochromatin and euchromatin (see examples below) plus DNA repeats that are characteristic of heterochromatin. The latter include different satellite DNA families of highly repetitive tandem repeats. Satellite DNAs are common at centromeres and include: alphoid DNA, with a 171 bp α repeat unit (found at all human centromeres); a 68 bp β repeat unit at the centromeres of the acrocentric chromosomes plus chromosomes 1, 9, and Y; plus different other satellite DNAs with comparatively small repeat units.

Like most heterochromatin DNA, the DNA of centromeric heterochromatin is very poorly conserved between species. Telomeric heterochromatin is the exception. It is based on TTAGGG repeats (that extend over lengths of 5–15 kb at the chromosome ends); the TTAGGG telomere repeat sequence is conserved throughout vertebrates and is highly similar to the telomere repeats of many invertebrates and plants.

Transposon-derived repeats in the human genome

Different classes of very highly repetitive DNA occur in the human genome in an interspersed fashion rather than as tandem repeats. They originated during evolution from **transposons** (mobile DNA elements that were able to migrate from one location in the genome to another). Examples are commonly found within genes, usually in introns and sometimes in exons.

The vast majority of the existing transposon-derived repeats in the human genome can no longer transpose—they lost key sequences or picked up inactivating mutations during evolution and so are regarded as 'transposon fossils'—but a small minority are able to transpose. Transposition is thought to be important in the birth of new exons and new regulatory sequences in genomes, and in the formation of novel hybrid genes.

About 6% of the interspersed repeats in euchromatin originated from DNA transposon families that transpose by a cut-and-paste mechanism. The great majority, however, arose from **retroposons** (or retrotransposons) that transpose through an RNA intermediate. At least 40% of the DNA in the human genome therefore originated by copying RNA using cellular reverse transcriptases—retroposons can be transcribed and the resulting RNA converted by reverse transcriptase to make a cDNA copy that integrates elsewhere in the genome (the same principle as shown in the figure in Box 2.4). As illustrated in **Figure 2.14** and described below, there are three major classes of retroposon-derived repeats in the human genome:

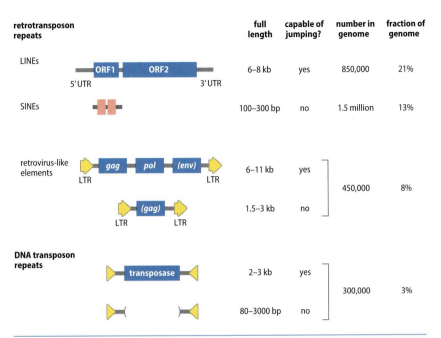

retrotransposon repeats		full length	capable of jumping?	number in genome	fraction of genome
LINEs		6–8 kb	yes	850,000	21%
SINEs		100–300 bp	no	1.5 million	13%
retrovirus-like elements		6–11 kb	yes	450,000	8%
		1.5–3 kb	no		
DNA transposon repeats		2–3 kb	yes	300,000	3%
		80–3000 bp	no		

Figure 2.14 Transposon-based repeats in the human genome. Retrotransposons are mobile elements that can potentially use a copy-and-paste transposition via an RNA intermediate. That is, an RNA transcript is converted into a cDNA by a reverse transcriptase, and the cDNA copy integrates elsewhere in the genome. The human genome has large numbers of retrotransposon repeats belonging to the three major classes shown. Only a very small percentage of the retrotransposon repeats are actively jumping (= transposing; most are truncated repeats or inactivated full-length repeats). Actively transposing LINE elements and retrovirus-like elements are said to be autonomous (able to jump independently because they carry a reverse transcriptase). SINEs such as the Alu repeat are non-autonomous: they need a reverse transcriptase to be supplied (for example by a neighboring LINE repeat). DNA transposons use a cut-and-paste transposition, but there are few in the human genome and only a small number can jump (many lack a transposase). Retrovirus-like elements and DNA transposons have repeats at their ends: long terminal repeats (LTR) orientated in the same direction in the retrovirus-like elements, and short repeats orientated in opposite directions at the termini of DNA transposons. (Adapted from The International Human Genome Sequencing Consortium [2001] *Nature* 409:860–921. With permission from Macmillan Publishers Ltd.)

- *LINES (long interspersed nuclear elements).* Full-length LINES are 6–8 kb long and can encode a reverse transcriptase, but many LINE repeats are truncated and the average size is close to 1 kb. There are three distantly related LINE families in the human genome, of which the most numerous is the LINE-1 (also called L1) family. The only human LINE elements that are currently capable of transposition are a small subset (about 80–100 copies) of the full-length LINE-1 repeats.

- *SINES (short interspersed nuclear elements).* Full-length SINEs range from 100 to 300 bp in length. About 70% of SINES belong to the Alu repeat family, which has close to 1.5 million copies. The Alu repeats are primate-specific and seem to have evolved from cDNA copies of 7SL RNA (a component of the signal recognition particle), which has an internal promoter sequence. Alu sequences are often transcribed (by adjacent promoters) but cannot make proteins. Nevertheless, some Alu repeats can transpose and rely on neighboring elements,

Figure 2.15 Retrotransposons can mediate exon shuffling. Exon shuffling can be performed with retrotransposons such as actively transposing members of the LINE1 (L1) sequence family, as shown here. Full-length L1 repeats have promoters and can be transcribed, but they have weak poly(A) signals (pA at top of dashed line) and so transcription often continues past such a signal until another downstream poly(A) signal is reached (for example after exon 3 (E3) in gene A). The resulting RNA copy contains a transcript not just of L1 sequences but also of a downstream exon (in this case E3). The L1 reverse transcriptase machinery can then act on the extended poly(A) sequence to produce a hybrid cDNA copy that contains both L1 and E3 sequences. Subsequent transposition into a new chromosomal location may lead to the insertion of exon 3 into a different gene (gene B).

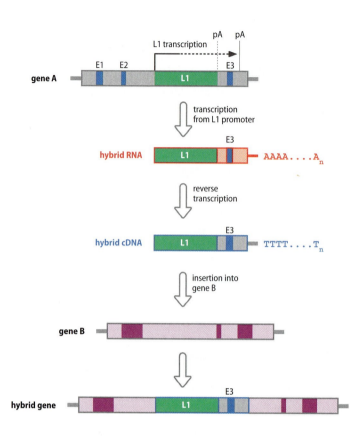

such as LINE elements, to produce the reverse transcriptase required for making cDNA copies.

- *Retrovirus-like LTR elements.* Full-length retrovirus-like elements (sometimes called human endogenous retroviruses or HERVs) are 6–11 kb long. In addition to containing long terminal repeats (LTRs) they may contain sequences resembling the key retroviral genes, including the *pol* gene that encodes reverse transcriptase, but there is little evidence of actively transposing human HERVs.

The huge contribution made by retrotransposition means that euchromatin is studded with interspersed highly repetitive DNA sequences that have been postulated to increase genetic novelty in different ways, including forming novel exons and regulatory sequences, and mediating **exon shuffling** between genes (**Figure 2.15**).

Summary

- Genes are transcribed to make RNA. Protein-coding genes make an mRNA that is decoded to make a polypeptide. RNA genes make a functional noncoding RNA.

- An mRNA contains a central coding sequence, flanked by noncoding untranslated regions that contain regulatory sequences.

- In eukaryotes, the DNA sequence corresponding to the coding sequence of an mRNA is often divided into exons that are separated by noncoding intervening sequences called introns.

- The primary RNA transcript, an RNA copy of both exons and introns, is cleaved at exon–intron boundaries. Transcribed intron sequences are discarded; transcribed exon sequences are spliced together.

- Specialized end sequences—a 5′ cap sequence and a 3′ poly(A) sequence—protect the ends of the mRNA and assist in transfer to the cytoplasm to engage with ribosomes.

- The coding sequence of an mRNA is translated using groups of three nucleotides (codons) to specify individual amino acids that are bonded together to make a polypeptide.

- Introns are also found in noncoding DNA, both in some RNA genes and in many DNA sequences that make untranslated regions in mRNAs.

- Noncoding RNAs perform many different functions in cells, but very many are gene regulators. Some regulatory RNAs are ubiquitous and perform general functions. Other RNAs regulate certain genes only, and are expressed in certain cell types.

- The human nuclear genome is composed of 24 different types of very long linear DNA molecules, one each for the 24 different types of human chromosomes (1–22, X, and Y). It has about 21,000 protein-coding genes and many thousands of RNA genes.

- Our mitochondrial genome consists of one type of small circular DNA molecule that is present in many copies per cell. It has 37 genes that make all the rRNAs and tRNAs needed for protein synthesis on mitochondrial ribosomes plus a few proteins involved in oxidative phosphorylation.

- The great majority of the genome consists of poorly conserved DNA sequences and only about 10% of genome sequences are thought to be under selective constraint to maintain function.

- Just over 1.2% of the nuclear genome is decoded to make proteins. These sequences have mostly been highly conserved during evolution—for each human protein, recognizably similar proteins exist in many other organisms.

- Other functionally constrained sequences include RNA genes and regulatory sequences, but they are often more rapidly evolving than protein-coding sequences.

- Repetitive DNA sequences are very common in the human genome. They include both tandem repeats (often sequential head-to-tail repeats) and dispersed repeats.

- Tandem repeats may be found within genes and coding sequences, and whole genes can be duplicated several times to produce a clustered gene family. Other gene families are made up of gene copies that are dispersed across two or more chromosomes.

- Gene families often contain defective gene copies (pseudogenes and gene fragments) in addition to functional genes.

- Dispersed gene copies often arise in evolution from RNA transcripts that are copied by a reverse transcriptase to make a complementary DNA that integrates randomly into chromosomal DNA (retrotransposition).

- DNA sequence lying outside exons is largely composed of repetitive sequences, including highly repetitive interspersed repeats such as Alu repeats. They originated by retrotransposition (DNA copies were made of RNA transcripts that then integrated into the genome). Very few of the repeats are currently able to transpose.

- The DNA of centromeres and telomeres is largely composed of very many tandemly repeated copies of short sequences.

- Gene and exon duplication has been a driving force during genome evolution. Novel genes and exons are occasionally produced by tandem duplication events. Novel exons and novel regulatory sequences can also be formed by retrotransposition.

Questions

Help on answering these questions and a multiple-choice quiz can be found at www.garlandscience.com/ggm-students

1. The sequence at the beginning of a human protein-coding gene is shown in **Figure Q2–1**. The sequence shown in capital letters is exon 1, and the ATG triplet shown in bold is translated to give the initiation codon of an mRNA. What is the number of the nucleotide that is the first to be transcribed into the mRNA sequence?

```
1   gtcagggcag agccatctat tgcctACATT TGCTTCTGAC ACAACTGTGT TCACTAGCAA


61  CCTCAAACAG ACACCATGGT GCACCTGACT CCTGAGGAGA AGTCTGCCGT TACTGCCCTG


121 TGGGGCAAGG TGAACGTGGA TGAAGTTGGT GGTGAGGCCC TGGGCAGgtt ggt
```

Figure Q2–1

2. The sequence in **Figure Q2–2** is from a central exon within a gene, with the exon nucleotides shown in bold capital letters flanked by conserved dinucleotides (underlined) from the flanking intron sequences. Translate the exon sequence in all three forward-reading frames. Is the exon likely to be coding DNA?

agAACCAGAGCCACTAGGCAGTCTTCGGACTACCGAGAGAGCCCCGTTTAAGTGCTGGATCGAgt

Figure Q2–2

3. Polypeptides are composed of amino acids that become covalently joined by peptide bonds. Illustrate how a peptide bond is formed by a condensation reaction between two amino acids— NH_2-CH(R^1)-COOH and NH_2-CH(R^2)-COOH—by giving the chemical reaction (continue to use different colors for the chemical groups that originate from the original two amino acids).

4. Describe the two different phosphodiester bonds in an mRNA and their functions.

5. A natural form of chemical crosslinking is often important in protein structure. Describe the mechanism and explain why it is important.

6. The nuclear and mitochondrial DNAs of human cells differ widely in many properties. Give four examples.

7. Gene families have arisen in evolution by mechanisms that copy DNA sequences. Give three examples of these mechanisms.

8. What are pseudogenes? Explain how they originate and explain how some pseudogenes are functionally important.

Further Reading

More detailed treatment of much of the subject matter in this chapter, including a detailed account of human genome organization, gene evolution, and the Human Genome Project, can be found in the following.

Strachan T & Read AP (2010) Human Molecular Genetics, 4th ed. Garland Science.

Protein-Coding Genes and Protein Structure

Agris PF et al. (2007) tRNA's wobble decoding of the genome: 40 years of modification. *J Mol Biol* 366:1–13; PMID 17187822.

Preiss T & Hentze MW (2003) Starting the protein synthesis machine: eukaryotic translation initiation. *BioEssays* 25:1201–1211; PMID 14635255.

Whitford D (2005) Protein Structure and Function. John Wiley & Sons.

RNA Genes and Regulatory RNA

Amaral PP et al. (2008) The eukaryotic genome as an RNA machine. *Science* 319:1787–1789; PMID 18369136.

Memczak S et al. (2013) Circular RNAs are a large class of animal RNAs with regulatory potency. *Nature* 495:333–338; PMID 23446348.

Poliseno L et al. (2010) A coding-independent function of gene and pseudogene mRNAs regulates tumour biology. *Nature* 465:1033–1038; PMID 20577206.

Siomi MC et al. (2011) PIWI-interacting small RNAs: the vanguard of genome defense. *Nat Rev Mol Cell Biol* 12:246–258; PMID 21427766.

Human Genome: Analysis and Internet Resources

Djebali S et al. (2012) Landscape of transcription in human cells. *Nature* 489:101–108; PMID 22955620.

ENCODE explorer (a resource with summaries of all facets of the ENCODE Project plus a compilation of ENCODE publications) is available at *Nature ENCODE* (http://www.nature.com/ENCODE).

ENCODE Project Consortium (2012) An integrated encyclopedia of DNA elements in the human genome. *Nature* 489:57–74; PMID 22955616.

GENCODE Project statistics (with useful statistics on the numbers of human genes and transcripts) are available at http://www.gencodegenes.org/stats.html.

Genome browsers—see Box 2.3.

User's Guide to the Human Genome *Nat Genet* 35 supplement no. 1, September 2003. Available at http://www.nature.com/ng/journal/v35/n1s/index.html.

Human Genome: Organization and Evolution

Bailey JA et al. (2002) Recent segmental duplications in the human genome. *Science* 297:1003–1007; PMID 12169732.

Conrad B & Antonorakis SE (2007) Gene duplication: a drive for phenotypic diversity and cause of human disease. *Annu Rev Genomics Hum Genet* 8:17–35; PMID 17386002.

Konkel MK & Batzer MA (2010) A mobile threat to genome stability: the impact of non-LTR retrotransposons upon the human genome. *Semin Cancer Biol* 20:211–221; PMID 20307669.

Long M et al. (2013) New gene evolution: little did we know. *Annu Rev Genet* 47:307–333; PMID 24050177.

Mills RE et al. (2007) Which transposable elements are active in the human genome? *Trends Genet* 23:183–191; PMID 17331616.

Muotri AR et al. (2007) The necessary junk: new functions for transposable elements. *Hum Molec Genet* 16:R159–R167; PMID 17911158.

Pink RC et al. (2011) Pseudogenes: pseudo-functional or key regulators in health and disease. *RNA* 17:792–798; PMID 21398401.

Ponting CP & Hardison RC (2011) What fraction of the human genome is functional? *Genome Res* 21:1769–1776; PMID 21875934.

Vinckenbosch N et al. (2006) Evolutionary fate of retroposed gene copies in the human genome. *Proc Natl Acad Sci USA* 103:3220–3225; PMID 16492757.

Principles Underlying Core DNA Technologies

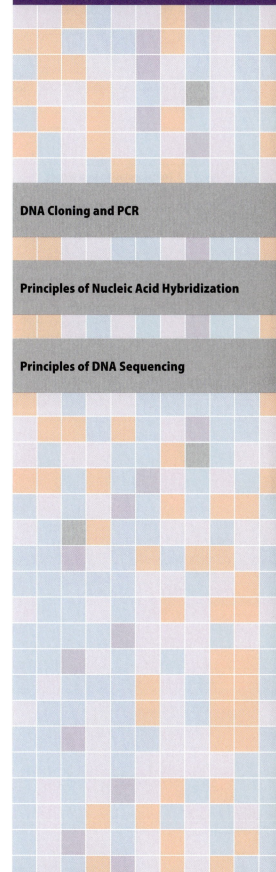

CHAPTER 3

- DNA Cloning and PCR

- Principles of Nucleic Acid Hybridization

- Principles of DNA Sequencing

Defining the genetic basis of disease requires the analysis of DNA and sometimes chromosomes. The vast majority of our genetic material is organized as immensely long DNA molecules, and changing just a single nucleotide out of the more than 3000 million nucleotides in our genome may cause disease. Sophisticated technologies have been developed to purify and manipulate genes, enabling studies on how they work and how they can be used therapeutically, and providing ways of analyzing mutations and studying the molecular basis of disease. We will outline how these approaches are used in later chapters. We describe here just the core technologies for purifying and analyzing DNA sequences.

Imagine that we wish to analyse or manipulate a single human exon or gene. We can readily isolate DNA from human cells but a single coding sequence exon, averaging just 150 bp, is a tiny fraction of the DNA, representing 1/20,000,000 of the genome. Many full-length genes are also extremely small components of the genome.

To overcome this difficulty, two quite different approaches are used. One way is to *selectively amplify* just the small piece of DNA that we are interested in, making many copies of just that one DNA sequence so that we get enough DNA to work with and analyze. The alternative approach is to use some method to *specifically recognize* a short sequence of interest so that we can track it and study it.

To selectively amplify a short region of DNA sequence we use DNA polymerases to make multiple copies of just the desired sequence, either within cells (DNA cloning) or simply in a test tube using the polymerase chain reaction (PCR). To specifically recognize a DNA sequence we use the principle of nucleic acid hybridization as described in Section 3.2.

The ultimate way of tracking changes in genes and DNA sequences is to sequence the nucleotides in a DNA sample. DNA sequencing used to be expensive, time-consuming, and restricted in scope, but all that has changed as a result of recent rapid technological advances. We consider the principles involved in DNA sequencing in Section 3.3, but we describe the most recent DNA sequencing techniques in Chapter 11. We outline how the principles described here are applied in practice in the context of different medical applications in various later chapters.

3.1 DNA Cloning and PCR

DNA cloning: fractionating and purifying DNA by transforming cells with recombinant DNAs

Cloning DNA in cells is a way of purifying DNA sequences. Large amounts (many identical copies) of a DNA sequence can be prepared so that the

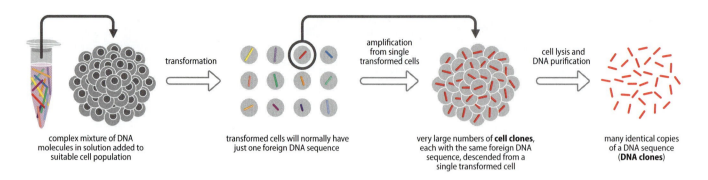

complex mixture of DNA molecules in solution added to suitable cell population

transformation

transformed cells will normally have just one foreign DNA sequence

amplification from single transformed cells

very large numbers of **cell clones**, each with the same foreign DNA sequence, descended from a single transformed cell

cell lysis and DNA purification

many identical copies of a DNA sequence (**DNA clones**)

Figure 3.1 Transformation as a way of fractionating a complex sample of DNA fragments. The key point is that transformation is selective: when a cell is transformed, it usually picks up a single DNA molecule from the environment, and so different fragments are taken up by different cells. (For simplicity, the figure shows only the DNA sequences that are to be cloned—in practice they would be joined to a vector molecule to make a recombinant DNA that is often circular.) Cell clones can form by cell division from a single transformed cell and be propagated to produce a large number of cells with an identical foreign DNA sequence that can be purified after the cells have been broken open.

sequence can be studied or put to some use. The cells involved are typically bacterial cells or, less frequently, yeast cells. The process initially involves treating the cells in some way so as to allow transfer of the DNA molecules that we wish to clone into the cells, a process known as **transformation**. In each case the DNA to be cloned is covalently joined (ligated) to some **vector** DNA sequence that will help it replicate within the host cells, as detailed below.

The joining of DNA fragments to vector molecules results in the formation of an artificial **recombinant DNA** that may be linear (as in the case of cloning artificial chromosomes in yeast), or circular, as in cloning DNA in bacteria (which are more readily transformed using small circular DNA molecules). There is normally some kind of selection or screening system that helps identify those cells that have been successfully transformed and that contain recombinant DNA.

The transformation process is selective: when foreign DNA does get into a cell, just a *single* DNA molecule is usually taken up by a cell. If a cell population is presented with a mixture of different foreign DNA fragments, therefore, different DNA fragments will be randomly allocated to different cells during transformation. That is, the population of cells serves as a sorting office that can efficiently fractionate a complex mixture of DNA fragments (**Figure 3.1**).

Amplification

Bacterial cloning systems offer the chance of making large quantities of a cloned DNA. That is, the inserted DNA is amplified to very high copy numbers as indicated in the images on the right of Figure 3.1. That is possible for two reasons. First, a single bacterium containing a cloned DNA can rapidly divide and eventually produce a huge number of identical bacterial cell clones, each with the same foreign DNA sequence. Secondly, as we will see below, some vector molecules can replicate *within* a bacterial cell to reach quite high copy numbers; if they have a foreign DNA sequence covalently linked to them, that too will be amplified within the cell (**Figure 3.2B**). We consider some of the details below.

Vector molecules

Fragments of human DNA would not normally be able to replicate if transferred into bacterial cells or yeast cells. To replicate within cells, the DNA molecules need a suitable origin of replication, a DNA sequence that will initiate DNA replication in that cell type (molecules like this are known as *replicons*). A convenient solution is to take advantage of replication origins in DNA molecules that naturally replicate within the host cells.

For cloning in bacteria, extrachromosomal replicons are typically used that replicate independently of the bacterial chromosome. Two useful

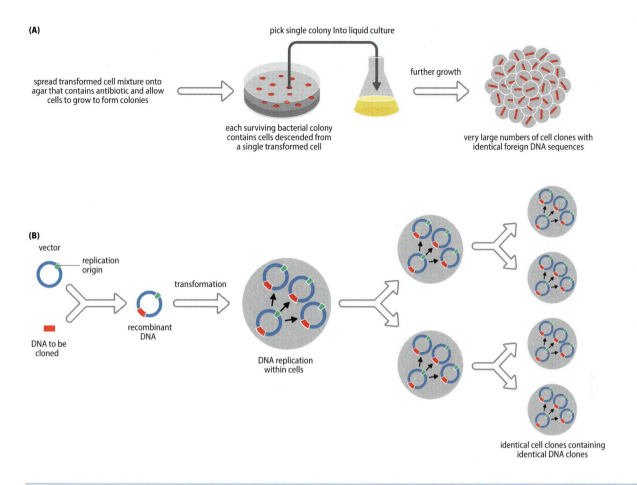

Figure 3.2 DNA cloning in bacterial cells: DNA copy number amplification and separation of clones from different transformed cells. (A) For a cell transformed by a recombinant DNA, an increase in cell number leads to a proportional expansion in copy number of that recombinant DNA. Growth occurs originally in a solid medium after the transformed cells have been spread out on a plate of agar containing antibiotics. Individual cells will be physically dispersed to different parts of the plate, and individual surviving cells then go through several rounds of cell division *in situ* to form *separate* visible colonies. An individual colony can be picked and allowed to go through a second round of amplification by growth in liquid culture. For simplicity, the cloned DNA fragments are shown in the absence of the vector molecule. (B) Vectors have their own replication origin and can replicate within a bacterial cell *independently* of the host chromosome, often replicating much more frequently than the host cell chromosome. For simplicity, the illustration here shows a very modest threefold amplification of the recombinant DNA, but some plasmids allow amplification to 100 copies or more in bacterial cells, leading to a large increase in copy number of the DNA clones.

sources are **plasmids** (small circular double-stranded DNAs that can replicate to high copy numbers in some cases) and bacteriophages (bacterial viruses)—see **Table 3.1** for examples of common cloning vectors. For some purposes, such as cloning very large DNA fragments in yeast, advantage is taken of a chromosomal replication origin.

To be useful as a cloning vector the original plasmid, bacteriophage, or other replicon needs to be genetically modified so that we can efficiently join a foreign DNA to it (as described below) and so that transformed cells can easily be recognized. When cloning is carried out in bacteria, for example, the vector will have been genetically engineered to contain a gene that confers resistance to some antibiotic that the host cells are sensitive to. After transformation, the cells are grown on agar containing the antibiotic; untransformed cells die but transformed cells survive. Because some cells are transformed by naked vector DNA (lacking other DNA), screening systems are often also devised to ensure that cells with recombinant DNA can be identified.

Table 3.1 Some classes of cloning vectors.

VECTOR	HOST CELL	MAXIMUM CLONING CAPACITY	COMMENTS
Plasmid	*E. coli*	5–10 kb	replicate independently of host chromosome and can reach high copy numbers; widely used for cloning small DNA fragments
Bacterial artificial chromosome (BAC)	*E. coli*	~200 kb	a modified plasmid vector with tight constraint on copy number that allows large fragments to be stably propagated
Yeast artificial chromosome (YAC)	*Saccharomyces cerevisiae*	>1 Mb	recombinants are effectively small linear chromosomes that consist mostly of human or other foreign DNA

Physical clone separation

How can cells that have taken up different DNA fragments be separated from each other? That relies on the formation of physically separated cell colonies. After transformation of bacterial cells, for example, aliquots of the cell mixture are spread over the surface of antibiotic-containing agar in Petri dishes (plating out); successfully transformed cells should grow and multiply; if the plating density is optimal, they form well-separated cell colonies (see Figure 3.2A). Each colony consists of identical descendant cells (cell clones) that originate from a single transformed cell and so the cell clones each contain the same single foreign DNA molecule.

An individual well-separated cell colony can then be physically picked and used to start the growth of a large culture of identical cells all containing the same foreign DNA molecule, resulting in very large amplification of a single DNA sequence of interest (Figure 3.2A). Thereafter, the cloned foreign DNA can be purified from the bacterial cells.

Making recombinant DNA

To make recombinant DNA, each DNA fragment of interest needs to be covalently joined (ligated) by a DNA ligase to a vector DNA molecule to form a **recombinant DNA** (which will then be transported into a suitable host cell, often a bacterial cell or a yeast cell). Before that is done, there is a need to prepare the DNA of interest and the vector DNA so that they can be joined efficiently, and there is a need to ensure that the recombinant DNAs are of optimal size.

To clone DNA in bacterial cells, we normally need to use relatively small DNA fragments. When DNA is isolated from the cells of complex organisms, the immensely long nuclear DNA molecules are fragmented by physical shearing forces to give an extremely heterogeneous collection of still rather long fragments with heterogeneous ends. The long fragments need to be reduced to pieces of a much smaller, manageable size with more uniform end sequences to facilitate ligation.

Recombinant DNA technology was first developed in the 1970s. The crucial breakthrough was to exploit the ability of restriction endonucleases to cut the DNA at *defined* places. As a result, the DNA could be reduced to small well-defined fragments with uniform end sequences that could be easily joined by a DNA ligase to similarly cut vector molecules (Box 3.1). Note that whereas most recombinant DNAs are circular, as shown in Box 3.1, sometimes very large pieces of DNA are cloned in yeast cells and here the recombinant DNA is a linear DNA molecule that is usually called a yeast artificial chromosome (YAC).

BOX 3.1 Restriction Endonucleases: From Bacterial Guardians to Genetic Tools.

The natural role of restriction endonucleases: host cell defense

Restriction endonucleases (also called restriction nucleases) are a class of bacterial enzyme that recognize specific short sequence elements within a double-stranded DNA molecule and then cleave the DNA on both strands, either within the recognition sequences or near to them.

The natural purpose of these enzymes is to protect bacteria from pathogens, notably bacteriophages (viruses that kill bacteria). They can disable the invading pathogen by selectively cutting the pathogen's DNA into small pieces. To ensure that its own genome is unaffected, the host cell produces a matching DNA methyltransferase that methylates its own DNA so that it is protected from subsequent cleavage by the restriction nuclease.

For example, restriction nuclease *Eco*RI from the *Escherichia coli* strain RY13 specifically recognizes the sequence GAATTC and cleaves DNA strands within this recognition sequence (called a **restriction site**). The same bacterial strain also initially produces an *Eco*RI methyltransferase that is used to modify its own genome: it recognizes the sequence GAATTC and methylates the central adenosine on both DNA strands. The *Eco*RI restriction nuclease cannot cleave at previously methylated GAATTC sequences within the bacterial genome but will cleave at unmethylated GAATTC sequences in the DNA of invading pathogens.

Restriction nucleases as molecular genetic tools

There are different classes of restriction nucleases, but type II restriction nucleases are widely used in manipulating and analyzing DNA. They recognize short sequence elements that are typically palindromes (the 5′→3′ sequence is the same on both strands, as in the sequence GAATTC); they then cleave the DNA either within or very close to the recognition sequence. Cleavage often occurs at asymmetric positions within the two strands to produce fragments with overhanging 5′ ends (**Figure 1**) or overhanging 3′ ends.

Under appropriate conditions, it is possible to use a restriction nuclease to cut complex genomic DNA into thousands or millions of fragments that can then be individually joined (*ligated*) using a DNA ligase to a similarly cut vector molecule to produce recombinant DNA molecules (**Figure 2**). For cloning DNA in bacterial cells, vector molecules are often based on circular plasmids that have been artificially engineered so that they contain unique restriction sites for certain restriction nucleases. The recombinant DNA molecules can then be transferred into suitable host cells and amplified.

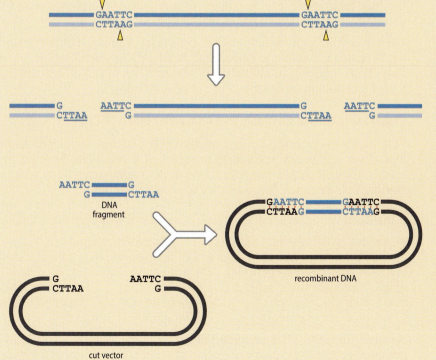

Figure 1 Asymmetric cutting of double-stranded DNA by the restriction nuclease *Eco*RI. Note that the underlined AATT sequence is an example of an overhanging 5′ end.

Figure 2 Formation of recombinant DNA. In this example, the vector has been cut at a unique *Eco*RI site to produce 5′ ends with an overhanging AATT sequence, and the DNA fragment to be cloned has the same 5′ AATT overhangs, having also been produced by cutting with *Eco*RI. The AATT overhangs are examples of *sticky ends* because they can hydrogen bond to other fragments with the same overhang (as shown in the recombinant DNA; hydrogen bonds are shown as vertical red lines), thus facilitating intermolecular interactions.

DNA libraries and the uses and limitations of DNA cloning

Once DNA cloning was established it was soon used to make **DNA libraries**; that is, collections of DNA clones representing all types of DNA sequence in a complex starting material.

DNA isolated from white blood cells, for example, provides a complex genomic DNA that can be cut into many pieces and attached to vector DNA molecules. The resulting mixture of different recombinant DNA molecules is used to transform bacteria to produce very many different clones, a genomic DNA library. A good genomic DNA library would have so many different DNA clones that there was a good chance that the library would include just about all the different DNA sequences in the genome.

An alternative was to make gene-centred DNA libraries. Until recently, it was imagined that the vast majority of human genes made proteins, and an obvious starting point was mRNA. RNA cannot be cloned, however. Instead, DNA copies needed to be made by using a specialized DNA polymerase, a reverse transcriptase that naturally copies a single-stranded RNA template to make a **complementary DNA** (**cDNA**) copy. Once the cDNA strand has been made, the original RNA is destroyed by treatment with ribonuclease and the copied DNA strand is copied in turn to make a complementary DNA, thereby making double-stranded cDNA.

Total double-stranded cDNA isolated from cells could then be used to make a cDNA library. But because different genes are expressed in different cell types, the range of DNA clones in a cDNA library could vary according to whether the cDNA originated from white blood cells or brain cells, for example. By comparing cDNA clones with the respective genomic DNA clones, exon–intron organizations could be determined.

DNA cloning started a revolution in genetics. It prepared the way for obtaining panels of DNA clones representing all the sequences in the genome of organisms, and that in turn made genome projects possible to obtain the complete sequence of genomic DNA in a variety of organisms. Once that was done, the structure of genes could be determined, paving the way for comprehensive studies to analyse gene expression and to determine how individual genes work.

There is a drawback: cloning DNA in cells is laborious and time-consuming. It is also not suited to performing rapid parallel amplifications of the same DNA sequence in multiple different samples of DNA. That required a new technology, as described in the next section.

The basics of the polymerase chain reaction (PCR)

PCR, a cell-free method for amplifying DNA, was first developed in the mid-1980s and would revolutionize genetics. It was both very fast and readily allowed parallel amplifications of DNA sequences from multiple starting DNA samples. If you wanted to amplify each exon of the β-globin gene from blood DNA samples from 100 different individuals with β-thalassemia, a single person could now do that in a very short time.

PCR relies on using a heat-stable DNA polymerase to synthesize copies of a small predetermined DNA segment of interest within a complex starting DNA (such as total genomic DNA from easily accessed blood or skin cells). To initiate the synthesis of a new DNA strand, a DNA polymerase needs a single-stranded oligonucleotide **primer** that is designed to bind to a *specific* complementary sequence within the starting DNA.

For the primer to bind preferentially at just one desired location in a complex genome, the oligonucleotide often needs to be about 20 nucleotides long or more and is designed to be able to base pair perfectly to its intended target sequence (the strength of binding depends on the number of base pairs formed and the degree of base matching).

To allow the primer to bind, the DNA needs to be heated. At a high enough temperature, the hydrogen bonds holding complementary DNA stands together are broken, causing the DNA to become single stranded. Subsequent cooling allows the oligonucleotide primer to bind to its perfect complementary sequence in the DNA sample (**annealing** or **hybridization**). Once bound, the primer can be used by a suitably heat-stable DNA polymerase to synthesize a complementary DNA strand.

In PCR, two primers are designed to bind to complementary sequences on opposing DNA strands, so that copies are made of both DNA strands. The primers are designed to be long enough for them to bind specifically to sequences that closely flank the DNA sequence of interest in such a way that the direction of synthesis of each new DNA stand is toward the sequence that is bound by the other primer. In further cycles of DNA denaturation, primer binding, and DNA synthesis, the previously synthesized DNA strands become targets for binding by the other primer, causing a chain reaction to occur (**Figure 3.3**).

The end result is that millions of copies can be made of just the desired DNA sequence of interest within the complex starting DNA. By amplifying the desired sequence we can now study it in different ways—by directly sequencing the amplified DNA, for example.

PCR is also very sensitive and robust and can successfully amplify DNA fragments from tiny amounts of tissue samples that may have been badly degraded—and even from single cells. As a result, there have been numerous applications in forensic and archaeological studies, and PCR is robust enough to allow the analysis of tissue samples that have been fixed in formalin. PCR can also be used to analyse RNA transcripts. In that case the RNA transcripts are first converted into cDNA with reverse transcriptase (the process is called reverse transcription-PCR or RT-PCR).

Quantitative PCR and real-time PCR

In routine PCR, all that is required is to generate a detectable or usable amount of product. However, for some purposes there is a need to quantitate the amount of product. There are different types of **quantitative PCR**. Some are variants of routine PCR and use standard PCR machines to give a relative quantitation of a sequence of interest within test samples and controls; others need specialized PCR machines and the quantity measured can be an absolute number of copies.

In Chapter 11 we describe different diagnostic DNA screening methods that use PCR to get relative quantitation. Fluorescently labeled PCR products from the exponential phase of the PCR reaction (**Figure 3.4**) are removed and analyzed to measure the ratio of the fluorescence exhibited by the PCR product from a test sample (one that is associated with disease or is suspected as being abnormal) and the fluorescence exhibited by the PCR product from a control sample. The basis of the quantitation is that during the exponential phase the amount of PCR product is proportional to the amount of target DNA sequence in the input DNA.

Real-time PCR is a form of quantitative PCR that is performed in specialized PCR machines, and it can provide absolute quantitation (the absolute number of copies) and also relative quantitation. Instead of waiting for the end of the reaction, the quantitation is performed while the PCR reaction is still progressing: the amplified DNA is detected as the PCR reaction proceeds in real time within the PCR machine. Important applications are found in profiling gene expression (using RT-PCR) and also in assays for altered nucleotides in DNA, as detailed in Chapter 11.

Figure 3.3 The polymerase chain reaction (PCR). The reaction usually consists of about 25–30 cycles of (a) DNA denaturation, (b) binding of oligonucleotide primers flanking the desired sequence, and (c) new DNA synthesis in which the desired DNA sequence is copied and primers are incorporated into the newly synthesized DNA strands. Numbers in the vertical strips to the left indicate the origin of the DNA strands, with original DNA strands represented by 0 and PCR products by 1 (made during first cycle), 2 (second cycle), or 3 (third cycle). The first cycle will result in new types of DNA product with a fixed 5′ end (determined by the primer) and variable 3′ ends (extending past the other primer). After the second cycle, there will be two more products with variable 3′ ends but also two desired products of fixed length (shown at the left by filled red squares) with both 5′ and 3′ ends defined by the primer sequences. Whereas the products with variable 3′ ends increase arithmetically (amount = $2n$, where n is the number of cycles), the desired products initially increase exponentially until the reaction reaches a stationary phase as the amount of reactants becomes depleted (see Figure 3.4). After 25 or so cycles, the desired product accounts for the vast majority of the DNA strands.

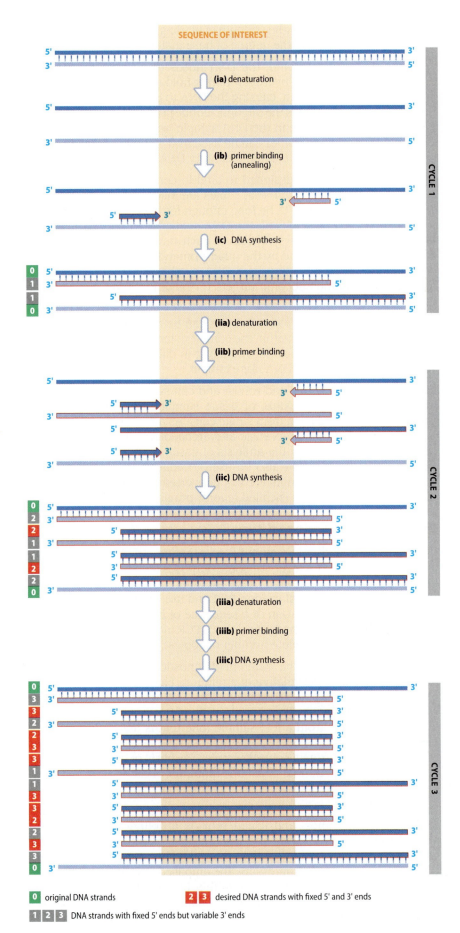

3.2   Principles of Nucleic Acid Hybridization

In a double-stranded DNA molecule, the hydrogen bonds between paired bases act as a glue that holds the two complementary DNA strands together. Two hydrogen bonds form between A and T in each A–T base pair, and three hydrogen bonds hold G and C together in each G–C base pair (see Figure 1.4). A region of DNA that is GC-rich (having a high proportion of G–C base pairs) is therefore more stable than a region that is AT-rich.

Each hydrogen bond is individually weak, but when base matching extends over many base pairs, the cumulative strength of the hydrogen bonds becomes quite strong. (Velcro® uses the same principle: a single Velcro hook and loop attachment is very weak, but thousands of them make for a strong fastening system.)

Double-stranded DNA can be subjected to different treatments that result in breaking of the hydrogen bonds so that the two DNA strands are separated (**denaturation**). For example, if we heat the DNA to a high enough temperature or expose it to strong concentrations of a highly polar molecule such as formamide or urea, the hydrogen bonds break and the two complementary DNA strands separate. Subsequent gradual cooling allows the separated DNA strands to come together again, re-forming the base pairs in the correct order to restore the original double-stranded DNA (**Figure 3.5A**).

Formation of artificial heteroduplexes

The association of any two complementary nucleic acid strands to form a double-stranded nucleic acid is known as nucleic acid **hybridization** (or annealing). Under experimental conditions, two single nucleic acid strands with a high degree of base complementarity can be allowed to hybridize to form an artificial duplex. For example, if we mix cloned double-stranded DNA fragments that come from two different sources but have high levels of sequence identity, and then heat the mixture to disrupt all hydrogen bonding, all the millions of molecules of double-stranded DNA in the DNA from each of the two sources will be made single-stranded (Figure 3.5B).

Now imagine allowing the mixture to cool slowly: two different types of DNA duplex can form. First, a proportion of the single-stranded DNA molecules will base pair to their original partner to reconstitute the original DNA strands (homoduplexes). But in addition, sometimes a single-stranded DNA molecule will base pair to a complementary DNA strand in the DNA from the other source to form an artificial **heteroduplex** (see Figure 3.5B). (Note that we will use the term heteroduplex to cover all artificial duplexes in which base pairing is not perfect across the lengths of the two complementary strands. In the example in Figure 3.5B there is perfect base matching over the length of the small DNA strands but much of the blue strands remains unpaired. Very rarely, complementary DNA strands from two different sources might be generated that have both identical lengths and perfect base matching—if so, they could form artificial homoduplexes.)

The formation of artificial duplexes, almost always heteroduplexes, is the essence of the nucleic hybridization assays that are widely used in molecular genetics. For convenience we have illustrated cloned double-stranded DNAs in Figure 3.5B. But as we will see below, the starting sources of nucleic acids often include RNA (usually already single-stranded) or synthetic oligonucleotides as well as DNA. Often, too, one or both starting nucleic acids are complex mixtures of fragments, such as total RNA from

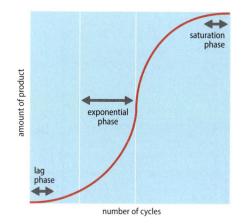

Figure 3.4 Different phases in a PCR reaction. After a lag phase, the amount of PCR product increases gradually at first. In the exponential phase, beginning after about 16–18 cycles and continuing to approximately the 25th cycle, the amount of PCR product is taken to be proportional to the amount of input DNA; quantitative PCR measurements are made on this basis. With further cycles, the amount of product increases at first but then tails off as the saturation phase approaches, when the reaction efficiency diminishes as reaction products increasingly compete with the remaining primer molecules for template DNA.

(A)

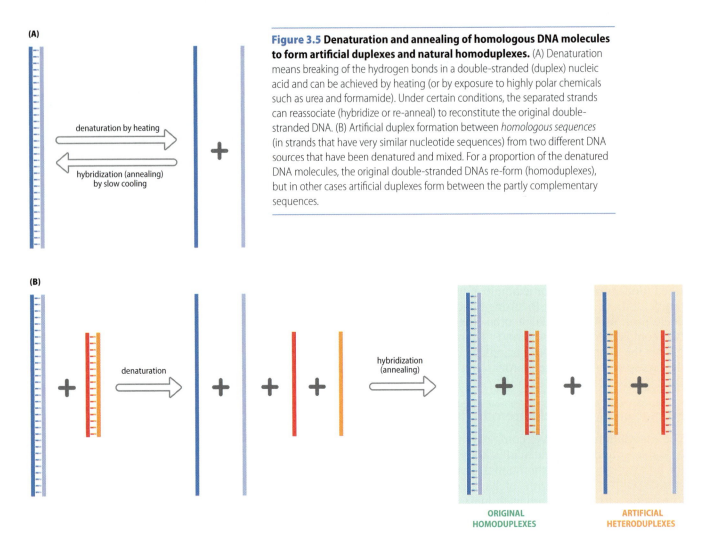

Figure 3.5 Denaturation and annealing of homologous DNA molecules to form artificial duplexes and natural homoduplexes. (A) Denaturation means breaking of the hydrogen bonds in a double-stranded (duplex) nucleic acid and can be achieved by heating (or by exposure to highly polar chemicals such as urea and formamide). Under certain conditions, the separated strands can reassociate (hybridize or re-anneal) to reconstitute the original double-stranded DNA. (B) Artificial duplex formation between *homologous sequences* (in strands that have very similar nucleotide sequences) from two different DNA sources that have been denatured and mixed. For a proportion of the denatured DNA molecules, the original double-stranded DNAs re-form (homoduplexes), but in other cases artificial duplexes form between the partly complementary sequences.

cells or fragments of total genomic DNA. Like cloned DNA, the starting nucleic acids are usually isolated from millions of cells, and so individual sequences are normally present in many copies, often millions of copies.

Hybridization assays: using known nucleic acids to find related sequences in a test nucleic acid population

Nucleic acid hybridization assays exploit the specificity of hybridization. Two single polynucleotide (DNA or RNA) or oligonucleotide strands will form a stable double-stranded hybrid (duplex) only if there is a significant amount of base pairing between them. The stability of the resulting duplex depends on the extent of base matching. Assay conditions can be chosen to allow perfectly matched duplexes only, or to allow degrees of base mismatching.

Hybridization assays can be performed in many different ways, with multiple applications in both research and diagnostics. But there is a common underlying principle: a *known*, well-characterized population of nucleic acid molecules or synthetic oligonucleotides (the **probe** population) is used to interrogate an imperfectly understood population of nucleic acids (the test sample). To do that as required, both nucleic acid populations must be separated into single strands and then mixed so that single probe strands can form artificial duplexes with complementary strands in the test sample.

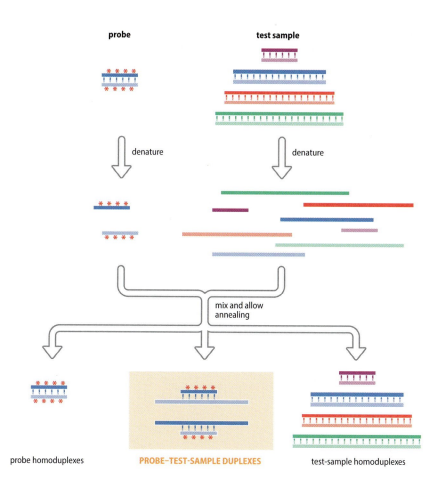

probe   test sample

denature   denature

mix and allow
annealing

probe homoduplexes   **PROBE–TEST-SAMPLE DUPLEXES**   test-sample homoduplexes

Figure 3.6 Heteroduplex formation in a nucleic acid hybridization assay. A defined probe population of known nucleic acid or oligonucleotide sequences and a test nucleic acid sample population are both made single-stranded (as required), then mixed and allowed to anneal. Many of the fragments that had previously been base paired in the two populations will reanneal to reconstitute original homoduplexes (bottom left and bottom right). In addition, new artificial duplexes will be formed between (usually) partly complementary probe and test-sample sequences (bottom centre). The hybridization conditions can be adjusted to favor formation of the novel duplexes. In this way, probes can selectively bind to and identify closely related nucleic acids within a complex nucleic acid population. In this example, the probe has been labeled in some way (red asterisks) but in some hybridization assays it is the test sample nucleic acid that is labeled.

Because the object of a hybridization assay is to use the probe to identify complementary or partly complementary test-sample strands, the probe–test-sample duplexes need to be labeled in some way so that they can be identified. To do that, either the probe or the test sample needs to be labeled at the outset (**Figure 3.6** gives one approach).

Using high and low hybridization stringency

A hybridization assay can be used to identify nucleic acid sequences that are distantly related from a given nucleic acid probe. We might want to start with a DNA clone from a human gene and use that to identify the corresponding mouse gene. The human and mouse genes might be significantly different in sequence, but if we choose a long DNA probe and reduce the stringency of hybridization, stable heteroduplexes can be allowed to form even though there might be significant base mismatches (**Figure 3.7A**).

Conversely, we can choose hybridization conditions to accept only perfect base matching. If we choose an oligonucleotide probe we can use a high hybridization stringency so that the only probe–test duplexes that can form are ones that contain exactly the same sequence as the probe (Figure 3.7B). That can happen because a single mismatch out of, say, 18 base pairs can make the duplex thermodynamically unstable. Oligonucleotides can therefore be used to identify alleles that differ by a single nucleotide (allele-specific oligonucleotides).

Two classes of hybridization assay

There are many types of hybridization assay, but they all fall into two broad classes. In one case the probe molecules are labeled and the

Figure 3.7 Using low or high hybridization stringency to detect nucleic acid sequences that are distantly related or show perfect base matching with a given probe. In any hybridization assay we can control the degree of base matching between complementary strands in the probe and test sample. If, for example, we increase salt concentrations and/or reduce the temperature, we lower hybridization stringency. (A) In some circumstances a long probe strand can form a thermodymically stable duplex with a comparable but distantly related strand within the test DNA (or RNA), even though there might be significant base mismatching. (B) Alternatively, we can use high temperatures and low salt concentrations to achieve high hybridization stringency that might allow only perfect base matching. That is most easily achieved with a short oligonucleotide probe and allows assays to discriminate between alleles that differ at a single nucleotide position.

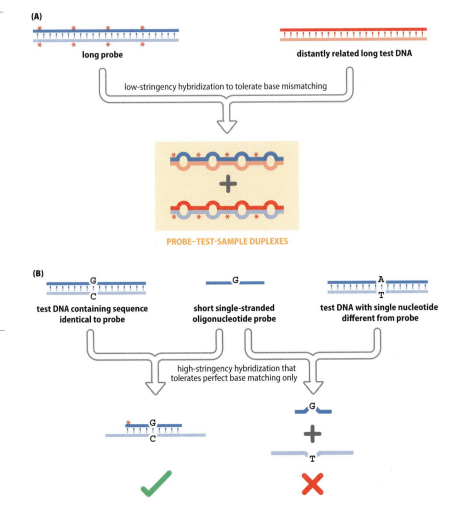

test-sample molecules are unlabeled, as in Figure 3.6. In that case, the probe is often a single type of cloned DNA and it is usually labeled by using a polymerase to synthesize complementary DNA or RNA strands in the presence of one or more fluorescently labeled nucleotides (**Box 3.2**). The alternative type uses unlabeled probe molecules and it is the test-sample molecules that are labeled (see below).

The point of using labeled nucleic acids in a hybridization assay is to allow probe–test sample heteroduplexes to be identified. But how can we distinguish between the label in these duplexes and the label in the original labeled probe or labeled test-sample DNA? The answer is to immobilize the unlabeled nucleic acid population on a solid support (often plastic, glass, or quartz) and expose it to an aqueous solution of the labeled nucleic acid population. When labeled nucleic acid strands hybridize to complementary sequences on the solid support, they will be physically bound to the support, but labeled molecules that do not find a partner on the support or that stick nonspecifically can be washed off. That leaves behind the complementary partners that the assay is designed to find (**Figure 3.8**).

Hybridization assays are used for different purposes (**Table 3.2** gives some examples). For decades, almost all hybridization assays used a homogeneous labeled probe (often, a single type of DNA clone) to search for related sequences in an immobilized complex test nucleic acid sample (see Figure 3.6 and the left part of Figure 3.8). As described in the next section, microarray-based hybridization assays have become very

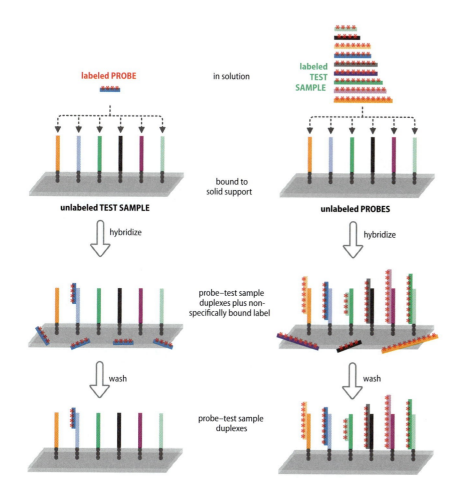

Figure 3.8 The two fundamental classes of hybridization assay and the use of solid supports to capture labeled probe–test sample duplexes. In both assays the unlabeled nucleic acid (or oligonucleotide) population is bound to a solid support and denatured, and is exposed to an aqueous solution of the labeled nucleic acid (or oligonucleotide) population that has also, as required, been denatured. Single-stranded molecules in the labeled population can hybridize to complementary sequences in the unlabeled population and so become bound to the solid support. Other labeled sequences that have not bound, or have bound nonspecifically at incorrect locations on the support, can be washed off. The bound labeled nucleic acids can then be studied and are sometimes retrieved by washing at higher temperatures to break the hydrogen bonds connecting them to their unlabeled partner strands on the supports. In the past, most hybridization assays used the labeled probe/immobilized test sample scheme shown on the left (see Table 3.2), but microarray hybridization assays use the labeled test-sample/immobilized probe scheme shown on the right (see Figure 3.9).

popular; they use unlabeled complex probe populations bound to a surface to interrogate a labeled test sample (see the right part of Figure 3.8 for the principle, and the next section for some detail).

Microarray hybridization: large-scale parallel hybridization to immobilized probes

Innovative and powerful hybridization technologies developed in the early 1990s permit numerous hybridization assays to be conducted simultaneously on a common sample under the same conditions. A DNA or oligonucleotide microarray consists of many thousands or millions of different unlabeled DNA or oligonucleotide probe populations that have been fixed to a glass or other suitable surface within a high-density grid format. Within each grid square are millions of identical copies of just one probe (a grid square with its probe population is called a *feature*). For

Table 3.2 Popularly used hybridization assays.

PROBE AND TEST SAMPLE LABELING	HYBRIDIZATION METHOD	APPLICATIONS	EXAMPLES
Labeled probe and unlabeled test sample (Figures 3.6 and 3.7A)	Southern blot	looking for medium-sized changes (hundreds of base pairs to several kilobases) in genes/DNA in test sample	Box 11.1 Figure 1
	tissue *in situ*	tracking RNA transcripts in tissues and embryos	
	chromosome *in situ*	studying large-scale changes using fixed chromosomes on a slide as the test sample	Figures 10.6A, 11.7B, 11.8
Unlabeled probe and labeled test sample (Figure 3.8)	microarray comparative genome hybridization	genomewide search for large (megabase-sized) changes in the DNA	Figure 11.5
	microarray-based expression profiling	simultaneously tracking expression of very many genes; cancer profiling	Figure 10.19

BOX 3.2 Labeling of Nucleic Acids and Oligonucleotides.

Hybridization assays involve the labeling of either the probe or the test-sample population. Usually this involves making labeled DNA copies of a starting DNA or RNA with a suitable DNA polymerase in the presence of the four precursor deoxynucleotides (dATP, dCTP, dGTP, and dTTP). In the case of a starting RNA, a specialized DNA polymerase, a reverse transcriptase, uses the RNA as a template for making a complementary DNA copy. For some purposes, labeled RNA copies are made of a starting DNA using a RNA polymerase and the four precursor ribonucleotides (ATP, CTP, GTP, and UTP).

Whichever procedure is used, particular chemical groups (labels) are introduced into the DNA or RNA copies and can be specifically detected in some way. Often, at least one of the four nucleotide precursors has been modified so that it has a label attached to the base; alternatively, labeled oligonucleotide primers are incorporated.

Unlike DNA or RNA, oligonucleotides are chemically synthesized by the sequential addition of nucleotide residues to a starting nucleotide that will be the 3' terminal nucleotide. Amine or thiol groups can be incorporated into the oligonucleotide and can then be conjugated with amine-reactive or thiol-reactive labels.

Different labeling systems can be used (**Table 1**). Fluorescent dyes—such as derivatives of fluorescein—are popular; they can be detected readily because they emit fluorescent light of a defined wavelength when suitably stimulated. Some other labels are detected by specific binding to an antibody or to a very strongly interacting protein (see Table 1). In these cases, the detecting protein is usually conjugated to a fluorescent group (**fluorophore** or **fluorochrome**) or to an enzyme, such as alkaline phosphatase or peroxidase, that can permit detection via colorimetric assays or chemical luminescence assays.

LABELING SYSTEM	EXAMPLES OF LABELS	LABEL DETECTION
Fluorescence	FITC (fluorescein isothiocyanate)	using laser scanners/fluorescence microscopy
Antibody detection	digoxigenin (a steroid found in *Digitalis* plants)	via a digoxigenin-specific antibody that is coupled to a fluorophore or suitable enzyme
Specific protein interaction	biotin (= vitamin B₇)	via streptavidin (a bacterial protein with an extraordinarily high affinity for biotin) that has been conjugated to a fluorophore or enzyme

Table 1 Popular systems for labeling nucleic acids.

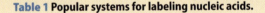

example, oligonucleotide microarrays often have a 1.28 cm × 1.28 cm surface that contain millions of different features, each occupying about 5 or 10 μm² (**Figure 3.9**).

A test sample—an aqueous solution containing a complex population of fluorescently labeled denatured DNA or RNA—is hybridized to the different probe populations on the microarray. After a washing step to remove nonspecific binding of labeled test-sample molecules to the array, the remaining bound fluorescent label is detected with a high-resolution laser scanner. The signal emitted from each feature on the array is analyzed with digital imaging software that converts the fluorescent hybridization signal into one of a palette of colors according to its intensity (see Figure 3.9).

Because the intensity of each hybridization signal reflects the number of labeled molecules that have bound to a feature, microarray hybridization is used to *quantitate* different sequences in complex test-sample populations such as different samples of genomic DNA or total cellular RNA (or cDNA). Frequent applications include quantifying different transcripts (expression profiling) and also scanning genomes to look for large-scale deletions and duplications, as described in Chapter 11.

3.3 Principles of DNA Sequencing

DNA sequencing is the ultimate DNA test. Until quite recently, Sanger dideoxy DNA sequencing was the predominant method. It relies on amplifying individual DNA sequences. For each amplified DNA, nested sets of labeled DNA copies are made and then separated according to size by gel electrophoresis.

In the last few years completely different technologies have allowed massively parallel DNA sequencing. No attempt is made to obtain the sequence of just a purified DNA component; instead, millions of DNA fragments present in a complex DNA sample are simultaneously sequenced without the need for gel electrophoresis.

Dideoxy DNA sequencing remains widely used for investigating specific DNA sequences, for example testing whether individuals have mutations in a particular gene. What the newer DNA sequencing technologies offer is a marked increase in sequencing capacity and the ability to sequence complex DNA populations, such as genomic DNA sequences, very rapidly. As a result of fast-developing technology, the running costs of DNA sequencing are plummeting, and very rapid sequencing of whole genomes is quickly becoming routine.

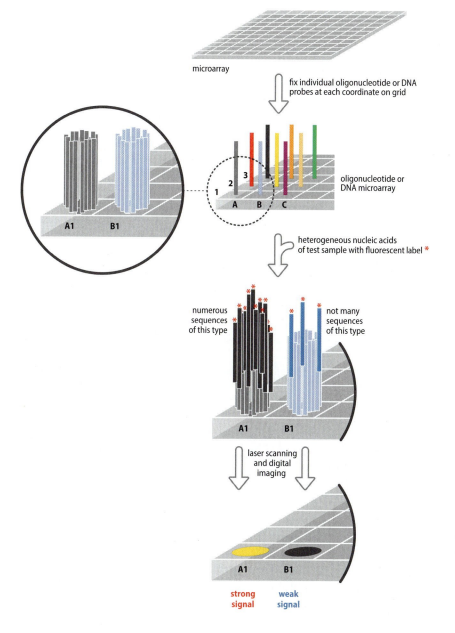

Figure 3.9 Principle of microarray hybridization. A microarray is a solid surface on which molecules can be fixed at specific coordinates in a high-density grid format. Oligonucleotide or DNA microarrays have thousands to millions of different synthetic single-stranded oligonucleotide or DNA probes fixed at specific predetermined positions in the grid. As shown by the expanded item enclosed within dashed lines, each grid square will have many thousands of identical copies of a single type of oligonucleotide or DNA probe (a *feature*). An aqueous test sample containing a heterogeneous collection of labeled DNA fragments or RNA transcripts is denatured and allowed to hybridize with the probes on the array. Some probes (for example the A1 feature) may find numerous complementary sequences in the test population, resulting in a strong hybridization signal; for other probes (for example the B1 feature) there may be few complementary sequences in the test sample, resulting in a weak hybridization signal. After washing and drying of the grid, the hybridization signals for the numerous different probes are detected by laser scanning, giving huge amounts of data from a single experiment. (For ease of illustration, we show test-sample nucleic acids with end labels, but sometimes they contain labels on internal nucleotides.)

Dideoxy DNA sequencing

Like PCR, dideoxy DNA sequencing uses primers and a DNA polymerase to make DNA copies of specific DNA sequences of interest. To obtain enough DNA for sequencing, the DNA sequences are amplified by PCR (or sometimes by cloning in cells). The resulting purified DNAs are then sequenced, one after another, in individual reactions. Each reaction begins by denaturing a selected purified DNA. A single oligonucleotide primer is then allowed to bind and is used to make labeled DNA copies of the desired sequence (using a provided DNA polymerase and the four dNTPs).

Instead of making full-length copies of the sequence, the DNA synthesis reactions are designed to produce a population of DNA fragments sharing a common 5′ end sequence (defined by the primer sequence) but with variable 3′ ends. This is achieved by simultaneously having the standard dNTP precursors of DNA plus low concentrations of ddNTPs, dideoxynucleotide analogs that differ from a standard deoxynucleotide only in that they lack an OH group at the 3′ carbon of the sugar as well as at the 2′ carbon (**Figure 3.10A**).

DNA synthesis continues smoothly when dNTPs are used, but once a dideoxynucleotide has been incorporated into a growing DNA molecule,

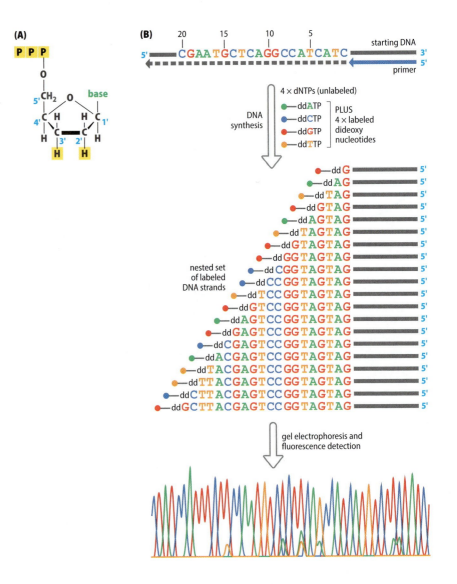

Figure 3.10 Principle of dideoxy sequencing. (A) Generalized structure of a 2′,3′ ddNTP. The sugar is *dideoxyribose* because the hydroxyl groups attached to both carbons 2′ and 3′ of ribose are each replaced by a hydrogen atom (shown by shading). (B) In dideoxy sequencing reactions, a DNA polymerase uses an oligonucleotide primer to make complementary sequences from a purified single-stranded starting DNA. The sequencing reactions include ddNTPs, which compete with the standard dNTPs to insert a chain-terminating dideoxynucleotide. Different labeling systems can be used, but it is convenient to use labeled ddNTPs that have different fluorescent groups according to the type of base, as shown here. The DNA copies will have a common 5′ end (defined by the sequencing primer) but variable 3′ ends, depending on where a labeled dideoxynucleotide has been inserted, producing a nested set of DNA fragments that differ by a single nucleotide in length. A series of nested fragments that differ incrementally by one nucleotide from their common 5′ end are fractionated according to size by gel electrophoresis; the fluorescent signals are recorded and interpreted to produce a linear base sequence such as the example given at the bottom.

chain extension is immediately terminated (the dideoxynucleotide lacks a 3'-OH group to form a phosphodiester bond). To keep the balance tilted toward chain elongation, the ratio of each ddNTP to the corresponding dNTP is set to be about 1:100, so that a dideoxynucleotide is incorporated at only about 1% of the available nucleotide positions.

If we consider competition between ddATP and dATP in the example in Figure 3.10B, there are four available positions for nucleotide insertion: opposite the T at nucleotide positions 2, 5, 13, and 16 in the starting DNA. Because the DNA synthesis reaction results in numerous DNA copies, then by chance some copies will have a dideoxyA incorporated opposite the T at position 2, some will have a dideoxyA opposite the T at position 5, and so on. Effectively, chain elongation is *randomly* inhibited, producing sets of DNA strands that have a common 5' end but variable 3' ends.

Fluorescent dyes are used to label the DNA. One convenient way of doing this, as shown in Figure 3.10B, is to arrange matters so that the four different ddNTPs are labeled with different fluorescent dyes. The reaction products will therefore consist of DNA strands that have a labeled dideoxynucleotide at the 3' end carrying a distinctive fluorophore according to the type of base incorporated.

All that remains is to separate the DNA fragments according to size by electrophoresis (Box 3.3) and to detect the fluorescence signals. In modern dideoxy sequencing, as the DNA fragments migrate in the gel they pass a laser that excites the fluorophores, causing them to emit fluorescence at distinct wavelengths. The fluorescence signals are recorded and an output is provided in the form of intensity profiles for the differently coloured fluorophores, as shown at the bottom of Figure 3.10B.

Dideoxy DNA sequencing is disadvantaged by relying on gel electrophoresis (slab polyacrylamide gels were used initially; more modern machines use capillary electrophoresis (see Box 3.3)). Because gel electrophoresis is not suitable for handling large numbers of samples at a time, dideoxy sequencing has a limited sequence capacity. It is therefore not well suited to genome sequencing (although it has been used in the past to obtain the first human genome sequences). In modern times it is often used for analyzing variation over small DNA regions, such as regions encompassing individual exons.

Massively parallel DNA sequencing (next-generation sequencing)

In the early to mid-2000s new sequencing-by-synthesis methods were developed that could record the DNA sequence while the DNA strand is being synthesized. That is, the sequencing method was able to monitor the incorporation of each nucleotide in the growing DNA chain and to identify which nucleotide was being incorporated at each step.

The new sequencing technologies, often called next-generation sequencing (NGS), represent a radical step-change in sequencing technology. Standard dideoxy sequencing is a highly targeted method requiring the purification of specific sequences of interest that are then selected to be sequenced, one after another. By contrast, massively parallel DNA sequencing is indiscriminate: all of the different DNA fragments in a complex starting DNA sample can be *simultaneously* sequenced without any need for gel electrophoresis. The difference in sequencing output is therefore vast.

As listed in Table 3.3, various NGS technologies are commercially well established. Some of them require amplification of the starting DNA; others rely on unamplified starting DNA ('single-molecule sequencing').

BOX 3.3 Slab Gel Electrophoresis and Capillary Electrophoresis for Separating Nucleic Acids According to Size.

Nucleic acids carry numerous negatively charged phosphate groups and will migrate toward the positive electrode when placed in an electric field. By arranging for them to migrate through a porous gel during electrophoresis, nucleic acid molecules can be fractionated according to size. The porous gel acts as a sieve: small molecules pass easily through the pores of the gel, but larger fragments are impeded by frictional forces.

Standard gel electrophoresis with agarose gels allows the fractionation of moderately large DNA fragments (usually from about 0.1 kb to 20 kb). Pulsed-field gel electrophoresis can be used to separate much larger DNA fragments (up to megabases long). It uses specialized equipment in which the electrical polarity is regularly changed, forcing the DNA molecules to alter their conformation periodically in preparation for migrating in a different direction. Polyacrylamide gel electrophoresis allows the superior resolution of smaller nucleic acids (it is usually used to separate fragments in size ranges up to 1 kb) and is used in dideoxy DNA sequencing to separate fragments that differ in length by just a single nucleotide.

In slab gel electrophoresis, individual samples are loaded into cut-out wells at one end of a solid slab of agarose or polyacrylamide gel. They migrate in parallel lanes toward the positive electrode (**Figure 1**). The separated nucleic acids can be detected in different ways. For example, after the end of an electrophoresis run, the gels can be stained with chemicals such as ethidium bromide or SYBR green that bind to nucleic acids and fluoresce when exposed to ultraviolet radiation. Sometimes the nucleic acids are labeled with fluorophores before electrophoresis, and during electrophoresis a recorder detects the fluorescence of individual labeled nucleic acid fragments as they sequentially pass a recorder placed opposite a fixed position in the gel.

The disadvantage of slab gel electrophoresis is that it is labor-intensive. The modern trend is to use capillary gel electrophoresis, which is largely automated. Fluorescently labeled DNA samples migrate through individual long and very thin tubes containing polyacrylamide gel, and a recorder detects fluorescence emissions as samples pass a fixed point (**Figure 2**). Modern dideoxy DNA sequencing uses capillary electrophoresis, as do many different types of diagnostic DNA screening methods that we outline in Chapter 11.

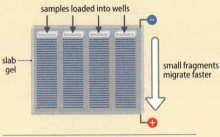

Figure 1 Slab gel electrophoresis.

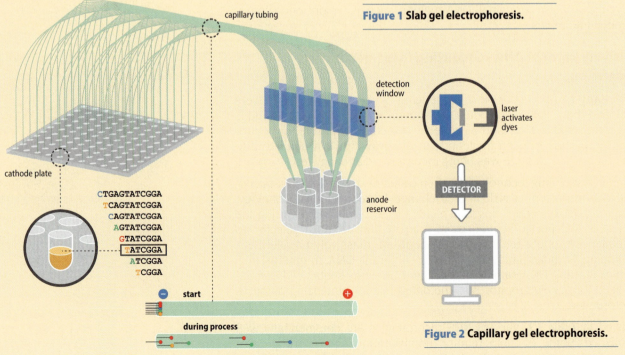

Figure 2 Capillary gel electrophoresis.

TECHNOLOGY CLASS	SEQUENCING MACHINE	READ LENGTH (NUCLEOTIDES)	READS PER RUN	RUN TIME
Conventional (chain termination sequencing)	ABI prism 3730 Sanger dideoxy sequencing	400–900	96	20 minutes to 3 hours
Massively parallel sequencing of PCR-amplified DNAs	Roche/454 pyrosequencer	400–600	1 million	7 hours
	Illumina/Solexa HiSeq 2000	150 × 2	many hundreds of millions	2 days to 10 days
	ABI SOLiD 4	35–75	hundreds of millions	7 days
	Life Technologies Ion Torrent	200	5 million	4 hours
Massively parallel sequencing of unamplified (single-molecule) DNAs	Pacific Biosciences SMRT (single-molecule real time) sequencing	~3000	up to 75,000	1 hour

They vary in different parameters, such as read lengths (the length of DNA sequence generated per starting DNA), run lengths, and the number of different DNA sequences that can be conducted in parallel.

Table 3.3 Major characteristics of some commercially available DNA sequencing technologies.

By comparison with the standard Sanger dideoxy sequencing, the NGS methods generally have high intrinsic error rates in base calling but the final reported sequences are much more accurate than the initial reads (after quality filtering and comparison of multiple sequence reads). And, importantly, they have significantly cheaper running costs per base (for example, the cost of sequencing with the HiSeq2000 machine is 1/34,000 of the cost when using Sanger dideoxy sequencing on an ABI3730 machine). Note, however, that the HiSeq2000 machine and other similar machines are not suited to low-capacity sequencing; they are economic only when processing very large numbers of different DNA sequences.

A variety of additional single-molecule sequencing technologies are currently also being piloted, and sequencing capacity is likely to be increased in the near future, with yet further decreases in sequencing costs. We will describe two widely used massively parallel DNA sequencing technologies in Chapter 11, but readers interested in the chemistry of these new technologies are advised to consult recent reviews such as the Mardis (2013) review listed in Further Reading.

NGS technologies are well suited to sequencing whole genomes and have sparked various projects to sequence the genomes of different individuals, as described in later chapters. They are also being used to sequence whole transcriptomes from various normal and abnormal tissues (by converting total RNA to cDNA). We consider medical applications of NGS technology in later chapters.

Summary

- In complex genomes, an individual gene, exon, or other sequence of interest is often a tiny fraction of the genome. To study a specific short DNA sequence like this either we must first *purify* it (by artificially increasing its copy number by replicating it with a DNA polymerase), or use some method that specifically *tracks* the sequence of interest.

- Making multiple copies of a DNA sequence can be done within cells (DNA cloning), or by using a cell-free system, the polymerase chain reaction (PCR).

- In cell-based DNA cloning, the DNA sequence of interest is first attached to a vector molecule that can self-replicate in a suitable host cell (often a bacterial cell). Vector molecules are modified DNAs that can readily replicate in the host cell, such as small circular plasmids or different types of bacteriophage.

- Restriction nucleases are used to cut large DNA molecules, such as chromosomal DNAs, into small pieces of discrete sizes that can easily be joined to similarly cut vector molecules, producing recombinant DNA molecules.

- Recombinant DNA molecules can be induced to enter a suitable host cell (transformation). Transformation is selective: each transformed cell has normally taken up a *single* DNA molecule. A transformed bacterial cell can multiply many times, and large numbers of identical copies of the recombinant DNA are produced.

- A DNA library is a bank of DNA clones that collectively include many different DNA sequences representing a complex starting population of genomic DNA or cDNA copies of a complex RNA population.

- In PCR, a DNA sequence of interest can be copied many times from a complex source of DNA by *in vitro* DNA synthesis. Specific oligonucleotide primers are designed to bind to the starting DNA at positions flanking the sequence of interest and then used to make DNA copies that can themselves serve as templates for making further copies, rapidly increasing the copy number of the sequence of interest.

- Nucleic acid hybridization is the key method used to track a DNA or RNA sequence of interest. The method relies on the specificity of base pairing—if two different nucleic acids are related in sequence, they may be able to form an artificial duplex that is stable under selected experimental conditions.

- To perform nucleic acid hybridization, a test nucleic acid population with some sequence of interest is made single-stranded (denatured) and mixed with a probe population of known denatured nucleic acids. The object is to identify heteroduplexes in which a single-stranded sequence of interest in the test sample has formed a stable hybrid with a known sequence within the probe population.

- In many types of nucleic acid hybridization, a homogeneous labeled probe population is used to identify related sequences in an unlabeled test population that is typically bound to a solid surface.

- In microarray hybridization, many thousands of unlabeled oligonucleotide probe populations are attached to a solid surface in a regular grid formation and hybridized in parallel with a labeled test nucleic acid population provided in solution. According to the amount of labeled DNA bound to each type of oligonucleotide, it is possible to quantitate specific sequences that are complementary to each of the different probes.

- In DNA sequencing, DNA samples are made single-stranded and an DNA polymerase is used to synthesize a complementary DNA in a way that provides a read-out of the base sequence.

- In standard dideoxy DNA sequencing, selected individual DNA samples are sequenced. The DNA synthesis step uses a mixture of normal and chain-terminating nucleotides, producing a nested set of fragments that differ incrementally by one nucleotide and that can be separated by gel electrophoresis.

- In massively parallel DNA sequencing (next-generation sequencing), a complex population of very many (often millions of) DNA templates are sequenced simultaneously and indiscriminately. There is no gel electrophoresis. Instead, the methods rely on being able to monitor which of the four nucleotides is being incorporated at each step in synthesizing the cDNA.

Questions

Help on answering these questions and a multiple-choice quiz can be found at www.garlandscience.com/ggm-students

1. What are the two broad classes of experimental method used to amplify (increase the copy number of) a DNA sequence of interest? Explain the essential difference between the two approaches.

2. What is a restriction endonuclease and what is its natural function?

3. Restriction nucleases often recognize palindromic DNA sequences and often cut the recognition sequence asymmetrically. For example, the restriction endonuclease EcoRI cleaves the sequence GGATCC by breaking the bond connecting the two guanines in the recognition sequence. What is meant by a palindromic DNA sequence, and why does asymmetric cutting of a palindromic DNA sequence generate fragments with 'sticky ends'?

4. When compared with cloning DNA in cells, PCR has multiple advantages as a method of amplifying DNA. Describe four advantages.

5. When compared with cloning DNA in cells, PCR has two major disadvantages. What are they?

6. Labeling of nucleic acids is an important way of tracking them in different types of reaction, such as in DNA hybridization, in DNA sequencing, and in quantitative PCR. That typically involves incorporating nucleotides that have been labeled in some way into the nucleic acid. The nucleotides are usually labeled by covalently attaching to them a specific chemical group that can be detected in some specific way. Two popular methods involve attaching some type of fluorescent dye or a biotin group. Explain how these labels can be specifically detected.

7. In a hybridization assay what is meant by a probe, and what is the point of a hybridization assay?

8. According to the conceptual design, there are 'standard' and reverse hybridization assays. How do they differ, and in what class would you place microarray hybridization?

9. In some hybridization assays, the hybridization stringency is deliberately designed to be low; other hybridization assays depend on very stringent hybridization. What is meant by hybridization stringency, how can high and low stringency be achieved, and in what circumstances would very high and low hybridization stringencies be required for a hybridization assay?

10. What function do dideoxynucleotides have in Sanger DNA sequencing?

11. Many types of DNA sequencing involve a DNA synthesis reaction in which labeled nucleotides are incorporated. Some of these methods also use some form of gel electrophoresis, but others are automated methods that do not use gel electrophoresis. When employed in DNA sequencing, what is the purpose of gel electrophoresis? Why is it not needed in some DNA sequencing methods?

Further Reading

DNA Cloning and PCR

Arya M et al. (2005) Basic principles of real-time quantitative PCR. *Expert Rev Mol Diagn* 5:209–219; PMID 15833050.

Brown TA (2010) Gene cloning and DNA analyses. An Introduction, 6th ed. Wiley-Blackwell.

McPherson M & Moller S (2006) PCR. The Basics, 2nd ed. Taylor & Francis Group.

Microarray-Based Hybridization

Geschwind DH (2003) DNA microarrays: translation of the genome from laboratory to clinic. Lancet Neurol 2:275–282; PMID 12849181.

Massively Parallel DNA Sequencing

Liu L et al. (2012) Comparison of next-generation sequencing systems. *J Biomed Biotechnol* 2012; PMID 22829749. [Comparison of performance features of commercially available platforms that sequence amplified DNA.]

Mardis ER (2013) Next-generation sequencing platforms. *Annu Rev Anal Chem* 6:287–303; PMID 23560931. [Detailed description of the chemistries of various current methods.]

Tucker T et al. (2009) Massively parallel sequencing: the next big thing in genetic medicine. *Am J Hum Genet* 85:142–154 ; PMID 19679224.

Principles of Genetic Variation

As explained in Section 1.3, each sperm cell and each egg cell is genetically unique. Every one of us arose from a single cell that formed by the fusion of a unique sperm and a unique egg cell.

Genetic variation describes differences between the DNA sequences of individual genomes. Because each of us has two nuclear genomes (a paternal genome and a maternal genome), genetic variation occurs within as well as between individuals. At any genetic **locus** (DNA region having a unique chromosomal location) the maternal and paternal **alleles** normally have identical or slightly different DNA sequences (we are said to be **homozygotes** if the alleles are identical, or **heterozygotes** if they differ by even a single nucleotide).

In addition to the genetic variation that we inherit, and that is present in all our cells (**constitutional** variation), DNA changes occur in the DNA of our cells throughout life (*post-zygotic* or *somatic* genetic variation). Most of them occur in a rather random fashion and cause small differences in the DNA within different cells of our body. However, programmed, cell-specific DNA changes occur in maturing B and T cells that allow each of us to make a wide range of different antibodies and T-cell receptors.

Individuals differ from each other mostly because our DNA sequences differ, but genetic variation is not the only explanation for differences in **phenotype** (our observable characteristics). A fertilized egg cell can split in two in early development and give rise to genetically identical twins (monozygotic twins) that nevertheless grow up to be different: although hugely important, genetic variation is not the only influence on phenotype. During development, additional effects on the phenotype occur by a combination of stochastic (random) factors, differential gene–environment interactions and epigenetic variation that is not attributable to changes in base sequence. We consider epigenetic effects and environmental factors in Chapter 6 when we examine how our genes are regulated.

In this chapter we look at general *principles* of human genetic variation and how variation in DNA relates to variation in the sequences of proteins and noncoding RNAs. We are not concerned here with the very small fraction of genetic variation that causes disease. That will be covered in later chapters, especially in Chapter 7 (where we look primarily at genetic variation in relation to monogenic disorders), Chapter 8 (genetic variation in relation to complex inherited diseases) and Chapter 10 (genetic variation and cancer).

In Section 4.1 we consider the origins of DNA sequence variation. DNA repair mechanisms seek to minimize the effects of DNA sequence variation, and in Section 4.2 we outline the different DNA repair mechanisms that work in our cells. We take a broad look at the extent of human genetic variation in Section 4.3 and at the different forms in which this variation manifests. In Section 4.4 we deal with functional genetic variation. Here, we examine in a general way how variation in the sequences of protein products is determined both by genetic variation and by post-transcriptional modification. In this section we also deal with aspects of population genetics that relate to the spread of advantageous DNA variants through human populations (but the population genetics of harmful disease-associated DNA variants is examined in Chapter 5).

Genetic variation is most highly developed in genes that work in recognizing foreign, potentially harmful molecules which have been introduced into the body. These molecules are sometimes under independent genetic control because they originate from another organism, as in the case of molecular components of microbial pathogens, and ingested plant and fungal toxins. When that happens, two types of Darwinian **natural selection** may sometime oppose each other, one that operates on us and one that operates on the organism that poses a potential threat to us. (We describe different types of natural selection later in this chapter and in following chapters. For now, consider natural selection as the process whereby some allele or combination of alleles determines a phenotype that may confer an increased or reduced chance of survival and reproductive success, with a resulting increase in frequency of favorable allele(s) in the population or reduced frequency of disadvantageous alleles.

Take the example of invading microbial pathogens. Natural selection works to maximize genetic variation in the frontline immune system genes that help us recognize **antigens** on the invaders. Some genetic variants in these genes will be more effective than others; accordingly, some individuals in the population will be more resistant than others to the potential harmful effects of specific microbial pathogens. But natural selection also works on the microbial pathogens to maximize genetic diversity of external molecules in an effort to escape detection by our immune defense systems.

As described in Section 4.5, the frontline genes in our immune system defenses need to recognize a potentially huge number of foreign antigens. Here, we describe the basis for the quite exceptional variability of some human leukocyte antigen (HLA) proteins and the medical significance of this variability. We also consider how exceptional post-zygotic genetic variation is created at our immunoglobulin and T-cell receptor loci so that an individual person can make a huge number of different antibodies and T-cell receptors.

As a result of natural selection too, some of the enzymes that work in breaking down complex molecules in the food and drink that we consume have developed quite high levels of polymorphism. Over very many generations they have had to deal with occasional ingestion of harmful molecules, such as plant and fungal toxins. Because they also handle the metabolism of artificial drugs, genetic variation in the same enzymes explains the wide variation between individuals in the ways in which we metabolize and respond to drugs. We consider this subject—pharmacogenetics—within the context of drug treatment in Section 9.2.

4.1   Origins of DNA Sequence Variation

Underlying genetic variation are changes in DNA sequences. **Mutation** describes both a process that produces altered DNA sequences (either a change in the base sequence or in the number of copies of a specific DNA sequence) and the outcome of that change (the altered DNA sequence). As events, mutations can occur at a wide variety of levels, and can have different consequences. They may contribute to a normal phenotype (such as height) or to a disease phenotype. They may have no obvious effect on the phenotype, or, very rarely, they may have some beneficial effect.

Mutations originate as a result of changes in our DNA that are not corrected by cellular DNA repair systems. The DNA changes are occasionally induced by radiation and chemicals in our environment, but the great majority arise from endogenous sources. The latter include spontaneous errors in normal cellular mechanisms that regulate chromosome segregation, recombination, DNA replication, and DNA repair and also spontaneous chemical damage to DNA.

Mutations are inevitable. They may have adverse effects on individual organisms, causing aging and contributing to many human diseases. But they also provide the raw fuel for natural selection of beneficial adaptations that allow evolutionary innovation and, ultimately, the origin of new species.

Genetic variation arising from endogenous errors in chromosome and DNA function

Natural errors in various processes that affect chromosome and DNA function—chromosome segregation, recombination, and DNA replication—are important contributors to genetic variation. That happens because no cellular function can occur with 100% efficiency—mistakes are inevitable. Endogenous errors in the above processes may often not have harmful consequences, but some of them make important contributions to disease. We examine in detail how they can cause disease in Chapter 7; in this section we take a broad look into how they affect genetic variation in general.

DNA replication errors

General errors in DNA replication are unavoidable. Each time the DNA of a human diploid cell replicates, 6 billion nucleotides need to be inserted in the correct order to make new DNA molecules. Not surprisingly, DNA polymerases very occasionally insert the wrong nucleotide, resulting in mispaired bases (a base mismatch; the likelihood of such an error simply reflects the relative binding energies of correctly paired bases and mispaired bases).

In the great majority of cases, the errors are quickly corrected by the DNA polymerase itself. The major DNA polymerases engaged in replicating our DNA have an intrinsic 3′→5′ exonuclease activity with a *proofreading function*. If, by error, the wrong base is inserted, the 3′→5′ exonuclease is activated and degrades the newly synthesized DNA strand from its 3′ end, removing the wrongly inserted nucleotide and a short stretch before it. Then the DNA polymerase resumes synthesis again. If mispaired bases are not eliminated by the DNA polymerase, a DNA mismatch repair system is activated (as explained below).

Another type of DNA replication error commonly occurs within regions of DNA where there are short tandem oligonucleotide repeats. If, for example, the DNA polymerase encounters a 30-nucleotide sequence with 15 sequential repeats of AT dinucleotide or 10 sequential repeats of the CAA trinucleotide, there will be an increased chance that during DNA replication a mistake is made in aligning the growing DNA strand with its template strand. A frequent result is that the template strand and newly synthesized strand pair up out of register by one (or sometimes more) repeat units, causing **replication slippage**, as detailed below. Errors like this are also often repaired successfully by the DNA mismatch repair system.

Although the vast majority of DNA changes caused by DNA replication errors are identified and corrected, some persist. That happens because although we have many very effective DNA repair pathways, DNA repair is also not 100% effective: unrepaired changes in DNA sequence are an important source of mutations. However, the great majority of the mutations introduced in this way do not cause disease: almost 99% of the genome is noncoding DNA, many mutations in coding DNA do not change an amino acid, and functional noncoding DNA sequences (such as those that specify noncoding RNAs) can often tolerate many kinds of sequence change without compromising the function of the RNA.

Chromosome segregation and recombination errors

Errors in chromosome segregation result in abnormal gametes, embryos, and somatic cells that have fewer or more chromosomes than normal and so have altered numbers of whole DNA molecules. Changes in chromosomal DNA copy number are not uncommon. If they occur in the germ line they often cause embryonic lethality or a congenital disorder (such as Down syndrome, which is commonly caused by an extra copy of chromosome 21), but changes in copy number of sex chromosomes are more readily tolerated. In somatic cells, changes in chromosomal DNA copy number are a common feature of many cancers.

Various natural errors can also give rise to altered copy number of a specific sequence within a DNA strand that may range up to megabases in length. That can occur by different recombination (and recombination-like) mechanisms in which non-allelic (but often related) sequences align so that chromatids are paired with their DNA sequences locally out of register. Subsequent crossover (or sister chromatid exchange) produces chromatids with fewer or more copies of the sequences. The ensuing duplication or deletion of sequences may, or may not, have functional consequences—we cover the mechanisms and how they can result in disease in Chapter 7.

Various endogenous and exogenous sources can cause damage to DNA by altering its chemical structure

DNA is a comparatively stable molecule. Nevertheless there are constant threats to its integrity, causing breakage of covalent bonds within DNA or inappropriate bonding of chemicals to DNA. Most of the damage originates spontaneously within cells (normal cellular metabolism generates some chemicals that are harmful to cells). A minority of the damage is induced by external sources.

Chemical damage to DNA can involve the cleavage of covalent bonds in the sugar–phosphate backbone of DNA, causing single-strand or double-strand breaks (**Figure 4.1A**). Alternatively, bases are deleted (by cleavage of the N-glycosidic bond connecting a base to a sugar; Figure 4.1B), or they are chemically modified in some way.

Many base modifications involve replacing certain groups on bases, or adding a chemical group: a methyl or larger alkyl group, or some other large chemical (Figure 4.1C). Sometimes base modification involves the formation of covalent bonds between two bases (**crosslinking**); the bases may be on the same strand (Figure 4.1Di) or on complementary DNA strands (Figure 4.1Dii). Chemically modified bases may block DNA or RNA polymerases, and cause base mispairing; if not repaired, they may induce mutations.

Endogenous chemical damage to DNA

Most of the chemical damage to our DNA arises spontaneously and is unavoidable. Every day, under normal conditions, around 20,000–100,000

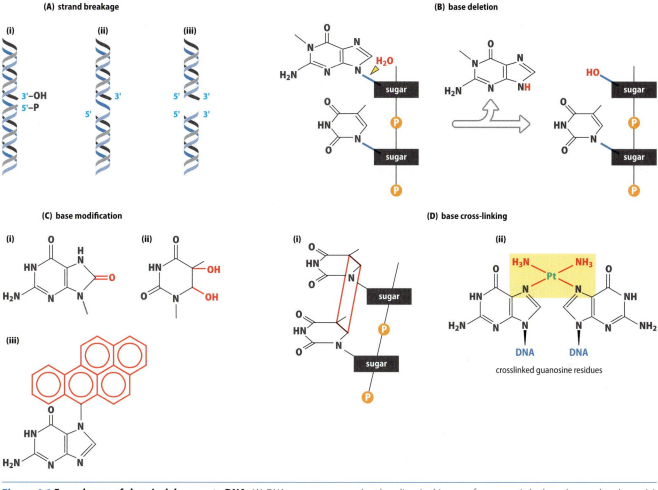

Figure 4.1 Four classes of chemical damage to DNA. (A) *DNA strand breakage.* A single strand may be broken by simple cleavage of a phosphodiester bond (i) or by a more complex single-strand break (ii) in which the ends are damaged and sometimes one or more nucleotides are deleted. Double-strand DNA breaks (iii) occur when both strands are broken at sites that are in very close proximity. (B) *Base deletion.* Hydrolysis cleaves the covalent N-glycosidic bond (shown in deep blue) connecting a base to its sugar. (C) *Base modification.* Altered bonding or added chemical groups are shown in red. Examples are: 8-oxoguanine (i), which base pairs to adenine and so induces mutations; thymidine glycol (ii), which is not a mutagen but blocks DNA polymerase; and a DNA adduct (iii) formed by covalent bonding, in this case of an aromatic hydrocarbon such as benzo(*a*) pyrene to N7 of a guanine residue. (D) *Base cross-linking.* This involves the formation of new covalent bonds linking two bases that may be on the same DNA strand (an *intrastrand cross-link*) or on complementary DNA strands (an *interstrand cross-link*). The former include cyclobutane pyrimidine dimers: linked carbon atoms 4 and 5 on adjacent pyrimidines on a DNA strand (i). This is the most prevalent form of damage incurred by exposure to UV radiation from the sun. The anti-cancer agent cisplatin, $(NH_3)_2PtCl_2$, causes interstrand cross-links by covalently bonding the N7 nitrogen atoms of guanines on opposite strands (ii).

lesions are generated in the DNA of each of our nucleated cells. Three major types of chemical change occur, as listed below. Hydrolytic and oxidative damage are particularly significant, breaking various covalent bonds in the nucleotides of DNA (**Figure 4.2**).

- *Hydrolytic damage.* Hydrolysis is inevitable in the aqueous environment of cells. It can disrupt bonds that hold bases to sugars, cleaving the base from the sugar to produce an abasic site—loss of purine bases (depurination) is particularly common (see Figure 4.1B). Hydrolysis also strips amino groups from some bases (deamination), leaving a carbonyl (C=O) group. Cytosines are often deaminated to give uracil, which base pairs with adenine (see the left part of Figure 4.5); adenine is occasionally deaminated to produce hypoxanthine, which effectively behaves like guanine by base pairing with cytosine.

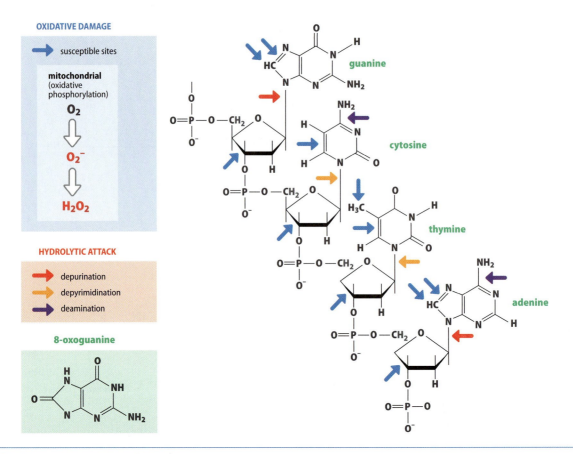

Figure 4.2 DNA sites that are susceptible to spontaneous hydrolytic attack and oxidative damage. Each day every nucleated human cell loses 5000 or more purines (A and G) and about 300 pyrimidines (C and T) as a result of hydrolytic attack. The N-glycosidic bond connecting a base to its deoxyribose sugar is cleaved by H_2O (see Figure 4.1B for the reaction). In deamination, an amino group is replaced by an oxygen atom to give a carbonyl group. About 100–500 cytosines are replaced by uracil in each cell every day (see the left part of Figure 4.5); a smaller number of adenines are replaced by hypoxanthines. In addition, normal metabolism generates reactive oxygen species that cleave certain chemical bonds not just in bases but also in sugar residues, causing breakage of DNA strands.

- *Oxidative damage.* Normal cellular metabolism generates some strongly electrophilic (and therefore highly reactive) molecules or ions. The most significant are **reactive oxygen species** (**ROS**) formed by the incomplete one-electron reduction of oxygen, including superoxide anions (O_2^-), hydrogen peroxide (H_2O_2), and hydroxyl radicals ($OH^\bullet$). ROS are generated in different cellular locations and have important roles in certain intercellular and intracellular signaling pathways, but they mostly originate in mitochondria (where electrons can prematurely reduce oxygen).

  Endogenous ROS attack covalent bonds in sugars, causing DNA strands to break. They also attack DNA bases, especially purines (see Figure 4.2); many derivatives are produced from each base. Some of the base derivatives are highly mutagenic, such as 7,8-dihydro-8-oxo-guanine (also called 8-oxoguanine or 8-hydroxyguanine), which base pairs with adenine; others are not mutagenic but nevertheless block DNA and RNA polymerases (see Figure 4.1C).

- *Aberrant DNA methylation.* As detailed in Chapter 6, many cytosines in our DNA are methylated by methyltransferases. Cells also use *S*-adenosylmethionine (SAM) as a methyl donor in a non-enzymatic reaction to methylate different types of molecules, but sometimes SAM can inappropriately methylate DNA to produce harmful bases. Each day about 300–600 adenines in each nucleated cell are converted to 3-methyladenine, a cytotoxic base that distorts the double helix, disrupting crucial DNA–protein interactions.

Chemical damage to DNA caused by external mutagens

A minority of the chemical damage to our DNA is caused by external agents that can induce mutation (**mutagens**), including radiation and harmful chemicals in the environment. Ionizing radiation (such as X-rays and gamma rays, and so on) interacts with cellular molecules to generate ROS that break chemical bonds in the sugar–phosphate backbone, breaking DNA strands (see below). Non-ionizing ultraviolet radiation causes covalent bonding between adjacent pyrimidines on a DNA strand (see Figure 4.1Di).

Our bodies are exposed to many harmful environmental chemicals—in our food and drink, in the air that we breathe, and so on. Some chemicals interact with cellular molecules to generate ROS. Other chemicals covalently bond to DNA forming a DNA adduct that may be bulky, causing distortion of the double helix. Cigarette smoke and automobile fumes, for example, have large aromatic hydrocarbons that can bond to DNA (see Figure 4.1Ciii). Electrophilic alkylating agents can result in base cross-linking.

4.2   DNA Repair

Cells have different systems for detecting and repairing DNA damage, according to the type of DNA damage. Some types of DNA damage may be minor—the net effect might simply be an altered base. Others, such as DNA cross-linking, are more problematic: they may block DNA replication (the replication fork stalls) or transcription (the RNA polymerase stalls).

Different molecular sensors identify different types of DNA damage, triggering an appropriate DNA repair pathway. If the DNA lesion is substantial and initial repair is not effective, cell cycle arrest may be triggered that may be temporary (we consider this in the context of cancer in Chapter 11) or more permanent. In other cases, as often happens in lymphocytes, apoptosis may be triggered.

The DNA repair process occasionally involves a simple reversal of the molecular steps that cause DNA damage. That, however, is rare in human cells. Examples include the use of *O*-6-methylguanine DNA methyltransferase to reverse the methylation of guanine at the O-6 position, and the use of DNA ligase to repair broken phosphodiester bonds (*DNA nicks*).

Normally, DNA repair pathways do not directly reverse the damage process. Instead, according to the type of DNA lesion, one of several alternative DNA repair pathways is used. Most of the time, the repair needs to be made to one DNA strand only, but sometimes both strands need to be repaired, as in the case of interstrand cross-linking (see Figure 4.1Dii) and double-strand DNA breaks (see Figure 4.1Aiii).

Errors in DNA replication and chemical damage to DNA are a constant throughout life. Inevitably, however, some mistakes are made in repairing DNA, and there are also inherent weaknesses in detecting some base changes, as described below. Inefficiency in detecting and repairing DNA damage is an important contributor to generating mutation. We consider the health consequences of defective DNA repair in **Box 4.1**, located at the end of Section 4.2. Before then, we consider the major DNA repair mechanisms in the next two sections—see **Table 4.1** for a roadmap of how individual types of DNA change are repaired.

DNA DAMAGE/ALTERATION	DNA REPAIR MECHANISM	COMMENTS
Base mismatches caused by replication errors	mismatch repair (Figure 10.17)	detailed in Section 10.3
Small insertions/deletions due to replication slippage (Figure 4.8)		
Small scale, single base modification (oxidation, deamination, methylation, etc.)	base excision repair (BER) (Figure 4.3A)	for modified bases, a DNA glycosylase[a] cuts out a base to produce an abasic site. At all abasic sites the remaining sugar-phosphate is eliminated and the gap sealed by inserting the appropriate nucleotide
Single base deletion—an *abasic site*—resulting from hydrolysis		
Bulky, helix-distorting DNA lesions (large DNA adducts; DNA intrastrand cross-links, and so on)	nucleotide excision repair (NER) (Figure 4.3B)	involves removal and re-synthesis of a sequence of several nucleotides spanning the altered site
Single-stranded DNA breaks (other than DNA nicks)	variant of base excision repair	initiated by poly(ADP ribose) polymerase binding to cleavage site
Double-strand DNA breaks	homologous recombination (HR)-mediated DNA repair (Figure 4.4)	accurate DNA repair but needs an intact homologous DNA strand to use as a template (limited to post-replication in S phase or occasionally G_2)
	nonhomologous end joining (NHEJ)	doesn't rely on a template strand and so is available throughout cell cycle. Less accurate than HR-mediated DNA repair
DNA interstrand cross-links	HR and Fanconi anemia DNA repair pathway	some uncertainty about mechanism

Table 4.1 Frequent types of DNA damage and alteration, and how they are repaired in human cells. [a]Different DNA glycosylases are specific for different types of modified base, for example uracil DNA glycosylase or *N*-methylpurine DNA glycosylase.

Repair of DNA damage or altered sequence on a single DNA strand

DNA damage or an error in DNA replication usually results in one strand having a DNA lesion or a wrongly inserted base, but leaves the complementary DNA strand unaffected at that location. In that case, the undamaged complementary strand may be used as a template to direct accurate repair.

- *Base excision repair (BER).* This pathway is specifically aimed at lesions where a single base has either been modified or excised by hydrolysis to leave an *abasic site* (roughly 20,000 such events occur in each nucleated cell every day). To replace a modified base by the correct one, a specific DNA glycosylase cleaves the sugar–base bond to delete the base, producing an abasic site. For all abasic sites, the residual sugar–phosphate residue is removed by a dedicated endonuclease and phosphodiesterase. The gap is filled by a DNA polymerase (to insert the correct nucleotide) and DNA ligase (**Figure 4.3A**). Note that the same DNA repair machinery occasionally makes a more substantial *long-patch repair*, which involves replacing more than a single base.

- *Single-strand break repair.* Simple single-strand breaks—also called DNA nicks—are caused by breakage of a single phosphodiester bond and are common. They are easily reversed by DNA ligase. More complex breaks occur when oxidative attack causes deoxyribose residues to disintegrate. A type of base excision repair is then employed: strand breaks are rapidly detected and briefly bound by a sensor molecule, poly(ADP-ribose), that initiates repair by attracting suitable repair proteins to the site. The 3′ or 5′ termini of most SSBs are damaged and need to be restored. The gap is then filled by a DNA polymerase and DNA ligase.

- *Nucleotide excision repair (NER).* This mechanism allows the repair of bulky, helix-distorting DNA lesions. After the lesion is detected, the damaged site is opened out and the DNA is cleaved some distance away on either side of the lesion, generating an oligonucleotide of about 30 nucleotides containing the damaged site, which is discarded.

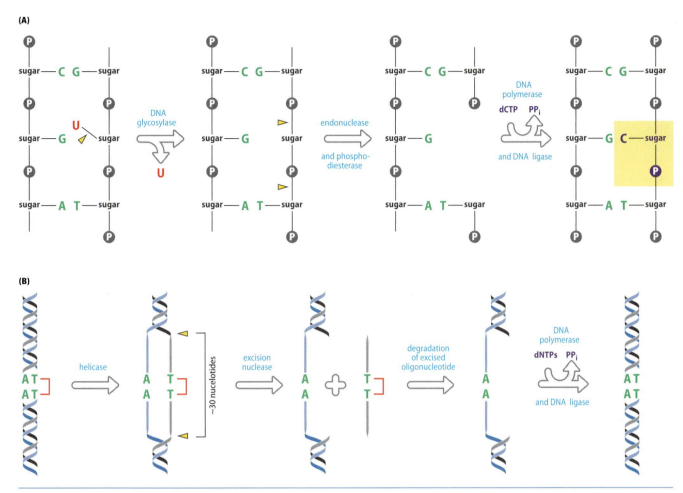

Figure 4.3 Base excision and nucleotide excision repair.
(A) *Base excision repair.* This pathway repairs modified bases and abasic sites produced by depurination or depyriminidation. Modified bases are first removed by a DNA glycosylase that cleaves the N-glycosidic bond that connects the base to the sugar, producing an abasic site. The cell has a range of different DNA glycosylases that are specific for common modified bases, such as 8-oxoguanine DNA glycosylase and uracil DNA glycosylase (as shown here). Abasic sites are usually repaired by excising the remaining sugar–phosphate residue (using a specialized endonuclease and a phosphodiesterase) and then inserting the correct nucleotide to match the undamaged complementary DNA strand. PP$_i$, pyrophosphate.
(B) *Nucleotide excision repair.* After identification of a bulky DNA lesion, this type of repair involves opening out the double helix containing the lesion over a considerable distance (using a helicase to unwind the DNA). Subsequently, an excision nuclease makes cuts on each side of the lesion on the damaged DNA strand, generating an oligonucleotide of about 30 nucleotides containing the damaged site, which is then degraded. The resulting gap is repaired by DNA synthesis using the undamaged strand as a template, and sealed by a DNA ligase.

Resynthesis of DNA is performed with the opposite strand as a template (see Figure 4.3B). The priority is to rapidly repair bulky lesions that block actively transcribed regions of DNA. A specialized subpathway, transcription-coupled repair, initiates this type of repair after detection of RNA polymerases that have stalled at the damaged site. Otherwise, an alternative global genome NER pathway is used.

- *Base mismatch repair.* This mechanism corrects errors in DNA replication. Errors in base mismatch repair are important in cancer and we describe this mechanism in Chapter 10.

Repair of DNA lesions that affect both DNA strands

Double-strand DNA breaks (DSBs) are normally rare in cells. They do occur naturally, however, and are necessary for specialized DNA rearrangements in B and T cells that maximize immunoglobulin and T-cell receptor diversity (as described near the end of this chapter).

DSBs also occur by accident, as a result of chemical attack on DNA by endogenous or externally induced ROS (but at much lower frequencies than SSBs). In these cases, DNA repair is required but can sometimes be difficult to perform. For example, when the two complementary DNA strands are broken simultaneously at sites sufficiently close to each other, neither base pairing nor chromatin structure may be sufficient to hold the two broken ends opposite each other. The DNA termini will often have sustained base damage and the two broken ends are liable to become physically dissociated from each other, making alignment difficult.

Unrepaired DSBs are highly dangerous to cells. The break can lead to the inactivation of a critically important gene, and the broken ends are liable to recombine with other DNA molecules, causing chromosome rearrangements that may be harmful or lethal to the cell. Cells respond to DSBs in different ways. Two major DNA repair mechanisms can be deployed to repair a DSB, as listed below; if repair is incomplete, however, apoptosis is likely to be triggered.

- *Homologous recombination (HR)-mediated DNA repair.* This highly accurate repair mechanism requires a homologous intact DNA strand to be available to act as a template strand. Normally, therefore, it operates after DNA replication (and before mitosis), using a DNA strand from the undamaged sister chromatid as a template to guide repair (**Figure 4.4**). It is important in early embryogenesis (when many cells are proliferating rapidly), and in the repair of proliferating cells after the DNA has replicated.

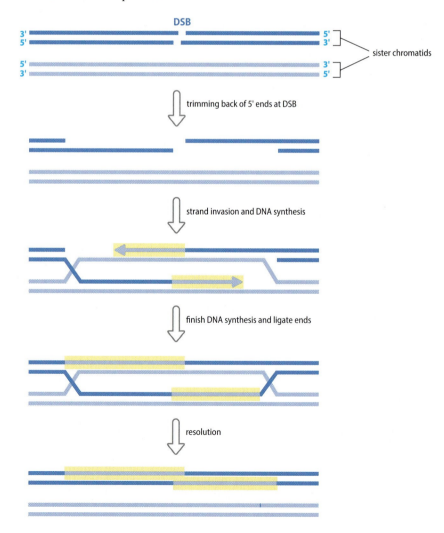

Figure 4.4 Homologous recombination-mediated repair of double-strand DNA breaks. The double-strand break (DSB) in the chromatid at the top is repaired using as a template the undamaged DNA strands in the sister chromatid (note that to make the mechanism easier to represent, the upper chromatid has, unconventionally, the 3′—→5′ strand placed above the 5′—→3′ strand). The first step is to cut back the 5′ ends at the DSB to leave protruding single-strand regions with 3′ ends. After strand invasion, each of the single-strand regions forms a duplex with an undamaged complementary DNA strand from the sister chromatid, which acts as a template for new DNA synthesis (shown by the arrows highlighted in yellow). After DNA synthesis, the ends are sealed by DNA ligase (newly synthesized DNA copied from the sister chromatid DNA is highlighted in yellow). The repair is highly accurate because for both broken DNA strands the undamaged sister chromatid DNA strands act as templates to direct the incorporation of the correct nucleotides during DNA synthesis.

- **Nonhomologous end joining** (**NHEJ**). No template strand is needed here because the broken ends are fused together. Specific proteins bind to the exposed DNA ends and recruit a special DNA ligase, DNA ligase IV, to rejoin the broken ends. Unlike HR-mediated DNA repair, NHEJ is, in principle, always available to cells. However, it is most important for the repair of differentiated cells and of proliferating cells in G_1 phase, before the DNA has replicated.

Repair of DNA interstrand cross-links

Cross-linking can occur between bases on complementary strands of a double helix, either as a result of endogenous metabolites or through exogenously supplied chemicals, notably many anti-cancer drugs (see the example of cisplatin in Figure 4.1Dii). Interstrand cross-links seem to be repaired by using a combination of NER, translesion synthesis, HR, and a complex of multiple different protein subunits that are encoded by genes mutated in Fanconi anemia (Fanconi anemia DNA repair pathway; see also Box 4.1).

Undetected DNA damage, DNA damage tolerance, and translesion synthesis

DNA damage may sometimes go undetected. For example, cytosines that occur within the dinucleotide CG are highly mutable as a result of inefficient DNA repair. In vertebrates, the dinucleotide CG is a frequent target for DNA methylation, converting the cytosine to 5-methylcytosine (5-meC). Deamination of cytosine residues normally produces uracil, which is efficiently recognized as a foreign base in DNA and eliminated by uracil DNA glycosylase. Deamination of 5-methylcytosine, however, produces a normal DNA base, thymine, that can sometimes go undetected as an altered base (**Figure 4.5**). As a result, C→T substitutions are the most frequent type of single-nucleotide change in our DNA.

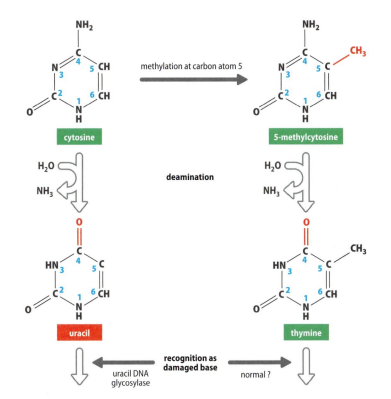

Figure 4.5 Why C→T mutations are so common in human DNA. Deamination of cytosine is a very common reaction in our cells and normally produces uracil, a base that is usually found in RNA, not DNA. Our cells are equipped with a special enzyme, uracil DNA glycosylase, that recognizes uracil residues in DNA and removes them as part of the base excision DNA repair pathway (Figure 4.3A). However, as in the DNA of other vertebrates, many of our cytosines are methylated at carbon atom 5. Deamination of 5-methylcytosine produces thymine, a base normally found in DNA. Although a stable CG base pair has been replaced by a TG base mismatch, the base mismatch may often escape detection by the base mismatch repair system (which focuses on DNA replication events). At the subsequent round of DNA replication the thymine will form a TA base pair, effectively producing a C→T mutation.

BOX 4.1 The Health Consequences of Defective DNA Damage Response/DNA Repair.

DNA damage accumulates in all of us throughout our lives. Inevitably, as we grow older, the incidence of somatic mutations increases, with consequences for increased risk of developing cancer and of declining efficiency in a variety of cellular processes, contributing to the aging process. More than 170 human genes are known to be involved in DNA damage responses and DNA repair (see Further Reading), and a wide variety of single-gene disorders have been described that result from germ-line mutations in genes that work in these pathways (Table 1 gives some examples).

As expected, increased susceptibility to cancer and accelerated aging are often found in these disorders, but a significant number have developmental abnormalities, and neurological features are very common. Although many cell types are regularly replaced, nondividing neurons are especially vulnerable. They have high oxygen and energy needs (with a resulting high frequency of oxidative damage), and they accumulate DNA damage over very long periods. In addition to various clinical features, cellular abnormalities are frequently seen, with respect to chromosome and genome instability as listed below.

Disease features

- *Cancer* (**C**) *susceptibility*. Not surprisingly, this is apparent in many inherited DNA repair deficiencies. Genome instability in mismatch repair deficiencies can induce cancer in highly proliferating tissues, notably intestinal epithelium. Individuals with xeroderma pigmentosum have little protection against UV radiation, and exposure to sunlight induces skin tumors (**Figure 1A**).

- *Progeria* (**P**). Some disorders have clinical features that mimic accelerated aging, notably individuals with Werner syndrome (Figure 1B), who prematurely develop gray hair, cataracts, osteoporosis, type II diabetes, and atherosclerosis, and generally die before the age of 50 as a result of cancer or atherosclerosis.

- *Neurological* (**N**) *features*. Neuronal death and neurodegeneration are common features. Individuals with ataxia telangiectasia experience cerebellar degeneration leading to profound ataxia and become confined to a wheelchair before the age

DNA REPAIR/DNA DAMAGE RESPONSE SYSTEM	SINGLE-GENE DISORDERS	DISEASE FEATURES[a]			
		C	P	N	I
Mismatch repair	hereditary nonpolyposis colorectal cancers (Lynch syndrome)	+	–	–	–
Nucleotide excision repair (NER)	xeroderma pigmentosum	+	–	+	–
NER (transcription-coupled repair)	Cockayne syndrome	–	+	+	–
	trichothiodystrophy	–	+	+	–
Single-strand break (SSB) repair	ataxia oculomotor apraxia 1	–	–	+	–
	spinocerebellar ataxia with axonal neuropathy I	–	–	+	–
Interstrand cross-link repair	Fanconi anemia	+	+	+	+
Double-strand break (DSB) repair (NHEJ)	Lig4 syndrome	+	–	+	+
	severe combined immunodeficiency	–	–	–	+
DNA damage signaling/DSB repair	ataxia telangiectasia	+	–	+	+
	Seckel syndrome	–	–	+	+
	primary microcephaly 1	–	–	+	–
Homologous recombination (HR)	Bloom syndrome	+	–	+	+
Telomere maintenance (TM)	dyskeratosis congenita	+	+	+	+
Base excision repair (BER) in mtDNA	spinocerebellar ataxia–epilepsy	–	–	+	–
	progressive external opthalmoplegia	–	–	–	–
HR, BER, TM	Werner syndrome	+	+	–	–

Table 1 Examples of inherited disorders of DNA repair/DNA damage responses. [a]C, cancer susceptibility; P, progeria; N, neurological features; I, immunodeficiency.

BOX 4.1 (continued)

of 10. Microcephaly is found in many disorders, sometimes accompanied by evidence of neurodegeneration and learning difficulties.

- *Immunodeficiency* (**I**). As described in Table 4.2 and Section 4.3, some proteins that work in DNA repair also function in specialized genetic mechanisms that occur exclusively in B and T lymphocytes. For example, the production of immunoglobulin and T-cell receptors requires components of the NHEJ repair pathway, and deficiency of these components typically results in hypogammaglobulinemia and lymphopenia or severe combined immunodeficiency.

Cell analyses revealing genome and chromosomal instability

The DNA of individuals with disorders of mismatch repair (described in Section 10.3) shows striking evidence of genome instability when short tandem repeat DNA polymorphisms known as microsatellite DNA are assayed. Cells from individuals with a DNA repair disorder quite often also show an increased frequency of spontaneous chromosome aberrations that can be characteristic of the disorder, as in the case of ataxia telangiectasia, Fanconi anemia, and Bloom syndrome (which shows very high levels of sister chromatid exchange).

Chromosome analyses can also provide a simple route to laboratory-based diagnosis. Fanconi anemia (which is characterized by the variable presence of assorted developmental abnormalities, plus progressive bone marrow failure and an increased risk of malignancy) can be caused by mutations in any one of at least 13 different genes that work to repair interstand cross-links, making DNA-based diagnosis difficult. Chromosome-based diagnosis is more straightforward: lymphocyte cultures are treated with diepoxybutane or mitomycin C—chemicals that induce DNA interstrand cross-links—and chromosomes are analyzed for evidence of chromatid breakage, which can produce characteristic abnormal chromosome formations (Figure 1C).

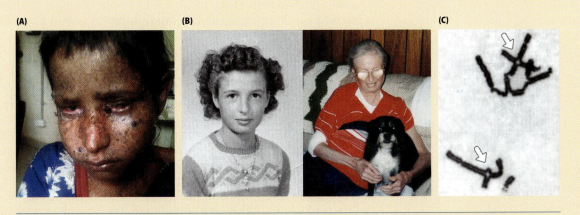

(A)   **(B)**   **(C)**

Figure 1 Examples of abnormal phenotypes in DNA-repair disorders. (A) Extensive skin cancer in xeroderma pigmentosum. (B) Accelerated aging in Werner syndrome—portraits of the same woman at age 13 (left) and age 56 (right). (C) Characteristic quadriradial and triradial chromosome formations in Fanconi anemia cells after treatment with mitomycin C. (A, courtesy of Himynameislax (CC BY-SA 3.0). B, from Hisama FM, Bohr VA, and Oshima J [2006] *Sci Aging Knowl Environ* 10:pe18. With permission from the AAAS (left) and the International Registry of Werner Syndrome (right). C, courtesy of Niall Howlett from Harney JA, Shimamura A and Howlett NG [2008] *Pediatr Health* 2:175–187. With permission from Future Medicine Ltd.)

Sometimes, DNA lesions may be identified but are not repaired before DNA replication (damage tolerance). For example, DNA lesions that block replication may be bypassed rather than repaired, and non-classical DNA polymerases are required to resume DNA synthesis past the damaged site (*translesion synthesis*). Subsequently, the gap in the daughter strand opposite the lesion is filled in; the lesion can be repaired later on, by using the daughter strand as a template in nucleotide excision repair. The non-classical DNA polymerases used in translesion synthesis exhibit

Table 4.2 Non-classical DNA-dependent mammalian DNA polymerases. The classical DNA-dependent DNA polymerases listed in Table 1.1 are mostly used generally in DNA synthesis and DNA repair and have very low error rates. By contrast, the non-classical polymerases listed here exhibit comparatively low fidelity of DNA replication and are reserved for specialized mechanisms, notably for bypassing damaged DNA and in B-cell-specific and T-cell-specific mechanisms used to make immunoglobulins and T-cell receptors. [a]Used to bypass an obstructive DNA lesion. [b]A specialized mechanism used in B cells to maximize the variability of immunoglobulins. [c]A specialized mechanism used in both B and T cells to maximize the variability of immunoglobulins and T-cell receptors, respectively.

DNA POLYMERASES	ROLES
Nonclassical DNA-dependent DNA polymerases	Mostly for bypassing damaged DNA, and/or for use in specialized genetic mechanisms in B and T cells
ζ (zeta), η (eta), Rev1	translesion synthesis[a] (to bypass obstructive DNA lesion)
ι (iota), κ (kappa)	translesion synthesis[a] plus other roles in DNA repair
ν (nu)	interstrand cross-link repair?
θ (theta)	translesion synthesis[a]; base excision repair; somatic hypermutation[b]
λ (lambda), μ (mu)	V(D)J recombination[c]; double-strand break repair; base excision repair
Terminal deoxynucleotide transferase	V(D)J recombination[c]

a low fidelity in DNA replication (Table 4.2). They have a higher success in incorporating bases opposite a damaged site, but they are prone to error by occasionally inserting the wrong base. As a result, replication forks are preserved, but at the cost of mutagenesis.

4.3   The Scale of Human Genetic Variation

Human genetic variation occurs by changes to our base sequence that can be classified into two categories. First, some changes to our DNA sequence do not affect the DNA content (that is, the number of nucleotides is unchanged). Quite often, for example, a single nucleotide is replaced by a different nucleotide. More rarely, multiple nucleotides at a time may be sent to a new location without net loss or gain of DNA content: these balanced translocations and inversions that result in chromosome breakage without net loss or gain of DNA may sometimes have no effect on the phenotype or be harmful.

A second class of DNA change causes a net loss or gain of DNA sequence: there is a change in *copy number* of a DNA sequence that can be large or small. The largest changes in copy number result from abnormal chromosome segregation, producing fewer or more chromosomes than normal, and therefore a change in the copy number of whole nuclear DNA molecules. They are almost always harmful: many result in spontaneous abortion and some give rise to developmental syndromes as described in Chapter 7. At the other end of the scale, the change in copy number can mean the deletion or insertion of a single nucleotide. In between are copy number changes that range from altered numbers of specific short oligonucleotide sequences to megabase lengths of DNA.

Overall, the most common DNA changes are on a small scale and involve only a single nucleotide or a very small number of nucleotides. Small-scale changes like this (sometimes called **point mutations**) often have no obvious effect on the phenotype; in that case they would be considered to be neutral mutations. That happens because 90% or so of our DNA is poorly conserved and of questionable value, and seems to tolerate small changes in DNA sequence without obvious effect.

DNA variants, polymorphism, and developing a comprehensive catalog of human genetic variation

Mutations result in alternative forms of DNA that are generally known as DNA **variants**. However, for any locus, if more than one DNA variant is common in the population (above a frequency of 0.01), the DNA variation is described as a **polymorphism**; DNA variants that have frequencies of less than 0.01 are often described as *rare variants*. (The 0.01 cutoff might seem arbitrary; it was initially proposed so as to exclude recurrent mutation.)

Our knowledge of human genetic variation comes from analyzing DNA from multiple individuals. Being diploid, each of us has two different nuclear genomes that we inherited from our parents, but to get a detailed view of human genetic variation we need to compare DNA from many different individuals. The draft human genome sequence (published in 2001) was a patchwork of DNA sequences from different individuals (so that sequences present in different regions of the genome came from different human donors). However, as next-generation sequencing technologies became available, personal genome sequencing projects were established; subsequently, larger population-based genome sequencing is providing a comprehensive catalog of human genetic variation (Box 4.2).

BOX 4.2 Personal and Population-Based Genome Sequencing.

The Human Genome Project involved sequencing DNA clones from different human cell lines; the resulting sequence was a patchwork of DNA sequence components derived from different individuals. Thereafter, various pilot projects obtained the genome sequence of single individuals (*personal genome sequencing*). More comprehensive projects are using next-generation sequencing (massively parallel DNA sequencing—see Section 3.3) to obtain genome sequences from thousands of individuals, taking us into an era of *population-based genome sequencing*.

One principal aim is to get a comprehensive catalog of normal human DNA variation by sequencing whole genomes from normal individuals. Another major goal is to correlate DNA variation with phenotype in an effort to identify genetic markers of disease, and in this case sequencing has been focused on **exomes** (in practice, the collective term for all exons in protein-coding genes) from individuals whose phenotypes have been well studied. Two of the first major population-based genome sequencing projects, the 1000 Genomes Project and the UK10K project, are described below. As we detail in Chapters 10 and 11, large-scale genome sequencing projects have also been launched to decode the genomes of tumors and of individuals with genetic disorders.

The 1000 Genomes Project (http://www.1000genomes.org/)

Initiated in 2008, the initial aim of sequencing the genome of 1000 individuals was subsequently upgraded to cover the genomes of 2500 individuals representing populations from Europe, East Asia, South Asia, West Africa, and the Americas. The genomes include those of several two-parent-and-child trios to assess the frequency of *de novo* mutations (by comparing the child's genome with that of its parents). See the 1000 Genomes Project Consortium (2010) paper under Further Reading for an interim publication.

The Wellcome Trust UK10K project (http://www.uk10k.org/)

This involves sequencing the genomes of 10,000 people in the UK whose phenotypes have been closely monitored; 4000 people are drawn from the TwinsUK study (in which participants have been studied for more than 18 years) and the ALSPAC (Avon Longitudinal Study of Parents and Children—also known as Children of the 90s) study, in which individuals have been followed since birth in 1991–92. The data for each group include extensive descriptions of their health and their development. The remaining 6000 people have a severe condition thought to have a genetic cause (including severe obesity, autism, schizophrenia, and congenital heart disease). Here the priority is exome sequencing.

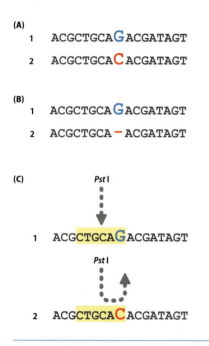

Figure 4.6 Classes of DNA variation affecting a single nucleotide position. (A) Single nucleotide variant (SNV) in which two variants differ by having a G or a C. (B) Insertion/deletion (*indel*) variation in which variant 1 has a G not present in variant 2. (C) Sometimes differences at a single nucleotide position can lead to the variable presence of a restriction site. Here variant 1 shown in (A) can be seen to have a recognition sequence (CTGCAG) for the restriction enzyme *Pst*I; in variant 2 the equivalent sequence (CTGCAC) will not be cleaved by *Pst*I. If the variants are common this would be an example of a *restriction fragment length polymorphism* (*RFLP*) that can conveniently be assayed by designing PCR primers to amplify the sequence containing it and then cutting the PCR product with *Pst*I.

Single nucleotide variants and single nucleotide polymorphisms

The most common type of genetic variation in the human genome is due to single nucleotide substitutions. For example, one variant may have a G at a defined position and another variant a C (**Figure 4.6A**). This type of variation produces single nucleotide variants (SNVs). If two or more alternative DNA variants of this type exceed a frequency of 0.01 in the population, the DNA variation is commonly described as **single nucleotide polymorphism** or **SNP** (pronounced "snip").

The pattern of SNV in the human genome is nonrandom for two reasons. First, there is a mechanistic reason: different DNA regions and different DNA sequences can undergo different mutation rates. For example, mitochondrial DNA has a much higher mutation rate than nuclear DNA (partly because of proximity to reactive oxygen species that are generated especially in mitochondria). And, as in other vertebrate genomes, there is an excess of C→T substitutions in the human genome (because the dinucleotide CG often acts as a methylation signal; C becomes 5-methylC, which is prone to deamination to give T—see Figure 4.5).

A second reason for nonrandom variation comes from our evolutionary ancestry. One might reasonably wonder why only certain nucleotides should be polymorphic and be surrounded by stretches of nucleotides that only rarely show variants. In general, the nucleotides found at SNP sites are not particularly susceptible to mutation (the germ-line single nucleotide mutation rate is about 1.1×10^{-8} per generation, roughly 1 nucleotide per 100 Mb, and SNPs are stable over evolutionary time). Instead, the alternative nucleotides at SNP sites mark alternative ancestral chromosome segments that are common in the present-day population. As described in Chapter 8, using SNPs to define ancestral chromosome segments is important in mapping genetic determinants of disease. We cover methods for assaying specific single nucleotide changes (including SNPs) in Chapter 11.

The imprecise cut-off between indels and copy number variants

Some point mutations create DNA variants that differ by the presence or absence of a single nucleotide, or by a relatively small number of nucleotides at a specific position. This is an example of *in*sertion/*del*etion variation or polymorphism (**indel** for short—see Figure 4.6B). A subset of SNPs or indels leads to the gain or loss of a restriction site, in which case cutting the DNA with the relevant restriction nuclease can generate restriction fragment length polymorphism (RFLP; see Figure 4.6C).

Strictly speaking, indels should be considered to be copy number variants (a person with a heterozygous deletion of even a single nucleotide at a defined position on a chromosomal DNA has one copy of that nucleotide, instead of the normal two copies). However, the modern convention is to reserve the term *indel* to describe deletions or insertions of from one nucleotide up to an arbitrary 50 or so nucleotides (chosen because many massively parallel DNA sequencing methods produce quite short sequences and are often not suited to detecting deletions or insertion of greater than 50 nucleotides). The term **copy number variation** (**CNV**) is most often applied to a change in copy number of sequences that result in larger deletions and insertions, usually more than 100 nucleotides up to megabases.

The frequency of insertion/deletion polymorphism in the human genome is about one-tenth the frequency of single nucleotide substitutions. Short insertions and deletions are much more common than long ones. Thus, 90% of all insertions and deletions are of sequences 1–10 nucleotides long, 9% involve sequences from 11 to 100 nucleotides, and only 1% involve sequences greater than 100 nucleotides. Nevertheless, because many of the last category involve huge numbers of nucleotides, CNV affects more nucleotides than single nucleotide substitutions.

Microsatellites and other polymorphisms due to variable number of tandem repeats

As described in Section 2.4, repetitive DNA accounts for a large fraction of the human genome. Tandem copies of quite short DNA repeats (1 bp to fewer than 200 bp) are common, and those with multiple tandem repeats are especially prone to DNA variation. A continuous sequence of multiple tandem repeats is known as an array. Different organizations are evident and the repeated sequences are classified as belonging to three classes, according to the total length of the array and genomic location:

- *satellite DNA* (array length: often from 20 kb to many hundreds of kilobases; located at centromeres and some other heterochromatic regions)

- *minisatellite DNA* (array length: 100 bp to 20 kb; found primarily at telomeres and subtelomeric locations)

- *microsatellite DNA* (array length: fewer than 100 bp long; widely distributed in euchromatin).

The instability of tandemly repeated DNA sequences results in DNA variants that differ in the numbers of tandem repeats (variable number of tandem repeats polymorphism, or VNTR polymorphism). Because microsatellite DNA arrays (usually called **microsatellites**) are frequently distributed within human euchromatin (roughly once every 30 kb) and are often highly polymorphic, they have been widely used in genetic mapping.

Microsatellites have very short repeats (one to four base pairs long), so microsatellite polymorphisms are sometimes also known as short tandem repeat polymorphisms (STRPs). Unlike SNPs (which almost always have just two alleles), microsatellite polymorphisms usually have multiple alleles (**Figure 4.7**). The variation in copy number arises as a result of **replication slippage**: during DNA replication, the nascent (newly synthesized) DNA strand slips relative to the template strand so that the two strands are slightly out of alignment—see **Figure 4.8**.

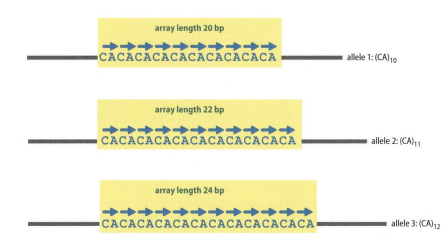

array length 20 bp
CACACACACACACACACACA   allele 1: (CA)$_{10}$

array length 22 bp
CACACACACACACACACACACA   allele 2: (CA)$_{11}$

array length 24 bp
CACACACACACACACACACACACA   allele 3: (CA)$_{12}$

Figure 4.7 Length polymorphism in a microsatellite. Here, a microsatellite locus is imagined to have three common alleles that differ in length as a result of having variable numbers of tandem CA repeats. See Figure 4.8 for the mechanism that gives rise to the variation in copy number.

(A) slippage causes insertion

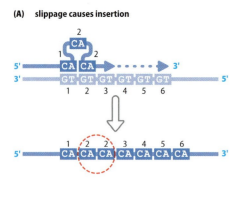

(B) slippage causes deletion

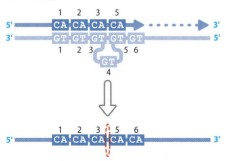

Figure 4.8 Microsatellite polymorphism results from strand slippage during DNA replication. The dark blue strand represents the synthesis of a new (nascent) DNA strand from the pale blue template DNA strand. During normal DNA replication, the nascent strand often partly dissociates from the template and then reassociates. When there is a tandemly repeated sequence, the nascent strand may mispair with the template strand when it reassociates, so that the newly synthesized strand has more repeat units (A) or fewer repeat units (B) than the template strand, as illustrated within the dashed red circles.

Individual microsatellite polymorphisms can be typed by first using PCR to amplify a short sequence containing the array, and then separating the PCR products according to size by polyacrylamide gel electrophoresis. They have been used extensively in both family studies and DNA profiling (to establish identity for legal and forensic purposes). However, unlike SNP assays, it is not easy to automate assays of micro-satellite polymorphisms.

Structural variation and low copy number variation

Until quite recently, the study of human genetic variation was largely focused on small-scale variation such as changes affecting single nucleotides and microsatellite polymorphisms. We now know that variation due to moderately large-scale changes in DNA sequence is very common. Such **structural variation** can be of two types: balanced and unbalanced.

In balanced structural variation, the DNA variants have the same DNA content but differ in that some DNA sequences are located in different positions within the genome. They originate when chromosomes break and the fragments are incorrectly rejoined, but without loss or gain of DNA. That can involve inversions and translocations that do not involve change in DNA content (**Figure 4.9**).

In unbalanced structural variation, the DNA variants differ in DNA content. In rare cases in which a person has gained or lost certain

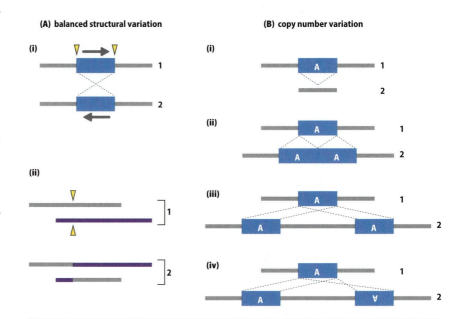

Figure 4.9 Structural variation and low copy number variation. The numbers 1 and 2 refer to alternative variants throughout. (A) Balanced structural variation involves large-scale changes that produce variants with the same number of nucleotides, including many inversions (i) and balanced translocations (ii). (B) Unbalanced structural variation includes unbalanced inversions and unbalanced translocations plus different types of low copy number variation (CNV). Copy number variants have different numbers of copies of a moderately long sequence (represented here by the box marked A). (i) CNV in which variants possess (1) or lack (2) a sequence, effectively a large-scale indel. This can result from an insertion (for example of a mobile element) or a deletion. (ii) CNV due to tandem duplication, effectively a large-scale VNTR that can sometimes have several copies, rather than just the one or two copies shown here. Sometimes additional insertion and inversion events can result in interspersed duplication with normal orientation of copies (iii) and interspersed duplication with inversion of a copy (iv).

chromosomal regions (as when a parent with a balanced reciprocal translocation passes one of the translocation chromosomes, but not the other, to a child), the gain or loss of substantial chromosomal segments often results in disease. Unbalanced structural variation also includes commonly occurring CNV in which the variants differ in the number of copies of a moderately long to very long DNA sequence. Some CNVs such as this contribute to disease, but very many CNVs are commonly found in the normal population.

Copy number variation can take different forms. One form is effectively simple insertion/deletion variation on a large scale in which DNA variants either lack or possess a specific sequence (see Figure 4.9Bi). Other forms result from tandem duplication that may be complicated by subsequent insertion and inversion events (see Figure 4.9Bii–iv). In some CNVs, the DNA sequence that varies in copy number can include part of a gene sequence or regulatory sequence and sometimes multiple genes. As a result, some CNVs are important contributors to disease, as described in later chapters.

Taking stock of human genetic variation

The data from population-based genome sequencing projects indicate that single nucleotide changes are the most common type of genetic variation, accounting for close to 75% of DNA changes. A study of 1092 individuals in 14 human populations by the 1000 Genomes Project Consortium (see Further Reading) reported a total of 38 million human SNPs (more than 1 per 100 nucleotides). However, the vast majority of these SNPs are rare in any population. Any one individual has just two haploid genomes and will be homozygous at many SNP loci. Personal genome sequencing shows that single nucleotide differences between the maternal and paternal genomes in one individual occur about once every 1000 nucleotides. Much of that variation falls outside coding sequences and mostly represents neutral mutations.

Structural variation is less common, accounting for close to one-quarter of mutational events, and is dominated by CNV. Because CNV often involves very long stretches of DNA, however, the number of nucleotides involved in CNVs significantly exceeds those involved in SNVs.

Various databases have been established to curate basic data on genetic variation in humans (and other species)—see **Table 4.3**.

4.4   Functional Genetic Variation and Protein Polymorphism

Thus far in this chapter, we have focused on the different types of human genetic variation at the DNA level, and their origins. But DNA variation has consequences with regard to gene products, and that can cause disease as described in later chapters. In a normal person, however, a

DATABASE	DESCRIPTION	WEBSITE
dbSNP	SNPs and other short genetic variations	http://www.ncbi.nlm.nih.gov/SNP/index.html
dbVar	genomic structural variation	http://www.ncbi.nlm.nih.gov/dbvar/
DGV		http://dgv.tcag.ca/
ALFRED	allele frequencies in human populations	http://alfred.med.yale.edu/alfred/index.asp

Table 4.3 General human genetic variation databases. Databases focusing on mutations that relate to phenotypes and disease are described in Chapters 7 and 8.

surprising number of genes are inactivated: each of us carries an average of about 120 gene-inactivating variants, with about 20 genes being inactivated in both alleles. That might seem alarming, but some genes do not carry out vital functions. For example, people with blood group O are homozygotes for an inactivating mutation in the *ABO* gene and fail to make a glycosyltransferase that transfers a carbohydrate group onto the H antigen. (The H-antigen is an oligosaccharide that is attached to certain lipids or proteins on the surface of some types of cell, notably red blood cells.) People with blood groups A, B, and AB do produce this enzyme and it transfers either an N-acetylgalactosamine (A allele) or a galactose (B allele) to the H-antigen. Effectively, therefore, people with blood group O make the H-antigen and people with other blood groups make a related antigen with an extra monosaccharide.

In this section we focus on variation at the level of gene products. That includes both functional genetic variation (usually DNA variation that causes a change in some functional gene product) and also post-transcriptional mechanisms that result in altered coding sequences. In Section 4.5 we focus on an important subset of functional variation: the extraordinary variation seen in some proteins that have important roles in the immune system. We consider how DNA variants contribute to pathogenesis in later chapters, notably Chapter 7.

Most genetic variation has a neutral effect on the phenotype, but a small fraction is harmful

The functional DNA variants that are primarily studied are those that have an effect on gene function, changing the structure of a gene product or altering the rate at which it is produced. Only a very small fraction of nucleotides in our DNA is important for gene function, however. Coding DNA sequences make up just over 1.2% of the genome. Some additional DNA sequences make functional noncoding RNAs; others regulate gene expression in some way—at the transcriptional, post-transcriptional, or translational level.

Estimating how much of the genome that is functionally important is not straightforward. One way is to determine the percentage of the genome that is subject to purifying selection to conserve functionally important sequences. That has traditionally been done by cross-species comparisons, but population-based human genome sequencing is also offering insights into functional constraint by enabling large-scale studies of within-species DNA variation. Current estimates suggest that perhaps a total of about 8–10% of the genome is under functional constraint. Mutation at any one of more than 90% or more of our nucleotides may have essentially no effect.

Even within the small target of sequences that are important for gene function, many small DNA changes may still have no effect. For example, many coding DNA mutations are silent: they do not change the protein sequence and would usually have no effect (unless they cause altered splicing—we show examples in Chapter 7). Single nucleotide changes in regulatory sequences or sequences that specify functional noncoding RNA sometimes also have no or very little effect. Because we know comparatively little about the functional significance of changes in noncoding RNA and regulatory sequences, we mostly focus here on coding DNA sequences. Of course, harmful mutations also occur in functional DNA, and very occasionally mutations have a beneficial effect.

Harmful mutations

A small fraction of genetic variation is harmful, and we consider the detail in other chapters, notably Chapter 7. Harmful mutations are subject to

purifying (negative) selection: people who possess them will tend to have lower reproductive success rates and the mutant allele will gradually be eliminated from populations over several generations. Harmful DNA changes include many different types of small-scale mutations, both in coding DNA (resulting in changes in amino acid sequence) and noncoding DNA (causing altered splicing, altered gene regulation, or altered function in noncoding RNA).

In addition, structural variation can often have negative effects on gene function. Genes may be inactivated by balanced structural variation if breakpoints affect how they are expressed. Copy number variation can lead to a loss or gain of gene sequences that can be harmful because the levels of some of our gene products need to be tightly controlled, as explained in later chapters.

Positive Darwinian selection and adaptive DNA changes in human lineages

Occasionally, a DNA variant has a beneficial effect on the phenotype that can be transmitted to offspring. DNA variants like this become prevalent through **positive selection** (individuals who possess the advantageous DNA variant may have increased survival and reproductive success rates; the DNA variant then increases in frequency and spreads throughout a population).

Positive selection has occurred at different times in human lineages. Some regions of the genome, notably the HLA complex (which contains genes involved in recognizing foreign antigens in host cells and activating T cells), have been subject to continuous positive selection since before the evolutionary divergence of humans and great apes. The selection involved here is thought to favour heterozygotes and has caused certain HLA proteins to be the most polymorphic of all our proteins—we provide details in Section 4.5. In addition, positive selection has been responsible for fostering different features that distinguish us from the great apes, notably human innovations in brain development and increased cognitive function. More recent adaptive changes have led to increased frequencies of certain DNA variants in different human populations.

The great majority of the selected variants seem to occur in noncoding regulatory DNA and result in altered gene expression. However, altered amino acids may sometimes confer functional advantages, as suggested in the case of *FOXP2*, a gene mutated in an autosomal dominant speech and language disorder (OMIM 602081). This gene is known to regulate the development of different brain regions, including Broca's area (which is involved in producing speech). Although *FOXP2* has been extremely highly conserved during evolution, it underwent mutations in human lineages, resulting in two amino acid changes that have been proposed to result from positive selection and may have contributed to language acquisition (**Figure 4.10**).

Adaptations to altered environments

After out-of-Africa migrations 50,000–100,000 years ago, modern humans settled in different regions and were exposed to different environments, including living in latitudes with low levels of sunlight, or living at high altitudes. Some adaptations also provide some protection against infectious diseases, notably malaria. As agriculture developed, DNA variants were selected in response to changes of diet (**Table 4.4** gives some examples).

Adaptations to local environments can involve down-regulating a physiological function, as in reduced skin pigmentation in Europeans.

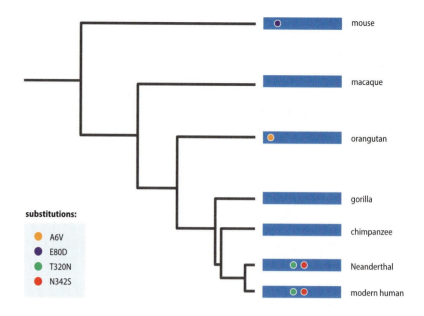

Figure 4.10 Human-specific amino acids in the FOXP2 protein: an example of positive selection fostering language acquisition? The *FOXP2* gene has been implicated in regulating different brain functions including motor control of the orofacial regions and vocalization. The gene has been very highly conserved: the chimpanzee, gorilla, and macaque have the same FOXP2 protein sequence, and the substitutions shown refer to changes from that common sequence. Thus, the mouse protein is identical except for one conservative substitution (an aspartate instead of glutamate at position 80, shown here) plus two indels in which the mouse sequence lacks one glutamine within a polyglutamine cluster and a span of 17 amino acids corresponding to amino acids 133–149 in the 732 amino acid human FOXP2 sequence (NCBI reference NP_683698.2). Given the extraordinary level of evolutionary conservation, it is very unusual to find two amino acid substitutions in the short evolutionary branch leading to modern humans and Neanderthals. Studies of mice genetically engineered to have the human-specific T320N and N342S substitutions suggest that these changes affect specifically cortico-striatal circuits (which are generally involved in developing motor and cognitive skills).

substitutions:
- 🟠 A6V
- 🟣 E80D
- 🟢 T320N
- 🔴 N342S

Ultraviolet (UV) radiation in sunlight is needed for a photolytic reaction that occurs in a deep layer of the dermis, and this is the principal source of vitamin D_3. Dark skins in equatorial populations protect skin cells from DNA damage caused by intense exposure to UV. Populations that migrated to northern latitudes were exposed to less UV, but the potentially reduced ability to make vitamin D_3 was offset by an adaptation that reduced the amount of melanin, maximizing UV transmission through skin. The most significant contributor was a nonsynonymous change in the *SLC24A5* gene, resulting in the replacement of alanine at position 111 by threonine (A111T). The SLC24A5 protein is a type of calcium transporter that regulates melanin production, and the A111T change results in defective melanogenesis. The A111T variant became fixed in European populations as a result of what is called a **selective sweep** (**Box 4.3**).

ALTERED ENVIRONMENT	ADAPTATION AND ITS EFFECTS	ASSOCIATED GENETIC VARIANTS
Reduced sunlight (low UV exposure)	decreased pigmentation; decreased melanin in skin allows more efficient transmission of the depleted UV to a deep layer of the dermis, where a photolytic reaction is needed to synthesize vitamin D_3	an *SLC24A5* variant (replacing the ancestral alanine at position 111 by threonine) is prevalent in European populations as result of a recent *selective sweep* (see Box 4.3)
High-altitude settlements (low O_2 tension)	in Tibetan[a] populations lowered hemoglobin levels and a high density of blood capillaries provide protection against hypoxia	variants in *EPAS1*, a key gene in the hypoxia response
Malaria-infested environments	alterations in red blood cell physiology, affecting transmission of the mosquito-borne parasites *P. falciparum* or *P. vivax* and conferring increased resistance to malaria	pathogenic mutations[b] in *HBB* or *G6PD* for *P. falciparum* malaria; inactivating *DARC* variants that do not express the Duffy antigen[c] in *P. vivax* malaria
Lifelong intake of fresh milk	persistence of lactase production in adults, allowing efficient digestion of lactose	the −13910T allele about 14 kb upstream of the lactase gene, LCT
High levels of dietary starch	increased production of enzyme needed to digest starch efficiently	high *AMY1A* copy number (Figure 4.11)

Table 4.4 Examples of genetic variants in adaptive evolution in human populations. Gene symbols denote genes as follows: *SLC24A5*, solute carrier family 24, member 5; *EPAS1*, endothelial PAS domain protein 1; *HBB*, β-globin gene; *G6PD*, glucose-6-phosphate dehydrogenase; *LCT*, lactase gene (converts lactose to galactose plus glucose); *AMY1A*, salivary α-amylase gene (converts starch into a mixture of constituent monosaccharides). [a]Andean populations show different anti-hypoxia adaptations. [b]Includes sickle-cell, thalassemia, and glucose-6-phosphate dehydrogenase deficiency mutations. [c]The Duffy antigen is a ubiquitously expressed cell surface protein that is required for infection of erythryocytes by *Plasmodium vivax*.

Adaptations to living in malaria-infested regions have often involved increased frequencies of harmful alleles associated with certain blood disorders, notably sickle-cell disease, thalassemia and glucose-6-phosphate dehydrogenase deficiency in the case of *Plasmodium falciparum* malaria. Heterozygotes with mutant alleles associated with these diseases exhibit small changes in phenotype that make them comparatively resistant to malaria. **Balancing selection** is involved: the mutant heterozygotes have higher rates of reproductive success than both mutant and normal homozygotes (**heterozygote advantage**)—we consider the details in Chapter 5.

The development of agriculture brought significant changes to the human diet. The domestication of wheat and rice led to high-starch diets, and the domestication of cows and goats led to lifelong consumption of fresh milk in some human populations. Adaptive responses to high-starch diets and extended milk consumption simply involved the increased production of enzymes required to metabolize starch or lactose (the major sugar in milk). The major enzyme that metabolizes starch, salivary α-amylase, is produced by *AMY1A*, and whereas our closest animal relative, the chimpanzee, has a single copy of the gene (and so a uniform diploid copy number of two) humans normally have multiple *AMY1A* genes. Individuals who take in a large amount of starch in their diets have a significantly higher *AMY1A* copy number and increased capacity to make salivary α-amylase than those who are used to low-starch diets (**Figure 4.11**).

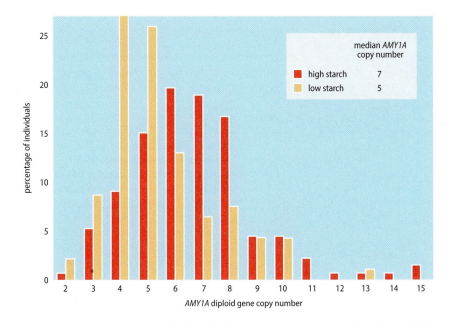

Figure 4.11 Recent acquisition of multiple genes encoding salivary α-amylase as an adaptation to high starch diets in human populations. The human *AMY1A* gene is normally present in multiple copies. The graph illustrates two points. First, the diploid copy number in humans is quite variable (between 2 and 15 from this data set) but generally high (chimpanzees have a single copy of this gene, giving a consistent diploid copy number of just 2). Secondly, individuals in populations that have high-starch diets have significantly more *AMY1A* gene copies than those in populations with low-starch diets. Salivary α-amylase hydrolyzes starch, and its activity is known to correlate with *AMY1A* copy number. The high *AMY1A* copy number is thought to be a beneficial adaptation that spread as human diets increasingly became rich in insoluble starch. (From Perry GH et al. [2007] *Nat Genet* 39:1256–1260; PMID 17828263. With permission from Macmillan Publishers Ltd.)

BOX 4.3 Selective Sweeps of Advantageous DNA Variants.

Positive selection for an advantageous DNA variant can leave distinctive signatures of genetic variation in the DNA sequence. Imagine a large population of individuals before positive selection occurs for some advantageous DNA variant on a region of, say, chromosome 22. If we were able to scan each chromosome 22 in the population before selection we might expect to find hundreds of thousands of different combinations of genetic variants (**Figure 1A**).

Now imagine that an advantageous DNA variant arises by mutation on one chromosome 22 copy and then gets transmitted through successive generations. If the advantageous variant is subject to strong positive selection, people who carry it will have significantly higher survival and reproductive success rates. As descendants of the original chromosome 22 copy carrying the variant become more and more

common, the selected DNA variant will increase in frequency to become a common allele (Figure 1B).

The entire chromosome 22 copy is not passed down as a unit: recombination will result in the replacement of some original segments by equivalent regions from other chromosome 22s. A short segment from the original chromosome 22 copy containing the favourable DNA variant and nearby 'hitchhiking alleles' will increase in prevalence in a **selective sweep** (Figure 1B), but the segment will be slowly reduced in size by recombination.

A genomic region that has been subject to a selective sweep will demonstrate extremely low heterozygosity levels. The genomic region on chromosome 15 that contains the *SLC24A5* locus in Europeans provides a good practical example—see **Figure 2**.

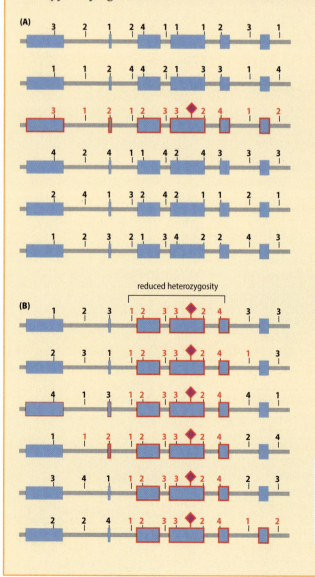

Figure 1 General effect of a selective sweep for an advantageous DNA variant. (A) Heterozygosity profile before selection. Imagine that an advantageous DNA variant has just occurred on a founder chromosome 22 (with genes outlined in red). Now imagine assaying genetic variation by using intronic and extragenic microsatellite markers, each with four common alleles (1 to 4), over each copy of chromosome 22 in the population. We might expect significant heterozygosity, as shown by the six representative chromosome 22s. (B) Heterozygosity profile after positive selection over many generations. Vertical transmission of the founder chromosome 22, recombination, and continued positive selection for the advantageous variant will result in an increased frequency of the advantageous DNA variant plus closely linked DNA variants, causing reduced heterozygosity for that chromosome segment.

BOX 4.3 (continued)

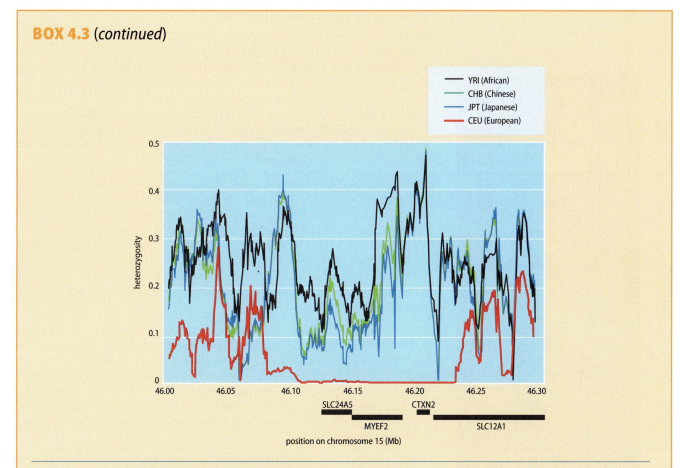

Figure 2 A strong selective sweep acting on an advantageous DNA variant in the *SLC24A5* gene in European populations. Heterozygosity levels in the region containing the *SLC24A5* gene on chromosome 15 were determined for a high density of common SNPs and averaged over 10 kb windows. The observed heterozygosity profiles for this chromosome region are unremarkable in African, Chinese, and Japanese populations. However, in the European population a strong selective sweep for a specific *SLC24A5* variant associated with reduced skin pigmentation has meant that almost all European chromosome 15s share a segment containing the favorable *SLC24A5* variant and hitchhiker alleles at the neighboring *MYEF2* and *CTXN2* loci. The result is a sharp decrease in heterozygosity for this chromosome region. (Adapted from Lamason RL, Mohideen MA, Mest JR et al. [2005] *Science* 310:1782–1786; PMID 16357253. With permission from the AAAS.)

Like other mammals, most of the world's human population are lactose intolerant: consuming any more than a small amount of milk in adult life causes abdominal pain, flatus, and diarrhea (the ability to digest lactose declines rapidly after weaning as levels of the enzyme lactase fall in the small intestine). In populations who had domesticated cows and goats, however, a cultural tradition developed of lifelong drinking of animal milk. Strong vertical transmission of this cultural practice led to selection for regulatory DNA variants that allowed lifelong expression of the lactase gene, *LCT* (lactase persistence). In each case mutations occur in a regulatory DNA region located about 14 kb upstream of the start codon.

Generating protein diversity by gene duplication and alternative processing of a single gene

Creating diverse forms of a protein can be achieved by different mechanisms, and in this section we consider two of them: gene duplication and post-transcriptional processing. We discuss in Section 4.5 additional mechanisms that are used in rather specialized cases in which protein diversity is of paramount importance to help us identify foreign pathogens.

Diversity through gene duplication

Gene duplication is common in the human genome and immediately offers the possibility of generating many different forms of a protein. For example, we produce a diverse repertoire of olfactory receptors as a result of gene duplication. The olfactory receptor gene family is our largest protein-coding gene family and comprises about 340 olfactory receptor (OR) genes plus about 300 related pseudogenes. The functional OR genes frequently have missense alleles and an average person is heterozygous at about one-third of the OR loci; we probably make about 450 or so variant olfactory receptors.

Most odorants bind to several OR variants: the precise identification of any one odor seems to depend on a combinatorial code of binding of different receptors, allowing us to potentially sense many thousands of different odors. There is, however, pronounced variation between individuals in the ability to detect specific odors, and the OR gene family demonstrates the greatest variation in gene content of any human gene family. In addition to the pseudogenes, alleles for deleterious mutations at functional gene loci are both common and highly variable between individuals (**Figure 4.12**).

Immunoglobulin and T-cell receptor genes provide the most extreme cases of multiple genes encoding variants of the same product. In germline DNA and in most somatic cells we have three immunoglobulin genes and four T-cell receptor genes. However, in maturing B and T cells, cell-specific rearrangements of the immunoglobulin and T-cell receptor genes, respectively, mean that each person has huge numbers of different gene variants for each of the immunoglobulin and T-cell receptor classes—we provide details in Section 4.5.

Having multiple extremely similar genes can raise the question of how to define an allele. Conventionally, a person has at most two alleles at an individual gene locus, but where duplicated genes make extremely similar gene products, a single individual might conceivably be viewed as having multiple alleles. For example, the duplicated *HLA-DRB1* and *HLA-DRB5* genes both make polymorphic HLA-DRβ chains, and a single person can make four different HLA-DRβ proteins. (Note that variants of the same HLA protein are traditionally known as alleles, irrespective of how rare they might be and whether they are encoded by different genes, as in the case of HLA-DRβ.)

Post-transcriptionally induced variation

Not all of the sequence variation in our proteins and functional non-coding RNAs is due to variation at the DNA level. Some variants are attributable to alternative post-transcriptional processing: a single allele at the DNA level can produce different variants (**isoforms**) of a protein or noncoding RNA. Alternative RNA splicing is very common and can lead to the skipping of exons from some transcripts and other forms of sequence difference. Specific nucleotides in transcripts can also occasionally be altered by processes known as RNA editing. We provide detail on these mechanisms in Chapter 6 within the context of understanding gene regulation.

4.5 Extraordinary Genetic Variation in the Immune System

Our most polymorphic genes deal with foreign molecules introduced into the body that are subject to independent genetic control. Complex

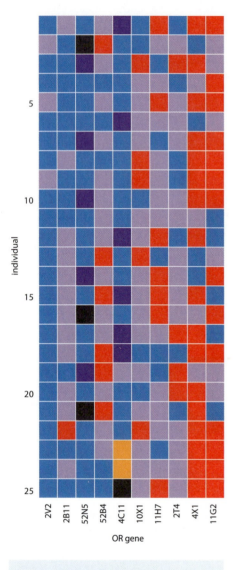

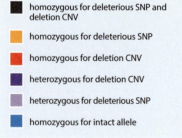

homozygous for deleterious SNP and deletion CNV

homozygous for deleterious SNP

homozygous for deletion CNV

heterozygous for deletion CNV

heterozygous for deleterious SNP

homozygous for intact allele

Figure 4.12 Genetic variation in some olfactory receptor (OR) genes. Genotype calls of ten OR genes, for which both an intact and inactive allele are present in the population. Each row represents an individual; every column represents a gene. (Data courtesy of Doron Lancet and Tsviya Olender, Department of Molecular Genetics, Weizmann Institute of Science, Rehovot.)

defense systems protect us from dangerous microbial pathogens and also from potentially harmful plant and fungal toxins in the food that we consume (by killing the microbe or the virus-infected cell and by metabolizing toxins so that they are rendered less dangerous and excreted).

The genes operating in the defense systems we use to recognize foreign molecules are polymorphic, or they undergo somatic rearrangements to produce different variants to enable them to deal with a great range of different microbes or toxins. Certain immune system genes show extraordinary genetic variation and are the subject of this section. In addition, a complex series of polymorphic enzymes is involved in metabolizing dietary components (which are effectively foreign molecules) and the genetic variation here underlies differences between individuals in how we respond to drugs. We discuss in Section 9.2 relevant aspects in a section on pharmacogenetics within the context of treating disease.

Pronounced genetic variation in four classes of immune system proteins

Our immune systems have a tough task. They are engaged in a relentless battle to protect us from potentially harmful microbial and viral pathogens. Not only must we be protected against a bewildering array of pathogens but, in addition, new forms of a pathogen can rapidly develop by mutation to provide new challenges to which we must continuously adapt.

Genes involved in identifying microbial pathogens are subject to constant positive selection to maximize diversity in the proteins involved in antigen recognition. That involves notably four types of proteins that belong to two broad classes as described below and in **Figure 4.13**.

- *Immunoglobulins.* Expressed on the surface of B cells in the bone marrow or secreted as soluble immunoglobulins (antibodies) by activated B cells, their main task is to recognize and bind specific foreign antigens. They can bind and neutralize toxins released by microbes, inhibit viruses from infecting host cells, and activate complement-mediated lysis of bacteria and phagocytosis.

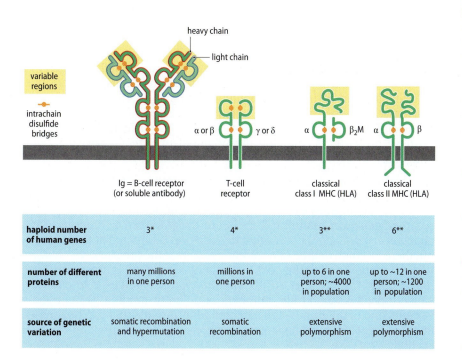

	Ig = B-cell receptor (or soluble antibody)	T-cell receptor	classical class I MHC (HLA)	classical class II MHC (HLA)
haploid number of human genes	3*	4*	3**	6**
number of different proteins	many millions in one person	millions in one person	up to 6 in one person; ~4000 in population	up to ~12 in one person; ~1200 in population
source of genetic variation	somatic recombination and hypermutation	somatic recombination	extensive polymorphism	extensive polymorphism

Figure 4.13 Extreme variation in four types of proteins that are used to recognize foreign antigens. Immunoglobulins (Igs), T-cell receptors and MHC (major histocompatibility complex) proteins are heterodimers with similar structures: globular domains (maintained by intrachain disulfide bridges) and N-terminal *variable regions* that bind foreign antigens (but otherwise have a conserved sequence, known as *constant regions* in Igs). They are cell surface receptors (except that Igs in activated B cells become secreted antibodies). Only a few human genes encode each protein chain, but nevertheless many different proteins are made because of special genetic mechanisms in B and T lymphocytes and because of selection for heterozygosity of HLA antigens. β_2M is β_2-microglobulin, the non-polymorphic light chain of class I HLA antigens. *It is estimated that we can make 10^7–10^8 different antibodies and close to the same number of different T-cell receptors. **See Figure 1 in Box 4.4 for genes encoding classical (and also non-classical) HLA antigens.

- *T-cell receptors.* Displayed on the surface of T cells, they work in cell-mediated immunity, along with proteins encoded by the major histocompatibility complex (MHC; known in humans as the HLA complex).

- *Class I MHC (HLA) proteins.* They are expressed on the surface of almost all nucleated cells and enable cytotoxic T lymphocytes to recognize and kill host cells that are infected by a virus or other intracellular pathogen.

- *Class II MHC (HLA) proteins.* They are displayed on the surface of very few types of cells, notably immune system cells that present foreign antigen to be recognized by helper T lymphocytes.

As illustrated in Figure 4.13, there is an extraordinary variety of each of the above types of proteins. The key point is that any *one* individual makes a huge variety of immunoglobulins and T-cell receptors (because of cell-to-cell genetic variation; different B cells in a single individual produce different immunoglobulins, and different T cells produce different T-cell receptors). By contrast, the extensive variety of MHC proteins is apparent at a *population* level. MHC proteins are highly polymorphic but a single person has a limited number of them, having at most two alleles at each of a small number of polymorphic MHC loci.

Random and targeted post-zygotic (somatic) genetic variation

As well as the genetic variation that we inherit from our parents, our DNA undergoes some changes as we develop from the single-celled zygote. Post-zygotic (somatic) genetic variation can involve mutations that occur randomly in all our cells. Although all our cells originate from the zygote, therefore, we are genetic **mosaics** who carry genetically different cells.

Much of the genetic variation between the cells of an individual is due to copy number variants, and mosaic patterns of copy number variations are a feature of human neurons. Small-scale mutations also arise post-zygotically that often have no functional consequences. Whereas an inherited mutation will appear in all of our nucleated cells, a somatic mutation will only be present in the cell in which it arose plus any cell lineages that arise by cell division from the progenitor cell. Some somatic mutations give rise to disease if they occur at an early stage in development or result in abnormal tumor cell populations (we consider mosaicism for pathogenic mutations in Box 5.3).

In addition to random somatic mutations, programmed changes are designed to occur in the DNA of maturing B and T cells; they are specifically targeted at immunoglobulin genes and T-cell receptor genes, respectively. These changes are positively selected so as to diversify our antibodies and our T-cell receptors so that we have huge repertoires of these proteins, maximizing the potential to detect different foreign antigens.

We inherit from each parent just three immunoglobulin genes (*IGH*, *IGK*, and *IGL*) and four T-cell receptor genes (*TRA*, *TRB*, *TRD*, and *TRG*). However, the immunoglobulin genes in maturing B cells and the T-cell receptor genes in maturing T cells are programmed to undergo DNA changes *in a cell-specific way*: genetic variation is targeted at these genes, but there is a high degree of randomness so that the precise DNA changes vary from one B cell to the next in the same individual, or from one T cell to the next. The net effect of these post-zygotic changes is to endow a single individual with huge numbers of different immunoglobulin gene variants and of different T-cell receptor gene variants that can be pressed into service. Four mechanisms are responsible, as described in the next section.

Other examples of positive selection for post-zygotic genetic variation occur in cancer cells; these can involve targeted somatic hypermutation as we explain in Chapter 10.

Somatic mechanisms that allow cell-specific production of immunoglobulins and T-cell receptors

Although a human zygote has a total of six immunoglobulin genes and eight T-cell receptor genes, somatic DNA changes in maturing B cells and T cells allow us to develop millions of different immunoglobulin (Ig) gene variants and millions of different T-cell receptor gene variants. Different mechanisms are involved, as described below.

Combinatorial diversity via somatic recombination

Each Ig and T-cell receptor gene is made up of a series of repeated gene segments that specify discrete segments of the protein, and different combinations of gene segments are used in protein production in different B cells or in different T cells of each individual. The different combinations of gene segments are made possible by somatic recombinations that occur in Ig genes in mature B cells and in T-cell receptor genes in mature T cells.

As an example, consider the gene segments that specify an Ig heavy chain. The variable region, which is involved in antigen recognition, is encoded by three types of repeated gene segment: V (encoding the first part of the variable region), D (diversity region), and J (joining region). The constant region defines the functional class of immunoglobulin (IgA, IgD, IgE, IgG, or IgM) and is encoded by repeated C gene segments (that have coding sequences split by introns). For each type of segment, the repeats are similar in sequence but nevertheless show some differences.

The first step in making an Ig heavy chain requires two sequential recombination events within the *IGH* gene of a maturing B cell. The end result is that one V gene segment, one D gene segment and one J gene segment are fused together to form a continuous VDJ coding sequence that will specify the variable region (**Figure 4.14**).

Once assembled, a VDJ coding unit activates transcription. RNA splicing fuses the VDJ transcript to transcribed coding sequences within the nearest C gene segments, initially Cμ (see Figure 4.14) and then, through alternative splicing, either Cμ or Cδ. The first immunoglobulins to be made by a B cell are membrane-bound IgM and then IgD. Subsequently, as B cells are stimulated by foreign antigen and helper T lymphocytes, they secrete IgM antibodies.

Later in the immune response, B cells undergo class-switching (also called isotype switching) to produce different antibody classes. Here, another type of somatic recombination positions an alternative C gene segment to be nearest to the J gene segments: either a Cγ, Cε, or Cα gene segment to produce respectively an IgG, IgE, or IgA antibody.

The key point about the somatic recombination events that diversify the variable regions is that they occur randomly in each maturing B or T cell, respectively (with the proviso that one each of the different repeated gene segments are brought together). That is, the genetic variation is produced by *cell-specific* recombinations. The $V_2D_3J_2$ combination in Figure 4.14 might occur in one maturing B cell, but a $V_4D_{21}J_9$ unit or a $V_{38}D_{15}J_4$ unit, for example, might be generated in neighboring B cells in the same person. The variable region of the T-cell receptor β chain is also formed by the same kind of VDJ recombination, but for both Ig light chains and T-cell receptor α chains a single VJ recombination is involved because their genes lack diversity gene segments.

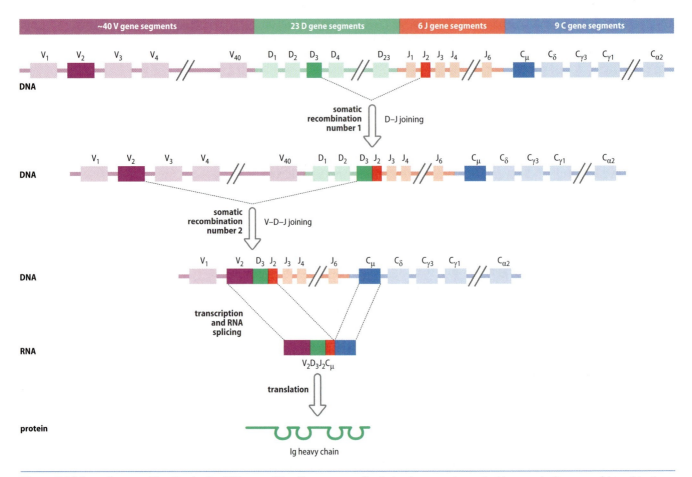

Figure 4.14 Somatic recombination in the *IGH* gene of B cells is used to make cell-specific immunoglobulin heavy chains. The human *IGH* gene has multiple but slightly different repeats for each of four types of gene segments: V (first part of variable region), D (diversity region), J (joining region), and C (constant region; although not shown here, each of the C gene segments has a coding sequence split by introns). An immunoglobulin heavy chain is made by bringing together coding sequences from one each of these four types of segments (shown here as filled boxes). Two sequential somatic recombinations produce first D–J joining, then a mature, functional VDJ coding sequence unit, which is effectively a large novel exon. In this example, the successful combination is $V_2D_3J_2$, but the choice of combinations is *cell-specific*. Once a functional VDJ exon has been assembled, transcription is initiated starting with this exon and RNA splicing joins the VDJ coding sequence to coding sequences in the *closest* constant (C) region gene segment, in this case $C\mu$. Another type of somatic recombination (known as *class switching*, but not shown here) can change the position of C gene segments so that other C gene segments can be used instead of $C\mu$ to give alternative classes of immunoglobulin.

Additional diversity generation

Two or three additional mechanisms are responsible for generating diversity in Igs and T-cell receptors as listed below. These mechanisms, together with V(D)J recombination, endow each of us with the potential of making many trillions of different antigen-binding sites, both for immunoglobulins and T-cell receptors. As required, individual B cells and T-cell receptors that successfully recognize foreign antigen are induced to proliferate to make identical clones with the same antigen specificity as the original cell.

- *Junctional diversity.* The somatic recombination mechanisms that bring together different gene segments in Ig or T-cell receptor genes variably add or subtract nucleotides at the junctions of the selected gene segments.

- *Protein chain combinatorial diversity.* Igs and T-cell receptors are heterodimers, and diversity is compounded by unique combinations of two unique protein chains. Note, however, that a B cell, for example, makes just one type of Ig. Although each diploid B cell has six Ig genes, only one of the two *IGH* alleles is (randomly) selected in each B cell to make a heavy chain (*allelic exclusion*) and only one of the four light chain genes is ever used (a combination of *light chain exclusion*—to select either a κ or λ light chain—plus allelic exclusion).

- *Somatic hypermutation.* This mechanism applies only to Igs and is used to further increase variability in the variable region after somatic recombinations have produced functional VDJ or VJ units. When B cells are stimulated by a foreign antigen, an activation-induced cytidine deaminase is produced by the activated B cell that deaminates cytidine to uridine. The uridines are variably repaired by base excision repair (see above), and the end result is that multiple nucleotides in the variable region are mutated.

MHC (HLA) proteins: functions and polymorphism

The HLA complex is the human major histocompatibility complex (MHC). The latter name came from the observation that certain MHC genes are the primary determinants in transplant rejection. That is an artificial situation: the normal function of MHC genes is to assist certain immune system cells, notably helping T cells to identify host cells that harbor an intracellular pathogen such as a virus.

Some MHC genes—called classical MHC genes—are extremely polymorphic. They are subject to positive selection to maximize genetic variation (people who are heterozygous for multiple MHC loci will be better protected against microbial pathogens and have higher reproductive success rates). The classical MHC proteins are deployed on the cell surface as heterodimers (see Figure 4.13). They serve to bind peptide fragments derived from the intracellular degradation of pathogen proteins and display them on the surface of host cells (**antigen presentation**) so that they can be recognized by T cells. Appropriate immune reactions are then initiated to destroy infected host cells. There are two major classes of classical MHC proteins, as detailed below.

Class I MHC proteins

Class I MHC proteins are expressed on almost all nucleated host cells. Their job is to help cytotoxic T cells (cytotoxic T lymphocytes; CTLs) to recognize and kill host cells that have been infected by a virus or other intracellular pathogen. When intracellular pathogens synthesize protein within host cells, a proportion of the protein molecules get degraded by proteasomes in the cytosol. The resulting peptide fragments are transported into the endoplasmic reticulum. Here, a newly formed class I MHC protein binds a peptide and is exported to the cell surface, where it is recognized by a CTL with a suitable receptor.

Because of cell-specific somatic recombinations (similar to those in Figure 4.14), individual CTLs make unique T-cell receptors that recognize *specific* class I MHC–peptide combinations. If the bound peptide is derived from a pathogen, the CTL induces killing of the host cell. Note that a proportion of normal host cell proteins also undergo degradation in the cytosol and the resulting self-peptides are bound by class I MHC proteins and displayed on the cell surface. But there is normally no immune response (starting in early fetal life, CTLs that recognize MHC-self peptide are programmed to be deleted, to minimize autoimmune responses).

Class II MHC proteins

Class II MHC proteins are expressed in professional antigen-presenting cells: dendritic cells, macrophages, and B cells. These cells also express class I MHC proteins but, unlike most cells, they make co-stimulatory molecules needed to initiate lymphocyte immune responses.

Whereas class I MHC proteins bind peptides from *endogenous* proteins (those made within the cytosol, such as a viral protein made after infection of that cell), class II MHC proteins bind peptides derived from *exogenous* proteins that have been transported into the cell (by endocytosis of a microbe or its products) and delivered to an endosome, where limited proteolysis occurs. The resulting peptide fragments are bound by previously assembled class II MHC proteins and transported to the cell surface so that a helper T lymphocyte with an appropriate receptor recognizes a specific class II MHC–peptide combination. (Helper T cells have critical roles in coordinating immune responses by sending chemical signals to other immune system cells.)

MHC restriction

T cells recognize a foreign antigen only after it has been degraded and become associated with MHC molecules (*MHC restriction*). A proportion of all normal proteins in a cell are also degraded, and the resulting peptides are displayed on the cell surface, complexed to MHC molecules. MHC proteins cannot distinguish self from nonself, and even on the surface of a virus-infected cell the vast majority of the many thousands of MHC proteins on the cell surface bind peptides derived from host cell proteins rather than from virus proteins.

The rationale for MHC restriction is that it provides a simple and elegant solution to the problem of how to detect intracellular pathogens—it allows T cells to survey a peptide library derived from the entire set of proteins in a cell *but only after the peptides have been displayed on the cell surface*.

MHC polymorphism

MHC polymorphism is pathogen-driven: strong selection pressure favors the emergence of mutant pathogens that seek to evade MHC-mediated detection. The MHC has evolved two counterstrategies to maximize the chance of detecting a pathogen. First, gene duplication has provided multiple MHC genes that make different MHC proteins with different peptide-binding specificities. Secondly, many of the MHC genes are extraordinarily polymorphic, producing the most polymorphic of all our proteins (**Table 4.5**).

The polymorphism of classical MHC proteins is focused on amino acids that form the antigen-binding pockets: different alleles exhibit different peptide-binding specificities. A form of long-standing balancing selection (also called overdominant selection) seems to promote MHC polymorphism. Heterozygosity is favored (presumably the ability to produce many different HLA proteins affords us greater protection against pathogens), and certain heterozygote genotypes seem to display greater fitness than others.

The balancing selection seems to have originated before the speciation event leading to evolutionary divergence from the great apes. HLA polymorphism is therefore exceptional in showing trans-species

Table 4.5 Statistics for the six most polymorphic HLA loci. Data were derived from the European Bioinformatics Institute's IMGT/HLA database (release 3.12.0 at 17 April 2013). The statistics for these and additional loci are available at http://www.ebi.ac.uk/ipd/imgt/hla/stats.html.

HLA GENE	-A	-B	-C	-DPB1	-DQB1	-DRB1
Number of alleles or DNA variants	2244	2934	1788	185	323	1317
Number of protein variants	1612	2211	1280	153	216	980

polymorphism: a human HLA allele may be more closely related in sequence to a chimpanzee HLA allele than it is to another human HLA allele. For example, human HLA-DRB1*0701 and HLA-DRB1*0302 show 31 amino acid differences out of 270 amino acid positions, but human HLA-DRB1*0701 and chimpanzee Patr-DRB1*0702 show only 2 differences out of 270.

The medical importance of the HLA system

The HLA system is medically important for two principal reasons. First, the high degree of HLA polymorphism poses problems in organ and cell transplantation. Secondly, certain HLA alleles are risk factors for individual diseases, notably many autoimmune diseases and certain infectious diseases; other HLA alleles are protective factors, being negatively correlated with individual diseases.

Transplantation and histocompatibility testing

After organ and cell transplantation, the recipient's immune system will often mount an immune response against the transplanted donor cells (the graft), which carry different HLA antigens from those of the host cells. The immune reaction may be sufficient to cause rejection of the transplant (but corneal transplants produce minimal immune responses—the cornea is one of a few immune privileged sites that actively protect against immune responses in several ways, including having a much reduced expression of class I HLA antigens).

Bone marrow transplants and certain stem cell transplants can also result in graft-versus-host disease (GVHD) when the graft contains competent T cells that attack the recipient's cells. GVHD can even occur when donor and recipient are HLA-identical because of differences in minor (non-HLA) histocompatibility antigens.

Immunosuppressive drugs are used to suppress immune responses after transplantion, but transplant success depends largely on the degree of HLA matching between the cells of the donor and the recipient. Histocompatibility testing (also called tissue typing) involves assaying HLA alleles in donor tissues so that the best match can be found for prospective recipients. The key HLA loci are the most polymorphic ones: *HLA-A, -B, -C, -DRB1, -DQB1*, and *-DPB1* (Table 4.5 and **Box 4.4**).

HLA disease associations

By displaying peptide fragments on host cell surfaces, HLA proteins direct T cells to recognize foreign antigens and initiate an immune response against cells containing viruses or other intracellular pathogens. Because HLA proteins differ in their ability to recognize specific foreign antigens, people with different HLA profiles might be expected to show different susceptibilities to some infectious diseases.

In autoimmune diseases, the normal ability to discriminate self antigens from foreign antigens breaks down, and autoreactive T cells launch attacks against certain types of host cells. Certain HLA antigens are very strongly associated with individual diseases, such as type I diabetes and rheumatoid arthritis; in general, genetic variants in the HLA complex are the most significant genetic risk factors that determine susceptibility to autoimmune diseases. Determining to what extent HLA variants are directly involved in the pathogenesis and how much is contributed by other variants that lie in the immediate vicinity of the HLA genes (and outside the HLA complex) is a major area of research—we consider HLA associations with individual diseases in some detail in Chapter 8.

BOX 4.4 HLA Genes, Alleles, and Haplotypes.

HLA genes

The HLA complex spans 3.6 Mb on the short arm of chromosome 6. The 253 genes in the complex include the 18 protein-coding HLA genes shown in **Figure 1**, ranging from *HLA-DPB1* (closest to the centromere) to *HLA-F*. Genes in the class I region make the heavy chain of class I HLA antigens (the non-polymorphic class I HLA light chain, β_2-microglobulin, is encoded by a gene on chromosome 15); the class II region has genes encoding both chains of class II HLA antigens. The intervening region does not contain any HLA genes, but it does contain multiple genes with an immune system function and is sometimes referred to as the class III region.

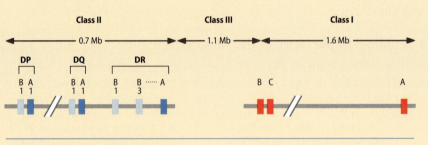

Figure 1 Classical (polymorphic) HLA genes within the HLA complex at 6p21.3. Genes in the class II HLA region encode α chains (dark shading) and β chains (pale shading) that pair up to form heterodimers within specific classes as indicated by horizontal bars above (DP, DQ, DR). Classical class I HLA genes encode a polymorphic class I α chain that forms a heterodimeric protein with the non-polymorphic β-microglobulin chain encoded by a gene on chromosome 15. Within the class I and class II HLA regions are several other non-polymorphic HLA genes and many HLA-related pseudogenes not shown here. The class III region includes certain complement genes. Some additional genes with an immune system function are found within the HLA complex plus some functionally unrelated genes such as the steroid 21-hydroxylase gene.

HLA alleles

Because of their extraordinary polymorphism, alleles of the classical, highly polymorphic HLA genes have been typed for many decades at the protein level (using serological techniques with panels of suitably discriminating antisera). The number of alleles that can be distinguished in this way is very high, for example 28 HLA-A alleles, 50 HLA-B alleles, and 10 HLA-C alleles (called Cw for historical reasons; the 'w' signifies workshop because nomenclature was updated at regular HLA workshops).

Serological HLA typing is still used when rapid typing is required, as in the case of solid organ transplants (in which the time between the chilling of an organ and the time it is warmed by having the blood supply restored needs to be minimized). However, much of modern HLA typing is performed at the DNA level, where very large numbers of alleles can be identified (see Table 4.5). The complexity means that the nomenclature for HLA alleles identified at the DNA level is quite cumbersome—see **Table 1** for examples.

HLA haplotypes

The genes in the HLA complex are highly clustered, being confined to an area that represents only about 2% of chromosome 6. Genes that are close to each other on a chromosome are usually inherited together because there is only a small chance that they will be separated by a recombination event occurring in the short interval separating the genes. Such genes are said to be *tightly linked* (we consider genetic linkage in detail in Section 8.1).

A **haplotype** is a series of alleles at linked loci on an *individual* chromosome; haplotypes were first used widely in human genetics with reference to the HLA complex. See **Figure 2** for how haplotypes are established by tracking the inheritance of alleles in family studies. Note that because the HLA genes are very closely linked, recombination within the HLA complex is rare.

NOMENCLATURE	MEANING
HLA-DRB1	an HLA gene (encoding the β chain of the HLA-DR antigen)
*HLA-DRB1*13*	alleles that encode the serologically defined HLA-DR13 antigen
*HLA-DRB1*13:01*	one specific HLA allele that encodes the HLA-DR13 antigen
*HLA-DRB1*13:01:02*	an allele that differs from *DRB1*13:01:01* by a synonymous mutation
*HLA-DRB1*13:01:01:02*	an allele that differs from *DRB1*13:01:01* by having a mutation outside the coding region
*HLA-A*24:09N*	a null allele that is related by sequence to alleles that encode the HLA-A24 antigen

Table 1 HLA allele nomenclature. For more details see http://hla.alleles.org/.

BOX 4.4 (*continued*)

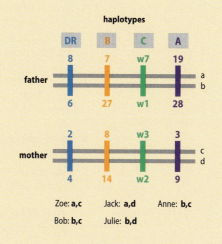

	alleles					haplotypes			
	HLA-DR	HLA-B	HLA-C	HLA-A		DR	B	C	A
father	6, 8	7, 27	w1, w7	19, 28					
mother	2, 4	8, 14	w2, w3	3, 9					
Zoe	2, 8	7, 8	w3, w7	3, 19					
Bob	2, 6	8, 27	w1, w3	3, 28					
Jack	4, 8	7, 14	w2, w7	9, 19					
Julie	4, 6	14, 27	w1, w2	9, 28					
Anne	2, 6	8, 27	w1, w3	3, 28					

Zoe: **a,c**   Jack: **a,d**   Anne: **b,c**

Bob: **b,c**   Julie: **b,d**

Figure 2 Deriving HLA haplotypes from family studies. Father, mother, and their three daughters, Zoe, Julie, and Anne, and two sons, Bob and Jack, have been tissue typed using serological reagents for four HLA antigens as shown at the left. By tracking which parental alleles have been passed on to individual children it is possible to deduce the parental HLA haplotypes. Father has one chromosome 6 with the HLA haplotype DR8, B7, Cw7, A19 (haplotype a) and another chromosome 6 with the HLA haplotype DR6, B27, Cw1, A28 (haplotype b). Similarly, mother has haplotypes c (DR2, B8, Cw3, A3) and d (DR4, B14, Cw2, A9). Father has transmitted haplotype a to Zoe and Jack, and haplotype b to Bob, Julie, and Anne. Mother has transmitted haplotype c to Zoe, Bob, and Anne, and haplotype d to Jack and Julie.

Summary

- The DNA in our cells accumulates changes over time (mutations) that usually have no significant effect on the phenotype.

- Some mutations adversely affect how genes work or are expressed; they can be associated with disease and because at least some people who carry them have a lower reproductive fitness they tend to be removed from populations (purifying selection).

- Very occasionally, a mutation may result in some benefit and may accumulate in frequency if it endows individuals with increased reproductive fitness (positive selection).

- Large-scale changes to DNA can result from abnormalities in chromosome segregation and recombination. Smaller-scale changes typically result from unrepaired errors in DNA replication or unrepaired chemical attacks on DNA.

- DNA is damaged within living cells and organisms by various types of chemical attack that break covalent bonds in DNA or form inappropriate covalent bonds with bases. One or both DNA strands may be broken, bases or nucleotides may be deleted, or inappropriate chemical groups may be covalently bonded to the DNA.

- Chemical attacks on DNA can arise from external agents (such as ionizing radiation or harmful chemicals), but mostly they originate spontaneously (DNA has certain chemically unstable bonds and chemical groups that are susceptible to attack by highly reactive chemicals produced naturally inside our cells).

- According to the type of chemical damage to DNA, different cellular pathways are used to repair a DNA lesion. Direct reversal of the chemical steps that cause damage is rare, and individual pathways often involve many molecular components.

- DNA variants often have low frequencies. More common variants that have a frequency of more than 0.01 are described as DNA polymorphisms.

- A single nucleotide variant (or polymorphism) involves the substitution of one nucleotide for another at a specific location. Nucleotide substitutions are nonrandom—for example, C→T substitutions are particularly common in vertebrate DNA.

- An indel is a site where variants differ by lacking or possessing one or a few nucleotides.

- Some DNA variants differ by having different numbers of copies of a tandemly repeated DNA sequence, producing length variation. Microsatellite variants are DNA sequences that show small length differences as a result of having fewer or more tandem copies of a simple repeat sequence with between one and four nucleotides.

- Structural variation results from large-scale changes in DNA. In balanced structural variation, the variants do not differ in DNA content. In unbalanced structural variation, there is substantial length variation between variants that often occurs as a result of copy number variation for a long nucleotide sequence.

- In population-based genome sequencing, whole diploid genomes from multiple individuals are sequenced, providing comprehensive data on human genetic variation.

- Recent positive selection for genetic variants in different human populations has allowed adaptation to different local environments and to major dietary changes.

- Gene duplication is the basis of our diverse repertoire of olfactory receptors.

- To identify foreign antigens efficiently, each of us makes an extraordinary variety of immunoglobulins and T-cell receptors. We inherit only three immunoglobulin and four T-cell receptor genes from each parent, but cell-specific somatic rearrangements in maturing B and T cells endow us with huge numbers of different immunoglobulin and T-cell receptor gene variants.

- Our most polymorphic proteins are produced by genes in the HLA complex (the human major histocompatibility complex). HLA proteins recognize and bind peptides from processed foreign antigens and present them on cell surfaces so that they can be recognized by specific T-cell receptors.

- The extreme polymorphism of HLA proteins means that recipients of tissue or organ transplants often mount strong immune responses to the foreign tissue. Tissue typing seeks to find reasonable matches between HLA antigens expressed by donor tissue and prospective recipients.

Questions

Help on answering these questions and a multiple-choice quiz can be found at www.garlandscience.com/ggm-students

1. Most of the constitutional variation in our DNA comes from endogenous sources. What are they?

2. A single-strand DNA break is a problem for the cell, but a double-strand DNA break is often an emergency. Explain why.

3. In what circumstances do cells use the nucleotide excision repair pathway?

4. Why should the repair of double-strand breaks in DNA be more accurate in cells after the DNA has replicated than before it has replicated?

5. With regard to human genomic DNA, what is meant by a DNA polymorphism and an allele, and how is the term allele used differently when it refers to genetic variation at the protein level?

6. The pattern of single nucleotide substitution in the human genome is not random: as in the genomes of other vertebrates, there is a marked excess of C→T substitutions. Why?

7. Single nucleotide polymorphisms (SNPs) are distributed across the genome but occur at only a small minority of nucleotide positions. Why should only certain nucleotides be polymorphic and be surrounded by stretches of nucleotide sequences that rarely show variants?

8. Explain the general concept of natural selection. What type of selection is involved when we talk about selection pressure to conserve sequences and their functions, and how does it operate?

9. What is positive selection? Give examples of individual genes in which positive selection appears to have led to an increase in heterozygosity, an increase in gene expression, or an increase in gene copy number.

10. The relationship between genetic variation at the level of genes and proteins is complicated and is not simply due to whether genetic variants in a single gene cause amino acid substitutions or not. What other factors are involved?

11. What is genetic mosaicism, and how widespread is it?

12. In human zygotes there are only six immunoglobulin genes, two alleles each at the *IGH*, *IGK*, and *IGL* loci, and yet each of us is able to make huge numbers of different antibodies. How is that possible?

Further Reading

DNA Damage and DNA Repair

Barnes DE & Lindahl T (2004) Repair and genetic consequences of endogenous DNA base damage in mammalian cells. *Annu Rev Genet* 38:445–476; PMID 15568983.

Ciccia A & Elledge SJ (2010) The DNA damage response: making it to safe to play with knives. *Mol Cell* 40:179–204; PMID 20965415. [An authoritative review on both DNA damage and repair, including detailed tabulation of frequencies of different types of DNA damage and of inherited disorders of DNA damage responses/DNA repair.]

Rass U et al. (2007) Defective DNA repair and neurodegenerative disease. *Cell* 130:991–1004; PMID 17889645.

Personalized and Population-Based Genome Sequencing

Levy S et al. (2007) The diploid genome sequence of an individual human. *PLoS Biol* 5:e254; PMID 17803354.

The 1000 Genomes Project Consortium (2010) A map of human genome variation from population scale sequencing. *Nature* 467:1061–1073; PMID 20981092.

The 1000 Genomes Project Consortium (2012) An integrated map of genetic variation from 1,092 human genomes. *Nature* 491:56–65; PMID 23128226.

Wheeler DA et al. (2008) The complete genome of an individual by massively parallel DNA sequencing. *Nature* 452:872–877; PMID 18421352.

Structural Variation, Copy Number Variation, and Indels

Alkan C et al. (2011) Genome structural variation and genotyping. *Nature Rev Genet* 12:363–376; PMID 21358748.

Conrad DF et al. (2010) Origins and functional impact of copy number variation in the human genome. *Nature* 464:704–712; PMID 19812545.

Mills RE et al. (2011) Mapping copy number variation by population-scale genome sequencing. *Nature* 470:59–65; PMID 21293372.

Mullaney JM et al. (2010) Small insertions and deletions (INDELs) in human genomes. *Hum Mol Genet* 19:R131–R136; PMID 20858594.

Human mutation rates

Campbell CD, Eichler EE (2013) Properties and rates of germline mutations in humans. *Trends Genet*. 29:575–584; PMID 23684843.

Lynch M (2010) Rate, molecular spectrum and consequences of human mutation. *Proc Natl Acad Sci USA* 107:961–968; PMID 20080596.

Functional Variation, Positive Selection and Adaptive Evolution

Fisher SE & Marcus GF (2006) The eloquent ape: genes, brains and the evolution of language. *Nature Rev Genet* 7:9–20; PMID 16369568.

Fu W & Akey JM (2013) Selection and adaptation in the human genome. *Annu Rev Genomics Hum Genet* 14:467-489; PMID 23834317.

Grossman SR et al. (2013) Identifying recent adaptations in large-scale genomic data. *Cell* 152:703–713; PMID 23415221.

Hulse AM & Cai JJ (2013) Genetic variants contribute to gene expression variability in humans. *Genetics* 193:95–108; PMID 23150607.

Olson MV (2012) Human genetic individuality. *Annu Rev Genomics Hum Genet* 13:1–27; PMID 22657391.

Post-Zygotic Genetic Variation

McConnell MJ et al. (2013) Mosaic copy number variation in human neurons. *Science* 342:632-637; PMID 24179226.

O'Huallachain M et al. (2012) Extensive genetic variation in somatic human tissues. *Proc Natl Acad Sci USA* 109:18018–18023; PMID 23043118.

Genetic Variation in the Immune System

Bronson PG et al. (2013) A sequence-based approach demonstrates that balancing selection in classical human leukocyte antigen (HLA) loci is asymmetric. *Hum Molec Genet* 22:252–261; PMID 23065702.

Murphy K (2011) Janeway's Immunobiology, 8th ed. Garland Science.

Parham P (2009) The Immune System. Garland Science.

Shiina T et al. (2009) The HLA genomic loci map: expression, interaction, diversity and disease. *J Hum Genet* 54:15–39; PMID 19158813.

Spurgin LG & Richardson DS (2010) How pathogens drive genetic diversity: MHC, mechanisms and misunderstandings. *Proc R Soc B* 277:979–988; PMID 20071384.

Single-Gene Disorders: Inheritance Patterns, Phenotype Variability, and Allele Frequencies

CHAPTER 5

Genes are functional units of DNA that make some product needed by cells, either the polypeptide chain of a protein or a functional noncoding RNA. In this chapter, however, we will consider genes very largely as abstract entities and we consider them within the context of single-gene disorders—diseases in which the genetic contribution is determined primarily by one gene locus. Although individually rare, single-gene disorders are important contributors to disease. Knowledge of single-gene disorders also provides a framework for understanding the more complex genetic susceptibility to common disease (which we cover in later chapters, notably Chapters 8 and 10).

We look first at the patterns of inheritance of single-gene disorders and provide an introductory basis for estimating disease risk according to the inheritance pattern (we provide more advanced disease risk calculations within the context of genetic counseling in Chapter 11).

We also consider how genes affect our observable characteristics. The term **phenotype** may be used broadly to describe the observable characteristics of a person, an organ, or a cell. But geneticists also use the word phenotype in a narrower sense to describe only those specific manifestations that arise in response to the differential expression of just one or a small number of genes. These manifestations may be harmful, and we can talk of a disease phenotype.

When the observable manifestations are not disease-associated we normally refer to a character or **trait**, for example blue eyes or blood group O. We can measure and record aspects of the phenotype, such as anatomical and morphological features, behaviour, or cognitive functions. Sophisticated laboratory procedures can be used to perform more extensive investigations at the physiological, cellular, and molecular levels.

Genetic variation—changes in the base sequence of our DNA—is the primary influence on the phenotype (identical twins are remarkably similar, after all). But it is not the only determinant of the phenotype: environmental factors make a contribution, too, as do **epigenetic** effects (which, unlike genetic mechanisms, are independent of the base sequence of DNA) and stochastic factors.

As we will see, there can be considerable complexity in the link between genetic variation and phenotype: even in single-gene disorders the phenotype is often variable in affected members of one family. Note that we do not deal with the molecular basis of single-gene disorders here; that will be covered in later chapters, notably Chapter 7.

We end the chapter by generally looking at the factors that affect allele frequencies in populations, and then focusing on frequencies of disease alleles (which are important practically for calculating disease risk for some types of single-gene disorder). And we explain why some single-gene disorders are common but others are rare.

Introduction: Terminology, Electronic Resources, and Pedigrees

The Basics of Mendelian and Mitochondrial DNA Inheritance Patterns

Uncertainty, Heterogeneity, and Variable Expression of Mendelian Phenotypes

Allele Frequencies in Populations

5.1   Introduction: Terminology, Electronic Resources, and Pedigrees

Background terminology and electronic resources with information on single-gene disorders

An individual gene or DNA sequence in our nuclear DNA has a unique chromosomal location that defines its position, its **locus** (plural **loci**). We can refer to the ABO blood group locus, for example, or the *D3S1563* locus (a polymorphic DNA marker sequence located on chromosome 3).

Alleles and allele combinations

In human genetics an **allele** means an individual copy of a gene or other DNA sequence that is carried at a locus on a *single* chromosome. Because we are diploid we normally have two alleles at any one chromosomal locus, one inherited on a chromosome passed down from mother (the maternal allele) and one inherited from father (the paternal allele). The term **genotype** describes the combination of alleles that a person possesses at a single locus (or at a number of loci). If both alleles are the same at an individual locus, a person is said to be **homozygous** at that locus and may be referred to as a homozygote. If the alleles are different, even by a single nucleotide, the person is said to be **heterozygous** at that locus, a heterozygote.

Although we are essentially diploid, men have two types of sex chromosomes, X and Y, which are very different in both structure and gene content. As a result, most DNA sequences on X do not have a direct equivalent (allele) on Y, and vice versa. Men are therefore **hemizygous** for such loci (because they normally only have one allele). Women normally have two alleles at each locus on the X chromosome.

Dominant and recessive phenotypes

In humans any genetic character is likely to depend on the expression of a large number of genes and environmental factors. For some, however, a particular genotype at a *single* locus is the primary determinant, being both necessary and sufficient for the character to be expressed under normal circumstances. Such characters are often said to be **Mendelian**, but that implies that a chromosomal locus is involved; a more accurate term is monogenic (which takes in to account both chromosomal loci and loci on mitochondrial DNA). Although collectively important, individual single-gene disorders are rare, and common genetic disorders depend on multiple genetic loci.

When a human monogenic disorder (or trait) is determined by a nuclear gene, the disorder (or trait) is said to be **dominant** if it is manifested in the heterozygote (who carries a normal allele and a mutant allele), or **recessive** if it is not. Sometimes two different phenotypes that result from mutations at a single gene locus can be simultaneously displayed by the heterozygote and are said to be **co-dominant**. For example, the AB blood group is the result of co-dominant expression of the A and B blood group phenotypes that are determined by different alleles at the *ABO* blood group locus. As described below, the inheritance of mitochondrial DNA is rather different, with important implications for associated phenotypes.

Electronic information on monogenic disorders

Various electronic resources provide extensive information on human single-gene disorders and characters (**Box 5.1**). *GeneReviews* provides excellent summaries for many of the more common single-gene disorders

BOX 5.1 Electronic Resources With Information on Human Single-Gene Disorders and Underlying Genes.

Some of the more comprehensive and stable resources are listed below. There are also many disease-specific databases; we describe some of these in Section 7.2.

Genecards (http://www.genecards.org). A gene-centered database, this contains about 50,000 automatically generated entries, mostly relating to specific human genes. It provides a large amount of biological information about each gene.

***GeneReviews*™** (http://www.ncbi.nlm.nih.gov/books/ NBK1116/; see PMID 20301295 for an alphabetic listing). This series of clinically and genetically orientated reviews of single-gene disorders is made available through NCBI's Bookshelf program. Individual reviews are assigned a PubMed identifier (**PMID**), an eight-digit number that in this case normally begins with 2030—for example, Huntington disease is at PMID 20301482. The series covers about 375 of the most common single-gene disorders, and for listed disorders there is more clinical information than in OMIM (see below).

OMIM (http://www.ncbi.nlm.nih.gov/omim). The Online Mendelian Inheritance in Man database is the most comprehensive single source of information on human Mendelian phenotypes and the underlying genes. Entries have accumulated text over many years, and the early part of an entry may often reflect historical developments rather than current understanding.

Each OMIM entry has an identifying six-digit number in which the first digit indicates the mode of inheritance. The initial convention for the first digit was: 1, autosomal dominant; 2, autosomal recessive; 3, X-linked; 4, Y-linked; and 5, mitochondrial. However, the distinction between autosomal dominant and autosomal recessive was discontinued for new entries after May 1994. After that date all new entries for autosomal traits and genes were assigned a six-digit number beginning with the number 6. See the review by McKusick (2007) for further details.

that are accessible through the widely used PubMed system for electronic searching of biomedical research literature. We therefore often provide the eight-digit PubMed identifier (**PMID**) for relevant *GeneReviews* articles on single-gene disorders. The Online Mendelian Inheritance in Man (**OMIM**) database is comprehensive, and we provide six-digit OMIM database numbers for some disorders.

Investigating family history of disease and recording pedigrees

The extent to which a human disorder has a genetic basis can often be established by taking a family history. Medical records may be available to health service professionals for some family members; details of deceased family members and others who may be difficult to contact may be obtained by consulting more accessible family members.

A **pedigree** is a graphical representation of a family tree that uses the standard symbols depicted in **Figure 5.1**. Generations are often labeled with Roman numerals that increase from top to bottom of the page (toward the youngest generation). Individuals within each generation are given Arabic numerals that increase from left to right. An extended family covering many generations may be described as a kindred. A family member through whom the family is first ascertained (brought to the attention of health care professionals) is known as the *proband* (also called propositus—feminine proposita) and may be marked with an arrow.

The term **sib** (sibling) is used to indicate a brother or sister, and a series of brothers and sisters is known as a sibship. According to the number

Figure 5.1 Pedigree symbols.

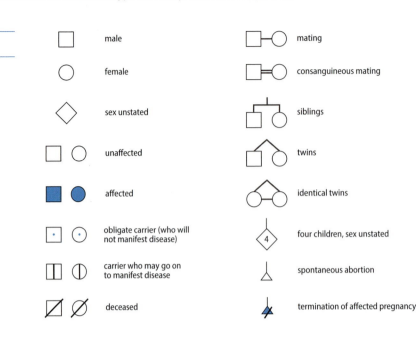

of steps in the pedigree that links two family members, they may be classified relatives of the first degree (parent and child; sibs); second degree (grandparent and grandchild; uncle/aunt and nephew/niece; half-sibs); third degree (first cousins), and so on. Couples who have one or more recent ancestors in common are said to be **consanguineous**.

5.2 The Basics of Mendelian and Mitochondrial DNA Inheritance Patterns

Mendelian characters are determined by chromosomal loci, either on an autosome (human chromosomes 1 to 22) or on a sex chromosome (X or Y). Females are diploid for all loci (they have 23 pairs of homologous chromosomes). Males are different. Like females they have two copies of each autosomal locus and of **pseudoautosomal** sequences (see below) at the tips of the sex chromosomes. However, they are hemizygous for the great majority of loci on the X and the Y (males have only one copy of the great majority of loci that are located on the X and the Y but outside the pseudoautosomal regions).

As a result of the above, there are five basic Mendelian inheritance patterns: autosomal dominant, autosomal recessive, X-linked dominant, X-linked recessive, and Y-linked (not Y-linked dominant or Y-linked recessive because males are never heterozygous for Y-linked sequences; the two Y chromosomes in rare XYY males are duplicates). In addition there is the unique pattern of inheritance of mitochondrial DNA mutations, which are substantial contributors to human genetic disease.

Autosomal dominant inheritance

A dominantly inherited disorder is one that is manifested in heterozygotes: affected persons usually carry one mutant allele and one normal allele at the disease locus. In autosomal dominant inheritance, the disease locus is present on an autosome (any chromosome other than the X or the Y), and so an affected person can be of either sex.

When an affected person has children with an unaffected person, each child would normally have a 50% chance of developing the disease (the affected parent can transmit either the mutant allele or the normal allele).

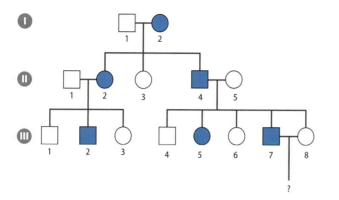

Figure 5.2 Pedigree showing autosomal dominant inheritance. Both sexes are affected and are equally likely to transmit the disorder. Affected individuals are typically heterozygotes (with one mutant allele and one normal allele) and usually have at least one affected parent. The question mark indicates the chance of having an affected child, which is 1 in 2 because one parent, III-7, is affected (there is a 50% chance of transmitting the mutant allele and a 50% chance of transmitting the normal allele).

Affected persons often have an affected parent (see a typical example of autosomal dominant inheritance in **Figure 5.2**).

Because the disorders are rare, affected individuals are almost always heterozygotes. Very occasionally, however, affected homozygotes are born to parents who are both affected heterozygotes. According to the effect of the mutation on the gene product, the affected homozygotes may show the same phenotype as the affected heterozygote. More commonly, affected homozygotes have a more severe phenotype than affected heterozygotes, as reported for conditions such as achondroplasia (PMID 20301331) and Waardenburg syndrome type I (PMID 20301703), or they have a much earlier age at onset of the disease, as in familial hypercholesterolemia (OMIM 143890).

In model organisms, a distinction is often made between different phenotypes seen in affected homozygotes and in affected heterozygotes (which are respectively called dominant and semidominant phenotypes in mice, for example). In human genetics, however, we refer to dominant phenotypes in affected heterozygotes simply because affected homozygotes are so rarely encountered.

Autosomal recessive inheritance

A person affected by an autosomal recessive disorder can be of either sex and is usually born to unaffected parents who are heterozygotes (the parents would be described as asymptomatic **carriers** because they carry one mutant allele without being affected). Affected individuals carry two mutant alleles at the disease locus, one inherited from each parent. Assuming that both parents of an affected child are phenotypically normal carriers, the chance that each future child born to these parents is also affected is normally 25% (the risk that one parent transmits the mutant allele is 1/2, so the risk that they both transmit the mutant allele to a child is $1/2 \times 1/2 = 1/4$).

Every one of us carries a single harmful allele at multiple loci associated with recessive phenotypes (carrying two such alleles can lead to disease, or even lethality in the prenatal period). When an autosomal recessive disorder is quite frequent, carriers will be common. In that case an affected child may often be born to two parents who carry different mutant alleles, and the affected individual with two different mutant alleles would be described as a **compound heterozygote** (**Figure 5.3A**).

Consanguinity

A feature of many recessive disorders, especially rare conditions, is that affected individuals often have two identical mutant alleles because the parents are close relatives; such couples are said to be *consanguineous*. In the example in Figure 5.3B the parents of the affected child in

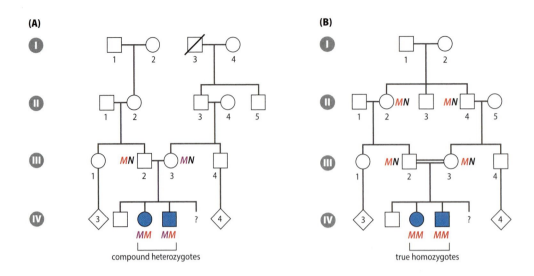

Figure 5.3 Pedigree showing autosomal recessive inheritance. (A) A pedigree for a common autosomal recessive disorder. The parents of the affected children in generation IV are carriers, with one normal allele (N) and one mutant allele (M). If they are not known to be related they might well have different mutant alleles (shown by pink or red M) and the affected children would be compound heterozygotes. From the pedigree alone, we would not know who carried mutant alleles in generations I and II. For each subsequent child of III-2 and III-3, the risk of being affected is 1 in 4, irrespective of sex (each parent has a 50% chance of transmitting the mutant allele, and the chance of inheriting both alleles is $1/2 \times 1/2 = 1/4$). (B) Involvement of consanguinity. Here we know that III-2 and III-3 are first cousins. They can be expected to be carriers, with one mutant allele (M) and one normal allele (N). We could infer that II-2 and II-4 inherited the same mutant allele (red M) from one parent (either I-1 or I-2). That means that III-2 and III-3 have the same mutant allele and their affected children will have inherited two identical mutant alleles and be true homozygotes. The chance that their fourth child will be affected (question mark) remains 1 in 4, irrespective of its sex.

generation IV are first cousins, and they will have 1/8 of their genes in common by genetic descent (**Box 5.2** shows how these calculations are made). The two parents, III-2 and III-3, have each inherited the same mutant allele ultimately from the same common ancestor (in this case, a common grandparent, either I-1 or I-2).

For rare disorders when there is doubt concerning the mode of inheritance, known parental consanguinity will strongly indicate autosomal recessive inheritance in a pedigree in which affected individuals have unaffected parents. But if consanguinity is not apparent, as in the pedigree in Figure 5.3A, alternative explanations are possible, as described below.

Disease-related phenotypes in carriers

Although carriers of an autosomal recessive disorder are considered asymptomatic, they may nevertheless express some disease-related trait that can distinguish them from the normal population. Take sickle-cell disease (OMIM 603903), for example. Affected individuals are homozygous for a β-globin mutation and produce an abnormal hemoglobin, HbS, that causes red blood cells to adopt a rigid, crescent (or sickle) shape. The sickle cells have a shorter life span that leads to anemia, and they can block small blood vessels, causing hypoxic tissue damage.

Carriers of the sickle-cell mutation are not quite asymptomatic. The sickle-cell allele produces a mutant β-globin that is co-dominantly expressed with the normal β-globin, and heterozygotes can have mild anemia (sickle-cell trait). However, under intense, stressful conditions such as exhaustion, hypoxia (at high altitudes), and/or severe infection, sickling may occur in heterozygotes and result in some of the complications associated with sickle-cell disease. Note that whereas sickle-cell disease is recessively inherited, the sickle-cell trait is expressed in the heterozygote and is therefore a dominant trait.

X-linked inheritance and X-chromosome inactivation

Before we go on to consider X-linked inheritance, we need to take account of mechanisms that compensate for the variable number of sex chromosomes in humans (and other mammals): females have two X chromosomes but males have one X and one Y.

Having different numbers of chromosomes usually has severe, often lethal consequences—the loss of just one of our 46 chromosomes is

BOX 5.2 Consanguinity and the Degree to Which Close Relatives Are Genetically Related.

Ultimately, all humans are related to one another, but we share the highest proportion of our genes with close family relatives. Mating between the most closely related family members (with 50% of their genes in common, such as parent/child, and sibs) is very likely to result in homozygotes for recessive disease and is legally prohibited and/or socially discouraged in just about all societies. Cousin marriages can, however, be quite frequent in some communities from the Middle East, parts of the Indian subcontinent and other parts of Asia. Because cousins share a significant proportion of their genes, the offspring of cousin marriages can have a high degree of homozygosity with increased chance of being affected by a recessive disorder.

Because a child's risk of being homozygous for a rare recessive allele is proportional to how related the parents are, it is important to measure consanguinity. When one person is a direct descendant of another, the proportion of genes they have in common is $(1/2)^n$, where n is the number of generational steps separating the two. This gives: parent–child, 1/2 of genes in common; grandparent–grandchild, $(1/2)^2 = 1/4$ of genes in common; greatgrandparent–greatgrandchild, $(1/2)^3 = 1/8$ of genes in common.

Calculating the coefficient of relationship

The **coefficient of relationship** is the proportion of alleles shared by two persons as a result of common genetic descent from one or more recent (definable) common ancestors (or, more loosely, the proportion of genes in common as a result of common genetic descent). To calculate this, one considers paths of genetic descent linking the two individuals through *each* common ancestor in a family. A single generational step in such a path reduces the shared genetic component from the common ancestor by 1/2.

Consider the example in **Figure 1**. I-2 has had three children, a brother and sister who are sibs because they also have a common father, I-1, and their half brother, II-5. Half-sibs, such as II-3 and II-5, have a single ancestor in common and so there is a single path connecting them to their common parent. So, the orange path connecting II-3 to II-5 via their common mother has two steps, making a contribution of $1/2 \times 1/2 = 1/4$ of genes in common.

I-1 and I-2 are common ancestors for the sibs in generation II, for the first cousins in generation III and for the second cousins in generation IV. To calculate the coefficient of relationship for relatives linked by two or more common ancestors, we need to calculate the contributions made by each path and then sum them. Thus, for the first cousins in generation III, the green path that links them through their common grandfather I-1 has four steps, making a contribution of $(1/2)^4 = 1/16$, and the orange path that links them through the common grandmother, I-2, also has four steps, making a contribution of $(1/2)^4 = 1/16$. Adding the two paths gives 2/16 or 1/8 of genes in common. More complicated inbreeding may mean that individuals have four or more recent common ancestors, but the principle is always the same: work out paths for each common ancestor and sum the contributions.

The **coefficient of inbreeding** is the probability that a homozygote has identical alleles at a locus as a result of common genetic descent from a recent ancestor. It is also the proportion of loci at which a person is expected to be homozygous because of parental consanguinity and is one-half of the coefficient of relationship of the parents. So, if the parents are first cousins, the coefficient of inbreeding is 1/16. Note that even quite highly inbred pedigrees result in relatively moderate coefficients of inbreeding.

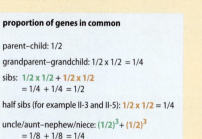

proption of genes in common

parent–child: 1/2

grandparent–grandchild: 1/2 x 1/2 = 1/4

sibs: 1/2 x 1/2 + 1/2 x 1/2
= 1/4 + 1/4 = 1/2

half sibs (for example II-3 and II-5): 1/2 x 1/2 = 1/4

uncle/aunt–nephew/niece: $(1/2)^3 + (1/2)^3$
= 1/8 + 1/8 = 1/4

first cousins: $(1/2)^4 + (1/2)^4$
= 1/16 + 1/16 = 1/8

first cousins once removed: $(1/2)^5 + (1/2)^5$
= 1/32 + 1/32 = 1/16

second cousins: $(1/2)^6 + (1/2)^6$
= 1/64 + 1/64 = 1/32

Figure 1 The proportion of genes in common between family members.

(A)

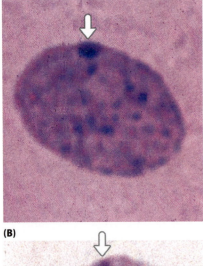

(B)

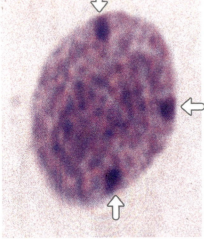

Figure 5.4 Barr bodies. (A) A cell from an XX female has a single inactivated X chromosome that forms a Barr body (arrow). (B) A cell from a 49,XXXXY male has one active X chromosome plus three inactivated X chromosomes that form Barr bodies (arrows). (Images courtesy of Malcolm Ferguson-Smith.)

lethal except for 45,X (Turner syndrome), and having an extra chromosome is usually lethal or results in a developmental syndrome such as trisomy 21 (Down syndrome). This happens because of problems with gene dosage: for some of our genes, the amount of gene product made must be tightly controlled (having one or three copies of these genes can be harmful because too little or too much product is made).

The sex difference regarding the Y chromosome is minimized by the conspicuous lack of genes on the Y. Most of the very few genes on the Y chromosome have male-specific functions, or they have an equivalent gene copy on the X (these X–Y gene pairs are largely concentrated at the tips of the sex chromosomes in the pseudoautosomal regions).

X-chromosome inactivation

Unlike the Y chromosome, the human X chromosome has many hundreds of important genes. To compensate for having different numbers of X chromosomes in males and females, a special mechanism is needed to silence genes on one of the two X chromosomes in each female cell (**X-inactivation**). That is to say, men are *constitutionally* hemizygous for most genes on the X chromosome, and women are *functionally* hemizygous for the same genes.

The X-inactivation mechanism is initiated after a cellular mechanism counts the number of X chromosomes in each cell of the early embryo. If the number of X chromosomes is two (or more), all except one of the multiple X chromosomes is inactivated. Each such X chromosome is induced to form a highly condensed chromosome that is mostly transcriptionally inactive, known as a Barr body (**Figure 5.4**). Note that some genes, including genes in the pseudoautosomal regions, escape inactivation (we consider the mechanism of X-inactivation in Chapter 6).

In humans the initial decision to inactivate one of the two X chromosomes is randomly made in the preimplantation embryo, beginning at around the eight-cell stage; some cells inactivate the paternal X and the remainder inactivate the maternal X. Once a cell has chosen which X to inactivate in the early embryo, that pattern of X-inactivation is continued in all descendant cells. Thus, a female who is heterozygous at a disease locus will be a genetic **mosaic**, containing cell clones in which the normal allele is expressed and clones in which the mutant allele is expressed. As described below, this has implications for the female phenotype in X-linked disorders.

X-linked recessive inheritance

In X-linked recessive disorders, affected individuals are mostly male, and affected males are usually born to unaffected parents. The mother of an affected male is quite often a carrier (and clearly so if she has affected male relatives). A distinguishing feature is that there is no male-to-male transmission because males pass a Y chromosome to sons (**Figure 5.5A**). However, a pedigree may *appear* to show male-to-male transmission when an affected man (with a condition such as hemophilia, for example) and a carrier woman produce an affected son (Figure 5.5B). The same parents could each potentially transmit a mutant X to produce an affected daughter.

In X-linked recessive disorders, female carriers with a single mutant allele can occasionally be quite severely affected and are known as manifesting heterozygotes. Because of X-inactivation, female carriers of an X-linked mutation are mosaics: some of their cells have the normal X chromosome inactivated and other cells have the mutant X inactivated, as seen most readily in skin disorders. Manifesting heterozygotes can occur by

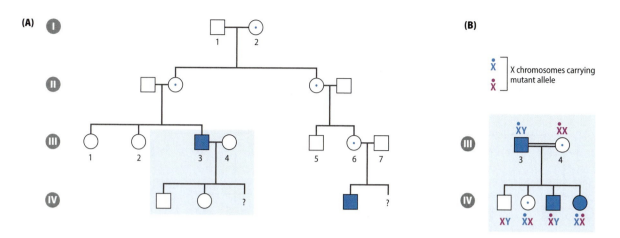

Figure 5.5 Pedigree showing X-linked recessive inheritance.
(A) Affected males in generations III and IV have inherited (via female carriers) a common mutant allele from I-2. For each child of a carrier mother such as III-6, the overall chance of being affected is 1 in 4, but this is sex-dependent: a son will have a 1 in 2 risk but a daughter will not be at risk (though she has a 1 in 2 risk of being a carrier). In the highlighted box, III-3 and III-4 have had two normal children, and the risk of having affected children would normally be very low (the father cannot transmit the mutant X allele to sons—he must transmit a Y—and any daughter will inherit a normal X from her mother). (B) The complication of inbreeding. Imagine that III-3 and III-4 were consanguineous and had the same mutant allele. We now have mating between an affected individual and a carrier, and there is a 1 in 2 chance that a child would be affected, irrespective of whether it was a boy or girl. The apparent male-to-male transmission is an illusion (the affected son has inherited a mutant allele from his mother, not from his father). The affected daughter is homozygous and although one mutant allele will be silenced in each of her cells by X-inactivation, she does not have a normal allele.

chance: most cells of a tissue that is critically important in disease development have an inactivated X carrying the normal allele.

Manifesting heterozygotes can occasionally occur because of nonrandom X-inactivation. That can happen when there is some advantage in inactivating the normal X chromosome instead of the mutant X chromosome. For example, an X-linked disorder may manifest in a woman who has an X-autosome translocation in which the breakpoint on the X is the cause of the disorder. If the X-autosome translocation chromosome were to be inactivated, neighboring autosomal genes would also be silenced, causing gene dosage problems and so the normal X is preferentially inactivated. Skewing of X-inactivation can often work in the other direction: some female carriers are asymptomatic because of nonrandom inactivation of the mutant X chromosome. We consider the mechanisms in Chapter 6.

X-linked dominant inheritance

As in autosomal dominant disorders, affected individuals with an X-linked dominant disorder can be of either sex, and usually at least one parent is affected. However, there are significantly more affected females than affected males, and affected females typically have milder (but more variable) expression than affected males.

The excess of affected females arises because there is no male-to-male transmission of the disorder. All children born to an affected mother (and an unaffected father) have a 50% chance of being affected, but an affected father with a single X chromosome will consistently have unaffected sons (they do not inherit his X chromosome), but his daughters will always be at risk because they will always inherit his affected X (**Figure 5.6A** gives an example pedigree).

The milder phenotype seen in affected females is a result of X-inactivation—the mutant allele is located on an inactivated X in a proportion of their cells. For certain X-linked dominant disorders, virtually

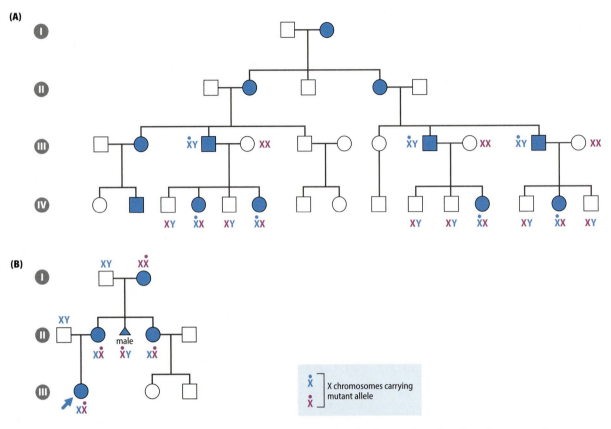

Figure 5.6 Pedigrees showing X-linked dominant inheritance. (A) Each child of an affected parent has a 1 in 2 chance of being affected. There is a 50% chance that an affected female can transmit a mutant X allele to sons and to daughters; the risk is the same, irrespective of the sex of the child. However, the risk to the children of an affected father depends crucially on the sex of the child: every son would be expected to be unaffected; instead, the risk of being affected is focused on daughters (as shown for the three affected males in generation III who each have had children—the father must pass on a Y chromosome to each son and so does not transmit the mutant X, but must transmit the mutant X chromosome to each daughter). (B) X-linked dominant inheritance with early male lethality. This example shows four affected females in a three-generation family with incontinentia pigmenti that was followed up after the birth of the affected granddaughter (arrowed). An affected male in generation II had spontaneously aborted. (Adapted from Minić S et al. [2010] *J Clin Pathol* 63:657–659; PMID 20591917. With permission from BMJ Publishing Group Ltd.)

all affected individuals are female: the phenotype is so severe in males that they die in the prenatal period, but the milder phenotype of affected females allows them to survive and reproduce. We illustrate this with the example of incontinentia pigmenti in Figure 5.6B; another disorder like this, Rett syndrome, is profiled in Section 6.3.

Pseudoautosomal and Y-linked inheritance

In female meiosis, the two X chromosomes recombine like any pair of homologous chromosomes; in male meiosis, however, recombination between the X and Y chromosomes is very limited. The X and Y are very different in size, organization, and gene content, and pairing between the X and Y at meiosis is very limited.

Despite their considerable differences, the X and Y nevertheless have some short gene-containing regions in common, notably the pseudo-autosomal regions that are located just before the telomere-associated repeats at the ends of both the short and long chromosome arms (**Figure 5.7**). The pseudoautosomal regions are distinctive because they are the only regions of the X and Y that can pair up during male meiosis and undergo recombination just like paired sequences do on homologous autosomal chromosomes (at each meiosis, there is an obligate X–Y crossover in the major pseudoautosomal region; recombination is less frequent in the minor pseudoautosomal region).

Outside the pseudoautosomal regions there is no recombination between the X and Y, and the remaining large central regions are X-specific and Y-specific regions. The X-specific region can engage in recombination in female meiosis, and sequences in this region can be transmitted to males or females; the Y-specific region is never involved in recombination and so is also called the *male-specific region*. With a few exceptions (see Figure 5.7), the sequences in the X-specific and male-specific regions are very different.

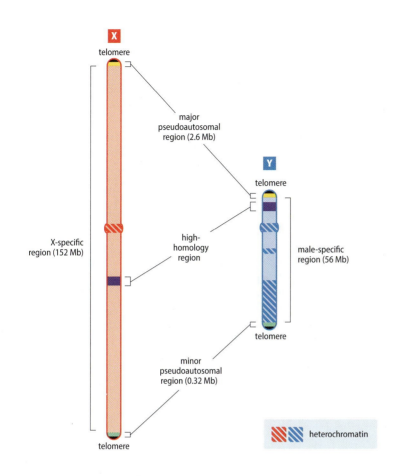

Figure 5.7 The human X and Y chromosomes: differences and major homology regions. The X and Y chromosomes differ greatly in size, heterochromatin content (much of the long arm of the Y is composed of heterochromatin), and DNA sequence. Colors indicate sequences that are X-specific (pink), Y-specific (blue), heterochromatic (hatched), or shared by the X and Y chromosomes only (other colors). The major pseudoautosomal regions on the X and Y (yellow) are essentially identical, as are the minor pseudoautosomal regions on the long arms (green). The pseudoautosomal regions are involved in X–Y pairing and recombination in male meiosis. The large central regions of the X and Y do not engage in recombination and are X-specific or Y-specific (also a *male-specific region* because it is not normally transmitted to females). As a result of an evolutionarily recent X–Y transposition event there is also roughly 99% sequence homology between sequences on Yp, close to the major pseudoautosomal region, and sequences at Xq21 (shown by purple boxes).

Pseudoautosomal inheritance

As a result of recombination in male meiosis, the individual X–Y gene pairs in the pseudoautosomal regions are effectively alleles. An individual allele in these regions can move locations between the X and Y chromosomes and so is neither X-linked nor Y-linked; instead, the pattern of inheritance resembles autosomal inheritance (**Figure 5.8**).

There are few genes in the pseudoautosomal regions, so that few pseudoautosomal conditions have been described. However, mutations in *KAL* can cause Kallmann syndrome (PMID 20301509), and the *SHOX* homeobox gene is a locus for two disorders. If one *SHOX* gene copy is damaged by mutation, the resulting heterozygotes have Leri–Weill dyschondreostosis (OMIM 127300). Homozygotes with mutations in both *SHOX* genes have a more severe condition, Langer mesomelic dysplasia (OMIM 249700).

Y-linked inheritance

The Y-specific region of the Y chromosome is a non-recombining male-specific region (see Figure 5.7). Population genetics dictates that non-recombining regions must gradually lose DNA sequences (as a way of deleting acquired harmful mutations because they cannot be removed by recombination). Over many millions of years of evolution, the Y chromosome has undergone a series of contractions and now has only 38% of the DNA present in the X (the X and Y are thought to have originated as a homologous pair of autosomes that began to diverge in sequence after one of them acquired a sex-determining region). As a result of DNA losses, the male-specific region of the Y chromosome has few genes and makes a total of only 31 different proteins, most of which are involved in male-specific functions.

Figure 5.8 Pseudoautosomal inheritance and X–Y recombination. (A) In this pedigree, affected individuals are heterozygous for a mutation in the major pseudoautosomal region, and the disorder shows dominant inheritance. Affected females carry the mutation on an X chromosome which can be passed to both sons and daughters. Affected males pass a Y chromosome containing the mutant allele to affected sons, but can also pass an X chromosome containing the mutant allele to affected daughters. This happens as a result of the obligatory X–Y recombination, which occurs within the major pseudoautosomal region. (B) When the crossover point occurs proximal to the mutant allele, the allele is transposed between the X and Y chromosomes.

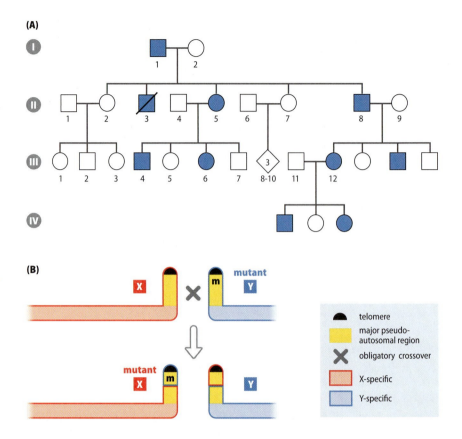

In Y-linked inheritance, males only should be affected and there should be exclusive male-to-male transmission. However, because of the lack of genes, Y-linked disorders are rare. Claims for some Y-linked traits, such as hairy ears (OMIM 425500), are now known to be dubious, but maleness is indisputably Y-linked. Interstitial deletions on the long arm of the Y chromosome are an important cause of male infertility (but infertile males are not normally able to transmit chromosomes unless conception is assisted by procedures such as intracytoplasmic sperm injection).

Matrilineal inheritance for mitochondrial DNA disorders

The mitochondrial genome is a small (16.5 kb) circular genome that has 37 genes (see Figure 2.11). It is much more prone to mutation than nuclear DNA, partly because of its proximity to reactive oxygen species (the mitochondrion is a major source of reactive oxygen species in the cell). As a result, mutations in mitochondrial DNA (mtDNA) are a significant cause of human genetic disease. Tissues that have a high energy requirement—such as muscle and brain—are primarily affected in mtDNA disorders.

Individuals with a mitochondrial DNA disorder can be of either sex, but affected males do not transmit the condition to any of their children. The sperm does contribute mtDNA to the zygote, but the paternal mtDNA is destroyed in the very early embryo (after being tagged by ubiquitin), and a father's mtDNA sequence variants are not observed in his children. That is, inheritance occurs exclusively through the mother (*matrilineal* inheritance). An additional, common feature of mitochondrial DNA disorders is that the phenotype is highly variable within families (**Figure 5.9**).

Variable heteroplasmy and clinical variability

Each cell contains multiple mitochondria, and there are often several hundred to thousands of mtDNA copies per cell. In some affected persons

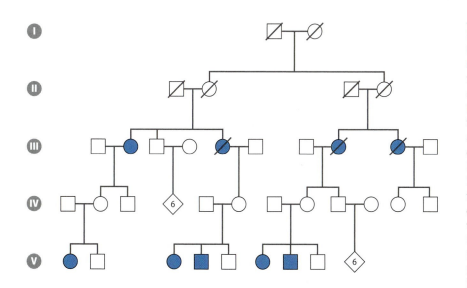

Figure 5.9 A pedigree illustrating matrilineal inheritance for a mitochondrial DNA disorder. Mitochondrial DNA disorders are transmitted by females only (because mtDNA originating from the sperm is quickly degraded in the early embryo). However, an affected female can pass on the condition to both sons and daughters. A common feature of mtDNA disorders is incomplete penetrance, as shown here by the absence of clinical phenotypes in several individuals who must be gene carriers, including three clear carrier females in generation IV, each of whom were born to an affected mother and went on to produce affected children of their own (the females in generations I and II might also have been expected to be carriers of the mutant mtDNA). *One* cause of this intrafamilial variability is variable heteroplasmy. The mutation here was shown to be a nucleotide substitution in the mitochondrial 12S rRNA gene that was associated with variable hearing loss. (From Prezant TR et al. [1993] *Nat Genet* 4:289–294; PMID 7689389. With permission from Macmillan Publishers Ltd.)

every mtDNA molecule carries the causative mutation (homoplasmy), but affected individuals frequently have cells with a mixed population of normal and mutant mtDNAs (**heteroplasmy**). The clinical features depend mostly on the proportion of mutant to normal mtDNA molecules in the cells of tissues with high energy requirements.

Although a human egg cell is haploid for nuclear DNA, it contains more than 100,000 mtDNA molecules. A heteroplasmic mother can give rise to children who differ widely from her and from each other in the ratio of mutant to normal mtDNA molecules in their tissues (variable heteroplasmy). As a result, there can be very significant clinical variability between affected members of the same family.

To explain rapid shifts in heteroplasmy that occur over only one generation, the mitochondrial genetic bottleneck hypothesis envisages that, during early development, germ-line cells pass through a bottleneck stage in which they contain very few mtDNA molecules. By chance, germ-line cells at this stage may have a much higher or much lower proportion of mutant mtDNA molecules than the somatic cells. As a result, a heteroplasmic mother could give rise ultimately to eggs with a much higher or much lower proportion of mutant mtDNA molecules than are present in her affected tissues.

Another contributor to variable heteroplasmy is the rapid evolution of a mtDNA variant within an individual. Mutant mtDNAs that have a large deletion or a large duplication can evolve rapidly, so that different tissues or even the same tissue at different times may show different distributions of the mtDNA variant.

5.3 Uncertainty, Heterogeneity, and Variable Expression of Mendelian Phenotypes

Section 5.2 dealt with the different modes of inheritance for phenotypes that are determined principally by single genes. Some complications were covered, including the effects of X-inactivation in females and hemizygosity in males, occasional differences between homozygotes and heterozygotes for autosomal dominant disorders, the occasional expression of disease symptoms in carriers of an autosomal recessive disorder, the mimicking of autosomal inheritance by genes in the pseudoautosomal regions of the X and Y chromosomes, and the unique features of mitochondrial DNA inheritance.

In this section we discuss broader complications that relate to uncertainty of mode of inheritance, and difficulties posed by heterogeneity in the links between DNA variation and phenotypes. In addition, we consider how affected individuals within a single family can show variable phenotypes for Mendelian disorders.

Difficulties in defining the mode of inheritance in small pedigrees

Many families are small and may have only a single affected person. If the disorder is rare and we do not know the underlying disease gene, how can we work out the mode of inheritance? Knowing the mode of inheritance is important in genetic counseling because it is on that basis that we calculate the risk of having a subsequent affected child. Until a disease gene has been identified and screened for mutations, however, the mode of inheritance inferred from examining the pedigree should be regarded simply as a working hypothesis.

Having a single affected child in a family with no previous history of a presumed genetic disorder might suggest the possibility of a recessive disorder, with a 1 in 4 risk that each subsequent child would be affected. Alternatively, it could be a dominant disorder and the affected individual could be a heterozygote. In that case, one parent carries the disease gene but does not display the phenotype, or the disorder is due to a *de novo* mutation (see below).

One possible way to work out the mode of inheritance is to study multiple families with the same disorder and calculate the overall proportion of affected children, (called the *segregation ratio*). But there are many difficulties with this approach. First, the disorder may be heterogeneous and be due to different genes in different families. Secondly, the total numbers of children who can be studied are often too small to get reliable estimates.

There are also problems in how the families are ascertained (that is, in finding the people and families who will be studied). Imagine trying to establish that a disorder is autosomal recessive. We could collect a set of families and try to show a segregation ratio of 1 in 4. However, there will be *ascertainment bias*: if there is no independent way of recognizing carriers, the families will be identified only through an affected child (families with two carrier parents and only unaffected children seems perfectly normal and not be included). By focusing on families who already have at least one affected child, the segregation ratio will inevitably be high. Ascertainment bias is most obvious for recessive conditions, but more subtle biases can distort the estimated ratios in any condition.

Happily, we are rapidly moving into an era in which underlying disease genes can quickly be found even for rare single-gene disorders. Rapid next-generation DNA sequencing is now being widely used to screen **exomes** (in practice, all exons of protein-coding genes) of affected individuals with the same condition. As described in later chapters, genes underlying some rare single-gene disorders have been successfully identified after sequencing exomes from only a very few unrelated individuals with the disorder.

New mutations and mosaicism

For a single-gene disorder, the observed incidence of mutant alleles in a defined population can be quite stable over time. A proportion of mutant alleles are transmitted from one generation to the next, and a proportion are lost because some individuals possessing mutant alleles do not transmit them. To keep the frequency of mutant alleles constant, new mutations make up for the loss of mutant alleles that are not transmitted to the next generation.

Persons who have a severe disorder usually do not reproduce or have a much reduced reproductive capacity (unless the disorder is not manifested until later in life). In severe autosomal recessive conditions, however, for each affected individual there are very many asymptomatic carriers who can transmit mutant alleles to the next generation. Because only a very small proportion of mutant alleles go untransmitted, the incidence of new mutation is low.

For severe dominant disorders, the mutant alleles are concentrated in affected individuals. If most individuals who carry the disease allele do not reproduce (because the disorder is congenital, say), the incidence of new mutation will be very high. If, however, there is a relatively late age at onset of symptoms, as with Huntington disease, individuals with the mutant allele may reproduce effectively, and the rate of new mutation may be very low.

For severe X-linked recessive disorders, the incidence of new mutation will also be quite high to balance the loss of mutant alleles when affected males do not reproduce. However, female carriers will usually be able to transmit mutant alleles to the next generation.

As a result of a new mutation, an affected person may be born in a family with no previous history of the disorder and would present as an isolated (*sporadic*) case. In rare disorders that have not been well studied, a sporadic case poses difficulty for calculating the risk that subsequent children could also be affected. The affected individual could be a heterozygote (as a result of *de novo* mutation, or the failure of the disorder to be expressed in one parent), but alternatively could be a homozygote born to carrier parents, or a hemizygous boy whose mother is a carrier of an X-linked recessive condition.

Post-zygotic mutations and mosaicism

Most mutations arise as a result of endogenous errors in DNA replication and repair. Mutations can occur during gametogenesis and produce sperm and eggs with a new mutant allele. In addition, *de novo* pathogenic mutations can also occur at any time in post-zygotic life. As a result of post-zygotic mutations, each individual person is a genetic **mosaic** with genetically distinct populations of cells that have different mutational spectra.

Post-zygotic mutations may result in somatic mosaicism that will have consequences only for that individual (**Box 5.3**). But certain post-zygotic mutations, often occurring comparatively early in development, may also result in **germ-line mosaicism**. A person who has a substantial proportion of mutant germ-line cells (a germ-line mosaic or gonadal mosaic) may not show any symptoms but will produce some normal gametes and some mutant gametes. The risks of having a subsequently affected child are much higher than if an affected child carries a mutation that originated in a meiotic division.

Heterogeneity in the correspondence between phenotypes and the underlying genes and mutations

There is no one-to-one correspondence between phenotypes and genes. Three levels of heterogeneity are listed below. As we will see below and in later chapters, both nongenetic factors (environmental and epigenetic) and additional genetic factors can also influence the phenotypes of single-gene disorders.

BOX 5.3 Post-Zygotic (Somatic) Mutations and Why We Are All Genetic Mosaics.

A pathogenic new mutation can be imagined to occur during gamete formation in an entirely normal person. Most mutations arise as a result of endogenous errors in DNA replication and repair, and although mutations do occur during gametogenesis and produce sperm and eggs with a new mutant allele, they can also occur at any time in post-zygotic life. As a result of post-zygotic mutations, each individual person is a genetic **mosaic** with genetically distinct populations of cells that have different mutational spectra.

Human mutation rates are around 10^{-6} per gene per generation, and so a person with a wild-type allele at conception has a roughly one in a million chance of transmitting it to a child as a mutant allele. In this case we are considering the chance of a mutation occurring in a lineage of germ-line cells from zygote to gamete, involving a series of about 30 cell divisions in females and several hundred divisions in males (about 400 by age 30 and increasing by about 23 per year because spermatogenesis continues through adult life).

Now consider post-zygotic mutations in somatic cell lineages. The journey from single-celled zygote to an adult human being involves a total of about 10^{14} mitotic cell divisions. With so many cell divisions, post-zygotic mutation is unavoidable—we must all be mosaics for many, many mutations. Having so many potentially harmful somatic mutations is usually not a concern because the number of cells that will fail to function correctly is normally very small. A cell will usually function normally after sustaining a harmful mutation in a gene that is not normally expressed in that cell type, and even if the cell does function abnormally as a result of mutation it might not give rise to many mutant descendants.

A person may be at risk of disease, however, if a mutated cell is able to give rise to substantial numbers of descendant cells that act abnormally (**Figure 1**). The biggest disease risk posed by somatic mutations is that they set off or accelerate a process that leads to cancer. As we describe in Chapter 10, cancers are unusual in that although they can be inherited, the biggest contribution to disease comes from somatic mutations.

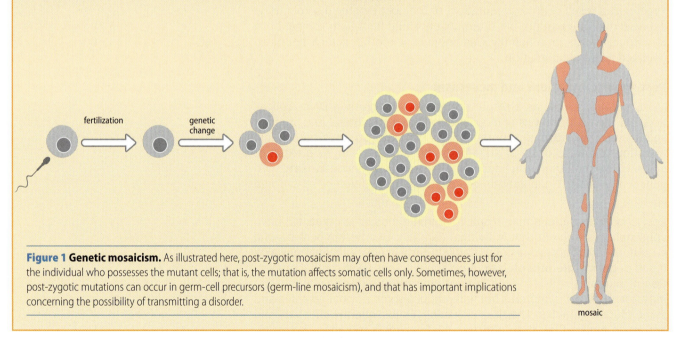

fertilization

genetic change

mosaic

Figure 1 Genetic mosaicism. As illustrated here, post-zygotic mosaicism may often have consequences just for the individual who possesses the mutant cells; that is, the mutation affects somatic cells only. Sometimes, however, post-zygotic mutations can occur in germ-cell precursors (germ-line mosaicism), and that has important implications concerning the possibility of transmitting a disorder.

Locus heterogeneity

The same clinical phenotype can often be produced by mutations in genes at two or more loci. The different genes often make related products that work together as a complex or in a common pathway; sometimes one gene is the primary regulator of another gene.

Locus heterogeneity explains how parents who are both affected with the same common recessive disorder produce multiple unaffected children. Recessively inherited deafness is the classic example (sensorineural hearing impairment mostly shows autosomal recessive inheritance, and deaf persons often choose to have children with another deaf person). If two deaf parents are homozygous for mutations at the same gene locus, one would expect that all their children would also have impaired hearing. If, instead, the parents are homozygous for mutations at two different recessive deafness loci, all their children would be expected to be double heterozygotes and have normal hearing (**Figure 5.10**).

As the underlying genes for single-gene disorders become known, it has become clear that very many conditions show locus heterogeneity. One might anticipate that many different genes contribute at different steps to broad general pathways (responsible for hearing or vision, for example). It is therefore unsurprising that autosomal recessive deafness or retinitis pigmentosa (hereditary retinal diseases with degeneration of rod and cone photoreceptors) can result from mutations in different genes.

More specific phenotypes can also be caused by mutations at any one of many different gene loci. Usher syndrome, for example, involves profound sensorineural hearing loss, vestibular dysfunction, and retinitis pigmentosa; autosomal recessive forms can be caused by mutations at any one of at least 11 different gene loci.

Bardet–Biedl syndrome (PMID 20301537) provides another illustrative example. It is a pleiotropic disorder (many different body systems and functions are impaired) and the primary features are: degeneration of light-sensitive cells in the outer regions of the retina (causing night blindness, tunnel vision, reduced visual acuity), learning disabilities, kidney disease, extra toes and/or fingers, obesity, and abnormalities of the gonads. Autosomal recessive inheritance is the typical inheritance pattern, and the disorder is caused by mutations in any of at least 15 genes, all involved in regulating how cilia function (**Figure 5.11**).

Allelic and phenotypic heterogeneity

Many different mutations in one gene can have the same effect and produce similar phenotypes. For example, β-thalassemia results from a deficiency of β-globin and can arise by any number of different mutations in the hemoglobin β chain (*HBB*) gene. Different mutations in a single gene can also often result in different phenotypes. That can arise in two ways: either different types of mutation somehow have different effects on how the underlying gene works—which we consider here—or some factors outside the disease locus have varying effects on the phenotype (described below).

Phenotype variation due to different mutations at a single gene locus may differ in degree (severe or mild versions of the same basic phenotype) or be extensive and result in rather different disorders. For example,

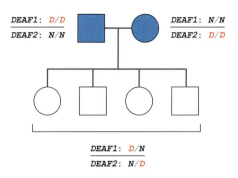

Figure 5.10 Locus heterogeneity explains why two parents with autosomal recessive deafness can consistently produce unaffected children. Imagine that the two parents are deaf because they have two mutant alleles at different autosomal recessive deafness loci, which we represent here as *DEAF1* and *DEAF2*. We represent normal alleles as *N* and deafness-associated alleles as *D*. In this case, sperm produced by the father would carry the *DEAF1*D* allele and the *DEAF2*N* allele, and eggs produced by the mother would carry the *DEAF1*N* allele and the *DEAF2*D* allele. All children would therefore be unaffected because they would be heterozygous at both loci. The normal phenotypes of each child result from complementation between normal alleles at the two loci. If, instead, both parents had autosomal recessive deafness caused by different mutations in the *same* gene, all their children would be expected to be deaf as a result of inheriting two mutant alleles at that locus.

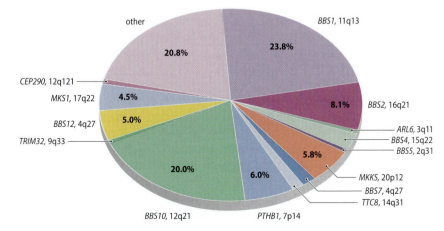

Figure 5.11 Extraordinary locus heterogeneity for Bardet–Biedl syndrome. Where shown, percentages indicate the proportion of mutant alleles that are attributable to indicated genes (whose subchromosomal locations are given below or adjacent to the gene symbol). Seven of the genes (*BBS1, BBS2, MKKS, PTHB1, BBS10, BBS12,* and *MKS1*) account for almost three-quarters of the identified pathogenic mutations in Bardet–Biedl syndrome. (Adapted from Zaghloul NA & Katsanis N [2009] *J Clin Invest* 119:428–437; PMID 19252258. With permission from the American Society for Clinical Investigation.)

CLASS OF DISORDER	DISORDER	INHERITANCE PATTERN[a]	OMIM NO.
Lipodystrophy	lipodystrophy, familial partial	AD	151660
	mandibulosacral dysplasia type A with lipodystrophy	AR	248370
Muscle/heart disease	limb girdle muscular dystrophy type IB	AD	159001
	Emery–Dreifuss muscular dystrophy type 2	AD	181350
	Emery–Dreifuss muscular dystrophy type 3	AR	181350
	congenital muscular dystrophy	AD	613205
	cardiomyopathy, dilated type IA	AD	150330
	Malouf syndrome (cardiomyopathy, dilated, with hypertrophic hypogonadism)	AR	212112
	heart–hand syndrome, Slovenian type	AD	610140
Neuropathy	Charcot–Marie–Tooth disease, type 2B1	AR	605588
Progeria	Hutchinson–Gilford progeria syndrome	AD	176670
	atypical Werner syndrome		
	atypical progeroid syndrome		

Duchenne and Becker muscular dystrophies (OMIM 310200 and 300376, respectively) represent severe and mild forms of the same type of muscular dystrophy and are both examples of dystrophinopathies (PMID 20301298). More extreme phenotype heterogeneity can result from mutations at some genes (see the example of the lamin A/C gene in **Table 5.1**).

Clinical phenotypes can also vary between affected members of the same family even although they have identical mutations. As we saw in Section 5.2, heteroplasmy can explain divergent phenotypes in family members affected by a mitochondrial DNA disorder. But single-gene disorders can also show intrafamilial variation in phenotype that may be due to genetic and nongenetic factors as described below.

Non-penetrance and age-related penetrance

The **penetrance** of a single-gene disorder is the probability that a person who has a mutant allele will express the disease phenotype. Dominantly inherited disorders, by definition, are manifested in heterozygotes and might be expected to show 100% penetrance. That might be true for certain dominant disorders. For many others, however, penetrance is more variable and the disorder can sometimes appear to skip a generation so that a person who must have inherited the disease allele is unaffected (**non-penetrance**—**Figure 5.12**).

Non-penetrance should not be viewed as surprising. Even in single-gene disorders—in which, by definition, the phenotype is largely dictated by the genotype at just one locus—other genes can play a part, as can epigenetic and environmental factors.

Variable age at onset in late-onset disorders

A disease phenotype may take time to manifest itself. If a disorder is present at birth, it is said to be congenital. In some disorders, however, there is a late age at onset so that the penetrance is initially very low but then increases with age. Age-related penetrance means a late onset of symptoms, and quite often the disease first manifests in adults.

The slow development of disease in adult-onset disorders may occur in different ways. Harmful products may be produced slowly but build up

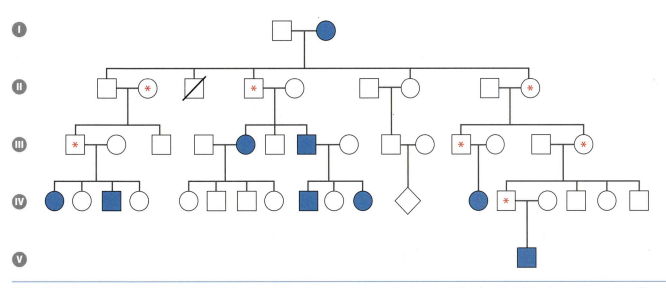

Figure 5.12 Non-penetrance in an autosomal dominant disorder.
Individuals with a red asterisk are asymptomatic disease gene carriers: they have inherited a mutant allele ultimately from the affected great-great-grandmother in generation I, but none of them expresses the disease phenotype. In this example, the disorder is evident only in individuals who have inherited a mutant allele from their father (in each case the unaffected individuals with a red asterisk inherited a mutant allele from their mother). As described in the text, an epigenetic mechanism known as *imprinting* can result in this type of parent-of-origin effect on the phenotype.

over time, for example. If pathogenesis involves a gradual process of cell death, it may take some time before the number of surviving cells drops to critically low levels that produce clinical symptoms. In hereditary cancers, a mutation is inherited at a tumor-suppressing gene locus and a second, somatic mutation is required to initiate tumor formation. The second mutation occurs randomly, but the probability of a second mutation increases with time and therefore with age.

Huntington disease (PMID 20301482) is a classic example of a late-onset single-gene disorder. In this case, mutant alleles produce an abnormal protein that is harmful to cells and especially toxic to neurons. The loss of neurons is gradual but eventually results in a devastating neurodegenerative condition. Huntington disease is highly penetrant. The onset of symptoms typically occurs in middle to late adult life, but juvenile forms are also known (**Figure 5.13**).

Age-at-onset curves for late-onset disorders are used in genetic counseling to calculate the chance that an asymptomatic person at risk of developing the disease carries the mutation. In Huntington disease an unaffected person who has an affected parent will have a 50% *a priori* risk that decreases with age (see Figure 5.13); if one is still free of symptoms by age 60, for example, the chance of developing the disease falls to less than 20%.

Variable expression of Mendelian phenotypes within families

Phenotypes resulting from mutation in mitochondrial DNA are highly variable because of the special mitochondrial property of heteroplasmy (see Section 5.2). Some types of Mendelian disorders, notably dominant phenotypes, are also prone to variable expression, and different family members show different features of disease (sometimes called variable expressivity—see **Figure 5.14** for an example pedigree). But, like non-penetrance, variable phenotype expression is occasionally seen in recessive pedigrees.

Non-penetrance can be regarded as an extreme endpoint of variable expression, and the factors that produce variable expression of

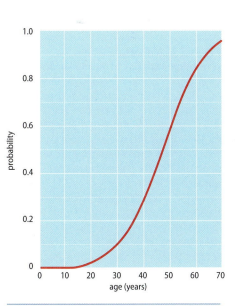

Figure 5.13 Age-related onset of Huntington disease. The curve shows the probability that an individual carrying a Huntington disease allele will have developed symptoms by a given age. (From Harper PS [2010] Practical Genetic Counselling, 7th ed. With permission from Taylor & Francis Group LLC.)

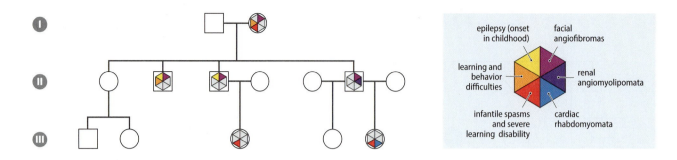

Figure 5.14 Variable phenotypes in a tuberous sclerosis family. Tuberous sclerosis is an autosomal dominant disorder caused by mutations in either the *TSC1* or *TSC2* gene. These two genes make two subunits of a tumor suppressor protein complex that works in a signaling pathway to regulate cell growth and proliferation. The disorder affects multiple body systems with characteristic tumor-like lesions in the brain, skin, and other organs, and is often associated with seizures and learning difficulties. However, as is evident in this family from the northeast of England, there can be considerable differences in *expressivity* of the disorder. (Pedigree information provided by Dr Miranda Splitt, Northern Region Genetics Service UK and Institute of Genetic Medicine, Newcastle University.)

Figure 5.15 Main explanations for phenotype variation in Mendelian disorders. (A) *Interfamilial variation in phenotype.* Unrelated individuals with the same Mendelian disorder may often have different mutations (red symbols) at the disease gene locus with different consequences for gene expression and disease. (B) *Intrafamilial variation in phenotype.* Affected members of a single family can be expected to have the same mutation at the disease gene locus but can nevertheless show differences in phenotype because of genetic or nongenetic factors. In the former case, the affected individuals may have different alleles at one or more modifier gene loci. Modifier genes make products that interact with the primary gene locus so as to modulate the phenotype, and different alleles of a modifier gene can have different effects. Alternatively, nongenetic factors can explain phenotype variation; an example is epigenetic regulation, in which the disease allele can be silenced by an altered chromatin conformation (green hatched box) or by variable exposure to an environmental factor (green circle) such as a specific virus or chemical during development *in utero*.

phenotypes within families are the same as those that result in non-penetrance. They include nongenetic factors—epigenetic regulation and environmental factors (**Figure 5.15B**) and stochastic factors. Additional genetic factors are also involved, notably modifier genes that regulate or interact with a Mendelian locus, affecting how it is expressed. Different alleles at a modifier gene locus may have rather different influences on the expression of the Mendelian locus (Figure 5.15B).

Imprinting

Certain phenotypes show autosomal dominant inheritance with parent-of-origin effects. Both sexes are affected, and the mutant allele can be transmitted by either sex but is expressed only when inherited from a parent of one particular sex. For some conditions, a mutant allele must be inherited from the father for the disease to be expressed (see Figure 5.12 for a possible pedigree). For other conditions, such as Beckwith–Wiedemann syndrome, the disease phenotype is expressed only if the disease allele is inherited from the mother (**Figure 5.16**).

The parent-of-origin effects are due to an epigenetic mechanism known as **imprinting**, which we describe in detail in Chapter 6. The mutant allele that is not expressed is often described as the imprinted allele. Beckwith–Wiedemann syndrome is said to be paternally imprinted, because paternally inherited alleles are not expressed.

Anticipation

Some disorders show consistent generational differences in phenotype. Disorders such as fragile X mental retardation syndrome, myotonic dystrophy, and Huntington disease are caused by unstable mutations (often called *dynamic mutations*) whose characteristics can change after they undergo DNA replication. As a result, the phenotype can vary between affected individuals in families but in a directional way; that is, it can be

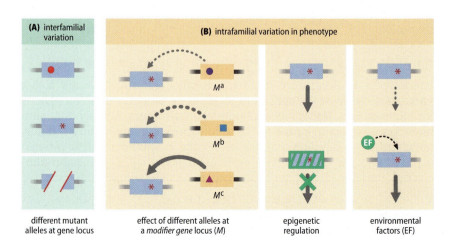

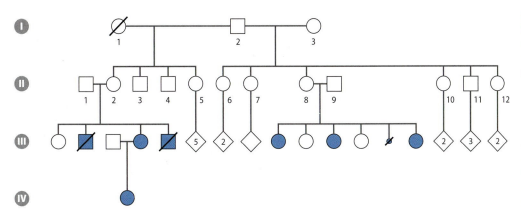

Figure 5.16 Parent-of-origin effect on the expression of an inherited disorder. This pedigree shows autosomal dominant Beckwith–Wiedemann syndrome (PMID 20301568), which manifests only when the underlying mutant allele is maternally inherited. The affected individuals in generation III must have inherited the mutant allele from their common grandfather I-2 but none of his 10 children in generation II have symptoms of disease, including two daughters, II-2 and II-8, who have gone on to have multiple affected children. (From Viljoen D & Ramesar R [1992] *J Med Genet* 29:221–225; PMID 1583639. With permission from BMJ Publishing Group Ltd.)

expressed at an earlier age and become increasingly severe with each new generation of affected individuals. This phenomenon is known as **anticipation** (Figure 5.17). We consider the molecular mechanisms in detail in Chapter 7.

5.4   Allele Frequencies in Populations

Genetic disorders that are comparatively common and serious have somehow avoided being eliminated by natural selection. This raises two questions: Are high mutation rates enough to explain why harmful disease alleles persist? And if so, why should some single-gene disorders be comparatively common but others very rare? In this section we are concerned primarily with allele frequencies and the factors that affect them.

The frequency of a single-gene disorder in a population relates to the frequency in the population of pathogenic alleles at the relevant disease locus (or loci). A high disease allele frequency might result if a gene were to be particularly susceptible to mutation. Genes that are very large, such as the dystrophin gene, or that contain many repetitive sequences that confer structural instability, might be expected to be especially prone to mutation.

Some of the most common single-gene disorders, such as sickle-cell anemia and the thalassemias, result from mutation in tiny genes—as we will see below, autosomal recessive disorders do not require high mutation rates to be common. Even in some autosomal dominant disorders, a high incidence of the disorder may not necessarily mean that the underlying gene loci have high mutation rates, as described below.

Some disorders may be caused by a *selfish mutation*. Achondroplasia (PMID 20301331) is a common single-gene disorder but is caused exclusively by mutation at just a single nucleotide, producing a highly specific change (glycine-to-arginine substitution at residue 380) in the FGFR3 (fibroblast growth factor receptor type 3) protein. The nucleotide that is altered is not thought to be highly mutable. Instead, the mutation may promote its own transmission: male germ-line cells that contain it may have a proliferative advantage and make a disproportionate contribution to sperm (so that there is a high allele frequency even although the mutation rate is not so exceptional). We consider selfish mutations in detail in Section 7.2.

We also need to explain why some single-gene disorders are common in some human populations but very rare in others. Cystic fibrosis is particularly common in northern European populations, for example, and sickle-cell anemia is especially frequent in tropical Africa but virtually absent from many other human populations.

Figure 5.17 A three-generation family affected with myotonic dystrophy. The degree of severity increases in each generation. The grandmother (right) is only slightly affected, but the mother (left) has a characteristic narrow face and somewhat limited facial expression. The baby is more severely affected and has the facial features of children with neonatal-onset myotonic dystrophy, including an open, triangular mouth. The infant has more than 1000 copies of the trinucleotide repeat, whereas the mother and grandmother each have about 100 repeats. (From Jorde LB, Carey JC & Bamshad MJ [2009] Medical Genetics, 4th ed. With permission from Elsevier.)

In all of these considerations, what do we mean by a human *population*? We could mean anything from a small tribe to the whole of humanity. An idealized population would be large with no barriers to random mating; as we will see below, some important principles in population genetics are based on this kind of population.

In practice, mating is often far from random because of different types of barriers. Geographic barriers can mean that people who live in locations that are remote or difficult to access form populations with limited genetic diversity and with distinctive allele frequencies. But even within single cities there are also many ethnic populations with distinctive allele frequencies. And, as we will see below, even within these populations, mating is not random.

Allele frequencies and the Hardy–Weinberg law

The frequency of an allele in a population can vary widely from one population to another. The concept of the **gene pool** (all of the alleles at a specific gene locus within the population) provides the reference point for calculating allele frequencies (which are often inaccurately represented in the literature as *gene frequencies*).

For a specific allele, say allele *A*1* at locus *A*, the **allele frequency** is the proportion of all the alleles in the population at locus *A* that are *A*1* and is given as a number between 0 and 1. Effectively, the allele frequency for *A*1* is the *probability* that an allele, picked at random from the gene pool, would be *A*1*.

The Hardy–Weinberg law

The Hardy-Weinberg law (or equilibrium, principle, theorem) provides a mathematical relationship between allele frequencies and genotype frequencies in an *idealized* large population where matings are random and allele frequencies remain constant over time.

Imagine that locus *A* has only two alleles, *A*1* and *A*2*, and that their respective frequencies are *p* and *q* (so that *p* + *q* = 1). The respective genotypes are combinations of two alleles at a time. To calculate the frequency of a genotype, we therefore first need to estimate the probabilities of picking first one specified allele from the gene pool (as the paternal allele, say), and then picking a second allele to be the maternal allele.

Imagine we pick *A*1* first (with a probability of *p*) and then we pick *A*1* again (with a probability of *p*). If the population is large, the two probabilities are independent events and so the joint probability of picking *A*1* first and then *A*1* again is the product of the two probabilities, namely p^2. This is the only way that we can arrive at the genotype *A*1.A*1*, whose frequency is therefore p^2 (**Figure 5.18**).

Now consider the genotype *A*1.A*2*. We can get this in two ways. One way is to first pick *A*1* (probability *p*) and then *A*2* (probability *q*), giving a joint probability of *pq*. But a second way is to pick *A*2* first (probability *q*) and then pick *A*1* (probability *p*), again giving a combined probability of *pq*. As a result, the frequency of the genotype *A*1.A*2* is 2*pq*.

In summary, in a suitably ideal population, the Hardy–Weinberg law gives the frequencies of homozygous genotypes as the square of the allele frequency, and the frequencies of heterozygous genotypes as twice the product of the allele frequencies. An important consequence is that if allele frequencies in a population remain constant from generation to generation, the genotype frequencies will also not change.

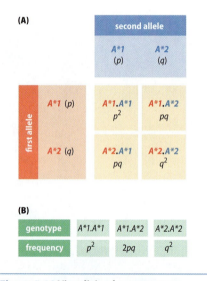

Figure 5.18 Visualizing how genotype frequencies are related to allele frequencies. In this example we consider a locus *A* that has two alleles, *A*1* and *A*2*, with respective frequencies *p* and *q*. Genotypes are unique combinations of *two* alleles, one from a father and one from a mother. (A) We can first construct a matrix of all possible pairwise allele combinations, whose frequencies are simply the products of the frequencies of the two alleles. (B) We then integrate any pairwise combinations that have the same two alleles (*A*1.A*2* is effectively the same as *A*2.A*1*) to get the frequencies of the three unique genotypes. Note that the Hardy–Weinberg law relates genotype frequencies to allele frequencies by a binomial expansion: $(p + q)^2 = p^2 + 2pq + q^2$ for two alleles (as shown here), $(p + q + r)^2$ for three alleles, $(p + q + r + s)^2$ for four alleles, and so on.

Applications and limitations of the Hardy–Weinberg law

The major clinical application of the Hardy–Weinberg law is as a tool for calculating genetic risk. In a family with a single-gene disorder, only one or two mutant alleles are normally found in the causative gene, but within a population there may be many different mutant alleles at the disease locus. To apply the Hardy–Weinberg law to single-gene disorders, all the different mutant alleles are typically lumped together to make one disease allele. That is, we envisage just two alleles according to their effect on the disease phenotype: a normal allele (N), with no effect on the phenotype, and a disease allele (D), which can be *any* mutant allele. If we assign frequencies of p for allele N and q for allele D, the genotype frequencies would be as follows: p^2 for NN (normal homozygotes), $2pq$ for ND (heterozygotes), and q^2 for DD (disease homozygotes).

Practical application of the Hardy–Weinberg law to single-gene disorders is largely focused on autosomal recessive disorders, where it allows the frequency of carriers to be calculated without having to perform relevant DNA tests on a large number of people (**Box 5.4**). Its utility depends on certain assumptions—notably random mating and constant allele

---

BOX 5.4 Using the Hardy–Weinberg Law to Calculate Carrier Risks for Autosomal Recessive Disorders.

Genetic counseling for autosomal recessive conditions often requires calculations to assess the risk of being a carrier. The proband who seeks genetic counseling is typically a prospective parent with a close relative who is affected. He/she is worried about the high risk of being a carrier and then about the risk that his/her spouse could also be a carrier.

The proband's chance of being a carrier can be calculated by using the principles of Mendelian inheritance, but the Hardy–Weinberg law is used to calculate the risk that his/her spouse could also be a carrier. If both parents were to be carriers, each child would have a 1 in 4 risk of being affected.

Take the specific example in **Figure 1**. The healthy proband (arrowed) has a sister with cystic fibrosis and is worried about the prospect that he and his wife might have a child with cystic fibrosis. His wife is Irish, and the Irish population has the highest incidence of cystic fibrosis in the world, affecting one birth in 1350.

The proband's parents can be presumed to be carriers, each with one normal allele N and one mutant allele M. Because the proband is healthy, he must have inherited one of three possible combinations of parental alleles: N from both parents (homozygous normal); N from father and M from mother (carrier); and M from father and N from mother (carrier). So from Mendelian principles, he has a risk of 2/3 of being a carrier (see Figure 1).

The risk that his wife is a carrier is the same as the probability that a person, picked at random from the Irish population, is a carrier. If we assign a frequency of p for the normal allele and q for the cystic fibrosis allele, the Hardy–Weinberg law states that the frequency of affected individuals will be q^2 and the frequency of carriers will be $2pq$. Because population surveys show that cystic fibrosis affects 1 in 1350 births in the Irish population, $q^2 = 1/1350$ and so $q = 1/\sqrt{1350}$, or $1/36.74 = 0.027$.

Since $p + q = 1$, the value of $p = 0.973$. The risk of the wife being a carrier ($2pq$) is therefore $2 \times 0.973 \times 0.027 = 0.0525$, or 5.25%. The combined risk that both the proband and his wife are carriers is $2/3 \times 0.0525 = 0.035$, or 3.5%.

For rare autosomal recessive disorders, the value of p very closely approximates 1, so the carrier frequency can be taken to be $2q$. However, if the disorder is especially rare, the chances that the prospective parents are consanguineous is much higher, making the application of the Hardy–Weinberg law much less secure.

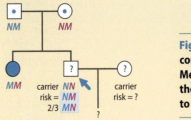

Figure 1 Using a combination of Mendelian principles and the Hardy–Weinberg law to estimate disease risk.

frequencies—that may not be strictly upheld. As described below, allele frequencies can change in populations, but the changes are often slow and in small increments, and often have minor effects in disturbing the Hardy–Weinberg distribution of genotypes. However, certain types of nonrandom mating can substantially upset the relative frequency of genotypes predicted by the Hardy–Weinberg law.

Nonrandom mating

In addition to geographical barriers to random mating, people also preferentially select mates who are similar to themselves in different ways. They may be members of the same ethnic group and/or sect, for example. Because breeding is less frequent between members from different communities, allele frequencies can vary significantly in the different communities. Geneticists therefore need to define populations carefully and calculate genetic risk by using the most appropriate allele frequencies.

Additional types of assortative mating occur. We also tend to choose a mate of similar stature and intelligence to us, for example. Positive assortment mating of this type leads to an increased frequency of homozygous genotypes and a decreased frequency of heterozygous genotypes. It extends to medical conditions. People who were born deaf or blind have a tendency to choose a mate who is similarly affected.

Inbreeding is a powerful expression of assortative mating that is quite frequent in some societies (see Box 5.2) and can result in genotype frequencies that differ significantly from Hardy–Weinberg predictions. Consanguineous mating results in an increased frequency of mating between carriers and a correspondingly increased frequency of autosomal recessive disease.

Ways in which allele frequencies change in populations

Allele frequencies can change from one generation to the next in different ways. Often changes in allele frequency are quite slow, but occasionally the composition of populations can change quickly, producing major shifts in allele frequency. Principal ways in which alleles change in the frequency of a population are listed below.

- **Purifying (negative) selection.** If a person affected by genetic disease is not likely to reproduce, disease alleles are lost from the population. This effect is much more pronounced in early-onset dominant conditions, in which—with the exception of non-penetrance—anyone with a mutant allele is affected by the time of puberty.

- *New mutations.* New alleles are constantly being created by the mutation of existing alleles. Some mutations produce new disease alleles by causing genes to lose their function or to function abnormally. There are numerous different ways in which a 'forward' mutation can cause a gene to lose its function, but a 'back mutation' (*revertant* mutation) that can restore the function of a nonfunctioning allele has to be very specific and so is comparatively very rare.

- *Influx of migrants.* If a population absorbs a large influx of migrants with rather different allele frequencies, then the overall gene pool will change.

- *Random sampling of gametes.* Only a certain proportion of individuals within a population reproduce. Out of all the alleles within the population, therefore, only those present in people who reproduce can be transmitted to the next generation. That is, a *sample* of the total alleles in the population is passed on and that sample is never exactly

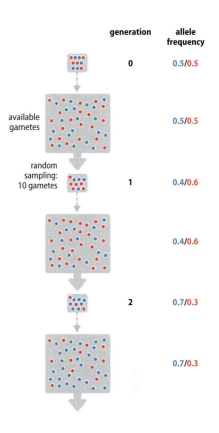

Figure 5.19 Random sampling of gametes in small populations can lead to considerable changes in allele frequencies. Small boxes represent gametes transmitted by reproducers to the next generation; large boxes represent all available gametes in the population. The comparative frequencies of the red and blue alleles can change significantly between generations when by chance the samples of transmitted gametes have allele frequencies that are rather different from the allele frequencies in the population. Such *genetic drift* is significant in small populations. (Adapted from Bodmer WF and Cavalli-Sforza LL [1976] Genetics, Evolution and Man. With permission from WH Freeman & Company.)

representative of the total population for purely statistical reasons. The smaller the size of a population, the larger will be the random fluctuations in allele frequency. This effect is known as **genetic drift** and in small populations it can cause comparatively rapid changes in allele frequencies between generations (**Figure 5.19**).

Population bottlenecks and founder effects

Genetic drift is most significant when population sizes are small. There have been several occasions during our evolution when the human population underwent a *population bottleneck*, a severe reduction in size before the reduced population (now with much less genetic variation) expanded again (**Figure 5.20A**). As a result, genetic variation in humans is very much less than in our nearest relative, the chimpanzee.

Another type of population reduction has periodically happened during human migrations, when a small group of individuals emigrated to form a separate colony. Again, the small population would represent a subset of the genetic variation in the original population and have different allele frequencies. Subsequent expansion of the founding colony would lead to a new population that continued to have limited genetic variation and distinctive allele frequencies reflecting those of the original settlers (a **founder effect**—see Figure 5.20B).

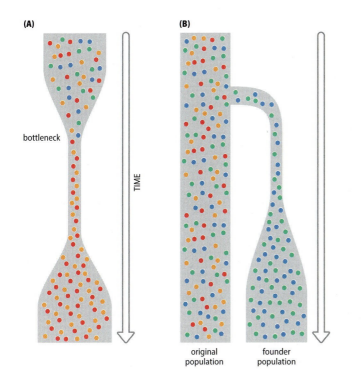

Figure 5.20 Altered allele frequencies after a population bottleneck and formation of a founder population. (A) In a population bottleneck, a severe reduction in the size of the population can lead to altered allele frequencies and much less genetic variation in the subset of the surviving population. Subsequent expansion will reestablish a large population but with reduced genetic variation compared with the time before the bottleneck. (B) A small group of individuals, with a subset of the genetic variation of the larger population, migrate to establish a separate colony (founder population) that can expand but continues to show different allele frequencies from the original foundation. In both images the vertical arrows indicate the passage of time.

If a founder colony happens to have an increased frequency of a disease allele, the new population that will descend from it can be expected to have an increased frequency of the disease. Various populations throughout the world have elevated frequencies of certain single-gene disorders as a result of a founder effect. In autosomal recessive disorders, the great majority of mutant alleles are found in asymptomatic carriers who transmit the mutant alleles to the next generation.

The Finnish and Ashkenazi Jewish populations have been particularly amenable to investigations of founder effects because of rapid recent population expansion, high education levels and very well developed medical services. After the introduction of agriculture from the Middle East in prehistoric times, Finland was one of the last regions of Europe to be populated, and the major expansion that led to the present population began only 2000–2500 years ago as migrants entered southern Finland. Thereafter, in the seventeenth century a second large population expansion began with the occupation of the uninhabited north of Finland.

Ashkenazi Jews (descended from a population that migrated to the Rhineland in the ninth century and from there to different countries in eastern Europe) and Sephardic Jews (primarily from Spain, Portugal, and north Africa) have been distinct populations for more than 1000 years. Until just a few hundred years ago, Ashkenazi Jews used to represent a minority of Jews, but they have undergone a rapid population expansion and now account for 80% of the global Jewish population. Founder effects have been documented in many other populations; see **Table 5.2** for examples.

A distinguishing feature of a founder effect is that affected individuals will usually have mutant alleles with the same ancestral mutation. For example, affected individuals in nine Amish families with Ellis–van Creveld syndrome were shown to be homozygous for the same pathogenic mutation in the *EVC* gene and for a neighboring nonpathogenic sequence change that is absent from normal chromosomes. In this case, genealogy studies were able to confirm a founder effect: all affected individuals could trace their ancestry to the same couple, a Mr Samuel King and his wife, who immigrated in 1774.

Mutation versus selection in determining allele frequencies

If we consider stable, large populations (so that migrant influx and genetic drift are not significant factors), the frequencies of mutant alleles (and

DISORDER AND INHERITANCE[a] (OMIM)	POPULATION	COMMENTS
Aspartylglucosaminuria; AR (208400)	Finnish	carrier frequency = 1 in 30
Ellis–van Creveld syndrome; AR (225500)	Amish, Pennsylvania	carrier frequency ≈ 1 in 8. Traced to a single couple who immigrated to Pennsylvania in 1774
Familial dysautonomia; AR (223900)	Ashkenazi Jews	carrier frequency = 1 in 30
Hermansky–Pudlak syndrome; AR (203300)	Puerto Ricans	thought to have been introduced by migrants from southern Spain
Alzheimer's disease type 3, early onset; AR (607822)	in remote villages in the Andes	all descended from a couple of Basque origin who settled in Colombia in the early 1700s
Huntington disease (HD); AD (143100)	in fishing villages around Lake Maracaibo, Venezuela	more people with HD here than in rest of world. About 200 years ago, a single woman with the HD allele bore 10 children. Many current residents of Lake Maracaibo can trace their ancestry and disease-causing allele back to this lineage
Myotonic dystrophy, type I, AD (160900)	in Saguenay–Lac-Saint-Jean, Quebec	prevalence of 1 in 500 (30–60 times more frequent than in most other populations). Introduced by French settlers

Table 5.2 Examples of single-gene disorders that are common in certain populations because of a founder effect.

[a]AR, autosomal recessive; AD, autosomal dominant.

genetic diseases) in a population are determined by the balance between two opposing forces: mutation and selection.

Purifying selection removes disease alleles from the population when a disorder causes affected individuals to reproduce less effectively than the normal population. The genetic term **fitness** (*f*) is applied here and is really a measure of reproductive success: it uses a scale from 0 to 1 to rank the capacity of individuals to reproduce and have children who survive to a reproductive age. Thus, a fitness of 0 (genetic lethal) means consistent failure to reproduce, and so mutant alleles are not transmitted vertically to descendants. Loss of mutant alleles from the population by purifying selection is balanced by the creation of new mutant alleles by fresh mutation, keeping constant the disease allele frequency in the population.

For autosomal dominant disorders, all people who have a disease allele might be expected to be affected (if we discount non-penetrance). Yet, according to the disorder, the fitness of individuals varies enormously. In many cases, affected individuals have severely or significantly reduced fitness. However, individuals affected by a late-onset disorder can have fitness scores that approach those of normal individuals—they are healthy in their youth and can reproduce normally (**Figure 5.21** gives some examples).

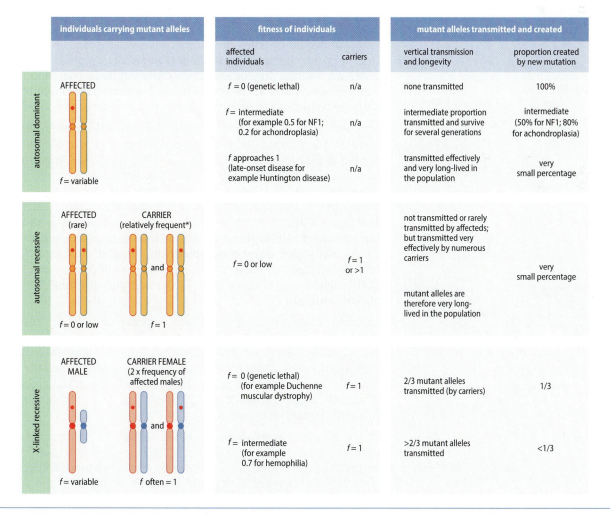

Figure 5.21 Fitness of individuals and mutant allele transmission/creation in single-gene disorders. Note that carriers of certain autosomal recessive disorders may have a higher fitness than normal individuals (*heterozygote advantage*—see following page).

For recessive disorders, mutant alleles are also found in carriers who have a single mutant allele. In autosomal recessive disorders, carriers vastly outnumber affected individuals. Recall the Hardy–Weinberg law that gives a ratio of $2pq$ (carriers) to q^2 (affecteds) = $2p/q \approx 2/q$ (p is very close to 1 for almost all recessive disorders). To take one example, cystic fibrosis occurs in roughly 1 in 2000 births in northern European populations, so $q^2 = 1/2000$. This gives $q \approx 1/45$ and $2/q \approx 2/(1/45) = 90$. That is, there would be about 90 carriers of cystic fibrosis for each affected individual in this population. Because carriers of autosomal recessive disease are normally asymptomatic, they are very effective at transmitting mutant alleles and so new mutations are rare in autosomal recessive disease.

For X-linked recessive disorders, there are two female carriers per affected male. This happens because mutant alleles that reside on an X chromosome get transferred by recombination between three types of X chromosome: a single X chromosome in males and two X chromosomes in females. If we discount manifesting heterozygotes and take an approximation that the fitness of carriers is close to 1, then for conditions in which affected males do not reproduce, natural selection removes 1 out of 3 mutant alleles from the population. Because the lost alleles are replaced by new mutant alleles, 1 out of 3 mutations are new mutations.

Heterozygote advantage: when natural selection favors carriers of recessive disease

We saw above how some populations have a particularly high incidence of a genetic disorder as a result of a founder effect. Another reason why a recessive condition may be especially common in one population is that under certain conditions a type of natural selection can favor a particularly high frequency of carriers.

Recall that natural selection works to eliminate disadvantageous alleles within the population (purifying selection) and also to promote an increase in frequency of advantageous alleles (positive selection). That occurs because natural selection works though the genetic *fitness* of individuals (their ability to reproduce and have children who survive to a reproductive age): disadvantageous alleles are alleles that reduce fitness; advantageous alleles increase fitness. But sometimes a disadvantageous allele can also simultaneously be an advantageous allele. A form of natural selection called **balancing selection** can cause a harmful disease allele to increase in frequency in a population because carriers of the mutant allele have a higher fitness than normal individuals (**heterozygote advantage**).

Sickle-cell anemia provides a classic example of heterozygote advantage. It is very common in populations in which malaria caused by the *Plasmodium falciparum* parasite is endemic (or was endemic in the recent past) but is absent from populations in which malaria has not been frequent. In some malaria-infested areas of west Africa, the sickle-cell anemia allele has reached a frequency of 0.15—far too high to be explained by recurrent mutation.

Sickle-cell heterozygotes have red blood cells that are inhospitable to the malarial parasite (which spends part of its life cycle in red blood cells). As a result, they are comparatively resistant to falciparum malaria. Normal homozygotes, however, frequently succumb to malaria and are often severely, sometimes fatally, affected. Heterozygotes therefore have a higher fitness than both normal homozygotes and disease homozygotes (who have a fitness close to zero because of their hematological disease).

Heterozygote advantage through comparative resistance to malaria has also been invoked for certain other autosomal recessive disorders that feature hemolytic anemia, such as the thalassemias and glucose-6-phosphate dehydrogenase deficiency. The high incidence of cystic fibrosis in northern European populations and Tay–Sachs disease in Ashkenazi Jews is also likely to have originated from heterozygote advantage, possibly through a greater resistance of carriers to infectious disease.

If continued over many generations, even a small degree of heterozygote advantage can be enough to change allele frequencies significantly (invalidating Hardy–Weinberg predictions that assume constant allele frequencies).

Distinguishing heterozygote advantage from founder effects

Diseases that are common in a population because of a founder effect typically originate from one (or occasionally two) mutant alleles. Most people in the population who carry mutant alleles can be expected to have the same ancestral mutation. Heterozygote advantage, by contrast, could be conferred by multiple different mutations of similar effect in the same gene.

If genealogical evidence is not available, strong support for a founder effect can still be obtained. DNA analyses may show that multiple individuals from the population have alleles with the same pathogenic mutation located within a common haplotype of nonpathogenic alleles at neighboring marker DNA loci. If, by contrast, it can be shown that there are multiple different disease alleles in the population or that the disease alleles are embedded in different haplotypes (suggesting different mutational events), heterozygote advantage is likely to apply. But sometimes it is difficult to distinguish between different possible contributions made by founder effects, heterozygote advantage, and even genetic drift when population sizes are very small.

Summary

- Some human disorders and traits are very largely determined by genetic variation at a single gene locus.

- Multiple members of a human kinship (extended family) may be affected by the same single-gene disorder as a result of genetic transmission of mutant alleles (individual versions of a gene at one locus) from one generation to the next.

- In dominantly inherited disorders, an affected person is usually a heterozygote—one allele at the disease locus is defective or harmful, but the other allele is normal.

- In recessive disorders, an affected person has defective alleles only at the disease locus. A person with one disease allele and one normal allele is usually an unaffected carrier who can transmit the harmful (mutant) allele to the next generation.

- A person with an autosomal recessive disorder may have two identical mutant alleles (a true homozygote) or two different mutant alleles (a compound heterozygote).

- Because the X and Y chromosomes have very different genes, men are hemizygous by having a single functional allele for most genes on these chromosomes.

- In X-linked recessive disorders, men are disproportionately affected (they have a single allele, but women with one mutant allele are usually asymptomatic carriers).

- One X chromosome is randomly inactivated in each cell of the early female embryo; descendant clonal cell populations have an inactivated maternal X or an inactivated paternal X. A female carrier of a mutant X-linked allele may be affected if the normal X has been inactivated in a disproportionately large number of cells.

- A genetic mosaic with a mixture of normal and mutant cells may be mildly affected but transmit the mutant allele to descendants who would have harmful mutations in each cell and be more severely affected.

- Some types of mutation are dynamically unstable and become more severe from one generation to the next (anticipation).

- Affected individuals in the same family can also show differences in phenotype because they have different alleles at some other genetic locus (modifier) that interacts with the disease gene locus.

- Our cells each contain many copies of mitochondrial DNA (mtDNA), and affected individuals in a family with a mtDNA disorder may be variably affected because of heteroplasmy (variable ratios of mutant to normal mtDNA copies per cell).

- In disorders of imprinting, individuals who have inherited a mutant gene may or may not be affected, depending on whether the mutant allele was inherited from the paternal or maternal line.

- There is no one-to-one correspondence between genes and phenotypes. Different mutations in the same gene can sometimes cause different disorders, and yet the same disorder is quite often caused by mutations in different genes.

- Some single-gene disorders are notably common in certain ethnic populations. For recessive disorders, a high carrier frequency may arise because asymptomatic carriers of the mutant allele have been reproductively more successful than individuals with two normal alleles (the single mutant allele may have given heterozygotes an advantage by providing greater protection against certain infectious diseases).

- Mutant alleles lost from the population (when individuals fail to reproduce) are balanced by new mutant alleles (created by fresh mutation). For recessive disorders and late-onset dominant disorders, comparatively few alleles are lost from the population and so fresh mutation rates are low. For a severe dominant disorder that manifests before puberty, the rate of fresh mutation may be very high.

- Allele frequencies can be calculated in populations by using the Hardy–Weinberg law, which gives the frequency of a homozygous genotype as the square of the allele frequency, and the frequency of a heterozygous genotype as twice the product of the allele frequencies.

Questions

Help on answering these questions and a multiple-choice quiz can be found at www.garlandscience.com/ggm-students

1. What is the likely inheritance pattern in the pedigree below?

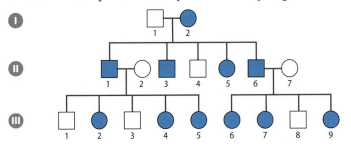

2. What is the likely inheritance pattern in the pedigree below? II-7 is pregnant again. What is the risk that the child will be affected?

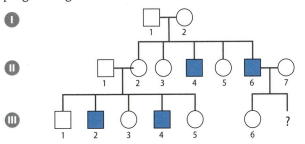

3. In X-linked recessive inheritance, women carriers may sometimes show disease symptoms. How does that happen?

4. What is meant by pseudoautosomal regions and pseudoautosomal inheritance?

5. Sickle-cell disease is an autosomal recessive disorder. Nevertheless, carriers of a sickle-cell disease mutation may not be asymptomatic. Explain why.

6. New mutations are rare in recessive disorders, but for severe, X-linked recessive disorders, about 1 in 3 mutations occur *de novo*. Why is that?

7. Give four alternative explanations for variable expression (including non-penetrance) in monogenic disorders.

8. The Hardy–Weinberg equilibrium is an important principle in population genetics. Explain what it means and how it can be usefully applied to Mendelian disorders.

9. An autosomal recessive disorder has a frequency of 1 in 3600 in a certain population. A woman who has two affected sibs with this condition has recently married an unrelated man from the same population group and is expecting a child by him. Deduce from a combination of the principles of Mendelian segregation and from the Hardy–Weinberg equilibrium what the risk is that any child that the couple have will be affected. [Hint: use Mendelian segregation to work out her risk of being a carrier, then use the Hardy–Weinberg equilibrium to work out what her prospective husband's risk of being a carrier is, and finally use Mendelian segregation to work out the chance of having an affected child if both are carriers.]

10. What is meant by balancing selection? How does it explain the high frequencies of certain recessive conditions?

Further Reading

Single-Gene Disorders

McCusick VA (2007) Mendelian inheritance in Man and its online version, OMIM. *Am J Hum Genet* 80:588–604; PMID 17357067.

Pagon RA et al. (eds) GeneReviews™. http://www.ncbi.nlm.nih.gov/books/NBK1116/; PMID 20301295 (see also Box 5.1).

General Mendelian Inheritance

Bennett RL et al. (2008) Standardized human pedigree nomenclature: update and assessment of the recommendations of the National Society of Genetic Counselors. *J Genet Counsel* 17:424–433; PMID 18792771.

Wilkie AOM (1994) The molecular basis of dominance. *J Med Genet* 31:89–98; PMID 8182727.

Zschocke J (2008) Dominant versus recessive: molecular mechanisms in metabolic disease. *J Inherit Metab Dis* 31:599–618; PMID 18932014.

X-Linked Inheritance and X-Inactivation

Franco B & Ballabio A (2006) X-inactivation and human disease: X-linked dominant male-lethal disorders. *Curr Opin Genet Dev* 16:254–259; PMID 16650755.

Mangs AH & Morris BJ (2007) The human pseudoautosomal region (PAR): origin, function and future. *Curr Genomics* 8:129–136; PMID 18660847.

Migeon BR (2007) Females are Mosaics. X Inactivation and Sex Differences in Disease. Oxford University Press.

Orstavik KH (2009) X chromosome inactivation in clinical practice. *Hum Genet* 126:363–373; PMID 19396465.

Allele Frequencies, Mosaicism, and Calculating Genetic Risk

Aidoo M et al. (2002) Protective effects of the sickle cell gene against malaria morbidity and mortality. *Lancet* 359:1311–1312; PMID 11965279.

Harper PS (2001) Genetic Counselling, 5th ed. Hodder Arnold.

Hartl D & Clark AG (2007) Principles of Population Genetics, 4th ed. Sinauer Associates.

Hurst LD (2009) Fundamental concepts in genetics: genetics and the understanding of selection. *Nature Rev Genet* 10:83–93; PMID 19119264.

McCabe LL & McCabe ER (1997) Population studies of allele frequencies in single gene disorders: methodological and policy considerations. *Epidemiol Rev* 19:52–60; PMID 9360902.

Van der Meulen MA et al. (1995) Recurrence risks for germinal mosaics revisited. *J Med Genet* 32:102–104; PMID 7760316.

Principles of Gene Regulation and Epigenetics

All our cells develop ultimately from the zygote; each nucleated cell in a person contains the same DNA molecules. However, only a subset of the genes in a cell are expressed, and that subset varies according to the type of cell. The global gene expression pattern of a cell dictates the form of a cell, how it behaves, and ultimately its identity—whether it will be a hepatocyte, for example, or a macrophage, or a sperm cell.

In Chapter 2 we outlined the basic details of gene expression. Here, we consider the processes that regulate how our genes are expressed. Different levels of gene regulation affect the production or stability of gene products: transcription, post-transcriptional processing (to make mRNA or a mature noncoding RNA), translation of mRNA, post-translational modification, folding of gene products, incorporation into a multisubunit functional molecule, and degradation of gene products.

We explore aspects of mRNA degradation and of protein folding in Chapter 7. In this chapter we are mostly concerned with gene regulation at the levels of transcription, post-transcriptional processing, and translation. Complex networks of interacting regulatory nucleotide sequences and proteins are involved, and these regulatory sequences can be divided into two broad classes, according to whether they are ***cis*-acting** or ***trans*-acting**, as described in **Box 6.1**.

Much of the control over how our genes are expressed is genetic: the control depends on the base sequence. Mutation of a *cis*-acting regulatory sequence often has consequences for just one allele of one or a few neighboring genes; mutation of genes specifying a *trans*-acting gene regulator may have consequences for the expression of multiple genes distributed across the genome.

Given that our cells all contain the same DNA, one might reasonably wonder how we could ever come to have different types of cells with distinct gene expression patterns. The answer must lie outside our DNA: additional factors are involved that do not depend on the DNA sequence. In every nucleated cell in our bodies, the global pattern of how DNA functions (and how its integrity is maintained) is determined by a combination of genetic controls (that depend on the DNA sequence) and **epigenetic mechanisms** (that are independent of the DNA sequence).

As explained below, epigenetic mechanisms can bring about differential changes in chromatin structure across chromosomal regions. Over certain short regions, the chromatin can be induced to form a very loose structure that allows a gene or genes in that short region to have access to transcription factors and to be expressed; in other regions, the chromatin is kept tightly condensed so that any genes in that region are not expressed.

BOX 6.1 *Cis*-Acting and *Trans*-Acting Gene Regulation.

A **cis-acting** regulatory DNA element is a DNA sequence whose function in gene regulation is limited to the *single* DNA molecule on which it resides. As an example, consider gene promoters. The promoter upstream of the human insulin gene on a paternally inherited chromosome 11 regulates the *paternal* insulin gene only, not the maternal insulin gene. See **Figure 1A, B** for a general illustration.

A *cis*-acting RNA sequence regulates the expression of the *single* RNA transcript that it is located on. These elements are often located in the untranslated regions in mRNA (Figure 1D), but individual splicing control sequences in pre-mRNA may be located fully within exons, fully within introns, or spanning exon-intron boundaries.

A **trans-acting** gene regulator is a regulatory protein or RNA molecule that is free to migrate by diffusion so as to recognize and bind specific short target nucleic acid sequences, thereby

regulating the expression of *both* alleles on distantly located genes (Figure 1C) or the RNAs they produce (Figure 1D). A *trans*-acting gene regulator often regulates *multiple* different genes that have target sequences that it can bind to, or RNA transcripts from multiple genes.

Not shown in Figure 1 are additional types of *cis*-regulation in which regulatory noncoding RNA acts on DNA or on nascent transcripts that remain attached to DNA. These types of gene regulation are particularly important in certain epigenetic phenomena such as X-chromosome inactivation or imprinting, and we consider them in the context of epigenetic gene regulation in Section 6.2.

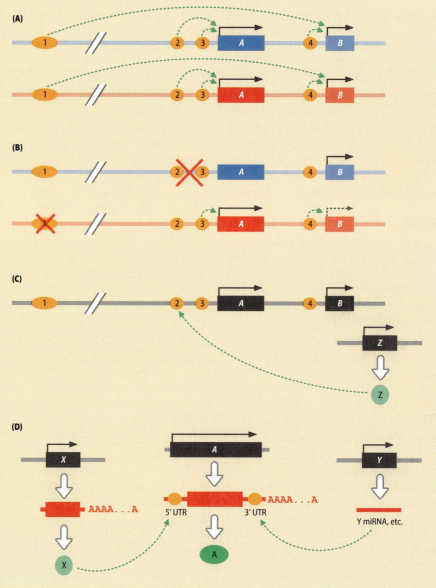

Figure 1 Examples of *cis*- and *trans*-acting regulation. (A, B) Regulation of neighboring genes *A* and *B* on homologous chromosomes by positive *cis*-acting regulatory elements 1 to 4 (orange ovals). Paternal DNA is shown in blue; maternal DNA is in pink. (A) Gene *A* is controlled by elements 2 and 3; gene *B* is controlled by proximal element 4 and remote element 1. (B) A mutation abolishing regulatory elements results in disrupted expression of a regulated gene on the same chromosome. Loss of paternal elements 2 and 3 inactivates the paternal *A* allele; loss of maternal element 1 severely reduces expression of the maternal *B* allele. (C) A remote gene *Z* (on another chromosome or the same one) produces a *trans*-acting regulatory protein Z that binds to regulatory element 2 on *both* paternal and maternal chromosomes (represented by a single, generic black chromosome). (D) *Cis*-acting regulatory elements in mRNA are often located in untranslated regions (UTRs). Expression of gene *A* is imagined here to be also regulated by *trans*-acting regulatory protein X and microRNA Y that bind to *cis*-acting elements in the 5' and 3' UTRs, respectively, of the mRNA.

Changes in the chromatin structure are largely dictated by changes in DNA methylation, histone modification, and the positioning of nucleosomes, but noncoding RNAs have important roles, too. Environmental factors can influence these processes, and stochastic factors also play a part.

We begin this chapter by looking at how genetic regulation governs how our genes are expressed. We then consider principles of epigenetic regulation. We end with a section on abnormal epigenetic regulation that results from abnormal chromosome inheritance, or from single-gene disorders in which mutations affect a gene involved in epigenetic regulation. We describe a very different type of epigenetic dysregulation that relates to protein folding in Chapter 7, and we examine epigenetic contributions to complex disease in Chapters 8 and 10.

6.1   Genetic Regulation of Gene Expression

As described in Figure 2.11, the mitochondrial genome is transcribed from fixed start points, generating large multigenic transcripts from each DNA strand, which are subsequently cleaved. By contrast, it is usual for nuclear genes to be transcribed individually, and transcription is regulated by genetic factors, as described in this section, and also by epigenetic factors, which we consider in Sections 6.2 and 6.3. Recently, however, geneticists have become increasingly aware of the importance of post-transcriptional controls, notably at the level of RNA processing and translation.

Promoters: the major on–off switches in genes

Along the lengths of each DNA strand of our very long chromosomal DNA molecules are **promoters**, *cis*-acting regulatory DNA sequences that are important in establishing which segments of a DNA strand will be transcribed. Each promoter is a collection of very short sequence elements that are usually clustered within a few hundred nucleotides from the transcription start site. For each DNA strand, transcription begins at fixed points on the DNA where the chromatin has been induced to adopt an 'open' structure (see below).

Nuclear genes are transcribed by three different types of RNA polymerases. A nucleolar RNA polymerase, RNA polymerase I, is dedicated to making three of the four different ribosomal RNAs (rRNAs) in our cytoplasmic ribosomes (the 28S, 18S, and 5.8S rRNAs). It transcribes clusters of about 50 tandem DNA repeats (each containing sequences for the 28S, 18S, and 5.8S rRNAs) on each of the short arms of chromosomes 13, 14, 15, 21, and 22. RNA polymerase II transcribes all protein-coding genes, genes that make long noncoding RNAs and some short RNA genes including many miRNA genes. RNA polymerase III transcribes tRNA genes, the 5S rRNA gene, and some other genes that make short RNAs.

None of the RNA polymerases acts alone; each is assisted by dedicated protein complexes. In the case of RNA polymerase II, for example, a core transcription initiation complex is formed by the sequential assembly of five multisubunit proteins (*general* transcription factors—see below) at specific sites on the DNA.

Some of the protein subunits of the transcription initiation complexes recognize and bind specific short DNA sequence elements of a promoter; others are recruited by binding to previously bound proteins. For a protein-coding gene, most core promoter elements are upstream of the start site, and the spacing of the elements is important.

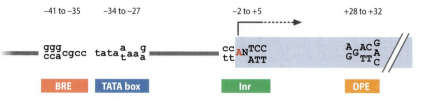

Figure 6.1 Consensus sequences for some core promoter elements often found in genes transcribed by RNA polymerase II. The TATA box is bound by the TATA-binding protein subunit of transcription factor IID. The initiator (Inr) element defines the transcription start site (the highlighted A) when located 25–30 bp from a TATA box. The downstream core promoter (DPE) element is only functional when placed precisely at +28 to +32 bp relative to the highlighted A of an Inr element. TFIIB binds to the BRE (TFII**B** **r**ecognition **e**lement) and accurately positions RNA polymerase at the transcription start site. However, none of these elements is either necessary or sufficient for promoter activity, and many active polymerase II promoters lack all of them. N represents any nucleotide. (Adapted from Smale ST & Kadonga JT [2003] *Annu Rev Biochem* 72:449–479; PMID 12651739. With permission from Annual Reviews.)

Figure 6.1 illustrates some important core promoter elements, but note that the composition of core promoter elements is highly variable and some promoters lack all the elements shown in Figure 6.1. We describe additional *cis*-acting elements in the next section.

Once the basal transcription apparatus has fully assembled, a component with DNA helicase activity is responsible for locally unwinding the DNA helix, and the activated RNA polymerase accesses the template strand.

Modulating transcription, tissue-specific regulation, enhancers, and silencers

As a metaphor for gene expression, imagine the output from a radio. The basal transcription apparatus described above would be the radio's on switch that is needed to get started. It is required in all cells and uses ubiquitous transcription factors that bind to *cis*-acting elements in the core promoter. But there is also the need for a volume control to amplify or reduce the signal, as required in different cell types or at different stages in a cell's life or development.

The role of the volume control is performed by additional circuits. First, additional, often non-ubiquitous, transcription factors bind to *cis*-acting regulatory elements other than those of the core promoter. These elements are sometimes distantly located from the gene they regulate, as described below. Then there are co-activator or co-repressor proteins that are recruited by bound transcription factors. In addition, diverse types of long noncoding RNAs regulate transcription, but because they often work in the epigenetic regulation of transcription we consider them in Section 6.2.

Enhancers, silencers, and insulator elements

DNA-binding sites for many transcription factors are often located within 1.5 kb of the transcriptional start site, and they can be located upstream or downstream of the transcription start site. For a gene with introns, these *cis*-acting regulatory elements are therefore often located just upstream of the gene or within the first intron. Sometimes, however, the regulatory elements may be a considerable distance from the promoter of the gene that they regulate.

In promoters the orientation and spacing of the sequence elements are important. Some additional *cis*-acting regulatory DNA elements modulate transcription in a way that is quite independent of their orientation. They belong to two classes: **enhancers**, which amplify transcription, and **silencers**, which repress transcription. They can be located close to a transcriptional start site, but quite often they are remote from their target gene (**Figure 6.2A**). To allow remote elements such as enhancers to work, the intervening DNA needs to be looped out—the proteins bound to the enhancer can now physically interact with proteins bound to the promoter of the target gene (see Figure 6.2B).

The long-distance action of elements such as enhancers needs to be targeted to the correct genes. To ensure that signals from regulatory elements do not affect genes other than the intended targets, another class

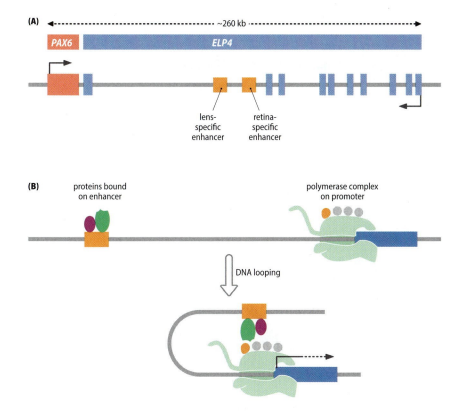

Figure 6.2 DNA looping brings proteins bound to remote enhancers or silencers into direct contact with proteins bound to the promoter. (A) Example of remote enhancers. The 30 kb *PAX6* gene that is mutated in aniridia (OMIM 106210) is known to be regulated by lens-specific and retina-specific enhancers (orange boxes) that are distantly located within a long intron of the neighboring *ELP4* gene. Vertical blue boxes represent *ELP4* exons (for the sake of clarity, *PAX6* is represented by a single box including both exons and introns). Exons and enhancer elements are not to scale. (B) DNA looping allows direct physical interactions between proteins bound to a distant *cis*-acting element (orange box) that regulates a gene (dark blue box) with some of the many proteins bound to the gene's promoter. For clarity, only the RNA polymerase is shown at the promoter.

of regulatory elements is needed to establish boundaries (boundary elements). Some boundary elements act as barriers between neighboring euchromatin and heterochromatin regions. Another type is an *insulator* that can block inappropriate interactions between enhancers and promoters by binding regulatory proteins, notably the CTCF regulator (see Figure 6.22B for an example).

Transcription factor binding

A protein **transcription factor** is a sequence-specific DNA-binding protein that binds at or close to *cis*-acting DNA sequences that are important in gene expression. In addition to ubiquitous general transcription factors that bind to core promoter elements, many other transcription factors bind to additional, often remote, *cis*-acting sequences, such as enhancers, as described below.

Although some genes need to be expressed in all cells ('housekeeping' genes), many are expressed only in specific tissues and/or at specific developmental stages, partly because the activity of the promoters relies on tissue-specific or developmentally regulated transcription factors that bind to non-core elements. Like other DNA-binding proteins, a transcription factor typically recognizes a specific short sequence (often four to nine nucleotides long) using a DNA-binding domain that contains some motifs that physically bind DNA, such as zinc fingers or leucine zippers (**Figure 6.3**).

For any one transcription factor there are from tens of thousands to hundreds of thousands of potential binding sites across the human genome, but only a tiny fraction are used, for two reasons. First, the binding site must be accessible (in a relaxed open chromatin conformation and away from direct contact with nucleosomes). Secondly, binding is *combinatorial*—different transcription factors work in concert by binding to adjacent recognition sequences.

Figure 6.3 Examples of DNA-binding motifs that are used by some transcription factors and other DNA-binding proteins. (A) The zinc finger motif involves the binding of a Zn^{2+} ion by four conserved amino acids (normally either histidine or cysteine) so as to form a loop (finger). They usually occur in clusters of sequential zinc fingers. The so-called C2H2 (Cys_2/His_2) zinc finger typically comprises about 23 amino acids, with neighboring fingers separated by a stretch of about seven or eight amino acids. The structure of a zinc finger may consist of an α-helix and a β-sheet held together by coordination with the Zn^{2+} ion, or of two α-helices, as shown here. In either case, the primary contact with the DNA is made by an α-helix binding to the major groove. (B) The leucine zipper is a helical stretch of amino acids rich in hydrophobic leucine residues, aligned on one side of the helix. These hydrophobic patches allow two individual α-helical monomers to join together over a short distance to form a coiled coil. Beyond this region, the two α-helices separate, so that the overall dimer is a Y-shaped structure. The dimer is thought to grip the double helix much like a clothes peg grips a clothes line. Leucine zipper proteins normally form homodimers but can occasionally form heterodimers (the latter provide an important combinatorial control mechanism in gene regulation).

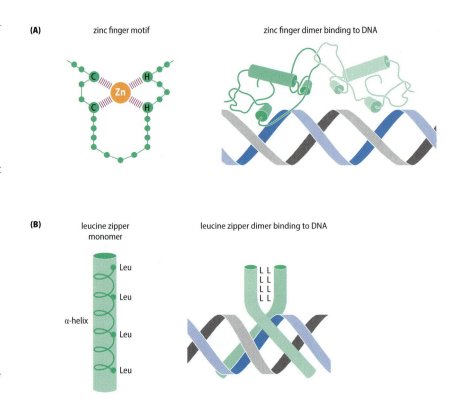

Figure 6.4 *Cis*-acting sequences that regulate RNA splicing. Pink boxes represent transcribed exon sequences. (A) The three fundamental RNA sequences that are engaged in the splicing mechanism. Bases in red are essentially invariant. Slashes indicate a gap that can vary in length from tens of nucleotides up to several hundred kilobases long in extreme cases. Spliceosomes contain several types of small nuclear RNA (snRNA), including U1 snRNA, which base pairs with the splice donor sequence, and U2 snRNA, which base pairs with the branch site sequence. (B) In addition to the splice donor (SD), splice acceptor (SA), and branch site (BS) sequences, other regulatory RNA sequences stimulate splicing (orange) or inhibit splicing (black). In this example, an exon has two exonic splice enhancers (ESE), and an exonic splice suppressor (ESS) and is flanked by introns that have intronic splice suppressors (ISS). The dashed lines indicate an alternative 3′ end to the exon, after the use of an alternative splice donor (sd) sequence that may be used instead of the usual splice donor sequence (SD) in some tissues or circumstances.

A transcriptional activation domain is present in transcription factors that stimulate transcription; transcriptional repressors often recruit specialized protein complexes to silence gene expression, as described in Section 6.2. Other proteins modulate transcription without binding to DNA. Instead, they work by protein–protein interactions that support other regulatory proteins (which bind DNA directly). There are two types: transcriptional *co-activators* (which enhance transcription) and *co-repressors* (which down-regulate transcription).

Genetic regulation during RNA processing: RNA splicing and RNA editing

Understanding the genetic control of splicing is important for understanding pathogenesis because mutations causing abnormal RNA splicing are a relatively common cause of disease. RNA editing is a less well understood form of RNA processing.

Regulation of RNA splicing

Like transcription, RNA splicing is subjected to different controls, and some splicing patterns are ubiquitous; others are tissue-specific. As illustrated in **Figure 6.4A**, three fundamental *cis*-acting regulatory RNA

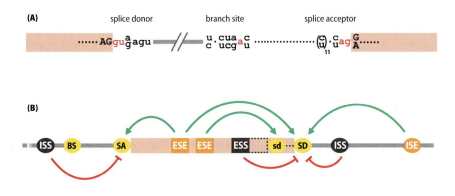

sequences are required for the basic splicing mechanism, which is performed by large ribonucleoprotein complexes known as spliceosomes. The **splice donor site** contains an invariant GU dinucleotide that defines the 5′ end of an intron at the RNA level. The **splice acceptor site** contains an invariant AG dinucleotide that defines the 3′ end of an intron at the RNA level and is embedded within a larger sequence that includes a preceding polypyrimidine tract. An additional control element, the branch site, is located very close to the splice acceptor; it contains an invariant A nucleotide and is responsible for initiating the splicing reaction. Note that the sequence surrounding the invariant GU and AG signals is variable—some splice sites are strong and readily used, whereas others are weak and used only occasionally.

Splicing is also regulated by two additional classes of short (often hexanucleotide) *cis*-acting regulatory RNA elements: splice enhancer sequences (which stimulate splicing) and splice suppressor sequences (which inhibit splicing). They are located close to splice junctions and can lie within exons or introns (Figure 6.4B). To help keep the spliceosome in place splicing enhancers bind SR proteins (so called because they have a domain based on repeats of the serine-arginine dipeptide). Splicing suppressors bind hnRNP proteins that are active in removing bound spliceosomes. Because different tissues and cell types can express different SR proteins and different hnRNP proteins, splicing patterns can vary between tissues.

Alternative splicing

More than 90% of human protein-coding genes undergo some kind of alternative splicing, when primary transcripts of a single gene are spliced in different ways (**Figure 6.5** gives some variations, and Figure 6.4B shows how they can be generated). Sometimes, a proportion of transcripts retain transcribed intronic sequence. **Exon skipping** occurs when one or more full-length exons are not represented in some transcripts. In other cases, there is some variability in the precise locations of exon–intron junctions, so that transcripts from one gene may have short or long versions of an exon.

Some of the variable transcripts may be functionally unimportant (splicing accidents must happen occasionally). Often, however, alternative splicing patterns show tissue specificity (so that, for example, one splice pattern is consistently produced in brain but another is found in liver), or there may be consistent differences in the use of specific splice patterns at different stages in development. By producing alternative products (**isoforms**) from individual genes, alternative splicing can increase functional variation.

Alternative isoforms may be retained in cells or they may be secreted or sent to different cellular compartments to interact with different molecules and perform different roles. For example, the –KTS isoforms of the WT1 Wilms tumor protein (**Figure 6.6A**) function as DNA-binding transcription factors, but the +KTS isoforms associate with pre-mRNA and may have a general role in RNA splicing. This pattern of alternative splicing has been conserved over hundreds of millions of years. The ERBB4 protein, a tyrosine protein kinase that is a member of the epidermal growth factor receptor family, has CYT1 and CYT2 isoforms that respectively possess or lack a binding site for the phosphatidylinositol-3-kinase signaling molecule (see Figure 6.6A).

Sometimes entirely different proteins are created from a common gene by alternative splicing. For example, alternative splicing of the *CDKN2A* gene produces two entirely different proteins which, nevertheless, have similar functions (see Figure 6.6B).

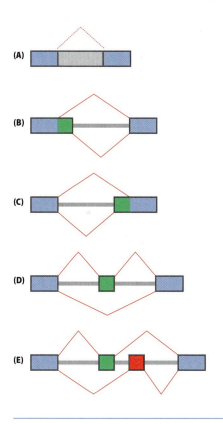

Figure 6.5 Types of alternative splicing event. (A) An intronic sequence (gray) is either excluded from a transcript or retained. (B, C) The use of alternative splice donor sites (B) or of alternative splice acceptor sites (C) results in the inclusion or exclusion of the sequences in green. (D) The exon in green may be either included or skipped (a *cassette exon*). (E) Alternative exons: the mature mRNA includes either the exon in green or the exon in red, but not both or neither. Blue boxes represent exons that are always included in the mature mRNA.

Figure 6.6 Examples of alternative splicing in human genes. (A) Alternative splicing producing variant proteins. Alternative splicing results in the variable presence of a 17 amino acid (17aa) peptide near the middle of the WT1 Wilms tumor protein and of a Lys-Thr-Ser tripeptide (KTS) between the third and fourth zinc finger (ZF) domains. Four different isoforms exist for the human ERBB4 protein. Just before the transmembrane (TM) domain there is the alternative presence of a 23-amino-acid peptide or a 13-amino-acid peptide (JM-a and JM-b isoforms, respectively). And within the tyrosine kinase (TK) domain is the variable presence of a 16-amino-acid peptide that has a binding site for phosphatidylinositol-3-kinase (CYT-1 isoforms have the peptide; CYT-2 isoforms lack it). (B) Alternative splicing of the *CDKN2A* gene produces two entirely different tumor suppressor proteins, p16-INK4A and p14-ARF, which work in cell cycle control. Exon 2, the one exon with coding sequence for both proteins, is translated in different reading frames.

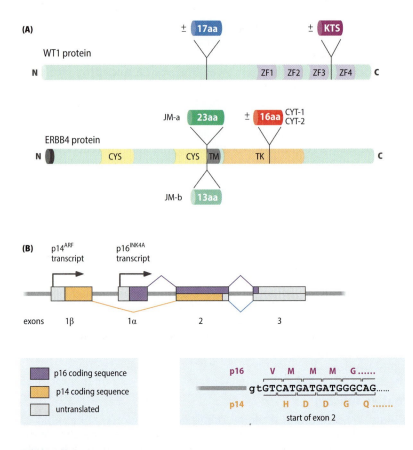

RNA editing

In some RNA transcripts, certain nucleotides naturally undergo deamination or transamination. When this happens in the coding sequences of mRNAs, the amino acid sequence of the protein will differ from that predicted by the genomic DNA sequence.

For example, certain adenines in some RNA transcripts are naturally deaminated to give the base inosine (I), which behaves like guanine (by base pairing with cytosine). In coding sequences, A→I editing is most commonly directed at CAG codons, which specify glutamine (Q). The resulting CIG codons behave like CGG and code for arginine (R), and so this type of RNA editing is therefore also called Q/R editing. Q/R editing is quite commonly found during the maturation of mRNAs that make neurotransmitter receptors or ion channels.

Some other types of RNA editing are known, including C→U editing (used in making apolipoprotein B mRNA, for example) and U→C editing (used in making mRNA from the *WT1* Wilms tumor gene). Once considered an oddity, RNA editing has recently been claimed to be much more widespread than previously thought, but its extent is still controversial, and its significance is unclear.

Translational regulation by *trans*-acting regulatory proteins

Regulation at the level of translation allows cells to respond more rapidly to altered environmental stimuli than is possible by altering transcription. According to need, stores of inactive mRNA may be held in reserve so that they can be translated at the optimal time. Controls are also exerted over where an mRNA is translated: some mRNAs are transported as ribonucleoprotein particles to specific locations within a cell; an example is tau mRNA, which is selectively localized to the proximal regions of axons rather than to dendrites.

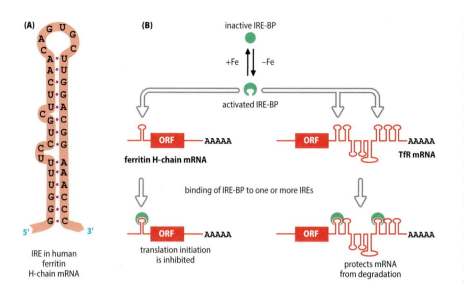

(A)

IRE in human
ferritin
H-chain mRNA

(B)

Figure 6.7 Iron-response elements in the ferritin and transferrin mRNAs. (A) Stem-loop structure of an iron-response element (IRE) in the 5′ untranslated region of the ferritin heavy (H)-chain mRNA. (B) When iron levels are low, a specific IRE-binding protein (IRE-BP) is activated and binds the IRE in the ferritin heavy-chain gene and also IREs in the 3′ untranslated region of the transferrin receptor (TfR) mRNA. Binding inhibits the translation of ferritin but protects the transferrin receptor mRNA from degradation, maximizing the production of transferrin receptor. When iron levels are high, the IRE-binding protein is inactivated, maximizing the production of ferritin and ecreasing the production of transferrin receptor. ORF (open reading frame) designates the central coding DNA of the mRNAs.

To control gene expression at the level of mRNA, *trans*-acting regulatory factors bind to specific *cis*-acting RNA elements in the untranslated regions of the mRNA. Single-stranded RNA is quite flexible (unlike DNA, which has a rather rigid structure), but typically it has a very high degree of secondary structure as a result of intra-chain hydrogen bonding (see Figures 2.4 and 2.6). RNA elements that bind protein are often structured as hairpins, as in the example of the iron-response element shown in **Figure 6.7A**.

As an example of translational regulation, consider how cells control the availability of two proteins involved in iron metabolism: ferritin (an iron-binding protein used to store iron in cells) and the transferrin receptor (which helps us absorb iron from the diet). When iron levels are low, the priority is to maximize the amount of iron that can be absorbed from the diet: transferrin receptor mRNA is protected from degradation so that it can make a protein product. Conversely, when iron levels are high, ferritin production is activated to store iron in cells. This happens without any change in the production of ferritin or transferrin receptor mRNAs. Instead, both these mRNAs have iron-response elements (IREs), which can be bound by a specific IRE-binding protein that regulates the production of protein from these mRNAs; the availability of this binding protein is also regulated by iron concentrations (see Figure 6.7B).

Post-transcriptional gene silencing by microRNAs

Trans-acting regulators such as the IRE-binding protein that work by binding to mRNA used to be viewed as quirky exceptions. The discovery of **RNA interference** (**Box 6.2**) and of the existence of tiny RNA regulators, notably microRNA, changed all that. MicroRNAs (miRNAs) are single-stranded regulatory RNAs that down-regulate the expression of target genes by base pairing to complementary sequences in their transcripts. Typically about 20–22 nucleotides long, they are formed by multiple processing events, including cleavages in the cytoplasm that are performed by the same endoribonucleases used in RNA interference.

An miRNA binds to any transcript that has a suitably long complementary sequence to form a stable heteroduplex (correct base pairing is important for the 'seed' sequence covering the first eight or so nucleotides from the 5′ end of the miRNA; some mismatches are tolerated when the remaining part of the miRNA pairs up). Because of the limited miRNA length and the tolerance of some base mismatches, a single type of miRNA can regulate transcripts from many different genes (**Figure 6.8**).

BOX 6.2 RNA Interference as a Natural Cell Defense Mechanism.

Long double-stranded RNA within the cells of higher organisms is not normal, and it can activate a defense mechanism called **RNA interference** (**RNAi**) that protects cells against viruses and transposable elements. Double-stranded RNA can be generated by viruses that have infected cells. It can also be made after transposons have been transcriptionally activated (transposons are often heavily methylated and transcriptionally inactive in animal cells; when inappropriately expressed, however, inverted copies can generate RNA transcripts from both DNA strands that will be complementary in sequence).

At the heart of RNAi are different endoribonucleases that work on double-stranded RNA. A cytoplasmic endoribonuclease called dicer cuts long double-stranded RNA asymmetrically into a series of short double-stranded RNA pieces known as **short interfering RNA** (**siRNA**)—see **Figure 1**. The siRNA duplexes are bound by different complexes that contain an Argonaute-type endoribonuclease (Ago) and some other proteins. Thereafter, the two RNA strands are unwound and one of the RNA strands is degraded by the Ago ribonuclease, leaving a single-stranded RNA, the guide strand, bound to the Argonaute complex. The complex is now activated, and the single-stranded RNA will guide it to its target by base pairing with complementary RNA sequences in the cells.

One type of Argonaute complex, shown in Figure 1, is the RNA-induced silencing complex (RISC). In this case, after the single-stranded guide RNA binds to a complementary long single-stranded RNA, the Ago enzyme will cleave the RNA, causing it to be degraded. Viral and transposon RNA can be inactivated in this way. Note that although mammalian cells have RNA interference pathways, the presence of long double-stranded RNA triggers an interferon response that causes *nonspecific* gene silencing and cell death.

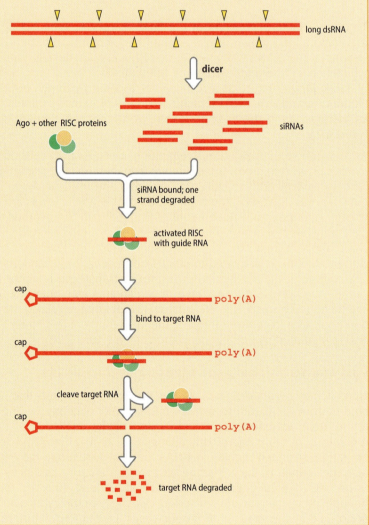

Figure 1 RNA interference using RNA-induced silencing complexes. In animal cells, unnatural long double-stranded RNA (dsRNA) is cleaved asymmetrically by an endoribonuclease called dicer that is specific for double-stranded RNA, generating short (20–22-nucleotide) pieces of RNA with overhanging 3′ ends called short interfering RNA (siRNA). The siRNAs are bound by protein complexes containing the Argonaute endoribonuclease (Ago) such as RNA-induced silencing complexes (RISC) shown here. The Ago ribonuclease attacks the bound siRNA, leading to the degradation of one RNA strand to leave one strand, the guide strand. The guide strand is so named because it can base pair to any other viral or transposon RNA with a complementary sequence: it guides the RISC complex to viral or transposon transcripts that can then be cleaved by the Ago endoribonuclease. The cleaved transcripts, lacking a protective cap or poly(A), are degraded by exonucleases.

We have several hundred miRNA genes, and they frequently show tissue-specific expression; many are important in early development, but miRNAs have been found to be important regulators in a whole range of different cellular and tissue functions. At least 50% of our protein-coded genes are thought to be regulated by miRNAs, and individual types of mRNA typically have recognition sequences for multiple miRNA regulators. Just

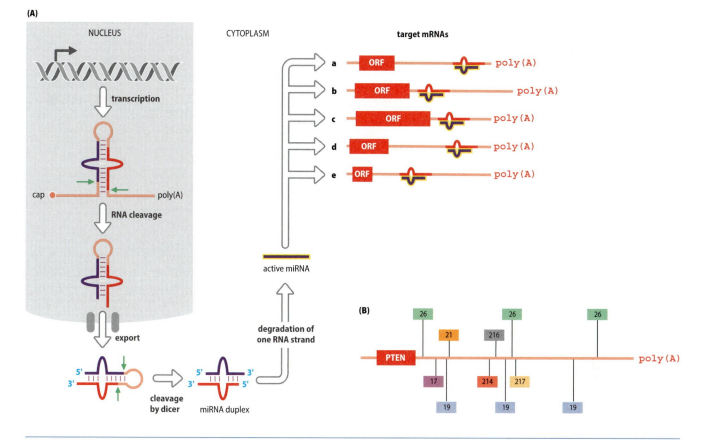

Figure 6.8 How microRNAs are produced and work in cells.
(A) miRNA genes are transcribed and cleaved in the nucleus to generate a stem-loop RNA that is exported to the cytoplasm and further cleaved asymmetrically by the endoribonuclease dicer to generate a miRNA duplex with overhanging 3′ ends. One strand of the duplex (called the passenger strand) is then cleaved and degraded, leaving the other strand (the guide strand) as a mature single-stranded miRNA. A typical human miRNA binds to and regulates transcripts produced by hundreds of different genes, and the vast majority of miRNA–target RNA heteroduplexes have imperfect base pairing. Shown here for illustration are five mRNAs produced from five different genes (a–e); the complementary sequence to which the miRNA binds is shown in red. ORF, open reading frame (= coding DNA). (B) An individual mRNA often has multiple miRNA-binding sites. The example here is the mRNA from the human *PTEN* tumor suppressor gene that has binding sites in the 3′ untranslated region for miRNAs belonging to seven miRNA families: three binding sites each for miR-19 and miR-26, and one each for miR-17, miR-21, miR-214, miR-216, and miR-217.

like protein transcription factors, miRNAs seem to be involved in complex regulatory networks, and they are subject to negative regulation by a wide range of RNA classes as described in the next section.

Currently, the details of how miRNAs bring about **gene silencing** to down-regulate the expression of target mRNAs are being worked out. Present evidence suggests that the degradation of mRNAs is enhanced and that translation is also arrested to some degree.

Repressing the repressors: competing endogenous RNAs sequester miRNA

Many of our pseudogenes are known to be transcribed. Some of them seem to have undergone purifying selection, indicating that they are functionally important. A landmark study published in 2010 provided the first real insights into how functional pseudogenes work: it showed that the human *PTEN* gene at 10q23 is regulated by a highly related processed pseudogene, *PTENP1*, located at 9p21.

PTEN makes a protein tyrosine phosphatase that is very tightly controlled (cells are very sensitive to even subtle decreases in abundance of this

Figure 6.9 Competing endogenous RNAs acting as miRNA sponges. In this example, a protein-coding gene *A* and a closely related pseudogene *ψA* produce RNA transcripts that have in common binding sites for certain miRNA classes. The pseudogene RNA can compete with the mRNA for binding by the same classes of miRNAs. Other RNA classes can also act in the same way, including certain long noncoding RNAs and circular RNAs.

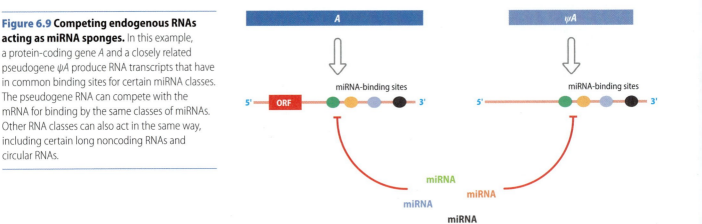

protein, and aberrant *PTEN* expression is common in cancers). *PTENP1* does not make a protein (one of the changes from the *PTEN* sequence disrupts the initiator methionine codon). It does, however, make a non-coding RNA that retains many of the miRNA-binding sites in the 3′ UTR of *PTEN* mRNA. The *PTENP1* RNA seems to regulate *PTEN* expression by binding to and sequestering miRNAs that would normally bind to the *PTEN* mRNA (**Figure 6.9** gives the principle).

Circular RNAs as abundant miRNA sponges

We now know that huge numbers of endogenous RNAs are involved in regulating the binding of miRNAs to their target RNAs. They do that by competing with the miRNA targets for binding to the relevant miRNAs (the competing endogenous RNAs have sometimes been called 'miRNA sponges' because they tend to soak up the available miRNAs). In addition to pseudogene RNAs, they include various long noncoding RNAs and circular RNAs that are surprisingly abundant (human fibroblast cells, for example, make more than 25,000 different circular RNAs). The circular RNAs largely overlap protein-coding sequences and so can contain sequences that correspond to the untranslated sequences of mRNAs, and therefore miRNA-binding sites. They are formed by head-to-tail splicing reactions that join first and last exon sequences—see the review by Tauli et al. (2013) under Further Reading.

6.2 Chromatin Modification and Epigenetic Factors in Gene Regulation

If different types of cells are programmed to express different sets of genes, but all nucleated cells in an individual originate from a single fertilized egg cell and have the same set of DNA molecules, how do we ever get different types of cells in the first place? How do uncommitted cells in the very early embryo give rise to increasingly specialized lineages of cells and then to different types of terminally differentiated cells?

The establishment of different cell lineages and cell differentiation involves programmes of altered gene expression during development that do not depend on the DNA sequence itself, but instead on epigenetic factors. In its broadest sense, **epigenetics** covers all phenomena that, without affecting the DNA sequence, can somehow produce heritable changes in how genomes function, both at the level of how genes are expressed and how sequences that determine chromosome functions, such as centromere sequences, work.

PHENOMENON	MECHANISM/COMMENTS
Epigenetic reprogramming in the early embryo	eggs and sperm are differentiated cells, and their genomes have different epigenetic marks. They combine to give a zygote whose genome is gradually reprogrammed to erase the great majority of the inherited epigenetic marks. By the blastocyst stage, the cells of the inner cell mass are pluripotent and will give rise ultimately to all cells of the body. Epigenetic marks are reestablished in the descendants of the cells of the inner cell mass to establish different cell lineages and permit cell differentiation
Epigenetic reprogramming in gametogenesis	readily detected as a wave of genomewide demethylation during germ cell development (erasing parental epigenetic marks) followed by comprehensive de novo DNA methylation to reset global patterns of DNA methylation and gene expression
Establishment of constitutive heterochromatin	centromere establishment relies on a specific histone H3 variant known as CENP-A
X-chromosome inactivation	silencing by a long noncoding RNA (XIST RNA) of almost all genes on one of the two X chromosomes in female cells
Genomic imprinting	silencing of one allele, asccording to parent of origin, at diverse gene loci (often organized in gene clusters) on different chromosomes
Position effects causing heterochromatinization	large-scale changes in DNA, causing genes to be relocated to a heterochromatin environment where they are silenced

Table 6.1 **Examples of epigenetic phenomena involving DNA and chromatin modification in mammalian cells.**

Epigenetic effects can be stably transmitted from one cell generation to the next, providing a form of cellular memory. For example, once a cell has been programmed to become a liver cell, daughter cells retain this programming so that they, too, are liver cells. There are also at least some instances in nature, notably in plants, in which epigenetic effects can be transmitted through meiosis, from one organism to subsequent generations. But epigenetic effects are also reversible.

Most epigenetic effects involve modifying DNA or histones, or changing the position of nucleosomes. The net effect is to alter the chromatin environment: the altered chromatin state causes changes in gene expression and can be propagated to daughter cells (**Table 6.1** gives examples).

Changes in chromatin structure producing altered gene expression

Depending on its chromatin environment, the properties of a DNA sequence can change. A functional gene that is embedded in highly condensed chromatin may not be accessible to transcription factors; the gene is said to be silenced. But if the chromatin structure is altered, adopting a more open, relaxed conformation, protein factors may be able to bind the promoter and related regulatory sequences to initiate transcription. Conversely, if a gene that is normally expressed were to be transposed by a translocation or inversion so that it now takes up residence within (or close to) a region of constitutive heterochromatin (permanently condensed heterochromatin), the gene would be silenced (an example of a **position effect**). Gene expression is therefore dependent not only on DNA sequence but also on chromatin structure.

Changes in chromatin structure often arise by specific types of chemical modification of the DNA strands and histones. For example, DNA is modified by adding methyl groups to a small percentage of the cytosines and by removing methyl groups from some of the methylated cytosines. Note that because methylated cytosines behave like cytosine and base pair with guanines, the base sequence is not considered to be altered. Extensive methylation of DNA sequences in vertebrates is generally characteristic of tightly packed chromatin; loosely structured chromatin ('open' chromatin) has low-level DNA methylation (**Figure 6.10**).

Figure 6.10 Altered chromatin states arise from DNA and chromatin modifications. Note that DNA methylation is associated with condensed chromatin, but different types of histone methylation are associated with open and condensed chromatin (see Table 6.2).

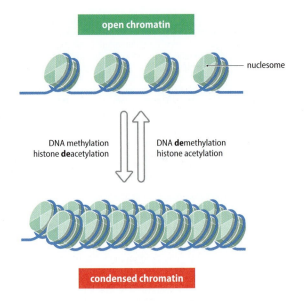

open chromatin

nucleosome

DNA methylation
histone **de**acetylation

DNA **de**methylation
histone acetylation

condensed chromatin

Histone modifications include different types of post-translational modification at specific amino acid positions on the different types of histone. Histone acetylation, for example, is associated with open chromatin, and histone deacetylation with condensed chromatin (see Figure 6.10).

Patterns of DNA methylation and histone modification that are heritable from one cell generation to the next are sometimes called **epigenetic marks** (or epigenetic settings). Many different enzymes are responsible for creating epigenetic marks. One class, called 'writers,' add chemical groups to modify DNA or histones covalently; 'erasers' work in the opposite direction to remove the chemical groups.

In addition, multiple different effector proteins—called 'readers'—are involved in binding to specific chemical groups on DNA or histones to interpret defined epigenetic marks. The readers may recruit additional factors to induce different changes in chromatin, such as chromosome compaction, or changes in nucleosome spacing and structure (**chromatin remodeling**). By adjusting the position of nucleosomes with respect to the DNA strand, promoters and other regulatory DNA sequences can become nucleosome-free, allowing access by transcription factors.

Nucleosome structure is also altered by histone substitution when standard histones in nucleosomes are substituted by minor histone variants that recruit regulatory factors. As described below, this can have different effects, such as activating transcription or defining centromeres. Although much of our knowledge of chromatin modification has come from studying patterns of DNA methylation and histone modification or substitution, numerous non-histone proteins and noncoding RNAs also have important roles in modifying chromatin structure.

Modification of histones in nucleosomes

Nucleosomes have 146 bp of DNA wrapped around a core of eight histone proteins, composed of two each of four different histone classes: H2A, H2B, H3, and H4. The histone proteins are positively charged (because of an excess of lysine and arginine residues) and have protruding N-terminal tails. The histone tails in **Figure 6.11A** are shown in isolation, but they can make contact with adjacent nucleosomes.

Each N-terminal histone tail shows a pattern of variable chemical modifications at specific amino acid positions. Individual amino acids in each

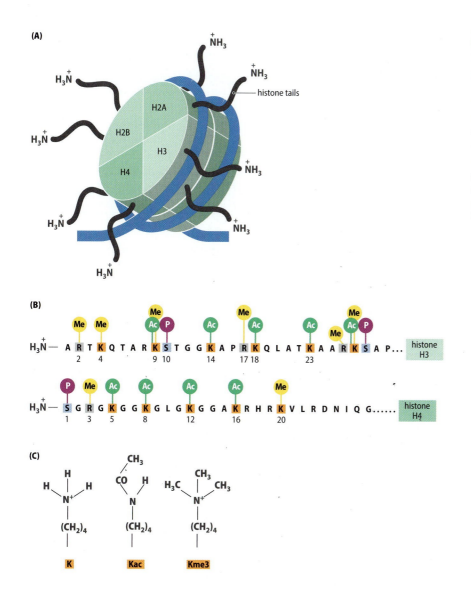

Figure 6.11 Histone modifications.
(A) Positively charged N-terminal histone tails protrude from nucleosomes and can associate with other nucleosomes (not shown). (B) Map of histone H3 and H4 tail modifications. Note that lysine at position 9 in H3 can be methylated (H3K9me) or acetylated (H3K9ac), but never both. Some lysine residues may have two or three methyl groups (see Table 6.2). (C) Examples of lysine modification, showing the standard side chain of lysine (K), an acetylated lysine (Kac), and a trimethylated lysine (Kme3).

tail may be methylated, acetylated, or phosphorylated (see Figure 6.11B), or subject to yet other types of modification, including tagging with ubiquitin. Particular types of amino acid are preferred targets for modifying the N-terminal histone tails: acetylation occurs only at lysine residues, phosphorylation mostly occurs at serines, but both lysine and arginine residues can be methylated.

Acetylation of lysines leads to loss of the positive charge (Figure 6.11C), and acetylated histone tails interact less well with neighboring nucleosomes than do the unacetylated histone tails. Histone acetylation therefore results in a more relaxed chromatin conformation. According to the specific amino acid position, lysines may also be modified to contain one, two, or three methyl groups, but in these cases the positive charge is retained on the side chain (see Figure 6.11C).

Histone modifications are performed by a series of different enzymes that are devoted to adding or removing a chemical group at specific amino acid positions. Thus, for example, there are multiple histone acetyl transferases (HATs) and histone deacetylases (HDACs), and suites of histone lysine methyltransferases (KMTs) and histone lysine demethylases (KDMs).

AMINO ACID	EUCHROMATIN					HETEROCHROMATIN	
	Promoters		Enhancers		Gene Bodies	Facultative	Constitutive
	Active	Inactive	Active	Inactive	Inactive		
H3K4	H3K4me2, H3K4me3		H3K4me1, H3K4me2				
H3K9	H3K9ac	H3K9me3		H3K9me2, H3K9me3	H3K9me2, H3K9me3	H3K9me2	H3K9me3
H3K27	H3K27ac	H3K27me3	H3K27ac			H3K27me3	
H4K12	H4K12ac					H4K12ac	H4K12ac
H4K20							H4K20me3

Table 6.2 **Examples of histone modifications characteristic of different chromatin states.**

The effect of modified histones and histone variants on chromatin structure

Histone modifications are 'read' by nonhistone proteins that recognize and bind the modified amino acids and then recruit other proteins to effect a change in chromatin structure. Proteins with a bromodomain recognize the acetylated lysines of nucleosomal histones, those with a chromodomain recognize methylated lysines, and different varieties of each domain can recognize specific lysine residues. Chromatin-binding proteins often have several domains that recognize histone modifications.

Certain individual types of histone modification are associated with open chromatin and transcriptional activation, or with condensed chromatin and transcriptional repression. For example, methylation of H3K4 (the lysine at position 4 on histone H3) is associated with open chromatin at the promoters of actively transcribed genes and at active enhancers (**Table 6.2**). By contrast, trimethylation of the lysine at position 9 on histone H3 (H3K9me3) is prominently associated with transcriptional repression, being widely found in constitutive heterochromatin and in inactive genes in euchromatin (see Table 6.2).

In addition to histone modification, core histone proteins can be replaced by minor variants, notably of histone classes 2A and 3 (the variants typically differ from the canonical histone by just a few amino acids). The minor histone variants are synthesized throughout interphase and are often inserted into previously formed chromatin by a histone exchange reaction powered by a chromatin remodeling complex. Once inserted, they recruit specific binding proteins to effect some change in the chromatin status for specific functions. A well-studied example is CENP-3A, a centromere-specific histone H3 variant that is responsible for assembling kinetochores at centromeres. **Table 6.3** gives other examples.

Modified histones and histone variants typically work together with DNA methylation and demethylation in regulating gene expression (**Figure 6.12**). H3K9me3 can bind heterochromatin protein 1, which in turn

CLASS	VARIANT	DESCRIPTION
H2A	H2AX	important in DNA repair and recombination (it is introduced at sites of double-strand breaks)
	H2A.Z	associated with the promoters of active genes. It also helps prevent the spread of silent heterochromatin and is important in maintaining genome stability
H3	H3.3	important in transcriptional activation
	CENP-A	a centromere-specific variant of H3. It is required for assembly of the kinetochore, to which spindle fibers attach

Table 6.3 **Examples of histone H2A and histone H3 variants.**

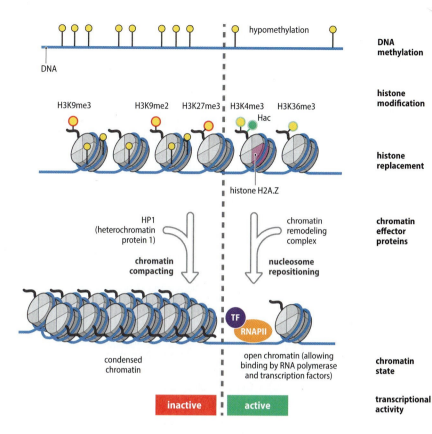

Figure 6.12 Contributions made by DNA and histone modification to different chromatin states in neighboring regions on a chromosome. For convenience, each nucleosome is shown with only one out of the eight N-terminal histone tails. Methylation modifications are shown by symbols filled in yellow—as small circles in DNA methylation, and as large circles in histone methylation; note that some histone methylations repress transcription (red outer lines), but other types are associated with transcription (green outer lines). The symbols Hac and filled green circle denote histone acetylation, which applies to multiple lysines on H3 and H4. Chromatin effector proteins such as heterochromatin protein 1 (HP1) or the repressive polycomb group protein complexes PRC1 and PRC2 are recruited to bind to specific histone modifications, often through the agency of long noncoding RNAs. Chromatin remodeling can involve creating nucleosome-free regions of DNA that allows the binding of transcription factors. RNAPII, RNA polymerase II.

recruits DNA methyltransferases that also serve to repress transcription. In turn, DNA methyltransferases and 5-meCG-binding proteins recruit histone deacetylases and appropriate histone methyltransferases to reinforce transcriptional repression.

The function of DNA methylation in mammalian cells

Whereas histones undergo very many different types of modification, the only covalent modification of DNA is methylation, which is essential for mammalian development. A principal function is in regulating gene expression—it stabilizes, or locks in, patterns of gene silencing so that transcription is suppressed in highly methylated regions of chromatin. Highly repetitive DNA sequences, such as satellite repeats in pericentromeric heterochromatin and dispersed transposons, are extensively methylated, but there is also significant—though more sporadic—methylation in the main body (exons and introns) of genes and in intergenic regions.

In the case of DNA methylation patterns, it is the extent of DNA methylation in the key *cis*-acting regulatory elements that distinguishes actively transcribing genes from silenced genes. Thus, the promoters and enhancers of actively transcribing genes are relatively free of DNA methylation. Along with characteristic histone modifications and histone variants such as H2A.Z and H3.3, such hypomethylated regions signal a locally open chromatin environment—transcription factors can gain access to and bind to their target sequences to stimulate transcription. Gene silencing is achieved when significant levels of DNA methylation in a gene's promoter and enhancer elements cause the chromatin to be condensed, denying access to transcription factors.

Like other epigenetic marks, global DNA methylation patterns vary between different cell types and different stages in development. In addition to a general role in gene expression silencing, DNA methylation has important roles in genomic imprinting and X-chromosome inactivation, and also in the suppression of retrotransposon elements. Retrotransposons are evolutionarily advantageous to genomes because they can insert varied DNA sequences at different locations in the genome, providing novel exon combinations in genes (Figure 2.15) and giving birth to new regulatory sequences and new exons. About 43% of the human genome is made up of retrotransposon repeats, but the number of actively transposing retrotransposons needs to be carefully regulated to prevent the genome from being overwhelmed. By suppressing transcription, DNA methylation acts as a necessary brake on excessive transposon proliferation.

DNA methylation: mechanisms, heritability, and global roles during early development and gametogenesis

In mammalian cells, DNA methylation involves adding a methyl group to certain cytosine residues, forming 5-methylcytosine (5-meC). The cytosines that are methylated occur within the context of a palindromic sequence, the CG dinucleotide (also called a CpG dinucleotide, where p represents phosphate).

The 5-meC base pairs normally with guanine (the methyl group is located on the outside of the DNA double helix, minimizing any effect on base pairing). It is recognized by specific 5-meCG-binding proteins that regulate chromatin structure and gene expression, as described below. In a somatic cell, about 70–80% of CG dinucleotides will have a methylated cytosine, but the pattern of methylation is variable across the genome and across genes (**Box 6.3**).

DNA methylation mechanism

DNA methylation is performed by DNA methyltransferases (DNMTs). The DNMT1 enzyme serves to maintain an existing DNA methylation pattern. During replication of a methylated DNA molecule, each parental DNA strand retains its pattern of methylated cytosines. The newly synthesized complementary DNA strand is formed by incorporating unmethylated bases, and so in the absence of any further DNA methylation the result would be a hemimethylated DNA (**Figure 6.13A**).

To maintain the original DNA methylation pattern, DNMT1 is normally present at the replication fork and methylates the nascent (growing) DNA strand. It methylates only those CG dinucleotides that are paired with a methylated CG on the opposing parental DNA strand. That is, the methylated parent DNA strands act as templates for copying the original methylation pattern (see Figure 6.13B). As a result, patterns of symmetric CG methylation can be faithfully transmitted from parent cell to daughter cells.

The enzymes DNMT3A and DNMT3B are *de novo* methyltransferases—they can methylate any suitable CG dinucleotide (see Figure 6.13A). They have important roles in epigenetic reprogramming when epigenetic marks are comprehensively reset across the genome, at two major stages. In each case, a wave of global DNA demethylation is followed by *de novo* DNA methylation that establishes a different methylation pattern.

DNA methylation in early development and gametogenesis

Major epigenetic reprogramming occurs in the early embryo. Egg cells and, notably, sperm cells have extensively methylated DNA (but rather different patterns of DNA methylation). Once a sperm has fertilized an egg, the introduced sperm genome (now within the male pronucleus) begins to undergo active DNA demethylation; after the male and female pronuclei fuse, global demethylation of the zygote begins and continues until

BOX 6.3 CpG Islands and Patterns of DNA Methylation Across the Genome and Across Genes.

DNA methylation is generally used to 'lock in' transcriptional inactivity in regions of our cells that do not require expression. Accordingly, heterochromatin and intergenic regions are subject to high levels of DNA methylation. Hypermethylation of some regions, such as pericentromeric heterochromatin, is important for genome stability; a significant decrease in methylation levels in these regions can lead to mitotic recombination and genome instability. Our genes are also subject to DNA methylation; however, by comparison with heterochromatin and intergenic regions, the DNA methylation is generally reduced and more variable.

As described in the text, DNA methylation in our cells is limited to cytosines and occurs within the context of the dinucleotide CG (because CG is the target for cytosine methylation). The resulting 5-methylcytosine can undergo spontaneous deamination to give thymidine (Figure 4.5), and during vertebrate evolution there has been a steady erosion of CG dinucleotides. As in other vertebrate genomes, therefore, the dinucleotide CG is notably under-represented in our DNA (41% of our genome is made up of G–C base pairs, giving individual base frequencies of 20.5% each for G and C; the expected frequency of the CG dinucleotide is therefore 20.5% × 20.5% = 4.2%, but the observed CG frequency is significantly less than 1%).

Within the sea of our CG-deficient DNA are nearly 30,000 small islands of DNA in which the CG frequency is the expected value but cytosine methylation is suppressed. Such **CpG islands** (or CG islands; the p signifies the phosphate connecting C to G) are often 1 kb or less in length and are notably associated with genes. Approximately 50% of CpG islands are located in the vicinity of known transcriptional start sites, as illustrated in **Figure 1**. A further 25% of CpG islands are found in the main gene body.

Note that CGIs associated with transcriptional start sites remain unmethylated even when the gene is not being transcribed. Whether a gene is silenced or expressed seems instead to be related to the methylation status of other CpGs that are often located up to 2 kb from transcriptional start sites in CpG island 'shores.'

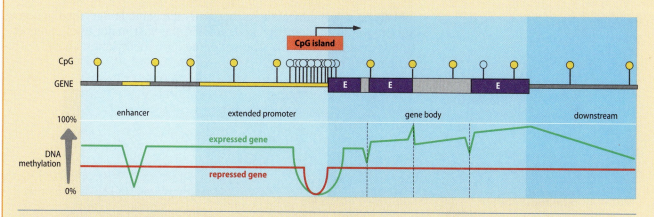

Figure 1 CpG density and DNA methylation levels across an idealized human gene. In this example we consider a gene that has single CpG island located in the vicinity of the transcriptional start site (marked by an arrow) and three exons (E). The CpG density across the gene is illustrated at the top, with open circles indicating unmethylated CpG dinucleotides and circles filled in yellow representing methylated CpG. If the gene is expressed, the gene body shows quite high DNA methylation levels, but the transcriptional start site and upstream enhancer are free of cytosine methylation, allowing access for *trans*-acting protein factors. Even when the gene is transcriptionally inactive, the cytosines remain unmethylated within the CpG island. (Adapted from Hassler MR & Egger G [2012] *Biochimie* 94:2219–2230. With permission from Elsevier Masson SAS.)

the early blastula stage in the preimplantation embryo (**Figure 6.14**). Then a wave of genome re-methylation occurs, coincident with initial differentiation steps that give rise to different cell lineages. Genome methylation is extensive in somatic cell lineages but moderate in trophoblast-derived lineages (which will give rise to placenta, yolk sac, and so on).

Significant epigenetic reprogramming also occurs during gametogenesis. The **primordial germ cells** that will give rise ultimately to gametes are

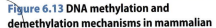

Figure 6.13 DNA methylation and demethylation mechanisms in mammalian cells. During replication of a DNA molecule containing methylated CG dinucleotides, the parental strand retains methylated cytosines, but the newly synthesized DNA incorporates unmodified cytosines. DNMT1 is usually available, and it specifically methylates any CG dinucleotides on the nascent strand that are paired with a methylated CG on the parental strand, regenerating the original methylation pattern (as detailed in bottom panel). If DNMT1 is not available, the hemimethylated DNA can give rise in a subsequent DNA replication to unmethylated DNA (*passive demethylation*). Unmethylated DNA can also be generated by an active demethylation process at certain stages in development (see Figure 6.14). DNMT3A and DNMT3B are used for *de novo* methylation at specific stages in development (see Figure 6.14).

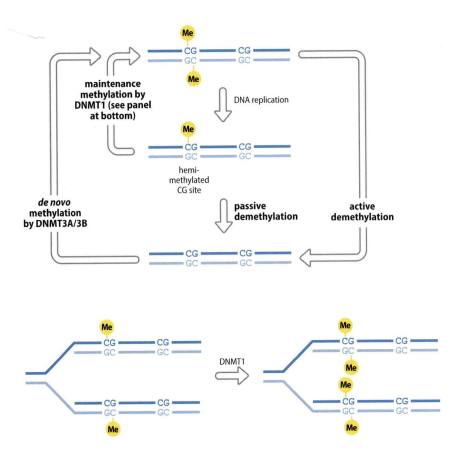

initially heavily methylated. As they enter the genital ridge, their genomes are progressively demethylated, erasing the vast majority of epigenetic settings (see Figure 6.14). Thereafter, *de novo* methylation allows epigenetic marks to be reset.

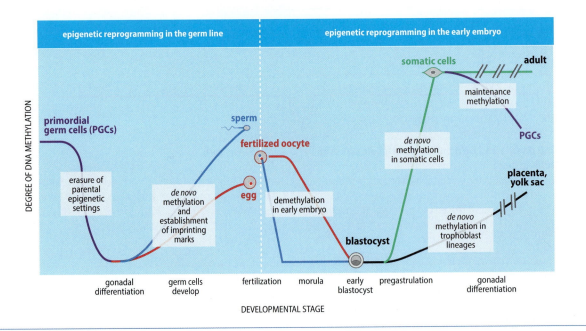

Figure 6.14 Changes in DNA methylation during mammalian development. Marked and often tissue-specific changes in overall methylation accompany gametogenesis and early embryonic development. Their causal role remains uncertain, although mice that are specifically unable to methylate sperm DNA are infertile. The horizontal time axis is necessarily abbreviated on the right-hand side of the figure leading toward birth and then adulthood (indicated by the use of double slashes).

Noncoding RNAs in epigenetic regulation

Diverse noncoding RNAs (ncRNAs) have important roles in gene regulation. They include tiny RNAs and long ncRNAs. The former are important in epigenetic regulation in many organisms, but their roles in mammalian epigenetic regulation are more limited. Thus, mammalian miRNAs are mostly focused on post-transcriptional gene silencing (but some repress key enzymes that are important in chromatin modification), and although endogenous siRNAs are important in the epigenetic regulation of centromeres in some organisms, it is not clear that they have a similar role in mammals (if they do, it might be limited to gametogenesis or early embryo development).

By contrast, mammalian long ncRNAs are very important in both genetic and epigenetic regulation. Long ncRNAs can work in epigenetic regulation by recruiting specific chromatin-modifying protein complexes to defined chromosomal loci; that can happen by mechanisms that do not rely on base pairing (**Figure 6.15D, E**). RNA is comparatively flexible, and through intrachain base pairing it can fold into complex secondary and tertiary structures (Figures 2.4 and 2.6), providing suitable surfaces and clefts to permit highly specific molecular interactions and binding of defined proteins.

We describe below how specific long ncRNAs have critical roles in epigenetic phenomena such as X-chromosome inactivation and imprinting. In particular, repressive polycomb protein complexes are often recruited by the ncRNAs to silence specific target genes.

Cis- and trans-acting regulation

Most of the epigenetic roles described for long ncRNAs involve *cis*-acting regulation (**Table 6.4**). In these cases, nascent transcripts of long ncRNA may act to tether chromatin-modifying complexes that act on neighboring genes. In some cases, a *cis*-acting ncRNA will silence several genes in a cluster (which sometimes involves making very long antisense transcripts, as described in Section 6.3). In the special case of X-chromosome inactivation, many hundreds of genes across the X chromosomes are silenced by a ncRNA (as described below).

Some mammalian long ncRNAs can work as *trans*-acting regulators. An example is HOTAIR (HOX antisense intergenic RNA), which is made by a gene in the *HOXC* homeobox gene cluster at 12q13. It acts as a scaffold that binds specific protein regulators at its 5′ and 3′ ends—respectively the PRC2 polycomb repressive complex and a complex containing the LSD1 eraser—and directs them to silence genes on other chromosomes, including multiple genes within the *HOXD* gene cluster on chromosome 2.

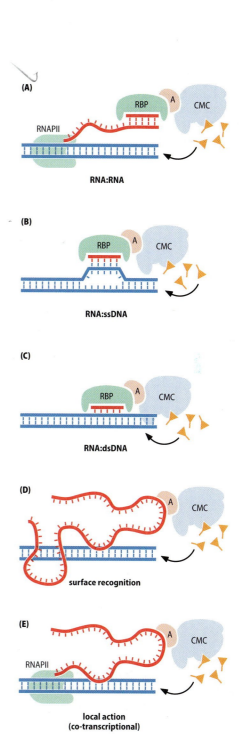

Figure 6.15 Depositing epigenetic signals using ncRNA. (A–C) Sequence-specific recognition models in which regulatory RNA base pairs with target sequences in (A) nascent transcripts, (B) one strand of an opened DNA double helix, or (C) a closed DNA double helix (forming a triple helix). (D, E) Recruitment of long ncRNA by (D) recognizing a target DNA sequence via complex surface-mediated interactions or (E) acting locally by being anchored to chromatin as a nascent transcript. Abbreviations are as follows: RNAPII, RNA polymerase II; RBP, RNA-binding protein; A, adaptor protein (hypothetical); CMC, chromatin-modifying protein complex; ssDNA, single-stranded DNA; dsDNA, double-stranded DNA. Orange flags indicate chemical groups that will bind to DNA or histones to act as epigenetic signals. (Adapted from Bonasio R et al. [2010] *Science* 330:612–616; PMID 21030644. With permission from AAAS.)

RNA	LOCUS OF ORIGIN	MODE OF ACTION	TARGET GENES	CHROMATIN-MODIFYING PROTEIN COMPLEX	HISTONE MODIFICATION[a]
Airn (108 kb)	in imprinted domain on mouse chromosome 17	*cis*	three neighboring genes on paternal chromosome 17	G9a	H3K9me
Kcnq1ot1 (91 kb)	in imprinted domain on mouse chromosome 7	*cis*	11 neighboring genes on paternal chromosome 7	G9a	H3K9me
				PRC2	H3K27me3
				PRC1	H2K119ub1[b]
Xist (17 kb)	at *Xic* on mouse X	*cis*	almost all genes on inactive X	PRC2	H3K27me3
HOTAIR (2 kb)	in *HOXC* homeobox gene cluster at 12q13	mostly *trans*	several genes in *HOXD* cluster at 2q31; many single genes across genome	PRC2	H3K27me3
				LSD1	H3K4me2/me3 demethylation

Table 6.4 Examples of mammalian long ncRNAs that cause gene silencing through identified chromatin-modifying repressive protein complexes. [a]See Table 6.2 on page 164 for histone modifications involving lysine methylation. [b]Monoubiquitylation at lysine 199 on histone H2A.

Genomic imprinting: differential expression of maternally and paternally inherited alleles

We are accustomed to the idea that in mammalian diploid cells both the paternal and maternal alleles are expressed (biallelic expression). For a significant proportion of our genes, however, only one of the two alleles is expressed—the other allele is silenced (monoallelic expression). That can occur in a random way so that in some cells of an individual the paternal allele of a gene is silenced, and in other cells the maternal allele is silenced (**Table 6.5**). However, for some genes a paternal allele is consistently silenced or a maternal allele is consistently silenced. That is, silencing of one allele occurs according to the parent of origin (**genomic imprinting**).

Because natural monoallelic expression occurs for a significant fraction of genes, the maternal and paternal genomes are not functionally equivalent in mammals. As a result, mammals, unlike several vertebrate species, cannot naturally reproduce by parthenogenesis (reproduction without fertilization: a haploid chromosome set duplicates within the oocyte). It is possible to manipulate mammalian eggs artificially to make a diploid embryo with two maternal genomes, but embryonic lethality always ensues: the maternal genome cannot by itself support development—both a maternal and a paternal genome are required. Parthenogenesis

CLASS	MECHANISMS	COMMENTS
Dependent on parent of origin	genomic imprinting	several genes are expressed only from paternally inherited chromosomes and several only from maternally inherited chromosomes
	X-inactivation in placenta	paternal X is always inactivated
Independent of parent of origin	X-inactivation in somatic cells[a]	random inactivation of most genes on either the paternal or maternal X (see Figure 6.18A)
	production of cell-specific Ig and T-cell receptors	each mature B or T cell makes Ig or T-cell receptor chains, using only one allele at a time. Once a functional chain is made by gene rearrangement at one randomly selected allele, a feedback mechanism inhibits further rearrangements (see Section 4.5)
	production of cell-specific olfactory receptors	each olfactory neuron expresses a single allele of just a single olfactory receptor (OR) gene (selected from several hundred OR genes) so that it fires in response to one specific odorant only. Depends on competition for a single monoallelic enhancer
	stochastic mechanisms	may be quite common

Table 6.5 Mechanisms generating natural monoallelic expression in mammals. [a]At least in eutherian mammals; in marsupials the paternal X is consistently inactivated.

fails in mammals because a subset of developmentally important genes are expressed only if inherited paternally; a different subset of genes are expressed only if inherited from a maternal egg.

As detailed below, imprinting patterns in genes are established by *cis*-acting regulatory sequences that also carry a type of imprint because they are methylated to very different extents in sperm DNA and egg DNA. The same differentially methylated region (DMR) can behave very differently when hypomethylated or extensively methylated.

A single allele can behave differently according to the parent of origin, but within an individual the pattern of transcriptional activity or inactivity is maintained through mitosis when somatic cells divide. The alleles are not intrinsically maternal or paternal, however: imprints need to be reversible. A man can inherit an allele from his mother that is inactive. But when he transmits that same allele in his sperm to the next generation the imprint needs to be erased so that the allele is reactivated (**Figure 6.16**).

Extent and significance of genome imprinting

Over 140 mouse genes have been experimentally shown to be imprinted genes, and a smaller number of human imprinted genes have been validated (a catalog of imprinted genes is available at http://igc.otago.ac.nz/).

Many known imprinted genes have a role in embryonic and placental growth and development, and a popular theory attributes imprinting to a conflict of evolutionary interest between mothers and fathers. Propagation of paternal genes would be favored if the offspring were all very robust, even at the expense of the mother (potentially, a man can father children by very many different mothers). Enhanced propagation of maternal genes, however, depends on the mother's being healthy enough to have multiple pregnancies.

Mammalian development is unusual in that the zygote gives rise to both an embryo and also extra-embryonic membranes, including the trophoblast (these act to support development, giving rise to the placenta). From the arguments above, paternal genes might be expected to promote the growth (and general robustness) of the fetus by maximizing the nutrients it can extract from the mother via the placenta. Paternal genes might therefore have a vested interest in supporting the development of the extra-embryonic membranes and placenta. Maternal genes, by contrast, might seek to limit the nutrient transfer so that it does not compromise the mother's health and future reproductive success. Some support for the paternal–maternal conflict theory comes from rare cases of uniparental diploidy in humans (**Figure 6.17**) and from artificially induced uniparental diploidy in mice.

The paternal–maternal conflict theory can only be a partial explanation of why imprinting evolved in mammals. Not all imprinted genes have roles in intrauterine growth, and not all are imprinted in the direction predicted by the parental conflict theory. The theory also doesn't explain why imprinting is tissue-specific for many genes. The insulin-like growth factor gene *IGF2*, for example, is maternally imprinted in many tissues but biallelically expressed in brain, adult liver, and so on, and the *UBE3A* gene, which is implicated in Angelman syndrome, is paternally imprinted in neurons but biallelically expressed in glial cells and other tissues.

Establishing sex-specific imprints by differential methylation

Imprinted genes are often found in clusters and under the control of a common differential methylation region known as an imprinting control

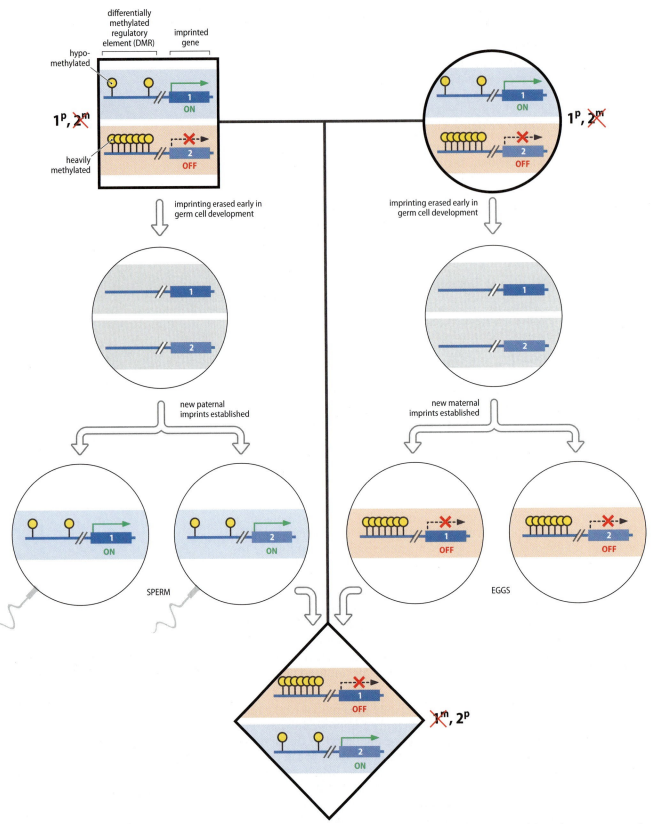

region (ICR). During germ cell development, parental imprints are erased. Thereafter, imprinting is established when sex-specific patterns of *de novo* DNA methylation are created within CG dinucleotides in the ICRs.

Although the sperm and egg have very different overall DNA methylation patterns, large-scale demethylation in the early embryo removes

Figure 6.16 Imprints can be reversed between generations. In this example, a differentially methylated regulatory element (DMR) is heavily methylated when inherited maternally and causes a neighboring imprinted gene to be silenced. The DMR is hypomethylated when inherited paternally, causing the gene to be expressed. The gene has two alleles, 1 and 2, and the man and woman at the top are heterozygotes: 1^p signifies that allele 1 was inherited on a paternal chromosome (pale blue shading); 2^m means that allele 2 was inherited on a maternal chromosome (pink shading). Red crosses indicate inactive alleles. During early germ cell development, rapid demethylation of DNA occurs, and paternal and maternal epigenetic settings are erased (see left part of Figure 6.14). The loss of parental imprints, including DNA methylation marks, means that the chromosomes mostly lose their parental epigenetic identities and are now represented in gray shading (middle of figure). Later in germ cell development, new imprints are established according to the sex of the individual. All sperm from the man either have an active allele 1 or an active allele 2; all the woman's eggs carry an inactive allele 1 or an inactive allele 2. Fertilization as shown can generate a child with the same genotype (1,2) as the parents, but the imprint has been reversed: allele 1 is now the inactive allele (having been maternally inherited) and allele 2 is functional.

the vast majority of the DNA methylation differences in the paternal and maternal genomes. An important exception is in the ICRs, where the sex differences in DNA methylation are retained in somatic cells (but will be erased in primordial germ cells).

Much of our knowledge of how ICRs regulate the expression of imprinted genes comes from studying certain imprinted gene clusters that are associated with developmental disorders in humans. We consider these mechanisms in the context of disorders of imprinting in Section 6.3; as we will see, long ncRNAs are also often associated with clusters of imprinted genes and are important in imprinting control.

X-chromosome inactivation: compensating for sex differences in gene dosage

As described in Section 7.4, a change in constitutional chromosome number (aneuploidy) is usually lethal, or it results in significant abnormalities. Chromosome loss is particularly damaging, and monosomy is

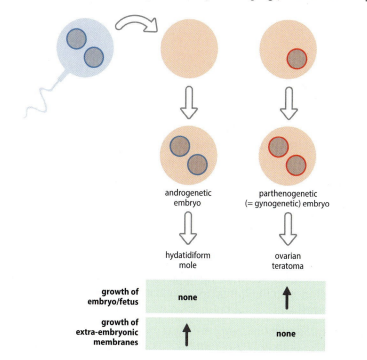

Figure 6.17 Uniparental diploidy and the divergent expression of paternal and maternal genomes. On rare occasions, a zygote is formed with a genome composed of paternal DNA only, producing an *androgenetic embryo* (this usually occurs when a diploid sperm fertilizes a faulty egg that lacks chromosomes, as shown here). Development produces an abnormal conceptus known as a hydatidiform mole, with widespread hyperplasia (overgrowth) of the trophoblast but no fetal parts. The reverse situation, where the zygotic genome is composed of maternal DNA only—producing a *pathenogenetic embryo*—gives rise to an ovarian teratoma that consists of disorganized embryonic tissues without the vital extra-embryonic membranes.

lethal in the early embryo with just one exception: loss of an X chromosome from a female embryo is viable and results in Turner syndrome (45,X).

Aneuploidies cause problems because of abnormal **gene dosage**. We have elaborate gene interaction systems, and the products of genes across the genome sometimes can work together in ways where the relative amounts of participating gene products need to be very tightly controlled (changes in the amounts of an individual component can wreak havoc with regulatory systems, for example). An average chromosome has multiple genes in which a change in copy number (producing abnormal gene dosage) is positively harmful. And yet there is one glaring difference that is somehow tolerated in mammals: females have two X chromosomes, but males have only one X chromosome and a Y chromosome.

Whereas Y-specific genes are rare and largely devoted to male-specific functions, the X chromosome has more than 800 protein-coding genes and many RNA genes that work in all kinds of important cell functions. As first proposed by Mary Lyon, a gene dosage compensation mechanism equalizes X-chromosome gene expression in male and female cells by causing one of the two X chromosomes in female cells to be heterochromatinized (**X-chromosome inactivation**).

X-chromosome counting and inactivation choices

Early in embryogenesis, our cells somehow count how many X chromosomes they contain, and then permanently inactivate all except one randomly selected X. At very early stages in development, both X chromosomes are active, but X-inactivation is initiated as cells begin to differentiate, occurring at the late blastula stage in mice, and most probably also in humans. Inactive X chromosomes remain in a highly condensed heterochromatic state throughout the cell cycle and can be seen as the Barr body (sex chromatin) on the periphery of the cell nucleus (see Figure 5.4).

The choice to inactivate the maternal or the paternal X is made randomly. But whichever of the parental X chromosomes is chosen for inactivation within a cell, that same X is inactivated in all daughter cells (**Figure 6.18A**). An adult female is thus a mosaic of cell clones, each clone retaining the pattern of X-inactivation that was established in its progenitor cell early in embryonic life. Figure 6.18B shows a striking example from the tortoiseshell cat.

X-inactivation is stable through mitosis but not across the generations. A woman's maternal X can equally well have been the active or inactive one in her mother, and has the same chance as her paternal X of being inactivated in her own cells.

XIST RNA and initiation of X-inactivation

Inactivation of a human X chromosome is initiated at an X-inactivation center (XIC) at Xq13. It then propagates along the whole length of the chromosome in what may be an extreme example of the tendency of heterochromatin to spread (we consider heterochromatin spreading in more detail in the context of disease in Section 6.3).

The transient pairing of the two XIC sequences is probably the mechanism by which the X chromosomes are counted. Within this region, the *XIST* gene encodes a 17 kb spliced and polyadenylated ncRNA, an **X**-**i**nactivation-**s**pecific **t**ranscript that is expressed only from the *inactive* X chromosome.

XIST is centrally involved in spreading heterochromatinization outward from the XIC: both *XIST* RNA and the Polycomb proteins that it recruits

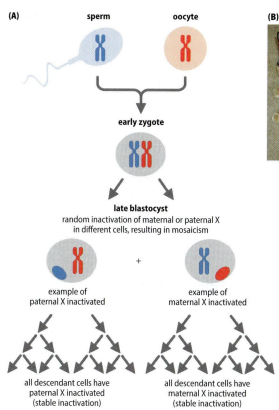

(A)

sperm   oocyte

early zygote

late blastocyst
random inactivation of maternal or paternal X
in different cells, resulting in mosaicism

example of
paternal X inactivated   +   example of
maternal X inactivated

all descendant cells have
paternal X inactivated
(stable inactivation)

all descendant cells have
maternal X inactivated
(stable inactivation)

(B)

Figure 6.18 X-chromosome inactivation.
(A) A randomly chosen X chromosome, either the maternal X or the paternal X, is inactivated in each cell of a 46,XX embryo. Once the choice is made, it is faithfully transmitted through all subsequent rounds of mitosis. Note that in the oogonia of a female, both X chromosomes are active; each has an equal chance of being passed on through the egg. (B) The calico (tortoiseshell and white) cat is heterozygous at an X-linked coat color locus. One allele specifies a black coat color, the other orange. The different color patches reflect clones in which different X chromosomes are inactivated. The white patches are the result of an unrelated coat color gene. Calico cats are always female, except for occasional XXY males. (Adapted from Migeon BR [1994] *Trends Genet* 10:230–235; PMID 8091502. With permission from Elsevier.)

seem to spread along the inactive X to initiate gene silencing along the length of the chromosome. As a result, the inactive X carries modifications typical of heterochromatin (H3K9me2, H3K9me3, H3K27me3, unmethylated H3K4, deacetylated H4, and frequent replacement of histone H2A by the macro-H2A histone variant). In differentiated cells that have already undergone X-inactivation, loss of *XIST* does not cause reactivation. That is, *XIST* is needed to establish X-inactivation but not to maintain it.

The mechanism of X-inactivation remains poorly understood. In addition to *XIST*, there are multiple other longer ncRNAs within the XIC. Several of them are known to have roles in the X-inactivation mechanism in mouse, but the organizations of the human XIC and its mouse counterpart, Xic, are rather different.

Escaping X-inactivation

A few genes on the X have active counterparts on the Y, notably in the terminal pseudoautosomal regions (but also in some other areas—see Figure 5.7). X-inactivation is therefore not a blanket inactivation of the entire chromosome, because no dosage compensation is needed for genes on the X that have functional equivalents on the Y. However, unlike in mouse, in which only a small number of genes escape X-inactivation and are not coated by *Xist* RNA, about 15% of genes on the human X somehow escape inactivation.

Epigenomes and dissecting the molecular basis of epigenetic regulation

The **epigenome**, the total collection of epigenetic settings across a genome, is a property of individual cell types and is both variable and flexible. Epigenetic marks such as cytosine methylation and histone modification patterns are not always associated with changes in transcriptional activity, and they are also not as stable as once thought.

Many can be faithfully propagated through mitosis from cell to daughter cells, but some others are much less stable. Cytosine methylation patterns within some coding sequences can change during the cell cycle, for example. Histone acetylation is very dynamic—sometimes a histone acetylation pattern can change only 2 hours after the acetyl group was first deposited.

We know that extensive reversal of epigenome settings is possible in mammals. A striking demonstration came with the birth of Dolly, a sheep that was the world's first cloned mammal. Dolly was produced by first removing the nucleus from a sheep oocyte and then transferring into the enucleated oocyte a nucleus taken from a fully differentiated uterine epithelial cell from another sheep. The cytoplasm of the oocyte reprogrammed—and effectively reversed—the epigenetic settings of the introduced nucleus, so that the nucleus mimicked the unspecialized nucleus of a zygote. Cellular differentiation is therefore clearly reversible in mammals.

The International Human Epigenetic Consortium (IHEC)

Our knowledge of the mechanics of epigenetic regulation remains limited. The IHEC (http://www.ihec-epigenomes.org/) was launched in 2010 to begin obtaining genomewide data on the epigenetic settings in human cells. An initial goal has been to characterize 1000 epigenomes in 10 years from different human tissues and individuals, with prospects for translating discoveries into improved human health.

The project has, however, been controversial. The epigenome is both variable between different tissues and cells and also plastic, being influenced by the environment, developmental stage, aging, and so on. The genome, by contrast, is a very consistent and stable property that shows very few differences between the nucleated cells in an individual, and even between different individuals of a species. We can therefore speak of a human genome with confidence, whereas a human epigenome is cell-specific and even within a cell it can vary according to circumstance.

There are also significant practical limitations and technical difficulties. Many cell types are not readily accessible for study (and the epigenetic settings in available surrogate cultured cells may not be faithful representatives of natural cells), and there is a clear need to improve the quality of antibody detection of many types of chromatin modification (which can be variable).

6.3 Abnormal Epigenetic Regulation in Mendelian Disorders and Uniparental Disomy

Abnormal regulation of how our genes and other functional DNA sequences work can arise in different ways. Changes in DNA sequence may affect how DNA sequences work without necessarily causing any great change in their chromatin environment. In Chapter 7 we look at how disease arises directly as a result of altered base sequences and copy number variation. In this section we are largely concerned with abnormal epigenetic regulation. That occurs in rare cases when two copies of the same chromosome are abnormally inherited from just one parent. In addition, some genetic disorders show abnormal epigenetic regulation.

Principles of epigenetic dysregulation

An abnormal epigenetic change (**epimutation**) at one or more loci can be the immediate cause of pathogenesis in Mendelian disorders that show abnormal epigenetic regulation. However, the *primary* event is often a genetic mutation at a defined locus that may be one of several types. It may be a gene that makes a protein or RNA that controls epigenetic modifications at other genes located elsewhere in the genome. It may be a *cis*-acting regulatory sequence that regulates epigenetic modifications of neighboring genes. In each case the epimutations determine the disease phenotype; because they lie downstream of a primary genetic event, they are often classified as *secondary epimutations* (**Figure 6.19**).

By contrast, *primary epimutations* can arise without any change to the base sequence. Instead, a chromatin state is reprogrammed, for example by some environmentally induced change, in a way that changes epigenetic controls (changes in metabolic factors may affect DNA methylation or histone modification states, and so on). Primary epimutations may be important in complex disease, and in Chapter 8 we examine the roles of both genetic and epigenetic factors in complex disease. Epigenetic

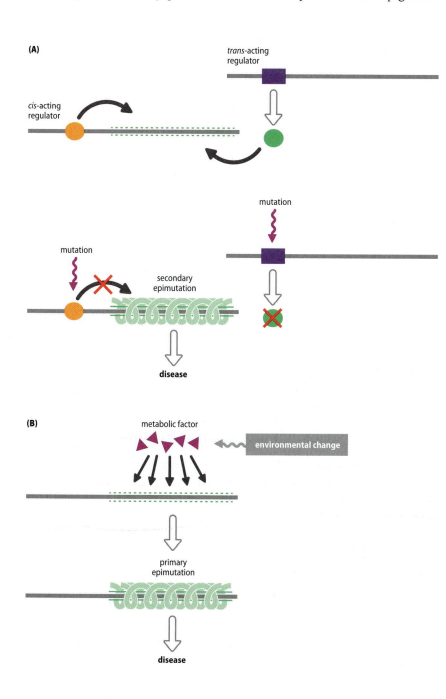

Figure 6.19 Primary and secondary epimutations. (A) Secondary epimutations arise through a change in chromatin state initated by mutation at a *cis*-acting or *trans*-acting epigenetic regulator. Alterations to the regulators cause a change in chromatin state—in this case from a transcription-permissive environment (upper panel) to a repressive heterochromatic environment (lower panel). (B) Primary epimutations can effect a change in chromatin state without any change in the base sequence. Here, we imagine that an environmental change has changed the concentration of some metabolic factor that is important in DNA or chromatin modification, causing the change in chromatin state.

factors have a particularly important role in cancer, and we consider these separately in Chapter 10. Here we focus on epigenetic dysregulation that arises through abnormal chromosome segregation or that is a feature of certain Mendelian disorders.

'Chromatin diseases' due to mutations in genes specifying chromatin modifiers

As detailed above, epigenetic marks are inscribed by a series of chromatin 'writers,' enzymes that methylate DNA or add different types of chemical group to defined amino acids on core histones. They can be recognized and bound by specific protein 'readers' or be removed by specific enzymes ('erasers').

Individual genes that produce a chromatin writer, eraser, or reader can potentially regulate very many different genes across the genome, and a mutation in a gene of this type may result in heritable abnormal chromatin organization at multiple loci. In some cases the disruption to normal gene regulation might be incompatible with life because many of the target genes of chromatin modifiers are important in early development. Some disorders, however, do arise through mutations at chromatin modifier loci. These so-called 'chromatin diseases' typically result in developmental disorders that can vary in phenotypes but are usually accompanied by mental retardation (Table 6.6). Affected individuals typically do not reproduce, usually presenting as sporadic (isolated) cases.

Rett syndrome: a classical chromatin disease

Rett syndrome (PMID 20301670) is a progressive X-linked neurodevelopmental disorder that affects girls almost exclusively. Development typically occurs normally for a period of 6–18 months, after which affected children regress rapidly for a while in language skills (often leading to loss of speech) and in motor skills, before the condition stabilizes over the long term.

During the period of regression, there is loss of purposeful use of the hands, and affected children go on to show stereotypical movements, notably characteristic hand-wringing. Growth retardation during this period results in microcephaly. Mental retardation can be severe in Rett syndrome, seizures are common, and autistic features are often seen. The condition varies in course and severity—some individuals never learn to talk or walk; others retain some abilities in these areas.

Table 6.6 Examples of chromatin diseases, disorders that arise from mutation in a chromatin modifier gene. [a]MR, mental retardation.

CLASS AND TYPE OF CHROMATIN MODIFIER		GENE	ASSOCIATED DISEASE (REFERENCE)	PHENOTYPE		
				Developmental	MR[a]	Other
Writers	DNA methyl transferase	DNMT3B	ICF syndrome (OMIM 242860)	facial anomalies	variable	immunodeficiency; centromeric instability
	histone acetyltransferase	CREBBP or EP300	Rubinstein–Taybi syndrome (PMID 20301499)	characteristic facial features; digit anomalies	yes	
Erasers	histone lysine demethylase	KDM5C	Claes–Jensen type of syndromic, X-linked mental retardation	variable—often mildly dysmorphic facial features; microcephaly	yes	
Readers	meCG-binding protein	MECP2	Rett syndrome (PMID 20301670)	see text on this page	variable	see text on this page
	chromatin remodeler	ATRX	α-thalassemia X-linked mental retardation syndrome (PMID 20301622)	cranial, facial, skeletal, and genital abnormalities; developmental delay and microcephaly	yes	thalassemia (ATRX regulates the α-globin genes among many others)

Classical Rett syndrome results from mutations in the X-linked *MECP2* gene that encodes the regulatory MECP2 protein (which recognizes and binds 5-methylcytosine and is especially important in neuron maturation). The phenotype is partly determined by the X-inactivation pattern: inactivating the normal X in a relatively high proportion of neurons would be expected to result in a particularly severe phenotype.

Failure to produce any functional MECP2 protein was initially expected to be lethal and would explain why affected males are so rarely seen. Affected males can occur as a result of a post-zygotic inactivating mutation, or have certain missense *MECP2* mutations and severe neonatal encephalopathy.

Disease resulting from dysregulation of heterochromatin

Epigenetic regulation causes distinctive patterns of heterochromatin and euchromatin to form in our cells. Heterochromatin is first formed at nucleation sites, consisting of either repetitive DNA or silencer elements, and can then expand across long distances on a chromosome, even a whole chromosome, as in X-chromosome inactivation. Heterochromatin spreading involves converting open chromatin to condensed transcriptionally silent chromatin and is facilitated by communication between nucleosomes.

To avoid silencing essential genes, cells have evolved different mechanisms to limit the spread of heterochromatin. One such mechanism depends on *barrier elements*, a type of insulator, that are able to protect genes from their surrounding environment. Barrier elements can include sequences that are selected to be comparatively nucleosome-free to provide a break in the nucleosome chain.

Altered heterochromatic states can impair normal gene expression in two quite different ways. Sometimes active genes are inappropriately exposed to heterochromatin controls and are silenced. An alternative form of dysregulation involves a reduction in heterochromatin and loss of gene silencing.

Inappropriate gene silencing

Aberrant heterochromatin regulation can silence genes that are normally meant to be expressed. A long-range **position effect** can mean that a gene is relocated to a position very close to constitutive heterochromatin (by a chromosome translocation or inversion, for example). In these cases, the boundary between euchromatin and heterochromatin can be reset, and the gene is silenced by heterochromatin spreading (**Figure 6.20**).

Some special types of mutation can also induce heterochromatin formation within or close to a gene and so silence it. That can happen in the case of very large expansions of noncoding triplet repeats that occur in some recessively inherited disorders such as fragile X-linked mental retardation and Friedreich's ataxia. We consider this type of abnormal heterochromatin within the context of disease due to unstable oligonucleotide repeat expansion in Section 7.3.

Heterochromatin reduction

A primary function of some tumor suppressor genes such as *BRCA1* is to maintain the integrity of constitutive heterochromatin. As described in Chapter 10, mutations in these genes can result in a loss of heterochromatin organization; reduced centromeric heterochromatin leads to mitotic recombination and genome instability.

Some mutations causing heterochromatin reduction and inappropriate gene activation also result in inherited disorders. A classic example is

Figure 6.20 Heterochromatin spreading after removal of a barrier element. *A barrier element* protects genes in the euchromatin region from being silenced by adjacent heterochromatin (shown here in green). Large-scale rearrangements, such as the inversion shown here, can relocate the protective barrier element so that it no longer separates gene *A* from a neighboring heterochromatic region. This allows heterochromatinization of what had been a euchromatic region, resulting in gene silencing (a *position effect*).

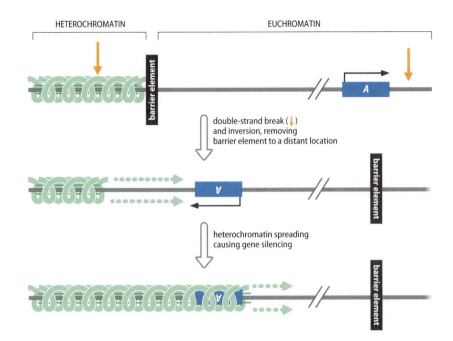

faciocapulohumeral dystrophy (FSHD), the third most common form of muscular dystrophy. This disorder occurs as a result of simultaneous inheritance of two genetic variants: one that causes a reduction in heterochromatin at an array of tandem macrosatellite repeats at 4q35, close to the telomere, and a second variant that creates a polyadenylation site close to the most telomeric of the repeats. The combination of the two variants allows inappropriate expression of the DUX4 transcription factor, which is normally silenced in somatic cells and has been thought to be toxic to muscle cells. One further complication is that the disorder is genetically heterogeneous: in the common FSHD1 form the reduction in heterochromatin is caused by copy number variation within the macrosatelite array at 4q35; in a second form, FSHD2, the reduction in heterochromatin is caused by mutation in a gene on chromosome 18 that regulates DNA methylation (**Box 6.4**).

Uniparental disomy and disorders of imprinting

Recall that the sperm and egg genomes each carry epigenetic marks that are rather different from each other. Although the great majority of the gametic epigenetic marks are then erased in the early embryo, remaining imprints are retained in our somatic cells. We have more than 100 classically defined imprinted genes, and many of them have important roles in early development. In some cases, the maternal allele is consistently silenced or preferentially silenced (that is, monoallelic expression in some cell types, but biallelic expression in others); for other genes, it is the paternal allele that is consistently or preferentially silenced.

Occasional cases of uniparental diploidy (producing an androgenetic or gynogenetic embryo as shown in Figure 6.17) can occur. They are invariably lethal in the early embryo (each embryo of this kind fails to express multiple imprinted genes needed for fetal development). Sometimes, however, abnormal regulation of imprinted genes is confined to genes on a single chromosome and results in a developmental disorder. This can arise by a change in DNA sequence at or near the imprinted gene locus or by abnormal epigenetic regulation of the imprinted gene that may result as a downstream effect, often because of other genetic changes.

Uniparental disomy arises when a zygote develops in which both copies of one chromosome originated either from the father or from the

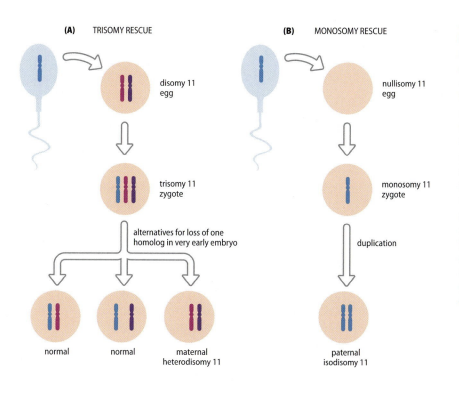

Figure 6.21 Uniparental disomy can arise by post-zygotic trisomy rescue or monosomy rescue. (A) Shown in the center is one type of trisomic zygote (with two maternal homologs plus one paternal chromosome, in this case chromosome 11). Trisomy 11 is lethal, but the trisomy can be corrected in the very early embryo by the loss of a chromosome 11 from one embryonic cell that then has a growth advantage and goes on ultimately to form an individual with the correct number of chromosomes. The disomic cell (and individual) may be normal (one paternal 11 plus one maternal 11) or have the two maternal chromosome 11 homologs (uniparental *heterodisomy*). (B) Monosomy rescue can occur by chromosome duplication, but in this case the result is uniparental *isodisomy* (the two chromosomes are identical, not homologs).

mother. It occurs most often after a trisomic conceptus is first formed with two chromosome homologs from one parent and a single chromosome copy from the other parent; loss of the latter chromosome very early in development results in a heterodisomy (**Figure 6.21A**). The alternative is monosomy rescue, which results in isodisomy (with two identical copies of a chromosome—see Figure 6.21B). As described in the next section, uniparental disomy can result in a disorder of development if the chromosome happens to contain imprinted genes that are important in development.

Abnormal gene regulation at imprinted loci

At an imprinted locus only one of the two parental alleles is consistently expressed (in at least some tissues), and alteration of the normal pattern of monoallelic expression can result in disease. Certain imprinted genes are important in fetal growth and development, so in these cases abnormal expression often results in recognizable developmental syndromes. Sometimes the normally expressed allele is not present or is defective, and deficiency of a gene product causes disease. In other cases, disease can be due to overexpression of a dosage-sensitive gene (**Table 6.7**).

Analysis of human imprinting disorders has helped us understand the underlying gene regulation. For example, a cluster of at least 10 imprinted genes in the subtelomeric 11p15.5 region has been well studied because of associations with Beckwith–Wiedeman syndrome (PMID 20301499) and many cases of Silver–Russell syndrome (PMID 20301568). The gene cluster has two different imprinting control regions, ICR1 and ICR2.

Figure 6.22A shows some key 11p15.5 genes regulated by ICR1 and ICR2. Both ICR1 and two nearby enhancer elements regulate *IGF2* (insulin growth factor type 2; paternally expressed) and *H19* (which makes a maternally expressed ncRNA). ICR2 regulates the *KCNQ1* gene (which makes a maternally expressed potassium channel), the *KCNQ1OT1* antisense RNA transcript (*KCNQ1* opposite strand transcript 1; paternally expressed), and *CDKN1C* (a suppressor of cell proliferation; maternally expressed).

BOX 6.4 Heterochromatin Reduction, Loss of Gene Silencing, and Digenic Inheritance in Facioscapulohumeral Dystrophy.

Facioscapulohumeral dystrophy (PMID 20301616) is dominantly inherited and has a highly variable phenotype. Muscle weakness can appear from infancy to late life (but typically in the second decade) and generally involves first the face and the scapulae followed by the foot dorsiflexors and the hip girdles. Typical features are striking asymmetry of muscle involvement from the left to right sides of the body (**Figure 1**), with sparing of bulbar, extraocular, and respiratory muscles.

Molecular studies show that the condition occurs through the simultaneous inheritance of two DNA variants. One of the DNA variants causes an epigenetic change reducing the amount of heterochromatin at a tandemly repeated macrosatellite sequence, D4Z4, located at 4q35 (**Figure 2A**; macrosatellite is used to describe non-centromeric satellite DNA in which the repeat unit is often from 200 bp to

several kilobases long and the array size is large). The second DNA variant results in the presence of a polyadenylation signal just downstream of the most telomeric D4Z4 repeat.

Individual D4Z4 repeat units are 3.3 kb long and each contains a copy of the *DUX4* retrogene, a heterochromatic gene that is silenced in somatic cells (but strongly expressed in germ cells of the testis, where it encodes a transcriptional activator). Normal individuals have long tandem arrays of D4Z4 repeats, with usually 11–100 repeats. The long arrays normally form heterochromatin by a mechanism that involves CpG methylation regulated by a protein made by the *SMCHD1* gene on chromosome 18 (Figure 2A).

FSHD1

In the most common form of the disease, affected individuals have 1–10 tandem D4Z4 repeats, and the contraction in the number of repeats can result in a significant reduction in the amount of heterochromatin at the D4Z4 array (Figure 2B). Reduced D4Z4 copy number is not by itself sufficient to cause disease: some normal individuals have 1–10 copies of D4Z4 (Figure 2C). An additional genetic variant is required, a polyadenylation site that lies close to the most telomeric D4Z4 repeat: affected individuals have the poly(A) signal, but normal individuals lack it.

The reduction in heterochromatin allows the internal *DUX4* genes within the repeat units to be transcribed. However, the *DUX4* transcripts are unstable if they are not polyadenylated, and in that case they are rapidly degraded. The additional presence of the polyadenylation site just downstream of the last repeat allows transcripts from the last *DUX4* sequence to undergo splicing and acquire the polyadenylation

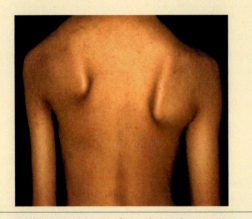

Figure 1 Striking asymmetry of muscle involvement in facioscapulohumeral dystrophy. Courtesy of the FSH Society.

Both ICR1 and ICR2 are activated when hypomethylated, and suppressed when extensively methylated. But they have opposite parental imprints: a maternally inherited chromosome 11 has a hypomethylated ICR1 and extensively methylated ICR2, but a paternal chromosome 11 has an extensively methylated ICR1 and a hypomethylated ICR2. They also use rather different control mechanisms (see Figure 6.22B).

Disease can result from significant changes in the methylation patterns or the base sequences of the ICRs and key imprinted genes. The most frequent cause of Silver–Russell syndrome is hypomethylation of ICR1 on both chromosome 11s so that both *IGF2* alleles are silenced. This can often happen by maternal 11 disomy or duplication of maternal 11p (see Table 6.7).

BOX 6.4 (continued)

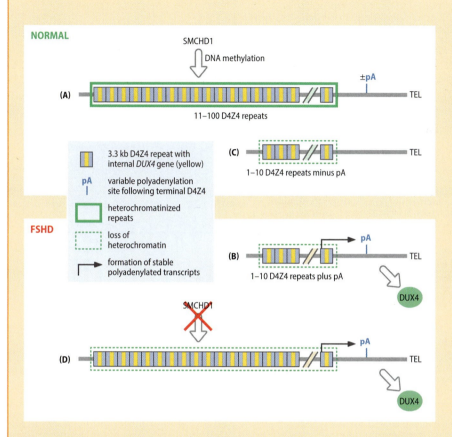

Figure 2 Heterochromatin reduction and inappropriate activation of the heterochromatic *DUX4* retrogene in FSHD. (A) A normal chromosome 4 with a heterochromatinized array of 11–100 D4Z4 repeats. A variable polyadenylation site is located adjacent to the last D4Z4 repeat. (B) In FSHD1, the D4Z4 array is reduced in size to 1–10 repeats, causing a marked decrease in heterochromatin; the downstream polyadenylation site is present, allowing both transcription and translation of the last *DUX4* sequence. (C) In the absence of the downstream polyadenylation signal, the *DUX4* sequence cannot produce stable transcripts even when the heterochromatin is decreased. (D) In FSHD2 the downstream polyadenylation sequence is present, together with a long D4Z4 array that nevertheless has decreased heterochromatin because of a failure to produce the SMCHD1 methylation regulator.

signal. The stable, polyadenylated transcripts go on to be processed and translated to produce DUX4.

FSHD2

In this case the reduction in heterochromatin does not occur by copy number variation: the affected individuals can have long arrays of D4Z4 repeats. Instead, mutation in the *SMCHD1* gene at 18p11 means that the long D4Z4 arrays are not heavily methylated, and the amount of heterochromatin declines (Figure 2D). That allows *DUX4* transcription; when the polyadenylation variant downstream of the terminal repeat is present, stable DUX4 transcripts are produced that allow the ectopic expression of DUX4.

Beckwith–Wiedemann syndrome is marked by fetal overgrowth. It can occur when ICR1 is extensively methylated on both chromosome 11s so that both *IGF2* alleles are expressed, causing excessive growth. It can also occur when ICR2 is hypomethylated on both chromosome 11s, causing silencing of both alleles of the growth-restricting gene *CDKN1C*. Paternal disomy 11 is a common cause.

Another well-studied imprinted gene cluster located at 15q11–12, which is associated with Angelman syndrome and Prader–Willi syndrome, has a bipartite imprinting control region that regulates all the genes in the cluster (**Box 6.5**).

Imprinting and assisted reproduction

Another aspect of imprinting disorders relates to concerns about apparently increased frequencies of these disorders in births in which assisted

DISORDER	DIAGNOSTIC CLINICAL FEATURES[a]	MOLECULAR BASIS OF IMPRINTING DISORDER	CAUSE	
			UPD[b]	Others include[c]
Pathogenesis due to underexpression of imprinted genes				
Prader–Willi syndrome	DD; low birth weight; hypotonia; hyperphagia	silencing/lack of active allele for imprinted genes at 15q11.2, including multiple *SNORD116* (*HB11-85*) snoRNA genes (see Box 6.5)	mat.15, ~25%	Δpat.15q11–13, ~70%
Angelman syndrome	DD (severe); no speech; epilepsy; ataxia	silencing/lack of active allele for imprinted *UBE3A* gene at 15q11.2 (see Box 6.5)	pat.15, ~5%	Δmat.15q11–13, ~75%
Silver–Russell syndrome	IUGR; faltering growth; short stature	silencing/lack of active allele for *IGF2* at 11p15.5 (see Figure 6.22) or for *MEST* (*PEG1*) at 7q31–32 (*IGF2* and *MEST* are maternally imprinted)	mat.11	loss of pat. ICR1 methylation, ~35–50%
			mat.7, ~8%	
Pathogenesis due to overexpression of imprinted genes and/or fetal growth promotion				
Beckwith–Wiedemann syndrome	macrosomia/overgrowth; macroglossia; umbilical defect	biallelic expression of *IGF2* at 11p15.5 (normally silenced on maternal 11) and/or biallelic expression of a ncRNA that suppresses a growth-restricting gene, *CDKN1C*; see Figure 6.22)	pat.11	loss of mat. ICR2 methylation, ~50%; gain of mat. ICR1 methylation, ~5%; *CDKN1C* mutation, ~5%
Transient neonatal diabetes	IUGR; neonatal diabetes with remission	biallelic expression of PLAGL1, a regulator of insulin secretion, at 6q24 (normally silenced on maternal 6)	pat.6	30%

Table 6.7 Examples of imprinted disorders of early childhood. [a]DD, developmental delay; IUGR, intrauterine growth retardation. [b]UPD, uniparental disomy; pat., paternal; mat., maternal. [c]ICR1, ICR2, imprinting control regions 1 and 2.

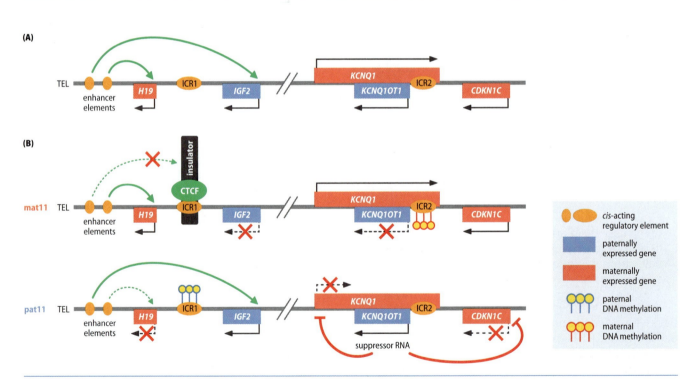

Figure 6.22 Imprinting control mechanisms in the 11p15 imprinted cluster. (A) Map of terminal genes and imprinting control regions (ICR1, ICR2) in the human 11p15 imprinted gene cluster. The gap (shown as slashes) between the insulin-like growth factor 2 gene (*IGF2*) and *KCNQ1* is about 600 kb and contains at least five other imprinted genes. Arrows indicate the direction of transcription. TEL, telomere. (B) Regulation of imprinted genes in the 11p15 cluster on maternal and paternal chromosome 11 (mat11 and pat11). ICR1 acts as an insulator. It is hypomethylated on mat11 and bound by the CTCF protein, which recruits other proteins to act as a barrier, blocking enhancer elements close to *H19* from activating the distant *IGF2* gene. On pat11, ICR1 is extensively methylated and is not bound by CTCF, allowing the enhancer elements to preferentially activate *IGF2* instead of *H19*. ICR2 is located in an intron of *KCNQ1* and acts as a promoter for the antisense RNA gene *KCNQ1OT1*, which encodes a *cis*-acting suppressor RNA. ICR2 is hypomethylated on pat11, allowing transcription of the *KCNQ1OT1* suppressor RNA, which inhibits the transcription of the neighboring genes *KCNQ1OT1* and *CDKN1C*. On mat11, ICR2 is extensively methylated, blocking transcription of *KCNQ1OT1* and allowing the expression of the neighboring genes.

BOX 6.5 Abnormal Imprinting Regulation in Angelman and Prader–Willi Syndromes.

Angelman syndrome (PMID 20301323) and Prader–Willi syndrome (PMID 20301505) are rare (1 in 20,000 live births) neurodevelopmental disorders that result from imprinting abnormalities in a cluster of imprinted genes at 15q11–q13. In Prader–Willi syndrome (PWS), affected individuals show mild intellectual disability and hyperphagia leading to obesity. Angelman syndrome phenotypes include severe intellectual disability and microcephaly, and affected individuals are prone to frequent laughter and smiling.

A single, bipartite imprinting control region regulates the whole cluster, which contains very many imprinted genes that are expressed only or preferentially on the paternal chromosome 15, including many small nucleolar RNA (snoRNA) genes, but just two imprinted genes that are prefentially expressed on the maternal chromosome 15 (**Figure 1**). The imprinted snoRNA genes seem to be located within introns of an extended transcription unit that includes a few exons making two proteins (SNURF and SNURPN) plus a large number of noncoding exons. By overlapping *UBE3A* and possibly *ATP10A*,

the very long transcripts might silence these two genes on paternal chromosome 15.

Angelman syndrome and PWS are primarily due to large deletions that remove the same 5 Mb region of DNA including the genes shown in Figure 1, either from maternal 15q11–q13 (Angelman syndrome) or paternal 15q11–q13 (PWS). This region is flanked by low-copy-number repeat sequences that make it inherently prone to instability, as shown in Figure 7.9C.

Angelman syndrome and PWS are both caused by genetic deficiency. In the former case, the key problem is loss or inactivation of the maternal *UBE3A* allele, which makes a ubiquitin-protein ligase (both *UBE3A* alleles are normally expressed in most tissues, but in neurons only maternal *UBE3A* is active). The PWS phenotype, however, is attributable to the deficient expression of different genes normally expressed only on the paternal chromosome 15, including *NDN*, which regulates adipogenesis, and *SNORD116/HBII-85* genes, which make a type of snoRNA that, in addition to its standard snoRNA role, also acts to regulate alternative splicing in some target genes.

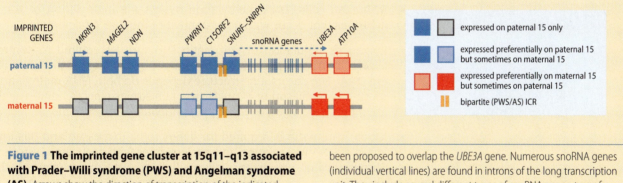

Figure 1 The imprinted gene cluster at 15q11–q13 associated with Prader–Willi syndrome (PWS) and Angelman syndrome (AS). Arrows show the direction of transcription of the indicated genes. The prominent long dashed blue arrow signifies that there is a long transcription unit with multiple noncoding exons that has been proposed to overlap the *UBE3A* gene. Numerous snoRNA genes (individual vertical lines) are found in introns of the long transcription unit. They include several different types of snoRNA genes, two of which are present in multiple copies: *SNORD116* (previously HBII-85) and *SNORD115* (previously HBII-52). ICR, imprinting control region.

reproductive technology (ART) has been employed. *In vitro* fertilization is now well accepted in economically advanced societies, where it accounts for 1–4% of births. Because early embryogenesis is a critical time for epigenetic regulation and is sensitive to environmental factors, ART might impose added stress on embryos that can result in altered epigenetic profiles.

Imprinting disorders are very rare, and statistical support for increased incidence in assisted conception is difficult to achieve. But studies in mice have shown that although intracytoplasmic sperm injection does introduce primary epimutations, they are normally corrected by epigenetic reprogramming in the germ line and are therefore not transmitted to subsequent generations.

Summary

- The regulation of gene expression is mostly genetic, and depends on the base sequence.

- *Cis*-acting regulatory sequences are located on the same DNA or RNA molecule as the sequences they regulate.

- *Trans*-acting gene regulators migrate in the cell to bind target sequences on DNA or RNA molecules. *Trans*-acting proteins bind to targets by using nucleic acid-binding domains; *trans*-acting RNAs often bind by base pairing.

- Gene promoters are composed of multiple short sequences. Core promoter elements are bound by ubiquitous transcription factors; other *cis*-acting regulatory elements are often bound by tissue-specific or developmental-stage-specific regulators.

- RNA splicing is largely dependent on recognition of *cis*-acting RNA sequence elements at splice junctions. Additional splice enhancer and splice suppressor sequences can be located in both introns and exons.

- Multiple different transcripts are produced for almost all of our genes and can give rise to alternative protein isoforms, increasing functional variation.

- Post-transcriptional regulation is often performed by tiny microRNAs. An individual miRNA binds to partly complementary sequences in untranslated regions of multiple target mRNA sequences and down-regulates expression.

- In human cells, DNA methylation means the methylation of some cytosines that occur within the context of the CpG dinucleotide.

- Epigenetic regulation of chromatin structure is required to control access by transcription factors to target sequences in the underlying DNA. The chromatin structure depends on both DNA methylation (CpG) and modification of nucleosomal histones (specific amino acids are methylated, acetylated, or phosphorylated, for example).

- 'Open' chromatin is required for genes to be expressed and has low levels of DNA methylation and nucleosomes with acetylated histones. Chromatin that has extensively methylated DNA and histones lacking acetyl groups is highly condensed, so that gene expression is suppressed.

- DNA methylation and histone modification patterns are prominent examples of epigenetic marks that can be stably inherited from one cell generation to the next. The appropriate chemical groups are added or removed by dedicated enzymes known, respectively, as 'writers' or 'erasers.'

- Methylated CGs and chemically modified amino acids on histones are bound and interpreted by specific proteins ('readers') that induce structural and functional changes in chromatin.

- Chromatin remodeling can involve altering nucleosome spacing at promoters and enhancers (to allow or deny access to transcription factors) and replacing standard histones by variants that promote some function, such as transcriptional activation.

- Long ncRNAs are important in epigenetic regulation in mammalian cells, often working as *cis*-acting regulators.

- In mammals, the DNA of sperm and egg cells is differentially methylated at certain *cis*-acting control sequences, causing differential regulation of maternal and paternal alleles of neighboring genes. Some genes are expressed on maternally inherited DNA only; others are expressed only if paternally inherited.

- X-inactivation means that one of the two X chromosomes in women (and female mammals) is heterochromatinized.

- Barrier elements separate heterochromatin from neighboring euchromatin regions. If they are deleted or relocated by inversions or translocations, the neighboring euchromatin region can be heterochromatinized, causing gene silencing (a position effect).

- Uniparental disomy means that a pair of homologous chromosomes has been inherited from one parent. Disease can result when both chromosomes carry one or more imprinted genes.

Questions

Help on answering these questions and a multiple-choice quiz can be found at www.garlandscience.com/ggm-students

1. In gene regulation, what are the similarities and essential differences between a promoter and an enhancer?

2. Three classes of *cis*-acting regulatory RNA element are important in the basic RNA splicing reaction carried out by spliceosomes. Where are they located, and what are their characteristics?

3. Alternative splicing of RNA transcripts must occasionally happen by accident. Quite often, however, the pattern of alternative splicing is thought to be functionally significant. What types of evidence suggest that alternative splicing can be functionally important?

4. The production of miRNAs uses many components of the cell's RNA interference machinery. What is the natural role of RNA interference in cells?

5. Give brief descriptions of the three principal molecular mechanisms that permit changes in chromatin structure required for epigenetic regulation.

6. A popular theory holds that gene imprinting evolved in mammals because of a conflict of evolutionary interest between fathers and mothers. What are the essential points of this theory, and why does it not explain all observations of imprinted genes in mammals?

7. Although patterns of DNA methylation can be stably inherited after cells divide, patterns of DNA methylation on a gamete transmitted by a parent to a child are altered in the child. Explain how this happens.

8. Rett syndrome is a classic example of a Mendelian chromatin disease in which pathogenesis occurs as a result of altered chromatin states at gene loci that are distinct from the one that is mutated. Explain how this happens. Why are individuals with Rett syndrome almost exclusively female?

Further Reading

Enhancers, Silencers, Insulators, and General Gene Regulation

Gaszner M & Felsenfeld G (2006) Insulators: exploiting transcriptional and epigenetic mechanisms. *Nat Rev Genet* 7:703–713; PMID 16909129.

Kolovos P et al. (2012) Enhancers and silencers—an integrated and simple model for their function. *Epigen Chromatin* 5:1; PMID 22230046.

Latchman DS (2010) Gene Control. Garland Science.

Alternative Splicing and RNA Editing

Kim E et al. (2008) Alternative splicing: current perspectives. *BioEssays* 30:38–47; PMID 18081010.

Tang W et al. (2012) Biological significance of RNA editing in cells. *Mol Biotechnol* 52:1–100; PMID 22271460.

MicroRNAs and Competing Endogenous RNAs

Baek D et al. (2008) The impact of microRNAs on protein output. *Nature* 455:64–71; PMID 18668037.

Memczak S et al. (2013) Circular RNAs are a large class of animal RNAs with regulatory potency. *Nature* 495:333–338; PMID 23446348.

Poliseno L et al. (2010) A coding-independent function of gene and pseudogene mRNAs regulates tumour biology. *Nature* 465:1033–1038; PMID 20577206.

Taulli R et al. (2013) From pseudo-ceRNAs to circ-ceRNAs: a tale of cross-talk and competition. *Nat Struct Mol Biol* 20:541–543; PMID 23649362.

Epigenetics in Gene Regulation (General)

Bonasio R et al. (2010) Molecular signals of epigenetic states. *Science* 330:612–616; PMID 21030644.

Portela A & Esteller M (2010) Epigenetic modifications and human disease. *Nat Biotechnol* 28:1057–1068; PMID 20944598.

Histone Modifications and DNA Methylation

Chen Z & Riggs AD (2011) DNA methylation and demethylation in mammals. *J Biol Chem* 286:18347–18353; PMID 21454628.

Deaton AM & Bird A (2011) CpG islands and the regulation of transcription. *Genes Dev* 25:1010–1022; PMID 21576262.

Smallwood SA & Kelsey G (2012) *De novo* DNA methylation: a germ cell perspective. *Trends Genet* 28:33–42; PMID 22019337.

Suganuma T & Workman JL (2011) Signals and combinatorial functions of histone modifications. *Annu Rev Biochem* 80:474–499; PMID 21529160.

Long Noncoding RNA in Genetic and Epigenetic Regulation

Batista PJ & Chang HY (2013) Long noncoding RNAs: cellular address codes in development and disease. *Cell* 152:1298–1307; PMID 23498938.

Brockdorff N (2013) Noncoding RNA and Polycomb recruitment. *RNA* 19:429–442; PMID 23431328.

Magistri M et al. (2012) Regulation of chromatin structure by long noncoding RNAs: focus on natural antisense transcripts. *Trends Genet* 28:389–396; PMID 22541732.

Mercer TR & Mattick S (2013) Structure and function of long noncoding RNAs in epigenetic regulation. *Nat Struct Mol Biol* 20:300–307; PMID 23463315.

Rinn JL & Chang HY (2012) Genome regulation by long noncoding RNAs. *Annu Rev Biochem* 81:145–166; PMID 22663078.

Genomic Imprinting, X-Inactivation, Heterochromatin Spreading

Barkess G & West AG (2012) Chromatin insulator elements: establishing barriers to set heterochromatin boundaries. *Epigenomics* 4:67–80; PMID 22332659.

Barlow DP (2011) Genomic imprinting: a mammalian epigenetic discovery model. *Annu Rev Genet* 45:379–403; PMID 21942369.

Pinter SF et al. (2013) Spreading of X chromosome inactivation via a hierarchy of defined Polycomb stations. *Genome Res* 22:1864–1876; PMID 22948768.

Sado T & Brockdorff N (2013) Advances in understanding chromosome silencing by the long non-coding RNA Xist. *Phil Trans R Soc B* 368:20110325; PMID 23166390.

Yang C et al. (2011) X-chromosome inactivation: molecular mechanisms from the human perspective. *Hum Genet* 130:175–185; PMID 21553122.

Wutz A (2011) Gene silencing in X-chromosome inactivation: advances in understanding facultative heterochromatin formation. *Nat Rev Genet* 12:542–553; PMID 21765457.

Epigenetic Dysregulation

De Waal E et al. (2012) Primary epimutations introduced during intracytoplasmic sperm injection (ISCI) are corrected by germline-specific epigenetic reprogramming. *Proc Natl Acad Sci USA* 109:4163–4168; PMID 22371603.

Demars J & Gicquel C (2012) Epigenetic and genetic disturbance of the imprinted 11p15 region in Beckwith–Wiedemann and Silver–Russell syndromes. *Clin Genet* 81:350–361; PMID 22150955.

Hahn M et al. (2010) Heterochromatin dysregulation in human diseases. *J Appl Physiol* 109:232–242; PMID 20360431.

Horsthemke B & Wagstaff J (2008) Mechanisms of imprinting of the Prader–Willi/Angelman region. *Am J Med Genet* 146A:2041–2052; PMID18627066.

Ishida M & Moore GE (2012) The role of imprinted genes in humans. *Mol Aspects Med* 34:826–840; PMID 22771538.

Lemmers RJ et al. (2012) Digenic inheritance of an *SCMD1* mutation and an FSHD-permissive D4Z4 allele causes facioscapulohumeral muscular dystrophy type 2. *Nat Genet* 44:1370–1374; PMID 20724583.

Lemmers RJLF et al. (2010) A unifying genetic model for facioscapulohumeral muscular dystrophy. *Science* 329:1650–1653; PMID 23143600.

Sahoo T et al (2008) Prader–Willi phenotype caused by paternal deficiency for the HBII-85 C/D box small nucleolar RNA cluster. *Nat Genet* 40:719–721; PMID 18500341.

Genetic Variation Producing Disease-Causing Abnormalities in DNA and Chromosomes

In Chapter 4 we outlined some basic principles of genetic variation. We covered the different types of genetic variation in our genome: both large-scale changes that make up structural variation, and point mutations. And we related how sequence variation at the level of gene product relates to sequence variation at the level of DNA. Here we begin to focus on the small fraction of human genetic variation that causes disease.

In this chapter we describe the different mechanisms that cause pathogenic DNA changes. The emphasis is on rare, highly penetrant variants that cause single-gene disorders and chromosome abnormalities, but we take a broad view of protein dysregulation in disease. In Chapter 8 we go on to consider variants of low penetrance that confer susceptibility to common complex disorders, and in Chapter 10 we examine how genetic variation, predominantly somatic mutations, contributes to cancer.

In Section 7.1 we give an overview of how genetic variation results in disease. According to the number of nucleotides that are changed (and the number of genes that are affected), we will consider the pathogenic changes as occurring at different levels (different mutation mechanisms can be involved, depending on the size of the DNA change). We describe in Section 7.2 the kinds of pathogenic mutations that typically involve changing just a single nucleotide (the most common type of mutation), or a very few nucleotides; they primarily affect the expression of a single gene. At a different level, as described in Section 7.3, various genetic mechanisms produce moderate- to large-scale mutations that affect from tens of nucleotides to a few megabases of DNA. Some very large-scale mutations and chromosome breaks produce recognizable changes at the chromosome level as detected by light microscopy. We consider these in Section 7.4 together with other types of chromosome abnormality that result from errors in chromosome segregation.

As initially considered in Chapter 5, the link between deleterious mutations and disease is often not straightforward. A deleterious mutation that might be expected to result in a single-gene disorder, according to the principles described in Chapter 5, may produce different degrees of disease severity, or no disease at all. Different factors—mosaicism, interactions with other gene loci, environmental factors, epigenetic factors—can result in variable phenotypes. In Sections 7.5 and 7.6 we consider the pathogenic effects of DNA variants, taking a general view of molecular pathology in Section 7.5, and focusing on altered protein folding and protein aggregation in Section 7.6. Finally, in Section 7.7 we consider the difficulties in correlating genotypes and phenotypes, and we describe how the phenotype of a monogenic disorder can be influenced by various factors including genetic variation at other gene loci and environmental factors.

CHAPTER 7

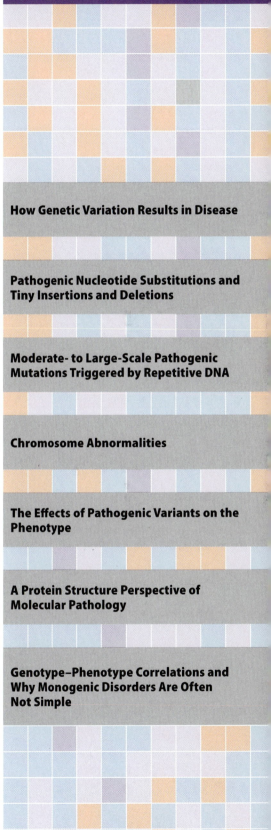

7.1 How Genetic Variation Results in Disease

The great majority of variation in our DNA appears to be without consequence. For the most part, that happens because just a small percentage of our genome is functionally important (the great majority of nucleotides within introns and extragenic DNA can be changed by small mutations without any obvious effect on the phenotype, and are tolerated). A second, and minor, reason is genetic redundancy: some genes are present in multiple, almost identical copies—an inactivating mutation in a single ribosomal RNA gene has no effect because each type of rRNA is made by hundreds of almost identical gene copies.

Pathogenic mutations do not occur haphazardly in our DNA. For example, single nucleotide substitutions, the most common type of pathogenic mutation, are not random: certain types of DNA sequence are more vulnerable to point mutation (mutation *hotspots*). And, as detailed below, different arrangements of repetitive DNA also predispose to different classes of mutation, including many large-scale DNA changes.

The genetic variation that causes disease may do so in two broad ways. First, it may cause a change in the sequence of the gene product. The result may be a loss of function: the mutant gene product may simply be incapable of carrying out its normal task (or may have a significantly reduced ability to work normally; a *hypomorph*). Alternatively, it may acquire an altered function (or occasionally a new function; a *neomorph*) that is harmful in some way (causing cells to die or to behave inappropriately). As we will see, loss of function and gain of function quite often involve a change in protein structure.

A second broad way in which mutations cause disease is by changing the *amount* of gene product made (**Figure 7.1**). Some small-scale pathogenic mutations in protein-coding genes result in unstable mRNAs and failure to make the normal gene product. Other mutations can result in altered gene copy number or adversely affect regulatory sequences controlling gene expression so that too little or too much product is made. That is a problem for a subset of our genes whose expression must be closely controlled, keeping quite tight limits on the amount of gene product made (as detailed below).

Although most pathogenic mutations affect individual genes, some mutations (and chromosome abnormalities) can simultaneously affect multiple genes. Large-scale deletions and duplications, for example, result in a simultaneous change in the copy number of multiple genes, with adverse effects. Additionally, a mutation in a regulatory gene can indirectly have consequences for how the many different target genes controlled by that gene are expressed.

7.2 Pathogenic Nucleotide Substitutions and Tiny Insertions and Deletions

Pathogenic single nucleotide substitutions within coding sequences

A single nucleotide substitution within a coding sequence has the effect of replacing one codon in the mRNA by another codon. There is, however, substantial redundancy in the genetic code: as explained below, all except two amino acids are specified by multiple codons (from two to six).

As a result of the redundancy in the genetic code, a mutated codon quite often specifies the same amino acid as the original codon. A coding

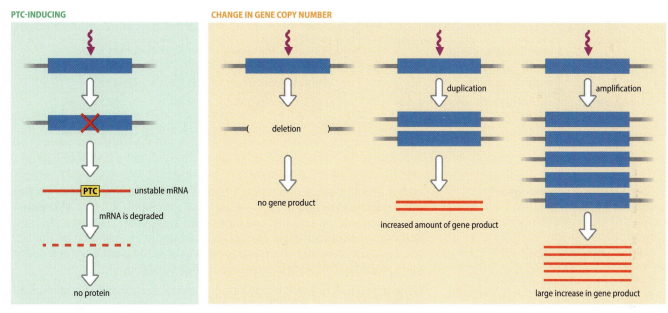

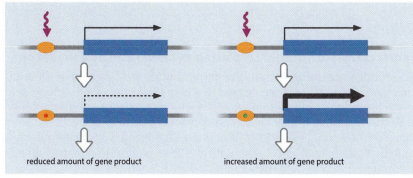

Figure 7.1 Classes of mutations that cause disease by altering the amount of a gene product. Some mutations that cause disease (or increase susceptibility to disease) do not change the sequence of a gene product; rather they change the amount of gene product that is made. Mutations that lead to premature termination codons (PTCs) often activate a pathway that leads to mRNA destruction (as explained in Box 7.1) and lack of gene product. Changes in gene copy number usually involve deletion and duplication events as detailed in Section 7.3, but gene amplification is common in many cancers, as explained in Chapter 10. Mutations in *cis*-acting regulatory sequences, such as enhancers and silencers, can lead to increased or reduced transcription.

sequence substitution such as this—one that does not change an amino acid—is known as a **synonymous** (or **silent**) **substitution**. Because there is no change in amino acid, no change in phenotype might be expected. Nevertheless, as discussed below, a minority of silent substitutions nevertheless cause disease, almost always by altering RNA splicing.

The alternative is a **nonsynonymous substitution**. There are different classes (**Table 7.1**), but the predominant one causes one amino acid to be replaced by another, a **missense mutation**. Missense mutations can sometimes have minimal effects on the phenotype, but they sometimes have adverse effects as described below.

Relative frequencies of silent and amino-acid-replacing substitutions

The relative frequencies of single nucleotide substitutions that are silent and those that cause an amino acid to be replaced (missense mutations) vary according to the base position in a codon. If we first consider the genes in nuclear DNA, 61 codons can specify an amino acid (**Figure 7.2**). If we take an average, two out of every three substitutions at the third base position in a codon are silent; by contrast, 100% of substitutions at the second base position and 184 out of 192 (about 96%) of substitutions of the first base are nonsynonymous.

Consider, for example, a G→A substitution that results in replacement of codon GG**G** by GG**A**. The genetic code shows that both the original GGG

CLASS	DEFINITION	EXAMPLE AND COMMENTS
Missense mutation	an amino-acid-specifying codon is replaced by a codon for a different amino acid	GGA (glycine) is replaced by CGA (arginine). The effect is greatest when the replacement amino acid has very different physiochemical properties
Stop–gain (nonsense) mutation	an amino acid-specifying codon is replaced by a premature stop codon	a G→T substitution causes GGA (glycine) to be replaced by UGA (stop). Results in unstable mRNA or production of a truncated protein (see Box 7.1)
Stop–loss mutation	a natural stop codon is replaced by an amino acid-specifying codon	UGA (stop) is replaced by GGA (glycine). Results in translational 'read-through' so that the first part of the 3′ untranslated region is translated and the protein has an extended C-terminus[a]

Table 7.1 Classes of nonsynonymous mutation. [a]The 3′ untranslated region will have termination codons in all three reading frames. As a result, in the case of a stop–loss mutation the 'read-through' past the normal stop codon usually picks up another termination codon quite quickly, and the extended C-terminus is often not very long. The effect on protein function may often not be great unless the extended C-terminus causes problems for protein folding or protein stability.

codon and the replacement GGA codon specify the amino acid glycine. If the substitution had been G→C or G→T (to give GG**C** or GG**U** codons), the altered codon again would have specified glycine. Like glycine, several other amino acids are exclusively, or largely, determined by the first two bases of a codon—there is flexibility in how the third base of a codon pairs with the 5′ base of the anticodon (*base wobble*).

Genetic redundancy at the first base position is responsible for silent substitutions in some arginine and leucine codons. Thus, codons **A**GA and **A**GG specify arginine, as do codons **C**GA and **C**GG (an A→C or C→A change at the first base position is silent in these cases). Similarly, codons **C**UA and **C**UG specify leucine, as do codons **U**UA and **U**UG.

Conservative substitution: replacing an amino acid by a similar one

Nucleotide substitutions that change an amino acid can have different effects. A major issue is the degree to which the replacement amino acid is similar to the original amino acid, based on properties such as polarity, molecular volume, and chemical composition (see below). Perhaps fewer than 30% have no, or very little, functional significance; the remainder are roughly equally split into those with weak to moderate negative effects on protein function, and those with strongly negative effects.

A nucleotide substitution that replaces one amino acid by another of the same chemical class is a **conservative substitution** and often has minimal consequences for how the protein functions. **Table 7.2** lists different chemical classes of amino acids plus some distinguishing features of individual amino acids (see Figure 2.2 for the chemical structures).

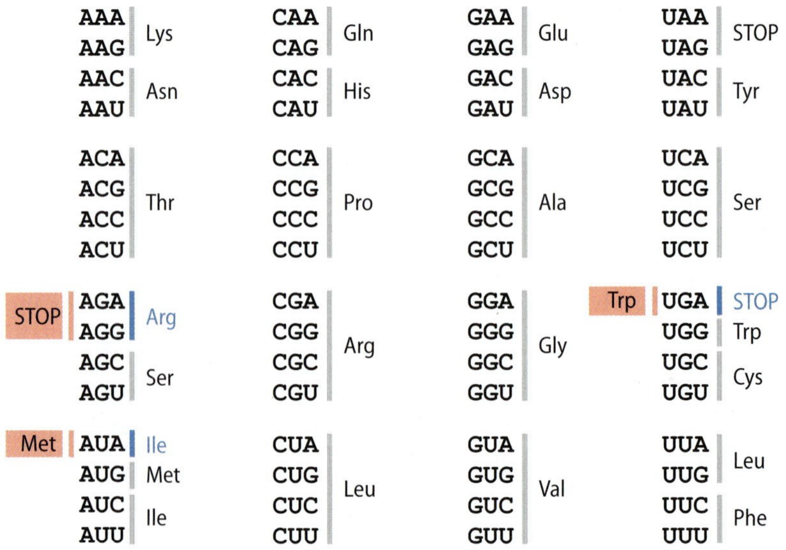

Figure 7.2 The genetic code. Pale gray bars to the right of the codons identify the 60 codons that are interpreted in the same way for mRNA from genes in nuclear and mitochondrial DNA. Four codons—AGA, AGG, AUA, and UGA—are interpreted differently. They are flanked on the right by blue bars showing the interpretation for nuclear genes, and on the left by pale pink bars showing how they are interpreted for genes in mtDNA. For nuclear genes, the 'universal' genetic code has 61 codons that specify 20 different amino acids, with different levels of redundancy from unique codons (Met, Trp) at one extreme, to as much as sixfold redundancy (Arg, Leu, Ser). The remaining three codons—UAA, UAG, and UGA—normally act as stop codons (according to the surrounding sequence context, however, UGA can occasionally specify a 21st amino acid, selenocysteine, and UAG can occasionally specify glutamine). For genes in mtDNA, 60 codons specify an amino acid and there are four stop codons (AGA, AGG, UAA, and UAG).

COMMON FEATURE OF SIDE CHAIN		AMINO ACIDS[a]	COMMENTS
Polar	Basic (positively charged)	Arg (R); Lys (K); His (H)	arginine and lysine have simple side chains with an amino ion ($-NH_3^+$); histidine has a more complex side chain with a positively charged imido group
	Acidic (negatively charged)	Asp (D); Glu (E)	simple side chains ending with a carboxyl ion ($-COO^-$)
	Amide group	Asn (N); Gln (Q)	simple side chains ending with a $-CONH_2$ group
	Hydroxyl group	Ser (S); Thr (T); Tyr (Y)	serine and threonine have short simple side chains with a hydroxyl group; tyrosine has an aromatic ring
	Polar with sulfhydryl (–SH) group	Cys (C)	disulfide bridges (–S–S–) can form between *certain* distantly spaced cysteines in a polypeptide and are important in protein folding
Nonpolar		Gly (G); Ala (A); Val V); Leu (L); Ile (I); Pro (P); Met (M); Phe (F); Trp (W)	glycine has the simplest possible side chain—a single hydrogen atom. Phenylalanine and tryptophan have complex aromatic side chains

Table 7.2 Amino acids can be grouped into six classes according to the chemical properties of their side chains. [a]See Figure 2.2 for structures of amino acids; for physicochemical properties of amino acids see http://www.ncbi.nlm.nih.gov/Class/Structure/aa/aa_explorer.cgi.

Nonconservative substitutions: effects on the polypeptide/protein

Replacing one amino acid by another belonging to a different chemical class may be expected to have more significant consequences. There are different considerations. One factor is whether the individual amino acid has an important role in the function of the protein. It might have a key role at the activation site of an enzyme, for example, or be a critical part of a specific recognition sequence used to bind some interacting molecule (a specific metabolite, protein, or nucleic acid sequence, for example), or have a side chain that needs to be chemically modified (such as by glycosylation or phosphorylation) for functional activity.

Additional factors include the potential effects on protein folding and protein structure (for a brief summary of protein structure, see Section 2.1 and Box 2.2 on page 29). Thus, for example, for thermodynamic reasons, globular proteins usually fold so that nonpolar, uncharged amino acids are buried in the interior and polar amino acids are on the outside, exposed to what is usually a hydrophilic aqueous environment; substitutions that change this pattern may induce incorrect protein folding.

Some amino acids are not tolerated in certain structural elements. Thus, for example, proline cannot be accommodated in an α-helix: if an amino acid is substituted by a proline the α-helix is disrupted. Conversely, certain amino acids have specific structural roles. Glycine, which has the smallest possible side chain (a single hydrogen atom), and proline (the only amino acid in which the side chain loops back to rejoin the polypeptide backbone) are important in allowing the polypeptide backbone to bend sharply. They often have important roles in protein folding—an extreme example is in the triple-stranded helical structure of collagens, in which glycines are required at about every third residue and prolines (and hydroxyprolines) are also extremely frequent.

Cysteine has a unique role in protein folding. The sulfhydryl (–SH) groups on *certain* distantly located cysteines on the same polypeptide may interact to form a disulphide bridge (–S–S–); this can be important in establishing globular domains (such as for the immunoglobulin superfamily proteins in Figure 4.13). Replacing either cysteine by any other amino acid breaks the intrachain disulphide bond; as a result, cysteine is the most conserved amino acid in protein evolution.

In addition to causing simple loss of normal function or incorrect protein folding, missense mutations can also result in some new protein property that is damaging to cells and tissues in some way, or alters their behavior. We consider this aspect in more detail when we discuss the effects of genetic variants in Section 7.5.

Mutations producing premature termination codons and aberrant RNA splicing

A natural termination (stop) codon in mRNA triggers the ribosome to dissociate from the mRNA, releasing the expected polypeptide. However, many pathogenic mutations cause an in-frame premature termination codon to be inserted into a coding sequence, either directly or indirectly.

Nonsense mutations are nonsynonymous substitutions that directly replace an amino-acid-specifying codon by a stop codon. For nuclear DNA, that means a substitution that produces one of three stop codons UAA, UAG, or UGA in the corresponding mRNA. Note that the genetic code for mitochondrial DNA is different—here, a UGA codon specifies tryptophan and both AGA and AGG act as stop codons instead of specifying arginine (see Figure 7.2).

A **frameshift mutation** may indirectly lead to a premature termination codon. If a sequence of coding nucleotides that is not a multiple of three is deleted or inserted, a shift in the translational reading frame results (see Box 2.1 Figure 1 on page 26 for the principle of frameshifts). If a different reading frame is used, an in-frame premature termination codon is quickly encountered. Frameshifts often involve deletions or insertions at the DNA level, but they can also occur from some mutations that produce altered splicing, such as intron retention.

At the DNA level, deletions or insertions of one or two nucleotides in coding DNA is a quite common cause of disease. In addition, some large genes are prone to large intragenic deletions that remove one or more exons, and they too can cause frameshifts, and transposons can occasionally accidentally insert into coding DNA (we consider large deletions and insertions like these in Section 7.3).

The net result of nonsense mutations and frameshifting mutations is that the mRNA is degraded by a mechanism known as nonsense-mediated decay, or it is translated to give a truncated protein (**Box 7.1**). Truncated proteins produced in this way can sometimes interfere with wild-type proteins produced from the normal allele and so affect their function (a *dominant-negative effect*). We consider the implications in Section 7.5.

Pathogenic splicing mutations

Disease-causing mutations that affect RNA splicing are common. The great bulk are in DNA sequences that specify *cis*-acting RNA elements regulating how a specific gene undergoes RNA splicing, and this will be the focus here. Note, however, that disease can occasionally be caused by mutations in genes encoding *trans*-acting regulators of splicing.

As illustrated in Figure 6.4A on page 124, fundamental *cis*-acting regulatory elements that control RNA splicing are located at or close to splice junctions. Point mutations in these sequences often have marked effects on RNA splicing, especially if they change highly conserved nucleotides, such as in GT (GU in RNA) and AG dinucleotides at the extreme 5′ end and 3′ end, respectively, of an intron. That can result in abnormal splicing patterns such as omission of an exon (**exon skipping**), or failure to splice out an intron (*intron retention*)—see **Figure 7.3**.

BOX 7.1 Nonsense-mediated decay as an mRNA surveillance mechanism.

Various RNA surveillance mechanisms monitor RNA integrity, checking for splicing accidents (such as when transcribed intron sequences are inappropriately retained in the mRNA—see Figure 7.3B2) and occasional errors in base incorporation during transcription. These errors frequently give rise to in-frame premature termination codons (PTCs) that could be dangerous—the aberrant transcripts could give rise to truncated proteins that might have the potential to interfere with the function of normal proteins.

To protect cells, an mRNA surveillance mechanism known as **nonsense-mediated decay (NMD)** degrades most mRNA transcripts that have an in-frame PTC. The primary NMD pathway is dependent on RNA splicing (single exon genes escape NMD because they do not undergo RNA splicing). Multisubunit protein complexes, called exon-junction complexes, are deposited shortly before the 3′ end of each transcribed exon during RNA splicing and remain bound at positions close to the exon–exon boundaries in mature RNA.

The first ribosome to bind and move along the mRNA displaces each of these complexes in turn before disengaging from the mRNA at the natural stop codon. If there is an in-frame PTC, however, the ribosome detaches from the mRNA at an early stage; some exon junction complexes remain bound to the RNA, which usually signals mRNA destruction. However, in-frame PTCs within or just before the last exon often escape NMD and are translated to give truncated proteins (Figure 1).

Nonsense mutations, frameshifting insertions and deletions, and certain splicing mutations (such as those that result in retention of an intron) can activate nonsense-mediated decay.

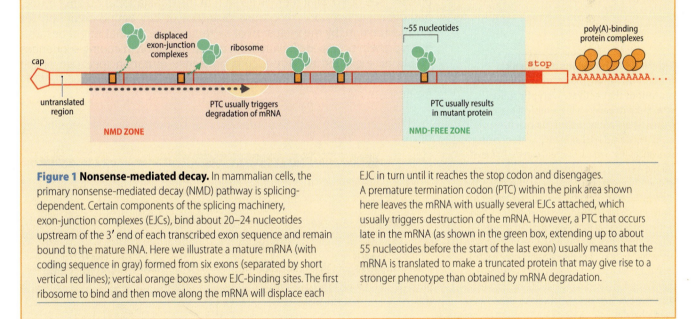

Figure 1 Nonsense-mediated decay. In mammalian cells, the primary nonsense-mediated decay (NMD) pathway is splicing-dependent. Certain components of the splicing machinery, exon-junction complexes (EJCs), bind about 20–24 nucleotides upstream of the 3′ end of each transcribed exon sequence and remain bound to the mature RNA. Here we illustrate a mature mRNA (with coding sequence in gray) formed from six exons (separated by short vertical red lines); vertical orange boxes show EJC-binding sites. The first ribosome to bind and then move along the mRNA will displace each EJC in turn until it reaches the stop codon and disengages. A premature termination codon (PTC) within the pink area shown here leaves the mRNA with usually several EJCs attached, which usually triggers destruction of the mRNA. However, a PTC that occurs late in the mRNA (as shown in the green box, extending up to about 55 nucleotides before the start of the last exon) usually means that the mRNA is translated to make a truncated protein that may give rise to a stronger phenotype than obtained by mRNA degradation.

Pathogenic mutations can also occur in additional *cis*-acting splice regulatory elements, including splice enhancer and splice silencer sequences in exons and introns (see Figure 6.4B). Mutations like this may be less readily identified as pathogenic mutations, and can explain why some synonymous substitutions are pathogenic (**Figure 7.4A**).

By chance, sometimes a sequence may be almost identical to a genuine splice donor or splice acceptor site (a *latent* or **cryptic splice site**), and changing a single nucleotide can cause it to become a novel splice site. Activation of cryptic splice sites produces truncated or extended exons (see **Figure 7.4B, C** for examples).

Abnormal splicing can have variable consequences. For a protein-coding gene, a loss of coding exon sequences (by exon skipping or exon

Figure 7.3 Splice-site mutations can cause exon skipping or intron retention. Exon sequences are enclosed within blue boxes and are represented as a series of dashes, with capital letters to indicate specified individual nucleotides. Intron sequences are represented as gray horizontal lines with lower-case letters to indicate key nucleotides. The dashed red lines indicate the positions where splicing will occur later *at the RNA level* to bring together transcribed exon sequences within RNA transcripts. (A) A normal situation with three exons separated by introns with conserved 5′ GT and 3′ AG terminal dinucleotides. (B1, B2) Possible outcomes of a G→T mutation at the conserved 3′ terminal nucleotide of intron 1, inactivating the splice acceptor site. The next available splice acceptor site (at the 3′ end of intron 2) could be used instead, causing skipping of exon 2 (B1) which will cause a frameshift if the number of nucleotides in exon 2 is not a multiple of three. An alternative outcome is to abandon splicing of intron 1 so that the intron 1 sequence is retained in mRNA, forming part of a large exon that also contains the sequences of the original exons 1 and 2 (B2).

Figure 7.4 Non-obvious pathogenic mutations: synonymous substitutions and mutations that activate a cryptic splice site. Exons (blue boxes) have nucleotides in capital letters; nucleotides of introns are in lower case. Dashed red lines indicate splicing of exon sequences that will occur later on *at the RNA level*. (A) The A→C mutation will lead to the replacement of one arginine codon (CGA) by another (CGC), but is nevertheless pathogenic because it changes the highlighted splice enhancer sequence (see Figure 6.4B) that occurs at the beginning of the exon. (B) A homozygous synonymous C→T substitution identified in exon 16 of the calpain 3 gene causes limb girdle muscular dystrophy (PMID 7670461). It replaces one glycine codon (GGC) by another (GGU), but simultaneously activates a cryptic splice site within this exon (GAGGGCAAAGGC), to become a functional splice donor. The resulting truncation of exon 16 (the last 44 nucleotides of the normal exon 16 are not included in RNA transcripts) causes a shift in the translational reading frame, and a premature termination codon is encountered early in exon 17. (C) Activation of a cryptic splice site within an intron that closely resembles the consensus sequence for a 3′ splice site (*splice acceptor*—see Figure 6.4A; the mutation produces the essential AG dinucleotide, as highlighted here). The resulting aberrant RNA splicing causes a sequence from the 3′ end of intron 1 to be incorporated with the sequence from exon 2 to extend exon 2, which can cause disease by introducing a frameshift or by disturbing protein structure or protein folding.

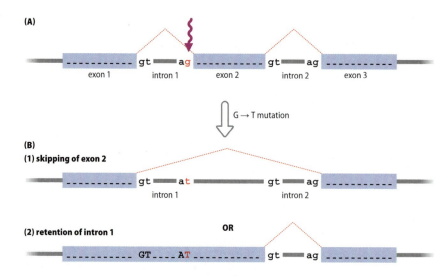

truncation) or a gain of coding exon sequences (by exon extension or intron retention) may result in a frameshift in the translational reading frame at the RNA level. In that case, the introduction of a premature termination codon might induce RNA degradation or produce a truncated protein (see Box 7.1). If there is no shift in the translational reading frame, pathogenesis may occur because of a loss of key amino acids, or, for exon extension, by the inclusion of extra amino acids that might destabilize a protein or impede its function. Intron retention in coding sequences would be expected to introduce a nonsense mutation (because of the comparatively high frequency of termination codons in all three reading frames).

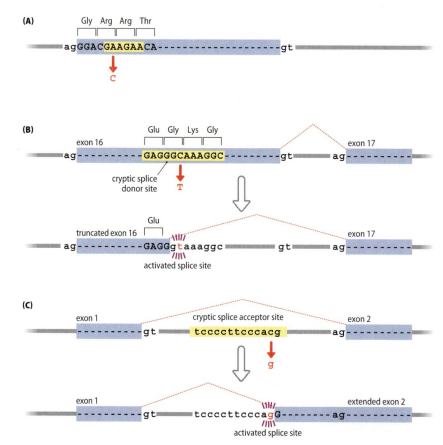

Genesis and frequency of pathogenic point mutations

As detailed in Section 4.1, single nucleotide substitutions often arise as a result of spontaneous chemical degradation of DNA that has not been repaired effectively. Some types of single nucleotide substitution are especially frequent in human DNA. C→T transitions are particularly common because the cytosine in CG dinucleotides is a hotspot for mutation in human (and vertebrate) cells. That happens because the CG dinucleotide is a target for cytosine methylation, and methylated cytosines tend to be deaminated to give thymines (see Figure 4.5 on page 89), which are not easily identified as altered bases by DNA repair systems.

Very small insertions and deletions are often produced by **replication slippage**, an error that typically occurs in DNA replication when a single nucleotide or short oligonucleotide is tandemly repeated. During DNA replication the nascent strand occasionally mispairs with the parent DNA strand (the mispairing is stabilized by base pairing between misaligned repeats on the two strands—see Figure 4.8, page 96, for the mechanism). Arrays with multiple tandem repeats are particularly susceptible to replication slippage; simply by chance, coding sequences occasionally have sequences with such repeats. For example, in about one out of four occasions, on average, two consecutive lysines in a protein are specified by the hexanucleotide AAAAAA. Any run of consecutive nucleotides of the same type means a significantly increased chance of replication slippage—in this case the daughter strands are liable to have five or seven A's, causing a frameshifting deletion or insertion.

Mutation rates in the human genome

Comprehensive genome sequencing in family members indicates that the genomewide germ-line nucleotide substitution rate is 10^{-8} per nucleotide per generation. That equates to about 30 *de novo* nucleotide substitutions on average in the 3 Gb haploid genomes inherited from each parent.

The mutation frequency varies across chromosomes and genes. Some gene-associated features make them more likely mutation targets. Genes are GC-rich and have a higher content of the CG dinucleotide; the cytosine in CG dinucleotides is a mutational hotspot, undergoing C→T transitions at a rate that is more than 10 times the background mutation rate.

The mitochondrial genome is extremely gene-rich, and the mutation rate is many times higher than in the nuclear genome. The mitochondrial genome is vulnerable, possibly because the great majority of reactive oxygen species are produced in mitochondria (see Section 4.1). Close proximity to these dangerous radicals results in much more frequent damage to the DNA, which is devoid of a protective chromatin coating, unlike nuclear DNA. Unrepaired DNA replication errors can also be significant in mtDNA.

Total pathogenic load

Only a small fraction of the novel changes that arise in the genomes we inherit from our parents is likely to be pathogenic, but because our parents are carriers of previously generated mutations our genomes contain many deleterious mutations. As yet there is no easy way to identify the total pathogenic load—all pathogenic mutations—within a genome.

The 1000 Genomes Project has shown that, depending on our ethnic background, each of us carries about 100 mutations that would be expected to result in loss of gene function (with an average of 20 genes that are homozygously inactivated), plus about 60 missense variants that severely damage protein structure. One prediction is that the average person might have over 400 damaging DNA variants. That might seem an impossibly high load of pathogenic mutations but many of these mutations are common variants in non-essential genes, such as the *ABO* blood group gene.

Effect of parental age and parental sex on germ-line mutation rates

Increased parental age often correlates with increased frequency of genetic disorders. We consider the maternal age effect in trisomy 21 in Section 7.4. For small-scale mutations there is often a higher frequency of *de novo* mutation in the male germ line, and paternal age effects can be apparent, as described below.

The frequency of *de novo* mutation can be expected to be high in gametes that have undergone many cell divisions since originating from the zygote through a **primordial germ cell** (the cells that are set aside in the early embryo to give rise to the germ line). That happens because DNA replication precedes each cell division, and mutations often arise as a result of uncorrected errors in DNA replication. Two meiotic cell divisions are required to form oocytes and sperm cells, but the number of preceding mitotic cell divisions required to produce the first meiotic cells is very different between the two sexes. All the egg cells that will be available to a woman are formed before birth. By contrast, after the onset of puberty in men, sperm are continuously being formed by the division of spermatogonial stem cells. The number of cell divisions required to produce gametes is therefore higher in men, and especially so in older fathers (**Figure 7.5**).

Paternal-age-effect disorders and selfish spermatogonial selection

A small group of exceptional congenital disorders occur spontaneously at remarkably high apparent rates, reaching 1 in 30,000 births for

Figure 7.5 Sex differences and paternal age differences in the number of cell divisions required to make gametes. The numbers shown represent the number of cell divisions that have been completed en route from a human zygote to gametes. Sperm and egg cells are generated as a result of two rounds of meiotic cell division (red arrows) preceded by multiple mitotic cell divisions (gray arrows). In females, all of the roughly 22 mitotic divisions required to get to the first meiotic cell are accomplished before birth, and indeed part of meiosis I has been completed by then but is then suspended until activated by ovulation. No matter how old mothers are, a total of about 24 cell divisions separate zygote from egg cells. Males are different because gametogenesis continues throughout adult life. About 30 cell divisions separate the zygote from the spermatogonial stem cell that is used to make the first sperm cells at the onset of puberty. From spermatogonial stem cell to gamete takes four mitotic divisions and then two meiotic divisions. So, at the onset of male puberty gametes are formed that have gone through 30 + 4 + 2 divisions = 36 cell divisions. Thereafter, spermatogonial stem cells divide every 16 days or so (or about 23 times per year). If we take an average age of 14 years, say, for the onset of male puberty, sperm from a 64-year-old man are formed by a process that has involved 34 + (23 x [64 − 14]) mitotic divisions plus two meiotic divisions, or a total of 1186 cell divisions.

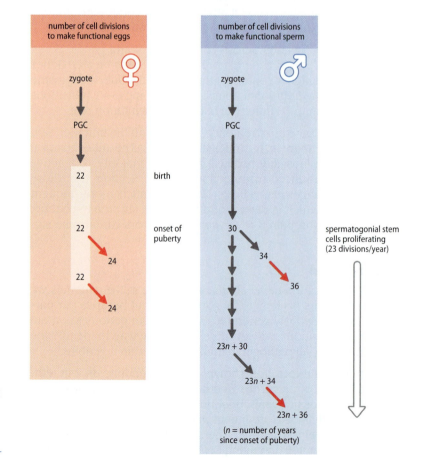

DISORDER	GENE REVIEWS PMID[a]	GENE	MUTATION/ AMINO ACID CHANGE	ESTIMATED BIRTH PREVALENCE FOR NEW MUTATIONS	PARENTAL ORIGIN OF MUTATION
Apert syndrome	20301628	FGFR2	p.Ser252Trp p.Pro253Arg	~1/65,000	100% paternal
Crouzon/Pfeiffer syndrome	20301628	FGFR2	>50 mutations	~1/50,000– 1/100,000	100% paternal
Achondroplasia	20301331	FGFR3	p.Gly380Arg[b]	1/30,000	100% paternal
Muenke syndrome	20301588	FGFR3	p.Pro250Arg	~1/30,000	100% paternal
Noonan syndrome	20301303	PTPN11	many mutations	~1/10,000	100% paternal

Table 7.3 Examples of paternal-age-effect disorders suggested to be associated with selfish spermatogonial selection. [a]PMID, PubMed Identifier. [b]The G380R change in the FGFR3 (fibroblast growth factor receptor 3) protein in achondroplasia is caused by either a G→C or a G→A change at nucleotide number 1138 in the reference cDNA sequence. For further information see the paper by Goriely & Wilkie (2012) in Further Reading.

achondroplasia, with marked paternal age effects and paternal germ-line transmission (see **Table 7.3** for some examples). That might suggest exceptional germ-line mutation rates (up to 1000-fold higher than the average rate). However, studies of mutation rates in sperm have suggested that the underlying mutations (which are missense mutations that change a single amino acid) do not occur at especially high frequencies. Instead, the mutations have been thought to result in mutant proteins that cause the dysregulation of spermatogonial stem cell behavior and confer a selective growth advantage on any spermatogonial stem cell that contains them. Stem cells containing these mutations might proliferate to reach high frequencies, explaining the paternal origin of many mutations and the paternal age effect. In each case the mutant proteins are fibroblast growth factor receptors or other proteins that work in the growth factor receptor–RAS signal transduction pathway. The underlying mutations belong to a class of gain-of-function mutations that we consider in Section 7.5.

Surveying and curating point mutations that cause disease

Sequencing of the human genome has intensified the search for disease-associated genetic variants, and various methods are used to establish whether or not a suspected genetic variant really does cause, or increase susceptibility to, disease. Point mutations are the most frequent contributors to disease, and data on known or suspected pathogenic point mutations have been curated in a variety of different databases.

According to circumstance, assessing whether an identified point mutation is pathogenic can be comparatively easy or difficult. We will return to explore this subject in detail in Chapter 11, when we consider diagnostic DNA approaches. For now, we give some brief points in the subsections below.

Point mutations in coding DNA

The vast majority of pathogenic point mutations have been recorded in protein-coding genes. For coding DNA, identifying some types of disease-causing mutation is comparatively easy. Nonsense mutations and insertions or deletions that change a known translational reading frame almost scream "Pick me!" at the investigator.

Making the correct call for missense mutations and small non-frameshifting deletions or insertions is harder. Evolutionary and population genetic studies can often help here: substitution or deletion of an amino acid is much more likely to be pathogenic if that specific amino acid has been strongly conserved during evolution. As detailed in Chapter 11, various computer programs can be used. One important application is to assess the likelihood of a missense mutation being pathogenic on the basis of predicted differences in the physicochemical properties of the original

Table 7.4 Examples of RNA genes that are mutated in single-gene disorders.

Table 7.4 Examples of RNA genes that are mutated in single-gene disorders. [a]More information on the loci can be accessed by web crosslinks from the loci listed in the Human Gene Nomenclature Database at www.genenames. org. [b]Identifier for PubMed database at www. ncbi.nlm.nih.gov/pubmed that gives details of relevant journal article. See also the review by Makrythanasis & Antonorakis (2012) under Further Reading.

RNA CLASS	LOCUS[a]	DISORDER	PMID[b]
miRNA	MIR96	Autosomal deafness, type 50 (OMIM 613074)	19363479
	MIR184	EDICT syndrome (OMIM 614303)	21996275
snRNA	RNU4ATAC	microcephalic osteodysplastic primordial dwarfism type 1 (OMIM 210710)	21474760
Long ncRNA	TERC	Type 1 Autosomal dominant dyskeratosis congenital (OMIM 127550)	11574891
		Susceptibility to anaplastic anemia (OMIM 614743)	12090986

amino acid and the substituted one. Querying databases of previously recorded mutations (see below) can also be very useful.

Point mutations in RNA genes and other noncoding DNA

Pathogenic point mutations in noncoding DNA are generally difficult to identify, with one exception: mutations at splice junctions, notably changes in the conserved GT and AG dinucleotides at the extreme ends of introns. Untranslated sequences and other regulatory sequences are often much less evolutionarily conserved than coding DNA, making mutations here much more difficult to evaluate. Identifying mutations within RNA genes is also not easy and is not helped by the comparatively poor evolutionary conservation of many RNA genes. Because RNAs have intricate secondary structure through extensive intra-chain hydrogen bonding (see Figures 2.4 and 2.6), mutations that affect secondary structure can have important consequences.

Other than mutations at splice junctions, point mutations in noncoding DNA might make a very small contribution to monogenic disorders but could be significant contributors to complex common diseases by causing important changes in gene expression. Only a handful of RNA genes have been implicated in single-gene disorders (see **Table 7.4** for examples), but various miRNAs are known to be important in the pathogenesis of other disorders, such as cancers.

Databases of human pathogenic mutations

Human mutation databases range from large general databases to more specific ones (**Table 7.5**). Locus-specific databases focus on a specific gene

Table 7.5 Examples of different types of mutation databases that curate disease-associated mutations. [a]An extensive list of locus-specific databases is maintained at http://www.centralmutations.org/Lsdb.php. The Human Genome Variation Society also maintains links to many other useful mutation databases at http://www.hgvs.org/dblist. For further background see Horaitis O & Cotton RG (2005) Current Protocols in Bioinformatics, Chapter 1, Unit 1; PMID 17893115; and Samuels ME & Rouleau G (2011) Nature Rev Genet 12:378–379; PMID 21540879.

DATABASE	DESCRIPTION	WEBSITE
Genomewide databases		
Human Gene Mutation Database	comprehensive data on germ-line mutations in nuclear genes associated with human inherited disease	http://www.hgmd.org
COSMIC	comprehensive catalog of somatic mutations in cancer	http://www.sanger.ac.uk/genetics/CGP/cosmic/
MITOMAP	mitochondrial genome database with prominent sections devoted to disease-associated mutations in mt-tRNA, mt-rRNA and mitochondrial coding and regulatory sequences	http://www.mitomap.org/
Locus-specific databases[a]		
Phenylalanine Hydroxylase Locus Knowledgebase	a list of mutations at the PAH locus, mostly centered on mutations causing phenylketonuria	http://www.pahdb.mcgill.ca
Databases by mutation category		
SpliceDisease Database	disease-associated splicing mutations	http://cmbi.bjmu.edu.cn/sdisease

(or sometimes genes) associated with an individual disorder. The submitted data include both pathogenic mutations and normal variants, and so the databases can be of help in evaluating whether newly identified mutations are likely to be pathogenic, as described in Chapter 11.

7.3 Moderate- to Large-Scale Pathogenic Mutations Triggered by Repetitive DNA

As detailed in Section 2.4, diverse types of DNA repeats account for a large amount of our genome, and various types of DNA duplication have been important in shaping our genome to allow the development of biological complexity. But there is a downside: DNA repeats also predispose DNA molecules to undergo changes that can cause disease, as a result of different mechanisms. We saw in the previous section that tandem repeats of a single nucleotide, dinucleotide, or other short oligonucleotides, are hotspots for replication slippage, and are a frequent cause of very small insertions and deletions. Larger-scale insertions and deletions are also very often facilitated by tandem DNA repeats or interspersed repeats, and they are the major topic of this section. (We discuss even larger changes to the genetic material, ones that are visible under the light microscope, within the context of chromosome abnormalities in Section 7.4.)

Most of the repeats involved in triggering large-scale mutations occur within introns or outside genes, but as we will see they can sometimes occur within coding sequences. The pathogenesis arises because the large-scale mechanisms either adversely change the structure or copy number of genes, or they adversely alter gene expression. We detail the principal genetic mechanisms giving rise to these changes in individual sections below; for now we will give a brief overview of the different types of DNA change.

Genes that have multiple large introns (which usually contain diverse types of repeats) are more prone to intragenic deletions. One or more exons can be eliminated, causing a shift in the translational reading frame if the *net* effect is to delete a number of coding nucleotides that is not a multiple of three (Figure 2C in Box 2.1 shows the general concept). Moderate- to large-scale insertions can also occur within a gene. That can involve the duplication of some of the exons in a gene (with the possibility of producing a translational frameshift, or an extended protein that might be unstable, not fold properly, or be functionally disadvantaged).

Quite often, large-scale pathogenic mutations are caused by exchanges between low-copy-number repeats that have very similar sequences (*homologous repeats*). Different families of low-copy-number repeats can be distinguished. They might be short interspersed sequences, or naturally duplicated sequences that can reach up to several hundreds of kilobases in length and contain multiple genes.

When two repeats with very similar sequences occur close to each other on the same chromosome arm, the high level of sequence identity between the repeats can lead to mispairing of chromatids. Non-allelic repeats can then pair up: repeat no. 1 on one chromatid might pair up with repeat no. 2 on the other chromatid. A subsequent recombination event occurring in the mispaired region produces a change in repeat copy number; the process is known as **non-allelic homologous recombination** (**NAHR**). Alternatively recombination occurs between homologous repeats on the same DNA molecule; an *intrachromatid recombination* such as this is also a form of NAHR. We describe different NAHR mechanisms below, show how they can generate insertions, deletions, and inversions, and give examples of how they cause disease.

Very occasionally, an insertion is caused by spontaneous random insertion of a DNA transposon or, more usually, a cDNA copy of a retrotransposon, such as an Alu repeat or a LINE1 (L1) repeat. (Only a small proportion of the Alu repeats and L1 elements are capable of transposition; see Figure 2.14.) Powerful transposon silencing mechanisms seek to minimize these events in germ-line cells, but they do happen occasionally; if a transposon copy inserts into coding DNA, normal gene expression can be disrupted.

A rather different type of insertion involves the expansion of arrays of certain tandem oligonucleotide repeats beyond normal limits, as described in the next two sections. That may mean modest expansions of tandem trinucleotide repeats within coding sequences, maintaining the translational reading frame. In addition, some tandem oligonucleotide repeats in noncoding DNA can expand to very high copy numbers, in ways that affect the expression of neighboring genes.

Pathogenic expansion of arrays of short tandem oligonucleotide repeats in coding DNA

Many of our genes have arrays of tandem trinucleotide repeats in coding sequences. Unlike tandem mononucleotide or dinucleotide repeats, changes in the number of trinucleotide repeats do not cause a translational frameshift, and small numbers of tandem trinucleotide repeats may be tolerated and are translated to produce proteins with the repetition of a single amino acid. Sometimes quite long runs of the same amino acid signify a functional domain, as in the case of polyglutamine and polyalanine, as described below.

Some of the tandem repeat arrays are polymorphic; others are not. In the former case the repeats may sometimes cause disease when the array expands in size above a critical number of repeats.

Pathogenic polyalanine expansion

Polyalanine regions are found in a small number of developmentally important transcription factors, where they serve as a flexible nonpolar linker or spacer between two folded domains. Individual polyalanine regions are often encoded by different alanine codons (notably GCG, GCT, or GCA). They are generally not polymorphic, and a very modest increase in the number of alanine codons can cause disease.

Nine different polyalanine expansion diseases are known, all congenital disorders of development except for a form of muscular dystrophy. According to the type of protein, the non-polymorphic polyalanine tract is normally 10–20 alanine residues long; pathogenesis occurs when the array is expanded by about a further 4–10 alanine residues on average (see Table 7.6 for some examples). The congenital polyalanine malformations are thought to occur by loss of the transcription factor function (expansion of the polyalanine array may induce aberrant protein folding or interfere with the correct localization of the protein within the cell).

Table 7.6 Examples of repeat expansion in polyalanine-associated diseases. Note that polyalanine tracts are not polymorphic and the level of expansion in disease is very modest. (Data from Brown and Brown [2004] *Trends Genet* 20:51–58; PMID 14698619.)

TYPE OF DISEASE	NUMBER OF ALANINE REPEATS	
	Normal	Disease
Oculopharyngeal muscular dystrophy (OPMD)	10	11–17
Blepharimosis, ptosis, and epicanthus inversus	14	22–26
Synpolydactyly type II	15	22–29
Hand–foot genital syndrome	18	24–26

Unstable expansion of CAG repeats encoding polyglutamine

Unlike polyalanine, polyglutamine (often called poly(Q)) is highly polar. Moderately long poly(Q) tracts are found in several proteins and are often involved in regulating the transcription of certain target genes. Unlike polyalanine, poly(Q) arrays are typically specified by just one codon (CAG); they show significant length polymorphism (possibly as a result of replication slippage).

After a certain critical length has been reached, poly(Q) arrays can expand to very significant lengths (**Table 7.7**). At the DNA level, the expanded alleles are meiotically and somatically unstable (unlike alleles encoding polyalanine expansions). That is, after a polyglutamine-coding sequence has reached a critical length, the extended array of repeats has a tendency to expand further and faster than the smaller arrays.

Unstable expansion of CAG repeats in coding sequences resembles the unstable expansion of various types of short oligonucleotide repeat in noncoding sequences, as described below and in the next section. They are sometimes described as **dynamic mutations** because the repeat length can increase from one generation to the next (and sometimes from mother cell to daughter cell in one individual). Increasing expansion of the repeats leads to increasing severity of the disease, as described below.

Four major types of late-onset neurodegenerative disorder result from unstable CAG repeat expansion producing an extended polyglutamine tract (see Table 7.7). In each case a buildup of toxic products produced by the mutant allele is thought to underlie the disease. The effect of toxic intracellular products is particularly marked in neurons (which are generally intended to be very long-lived and not replaced when lost). Depending on the disease, harmful consequences are apparent in different neural components. Motor neurons in the brainstem and spinal cord are principally affected in spinal bulbar muscular atrophy, for example, and neurons in different regions of the cerebellum are affected in the different forms of spinocerebellar ataxia (see Table 7.7).

The expansion mechanism remains unclear, and until recently the pathogenesis was thought to be due to toxic intracellular aggregates that were formed by proteins with excessively long polyglutamine tracts. Animal models now suggest that the pathogenesis might also involve toxic RNAs in some cases. For example, RNA transcripts containing expanded CAG repeats in the mouse homolog of the major Huntington disease gene, *HTT*, are neurotoxic under experimental conditions.

The question of toxic protein or toxic RNA is also relevant to Huntington disease-like 2, which until recently was thought to result from unstable expansion of a CTG repeat in the 5′ UTR of the *JPH3* gene. Now we know that an antisense transcript is produced that contains the complementary CAG repeat sequence. Translation of this RNA transcript might result in a similar polyglutamine pathogenesis to that in Huntington disease (**Figure 7.6**).

Table 7.7 Examples of repeat expansion in polyglutamine-associated diseases.
Polyglutamine arrays are polymorphic in the normal population, and can expand to quite high repeat numbers in disease in comparison with polyalanine expansions (see Table 7.6). This reflects differences in the polyalanine and polyglutamate expansion mechanisms. Multiple other forms of spinocerebellar ataxia (SCA) not shown here are also caused by polyglutamine expansion, including SCA1, SCA2, SCA3, SCA6, SCA7, and SCA17. The major SCA features are slowly progressive incoordination of gait, often associated with poor coordination of hands, speech, and eye movements, and atrophy of the cerebellum, but different regions within the cerebellum are affected in the different SCAs.

TYPE OF DISEASE	NUMBER OF GLUTAMINE REPEATS	
	Normal	Disease
Huntington disease	6–35	36–121
Spinal bulbar muscular atrophy (SBMA; Kennedy disease)	6–36	38–62
Dentatorubropallidoluysian atrophy	3–38	49–88
Spinocerebellar ataxia type 7 (SCA7)	7–18	38–200

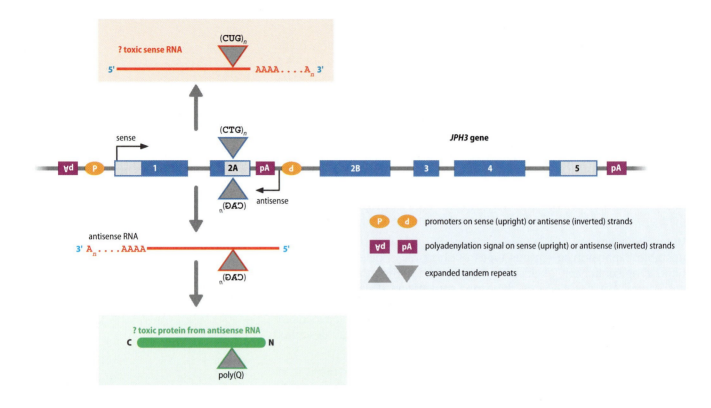

Figure 7.6 Bidirectional transcription at the *JPH3* locus associated with Huntington disease-like 2: toxic RNA or toxic protein?
Exon organization for sense transcripts of the *JPH3* gene is shown by boxes with a coding sequence represented by blue coloring and untranslated regions in pale gray. *JPH3* undergoes alternative splicing so that exon 2A, which contains the tandem CTG repeat, is variably used in transcripts. Some transcripts join exon 1 to exon 2A only, and a polyadenylation signal (pA) after exon 2A is used. Other transcripts omit exon 2A: they join exon 1 to exons 2B, 3, 4, and 5, and use the polyadenylation signal after exon 5. The disease was initially thought to be caused by toxic sense transcripts containing exon 2A with an expanded CUG repeat. However, it became clear that an antisense transcript with CAG repeats is also produced from the *JPH3* gene (using an internal promoter close to exon 2A). Translation of this transcript may cause disease by producing a toxic protein with an expanded poly(Q) region: in a mouse disease model, the Huntington-like pathogenesis remains after the ortholog of *JPH3* has been genetically engineered to express the antisense transcript only; PMID 21555070.

Pathogenic unstable expansion of short noncoding tandem repeats

Different types of short tandem noncoding repeats can also undergo unstable expansion in specific regions of the genome. Several are found in the promoter regions, untranslated sequences, or introns of genes; they are often composed of repeats with from three to six nucleotides and can result in disease when expanded. In some other cases the sequences are in gene-poor regions and do not cause disease.

Large-scale expansions of tandem oligonucleotide repeats underlie a proportion of chromosome abnormalities known as **fragile sites**, specific regions of the genome that appear as constrictions or gaps on the metaphase chromosomes prepared from cells treated with certain chemicals to partly inhibit DNA replication. Fragile sites can be classified as common (present in virtually all individuals and thought to be an intrinsic part of chromosome structure) or rare (occurring in a small percentage of individuals—this category includes fragile sites resulting from a large expansion of tandem noncoding repeats). Although fragile sites are often located in gene-poor regions, the rare fragile sites include three examples—FRAXA (Xq27.3), FRAXE (Xq28), and FRA11B (11q23.3)—that are located within or close to a gene and disturb its expression, causing disease, as detailed for FRAXA below. Many other examples of large-scale unstable expansion of short noncoding tandem repeats cause disease without producing fragile sites (**Table 7.8**).

For some noncoding tandem repeats that can undergo unstable expansion to cause disease, an intermediate stage can be identified. After a modest expansion in the normal number of repeats, a normal allele can change to a **premutation** in which the expanded array of repeats is unstable but usually not big enough to cause the disease associated with very large-scale expansion. For example, alleles with from 55 to close to 200 CGG repeats in the 5′ UTR of the *FMR1* gene are premutations for fragile X syndrome (they do not cause this disease but are unstable and

REPEAT UNIT	LOCATION	ASSOCIATED DISEASE	COPY NUMBER[a]		MECHANISM
			Normal	Disease	
GAA	intron 1 of *FXN* gene at 9q21	Friedreich ataxia (PMID 20301458)	6–32	200–1700	loss of function
CGG	5′ UTR of *FMR1* gene; also causes the fragile site FRAXA at Xq28	fragile X syndrome (with intellectual disability) (PMID 20301558)	5–54	>200 (up to several thousand)	
CTG	3′ UTR of *DMPK* gene at 19q13	myotonic dystrophy type I (PMID 20301344)	5–37	50–10,000	mutant RNA is toxic to cells or has a *trans*-dominant effect (see Box 7.2)
CCTG	intron of *CNBP* gene at 3q21	myotonic dystrophy type II (PMID 20301639)	10–26	75–11,000	
ATTCT	intron 9 of *ATXN10* gene at 22q13	spinocerebellar ataxia type 10 (PMID 20301354)	10–29	500–4500	
GGGGCC	noncoding region of the *C9ORF72* gene at 9p21	frontal dementia and/or amyotrophic lateral sclerosis (PMID 20301623)	0–23	up to 1600	

can quickly expand from one generation to the next to reach the critical number of more than 200 repeats associated with fragile X syndrome).

Individuals with very high repeat numbers are more severely affected and develop symptoms at an earlier age than those individuals who have a number of repeats that falls at the lower end of the disease range. Because the mutation is dynamic, transmission down successive generations in a family can lead to mutant alleles with an increasing repeat number and increasing disease severity in successive generations (which has been referred to as *anticipation*—see Figure 5.17 on page 137).

In some of the associated disorders the pathogenesis arises because the expanded repeats block the expression of a protein-coding gene; in other cases the problem appears to be a toxic or *trans*-dominant RNA (see Table 7.8 and **Box 7.2**).

Pathogenic sequence exchanges between chromatids at mispaired tandem repeats

Many human genes and gene regions have significant arrays of long tandemly repeated DNA sequences. This can include the repetition of exons, whole genes, and even multiple genes (see Figure 2.12 on page 45 for examples). Tandem repeats within genes or spanning coding sequences can predispose to a type of non-allelic homologous recombination that can cause disease.

Normally, recombination between homologous chromosomes occurs after the chromosomes have paired up with their DNA sequences in perfect alignment. However, local misalignment of the paired chromosomes is more likely to occur in regions where there are highly similar tandem repeats—the DNA molecules of the two chromatids can line up out of register. That is, the alignment is staggered and one or more repeats on each chromatid do not pair up with their normal partner repeat on the other chromatid.

A subsequent recombination within the mismatched sequences is known as **unequal crossover** (**UEC**) and results in one chromatid with an insertion (more tandem repeats) and one with a deletion (fewer tandem repeats) An equivalent process can also occur between sister chromatids, an **unequal sister chromatid exchange** (**UESCE**) (**Figure 7.7**). UEC and UESCE cause reciprocal exchanges between misaligned chromatids: one chromatid gains an extra DNA sequence, and the other loses an equivalent sequence. Disease may result from a change in gene copy number, or through the formation of hybrid genes that lack some of the functional gene sequence.

Table 7.8 Examples of pathogenic expansion of noncoding short tandem repeats.
[a]Individuals can sometimes have unstable *premutation alleles* with a number of repeats that is intermediate between the maximum found on normal alleles and the minimum found in alleles associated with the disease resulting from full-scale expansion (see the example of the *FMR1* gene in Box 7.2).

BOX 7.2 *FMR1*-related disorders and myotonic dystrophy: different pathogeneses due to unstable noncoding tandem repeats.

FMR1-related disorders: epigenetic gene silencing, loss of protein function, and possible toxic RNA

Normal individuals have between 5 and 54 tandem copies of a CGG repeat in the 5′ UTR of the *FMR1* gene at Xq27.3. The *FMR1* gene makes a regulatory RNA-binding protein that is important for proper synapse development and function (it works by binding certain target mRNAs, inhibiting translation).

Expansion to more than 200 and up to several thousand CGG repeats is known as a *full mutation*. Individuals with this type of expansion reveal a fragile site (FRAXA) at Xq27 (**Figure 1**) and develop the fragile X syndrome. The most conspicuous feature is intellectual disability (often profound in affected males, but mild in affected females); behavioral abnormalities are also common. The pathogenesis in fragile X syndrome results from loss of function—the expanded CGG repeats are targets for cytosine methylation, resulting in hypermethylation of the region

and aberrant methylation in the upstream regulatory region, causing epigenetic silencing of *FMR1*. Point mutations also occasionally cause fragile X syndrome by inactivating *FMR1* with loss of protein function.

Alleles with 55–200 CGG repeats are *premutation* alleles—they do not cause the classical fragile X syndrome but they are unstable and can become full mutations when transmitted to the next generation. Although carriers of a premutation allele do not have fragile X syndrome, they have been proposed to be at risk of different diseases not obviously related to fragile X syndrome, notably the neurodegenerative condition FXTAS (fragile X tremor-ataxia syndrome). In premutation individuals the *FMR1* gene is expressed fully, and in addition to *FMR1* sense transcripts, antisense transcripts are produced from the same region. Toxic RNAs have been proposed to underlie the pathogenesis of FXTAS—see the reviews by Santoro et al. (2012) and Hagerman & Hagerman (2013) under Further Reading.

Myotonic dystrophy: nuclear inclusions of mutant RNAs causing misregulation of alternative splicing

Myotonic dystrophy is a multisystem disorder that notably includes progressive muscle deterioration and myotonia (an inability to relax muscles after contraction). There are two types, resulting from the unstable expansion of different types of noncoding repeat (see Table 7.8) in two rather different genes: *DMPK* (a protein kinase gene) and *ZNF9* (which makes an RNA-binding protein).

If translated, the different types of repeat (CTG in type 1 and CCTG in type 2) would produce very different polypeptide sequences. However, at the RNA level the two types of repeat have a CUG sequence in common, and this is the sequence that is thought to be important in pathogenesis. Animal studies suggest that the pathogenesis is exclusively due to an expression product and is quite unrelated to the gene containing the expanded repeat (as shown by constructing a mouse model of the disease simply by inserting an expanded CTG repeat into a host actin gene).

Instead of being transported to the cytoplasm, the mutant RNAs accumulate within the nucleus. The expanded repeats within the RNAs form very large hairpins that bind certain RNA-binding proteins, and

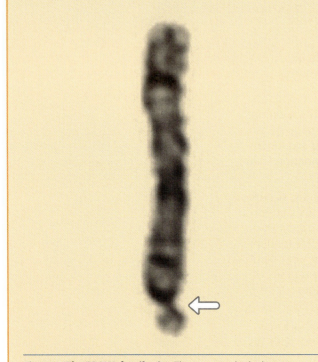

Figure 1 The FRAXA fragile site. The arrow marks the location of the FRAXA fragile site, which can be demonstrated to occur as a constriction at Xq27 in individuals with fragile X syndrome. (Courtesy of Graham Fews, West Midlands Regional Genetics Laboratory, UK.)

BOX 7.2 (continued)

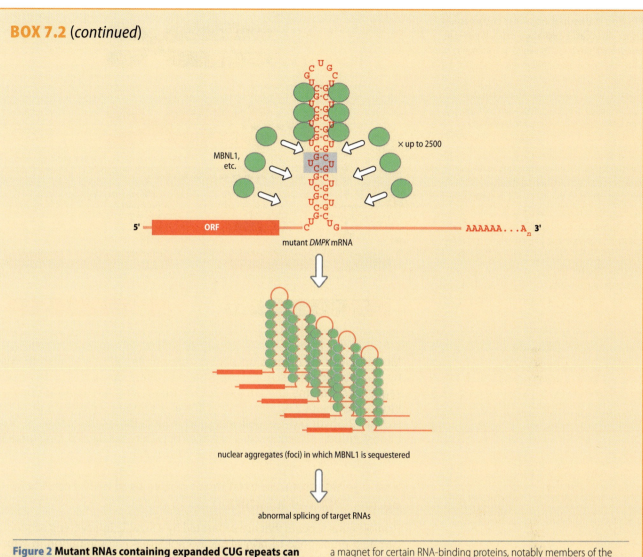

Figure 2 Mutant RNAs containing expanded CUG repeats can bind and inactivate muscleblind regulators of alternative splicing. In type I myotonic dystrophy mutant *DMPK* mRNA is retained in the nucleus. The massive CUG repeat expansion in the 3′ UTR causes a giant hairpin to form, stabilized by G–C base pairs between opposed CUG repeats (such as the example boxed in gray). The hairpin acts as a magnet for certain RNA-binding proteins, notably members of the muscleblind family, such as MBNL1, that regulate alternative splicing of certain target mRNAs. The mutant RNAs aggregate to form foci in the nucleus that bind and sequester MBNL1 proteins, stopping them from interacting with their normal targets.

the mutant RNAs form small aggregates that are seen to be discrete *foci* (or *inclusions*). In particular, proteins with a preference for binding CUG motifs are selectively bound, notably MBNL1 and other members of the muscleblind family of splicing regulators (named because of their homology to *Drosophila* proteins required for terminal differentiation of muscle and photoreceptor cells; **Figure 2**).

Because they bind in large quantities to the expanded CUG repeats, MBNL1 and related proteins are effectively sequestered within the mutant RNA foci—they cannot get close to their normal target mRNAs and so

lose their ability to function normally. An additional effect is that the mutant RNAs or nuclear aggregates seem to induce the increased production of another protein that binds to CUG repeats called CUGBP1 (CUG-binding protein 1). Like MNBL1, CUGBP1 is a regulator of alternative splicing. The combined effect of reducing the availability of the muscleblind family of splicing regulators and up-regulating the CUGBP1 splicing regulator causes misregulation of alternative splicing of multiple target mRNAs including some key muscle proteins and various other proteins, causing disease.

Figure 7.7 Unequal crossover and unequal sister chromatid exchange cause deletions and insertions. (A) Mispairing of chromatids with two very similar tandem repeats (1, 2). The very high sequence identity between these repeats can facilitate misalignment between the DNA of aligned chromatids so that repeat 1 on one chromatid aligns with repeat 2 on another chromatid. The misaligned chromatids can be on non-sister chromatids of homologous chromosomes, such as chromatids *b* and *c*, in which case recombination (large orange X) in the misaligned region results in an unequal crossover (B). Alternatively, there can be a recombination-like unequal sister chromatid exchange (small orange X) between misaligned repeats on sister chromatids (*c*, *d*) of a single chromosome (C). In either case, the result is two chromatids, one with three repeat units and the other with a single repeat unit (a hybrid of sequences 1 and 2—shown here as 1/2 or 2/1).

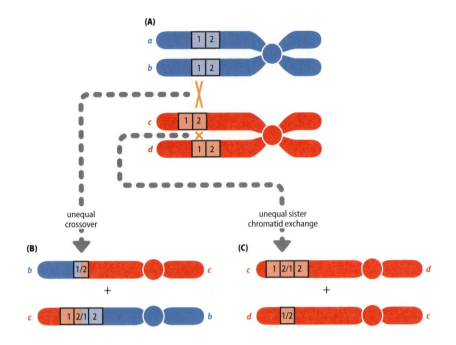

Misalignment of repeats on paired chromatids can also cause disease by a *nonreciprocal* sequence exchange. Here, one of the interacting sequences remains unchanged, but the other is mutated (**gene conversion**—see **Figure 7.8**). See **Box 7.3** for a common single-gene disorder, steroid 21-hydroxylase deficiency, in which the pathogenesis is due almost entirely to sequence exchanges between misaligned long tandem repeats, resulting in deletion or gene conversion.

Pathogenic sequence exchanges between distant repeats in nuclear DNA and in mtDNA

Homologous repeats that are separated by a sizable intervening sequence can also predispose to non-allelic homologous recombination in which sequence exchanges occur between misaligned repeats. The repeats may be **direct repeats** (oriented in the same 5′→3′ orientation); in that case, the intervening sequence between the repeats can be deleted or duplicated. That may mean loss or duplication of exons, which can be frameshifting mutations, or loss or duplication of multiple genes (both of which can be pathogenic, as described below). Exchange between *inverted repeats* (repeats oriented in opposite 5′→3′ directions) can also

Figure 7.8 Principle of gene conversion. Gene conversion is a *nonreciprocal* sequence exchange between two related sequences that may be alleles or non-allelic (such as misaligned repeats). Sequence information is copied from one of the paired sequences (the donor sequence) to replace an equivalent part of the other sequence (the acceptor sequence). The size of the sequence that is converted—the conversion tract—is often a few hundred nucleotides long in mammalian cells. See Chen et al. (2007) under Further Reading for details on possible gene conversion mechanisms.

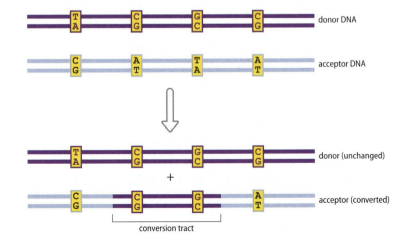

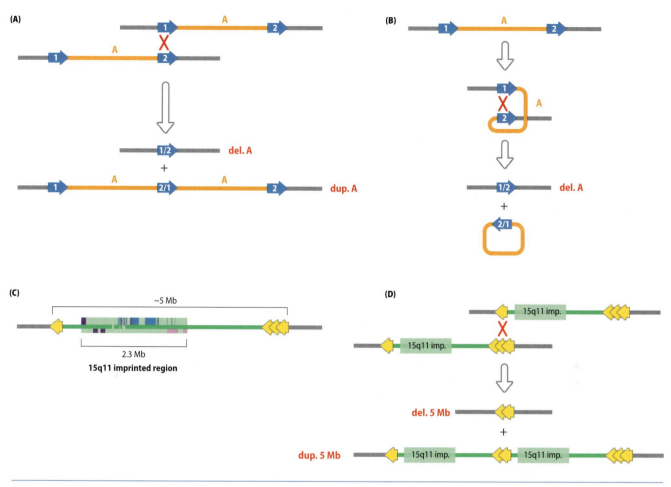

Figure 7.9 Deletion/duplication events due to non-allelic homologous recombination between low-copy-number direct repeats. Block arrows indicate highly similar low-copy-number *direct repeats* (oriented in the same direction). (A) DNA sequence A is flanked by low-copy-number repeats 1 and 2 that have identical or highly similar nucleotide sequences. Chromatid misalignment can occur so that repeat 1 on one chromatid pairs with repeat 2 on the other chromatid, and subsequent crossover can result in a chromatid with a deletion of sequence A (del. A) and one with two copies of sequence A (dup. A). (B) An intrachromatid 'recombination' between direct repeats on the same DNA molecule can also produce a deletion of sequence A. If the other product, a circular DNA containing sequence A, lacks a centromere, it will be lost after cell division. (C, D) A practical example. In (C) a low-copy-number repeat family (block arrows) flanking a roughly 5 Mb region of DNA contains a 2.3 Mb imprinted gene cluster at 15q11 (shown in detail in Figure 1 in Box 6.5). (D) Sequence exchanges between mispaired repeats flanking the 15q11 imprinted region predispose to frequent roughly 5 Mb deletions (del. 5 Mb), causing Angelman syndrome (if on maternally inherited chromosome 11) or Prader–Willi syndrome (if on paternal chromosome 11). Carriers of the reciprocal 5 Mb duplication (dup. 5 Mb) have a less severe combination of mental retardation and behavioral difficulties.

cause inversion of the intervening sequence; this can also be pathogenic, as illustrated in the last subsection below.

Chromosome microdeletions and microduplications

Just as with mispairing of tandem repeats, distantly spaced direct repeats can mispair when chromatids are aligned within homologous chromosomes. Subsequent crossover at mismatched direct repeats results in deletions or duplications of the intervening sequence (**Figure 7.9A**). An equivalent type of exchange can occur between mispaired short direct repeats on the same DNA strand (a form of intrachromatid recombination), and this can also lead to deletion (**Figure 7.9B**).

As an example, consider recurrent *de novo* deletions that cause Angelman syndrome (AS) and Prader–Willi syndrome (PWS). As detailed in Box 6.5, imprinted genes associated with AS and PWS map to an imprinted gene cluster at 15q11. A 5 Mb DNA segment spanning the AS–PWS critical

BOX 7.3 Steroid 21-hydroxylase deficiency, a disorder caused by gene–pseudogene sequence exchanges.

Steroid 21-hydroxylase is a cytochrome P450 enzyme needed by the adrenal gland to produce the glucocorticoid hormone cortisol and aldosterone (which regulates sodium and potassium levels). Genetic deficiency in this enzyme is much the most common cause of congenital adrenal hyperplasia. In classical (congenital) forms of the disorder, excessive adrenal androgen biosynthesis results in virilization of affected individuals, so that girls are often born with masculinized external genitalia. Classically affected individuals may have the 'simple-virilizing' form of the disorder, but some also excrete large amounts of sodium in their urine, which leads to potentially fatal electrolyte and water imbalance ('salt-wasting' phenotype).

Steroid 21-hydroxylase is encoded by a gene, *CYP21A2*, located in the class III region of the HLA complex. *CYP21A2* resides on an approximately 30 kb segment of DNA that is tandemly duplicated, with about 98% sequence identity between the tandem 30 kb repeats. As a result, there is a closely related copy of the *CYP21A2* gene sequence on the other repeat, a pseudogene called *CYP21A1P*. The pseudogene has multiple deleterious mutations distributed across its length (**Figure 1A**) and does not make a protein.

The very high sequence identity between the tandem repeats makes paired chromatids liable to local misalignment: a repeat containing the functional *CYP21A2* gene on one chromatid mispairs with the

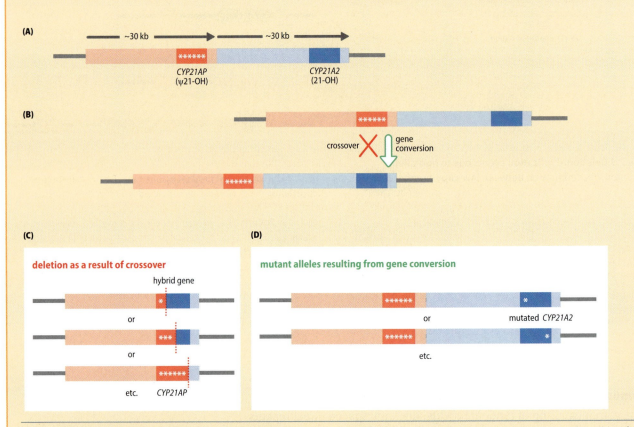

Figure 1 Tandem duplication of repeats containing the *CYP21A2* gene and a closely related pseudogene predispose to pathogenic sequence exchanges. (A) The steroid 21-hydroxylase gene, *CYP21A2*, and a closely related pseudogene, *CYP21AP* (with multiple inactivating mutations shown by asterisks), are located on tandemly duplicated repeats about 30 kb long that also contain other genes (not shown here for the sake of simplification). (B) Mispairing of non-allelic repeats is facilitated by the 98% sequence identity between them. Subsequent reciprocal exchange (recombination, shown by X) or a nonreciprocal exchange (gene conversion) can result in a loss of functional sequence. (C) Unequal crossover can produce chromosomes with a single 30 kb repeat that may contain a hybrid gene, part *CYP21AP* and part *CYP21A2*, or just the *CYP21AP* pseudogene (dashed vertical lines indicate the crossover point). (D) Gene conversion replaces a segment of the normal *CYP21A2* gene by the equivalent sequence from the pseudogene. The conversion tract is often only a few hundred nucleotides long and can occur at different regions, introducing a copy of an inactivating mutation from any part of the pseudogene.

BOX 7.3 (*continued*)

MUTATION CLASS AND LOCATION	NORMAL 21-OH GENE SEQUENCE (*CYP21A2*)	PATHOGENIC POINT MUTATION	EQUIVALENT *CYP21AP* PSEUDOGENE SEQUENCE
Intron 2, splicing mutation	CCCA**C**CCTCC	CCCA**G**CCTCC	CCCA**G**CCTCC
Exon 3, deletion of 8 nucleotides within codons 111–113	**GGA GAC TAC** TCx Gly Asp Tyr Ser	G.. ... ... TCx **Val**	G.. ... ... TCx
Exon 4, missense: I173N	ATC A**T**C TGT Ile Ile Cys	ATC A**A**C TGT Ile **Asn** Cys	ATC A**A**C TGT
Exon 6, multiple missense mutations (codons 237–240)	A**T**C G**T**G GAG A**T**G Ile Val Glu Met	A**A**C G**A**G GAG A**A**G **Asn Glu** Glu **Lys**	A**A**C G**A**G GAG A**A**G
Exon 7, missense: V282L	CAC **G**TG CAC His Val His	CAC **T**TG CAC His **Leu** His	CAC **T**TG CAC
Exon 8, nonsense: Q319X	CTG **C**AG GAG Leu Gln Glu	CTG **T**AG GAG Leu **STOP**	CTG **T**AG GAG
Exon 8, missense: R357W	CTG **C**GG CCC Leu Arg Pro	CTG **T**GG CCC Leu **Trp** Pro	CTG **T**GG CCC

Table 1 Pathogenic point mutations in the steroid 21-hydroxylase gene are copied from a closely related pseudogene. The gene conversion tract (the region copied from the pseudogene sequence) is usually no longer than a few hundred nucleotides—see Collier S et al. (1993) *Nat Genet* 3:260–265; PMID 8485582.

repeat containing the *CYP21AP* pseudogene on the other chromatid (Figure 1B).

In more than 99% of cases with 21-hydroxylase deficiency, mispairing between the two repeats and subsequent sequence exchanges is thought to be responsible for pathogenesis. About 75% of the mutations that cause disease are roughly 30 kb deletions caused by unequal crossover or unequal sister chromatid exchange. If the crossover point occurs between gene and pseudogene, a single nonfunctional hybrid 21-hydroxylase gene results; or if it is located just beyond the paired gene and pseudogene, it leaves just the 21-hydroxylase pseudogene (Figure 1C).

The remaining 25% or so of pathogenic mutations are point mutations, but in the vast majority of cases the point mutation is introduced into the *CYP21A2* gene by a gene conversion event that copies a sequence containing a deleterious mutation from the pseudogene, replacing the original sequence (Figure 1D and **Table 1**).

See the Chen et al. (2007) reference under Further Reading for other human diseases in which gene conversion is commonly involved in pathogenesis.

region and extending into 15q13 is prone to deletion and duplication as a result of recombination between low-copy-number repeats flanking the region (yellow arrowheads in **Figure 7.9C**). Deletions of the 5 Mb region are the principal cause of AS and PWS (according to the parent of origin—see Table 6.7 on page 184), but the reciprocal duplication of the 5 Mb region results in a milder phenotype (**Figure 7.9D**).

Like many other megabase-sized deletions and duplications, the common 5 Mb deletions found in AS and PWS are not detectable using standard light microscopy of stained chromosome preparations. Accordingly, these subcytogenetic changes are often known as chromosome *micro-deletions* or *microduplications*. The latter can be pathogenic if the

Table 7.9 Direct repeats are hotspots for pathogenic deletions in mtDNA. [a]See Figure 2.11 on page 43 for the mtDNA gene map. The pairs of direct repeats are identical except for those causing the 7664 bp deletion (as indicated in bold). The 4977 bp deletion is a particularly common deletion.

DELETION SIZE (BP)	SEQUENCE AND LOCATION OF REPEATS (IN INDICATED GENES)[a]	
	Repeat 1	Repeat 2
4420	AACAACCCCC 10942–10951 (*ND4*)	AACAACCCCC 15362–15371 (*CYTB*)
4977	ACCTCCCTCACCA 8470–8482 (*ATP8*)	ACCTCCCTCACCA 13447–13459 (*ND5*)
7521	AGGCGACC 7975–7982 (*CO2*)	AGGCGACC 15496–15503 (*CYTB*)
7664	CCTCC**G**TAGACCTAACC 6325–6341 (*CO1*)	CCTCCT**A**GACCTAACC 13989–14004 (*ND5*)
7723	TCACAGCCC 6076–6084 (*CO1*)	TCACAGCCC 11964–11972 (*ND4*)

duplicated region contains dosage-sensitive genes—we consider the effects on gene expression and the clinical impact in Section 7.5.

Deletions resulting from direct repeats in mtDNA

More than 120 different (single) large-scale mtDNA deletions are recorded in the MITOMAP database as being associated with disease. In addition to being involved in classical mitochondrial disorders, deletions in mtDNA may contribute to many common diseases, notably Parkinson disease, and to the aging process. Note, however, that most mtDNA deletions are sporadic and are not transmitted to offspring.

In most cases a mtDNA deletion is flanked by short direct repeats that are often perfect copies (see **Table 7.9** for examples). Physical association of the larger direct repeats promotes the formation of deletions, either during mtDNA replication (when the single-stranded repeats are exposed), or more probably during the repair of double-strand breaks.

The 4977 bp deletion is flanked by two large (13 bp) direct repeats. It is the most frequent deletion associated with mtDNA disease, and is also the most frequent somatic mtDNA mutation—hence its name the 'common deletion.'

Intrachromatid recombination between inverted repeats

Inverted repeats on a single chromatid can also mispair by looping out the intervening sequence. Subsequent recombination at the mispaired sequences will produce an inversion of the intervening sequence (**Figure 7.10A**). An inversion like this can cause disease, for example by relocating part of a gene, by disrupting the gene, or by separating a gene from important *cis*-acting control elements.

An instructive example is provided by hemophilia A: in about 50% of cases the cause is a large inversion that disrupts the *F8* gene, which makes blood clotting factor VIII. A low-copy-number repeat, *F8A1*, within intron 22 of the *F8* gene can mispair with either of two very similar repeat sequences, *F8A1* and *F8A2*, that are located upstream of *F8* and in the opposite 5′→3′ orientation to *F8A1*. Subsequent recombination between mispaired *F8A* repeats produces an inversion of about 500 kb of intervening sequence, causing disruption of the *F8* gene (**Figure 7.10B**).

7.4   Chromosome Abnormalities

Many large-scale changes to our DNA sequences that cause diseases are more readily studied at the level of chromosomes, as are changes

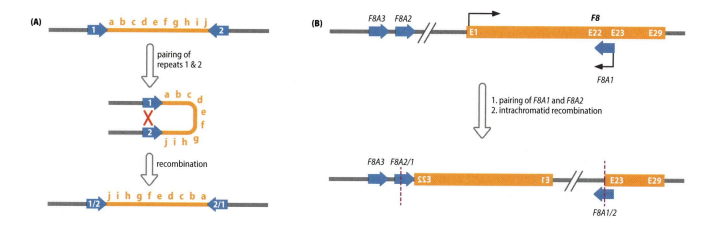

Figure 7.10 Intrachromatid recombination between inverted repeats produces inversions and is a common cause of hemophilia A. (A) Inverted repeats 1 and 2 on the same DNA strand can mispair by inducing looping of the intervening DNA. Subsequent recombination within the paired repeats produces hybrid repeat sequences (1/2 and 2/1) and inversion of the intervening DNA. (B) In about 50% of cases with hemophilia A, the mutation is a large inversion that disrupts the blood clotting factor VIII gene (*F8*). The 191 kb *F8* gene has 29 exons, and within the large intron 22 is a small gene, *F8A1*, that is transcribed from the opposite strand. *F8A1* is a member of a family of low-copy-number repeats that includes two closely related sequences, *F8A2* and *F8A3*, located upstream of *F8*. Mispairing between either of these repeats and *F8A1* can induce an inversion, by looping out of the intervening DNA to allow recombination between the mismatched repeats (such as *F8A1* and *F8A2*, as shown here; the dashed red vertical lines mark the boundaries of the inversion). The resulting inversion disrupts the *F8* gene, splitting it into two oppositely oriented fragments, one containing exons 1–22 and the other from exons 23–29.

in chromosome copy number resulting from errors in chromosome segregation. In standard cytogenetic **karyotyping** suitable metaphase or prometaphase chromosome preparations are chemically stained to reveal a pattern of alternating light and dark bands under light microscopy, which are examined to identify chromosome abnormalities. See **Box 7.4** for some details of the techniques and relevant nomenclature. Alternative approaches use DNA hybridization to identify or screen for abnormalities, and we detail these in Chapter 11.

According to their distribution in cells of the body, chromosome abnormalities can be classified into two types. A **constitutional** abnormality is present in all nucleated cells of the body and so must have been present very early in development. It can arise as a result of an abnormal sperm or egg, through abnormal fertilization, or through an abnormal event in the very early embryo. A somatic (or acquired) abnormality is present in only certain cells or tissues of a person, who is therefore a genetic **mosaic** (by possessing two populations of cells with altered chromosome or DNA content, each deriving from the same zygote).

Chromosomal abnormalities, whether constitutional or somatic, can be subdivided into two categories: structural abnormalities (which arise through chromosome breakage events that are not repaired) and numerical abnormalities (changes in chromosome number that often arise through errors in chromosome segregation). See **Table 7.10** for a guide to the nomenclature of human chromosome abnormalities.

Structural chromosomal abnormalities

As detailed in Section 4.1, abnormal chromosome breaks (caused by double-strand DNA breaks) occur as a result of unrepaired damage to DNA or through faults in the recombination process. Chromosome breaks that occur during the G2 phase (after the DNA has replicated) are really *chromatid* breaks: they affect only one of the two sister chromatids. Those that occur during the G1 phase that are not repaired by S phase (when

BOX 7.4 Human chromosome banding and associated nomenclature.

Chromosome preparation and chromosome banding methods

To study chromosomes under the light microscope, the chromosomes must be suitably condensed—metaphase (or prometaphase) chromosome preparations are required. A peripheral blood sample is taken and separated white blood cells are stimulated to divide by using a mitogen such as phytohemagglutinin. The white blood cells are grown in a rich culture medium containing a spindle-disrupting agent (such as colcemid) to maximize the number of metaphase cells (cells enter metaphase but cannot progress through the rest of M phase). Prometaphase preparations can also be obtained; they have slightly less-condensed chromosomes, making analysis easier.

Chromosome banding involves treating chromosome preparations with denaturing agents; alternatively they are digested with enzymes and then exposed to a dye that can bind to DNA. Some dyes preferentially bind to AT-rich sequences; others bind to GC-rich sequences. The dyes show differential binding to different regions across a chromosome that will reflect the relative frequencies of AT and GC base pairs.

The most commonly used method in human chromosome banding is **G-banding**. The chromosomes are treated with trypsin and stained with Giemsa, which preferentially binds AT-rich regions, producing alternating dark bands (Giemsa-positive; AT-rich) and light bands (Giemsa-negative; GC-rich). Because genes are preferentially associated with GC-rich regions, dark bands in G-banding are gene-poor; light bands are gene-rich.

Human chromosome and chromosome banding nomenclature

Human chromosome nomenclature is decided periodically by the International Standing Committee on Human Cytogenetic Nomenclature; see under Further Reading for the most recent ISCN report published in 2013. The nomenclature assigns numbers 1–22 to the autosomes according to perceived size, and uses the symbols p and q to denote, respectively, the short and long arms of a chromosome. Depending on the position of the centromere, chromosomes are described as *metacentric* (centromere at or close to the middle of the chromosome), *submetacentric* (centromere some distance from the middle and from telomeres), or *acrocentric* (centromere close to a telomere).

Each chromosome arm is subdivided into a number of regions, according to consistent and distinct morphological features (depending on the size of the

chromosome arm, there may be from one to three regions). Each region is in turn divided into bands, and then into sub-bands and sub-sub-bands, according to the banding resolution (**Figure 1**).

The numbering of regions, bands, sub-bands, and sub-sub-bands is done according to relative proximity to the centromere. If a chromosome arm has three regions, the region closest to the centromere would be region 1, and the one closest to the telomere would be region 3. For example, the band illustrated in Figure 1A would be known as 4q21 for these reasons: it is located on the long arm of chromosome 4 (= 4q); it resides on the second (of three regions) on this chromosome arm (= 4q2); within this region it is the nearest band (band 1) to the centromere. The last two digits of band 4q21 are therefore pronounced two-one (not twenty one) to mean region two, band one. Similarly in Figure 1C the numbers following 4q in the sub-sub-band 4q21.22 are pronounced two-one-point-two-two.

Note also that in chromosome nomenclature the words **proximal** and **distal** are used to indicate the relative position on a chromosome with respect to the centromere. Thus, proximal Xq means the segment of the long arm of the X that is closest to the centromere. Similarly, distal 3p means the portion of the short arm of chromosome 3 that is most distant from the centromere (= closest to the telomere).

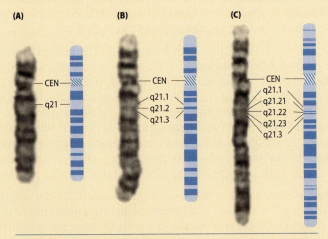

Figure 1 Chromosome banding resolutions can resolve bands, sub-bands and sub-sub-bands. G-banding patterns for human chromosome 4 (with accompanying ideogram to the right) are shown at increasing levels of resolution. The levels correspond approximately to (A) 400, (B) 550, and (C) 850 bands per haploid set, allowing the visual subdivision of bands into sub-bands and sub-sub-bands as the resolution increases. CEN, centromere. For an example of a full set of banded chromosomes, see Figure 2.8. (Adapted from Cross I & Wolstenholme J [2001] in Human Cytogenetics: Constitutional Analysis, 3rd ed (Rooney DE, ed). With permission from Oxford University Press.)

TYPE OF ABNORMALITY	EXAMPLES	EXPLANATION/NOTES
Numerical		
Triploidy	69,XXX, 69,XXY, 69,XYY	a type of polyploidy
Trisomy	47,XX,+21	gain of a chromosome is indicated by +
Monosomy	45,X	a type of aneuploidy; loss of an autosome is indicated by –
Mosaicism	47,XXX/46,XX	a type of mixoploidy
Structural		
Deletion	46,XY,del(4)(p16.3)	terminal deletion (breakpoint at 4p16.3)
	46,XX,del(5)(q13q33)	interstitial deletion (5q13–q33)
Inversion	46,XY,inv(11)(p11p15)	paracentric inversion (breakpoints on same arm)
Duplication	46,XX,dup(1)(q22q25)	duplication of region spanning 1q22 to 1q25
Insertion	46,XX,ins(2)(p13q21q31)	a rearrangement of one copy of chromosome 2 by insertion of segment 2q21–q31 into a breakpoint at 2p13
Ring chromosome	46,XY,r(7)(p22q36)	joining of broken ends at 7p22 and 7q36
Marker	47,XX,+mar	indicates a cell that contains a *marker chromosome* (an extra unidentified chromosome)
Reciprocal translocation	46,XX,t(2;6)(q35;p21.3)	a balanced reciprocal translocation with breakpoints at 2q35 and 6p21.3
Robertsonian translocation (gives rise to one derivative chromosome)	45,XY,der(14;21)(q10;q10)	a balanced carrier of a 14;21 Robertsonian translocation. q10 is not really a chromosome band, but indicates the centromere; der is used when one chromosome from a translocation is present
	46,XX,der(14;21)(q10;q10),+21	an individual with Down syndrome possessing one normal chromosome 14, a Robertsonian translocation 14;21 chromosome, and two normal copies of chromosome 21

Table 7.10 Nomenclature of chromosome abnormalities. This is a short nomenclature; a more complicated nomenclature is defined by the ISCN (2013) report that allows complete description of any chromosome abnormality; see Schaeffer et al. (2013) under Further Reading.

the DNA replicates) become chromosome breaks (both sister chromatids are affected). A cell with highly damaging chromosome breaks may often be removed by triggering cell death mechanisms; if it survives with unrepaired breaks, chromosomes with structural abnormalities can result.

Errors in recombination that produce structural chromosome abnormalities can occur at meiosis. Paired homologs are normally subjected to recombination mechanisms that ensure the breakage and rejoining of non-sister chromatids, but if recombination occurs between mispaired homologs, the resulting products may have structural abnormalities. Intrachromatid recombination can also be a source of structural abnormalities.

A form of somatic recombination also occurs naturally in B and T cells in which the cellular DNA undergoes programmed rearrangements to make antibodies and T cell receptors. Abnormalities in these recombination processes can also cause structural chromosomal abnormalities that may be associated with cancer.

Structural chromosome abnormalities are often the result of incorrect joining together of two broken chromosome ends. Different mechanisms are possible, as detailed in the following subsections.

Large-scale duplications, deletions, and inversions

Moderately large-scale duplications and deletions within a chromosome arm can occur as a result of exchanges between mispaired chromatids of the types described in Section 7.3, but larger changes can occur when breaks occur in both arms of a chromosome. If a single chromosome sustains two breaks, incorrect joining of fragments can result in

Figure 7.11 Stable outcomes after incorrect repair of two breaks on a single chromosome. (A) Incorrect repair of two breaks (orange arrows) occurring in the same chromosome arm can involve loss of the central fragment (here containing hypothetical regions e and f) and rejoining of the terminal fragments (deletion), or inversion of the central fragment through 180° and rejoining of the ends to the terminal fragments (called a paracentric inversion because it does not involve the centromere). (B) When two breaks occur on different arms of the same chromosome, the central fragment (encompassing hypothetical regions b to f in this example) may invert and rejoin the terminal fragments (pericentric inversion). Alternatively, because the central fragment contains a centromere, the two ends can be joined to form a stable ring chromosome, while the acentric distal fragments are lost. Like other repaired chromosomes that retain a centromere, ring chromosomes can be stably propagated to daughter cells.

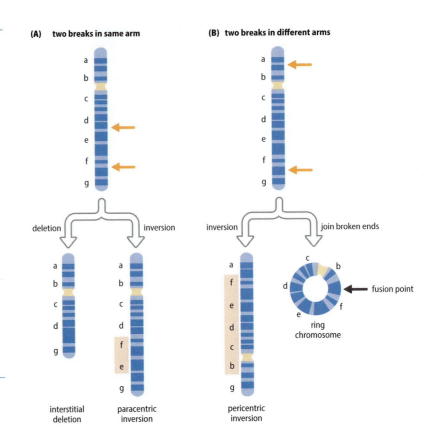

chromosome material being lost (deletion), switched round in the reverse direction (inversion), or included in a circular chromosome (a ring chromosome) (**Figure 7.11**).

Structurally abnormal chromosomes with a single centromere can be stably propagated through successive rounds of mitosis. However, any repaired chromosome that lacks a centromere (an *acentric* chromosome) or possesses two centromeres (a *dicentric* chromosome) will normally not segregate stably at mitosis, and will eventually be lost.

Chromosomal translocations

If two different chromosomes each sustain a single break, incorrect joining of the broken ends can result in the movement of chromosome material between chromosomes (**translocation**). A *reciprocal translocation* is the general term used to describe an exchange of fragments between two chromosomes (**Figure 7.12A**). If an acentric fragment from one chromosome (one that lacks a centromere) is exchanged for an acentric fragment from another, the products each have a centromere and are stable in mitosis. Structurally rearranged chromosomes like this that have a centromere are known as **derivative chromosomes**. Exchange of an acentric fragment for a centric fragment results in acentric and dicentric chromosomes that are normally unstable in mitosis (but see below for an exceptional class of translocations in which dicentric products are stable).

For a chromosomal translocation to occur, the regions containing the double-strand DNA breaks in the two participating chromosomes must be in very close proximity (permitting incorrect joining before double-strand DNA breaks can be repaired). The spatial distribution of chromosomes in the nucleus is not random, and chromosomes tend to occupy certain 'territories.' Chromosomes that tend to be physically closer to each other are more likely to engage in translocation with each other. For example, human chromosomes 4, 13, and 18 are preferentially located at the periphery of the nucleus and frequently translocate with each other but not with physically distant chromosomes localized in the interior of the nucleus. Specific types of translocations are common in certain cancers, and may reflect close physical association of the two chromosomal regions that participate in translocation.

One exceptional form of chromosome association occurs between the very small short arms of the five human acrocentric chromosomes (chromosomes 13, 14, 15, 21, and 22). Each of the short arms of these chromosomes has about 30–40 large tandem DNA repeats, each containing sequences for making three ribosomal RNAs: 28S, 18S, and 5.8S rRNAs; the five ribosomal DNA (rDNA) regions congregate at the nucleolus to produce these rRNAs. The close physical association of these five chromosome arms is responsible for a specialized type of translocation, called Robertsonian translocation or centric fusion, that involves breaks in the short arms of two different acrocentric chromosomes followed by exchange of acentric and centric fragments to give acentric and dicentric products (**Figure 7.12B**).

The acentric chromosome produced by a Robertsonian translocation is lost at mitosis without consequence (it contains just highly repetitive noncoding DNA plus rRNA genes that are also present at high copy number on the other acrocentric chromosomes). The other product is an unusual dicentric chromosome that is stable in mitosis: the two centromeres are in close proximity (centric fusion) and often function as one large centromere so that the chromosome segregates regularly. (Nevertheless, such a chromosome may present problems during gametogenesis.)

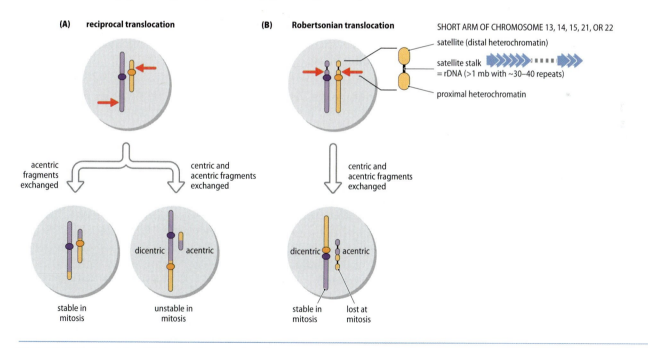

Figure 7.12 Reciprocal and Robertsonian translocations.
(A) Reciprocal translocation. The *derivative* chromosomes produced by the translocation are stable in mitosis when one acentric fragment is exchanged for another, but when a centric fragment is exchanged for an acentric fragment, the derivative chromosomes are usually unstable. (B) Robertsonian translocation (centric fusion). This is a highly specialized reciprocal translocation in which exchange of centric and acentric fragments produces a dicentric chromosome that is nevertheless stable in mitosis. It occurs exclusively after breaks in the short arms of two of the acrocentric chromosomes 13, 14, 15, 21, and 22. As illustrated, the short arms of the human acrocentric chromosomes have a common structure. A region of distal heterochromatin (called a *satellite*) is joined to a proximal heterochromatic region by a thin *satellite stalk* made up of *ribosomal DNA* (rDNA), an array of tandem DNA repeats that each make three types of rRNA. Breaks that occur close to the centromere can result in a dicentric chromosome in which the two centromeres are so close that they can function as a single centromere. The loss of the small acentric fragment has no phenotypic consequences.

More complex translocations can involve multiple chromosome break-ages. Insertions typically require at least three breaks—often a fragment liberated by two breaks in one chromosome arm inserts into another break that may be located in another region of the same chromosome or another chromosome.

Isochromosomes

An additional rare class of structural abnormality can arise from recombination after an aberrant chromosome division. The product is a symmetrical **isochromosome** consisting of either two long arms or two short arms of a particular chromosome. Human isochromosomes are rare, except for i(Xq) and also i(21q), an occasional contributor to Down syndrome.

Chromosomal abnormalities involving gain or loss of complete chromosomes

Three classes of numerical chromosomal abnormalities can be distinguished: polyploidy, aneuploidy, and mixoploidy (**Table 7.11**).

Polyploidy

Three per cent of recognized human pregnancies produce a triploid embryo (**Figure 7.13A**). The usual cause is two sperm fertilizing a single egg (dispermy), but triploidy is sometimes attributable to fertilization involving a diploid gamete. With three copies of every autosome, the dosage of autosomal genes might be expected to be balanced, but triploids very seldom survive to term, and the condition is not compatible with life (but diploid/triploid mosaics can survive). The lethality in triploids may be due to an imbalance between products encoded on the X chromosome and autosomes, for which X-chromosome inactivation would be unable to compensate.

Tetraploidy (**Figure 7.13B**) is much rarer and always lethal. It is usually due to failure to complete the first zygotic division: the DNA has replicated to give a content of $4C$, but cell division has not then taken place as normal. Although constitutional polyploidy is rare and lethal, some types of cell

Table 7.11 Clinical consequences of numerical chromosome abnormalities. [a]In humans, the embryonic period spans fertilization to the end of the eighth week of development; fetal development then begins and lasts until birth.

ABNORMALITY	CLINICAL CONSEQUENCES
Polyploidy	
Triploidy (69,XXX or 69,XYY)	1–3% of all conceptions; almost never born live and do not survive long
Aneuploidy (autosomes)	
Nullisomy (missing a pair of homologs)	lethal at pre-implantation stage of embryonic development
Monosomy (one chromosome missing)	lethal during embryonic development
Trisomy (one extra chromosome)	usually lethal during embryonic[a] or fetal[a] stages, but individuals with trisomy 13 (Patau syndrome) and trisomy 18 (Edwards syndrome) may survive to term; those with trisomy 21 (Down syndrome) may survive beyond age 40
Aneuploidy (sex chromosomes)	
Additional sex chromosomes	individuals with 47,XXX, 47,XXY, and 47,XYY all experience relatively minor problems and a normal lifespan
Lacking a sex chromosome	while 45,Y is never viable, in 45,X (Turner syndrome), about 99% of cases abort spontaneously; survivors are of normal intelligence but are infertile and show minor physical diagnostic characteristics

are naturally polyploid in all normal individuals—for example, our muscle fibers are formed by recurrent cell fusions that result in multinucleate *syncytial* cells.

Aneuploidy

Normally our nucleate cells have complete chromosome sets (euploidy), but sometimes one or more individual chromosomes are present in an extra copy, or are missing (**aneuploidy**). In trisomy, three copies of a particular chromosome are present in an otherwise diploid cell, such as trisomy 21 (47,XX,+21 or 47,XY,+21) in Down syndrome. In monosomy a chromosome is lacking from an otherwise diploid state, as in monosomy X (45,X) in Turner syndrome. Cancer cells often show extreme aneuploidy, with multiple chromosomal abnormalities.

Aneuploid cells arise through two main mechanisms. One mechanism is **nondisjunction**, in which paired chromosomes fail to separate (*disjoin*) during meiotic anaphase I and migrate to the same daughter cell, or sister chromatids fail to disjoin at either meiosis II or mitosis. Nondisjunction during meiosis produces gametes with either 22 or 24 chromosomes, which after fertilization with a normal gamete produce a trisomic or monosomic zygote (**Figure 7.14**). If nondisjunction occurs during mitosis, the individual is a mosaic with a mix of normal and aneuploid cells.

Anaphase lag also results in aneuploidy. If a chromosome or chromatid is delayed in its movement during anaphase and lags behind the others, it may fail to be incorporated into one of the two daughter nuclei. Chromosomes that do not enter a daughter cell nucleus are eventually degraded.

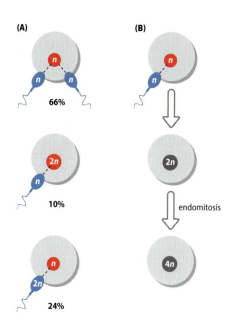

Figure 7.13 Origins of triploidy and tetraploidy. (A) Origins of human triploidy. Dispermy (top) is the principal cause, accounting for 66% of cases. Triploidy is also caused by diploid gametes that arise by occasional faults in meiosis, such as nondisjunction (see Figure 7.14); fertilization of a diploid ovum (middle) and fertilization by a diploid sperm (bottom) account for 10% and 24% of cases, respectively. (B) Tetraploidy involves normal fertilization and fusion of gametes to give a normal zygote. Subsequently, however, tetraploidy arises when DNA replicates without subsequent cell division (*endomitosis*).

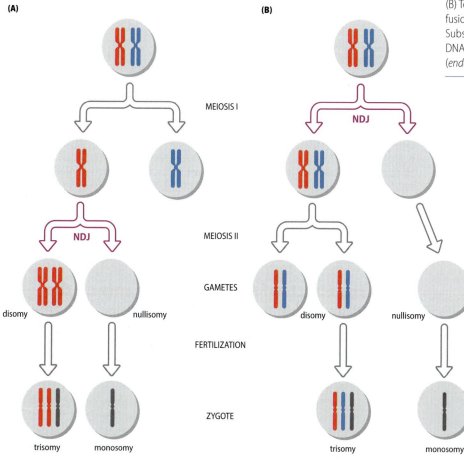

Figure 7.14 Meiotic nondisjunction and its consequences. In meiotic nondisjunction (NDJ) a pair of homologous chromosomes fail to disjoin and migrate to the same daughter cell, instead of being individually allocated to each daughter cell. That can happen at the second meiotic division (A) or at the first meiotic division (B). The result is that a person produces abnormal gametes, with either no copies of a chromosome (nullisomy) or two copies of a chromosome (disomy). In the latter case, the two chromosome copies may originate from a single parent (for NDJ at meiosis I) or individually from each parent (for NDJ at meiosis I). Fertilization with a normal gamete containing one copy of the chromosome (shown in gray) results in trisomy or monosomy.

Having the wrong number of chromosomes has serious, usually lethal, consequences (Table 7.11). Even though the extra chromosome 21 in a person with trisomy 21 (Down syndrome) is a perfectly normal chromosome, inherited from a normal parent, its presence causes multiple abnormalities that are present from birth (congenital). Embryos with trisomy 13 or trisomy 18 can also survive to term but both result in severe developmental malformations, respectively Patau syndrome and Edwards syndrome. Other autosomal trisomies are not compatible with life. Autosomal monosomies have even more catastrophic consequences than trisomies and are invariably lethal at the earliest stages of embryonic life. We consider in Section 7.5 how gene dosage problems in aneuploidies result in disease.

Maternal age effects in Down syndrome

In principle, the nondisjunction that causes a gamete to have an extra copy of chromosome 21 could occur at either meiotic division in spermatogenesis or oogenesis, but in practice in about 70% of cases it occurs at meiosis I in the mother. This is almost certainly a consequence of the extremely long duration of meiosis I in females (it begins in the third month of fetal life but is arrested and not completed until after ovulation). That means this one meiotic division can take decades; by contrast, male meiosis occurs continuously in the testes from puberty to old age. Not only is there a sex difference in the origin of the extra chromosome 21 but there is also a very significant *maternal age effect*. Thus the risk of having a Down syndrome child in a 20-year-old pregnant woman is about 1 in 1500, but that increases to about 1 in 25 for a 45-year-old woman.

Mixoploidy

Mixoploidy means having two or more genetically different cell lineages within one individual. Usually, the different cell populations arise from the same zygote (*mosaicism*). More rarely, a person can have different cell populations that originate from different zygotes and is described as a **chimera**; spontaneous chimerism usually arises by the aggregation of fraternal twin zygotes or immediate descendant cells within the very early embryo. Abnormalities that would otherwise be lethal (such as triploidy) may not be lethal in mixoploid individuals.

Aneuploidy mosaics, with a proportion of normal cells and a proportion of aneuploid cells, are common. This type of mosaicism can result when nondisjunction or chromosome lag occurs in one of the mitotic divisions of the early embryo (any monosomic cells that are formed usually die). Polyploidy mosaics (such as human diploid/triploid mosaics) are occasionally found. As the gain or loss of a haploid set of chromosomes by mitotic nondisjunction is extremely unlikely, human diploid/triploid mosaics most probably arise by fusion of the second polar body with one of the cleavage nuclei of a normal diploid zygote.

7.5 The Effects of Pathogenic Variants on the Phenotype

Sections 7.2 to 7.4 described how disease-causing genetic changes arise in DNA and chromosomes. Here we consider the effects of the pathogenic variants on the phenotype. There are two major considerations. The first is the effect of a pathogenic variant on gene function. The simplest situation occurs when the variant affects how a single gene functions, which we consider in the section below. Some variants, however, simultaneously affect, directly or indirectly, how multiple genes work. A large-scale mutation, for example, can directly change the sequence or copy number of multiple neighboring genes. In addition, a simple mutation in a regulatory gene can indirectly affect the expression of multiple target genes, and can potentially cause complex phenotypes. A second consideration is the extent to which normal copies or different copies of the mutant gene are also available. For diploid nuclear genes, that means considering how a mutant allele and a normal allele work in the presence of each other, or estimating the combined effect of two mutant alleles. The situation for mitochondrial DNA mutants is quite different, and rather unpredictable, partly because we have so many copies of mtDNA in each cell, and partly because, unlike nuclear DNA, mtDNA replication is not governed by the cell cycle (Box 7.5).

Another consideration concerns interacting factors. The affected genes do not work in isolation: many factors—genetic (other loci), epigenetic, and environmental—affect the extent to which an individual pathogenic variant affects the phenotype. The situation becomes even more complicated when multiple genetic variants at different loci are involved—we examine this in Chapter 8, within the context of complex disease.

Mutations affecting how a single gene works: loss-of-function and gain-of-function

Mutations that affect how a single gene works can have quite different effects on how the gene makes a product. Some affect expression levels only, often causing complete failure to express the normal gene product, or causing a substantial reduction in expression. Mutations that increase gene copy number and occasional activating point mutations result in overexpression (which can be a problem for certain dosage-sensitive genes). Other mutations can result in an altered gene product that lacks the normal function, or that has an altered or new function.

BOX 7.5 The basis of disease variability due to mutations in mitochondrial DNA.

Human mitochondrial DNA (mtDNA), which is transmitted exclusively by females, is a circular 16.5 kb DNA that makes 13 of the 92 different polypeptides needed for the protein complexes of the mitochondrial oxidative phosphorylation system, plus 24 different RNAs needed for translating mitochondrial mRNAs (for the mtDNA gene map, see Figure 2.11 on page 43). Unlike nuclear DNA, mtDNA molecules are present in very large numbers in each cell (from several hundreds to many thousands of copies), and they replicate independently of the cell cycle ('relaxed DNA replication').

Because mtDNA molecules are present in so many copies per cell, multiple mutant mtDNA molecules can coexist with multiple wild-type molecules (**heteroplasmy**). And because DNA replication is relaxed, and clonal expansion of mutant DNAs is variable, the proportion of mutant to wild-type mtDNA molecules can vary substantially in different cells of a single individual, and between related individuals who have the same mtDNA mutation. That leads to very significant clinical variability.

The frequency of pathogenic mtDNA mutations in the population is high (about 1 in 200 people), but the majority of these people will not be affected, mostly because of low levels of heteroplasmy. Most mtDNA mutations do not cause a biochemical phenotype until the mutant mtDNAs reach a certain proportion of the total mtDNA, a threshold level that can vary according to the nature of the mutation (Table 1).

MUTATION CLASS	THRESHOLD LEVEL (%) FOR PHENOTYPE
Point mutations in mt-tRNA genes (in general)	>90
mtDNA deletions (see end of Section 7.3)	50–60
m.5545C→T in the mitochondrial *MT-TW* gene (encoding the mt-tRNA for tryptophan)	<25

Table 1 Examples of how threshold levels required for mtDNA mutations to express a phenotype can vary according to the mutation. Note that because mtDNA replication is not tied to the cell cycle, clonal expansion of pathogenic mtDNA mutations (notably mtDNA deletions) in post-mitotic cells such as neurons and muscle cells is very common in aging and disease (where the effect is to reduce the energy output of the cells).

One broad way of classifying the overall effect of a mutation is to consider whether the mutation results in a loss of function or a gain of function, as described below.

Loss-of-function mutations

Loss-of-function mutations mean that the final gene product is simply not produced by the mutated DNA molecule, or it is made but does not work properly (because very little of it is produced or it is nonfunctional). A severe mutation might cause the deletion of an entire gene, and would be an example of a **null allele**. Other mutations can be null alleles if they have a comparable phenotypic effect. For protein-coding genes, that can mean mutations that somehow introduce premature termination codons, which frequently activate pathways to degrade the mRNA so that no protein is made (see Box 7.1). They include nonsense mutations and frameshifting mutations at the DNA level, and also RNA splicing mutations that cause a translational frameshift *at the RNA level*, as a result of exon skipping or intron retention (see Figure 7.3), or through exon truncation or exon extension (see Figure 7.4).

Some pathogenic mutations (such as some missense mutations) allow a gene product to be made but not function at all, or not very well. Specific amino acids can have critical functional roles (including in post-translational processing, such as glycosylation or phosphorylation—see Table 2.2). Others have critical structural roles (many cysteines participate in disulphide bonding; many glycines allow flexible bending of the polypeptide backbone). Other amino acids may have important but not absolutely essential roles so that they can be replaced by chemically similar or different amino acids with different consequences for functional activity.

Gain-of-function mutations

Gain-of-function mutations often give rise to products that are positively harmful in some way. They are common in cancers, but in inherited disorders they are often much less common than loss-of-function mutations. An outstanding exception to this general rule is provided by the paternal age effect disorders that we previously considered in Table 7.3. Here, different missense mutations can activate members of the growth factor receptor–RAS signal transduction pathway and have been believed to confer a selective growth advantage on spermatogonial stem cells, causing clonal expansion—see **Figure 7.15** for the example of mutations in the *FGFR3* (fibroblast growth factor receptor 3) gene, causing bone dysplasia.

Some gain-of-function mutations produce a radically altered product. We described in Section 7.3 the case of unstable dynamic mutations that give rise to toxic proteins and toxic or *trans*-dominant RNAs.

Figure 7.15 Pathogenic amino acid substitutions in the FGFR3 protein due to gain-of-function missense mutations that are likely to confer a selective growth advantage on spermatogonial stem cells. The fibroblast growth factor receptor 3 (FGFR3) protein has three immunoglobulin-like domains (IgI, IgII, and IgIIIa/c) in the extracellular region, a single hydrophobic transmembrane (TM) domain, and two cytoplasmic tyrosine kinase domains (TK1, TK2). According to the color scheme, the indicated amino acid substitutions shown below cause a type of bone dysplasia syndrome. They arise from gain-of-function mutations that have been hypothesized to confer a selective advantage on spermatogonial stem cells containing the mutant allele. AN, acanthosis nigricans. (Adapted from Goriely A & Wilkie AO [2012] *Am J Hum Genet* 90:175–200. With permission from Elsevier.)

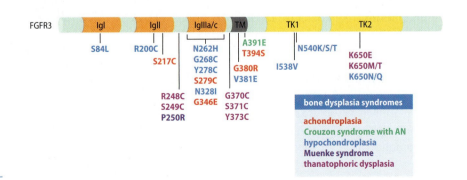

Most gain-of-function mutations do not produce a radically new product. Instead, they tend to make products that are expressed inappropriately in some way—such as in the wrong cell type, at the wrong time, or in response to the wrong signal. For example, the normal gene product might be an ion channel that is meant to be open only under certain conditions but when mutated is permanently open. Or it might be a cell-surface receptor that is inappropriately expressed, so that a particular signaling pathway is available when it should not be. As a result, the ectopically or inappropriately expressed product is able to interact with other cell components that it normally might not interact with, causing disease.

Sometimes it is a question of altered specificity in some system that is meant to protect cells from a potentially dangerous enzyme or protein; a mutant product may act on a different substrate, with catastrophic consequences. A good example is the Pittsburgh variant of α_1-antitrypsin (α_1-AT), a member of the serpin family of serine protease inhibitors. Serine proteases cleave target proteins at specific sites (they get their name because their function is critically dependent on a serine residue at the active site).

Serpins such as α_1-AT work as molecular mousetraps, offering themselves as bait to entrap and destroy certain, potentially dangerous, serine proteases. Thus, an important function of α_1-antitrypsin (α_1-AT) is to protect normal tissues from high levels of elastase, a protease expressed by neutrophils during inflammatory responses (high elastase levels trigger a compensatory increased α_1-AT production to suppress elastase). Elastase cleaves a specific peptide bond in the α_1-AT backbone that activates a radical conformational change in the α_1-AT molecule, leading to elastase destruction (**Figure 7.16A**). In the Pittsburgh variant a missense mutation changes the specificity of α_1-AT. The mutant α_1-AT now attacks

Figure 7.16 A missense mutation changes α1-antitrypsin from an elastase inhibitor to a thrombin inhibitor, causing a fatal bleeding disorder. (A) α_1-Antitrypsin (α_1-AT) acts as a molecular mousetrap to ensnare and destroy certain serine proteases, such as elastase, that cleave a certain peptide bond in its reactive center loop (yellow). The filled blue circles indicate the central carbon atom of key amino acids, including the active serine of elastase (with red side chain), plus Met 358 (with green side chain) and Ser 359 of α_1-AT. Elastase cleaves the peptide bond joining Met 358 to Ser 359 and initiates the reaction when the side chain of its active serine forms a bond with the central carbon of Met 358. After cleavage, the reactive loop of α_1-AT (yellow) snaps back into the main β-sheet (red ribbon with arrows), flinging the captured elastase (still bonded to Met 358) to the opposite end of the α_1-AT molecule) in a process that radically distorts the structure of the protease, inhibiting its function. (B) Normally, α_1-AT can inhibit elastase, which cleaves the peptide bond between Met 358 and Ser 359. In the Pittsburgh variant of α_1-AT, the methionine at position 358 has been replaced by arginine and the new Arg 358–Ser 359 peptide bond is now recognized and cleaved by thrombin, a protease that acts in blood clotting. As a result, α_1-AT-Pittsburgh acts as an anti-thrombin agent, resulting in a fatal bleeding disorder.

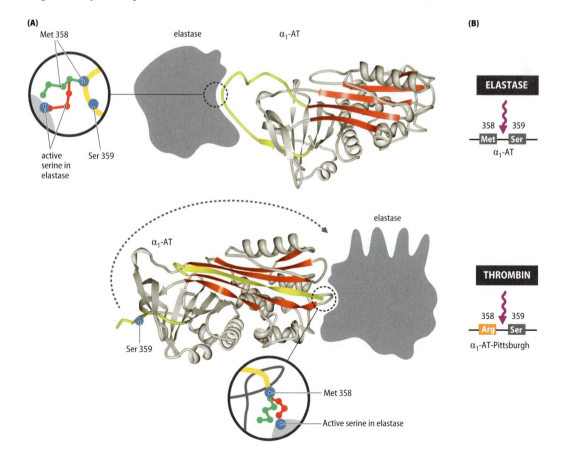

a different serine protease, the blood clotting factor thrombin, causing a lethal bleeding disorder (**Figure 7.16B**).

As we will see in Chapter 10, gain-of-function mutations are particularly common in cancer; many of these arise from chromosomal translocations and other rearrangements that create chimeric genes. Chromosomal translocations cause major problems in meiosis and so they are rarely responsible for inherited disease. But cancers arise from a mitotic division of mutant somatic cells in which the chromosomal rearrangements can be readily propagated from mother cell to daughter cells.

The effect of pathogenic variants depends on how the products of alleles interact: dominance and recessiveness revisited

Most of the genes in our diploid cells are present in two copies—one inherited from the mother and one from the father. For heterozygotes, therefore, we need to consider how the expression of a pathogenic variant (mutant allele) might affect the normal allele. To what extent might the effect of the mutant allele be reduced by—or compensated for by—having a normal allele? And, secondly, can a mutant allele have an adverse effect on how the normal allele is expressed?

Loss-of-function versus gain-of-function mutations in recessive and dominant disorders

Recall that in dominant conditions the disease phenotype is somehow expressed in the heterozygote, but in recessive conditions heterozygotes are unaffected. In autosomal recessive conditions, therefore, the disorder is expressed only when both alleles are pathogenic variants. Heterozygous carriers cannot have a gain-of-function mutation (otherwise they would be expected to be affected) and so have one loss-of-function mutation, and affected individuals have two alleles with loss-of-function mutations.

Dominant conditions are less uniform. There is one mutant allele and, according to the disorder, the mutant allele may be a gain-of-function or a loss-of-function mutation. It is easy to imagine how a gain-of-function mutation might work in a heterozygote if we think of it as being positively harmful, causing damage even if the other allele is pumping out the correct product. Think of the toxic products produced by many unstable oligonucleotide repeat expansions—the presence of functional product made by the normal allele is not going to stop the toxic effects of the mutant allele. (Potential gene therapies that rely on providing normal alleles could never work; instead we would have to provide some way of inhibiting the harmful effects of the mutant allele.)

But how does just one loss-of-function mutation cause a dominant disorder? Why is the normal product made by the unaffected allele not enough? In a few cases, such as for imprinted loci (see Section 6.3), only one of the two alleles is normally expressed, and the unaffected allele happens to be the one that is silenced. Thus, for example, a single loss-of-function mutation of the maternal *UBE3A* allele is enough to cause Angelman syndrome—the paternal *UBE3A* allele is not expressed (at least, not in the brain).

In most cases, both alleles of a diploid gene locus are normally expressed, but a single loss-of-function mutation can nevertheless cause disease if the gene concerned shows exceptional dosage sensitivity (**Box 7.6**). (Note that for all dominantly inherited disorders caused by a loss-of-function mutation, the phenotype is recessive at the cellular level: the normal phenotype can be restored by introducing a functioning allele or by reactivating a silenced allele.)

BOX 7.6 Dosage-sensitive genes and haploinsufficiency.

A small number of our genes are not essential (carriers of blood group O, for example, have two inactive alleles at the *ABO* gene locus without any harmful effects). For very many of our single-copy genes, however, homozygous inactivation is a problem, and complete absence of a gene product often results in disease (or is lethal). The amount of product made by most genes can nevertheless show very significant variation without harmful effects, provided it is above some critically low level.

For a diploid locus, an occasional gene duplication—resulting in a total of three gene copies and an expected 50% increase in gene product—often makes no obvious difference to the phenotype. According to circumstances, overproduction can even be advantageous: Western populations that traditionally have diets rich in starch have many copies of the starch-processing α-amylase gene, *AMY1A* (see Figure 4.11).

Similarly, for many genes, a reduction to 50% of gene product is often inconsequential. In recessive loss of function, one null allele at a diploid locus typically causes no harm (provided the other allele is working normally). Even if the second allele also has a mutation that leaves it only partly functional (a hypomorph), so that the combined output of the two alleles is, say, 30% of the averaged normal amount of gene product, there may be little evidence of pathogenesis (**Figure 1**). For these genes, the product levels need to drop to rather low levels before disease becomes apparent. (This particularly applies to products such as enzymes that can be used over and over again.)

A minority of our genes are especially **dosage-sensitive**—the amount of product that is made is critically important. Changes in gene copy number (gene *dosage*) can cause disease by changing the amount of gene product beyond normal limits, and certain types of point mutation can have the same effect by reducing or amplifying gene expression. Disease can occur when too much of a product is made, and sometimes that happens when the increase is only 50% above the normal amount. For example, as described below, type 1A Charcot–Marie–Tooth disease (a hereditary motor and sensory neuropathy) is often caused by a duplication that gives rise to three copies of the *PMP22* gene, or by point mutations that lead to the overexpression of *PMP22*.

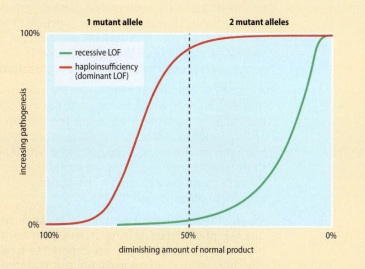

Figure 1 The relationship of disease susceptibility to diminishing amounts of gene product in dominant and recessive disorders due to loss-of-function mutations. The vertical dotted line marks the point at which an individual has a fully functional normal allele and a null allele (such as a gene deletion) and might be expected to make about 50% of the normal amount of gene product. If the allele is not null but partly functional, the amount of product made increases (as shown on left). If the second allele also has a loss-of-function (LOF) mutation, the amount of gene product made diminishes, depending on the severity of that mutation (as shown on the right). For dosage-sensitive genes, the reduction from 100% to 50% gene product is sufficient to result in disease (haploinsufficiency). Increasing the amount of gene product (by having three gene copies) can also induce disease. In recessive disorders, the disease is normally manifested only when both alleles have loss-of-function mutations; pathogenesis increases rapidly as the amount of normal product made approaches zero. There is some variability in recessive diseases—for some disorders, heterozygotes may show limited pathogenesis. Note that the idealized curves shown here do not take into account other factors such as modifier genes or environmental factors.

BOX 7.6 (*continued*)

More commonly, disease is due to a loss-of-function mutation in one allele of a dosage-sensitive gene (the other allele can be normal and expressed). If the mutant allele is a null, such as a gene deletion, the amount of normal gene product might be expected to be reduced by about 50% (but pathogenesis can sometimes be observed when the mutant allele retains partial function). Because heterozygotes are affected, this type of loss of function is dominantly inherited, and is known as **haploinsufficiency** (see Figure 1). Heterozygotes for a loss-of-function mutation in a dosage-sensitive gene are rare, and so having two loss-of-function alleles—usually by being a compound heterozygote—is extremely rare. When observed, the phenotype is often slightly more severe than that for a heterozygote.

Dosage-sensitive genes typically make products that need to be calibrated against the level of some other interacting or competing gene product. In many cases, the products have roles in quantitative signaling systems or other situations in which precisely defined ratios of the products of different genes are important for them to work together effectively. Genes that regulate other genes are likely candidates: they might do so by making transcription factors, signaling receptors, splicing regulators, or chromatin modifiers, for example. Or the different gene products may be antagonistic, competing with each other to ensure that some critical reaction is carried out that is important in development or metabolism. Because chromosomes usually have multiple dosage-sensitive genes, constitutional aneuploidies are often lethal, but some are viable (see Table 7.11 on page 218).

Striking loss of function produced by dominant-negative effects in heterozygotes

In heterozygotes a null allele causing a loss of function does not normally affect the function of the normal allele. Sometimes, however, a mutation results in a mutant protein that cannot perform the function of the normal protein and also inhibits the function of wild-type protein produced by the normal allele. A mutant protein that antagonizes the wild-type protein produced from the normal allele is sometimes known as an *antimorph*, and provides an example of a **dominant-negative effect**.

A common example occurs when the normal protein is part of a multimer that is inactive if it incorporates any of the mutant protein. Imagine the simplest possible case, when the multimer is a homodimer. A heterozygote for a null allele might be expected to make 50% of the normal homodimer. However, a heterozygote who makes equal quantities of the normal monomer and a mutant monomer (which can only form inactive dimers) might be expected to make only 25% of the normal amount of functional dimer (**Figure 7.17A**).

Disorders of structural proteins that work as multimers often provide disease phenotypes arising from dominant-negative proteins, including examples in osteogenesis imperfecta (collagens), Marfan syndrome (fibrillins), and epidermolysis bullosa (keratins). As an illustration, consider collagens in osteogenesis imperfecta (also called brittle bone disease; PMID 20301472). Collagens are initially synthesized as procollagens in which three polypeptide chains are wound round each other, beginning at the C-terminal end, to form a stiff, rope-like triple helix. Each helical polypeptide chain undergoes a remarkable one turn every three amino acids, which is made possible because collagen polypeptides have a unique structure based on tandem repeats of a three-residue sequence with the general formula Gly–X–Y. The glycines, with a single hydrogen atom in the side chain, provide the flexibility; X and Y are often (but not always) proline and hydroxyproline, respectively, which have unique side chains that loop back and connect to the polypeptide backbone, stabilizing the helical structure.

In type I procollagen, the precursor of the most common type of collagen, two of the three polypeptide chains are made by the *COL1A1* gene and the third is made by another gene, *COL1A2*. A null mutation in *COL1A1* (or *COL1A2*) results in a mild form of osteogeneis imperfecta—in each case the amount of procollagen made might be expected to decrease to 50% (**Figure 7.17B**). By contrast, a mutation that replaces a glycine in a collagen polypeptide by any other amino acid usually has dominant-negative effects: it causes abnormal packing of the collagen polypeptides into a triple helix for any collagen molecule into which it is incorporated. Such a mutation in the *COL1A1* gene causes a severe form of osteogenesis imperfecta, the type IIA form, because the amount of procollagen produced would be expected to decrease to just 25% (see Figure 7.17B).

The antagonistic effect of an antimorph in preventing wild-type protein from performing its function might be considered a gain of function, but as shown in Figure 7.17B the net effect is to create a greater loss of function than a null allele.

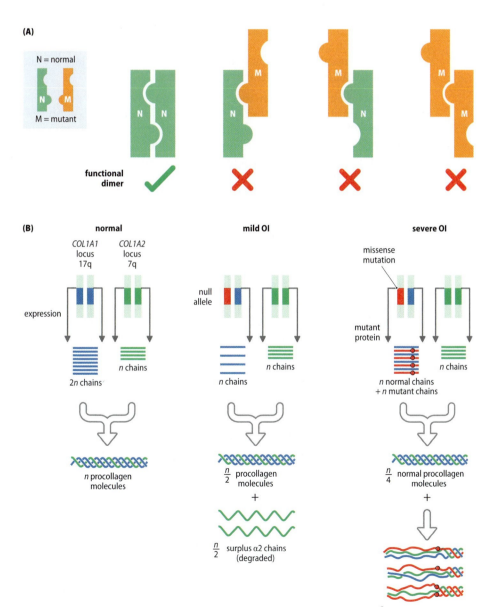

Figure 7.17 Dominant-negative effects: when a heterozygous missense mutation is more harmful than a null allele. (A) A hypothetical example: the product of a gene forms a homodimer, and the mutated allele produces a protein at normal quantities that can only form nonfunctional dimers. As a result, only one-quarter of the normal amount of functional dimers is made (there are two ways of forming nonfunctional heterodimers and one type of nonfunctional mutant homodimer). (B) A clinical example: a dominant-negative mutation causing a severe type of osteogeneis imperfecta (OI). Two *COL1A1*-encoded polypeptides and one *COL1A2*-encoded polypeptide are required to make a triple-helical type I procollagen. A null allele at *COL1A1* simply reduces the amount of type I procollagen by half and results in a mild form of OI. However, mutations that replace a structurally important glycine with any other amino acid usually have strong dominant-negative effects because they disrupt the packing of the three chains in the triple helix. The mutant *COL1A1*-encoded polypeptide is included within three-quarters of the type I procollagen molecules made, making them nonfunctional. With only 25% of the normal type I procollagens, affected individuals develop a severe type of OI.

Gain-of-function and loss-of-function mutations in the same gene produce different phenotypes

Phenotypes arising from a loss of function are typically associated with mutational heterogeneity (there are many different ways of inactivating a gene—frameshifting mutations, nonsense mutations, major splice-site mutations, missense mutations, and whole gene deletions). Gain-of-function mutations are less common in inherited disorders, and the resulting phenotypes are typically associated with mutational homogeneity. They may be a class of unstable oligonucleotide expansions, for example, or specific activating missense mutations, or a mutation that results in overexpression, but not a great range of different types of mutation.

Gain-of-function mutations and loss-of-function mutations in the same gene quite often produce very different phenotypes. In some cases, an inherited disorder is caused by the loss-of-function mutation, but a gain-of-function mutation in the same gene produces a cancer. For example, loss-of-function mutations in the *RET* gene (which makes a tyrosine kinase) result in susceptibility to Hirschsprung's disease (OMIM 142623) in which affected individuals have a congenital absence of ganglia in the gut. But different, very specific, kinds of activating missense mutations in the same gene result in different types of cancer: either medullary thyroid carcinoma or multiple endocrine neoplasia types 2A or 2B.

Loss-of-function mutations in the androgen receptor gene (*AR*) cause androgen-insensitivity syndrome (also called testicular feminization syndrome; PMID 20301602). Affected individuals have a 46,XY karyotype but because their androgen receptors do not work normally the end organs are insensitive to androgens, resulting in an X-linked recessive form of pseudohermaphroditism. Exon 1 of the *AR* gene also happens to have a tandem CAG repeat that can undergo unstable expansion to produce androgen receptor proteins with an expanded polyglutamine tract. The resulting proteins (and possibly RNA transcripts) are toxic to vulnerable cells and lead to spinal and bulbar muscular atrophy (also called Kennedy's disease; PMID 20301508). In this case there is degeneration of lower motor neurons affecting certain muscles in the arms and legs and also some muscles in the face and throat (bulbar muscles).

Another case of divergent gain-of-function and loss-of-function phenotypes in one gene is provided by the *PMP22* gene, which makes a peroxisomal membrane protein. Because the great majority of pathogenic mutations are duplications or deletions of a 1.4 Mb region at 17p11.2 that contains multiple genes in addition to *PMP22*, we consider the pathogenesis in the next section.

Multiple gene dysregulation resulting from aneuploidies and point mutations in regulatory genes

Some genes produce regulatory proteins or RNAs that regulate many different target genes. Examples include genes encoding master transcription factors and splicing regulators during development and genes involved in global epigenetic regulation, such as those that make chromatin modelers and DNA methyltransferases. These genes are typically dosage-sensitive, and heterozygotes with a loss-of-function mutation can often show complex phenotypes (see Table 6.6 on page 178 for some examples).

Whole-chromosome aneuploidies and large-scale intrachromosomal deletions and duplications (*segmental aneuploidies*) also directly affect multiple genes simultaneously, in this case by changing their copy number. However, because most genes are comparatively insensitive to dosage effects, the phenotype in affected heterozygotes is due to the combined effects of a comparatively small number of dosage-sensitive genes.

When whole chromosomes are involved, the number of dosage-sensitive genes can, however, be high, and reducing the gene copy number by 50% could be expected to have severe consequences. The cumulative effect of deleting one copy of each dosage-sensitive gene across a whole chromosome is simply too much to support embryonic or fetal development, and monosomies are almost always lethal.

One monosomy can sometimes be viable. The 45,X genotype leads to spontaneous abortion in about 99% of cases but occasionally leads to Turner syndrome, a comparatively mild condition in which women are short with certain minor physical abnormalities (such as webbed necks and low-set ears) and are sterile because of gonadal dysfunction. X-chromosome inactivation means that women are normally functionally hemizygous for most X-linked genes, but a few genes on the X chromosome, including genes in the pseudoautosomal regions, are not subject to X-inactivation, however.

Providing an extra gene copy for dosage-sensitive genes might be expected to have less harmful effects, but across a whole chromosome the combined effects of dosage imbalance in multiple genes means that most autosomal trisomies are also lethal. One might expect, however, that chromosomes with few genes might have correspondingly fewer dosage-sensitive genes; the three autosomal trisomies that are compatible with life—trisomies 13, 18, and 21—each involve chromosomes that have comparatively few genes.

Having extra sex chromosomes has far fewer ill effects than having an extra autosome because of X-inactivation (inactivating all X chromosomes except one) and the scarcity of genes on the Y chromosome. People with 47,XXX and 47,XYY karyotypes often function within the normal range, and in comparison with people with any autosomal trisomy, men with 47,XXY (Klinefelter syndrome) have relatively minor problems, notably hypogonadism and reduced fertility.

Segmental aneuploidies

Large-scale subchromosomal deletions and duplications also cause disease by simultaneously changing the copy number of multiple linked genes. If they occur in chromosomal regions that are constitutionally or functionally hemizygous, the functional copy number of some genes will be reduced to zero, so that no gene product is made at all. A profound effect can therefore be expected (surprisingly, for some of our gene loci, a complete absence of gene product does not result in a clinical phenotype; however, such genes are comparatively rare).

Males are constitutionally hemizygous for X- and Y-specific regions; large deletions in these regions therefore result in a complete absence of multiple gene products. The Y chromosome has few genes and they are very largely devoted to male-specific functions; accordingly, large deletions here are associated with azoospermia and infertility. Large deletions within the X chromosome in males are often lethal because of the high density of genes that perform a wide range of important functions. Certain regions such as Xp21 are comparatively gene-poor, however, and large deletions here can result in disease phenotypes. Occasionally, autosomal deletions also cause disease by reducing functional gene copy number to zero. Here, the phenotype results from the deletion of genes expressed from one allele only, as when an imprinted gene cluster at

15q11 is deleted (see Figure 7.9C, D), causing Angelman syndrome (when deletion occurs on the maternal chromosome 15) or Prader–Willi syndrome (deletion on paternal chromosome 15).

Large-scale deletions and duplications on autosomes can cause disease by changing the copy number of comparatively rare dosage-sensitive genes. Deletions reduce the functional gene copy number to one (so that disease results from haploinsufficiency in dosage-sensitive genes), and duplications increase gene copy number to three, resulting in the overexpression of multiple genes.

Contiguous gene syndromes

Subchromosomal deletions can be caused by occasional, random double-stranded breaks within DNA. The same clinical phenotype may be caused by deletions of differing size that eliminate the same key genes. Deletions that eliminate many genes often result in mental retardation because so many of our genes function in the developing brain.

Sometimes, the deletions result in a *contiguous gene syndrome* in which an affected individual has features of two or more quite different disorders. Males with a large deletion in Xp21 can have a combination of Duchenne muscular dystrophy, chronic granulomatous disease, retinitis pigmentosa, and mental retardation. Similarly, autosomal deletions spanning the *WT1* and *PAX6* genes at 11p13 result in WAGR syndrome (Wilm's tumor, aniridia, genitourinary abnormalities, and developmental delay) —much of the phenotype is due haploinsufficiency for *PAX6* and *WT1*.

Other clinical phenotypes typically result from recurring large deletions or duplications of very narrowly defined chromosome segments. In these chromosomal microdeletions and microduplications, the deletion or duplication is caused by recombination between low-copy-number flanking repeat sequences (**Table 7.12**).

Although large deletions and duplications can result in altered copy number for many genes, the phenotype might yet be due to a single gene. Deletion or duplication of a 1.4 Mb segment at 17p12 that contains about 15 genes is implicated in Charcot-Marie-Tooth type 1A (CMT1A) or hereditary neuropathy with liability to pressure palsies (HNPP; see Table 7.12). However, the pathogenesis is due to the *PMP22* gene—loss-of-function point mutations in *PMP22* produce the HNPP phenotype, and activating point mutations in *PMP22* result in the CMT1A phenotype.

Table 7.12 Examples of disorders resulting from recurrent large deletions and duplications brought about by recombination between low-copy-number repeats. [a]There are multiple *DAZ* genes on adjacent DNA repeats.

DISORDER	LOCATION	LENGTH OF RECOMBINING REPEATS	DELETION (Δ) OR DUPLICATION (DUP.) AND SIZE OF INTERVAL	KEY DISEASE LOCUS
Azoospermia type AZFc	Yq11.2	230 kb	Δ 3.5 Mb	*DAZ* family[a]
Angelman syndrome	15q11–q13	400 kb	paternal Δ 5 Mb	*UBE3A*
Prader–Willi syndrome			maternal Δ 5 Mb	*SNORD116*
Hereditary neuropathy with liability to pressure palsies	17p12	24 kb	Δ 1.4 Mb	*PMP22*
Charcot–Marie–Tooth type 1A			dup. 1.4 Mb	
DiGeorge syndrome/VCFS	22q11.2	225–400 kb	Δ 3 Mb or 1.5 Mb	*TBX1*
Smith–Magenis syndrome	17p11.2	175–250 kb	Δ 4 Mb	*RAI1*
Potocki–Lupski syndrome			dup. 4 Mb	
Williams–Beuren syndrome	7q11.2	300–400 kb	Δ 1.6 Mb	*ELN*
Sotos syndrome	5q25	400 kb	Δ 2 Mb	*NSD1*

7.6   A Protein Structure Perspective of Molecular Pathology

Until now we have looked at pathogenesis mostly from the perspective of altered gene expression or changes in protein sequence that either result in loss of function or produce a gain of function that does not necessarily involve a major change in protein structure. However, for many disorders the pathogenesis is due to major changes in protein *structure* that are typically induced by single nucleotide substitutions or other point mutations.

We briefly described elements of protein structure and folding in Section 2.1 and Box 2.2. In the next two sections we consider disease caused when proteins adopt altered structures and when changes in structure can predispose proteins to form aggregates that can cause disease. Understanding the basis of major changes in protein structure is important for understanding molecular pathology and developing novel therapeutic strategies.

Pathogenesis arising from protein misfolding

We have previously considered various aspects of how the expression of protein-coding genes is regulated to give a functional product, dwelling mostly on transcriptional, post-transcriptional, and translational control in Chapter 6, and touching on mRNA surveillance in Box 7.1. However, for proteins to function correctly they also need to fold properly to assume the correct three-dimensional conformation so that they can bind the appropriate interacting molecules. They need to function correctly within the right environment (in a hydrophilic environment, proteins fold up with hydrophobic amino acids located in the interior, and hydrophilic amino acids on the surface). Proteins also need to be able to interact correctly with other proteins when forming multimers.

Regulation of protein folding

Protein folding is not straightforward: there are various different paths that can be taken by an unfolded or partly folded protein to arrive at the final conformation, and natural errors in protein folding are common. Some proteins can fold correctly without help, but many proteins require assistance in folding from specific molecular chaperones such as Hsp60 or Hsp70 (chaperones are sometimes called *heat-shock proteins* and their names prefixed by *Hsp*, because their expression is drastically increased when cells are exposed to even moderate temperature increases, such as from 37°C to 42°C, that nevertheless cause an increase in protein misfolding).

Chaperones can help to fold both incompletely folded proteins and also some incorrectly folded proteins. However, when attempts to refold a protein are unsuccessful, the protein is shunted into a proteolytic pathway in which it is destroyed by the *proteasome*, a complex compartmentalized protease that is distributed in many copies throughout the cytosol.

Aberrant protein folding causing disease

Protein misfolding is a common cause of disease in many genetic disorders, such as cystic fibrosis and phenylketonuria, in which mutations that change a single amino acid are quite common. Thus, about 90% of individuals with cystic fibrosis have one or two copies of the p.Phe508del allele in which deletion of a single phenylalanine residue is sufficient to cause aberrant protein folding that cannot be rectified by chaperones. Whereas the normal protein would continue its normal journey to take up residence in the plasma membrane, the mutant protein is rapidly subjected to endoplasmic-reticulum-associated degradation.

Sometimes mutations destroy the ability to adopt highly specific structures that are required for the assembly of multimers such as collagens, fibrillins, and keratins. Collagens, for example, require the complex packaging of three collagen polypeptides into trihelical structures in which three individual collagen strands are wound round each other. As described above, missense mutations that replace glycines in collagen chains cause major packaging problems (see Figure 7.17B).

How protein aggregation can result in disease

The pathogenesis of many disorders, both monogenic and common diseases, involves the aggregation of proteins; this can produce soluble oligomers and large complexes that can be insoluble—the aggregated proteins are often manifested as cellular inclusions or pericellular deposits.

At present there is still some uncertainty about the significance of protein aggregates observed in many common diseases—are they a direct cause of disease, or are they more peripherally associated with pathogenesis? In the monogenic disorders there can be more certainty, although sometimes here the precise pathogenetic process is still not clear, as described in Section 7.3 for disorders involving the unstable expansion of polyglutamine repeats. In some monogenic disorders, however, the evidence for mutation-induced protein aggregation is clear; we give two examples below, one in which proteins aggregate to form extremely long protein fibers, and one in which the damage is done by protein aggregates in inclusion bodies within cells.

We end this section with the exceptional case of protein aggregation in prion diseases. Why are they exceptional? Well, they defy all the rules. The mutant proteins act as templates that are used to replicate the mutant protein shape. That kind of mechanism has sometimes been viewed as a type of epigenetic information transfer that does not involve nucleic acids (and so is quite distinct from the epigenetic mechanisms described in Section 6.3, which rely on heritable chromatin states). We consider the extent to which similar pathogenetic mechanisms might apply to other monogenic disorders and also to common neurodegenerative diseases such as Parkinson and Alzheimer disease.

Sickle-cell anemia: disruptive protein fibers

Normal adult hemoglobin is a tetramer with two α-globin chains and two β-globin chains. Individuals affected by sickle-cell anemia (PMID 20301551) are homozygous for a specific missense mutation that replaces a charged, hydrophilic glutamate residue at position 6 in the β-globin chain by a hydrophobic valine residue. The resulting mutant hemoglobin S (HbS) has a strong tendency to aggregate when deoxygenated, resulting in fibers composed of 14 long strands of HbS tetramers (**Figure 7.18A, B**). The fibers cause red blood cells to be deformed so that they are crescent-shaped, like a sickle. The much shorter life span of the abnormal sickle cells (10–20 days, in contrast with the normal 90–120 days) means that the body cannot replace dead red blood cells fast enough, resulting in anemia. The HbS fibers also block small blood vessels, causing hypoxic tissue damage.

The 14-strand structure is built on the lateral association of seven sets of paired HbS tetramer strands. The side chain of mutant valines on the β-globin chains of one HbS strand can interact with a complementary pocket on β-globin residues of the neighboring HbS tetramer. That type of bonding drives the formation of paired strands of HbS tetramers (**Figure 7.18C, D**) that then form the higher-order structure shown in Figure 7.18A, B by additional lateral associations.

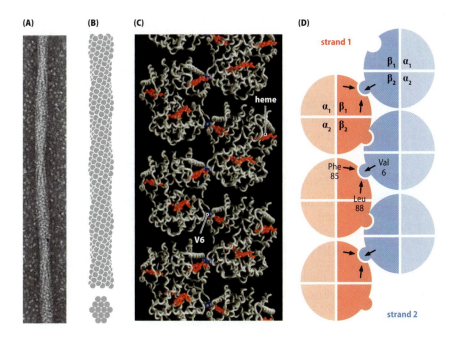

strand 1

β_1 α_1
β_2 α_2

heme

α_1 β_1
α_2 β_2

Phe 85 → ← Val 6

Leu 88

V6

strand 2

Figure 7.18 Aggregation of hemoglobin to form complex fibers in sickle-cell disease. (A, B) The 14-strand structure of deoxyhemoglobin S fibers including an electron micrograph in (A) of a stained fiber and the interpreted structure in (B), showing a lateral image at the top and a cross section at the bottom. The 14-strand structure is built from seven sets of paired strands. (C, D) The basic paired strand component of deoxyhemoglobin S fibers. (C) Structural model showing the mutant valine (V6; in blue color) located on the outside of the β-globin chains, facilitating lateral contact between β-globin chains on different HbS tetramers. Heme groups are shown in red. (D) Diagram illustrating how each double strand of hemoglobin tetramers is stabilized by lateral contacts involving the mutant valine on a β-globin chain from one strand interacting with a pocket formed between two helices on a β-globin chain on a Hb tetramer on the opposing strand. (A, B, From Dykes G, Crepeau RH & Edelstein SJ [1978] *Nature* 272:506–510. With permission from Macmillan Publishers Ltd; C, from Harrington DJ, Adachi K & Royer WE Jr [1997] *J Mol Biol* 272:398–407. With permission from Elsevier.)

α₁-Antitrypsin deficiency: inclusion bodies and cell death

α_1-Antitrypsin (α_1-AT) is made by the liver and secreted to regulate the levels of certain serine proteases such as elastase (which can be overproduced by neutrophils during inflammation and might damage sensitive tissue, such as the alveoli of lungs, if not kept in check—see Figure 7.16 above and the related text). α_1-Antitrypsin deficiency (PMID 20301692) is common in Caucasian populations in which two missense mutations are especially common: the mild PI*S allele (E264V) and the severe PI*Z allele (E342K).

Plasma concentrations of α_1-AT in ZZ homozygotes (about 15% of normal) and in SZ compound heterozygotes (about 40%) are not high enough to protect lungs from damage by elastase over a lifetime, especially in people who smoke. Affected individuals often develop emphysema, a form of chronic obstructive lung disease in which tissues needed to support the shape and function of the lungs are destroyed. The low plasma α_1-AT concentrations typically do not result from failure of liver cells to make any of the protein; instead the problem is a blockage in α_1-AT processing and secretion from liver cells.

The retained α_1-AT proteins can aggregate in the endoplasmic reticulum of hepatocytes to form intracellular inclusions (*inclusion bodies*) that can be readily recognized by using suitable stains and can be seen to contain bead-like polymerases in ZZ homozygotes (**Figure 7.19**). The inclusion bodies cause hepatocytes to die and can result in eventual cirrhosis of the liver, especially in ZZ homozygotes.

Seeding by aberrant protein templates

In Section 6.3 we explored epigenetic gene regulation in inherited disorders. That relies on heritable chromatin states rather than the DNA sequence, but yet another type of information that directs heritable changes in gene products is confined to the protein level. Prion diseases recently gained much public attention because of the danger to public health in eating meat from infected cattle ('mad cow disease'). Similar mechanisms allow the cellular spreading of protein aggregation in certain other neurodegenerative diseases such as amyotrophic lateral sclerosis, and Alzheimer and Parkinson disease (see **Box 7.7**).

(A)

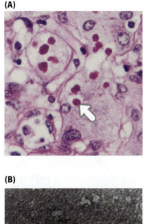

(B)

Figure 7.19 Intracellular inclusion bodies and protein aggregates in α1-antitrypsin deficiency. (A) Staining of hepatocytes with a periodic acid–Schiff stain reveals inclusion bodies as bright pink globules (arrowed). (B) Electron microscopy showing bead-like polymers of Z-type α_1-antitrypsin. (A, Courtesy of the National Society for Histotechnology; B, from Lomas DA, Finch JT, Seyama K et al. [1993] *J Biol Chem* 268:15333–15335. With permission from The American Society for Biochemistry and Molecular Biology.)

BOX 7.7 Prion diseases and prion-like neurodegenerative diseases: seeded spreading of protein aggregates between organisms and cells.

Prion diseases—also known as transmissible spongiform encephalopathies—are progressive, fatal, and incurable neurodegenerative disorders that affect humans and other animals, in which holes develop in brain tissues, giving them a sponge-like texture. The disease can be spread from one organism to another by ingesting or internalizing affected tissue. For example, consumption of affected tissue from cows with bovine spongiform encephalopathy has led to outbreaks of variant Creutzfeldt–Jakob disease (vCJD; also known in this specific instance as 'mad cow disease'). In addition to acquired prion protein disease, sporadic and hereditary forms exist. Creutzfeldt–Jakob disease (CJD), fatal familial insomnia, and Gerstmann–Straussler–Scheinker syndrome are dominantly inherited allelic disorders resulting from mutations in the *PRNP* prion protein gene at 20p13 (PMID 20301407).

In prion disease, a normal cellular form of prion (PrP^C) is misfolded into an abnormal conformation (PrP^{Sc}) that is rich in β-pleated sheets and is prone to aggregation (**Figure 1A**; the Sc superscript comes from scrapie, a sheep prion disease that was one of the first to be studied).

The most striking characteristic of PrP^{Sc} is that when it comes into contact with normal PrP^C proteins it can induce them to switch conformation so that they, too, adopt the PrP^{Sc} structure. Thus, if our cells are exposed to abnormal prion proteins from an infected animal or person, the abnormal foreign prion proteins will induce host PrP^C proteins to adopt the PrP^{Sc} structure (Figure 1B).

The abnormal prion protein structure can effectively self-propagate by a form of replication that has nothing to do with nucleic acid sequences. In that respect the disease mechanism resembles classical epigenetic mechanisms (which typically involve chromatin modifications). The abnormal prion proteins are infectious because the misfolded protein

can be acquired (by the ingestion of infected cells or tissue). Alternatively, prion proteins originate by a chance misfolding of a newly synthesized PrP^C protein in a sporadic case, or develop as a result of a genetic mutation (in which the mutant sequence has a greater propensity to misfold).

The brain is the main target of prion toxicity—neurons, being extremely long-lived and not effectively replaced, are especially vulnerable to toxic protein aggregates. How prions enter the body and infect brain cells is an interesting question. Somehow, the abnormal protein aggregates can get past mucosal barriers, survive the attentions of innate and adaptive immunity, pass across the blood–brain barrier, and spread to different brain cells. Infection can be efficient. vCJD has been transmitted to hemophiliacs who were treated with a factor VIII extract isolated from blood samples provided by donors who included subclinically infected blood donors. Growth hormone deficiency and infertility have also been treated in the past with growth hormones or fertility hormones recovered from human cadaveric pituitary glands, but because the pituitary extracts had been contaminated by infected human brain tissue, more than 160 treated people died of vCJD.

Amyloid diseases and prion-like neurodegenerative disease

Prion proteins are members of the family of amyloid proteins, which have a high content of β-sheets,

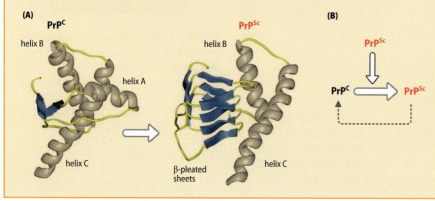

(A)

PrP^C

helix B

helix A

helix C

PrP^{Sc}

helix B

β-pleated sheets

helix C

(B)

PrP^{Sc}

PrP^C → PrP^{Sc}

Figure 1 Conversion of normal prion protein (PrP^C) to the aggregation-prone isoform PrP^{Sc}. (A) The normal PrP^C protein has three α-helices (A, B, and C in gray) and two very short β-strands (blue), but it can misfold to give the PrP^{Sc} isoform; because it has a very high content of β-strands that form a β-pleated sheet, the PrP^{Sc} isoform is susceptible to aggregation. (B) A PrP^{Sc} isoform that may have been induced by mutation, arisen spontaneously, or be derived from another person or an animal with prion protein disease (for example after blood transfusion or the ingestion of affected animal tissue) can induce a normal human PrP^C protein to change to the infectious PrP^{Sc} isoform. The latter can then induce other host PrP^C proteins to convert to PrP^{Sc}, spreading the disease between cells. (A, Adapted from Norrby E [2011] *J Intern Med* 270:1–14. With permission from John Wiley and Sons, Inc.)

BOX 7.7 (continued)

DISEASE	AMYLOID PROTEIN (PRECURSOR)
Non-neurodegenerative disorders	
Atherosclerosis	apolipoprotein A1
Rheumatoid arthritis	IAPP/amylin
Type 2 diabetes	serum amyloid A
Neurodegenerative disorders	
Alzheimer disease	β-amyloid/Aβ (APP); Tau
Amyotrophic lateral sclerosis (motor neuron disease)	SOD1
Frontotemporal lobar degeneration (FTLD)-tau[a]	tau
Huntington disease	huntingtin
Parkinson disease	α-synuclein
Prion protein diseases	PrP^{Sc} (PrP^{C})

Table 1 Examples of amyloid diseases. IAPP, islet amyloid polypeptide; APP, amyloid precursor protein; SOD1, superoxide dismutase 1. [a]The non-amyloid protein TDP-43 is also commonly found to be aggregated in FTLD.

making them prone to aggregation and the formation of elongated, unbranched amyloid fibrils. The spines of the fibrils consist of many-stranded β-sheets, arranged in a cross-β structure (**Figure 2**).

Amyloid proteins are frequently associated with disease (**Table 1**), and the aggregates may be extracellular (PrPSc; β-amyloid), nuclear (huntingtin), or cytoplasmic (SOD1; Tau; and synuclein, which forms

Lewy bodies). Amyloid protein aggregation is seen in some common diseases that are not associated with neurodegeneration, such as type 2 diabetes in which aggregates of serum amyloid A protein is found in the pancreatic islets of Langerhans. However, neurodegeneration is the most striking clinical characteristic of many amyloid diseases (see Table 1).

Neurodegenerative amyloid diseases, such as Alzheimer disease, Parkison disease, amyotrophic lateral sclerosis, and frontotemporal disease, resemble prion protein diseases in many ways and are sometimes classified as prionoid diseases. The direct involvement of the aggregated proteins in disease is supported from familial forms of these disorders in which mutations in the relevant gene promote the formation of amyloid protein, including mutations in *APP* (Alzheimer disease), *SNCA*, encoding α-synuclein (Parkinson disease), *SOD1* (amyotrophic lateral sclerosis), and *MAPT*, the microtubule-associated protein tau (frontotemporal dementia), for example.

There is no evidence from animal studies that the aggregated proteins in these disorders are infectious like prion proteins. But there is quite strong evidence that the pathogenesis resembles prion protein disease in two respects. First, like prion proteins, misfolded amyloid proteins in these disorders can induce the formation of the amyloid state in the normal proteins so that they aggregate. Secondly, for several of the disorders there is strong evidence for cell-to-cell spreading of the disorder; see the review by Polymenidou & Cleveland (2011) and the paper by Stöhr et al. (2012) in Further Reading.

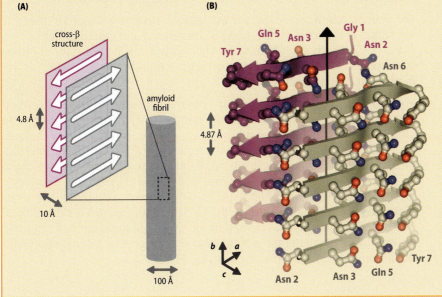

(A) cross-β structure

4.8 Å

amyloid fibril

10 Å

100 Å

(B) Gln 5 Asn 3 Gly 1 Asn 2 Tyr 7 Asn 6

4.87 Å

b a c

Asn 2 Asn 3 Gln 5 Tyr 7

Figure 2 Characteristics of amyloid fibers.
(A) Amyloid fibrils generally have a diameter of about 100 Å (10 nm) and are formed from the aggregation of proteins with a high β-sheet content. A characteristic property is the cross-β spine, a set of β-strands that run perpendicular to the axis of the fibril.
(B) Certain protein segments that are six to seven amino acids long bind to other copies with the same amino acid sequence to form two tightly interdigitating β-sheets. In the case of the prion protein, that sequence is GNNQQNY or Gly-Asn-Asn-Gln-Gln-Asn-Tyr, as shown here. (Adapted from Eisenberg D & Jucker M [2012] *Cell* 148:1188–1203. With permission from Elsevier.)

7.7 Genotype–Phenotype Correlations and Why Monogenic Disorders Are Often Not Simple

Assessing the effect of pathogenic variants on the phenotype is a component of the broader quest to understand genotype–phenotype correlations. If we know the genotype, to what extent can we predict the phenotype? This can be a difficult question to answer, even for monogenic disorders.

The effect of a pathogenic variant on the phenotype is not just dependent on the effect of the mutation on the ability of that allele to make its normal gene product and the interaction with the product made by the other allele. Genes from other loci can also influence the disease phenotype, as can environmental factors. It is now clear that monogenic disorders are often not as simple as sometimes described, and the division of genetic disorders into chromosomal, monogenic, and multifactorial disorders is a simplification.

The difficulty in getting reliable genotype–phenotype correlations

Interpreting the effect of a single pathogenic mutation, even in a well-defined fully characterized gene, is often not straightforward. Splicing mutations can be difficult to gauge; apparently harmless synonymous substitutions can be pathogenic; missense mutation effects are not easy to predict (loss-of-function, gain-of-function, or no clear effect?).

Even for nonsense and frameshifting mutations, the effects of the mutation may be hard to predict. Largely depending on where a premature termination codon is introduced within the mRNA, the effect may be to trigger mRNA destruction, with failure to make a protein, or to produce a mutant protein that may or may not have a gain of function (see Box 7.1). And who would have expected that deleting a single nucleotide in the 2.5 Mb dystrophin gene could produce the severe Duchenne form of muscular dystrophy, while deleting a 1 Mb region containing that same nucleotide along with a million others (including many coding exons) would result in a much milder form of muscular dystrophy? As we describe in Box 9.6 on page 360, part of the explanation lay in the differential effects of frameshifting and in-frame deletions, and the observation prompted a novel RNA therapy for Duchenne muscular dystrophy.

For individuals affected by an autosomal recessive disorder there is an added complication: the need to assess the combined effect of two mutant alleles. In such situations, the mutant alleles are typically loss-of-function mutations and the degree of overall residual function is the major determining factor.

For some disorders, such as many enzyme deficiencies, there is a good correlation between product levels and severity of the phenotype. In steroid 21-hydroxylase deficiency, for example, individuals with non-classical forms (later onset, mild) typically have 10–15% of residual enzyme activity, whereas in classical forms (congenital, severe) there is from about 2% residual enzyme activity, which usually manifests as a 'simple-virilizing' phenotype, to 0% enzyme activity in the most severe ('salt-wasting') form (see the beginning of Box 7.3 for the clinical phenotypes). Phenotypes due to deficient X-linked hypoxanthine guanine phosphoribosyltransferase activity also show significant correlation with the amount of residual enzyme activity (**Figure 7.20**).

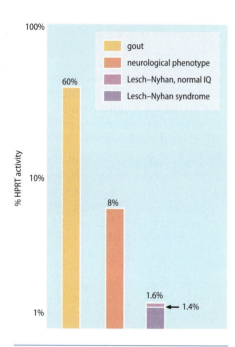

Figure 7.20 Different threshold levels for different phenotypes resulting from loss of activity of hypoxanthine guanine phosphoribosyltransferase (HPRT). Loss-of-function mutations in the X-linked *HPRT* gene can result in gout (which becomes manifest at less than 60% of normal HPRT activity). If the HPRT activity falls to below 8%, additional neurological features can begin to develop and are manifested as clumsiness and involuntary movements such as migrating contractions (chorea), and twisting and writhing (athetosis). A decrease in HPRT activity to less than 1.4% results in full Lesch–Nyhan syndrome (with choreathetosis and additional spasticity, self-mutilation, and mental retardation), but with an HPRT activity of about 1.4–1.6%, individuals with Lesch–Nyhan syndrome can have normal intelligence.

Exceptional versus general reasons for poor genotype–phenotype correlations

For many monogenic disorders, genotype–phenotype correlations can be extremely complicated. Sometimes affected individuals who have identical mutant alleles (affected members of the same family; genotyped affected individuals within a population) show remarkable differences in phenotype.

We have already considered some exceptional factors that contribute to poor genotype–phenotype correlations in some Mendelian disorders: epigenetic factors (notably, parent-of-origin effects, as described in Section 6.3); dynamic mutations (due to an unstable expansion of oligonucleotide repeats that can cause intergenerational differences in phenotype); and mosaicism (including differential X-chromosome inactivation that results in different effects for an X-linked mutation in women).

Disorders caused by mitochondrial mutations also show particularly poor genotype–phenotype correlation. Here, there is the exceptional problem of multiple copies of mtDNA per cell, so that cells can be **homoplasmic** for a mutant allele (all mtDNA copies carry the mutation) or **heteroplasmic** (a mix of normal and mutant mtDNA molecules). Because egg cells typically contain more than 100,000 mtDNA molecules, every child of an affected heteroplasmic mother will inherit at least some mutant mtDNA molecules, but the proportion is difficult to predict and the ratio of mutant to normal mtDNA copies can change over time. As a result, mutations in mtDNA can have low penetrance and rather unpredictable effects.

Modifier genes and environmental factors: common explanations for poor genotype–phenotype correlations

In addition to the exceptional factors described above, two *general* factors can explain differences in the phenotype of a monogenic disorder in affected members of the same family (who can be expected to have identical mutations in the case of a monogenic disorder) and in affected individuals within a population who have been revealed to have identical genotypes.

One of these is genetic variation at other loci. Genes that interact with the disease locus to modify the disease phenotype are known as **modifier genes** (the interaction between a disease locus and a modifier gene locus is called *epistasis*). Different alleles at a modifier locus can have different effects on a disease phenotype—they may sometimes have a protective effect (resulting in a milder disease phenotype) or an aggravating effect (inducing a more severe phenotype). The second general factor that influences a disease phenotype comes from the environment, as described below.

Modifier genes: the example of β-thalassemia

Until recently, modifier genes were not easy to identify directly in humans. Instead, heavy reliance was often placed on carrying out various types of analyses in animal disease models, which could then suggest candidate modifier genes for human diseases. Here we consider how modifier genes can affect the phenotype of a well-studied blood disorder, β-thalassemia.

Individuals with β-thalassemia have a genetic deficiency in β-globin, a component of hemoglobin. Although monogenic, this disorder is far from simple. It is usually autosomal recessive, but in occasional individuals the phenotype is dominant (one allele is normal; the other has an exceptional gain-of-function mutation). Although mutation in the β-globin gene, *HBB*, is the predominant factor in causing the disease, affected individuals with identical *HBB* alleles can show very significant differences in phenotype. Genetic variation at several modifier loci is also very important.

Adult hemoglobin is a tetramer with two α-globin chains and two β-globin chains; the synthesis of α- and β-globin chains is normally tightly regulated to ensure a 1:1 production ratio. However, when mutation in *HBB* results in a reduced production of β-globin chains, there will be a relative excess of α-globin chains. The excess α-globin monomers, present at high concentration, aggregate and precipitate, causing the death of early hemoglobin-producing cells in the bone marrow, and ineffective production of red blood cells. Those red blood cells that reach the peripheral blood also contain excess α-globin, which induces the formation of inclusion bodies and increased production of reactive oxygen species, leading to membrane damage and hemolysis. Because the anemia that results from lower numbers of red blood cells is life-threatening, current therapy is based largely on blood transfusions.

Genetic variation at other globin loci can affect the clinical severity of β-thalassemia. Thus, a mutation causing a reduced output of α-globin chains reduces the globin chain imbalance and allows the production of more red blood cells. Normal individuals usually have two tandemly repeated α-globin genes (*HBA1* and *HBA2*) on each chromosome 16, but as a result of unequal crossover the number of copies of the α-globin gene can vary: 0 (–); 1 (–α); 2 (αα); 3 (ααα); or 4 (αααα). Large numbers of α-globin genes can further add to the excess of α-globin chains that results from reduced β-globin production (**Figure 7.21**). As evidence of the modifier effect, individuals who are heterozygous for a null β-thalassemia (β⁰) allele but have a total of six or more α-globin genes can have a disease phenotype that resembles homozygous β-thalassemia).

The β-thalassemia phenotype is also modified by genetic variants that control the production of hemoglobin F (HbF), which is composed of two α-globin chains and two γ-globin chains. HbF is the dominant hemoglobin made during the fetal period (its high O$_2$-binding capacity makes it suited to working at the fetal stage), but there is a rapid decrease in HbF production at birth, and although it is still present at significant levels in infants it usually accounts for less than 1% of hemoglobin in adults. However, HbF can account for 10–40% of the hemoglobin in rare affected individuals with hereditary persistence of fetal hemoglobin, and there is significant variation between normal individuals in HbF levels. By forming more HbF, γ-globin polypeptides compensate for reduced production of β-globin. The elevated HbF levels in infants is thought to be protective, explaining the delayed onset of symptoms in β-thalassemia, and comparatively high HbF levels at later stages may be partly protective.

Many of the complications of the disease are also modified by genetic variation at other loci (see Figure 7.21). There are also differences in the patterns of adaptation to anemia at different ages, and environmental factors, notably exposure to malaria, can also modify the phenotype.

Environmental factors influencing the phenotype of genetic disorders

In some disorders, expression of the disease phenotype depends very significantly on environmental factors that may act at different levels: at a distance (external radiation sources); by direct exposure of our cells to harmful or potentially harmful chemicals that we ingest in food and drink or inhale (such as tobacco smoke or atmospheric pollution); and by contact with microbes and toxins.

Environmental factors are especially important in triggering cancers, as described in Chapter 10. Environmental factors are also very important in

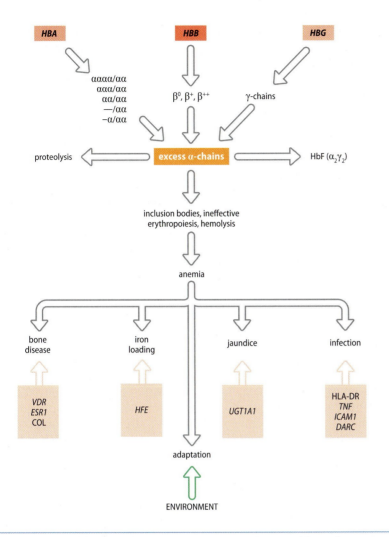

Figure 7.21 Multiple factors determine the β-thalassemia phenotype. The basic pathology of β-thalassemia results from excess α-globin chains damaging red blood cell precursors and red blood cells. Depending on the mutations at the disease locus *HBB*, there can be different levels of decrease in β-chain production (β^0, null; β$^+$, partial function; and so on) with a direct effect on the amount of excess α-globin chains. The phenotype can also vary as a result of variation in α-globin copy number, variation in the ability to produce HbF after birth (which uses up variable amounts of α-chains), and, possibly, through different rates of removal of α-chains by proteolysis. The many complications of the resulting anemia can also be modified by genetic variability, including variation in the genes listed at the bottom. Genes or gene loci are: *HBB*, β-globin; HBA, α-globin loci; HBG, γ-globin loci; *VDR*, vitamin D receptor; *ESR1*, estrogen receptor-1; COL, collagen loci; *HFE*, locus for hereditary hemochromatosis; *UGT1A1*, UDP glucuronyltransferase involved in bilirubin metabolism; HLA-DR, major histocompatibility complex loci; *TNF*, tumor necrosis factor; *ICAM1*, intercellular adhesion molecule-1; *DARC*, Duffy antigen receptor for chemokines. (Adapted from Weatherall D [2010] *Nature Med* 16:1112–1115. With permission from Macmillan Publishers Ltd.)

common diseases, whether at the earliest stages of development (factors in the uterine environment) or at later stages (such as exposure to chemicals and microbes). We consider some aspects in Chapter 8.

Environmental factors are also known to be important in some single-gene disorders. We illustrate the example of how dietary factors can influence disease with reference to phenylketonuria in **Box 7.8**. In Chapter 9 we also consider how differential sensitivity to drugs can influence other monogenic disorders, within the broader context of pharmacogenetics.

BOX 7.8 Phenylketonuria as an inborn error of metabolism, a multifactorial condition, and an embryofetopathy.

The first genetic disorders that were investigated at the molecular level were *inborn errors of metabolism*. Affected individuals lacked a single enzyme that catalyzed one step in a metabolic pathway (usually consisting of a series of enzyme-catalyzed steps in which the product of one step becomes the substrate for the next step). Deficiency in one such enzyme would cause a metabolic block (**Figure 1A**). The resulting buildup of the substrate proximal to the block might drive an alternative pathway (red arrow). By analyzing blood and urine samples, pioneers in the field were able to obtain molecular clues as to the cause of a genetic disorder many decades before we knew about DNA structure and were able to study genes.

Phenylketonuria (PMID 20301677) was one of the earliest inborn errors of metabolism to be studied; it results from a deficiency of phenylalanine hydroxylase, a liver enzyme that converts phenylalanine to tyrosine (Figure 1B). Genetic deficiency in this enzyme results in elevated levels of phenylalanine (hyperphenylalaninemia) that can be

sub-clinical (120–600 µmol/liter) or result in mild or classical phenylketonuria (600–1200 µmol/liter or more than 1200 µmol/liter, respectively, in untreated individuals).

Elevated phenylalanine concentrations drive the production of phenylketone derivatives (see Figure 1B), which are excreted. The clinical symptoms of phenylketonuria are largely due to the toxic effects of very high phenylalanine levels in the brain—untreated children show progressively impaired brain development, leading to severe intellectual disability and various other symptoms including behavioral problems.

The standard treatment is really a form of prevention. Infants identified as having very high levels of blood phenylalanine are placed on a low-phenylalanine diet that is generally successful (although there may be problems with compliance in later years). The low-phenylalanine diet works because phenylketonuria is really a multifactorial disorder—two factors are absolutely required for the disease to manifest itself: a genetic factor (mutations at the *PAH* locus

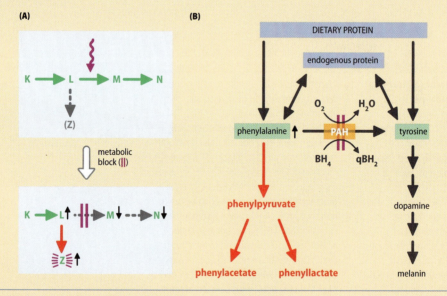

Figure 1 Metabolic blocks: principles and the example of phenylketonuria. (A) Principle of a metabolic block. Metabolites K, L, M, and N are linked by a series of enzyme-catalyzed reactions (green arrows) in which the product of an enzyme step serves as the substrate for the next enzyme. Here, as a result of genetic deficiency, there is a lack of the enzyme that converts L to M, leading to low concentrations of M, with a knock-on effect for the next step and a decreased concentration of N. The substrate L, proximal to the block, increases in concentration and that may lead to excessive production of metabolite Z, which becomes a biomarker of the disease.

(B) Phenylalanine is converted to tyrosine by phenylalanine hydroxylase (PAH), which requires the cofactor tetrahydrobiopterin (BH_4). When mutations cause homozygous deficiency in PAH, the conversion of phenylalanine to tyrosine is blocked (double magenta bar). As a result, high levels of phenylalanine build up (hyperphenylalaninemia), driving the production of new phenylalanine metabolites (shown by red arrows). Three phenylketones are produced (phenylpyruvate, phenylacetate, and phenyllactate) and are excreted. Deficiency of different genes involved in BH_4 metabolism can also cause hyperphenylalaninemia.

BOX 7.8 (*continued*)

causing homozygous deficiency of phenylalanine hydroxylase) and an environmental factor (normal L-phenylalanine levels in dietary protein).

Phenylketonuria is classified as a monogenic disorder only because the vast majority of us are exposed to the environmental factor. Because affected sibs can show significant differences in clinical phenotype, modifier genes are also likely to be involved. The phenotype could be influenced by genetic variation in different processes (such as protein degradation, phenylalanine transport and disposal, transport of phenylalanine across the blood–brain barrier, brain sensitivity to phenylalanine toxicity).

Very high levels of phenylalanine in phenylketonuria can be teratogenic and can result in *embryofetopathy*. A homozygous mother (who might nevertheless have a mild phenotype that could go unrecognized) can have heterozygous offspring who go on to develop mental retardation. During pregnancy the placenta naturally selects for higher concentrations of amino acids; as a result, phenylalanine levels may double in fetal blood, causing serious damage to brain and some other organ systems during development. Again, this can be prevented or ameliorated if the expectant mother is placed on a low-phenylalanine diet from the earliest stages of pregnancy.

Summary

- A small fraction of genetic variation causes disease by changing the expression of gene products (by altering gene copy number, or through point mutations) or by changing the sequence of a gene product.

- The genetic code is redundant (most amino acids can be specified by multiple different codons) and nearly universal (mitochondria use a slightly different code).

- Synonymous single nucleotide substitutions (silent mutations) replace one codon by another without changing the amino acid. They occasionally cause disease by altering RNA splicing.

- Nonsynonymous substitutions replace an amino-acid-specifying codon by a codon that specifies another amino acid (missense mutation) or by a stop codon (nonsense mutation).

- A missense mutation is likely to be pathogenic if the replacement amino acid is physicochemically very different from the original amino acid.

- Splicing mutations often alter important splice junction sequences. Additional splicing mutations change other important splice regulatory sequences within exons and introns, or activate a cryptic splice site to make a novel splice site.

- Insertions or deletions can produce a shift in the translational reading frame if the resulting number of nucleotides in a coding sequence is no longer a multiple of three. Frameshift mutations usually introduce an in-frame premature stop codon. RNA splicing mutations can also cause a translational frameshift at the RNA level.

- An in-frame premature termination codon often signals degradation of the mRNA (nonsense-mediated decay), but if the premature stop codon is close to the normal stop codon a truncated protein is usually produced that may sometimes result in a more severe phenotype.

- The CG dinucleotide is often a target for cytosine methylation and is a hotspot for C→T mutations because the resulting 5-methylcytosine is prone to deamination to give a thymine.

- Runs of small tandem repeats are prone to replication slippage and often cause frameshifting insertions and deletions.

- Long arrays of certain tandem oligonucleotide repeats can undergo unstable expansion. Such dynamic mutations show meiotic and mitotic instability and can result in expression products that are toxic to cells or otherwise have positively harmful effects.

- Non-allelic homologous recombination usually means a sequence exchange that occurs after pairing of non-allelic repeats that have highly similar sequences.

- Reciprocal exchange between mispaired tandem repeats (unequal crossover) or between distantly spaced repeats can result in changes in copy number, resulting in deletions or duplications. In alternative nonreciprocal exchanges the sequence of one copy is replaced in part by the sequence of another copy (gene conversion).

- Exchange between inverted repeats on the same strand can produce pathogenic inversions.

- Breaks in the DNA of one chromosome can result in subchromosomal deletions, inversions, and also ring chromosomes (formed after a chromosome has lost a terminal segment on each arm and the two broken ends join up).

- Translocations occur when two chromosomes undergo breakages and then exchange fragments. A balanced translocation means that there has been no obvious net loss of DNA.

- Aneuploidy involves the gain or loss of whole chromosomes. The effects on the phenotype are due to a minority of genes that are especially dosage-sensitive.

- Loss-of-function mutations can result in complete absence or complete functional inactivity (a null allele), reduced expression, or an altered product with reduced functional activity.

- Haploinsufficiency means that a loss of function of one allele causes a phenotype in the presence of a working normal allele.

- Dominant-negative mutations result in a mutant gene product that somehow impairs the activity of the normal allele in a heterozygote.

- Gain-of-function mutations have a phenotypic effect in the presence of a normal allele that cannot be compensated for by producing more of the normal gene product.

- Loss-of-function and gain-of-function mutations in the same gene can result in different phenotypes.

- Different components of a phenotype may be manifested at different threshold levels of gene function.

- Some proteins, notably prion proteins, can misfold to give a structure that is prone to self-aggregation and can induce other normally folded versions of the protein to misfold, seeding protein aggregation and thus causing disease.

• Predicting the phenotype from the genotype is often difficult, even for a monogenic disorder. The effect of some types of mutation can be difficult to predict, and the phenotype is often influenced by genetic differences at other gene loci (modifier genes) and by environmental factors.

Questions

Help on answering these questions and a multiple-choice quiz can be found at www.garlandscience.com/ggm-students

1. In the sequence below, the blue nucleotides represent an exon containing coding DNA near the beginning of a large gene, and green lines and letters are the flanking intron sequence. Nine mutations are shown: single nucleotide changes below (1–7) and two large-scale mutations above (8, 9). Red dashes indicate deleted nucleotides. Which mutation class does each belong to? Comment on the likely effect of each.

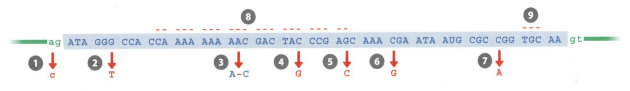

2. What is meant by the term dynamic mutations?

3. What are the different characteristics of pathogenic unstable expansion of oligonucleotide repeats in coding DNA and noncoding DNA?

4. In the example below, there has been mispairing of chromatids during meiosis so that a single-copy protein-coding gene, A, and a closely homologous pseudogene, ψA, with multiple inactivating mutations (red asterisks) pair up. Illustrate two types of sequence exchange between these mispaired repeats that can give rise to aberrant sequences causing disease.

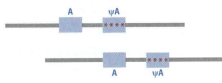

5. In the figure below, two very highly homologous repeats, 1 and 2, occur on one chromatid in the same orientation (direct repeats, shown on the left) or in opposite orientations (inverted repeats, shown on the right). The intervening DNA, shown as an X with a green start sequence and a red end sequence, can loop out, allowing the repeats to align in the same orientation and to engage in intrachromatid recombination. Illustrate how this happens and explain the consequences for each case.

6. What is meant by a loss-of-function mutation and a gain-of-function mutation? To what extent do they give rise, respectively, to recessive and dominant phenotypes?

7. Abnormal protein structure is a key cause of disease in certain single-gene disorders. Give three examples of diseases like this, and a brief outline of how the abnormal protein structure causes disease.

8. Genotype–phenotype correlations can be poor for many monogenic disorders. That may be due to different genetic and nongenetic contributing factors. For each of the contributing factors listed below, give an example of a disorder (or class of disorders) in which they have a prominent effect and illustrate how the effect causes phenotype variability:
   • modifier genes
   • cellular mosaicism
   • epigenetic effects
   • environmental factors.

Further Reading

Amino Acid Substitutions and Silent Mutations

Betts MJ & Russell R (2007) Amino-acid properties and consequences of substitutions. In Bioinformatics for Geneticists, 2nd ed (Barnes MR ed.), pp 311–341. Wiley-Blackwell.

Boyko AR et al. (2008) Assessing the evolutionary impact of amino acid mutations in the human genome. *PLoS Genet* 4:e100083; PMID 18516229.

Grantham R (1974) Amino acid difference formulae to help explain protein evolution. *Science* 185:862–864; PMID 4843792. [Gives a matrix that quantifies the effect of amino acid substitutions.]

Ng PC & Henikoff S (2006) Predicting the effects of amino acid substitutions on protein function. *Annu Rev Gen Hum Genet* 7:61–80; PMID 16824020.

Sauna ZE & Kimchi-Sarfaty C (2011). Understanding the contribution of synonymous mutations to human disease. *Nature Rev Genet* 12:683–691; PMID 21878961.

Nonsense-Mediated Decay (NMD)

Bhuvanagiri M et al. (2010) NMD: RNA biology meets human genetic medicine. *Biochem J* 430:365–377; PMID 20795950.

Popp MW & Maquat LE (2013) Organizing principles of mammalian nonsense-mediated mRNA decay. *Annu Rev Genet* 47:139–165; PMID 24274751.

Splicing and Regulatory Mutations

Cartegni L et al. (2002) Listening to silence and understanding nonsense: exonic mutations that affect splicing. *Nature Rev Genet* 3:285–298; PMID 11967553.

Jarinova O & Ekker M (2012) Regulatory variations in the era of next-gen sequencing: implications for clinical molecular diagnostics. *Hum Mutat* 33:1021–1030; PMID 22431194.

Sterne-Weiler T et al. (2011) Loss of exon identity is a common mechanism of human inherited disease. *Genome Res* 21:1563–1571; PMID 21750108.

Wang GS & Cooper TA (2007) Splicing in disease: disruption of the splicing code and the decoding machinery. *Nature Rev Genet* 8:749–761; PMID 17726481.

Mutation Rates, Mutation Load, and Noncoding DNA Mutations

Conrad DF et al. (2011) Variation in genome-wide mutation rates within and between human families. *Nature Genet* 43:712–714; PMID 21666693.

Goriely A & Wilkie AOM (2012) Paternal age effect mutations and selfish spermatogonial selection: causes and consequences for human disease. *Am J Hum Genet* 90:175–200; PMID 22325359.

Keightley PD (2012) Rates and fitness consequences of new mutations in humans. *Genetics* 190:295–304; PMID 22345605.

MacArthur DG et al. (2012) A systematic survey of loss-of-function variants in human protein-coding genes. *Science* 335:823–828; PMID: 22344438.

Makrythanasis P & Antonorakis S (2013) Pathogenic variants in non-protein-coding sequences. *Clin Genet* 84:422–428; PMID 24007299.

Xue Y et al. (2012) Deleterious- and disease-allele prevalence in healthy individuals: insights from current predictions, mutation databases, and population-scale resequencing. *Am J Hum Genet* 91:1022–1032; PMID 23217326.

Pathogenic Unstable Expansion of Short Tandem Repeats

Bañez-Coronel M et al. (2012) A pathogenic mechanism in Huntington's disease involves small CAG-repeated RNAs with neurotoxic activity. *PLoS Genet* 8:e1002481; PMID 22383888.

Hagerman R & Hagerman P (2013) Advances in clinical and molecular understanding of the FMR1 premutation and fragile X-associated tremor/ataxia syndrome. *Lancet Neurol* 12:786–798; PMID 23867198.

Klein AF et al. (2011) Gain of RNA function in pathological cases: focus on myotonic dystrophy. *Biochimie* 93:2006–2012; PMID 21763392.

Orr HT & Zoghbi HY (2007) Trinucleotide repeat disorders. *Annu Rev Neurosci* 30:575–621; PMID 17417937.

Santoro MR et al. (2012) Molecular mechanisms of fragile X syndrome: a twenty year perspective. *Annu Rev Pathol Mech Dis* 7:219–245; PMID 22017584.

Wojciechowska M & Krzyzosiak WJ (2011) Cellular toxicity of expanded RNA repeats: focus on RNA foci. *Hum Mol Genet* 20:3811–3821; PMID 21729883.

Gene Conversion and Non-Allelic Homologous Recombination

Chen JM et al. (2007) Gene conversion: mechanisms, evolution and human disease. *Nature Rev Genet* 8:762–775; PMID 17846636.

Liu P et al. (2012) Mechanisms for recurrent and complex human genomic rearrangements. *Curr Opin Genet Dev* 22:1–10; PMID 22440479.

Chromosome Nomenclature and Abnormalities

Nagaoka SI et al. (2012) Human aneuploidy: mechanisms and new insights into an age-old problem. *Nature Rev Genet* 13:493–504; PMID 22705668.

Roukos V et al. (2013) The cellular etiology of chromosome translocations. *Curr Opin Cell Biol* 25:357–364; PMID 23498663.

Schaffer LG, McGown-Jordan J, Schmid M (eds) (2013) ISCN 2013: An International System for Human Cytogenetics Nomenclature. Karger.

Gain-of-Function Mutations

Carrell RW & Lomas DA (2002) α_1-antitrypsin deficiency—a model for conformational diseases. *N Eng J Med* 346:45–54; PMID 11778003. [Includes the Pittsburgh gain-of-function mutation.]

Lester HA & Karschin A (2000) Gain-of-function mutants: ion channels and G protein-coupled receptors. *Annu Rev Neurosci* 23:89–125; PMID 10845060.

Mitochondrial DNA and Disease

Greaves LC et al. (2012) Mitochondrial DNA and disease. *J Pathol* 226:274–286; PMID 21989606.

Schon EA et al. (2012) Human mitochondrial DNA: roles of inherited and somatic mutations. *Nature Rev Genet* 13:878–809; PMID 23154810.

Haploinsufficiency and Molecular Basis of Genetic Dominance

Veitia RA & Birchler JA (2010) Dominance and gene dosage balance in health and disease: why levels matter! *J Pathol* 220:174–185; PMID 19827001.

Wilkie AOM (1994) The molecular basis of genetic dominance. *J Med Genet* 31:89–98; PMID 8182727.

Protein Misfolding and Protein Aggregation in Disease

Eisenberg D & Jucker M (2012) The amyloid states of proteins in human diseases. *Cell* 148:1188–1203; PMID 22424229.

Gregersen N et al. (2006) Protein misfolding and human disease. *Annu Rev Genomics Hum Genet* 7:103–124; PMID 16722804.

Polymenidou M & Cleveland DW (2011) The seeds of neurodegeneration: prion-like spreading in ALS. *Cell* 147:498–508; PMID 22036560.

Stöhr J et al. (2012) Purified and synthetic Alzheimer's amyloid beta (Ab) prions. *Proc Natl Acad Sci USA* 109:11025–11030; PMID 22711819.

Westermark GT & Westermark P (2010) Prion-like aggregates: infectious agents in human disease. *Trends Mol Med* 16:501–507; PMID 20870462.

Modifier Genes

Drumm ML et al. (2012) Genetic variation and clinical heterogeneity in cystic fibrosis. *Annu Rev Pathol* 7:267–282; PMID 22017581.

Hamilton BA & Yu BD (2012) Modifier genes and the plasticity of genetic networks in mice. *PLoS Genet* 8:e1002644; PMID 22511884.

Sankaran VG et al. (2010) Modifier genes in Mendelian disorders: the example of hemoglobin disorders. *Ann NY Acad Sci* 1214:47–56; PMID 21039591.

Identifying Disease Genes and Genetic Susceptibility to Complex Disease

Until recently, the molecular identification of rare genes for monogenic disorders was laborious, often consuming many years of painstaking effort. Now, in the era of massively parallel DNA sequencing (next-generation sequencing), it is becoming routine. We cover the principles in Section 8.1. Some difficulties remain, however, because for some single-gene disorders the disease phenotypes do not have a very well-defined, distinctive pathology. And a good deal of follow-up work will be needed to dissect out all the factors in monogenic diseases, which are sometimes rather complex, as described in Section 7.7.

The next big challenge has been to identify genes underlying complex (multifactorial) diseases, in which there is no obviously predominant disease locus (at least, not to the extent found in monogenic disorders). Instead, expression of the disease phenotype may be dependent on a few genes (oligogenic disorders) or many genes (polygenic disorders), with variable (and sometimes very strong) contributions from environmental factors. We cover the background to complex disease and polygenic theory in Section 8.2.

The genetic contribution to complex diseases differs according to the disease and between populations, and its overall impact can vary within a single population, depending on changeable environmental conditions. Investigations into the genetic susceptibility to multifactorial disease began decades ago but had limited success until the mid-2000s. More recently, however, many DNA variants have been identified to confer susceptibility to complex diseases (genetic risk factors) and some have been shown to lower disease susceptibility (protective factors). We outline the different general approaches used to uncover the genetic susceptibility to complex diseases in Section 8.2.

In Section 8.3 we consider the progress that has been made in understanding the genetic contribution to complex disease, and we set that in the context of approaches to understand gene–environment interactions and the contribution made by epigenetic factors. We consider investigations of common cancers separately in Chapter 10.

8.1 Identifying Genes in Monogenic Disorders

The identification of genes underlying monogenic disorders began with a very few exceptional cases in the 1970s and early 1980s; the genes were able to be identified through a known protein product (functional cloning), or as a result of huge enrichment of corresponding mRNAs in certain cell types. For example, hemophilia A results from a deficiency of blood clotting factor VIII, and enough of the factor VIII protein was purified from pig blood to obtain a partial amino acid sequence. By consulting the genetic code it was possible to design degenerate oligonucleotides (a panel of different but related oligonucleotides to cover all codon possibilities) for an optimal sequence of amino acids. The resulting oligonucleotides were used as probes to screen DNA libraries, identifying first homologous cDNA clones and eventually human gene clones. The genes for α- and β-globin, which are mutated in α- and β-thalassemia, were also readily purified because they are expressed to give most of the mRNA in precursors to red blood cells.

Candidate gene approaches could occasionally be used to identify a human disease gene. Knowledge of the biology of the condition would suggest plausible gene candidates, for example. Other candidate genes often came from observed similarities between the disease phenotype and a highly related phenotype in humans or animals. After the *FBN1* fibrillin gene was shown to be a locus for Marfan syndrome, for example, the related *FBN2* fibrillin gene was investigated and shown to be mutated in a very similar disorder, congenital contractural arachnodactyly. And after the *Sox10* gene was identified as the locus for the *Dominant megacolon* (*Dom*) phenotype in mouse, the human *SOX10* was quickly shown to be mutated in Waardenburg–Hirschsprung disease (*Dom* is a model of Hirschsprung disease, but the mutant mice also have pigmentary abnormalities reminiscent of Waardenburg syndrome.)

Position-dependent strategies

More general strategies were developed to identify disease genes for which little was known about the kind of gene product they might make. They relied on first establishing a chromosomal position for the disease gene. After that, effort could be concentrated on finding genes that map in that region and testing them to see whether they were significantly mutated in the disease of interest.

The first position-dependent strategies involved hugely laborious *positional cloning*: DNA clones had to be mapped to that chromosomal region, and then painstaking efforts were made to find gene sequences which then had to be characterized in detail to yield candidate genes. However, as the genome began to be deciphered, all of our protein-coding genes began to be mapped to specific subchromosomal regions and to be studied. That ushered in comparatively simple *positional candidate* strategies for identifying disease genes. Genes in the subchromosomal region of interest could now be identified by consulting gene, genome, and literature databases, further studied as required, and then prioritized for mutation screening, according to their known characteristics.

Position-dependent strategies require previous evidence suggesting a promising subchromosomal location for a disease gene of interest, but for decades that had been problematic. Disorders that seemed to show X-linked inheritance did at least narrow the search to just one chromosome—but that still left 155 Mb of DNA to examine for the gene of interest. As described below, more precise locations can be obtained from disease-associated chromosomal abnormalities, or by screening for large-scale mutations, but abnormalities like these are rare.

A more general strategy for finding subchromosomal locations for disease genes became possible using genetic mapping. That began by mapping polymorphic DNA markers to specific chromosomes and thence to subchromosomal regions, paving the way for a human genetic map. Once that had been achieved, the genes underlying a wide range of monogenic disorders could be mapped to subchromosomal locations by linkage analyses (as explained below). Ultimately, detailed gene maps were obtained for each chromosome; they were enormously helpful for positional candidate approaches with which to identify disease genes.

The final step: mutation screening

Whichever method is used to identify candidate disease genes, the next step is to prioritize some of them for testing; computational approaches can help here—see Piro & Di Cunto (2012) under Further Reading. Promising candidate genes are tested for evidence of disease-associated mutations (by testing multiple unrelated affected individuals plus normal controls). For a highly penetrant dominant disorder, pathogenic mutations should normally be found in affected individuals only; for a recessive disorder, the pathogenic mutations will occasionally be found in normal individuals (who might then be suspected to be heterozygous carriers).

A tiny number of RNA genes have been identified as loci for monogenic disorders (see Table 7.4 on page 200); the vast majority of pathogenic mutations have been identified in protein-coding genes. Loss-of-function mutations are expected in recessive conditions, and in dominant disorders resulting from haploinsufficiency. They are relatively easy to identify in protein-coding genes because they often occur in coding DNA sequences (with frequent changes to the specified protein sequence) and/or cause altered splicing (often at positions at or close to exon–intron boundaries). Gain-of-function mutations often involve specific missense mutations (which would not be expected in controls).

Mutation screening typically involves amplifying individual exons and the immediately surrounding intronic sequence from a candidate gene and sequencing them to identify mutations associated with disease. As we will see in the final part of Section 8.1, genomewide gene or exon sequencing can often dispense with the time-consuming need to first identify candidate disease genes.

Linkage analysis to map genes for monogenic disorders to defined subchromosomal regions

Genetic markers (polymorphic loci) from across the genome can be used to track the inheritance of a gene by using **linkage analysis**. Different types of linkage analysis can be carried out, but success usually depends on having suitably informative families with multiple affected individuals. Because human family sizes are generally very small, multiple different families need to be investigated. Hundreds of genetic markers are needed, from defined locations distributed across the genome. That became possible with the development of human genetic maps.

Human genetic maps

Long before human genetic mapping began, genetic maps had been available for many model organisms. They were obtained by mapping gene mutations that cause readily identifiable phenotypes. In *Drosophila*, for example, crosses can easily be set up to breed mutant white-eyed flies with flies that have abnormal curly wings; the progeny are then examined to see if the two mutant phenotypes segregate together or not. In humans, however, that kind of approach could never be applied—a different strategy was needed.

Instead of having a genetic map based on gene mutations, the solution to making a human genetic map was to identify *general* DNA variants (that mostly do not map within coding sequences, and that usually had no known effects on the phenotype). Different types of variants were identified and mapped to specific locations in our genome, beginning with restriction fragment length polymorphisms (RFLPs) that created or destroyed a restriction site (see Figure 4.6), followed by microsatellite polymorphisms that varied in the copy number of short tandem repeats (see Figure 4.7).

The first comprehensive map of human genetic markers (polymorphisms) did not appear until 1994. Based on microsatellite and restriction site polymorphisms, it had a marker spacing of just over one marker per megabase of DNA. Microsatellite markers have the advantage that they are highly polymorphic (with multiple alleles, each having a different number of copies of the repeat), whereas restriction site polymorphisms often have just two alleles.

The most recent maps are based on single nucleotide polymorphisms (SNPs). They also have limited polymorphism, with often just two alleles. But they have two very strong advantages: they are extremely abundant in the human genome, and they are amenable to automated typing. By October 2008 the International HapMap Project had defined 3.9 million unique human SNPs, giving an extremely high resolution human genetic map (on average, one SNP per 700 bp).

Data on individual SNPs can be accessed at the dbSNP database (http://www.ncbi.nlm.nih.gov/SNP). Identifying reference numbers are composed of a seven-digit to nine-digit number prefixed by rs, such as rs1800588 (rs is an abbreviation for reference SNP). The database can also be queried with a gene symbol to find SNPs in a specific gene; the resulting data can be filtered progressively to get human SNPs, and then SNPs that are of clinical interest or that have been recorded in the corresponding locus-specific database.

Principle of genetic linkage

One fundamental principle underlies genetic linkage: alleles at very closely neighboring loci on a DNA molecule are co-inherited because the chance that they are separated by recombination is very low. By extension, alleles at distantly spaced loci on the same DNA molecule are much more likely to be separated by recombination at meiosis. (During human meiosis, chromosomes are often split by recombination into between two and seven segments—see **Figure 8.1** for an example.)

A **haplotype** is a series of alleles at two or more neighboring loci on a *single* chromosomal DNA molecule. In human genetics, the term was first widely used within the context of the HLA system (readers who are unclear about haplotypes might wish to have a look at Figure 2 in Box 4.4 on page 113 to see how haplotypes are interpreted after obtaining genotypes from multiple family members).

We illustrate the principle of a *disease haplotype* in **Figure 8.2** within the context of a gene for an autosomal dominant disorder. The marker loci flanking the disease locus in Figure 8.2 are imagined to be very close to the disease locus. Recombination within this region would be extremely rare—the marker loci would be said to be *tightly linked* to the disease locus.

In Figure 8.2B the disease haplotype is transmitted unchanged through four meioses: from the grandfather (I-1) to his affected son and daughter, and then to the two affected grandchildren (III-2 and III-3). For marker loci that are increasingly distant from the disease locus, the incidence of recombination between disease locus and marker locus becomes progressively greater.

As the distance separating a marker locus and the disease locus on the same chromosome increases, a point will be reached where the chance of recombination between the two loci would equal the chance of no recombination between them. The marker would then be said to be *unlinked* to the disease locus (it would be no different from a marker on a different chromosome, for which an allele would have a 50% chance of segregating with disease, just by chance).

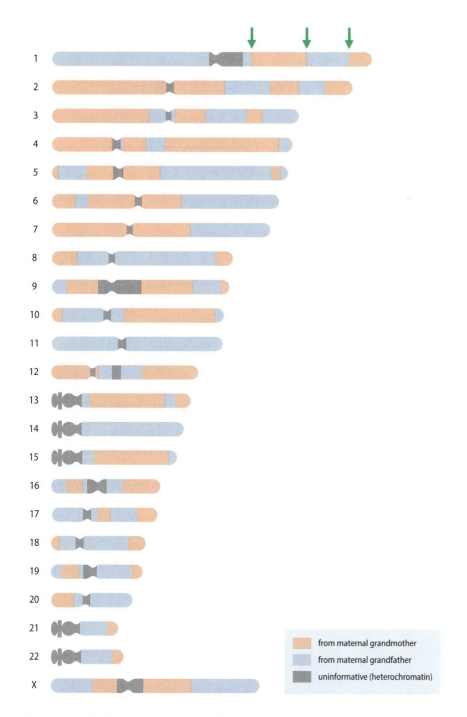

from maternal grandmother

from maternal grandfather

uninformative (heterochromatin)

Human meiotic recombination frequencies

For any two loci, the chance of recombination is a measure of the distance between them. Loci that are separated by recombination in 1% of meioses are said to be 1 *centimorgan* (*cM*) apart. Genetic distances are related to physical distances, but not in a uniform way: there is a rough correspondence between a genetic distance of 1 cM in humans and a physical distance of close to 1 Mb of DNA, but there are considerable regional variations across chromosomes.

Recombination is much more common at subtelomeric regions than in the middle of chromosome arms, for example, and is much less frequent in heterochromatic regions. At higher resolution, the majority (60% or more) of crossovers occur at a number of short hotspots, about 1–2 kb long, across the genome.

Figure 8.2 Inheritance of a disease haplotype in an autosomal dominant disorder. (A) The disease locus, highlighted by the yellow box, can be imagined to have two alleles: D (disease) and N (normal). Here we also consider alleles at four marker loci that are physically located very close to the disease locus, for example within 0.5 Mb. The disease haplotype in the affected individual is defined by the sequence of alleles at consecutive neighboring marker loci that could be represented here as 2–3–1–4 (when read in the proximal to distal direction). (B) The haplotypes in (A) belong to the affected grandfather (I-1) in this pedigree. The highlighted disease haplotype (**a**) is transmitted without change to affected individuals in generations II and III.

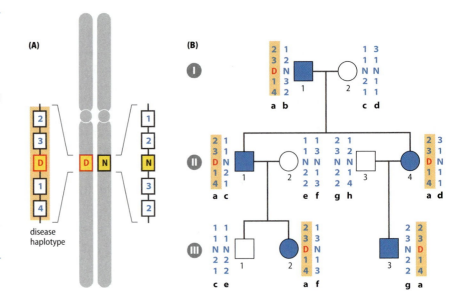

Recombination frequencies in human meiosis also show significant sex differences. Using dense genomewide SNP mapping in nuclear families, two large studies (by Cheung et al. [2007] and Coop et al. [2008]—see under Further Reading) looked at 728 and 557 meioses respectively; the overall mean scores averaged across the two studies was 25.2 crossovers in male meiosis and 39.1 crossovers in female meiosis. But there is also variation between individuals, and even between individual meioses within a single individual (as shown in **Figure 8.3**). Overall, therefore, there is no one correct human genetic map length: for any one subchromosomal region, the correspondence between the physical (DNA sequence) length and genetic map length will vary from one meiosis to another.

Standard genomewide linkage analyses

To map disease genes to specific chromosomal regions, genomewide linkage analyses can be used. Usually, several hundred genetic markers from defined loci across the whole genome are genotyped in family members with the disease. The results may show some marker loci that

Figure 8.3 Individual differences in the numbers of recombinations per meiosis. Each dot represents the number of recombinations identified in an individual meiosis; each vertical line of dots represents the scores determined in each of multiple meioses in a single male or female, as shown. The number of recombinants per meiosis was determined by genotyping individuals in families with 6324 SNP markers. (From Cheung VG et al. [2007] *Am J Hum Genet* 80:526–530; PMID 17273974. With permission from Elsevier.)

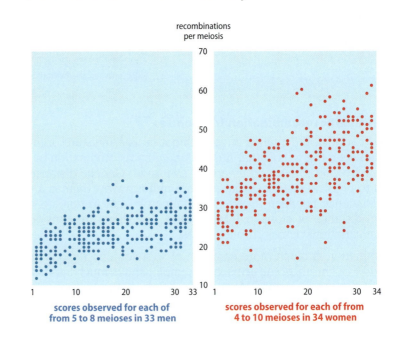

are tightly linked to the disease, thereby indicating a subchromosomal location for the disease gene.

Using, say, 400 markers for genomewide linkage analyses would give a marker density of one every 7–8 Mb or so. Given that our chromosomes are about 50–250 Mb in length and are split by meiotic recombination into usually only two to seven segments (see Figure 8.1), there is a high chance that one or more of a 400 genomewide marker set will be sufficiently close to the disease locus for a marker allele to co-segregate with a disease allele.

The segregation of alleles from each marker locus is followed through a suitably large number of informative meioses (see **Figure 8.4** for examples of informative and uninformative meioses). In practice that means having access to samples from multiple affected and unaffected members usually drawn from several families.

In an idealized situation, **recombinants** and non-recombinants can be clearly identified. In the autosomal dominant pedigree in **Figure 8.5**, the affected individual in generation II is heterozygous for a disease allele and he is also heterozygous for a marker, having a 1,2 genotype. In the highly unusual circumstances shown in this figure it is possible to identify recombinants and non-recombinants unambiguously. In practice, recombinants often cannot be identified unambiguously (linkage studies often use families that do not have such an ideal structure, and key meioses may often be uninformative).

To get round the difficulties of identifying recombinants in human genetic mapping, sophisticated computer programs are needed. They do not attempt to identify individual recombinants. Instead, their job is to survey the linkage data and then calculate alternative probabilities for linkage and nonlinkage. They then express the ratio of these probabilities as a logarithmic value called a **lod score**, as described in **Box 8.1**. Programs such as these are dependent on previous information on the mode of inheritance, disease gene frequency, and the penetrance of the genotypes at the disease locus (that is, the frequency with which the genotypes manifest themselves in the phenotype). For monogenic disorders, the mode of inheritance and disease gene frequency often do not present much difficulty; penetrance can be a more difficult problem.

Linkage can theoretically be achieved with 10 informative meioses, but in practice linkage analysis is rarely successful when there are fewer than 20 or so meioses (see Box 8.1). A major confounding problem in linkage analysis is locus heterogeneity, when the same disease in different families under study may be caused by different genes—it is important to try to study families with extremely similar disease phenotypes.

After obtaining evidence of linkage, crossover points are deduced to identify a minimal subchromosomal region for a disease gene (**Figure 8.6**).

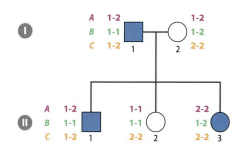

Figure 8.4 Informative and uninformative meioses. Let us assume full penetrance and autosomal dominant inheritance of the phenotype in the pedigree shown here. The disease allele has been transmitted from the father to the son and one daughter (II-3). Genotypes for three unlinked marker loci, A, B, and C, are shown by the respective colored figures. Consider marker A. For the affected son it is impossible to tell which parent contributed allele 1 and which contributed allele 2, but it is possible to infer that each parent contributed an allele 1 to II-2 and an allele 2 to II-3. Marker B is completely uninformative here because I-1 is homozygous and it is impossible to tell which of the two paternal alleles 1 was transmitted to each child. Marker C is informative in each case: the father transmitted allele 1 to his affected son, but transmitted allele 2 to both daughters.

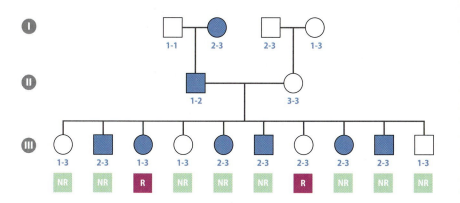

Figure 8.5 Unambiguous identification of recombinants and non-recombinants in an idealized pedigree. Members of this autosomal dominant pedigree have been typed for a marker that has three alleles (1, 2, and 3). The available data suggest that allele 2 of the marker is segregating with the disease. In that case, 8 out of the 10 children in generation III are non-recombinants (NR): either they have inherited both the disease and allele 2 from their affected father, or they have inherited paternal allele 1 and are unaffected. The other two are recombinants (R): either paternal allele 1 has segregated with disease, or paternal allele 2 is associated with a normal allele at the disease locus.

BOX 8.1 Lod scores and statistical evidence for linkage.

Computer-based linkage analysis programs calculate two alternative probabilities: (i) the likelihood of the marker data, given that there is linkage between the marker and the disease locus at specified recombination fractions; and (ii) the likelihood of the marker data, assuming that the marker is unlinked to the disease locus. The *likelihood ratio* is the ratio of likelihood (i) and likelihood (ii), and provides evidence for or against linkage.

The convention is to use the logarithm of the likelihood ratio, called the **lod score** (logarithm of the odds). Individual lod scores are calculated for a defined recombination fraction (θ), and so for each marker the computer programs provide a table of lod scores for different recombination fractions ($\theta = 0$, 0.10, 0.20, 0.30, and so on). The reported recombination fraction is chosen to be the one where the lod score (Z) is at a maximum (Z_{max}).

A lod score of +3 is normally taken to be the *threshold* of statistical significance for linkage between two loci. It means that the likelihood of the data given that the two loci are linked is 1000 times greater than if it is unlinked ($\log_{10}1000 = 3$). Linkage between two loci (say, a disease locus and a marker locus) can theoretically be achieved with 10 informative meioses. At each informative meiosis there are two choices: the two loci are linked (a specific marker allele segregates with disease), or they are not linked (the marker allele does not segregate with disease). If the same marker allele segregates with disease in each of 10 informative meioses in a large pedigree, the odds of that happening by chance are 2^{10} to 1 against, or just over 1000:1 against. In practice, because of poor family structures and some uninformative meioses, 20 or more meioses are often needed for linkage to be successful; DNA samples are usually needed from affected and unaffected members of multiple families.

The ratio of 1000:1 might seem overwhelming evidence in favor of linkage, but it is required to offset the inherent improbability of linkage. With 22 autosomes, two randomly chosen loci are unlikely to be on the same chromosome. Even if the two loci are on the same chromosome, however, they may be well separated, and so unlinked. Factoring in both of these observations, the prior odds are about 50:1 *against* linkage, or 1:50 in favor of linkage. That means we need pretty strong evidence from linkage analysis data to counteract the low starting probability. A likelihood ratio of 1000:1 in favor of linkage multiplied by a prior odds of 1:50 in favor of linkage gives a final odds of only 20:1 in favor of linkage. That is, a single lod score of 3 is not proof of linkage; there is a 1 in 20 chance that the loci are not linked.

Higher lod scores provide greater support for linkage. A lod score of 5 is 100 times more convincing than a lod score of 3. In practice, therefore, genomewide claims for linkage based on a single lod score less than 5 should be treated as provisional evidence for linkage. However, significant lod scores may often be obtained for several markers clustered in one subchromosomal region; if so, the combined data provide strong evidence of linkage. See Table 1 for the example of a dominantly inherited skin disorder, Hailey–Hailey disease (OMIM 169600), in which four neighboring markers in the 3q21–q24 interval show significant evidence of linkage.

The threshold for excluding linkage is a lod score of –2. *Exclusion mapping* can be helpful in excluding a candidate gene of interest, and in genomewide studies the exclusion of a substantial fraction of the genome can direct extensive analysis of the remaining regions.

MARKER	LOD SCORE (Z) AT				MAXIMUM LIKELIHOOD ESTIMATES	
	0.00	0.10	0.20	0.30	Z_{max}	AT $\theta =$
D3S1589	–0.99	2.29	1.90	1.14	2.29	0.09
D3S1587	4.54	3.80	2.83	1.73	4.54	0.00
D3S1292	2.62	4.98	3.84	2.41	5.32	0.04
D3S1273	3.36	5.52	4.12	2.54	6.10	0.03
D3S1290	–2.81	3.83	3.05	1.94	3.90	0.07
D3S1764	–8.62	2.21	2.06	1.38	2.26	0.13

Table 1 Pairwise lod scores for Hailey–Hailey disease and markers at 3q21–q24. The descending order of markers is from proximal to distal. Analyses were carried out in six disease families. The underlying disease gene was subsequently found by positional cloning to be *ATP2C1* and to map just proximal to *D3S1587*. (Data from Richard G et al. [1995] *J Invest Dermatol* 105:357–360.)

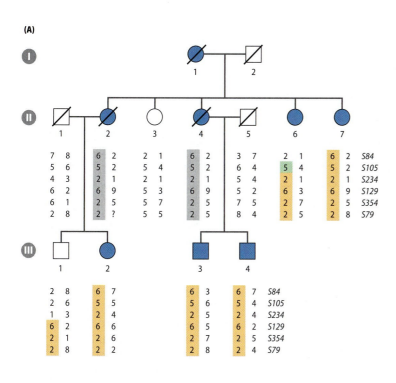

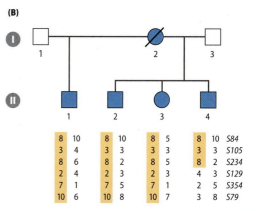

Figure 8.6 Defining the minimal candidate region by inspection of haplotypes. The two pedigrees show a dominantly inherited skin disorder, Darier–White disease, that had previously been mapped to 12q. The 12q marker haplotype segregating with disease is highlighted by orange shading. Gray boxes mark inferred haplotypes in deceased family members. In pedigree A the recombination in II-6 maps the disease gene distal to marker *D12S84* (abbreviated to *S84* in the figure); the *D12S105* marker is uninformative because I-1 was evidently homozygous for allele 5—compare the genotypes of II-3 and II-7. The recombination shown in III-1 suggests that the disease gene maps proximal to *D12S129*, but this requires confirmation (the interpretation depends on the genotypes of II-1 and II-2 being inferred correctly, and on III-1 not being a non-penetrant gene carrier). The recombination in II-4 in pedigree B provides the confirmation. The combined data locate the Darier gene to the interval between *D12S84* and *D12S129*. (Adapted from Carter SA et al. [1994] *Genomics* 24:378–382; PMID 7698764. With permission from Elsevier.)

Autozygosity mapping in extended inbred families

The term **autozygosity** means homozygosity for markers that are *identical by descent*—that is, the two alleles are copies of one specific allele that was transmitted to both parents from a recent common ancestor. In some societies, such as in the Middle East and parts of Asia, cousin marriages are quite common, and in extended inbred families there may be several individuals who are autozygous for an allele because of parental consanguinity. As illustrated in Box 5.2 on page 123, second cousins share respectively 1/32 of their genes, and so their children would be autozygous at 1/64 of all loci.

Homozygosity for a particular marker allele could be due to autozygosity. Alternatively, it might result from the inheritance of a second, independent copy of that allele that has been brought into the family at some stage; alleles such as this would be said to be *identical by state*. Homozygosity for a haplotype of marker alleles, however, is more likely to indicate autozygosity; if there are additional sibs who are homozygous for the same marker haplotypes, quite small consanguineous families can generate significant lod scores. Autozygosity mapping can therefore be a very efficient way of mapping a recessive monogenic disorder; see Goodship et al. (2000) under Further Reading for a successful application.

Chromosome abnormalities and other large-scale mutations as routes to identifying disease genes

Some affected individuals show a specific chromosome abnormality or other very large-scale mutation that can be detected quite readily (see below). Abnormalities such as these might occur coincidentally. That is, the abnormality might have nothing to do with disease (some parts of our genome do not contain critically important genes, and chromosome abnormalities affecting these regions might be found in a small percentage of normal people). Alternatively, the abnormality causes the disease by affecting the expression or structure of a gene or genes in that region. That would be more likely if the same DNA region were disrupted in two or more unrelated individuals with the same disease, or if the abnormality

Table 8.1 Examples of successful gene identification prompted by the identification of disease-associated chromosomal abnormalities. (PMID, PubMed identifier at http://www.ncbi.nlm.nih.gov/pubmed/ – see glossary.)

DISORDER	CHROMOSOME ABNORMALITY	COMMENTS	PUBLICATION PMID
Duchenne muscular dystrophy	an affected boy with a cytogenetically visible deletion at Xp21.3 and a woman with a balanced Xp21;21p12 translocation	positional cloning strategies identified genes within the deletion/ translocation breakpoint, finally implicating the giant dystrophin gene	2993910; 3001530
Sotos syndrome	a girl with a *de novo* balanced translocation with breakpoints at 5q35 and 8q24.1	the disease gene, *NSD1*, was found to be severed by the 5q35 breakpoint	11896389

occurred *de novo* in a sporadic case (that is, the affected individual has no family history of the disorder, and the abnormality is not present in either of the parents).

Chromosomal abnormalities occur rarely, so this approach can never be a general one for identifying disease genes. But it has been a useful for some disorders, notably dominantly inherited disorders with a severe congenital phenotype (because they normally cannot reproduce, affected individuals occur as sporadic cases, making it very difficult to carry out linkage analyses). Metaphase or prometaphase chromosome preparations can be prepared readily from blood lymphocytes, and then stained with DNA-binding dyes to reveal altered chromosome banding patterns that indicated a chromosome abnormality such as a translocation, deletion, inversion and so on (see Box 7.4 for the chromosome banding methodology).

Balanced translocations and inversions can be particularly helpful because, unlike large deletions, they may involve no net loss of DNA, and the underlying disease gene might be expected to be located at, or close to, a breakpoint that can readily be identified. See **Table 8.1** for some examples.

Genomewide screens for large-scale duplications and deletions are also available through comparative genome hybridization (CGH), a technique that we describe in detail in Chapter 10 within the context of genetic testing in Section 11.2 (pages 438–439). Briefly, CGH is a form of competition hybridization in which patient and normal control samples are labeled with different fluorophores (chemicals that fluoresce under ultraviolet light at specific wavelength ranges). They compete with each other to hybridize to large panels of DNA clones from across the genome. If the patient DNA has a deletion in a specific genomic region, its hybridization signal from that region will be reduced by 50%; or if there is a duplication, the hybridization signal will be increased by 50%.

Exome sequencing: let's not bother getting a position for disease genes!

Although many genes underlying monogenic disorders have been identified by the methods used above, some monogenic disorders have not been well studied because they are very rare, or because they are not readily identifiable. There may be difficulties, for example, in recognizing individual phenotypes within complex sets of overlapping phenotypes, such as intellectual disability. More than 7000 monogenic disorders are estimated to exist, and although the methods above have been very successful in gene identification, a new approach was needed to identify genes in the substantial proportion of monogenic disorders that are very rare or where there is difficulty in distinguishing the phenotype from related disorders.

The new approach? That would be massively parallel DNA sequencing (also called next-generation sequencing). (We give an introduction to massively parallel DNA sequencing in Section 3.3, and provide details on two popular methods in Chapter 11). This new approach offers a vast increase in sequencing capacity, and the time taken to sequence a human genome has fallen from several years to just a few days, and soon a few hours. And the expense has dropped drastically. The inevitable consequence has been an explosion of genome sequencing.

The ability to sequence whole genomes (and therefore all genes) both rapidly and cheaply means that disease genes can often be identified without any need to first find a chromosomal position for the disease gene. Because the vast majority of disease genes are protein-coding genes (see Table 7.4 for some rare examples of RNA disease genes) and given that the great majority (perhaps 85% or so) of currently known disease-causing mutations occur in the exons of protein-coding genes, whole-genome sequencing initially appeared a rather laborious and costly approach to identify a disease gene. However, all the exons of the protein-coding genes together account for only just over 30 Mb of our DNA; sequencing this fraction, just over 1% of the genome, appeared an easier and cheaper option than genome sequencing.

Exome is the collective term for all exons in the genome. Operationally, exome sequencing has largely focused on the exons of protein-coding genes (RNA genes have been a low priority, mostly because they are viewed as infrequent contributors to disease). Exome sequencing involves first capturing exons from the DNA of affected individuals, and then sequencing the captured DNA. In practice, exome capture is designed to capture exons with a little flanking intron sequence (to cover splice junctions) plus DNA sequences specifying some miRNAs; hybridization with a control set of cloned exon sequences allows capture of the desired exons, as shown in **Figure 8.7**.

To identify a gene for a rare disorder, sequencing of the exomes from just a few affected individuals is often sufficient. That is so because clearly deleterious mutations (frameshifting and nonsense mutations, and some types of nonconservative amino acid substitution) can often be easily identified in protein-coding sequences. Although each of us carries a surprising number of deleterious mutations like this, they are scattered throughout the genome and vary from individual to individual; however, unrelated individuals with the same single-gene disorder might be expected to often have causative mutations in the same gene. Where there is parental consanguinity, exome sequencing can sometimes be used to identify genes underlying an autosomal recessive condition by exome sequencing of affected members of a single family.

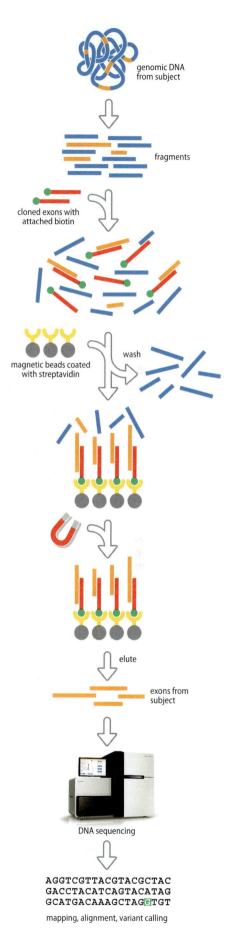

Figure 8.7 Exome capture and sequencing. A genomic DNA sample to be analyzed is randomly sheared and the fragments are used to construct a DNA library (library fragments are flanked by adapter oligonucleotide sequences; not shown). The DNA library is then enriched for exon sequences (orange rectangles) by hybridization to DNA or RNA baits that have been designed to represent human exon sequences (red rectangles). The baits have a biotin group (green circle) attached to their end. After capturing exon sequences from the test sample DNA by hybridization, the biotinylated baits can be selected by binding to magnetic beads (gray) coated with streptavidin (yellow goblet shapes)— streptavidin has an extremely high affinity for binding to biotin, and the streptatividin– biotin–exon complexes can be removed using a magnet. The captured exon sequences from the sample DNA are subjected to massively parallel DNA sequencing and the data are interpreted as described in the text. (Adapted from Bamshad MJ et al. [2011] *Nat Rev Genet* 12:745–755; PMID 21946919. With permission from Macmillan Publishers Ltd.)

DISEASE	TYPE OF DISORDER	EXOMES SEQUENCED FROM	INHERITANCE	OMIM NUMBER	UNDERLYING GENE	PUBLICATION PMID
Miller syndrome	congenital disorders of development	mostly unrelated sporadic cases	AR	263750	*DHODH*	19915526
Kabuki syndrome			AD	126064	*MLL2*	20711175
Schinzel–Giedion syndrome			AD	269150	*SETBP1*	20436468
Osteogenesis imperfect type VI	connective tissue disorder	affected sibs born to consanguineous parents	AR	613982	*SERPINF1*	21353196
Spastic paraplegia 30	early-onset neuromuscular disorder		AR	610357	*KIF1A*	21487076

Table 8.2 Examples of using exome sequencing for successful identification of genes causing monogenic disorders. For a recent review of different strategies for exome sequencing to identify disease genes, and of successful applications, see Gilissen C et al. (2012) *Eur J Hum Genet* 20:490–497. AD, autosomal dominant; AR autosomal recessive. (PMID, PubMed identifier at http://www.ncbi.nlm.nih.gov/pubmed/ – see glossary.)

Since its first successful application in identifying disease genes in 2009, exome sequencing has been dramatically successful in identifying genes underlying very rare autosomal recessive and congenital dominant disorders (neither of which is amenable to linkage analyses because of a lack of suitable families); see **Table 8.2** for examples. It has also been important in the case of extremely heterogeneous phenotypes—in one compelling study, 50 novel genes were identified for recessive cognitive disorders (**Figure 8.8**). Similar studies in the future are likely to gravitate towards whole genome sequencing as sequencing costs drop (the first DNA sequencing machines to allow whole genome sequencing for $1000 per genome appeared in January 2014).

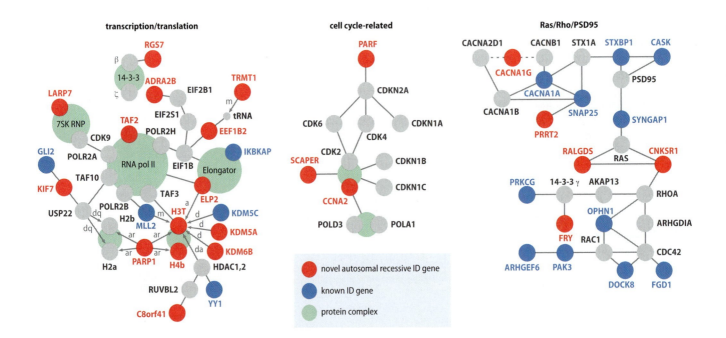

Figure 8.8 Known and novel genes for intellectual disability are part of protein and regulatory networks. In a single exome sequencing study, 50 novel genes were identified for recessive intellectual disability. The novel genes were predicted to encode components of protein and regulatory networks that had been implicated by studies of known genes for intellectual disability, including the transcriptional/translational network, the cell cycle-related network, and the Ras/Rho/PSD95 network. (Adapted from Najmabadi H et al. [2011] *Nature* 478:57–63; PMID 21937992. With permission from Macmillan Publishers Ltd.)

8.2   Approaches to Mapping and Identifying Genetic Susceptibility to Complex Disease

The polygenic and multifactorial nature of common diseases

It is convenient to divide genetic phenotypes into monogenic phenotypes (influenced predominantly by a single genetic locus) and polygenic phenotypes (multiple genetic factors determine the phenotype, each making a small contribution). Individual single-gene disorders are rare; common genetic disorders—such as type 1 and 2 diabetes, coronary artery disease, stroke, rheumatoid arthritis, Alzheimer disease, and so on—are polygenic.

In reality, the division between these two categories is not quite so clear cut. We may refer to single-gene disorders, for example, but in virtually no case is the disease phenotype entirely attributable to a single gene locus (see Figure 7.21 for the example of β-thalassemia). We can see the effects of other loci when the phenotype in a 'single-gene' disorder differs between affected members of the same family (see Figure 5.14). The difference in phenotype can be explained if affected family members (who would be expected to have the same disease allele or alleles) have different alleles at one or more *modifier loci*. The product of a modifier gene interacts with the disease allele in some way: it may regulate expression of the disease allele, or it may interact in the same pathway as the product of the disease allele so as to affect its function, for example (see Section 7.7 for examples of modifier genes).

What distinguishes a single-gene disorder is that, although there may be minor effects from variants at other genetic loci, rare genetic variants at a primary gene locus have a very great effect on the phenotype. By contrast, it is usual to think of a polygenic disorder as being one in which the genetic susceptibility to disease risk is not dominated by a primary locus where individual variants can have extremely strong effects on the phenotype.

There may be a predominant genetic locus in a true polygenic disease, but its effect would not normally be large: variants at that locus would not be expected to be necessary or sufficient to cause disease. However, as described later, there can also be a problem with heterogeneity: individual complex diseases can be a collection of related phenotypes, and in some of these rare variants can be of quite strong effect.

The phenotypes of polygenic diseases are also influenced by nongenetic factors, including environmental factors (which may sometimes work though epigenetic modification) and also chance (*stochastic*) factors. Environmental factors can sometimes have a strong influence in certain monogenic disorders, but their effects are very important right across the spectrum of polygenic disorders.

Because a combination of multiple genetic susceptibility loci and environmental factors is involved, common genetic diseases have long been considered to be *complex* or *multifactorial* diseases. Underlying polygenic theory is the idea that there is a *continuous susceptibility to the disease within the population*, and that disease manifests when a certain threshold is exceeded (Box 8.2).

Complexities in disease risk prediction

As we will see, some complex diseases can run in families. However, the tendency is much less marked than for simple monogenic diseases. A minority of families may have multiple affected members, and there may even be subsets of families that seem to show Mendelian or quasi-Mendelian transmission (such as in early-onset forms of Alzheimer disease,

BOX 8.2 Polygenic theory and the liability threshold concept to explain dichotomous traits.

Human traits can be divided into two classes. Some, like diseases, are *dichotomous* (you either have the trait or you do not). Others, such as height or blood pressure, are *continuous* (or *quantitative*) *traits*—everybody has the trait, but to differing degrees. To explain quantitative traits, polygenic theory envisages the additive contributions of variable alleles from multiple loci. Many alleles at the underlying **quantitative trait loci** (**QTL**) might have subtle differences (causing modest changes in expression levels of certain genes, for example); different combinations of alleles at multiple loci might then produce continuous traits (adult height, for example, is known to be governed by genetic variants at a minimum of 180 loci).

Polygenic theory can be extended to also explain dichotomous traits using the concept of a *liability threshold*. The idea is that there a continuous liability (susceptibility) to disease in the population, but only people whose susceptibility exceeds a certain threshold will develop disease. For each complex disease, the susceptibility curve for the general population will be a normal (bell-shaped) curve: most individuals will have a medium susceptibility, with smaller numbers of people having low to very low susceptibility or high to very high susceptibility (**Figure 1A**). Only a small percentage of the general population will have a susceptibility that exceeds the threshold so that they are affected by the disease (shown in red in Figure 1).

Close relatives of a person affected by a complex disease have an increased susceptibility to that disease (Figure 1B). The affected person must have a combination of different high-susceptibility genes; relatives will share a proportion of genes with the affected person and will therefore have an increased chance of having high-susceptibility genes.

By chance, some relatives may have only a few of the high-susceptibility genes, but others may have many high-susceptibility genes in common with the affected person. While disease susceptibility can show wide variation among first-degree relatives (parent and child, sibs), the overall effect is that the curve—and the median susceptibility (dashed vertical line)—is displaced to higher susceptibilities to the disease (Figure 1B). Because the threshold remains the same, more individuals are affected, and the relatives that are most closely related to an affected person are more likely to be affected.

Variability in the liability threshold for an individual disease

Thresholds of susceptibility to a complex disease are not absolutely fixed—they can show differences between the two sexes, for example. For many auto-immune disorders, women have significantly more disease risk than men. The reverse is true for some other conditions. For example, pyloric stenosis occurs in about 1 in 200 newborn males, but only in about 1 in 1000 newborn females. That is, a double threshold exists, one for females and one for males. The female threshold for pyloric stenosis is farther from the mean than that for the male. However, because it takes more deleterious genes to create an affected female, she has more genes to pass on to the next generation. Her male offspring are at a relatively high risk of being affected when compared with the population risk.

The threshold model accommodates environmental effects by postulating that such effects can reposition the threshold with respect to the genetic susceptibility. Protective environmental factors move the threshold to higher genetic susceptibilities; other environmental factors can increase the risk of disease by moving the threshold to lower genetic susceptibilities.

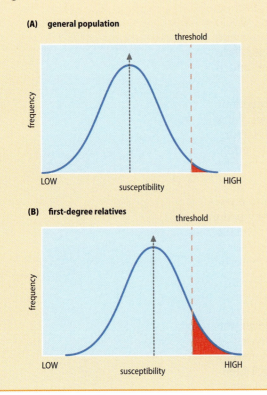

(A) general population

(B) first-degree relatives

Figure 1 Distribution of susceptibility to a complex disease in the general population and in first-degree relatives of an affected person.

diabetes, Parkinson disease, and various types of cancer). Frequently, however, there is little evidence of family history; quite commonly an affected individual appears as a sporadic case.

The calculation of disease risk for complex diseases is therefore often very different than for Mendelian disorders. In the latter case, disease risks are mostly based on theoretical calculations that remain stable (but there can sometimes be complications, notably because of low penetrance and variable expressivity).

For some complex diseases, we are beginning to accumulate information on major predisposing genetic and environmental factors; this may lead to a more informed measure of disease risk than has previously been possible. Traditionally, a lack of knowledge about the predisposing risk factors has meant that for complex diseases the disease risk has often been *empiric*: that is, it is based on observed outcomes in surveys of families, and is often *modified according to past incidence of disease*.

Take the following examples. If a couple with a cystic fibrosis child go on to have another two affected children, the risk of having an affected child is stable and remains 1 in 4 for each new pregnancy after that. By contrast, a couple who have had one baby with a neural tube defect would be quoted a specific recurrence risk for a subsequent pregnancy that would be based on observed frequencies in the population; if they do go on to have a second affected child, the disease risk for a subsequent pregnancy would now be substantially higher. The real disease risk would not have changed; the birth of the second affected child helps us to recognize that the parents carry more high-susceptibility genes than was apparent after the birth of their first affected child.

Difficulties with lack of penetrance and phenotype classification in complex disease

Researchers seeking to identify the genetic susceptibility in multifactorial disorders are confronted by multiple challenges. In the sections below we cover some of the technological difficulties that needed to be surmounted. Here we consider more intrinsic difficulties that are due to the general lack of penetrance, or to problems with defining and classifying disease phenotypes.

Recall that reduced penetrance can be a feature of some monogenic disorders, such as imprinting disorders. But, in general, DNA variants at a Mendelian disease locus are of very strong effect and are therefore highly penetrant. As a result, affected people typically have a disease-associated genotype; unaffected people do not. For complex disease, however, the picture is different. If we discount rare Mendelian subsets of complex disease, reduced penetrance is usually the standard, simply because multiple genes make small contributions to the phenotype. That is, a DNA variant that is strongly associated with a complex disease is often at best a **susceptibility factor**: its overall frequency should be significantly increased in affected individuals compared with controls, but normal people can quite often possess that variant, and affected people can quite often lack it.

Phenotype classification and phenocopies

In many complex diseases the phenotype can have many variable components. As detailed in Section 5.3, monogenic disorders can also have variable phenotypes, but here the high penetrance and the range of phenotypic features in multiple affected individuals reported in the literature allow us to be clear about which aspects of a person's phenotype are disease components, and which are not. In complex diseases, however, the situation is not so simple, and phenotype delineation and classification can be a major problem.

Some affected people, who do not have genotypes commonly associated with the disease, are **phenocopies** that have been wrongly classified as having the disease under study. For some phenocopies the disease is caused by different genetic factors from those that were expected to apply. For example, accurate diagnosis of Alzheimer disease has traditionally relied on post-mortem brain pathology. If we wish to study living patients, various clinical tests can be conducted, but a subsequent diagnosis of Alzheimer disease is often provisional ('*probably* Alzheimer disease'); post-mortem examination might subsequently show a different type of dementia, such as Lewy dementia, frontotemporal dementia and so on. For other phenocopies, the phenotype might have an environmental origin.

According to the condition, defining the disease phenotype can be straightforward or very challenging. At one extreme are conditions in which there is a very recognizable and rather specific phenotype. For example, in primary biliary cirrhosis, the most common autoimmune form of chronic liver disease, affected individuals have markedly similar phenotypes, and have signature autoantibodies that are specifically directed against the E2 subunit of the mitochondrial pyruvate dehydrogenase complex. At another extreme are some behavioral and psychiatric conditions, in which even classifying individuals as affected and unaffected can sometimes not be straightforward. The pathology might not be well defined and there can be heavy reliance on interviews (and subjective information). Here, clear diagnostic criteria are of paramount importance.

Deciding which aspects of a person's phenotype are components of a complex disease is not easy, and deciding how far we should lump together different, but clearly related, phenotypes is a significant issue in complex disease studies. If two first cousins have had different types of congenital heart malformation, for example, should we consider the phenotypes as independent occurrences, or lump them together and report two affected individuals in one family?

Estimating the contribution made by genetic factors to the variance of complex diseases

Phenotypes are determined both by genetic factors and by nongenetic factors (that are often described as environmental factors but include both stochastic and epigenetic factors). The *variance* (V) of a phenotype is a statistical term defined as the square of the standard deviation. The total variance of a phenotype, $V_P = V_G + V_E$ (where V_G is the genetic variance and V_E is the environmental variance). V_G is the sum of three components: additive genetic effects (the combined contributions from different loci), dominance effects (interaction of alleles at a single locus), and interaction effects (interaction between genes at different loci; the effect of genes at several loci may not be simply additive if they interact with each other).

The **heritability** (h^2) is that proportion of the variance that can be attributed to genetic factors—that is, $h^2 = V_G/V_P$—and has values ranging from 0 (no genetic factors involved) to 1 (exclusively due to genetic factors). As described below, the ratio of genetic to environmental involvement in the etiology of a disease varies according to the disease class.

Heritability has been important in the prediction of disease risk, but it is not a fixed property, and it describes a population, not an individual. More accurately, it describes the genetic contribution to variance *within a population and in a specific environment*. To estimate the heritability of a complex trait or disease, the incidence of disease is compared in genetically related individuals. Three types of study have been undertaken: family studies, adoption studies, and twin studies.

Family studies

Having a relative with a complex disease increases your risk of developing that disease. The **risk ratio**, the disease risk to a relative of an affected person divided by the disease risk to an unrelated person, is designated as λ, with a subscript that defines the relationship. For example, λ_S represents the comparative disease risk for a sib (brother or sister) of an affected individual. Unlike a monogenic disorder (for which the risks to family members are fairly precisely defined, according to simple theoretical calculations), the risks for complex diseases have quite often been empiric; that is, they have often been based on surveys of disease incidence in families.

The risk ratio (λ) is a measure of how important genetic factors are in the etiology of the disease. Not surprisingly, therefore, extremely high λ_S values are found in monogenic disorders. For example, in populations of western European origin, the general lifetime risk of cystic fibrosis is about 1 in 2000, but the risk to a fetus is 1 in 4 if the parents have had a previously affected child. In that case, $\lambda_S = 1/4$ divided by $1/2000 = 500$. For some complex diseases, λ_S values can be quite high (**Table 8.3**).

The drawback of using family studies to infer genetic factors is that family members would normally be exposed to some common environmental factors as a result of the shared family environment.

Adoption and twin studies

Adoption studies seek to separate the contributions of genetic and environmental factors. Children of affected individuals who were adopted at birth into an unaffected family might be expected to share genetic factors with their biological parents but be exposed to different environments. Studies compare the incidence of disease in adopted children with that in children raised in a family with affected relatives.

An alternative is twin studies that measure how often the twins are concordant (both have the disease) or discordant (only one is affected). Monozygotic twins are genetically identical but dizygotic twins, like any pair of sibs, share 50% of their genes. Regardless of zygosity, twins should be exposed to rather similar environment factors.

Table 8.4 lists observed concordance rates for monozygotic and dizygotic twins for various complex diseases. There are two major points to note. First, there are significant discordance rates between monozygotic twins—genetics is not everything! Secondly, it is clear that in certain diseases the concordance between monozygotic twins is much greater than that between dizygotic twins. Diseases in which genetic factors have a large role show a relatively high concordance in monozygotic twins and a much lower concordance rate in dizygotic twins: the greater this ratio, the greater the genetic contribution. Thus, for example, genetic factors would seem to be much more important in schizophrenia than in Parkinson disease (see Table 8.4).

The best type of study would be to look at monozygotic twins separated at birth and raised in different environments, but this occurs so rarely that statistically valid sample sizes cannot be obtained.

Variation in the genetic contribution to disorders

Heritability studies have indicated that genetic factors make different contributions to different diseases. Monogenic disorders are primarily determined by genetic factors (but in some cases environmental factors

DISEASE	RELATIVE RISK FOR SIB (λ_S)
Alzheimer disease (late-onset)	4
Autism spectrum disorder	6.5
Breast cancer, female	2
Crohn's disease	25
Multiple sclerosis	20
Schizophrenia	9
Type 1 diabetes	15
Type 2 diabetes	3

Table 8.3 Approximate relative risks for sibs compared with the general population for selected complex diseases, averaged from many studies.

DISEASE	CONCORDANCE (%)	
	In MZ twins	In DZ twins
Type 1 diabetes	42.9	7.4
Type 2 diabetes	34	16
Multiple sclerosis	25.3	5.4
Crohn's disease	37	10
Ulcerative colitis	7	3
Alzheimer disease	32.2	8.7
Parkinson disease	15.5	11.1
Schizophrenia	40.8	5.3

Table 8.4 Degree of concordance between twin pairs for select complex diseases, averaged from many studies.
MZ, monozygotic; DZ, dizygotic.

can make very significant contributions). By contrast, infectious diseases are primarily determined by environmental factors (exposure to the infectious agent). Here, however, host genetic factors also make a contribution, so that people vary in their susceptibility to an infectious disease and a small number of individuals may be disease-resistant.

For the great bulk of other complex diseases, genetic and environmental factors both make large contributions to the phenotype. In some complex diseases, such as schizophrenia, autism spectrum disorder, Alzheimer disease, type 1 diabetes, multiple sclerosis, and Crohn's disease, there is a strong genetic contribution; in others, such as Parkinson disease and type 2 diabetes, genetic factors seem to be less important.

The heritability of an individual disease varies from one population to another. It can also vary in the same population in response to a changing environment. Consider phenylketonuria. As detailed in Box 7.8, a deficiency of phenylalanine hydroxylase produces elevated phenylalanine and toxic by-products that can result in cognitive disability. In the recent past the disease was almost wholly due to genetic factors, and so the heritability was extremely high. In modern times, neonatal screening programs in many countries allow early detection and treatment using low-phenylalanine diets. Now, in societies with advanced health care, phenylketonuria results mostly from environmental factors that lead to failure to deliver the treatment (inefficiency in health care systems, reluctance of families to seek out treatment, non-compliance with the diet, and so on).

The incidence of some complex diseases is very dependent on environments that can change very significantly with time. Thus, the huge recent increase in type 2 diabetes (mostly as a result of increasingly unhealthy diets and lack of exercise) means that the heritability of this disorder in many populations is now much reduced when compared with just a few decades ago.

Linkage analyses to seek out genes for complex diseases

The linkage analyses used to map genes for monogenic disorders are said to be *parametric*—the data can be analysed only if a specific genetic model is assumed. The model needs to give details of certain key parameters: the mode of inheritance, disease gene frequency, and penetrance of disease genotypes. Parametric linkage analyses have been very successful in mapping genes for Mendelian disorders, but they have more limited applicability in complex disease (because of the general difficulty in providing all the required parameters).

Parametric linkage analyses in Mendelian subsets

Parametric linkage analyses are more readily applied when a complex disease shows very strong familial clustering. A near-Mendelian pattern may simply reflect the chance occurrence of several affected people in one family. If, by chance, most members of a family possess multiple genetic variants conferring high susceptibility to disease (not necessarily at the same loci in different affected family members), a single common disease-susceptibility allele (that has been transmitted in a Mendelian fashion within that family) might be enough to tip the balance past the susceptibility threshold.

For several complex diseases there are also subsets in which the disorder shows clear Mendelian inheritance. That is most obvious when there is dominant or quasi-dominant inheritance, as in early-onset forms of Alzheimer disease, Parkinson disease, diabetes, and various types of cancer. Large pedigrees such as the one shown in **Figure 8.9** offered an

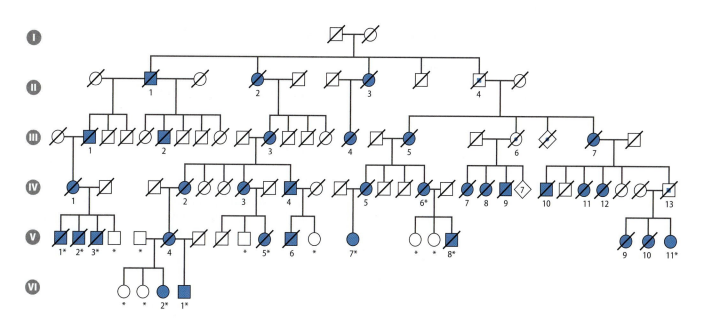

Figure 8.9 An exceptional pedigree showing dominantly inherited Alzheimer disease. Affected members of this pedigree had early-onset Alzheimer disease (average age at onset 46 ± 3.5 years). They were subsequently shown to have a mutation in the presenilin 1 gene. (From Campion D et al. [1995] *Neurology* 45:80–85; PMID 7824141. With permission from Wolters Kluwer Health.)

early way in to identify genes associated with a complex disease phenotype. In early-onset Alzheimer disease, for example, these approaches identified three disease genes: the amyloid precursor protein gene (*APP*), and two genes involved in processing the APP protein—we discuss the details in Section 8.3.

How might the phenotypes of a complex disease and one that segregates like a Mendelian disorder be so similar? The gene mutated in the Mendelian subset might also be a disease-susceptibility locus for the complex disease (with a rare, highly penetrant variant in the Mendelian subset, and a common variant of weak effect at the same locus in the complex disease). Or, if genes mutated in Mendelian subsets are not significant disease-susceptibility loci for a complex disease (as seems to be the case in Alzheimer disease), the common pathogenesis might suggest that at least different genes associated with the two forms of the disease are part of a common biological pathway or process.

Nonparametric and affected sib-pair linkage analysis

Nonparametric methods of linkage analysis do not require any genetic model to be stipulated, and so can generally be applied to analyzing complex disease. They rely on the principle that, regardless of the mode of inheritance, affected individuals in the same family would tend to share not just major disease-susceptibility genes but also the immediate chromosomal regions. That is, a major disease-susceptibility locus and a very closely linked marker would show a strong tendency to be co-inherited within affected individuals in the same family (because of the very low chance of recombination between the marker locus and the disease locus).

Nonparametric linkage studies occasionally use samples from all affected family members, but it is usually more convenient to simply use affected sib pairs. The aim here is to obtain genomewide marker data in affected sibs from multiple families and then identify chromosomal regions that have been shared more often than would be predicted by random Mendelian segregation. As sibs share 50% of their genes, affected sibs need to be studied in many families with the same complex disease. For marker loci that are not linked to a major disease susceptibility gene, sibs would be expected to share 50% of alleles on average (some sib pairs

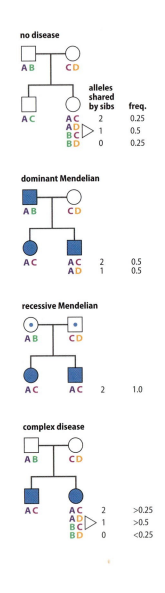

Figure 8.10 Principle of affected sib-pair analysis. By random segregation, any pair of sibs share 2, 1, or 0 parental haplotypes in the relative proportions 1:2:1. Pairs of sibs who are both affected by a dominant Mendelian condition must share the segment that carries the disease allele, and they may or may not (a 50:50 chance) share a haplotype from the unaffected parent. Pairs of sibs who are both affected by a recessive Mendelian condition necessarily share the same two parental haplotypes for the relevant chromosomal segment. For complex conditions, haplotype sharing greater than that expected to occur by chance may allow the identification of chromosomal segments containing susceptibility genes.

might share 2, 1, or 0 alleles by chance, but the *overall average* across all sets of sibs would be 1 allele in common). For marker loci close to a major disease susceptibility gene, affected sibs would be expected to share significantly more than 50% of alleles (see **Figure 8.10** for the principle).

Affected sib-pair analyses are comparatively easy to carry out (they need samples from just a few people per family) and robust (the method makes few assumptions). But there are inevitable limitations. Because any individual susceptibility factor is neither necessary nor sufficient for a person to develop a complex disease, the underlying genetic hypothesis is weaker than for Mendelian conditions. This means that finding statistically significant evidence for a disease susceptibility factor is going to be harder.

The calculations in **Table 8.5** show that under ideal conditions affected sib-pair analyses can be carried out with reasonable numbers of samples (typical studies use a few hundred sib pairs). But if the effects are weak, unfeasibly large numbers of samples are needed to detect them, and the studies will be defeated if there is a high degree of heterogeneity (individual susceptibility factors will operate in just a small proportion of families).

Genomewide nonparametric linkage scans require higher thresholds for statistical significance. Lod scores above 5.4 are considered highly significant evidence for linkage; scores between 3.6 and 5.4 are significant; and scores from 2.2 to 3.6 are suggestive. In practice, affected sib-pair analyses deliver typically modest lod scores that often do not reach statistically significant thresholds.

The review by Altmüller et al. (2001) under Further Reading reports an analysis of different nonparametric linkage studies for many diseases, and underscores the difficulties. Of ten genomewide linkage studies of schizophrenia analyzed in that review, four were unable to find any evidence of linkage, five found only suggestive evidence of linkage (lod scores between 2.2 and 3.5, but at many different regions on eight different chromosomes), and only one recorded more significant evidence of linkage. In addition to the lack of any great consistency in the results,

Table 8.5 Number of affected sib pairs needed to detect a disease susceptibility factor. Here p is the frequency of the disease susceptibility allele and q is the frequency of normal alleles at the disease susceptibility locus (so that $p + q = 1$). The relative risk of disease (γ) is a measure of how the disease risk changes when comparing persons with the susceptibility factor to those without. The calculations are based on the formulae derived by Risch & Merikangas in their 1996 paper (see Further Reading). [a]Really the number of affected sib pairs with 80% power to detect the effect. The take-home message is that unless the disease susceptibility factor is both quite common and confers a high disease risk, very many affected sib pairs are required to detect it.

RELATIVE RISK OF DISEASE (γ)	PROBABILITY OF ALLELE SHARING BY AFFECTED SIBS (Y)		NUMBER OF AFFECTED SIB PAIRS NEEDED TO DETECT THE EFFECT[a]	
	At $p = 0.1$	At $p = 0.01$	At $p = 0.1$	At $p = 0.01$
1.5	0.505	0.501	115,481	7,868,358
2.0	0.518	0.502	9162	505,272
2.5	0.536	0.505	2328	103,007
3.0	0.556	0.509	952	33,780
4.0	0.597	0.520	313	7253
5.0	0.634	0.534	161	2529

getting independent replication of significant results proved very difficult. But in some cases, such as an 8p21 location, the initial finding was replicated in multiple populations, eventually implicating an allele of the neuregulin gene, *NRG1*, as a risk factor (the gene was known to be involved in synaptic transmission).

Identifying the disease-susceptibility gene

Even if a significant candidate chromosome region can be identified for a complex disease susceptibility locus, finding the implicated gene is generally problematic—sibs share large chromosome segments, and so any candidate chromosome regions that can be found are very large (in Mendelian disorders, by contrast, the candidate chromosomal region can be progressively reduced by looking for rare recombinants between marker loci and the disease locus). Candidate gene approaches can be used, but to get closer to the disease susceptibility gene additional *linkage disequilibrium* mapping methods have sometimes been used. The same methods are regularly used in association analyses, and we will consider these in detail in the next section.

Despite the above difficulties, genomewide linkage studies have had a measure of success in mapping susceptibility genes in complex disease to specific candidate regions that would then allow subsequent gene identification using other approaches. In addition to the schizophrenia-associated *NRG1* allele mentioned above, successes include mapping of genes conferring susceptibility to age-related macular degeneration to the 1q32 region, and genes conferring susceptibility to Crohn's disease at the 16q11–16q12 region. Those advances allowed association analyses to be targeted to these regions as described below, ultimately allowing identification of the *CFH* (complement factor H) gene at 1q32 and the *NOD2* gene at 16q12 as novel disease-susceptibility factors.

The *CFH* gene had previously been well known, but the gene that came to be known as *NOD2* was identified only very shortly before being implicated in Crohn's disease; it would provide the first molecular insights into the pathogenesis of this disease. In Crohn's disease an abnormal immune response is directed against various *nonself* antigens in the gut, including harmless (and often beneficial) commensal bacteria; the resulting accumulation of white blood cells in the lining of the intestines produces chronic inflammation.

The *NOD2* gene was finally implicated in Crohn's disease by identifying three comparatively common variants: two missense mutations and, notably, a frameshift mutation that was presumed, however, to have a very weak effect (**Figure 8.11**). In one survey presumptive disease-causing mutations in *NOD2* were reported in 50% of 453 European patients. The three common mutations accounted for 81% of the mutations; homozygotes or compound heterozygotes for these mutations are not uncommon in Crohn's disease but are very rare in the normal population. A heterogeneous set of rare missense mutations were suggested to account for the remaining causal variants.

The NOD2 protein is now known to be part of the **innate immune system** (that produces the initial non-specific immune responses against pathogens). It has a C-terminal domain that recognizes a specific peptide motif found in a wide variety of bacterial proteins (see Figure 8.11). The three common DNA variants in Figure 8.11 seem to be partial loss-of-function mutations that impair the ability of NOD2 to recognize bacterial protein.

Gut flora (*microbiota*) include many microbes that are of active benefit to us. They help us derive additional energy through the fermentation of

Figure 8.11 The Crohn's disease susceptibility factor NOD2: common variants and expression in Paneth cells. (A) Domain structure of the 1040-residue NOD2 protein and corresponding location of common variants associated with Crohn's disease (in red). The 3020insC variant appears to be a mild frameshift mutation; it inserts a cytosine, causing a stop codon to be introduced at the next codon position (codon 1008), eliminating just the final 33 amino acids. Like 3020insC, the missense mutations G908R and R702W are located within or close to the LRR domain. Domains: CARD1, CARD2, caspase-activating recruitment domains; NOD, nucleotide-binding oligomerization domain; LRR, leucine-rich repeats domain. The LRR domain is now known to bind to specific breakdown products of peptidoglycan, a major component of bacterial cell walls. (B) The NOD2 protein is predominantly expressed in Paneth cells, specialized secretory epithelial cells found at the base of intestinal crypts (arrows show examples of staining with a specific anti-NOD2 antibody). Paneth cells secrete certain anti-microbial peptides, notably α-defensins. (B, From Ogura Y et al. [2003] *Gut* 52:1591–1597; PMID 14570728. With permission from the BMJ Publishing Group Ltd.)

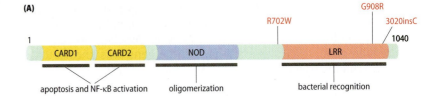

(A)

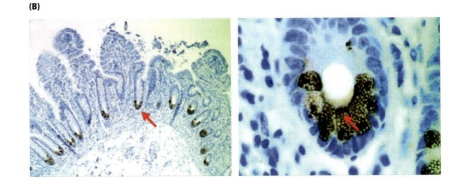

(B)

undigested carbohydrates, help us break down xenobiotics, and synthesize vitamins (B and K) for us. Although they are foreign microorganisms they are therefore tolerated by suppressing the usual innate immune responses. NOD2 works in this area by down-regulating those innate immune responses that require the Toll-like receptors; when NOD2 is impaired, a strong immune response is launched in response to the gut flora, causing inflammation.

The principle of allelic association

Linkage analyses have limited power to detect susceptibility factors in complex disease, and have very largely been supplanted by the alternative of association studies. Whereas linkage is a genetic phenomenon, a property of *loci* that is studied in *families*, **association** is essentially a *statistical* property that describes the joint occurrence of *alleles* (and/or *phenotypes*) in individuals within a *population*.

If allele *A*1 at locus *A* is found to be significantly more frequent in people affected by a specific complex disease than would be expected from the individual population frequencies of *A*1 and the disease gene, we would say that allele *A*1 is positively associated with disease, a disease-susceptibility allele. Conversely, if it were significantly less frequent in affected individuals, it would be viewed as a disease-resistance allele.

To investigate disease associations, **case-control studies** can be carried out in which genetic variants are genotyped in affected individuals (cases) and controls. Different methods can be used to measure the disease risk for each tested genetic variant, notably the **odds ratio**; that is, the odds of being affected when possessing a specific genetic variant divided by the odds of being affected when lacking the genetic variant (see **Table 8.6** for a worked example).

Significant associations between genetic variants and disease may be caused by nongenetic factors (as described below) or by genetic factors (in which case the genetic variant is directly involved in pathogenesis or is very closely linked to a disease-susceptibility allele). Unlike linkage, which works over long ranges in DNA molecules, genetic association works over very short distances only. Genomewide linkage analyses

HLA-CW6 STATUS	NUMBER OF CASES (WITH PSORIASIS)	NUMBER OF UNAFFECTED CONTROLS	ODDS OF BEING AFFECTED		ODDS RATIO
Present	900	328	900/328	→	(900/328) ÷ (100/672) = (900/328) × (672/100) = 18.44
Absent	100	672	100/672		

Table 8.6 A worked example of the odds ratio in case-control studies. The odds ratio is the odds of being affected when possessing a specific genetic variant divided by the odds of being affected when lacking the genetic variant. In this entirely hypothetical example, we imagine a case-control study of psoriasis in which 1000 affected individuals (cases) and 1000 unaffected controls have been genotyped for the HLA-Cw6 marker, giving the calculation in the final column.

require just a few hundred markers distributed across the genome, but genomewide association analyses typically need many hundreds of thousands of markers to find a marker allele that is both associated with—and very tightly linked to—the disease allele.

For many years, the technology simply was not available to carry out genomewide association studies. Instead, disease association analyses focused on testing candidate genes that were known or suspected to function in some biological pathway that appeared highly relevant to the disease under study (but after the human genome was first sequenced, it became clear that we had no or very little idea about the functions of at least one-third of our genes). And choosing candidate genes is not always straightforward (genes subsequently found to underlie some complex diseases would often not have been obvious candidates).

Despite the caveats above, candidate gene association studies have delivered some very significant successes in identifying disease-susceptibility genes. HLA genes were among the first genes to be tested (this was made possible at a very early stage by taking advantage of their exceptional polymorphism at the protein level). As described in **Box 8.3** and in Section 8.3, certain HLA variants have been identified to be the largest genetic contributors to a wide variety of important autoimmune disorders. Candidate gene association studies were also successful in identifying some non-HLA variants of large effect, such as *CTLA4* variants that confer susceptibility to Graves disease and type I diabetes mellitus, and the *APOE*ε4* allele that confers susceptibility to Alzheimer disease.

Linkage disequilibrium as the basis of allelic associations

Associations between genetic variants and disease can be caused by different factors, both genetic and nongenetic. In the former case, population substructure and history are important. As previously considered in Section 5.4, human populations within countries, regions, and cities are often *stratified* into different groups (organized along ethnic, cultural, and religious lines) whose members preferentially mate within the group rather than with members of another group. As a result of population **stratification**, different subgroups within a broad population often have significantly different frequencies of a genetic variant, and this can confound genetic analyses. To minimize problems arising from population stratification, association studies need controls with the same type of population ancestry as those with the trait being studied.

Genetic variants associated with disease might be directly involved in pathogenesis, or they may be tightly linked to a disease-susceptibility allele. In the latter case the haplotype containing the genetic variant and the disease susceptibility allele has a higher frequency than would be predicted from the individual frequencies of the genetic variant and susceptibility allele. This is an example of **linkage disequilibrium**, the nonrandom association of alleles at two or more loci.

As a concept, linkage disequilibrium describes *any* nonrandom association of alleles at different loci; in practice, the alleles are at very closely linked loci. For example, linkage disequilibrium is often evident for alleles

BOX 8.3 HLA associations with autoimmune disorders.

The human major histocompatibility complex (MHC) at 6p21.3 contains many genes that function in the immune system, notably HLA genes that make highly polymorphic cell surface proteins involved in cell-mediated immune responses (see Box 4.4 on page 112 for HLA nomenclature and a simplified HLA gene map). One of the main jobs of HLA genes is to signal the presence of virus-infected cells in our body and guide suitably discriminating T cells to initiate an immune response to kill the infected cells.

All proteins within our cells (whether of normal host origin or from intracellular pathogens such as viruses) undergo turnover, whereby the proteins are degraded to peptides within the proteasome. The resulting peptides are bound by newly synthesized HLA proteins and are then transported to the cell surface so that the HLA–peptide complex is recognized by a specialized T-cell receptor on the surface of T cells (**Figure 1A**).

Immune *tolerance* ensures that self peptides (originating from normal host proteins) do not normally trigger an immune response. At an early stage in thymus development, T cells with receptors that recognize self peptides bound to HLA are eliminated; thereafter T cells are normally focused on nonself ('foreign') peptides (such as those from pathogens). Different T cells in a person contain different T-cell receptors to maximize the chance that a nonself peptide presented by an HLA protein can be recognized. When that happens, T cells are activated to mount an immune response (see Figure 1).

Viruses readily mutate in an attempt to avoid triggering immune responses, and the number of potential foreign peptides is huge. This explains why T-cell receptors are genetically programmed, like antibodies, to be extraordinarily diverse (see Section 4.5). HLA proteins vary in their ability to present specific peptides for recognition and so they, too, are selected to be highly polymorphic.

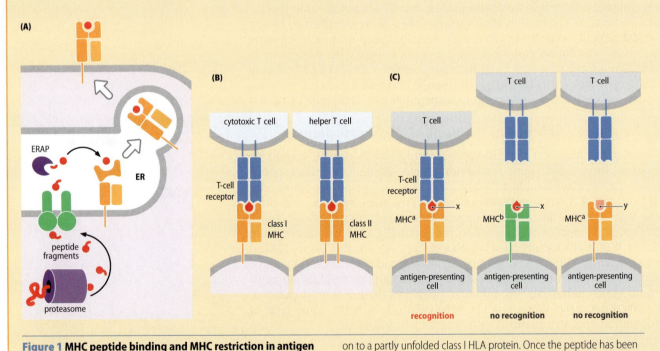

Figure 1 MHC peptide binding and MHC restriction in antigen presentation. (A) Class I MHC proteins (known as class I HLA proteins in human cells) serve to bind peptides and display them on the cell surface. The peptides are produced by the degradation of any protein synthesized within the cell (either a host cell protein or one made by a virus or other intracellular pathogen). Peptide fragments are produced within the proteasome and transported into the endoplasmic reticulum. Here they are snipped by an endoplasmic reticulum aminopeptidase (ERAP) to the proper size needed for loading on to a partly unfolded class I HLA protein. Once the peptide has been bound, the HLA protein completes its folding and is transported to the plasma membrane with the bound peptide displayed on the outside. (B) Receptors on cytotoxic T cells bind class I MHC–peptide complexes; those on helper T cells bind class II MHC–peptide complexes. (C) MHC restriction. T cells have cell-specific receptors that recognize a combination of a specific peptide and a specific MHC protein. (Adapted from Murphy K [2011] Janeway's Immunobiology, 8th ed. Garland Science.)

BOX 8.3 (*continued*)

DISORDER	CLASS OF HLA ANTIGEN	FREQUENCY OF HLA ANTIGEN		ODDS RATIO[a]
		Affecteds	Controls	
Ankylosing spondylitis	HLA-B27	>0.95	0.09	69.1
Celiac disease	HLA-DQ2 and -DQ8	0.95	0.28	15.4
Multiple sclerosis	HLA-DQ6	0.59	0.26	4.1
Narcolepsy	HLA-DQ6	>0.95	0.33	129.8
Psoriasis	HLA-Cw6	0.87	0.33	13.3
Rheumatoid arthritis	HLA-DR4	0.81	0.33	3.8
Type 1 diabetes	HLA-DQ8 and -DQ2	0.81	0.23	9.0
	HLA-DQ6	<0.1	0.33	0.22

Table 1 Examples of HLA disease associations in the Norwegian population. All except one of the associations shown here show an odds ratio greater than 1, indicating disease risk, but in type 1 diabetes HLA-DQ6 is a protective allele (carriers have less risk of the disease than the general population). (HLA antigen frequency courtesy of Erik Thorsby.) [a]See Table 8.6 for how odds ratios are calculated.

In autoimmune disorders there is a breakdown in the ability to distinguish self from nonself. As a result, cells in the body can be attacked by *autoantibodies* and by autoreactive T cells that inappropriately recognize certain host antigens (autoantigens). In diseases such as type 1 diabetes, rheumatoid arthritis, and multiple sclerosis, activated T cells kill certain populations of host cells (such as insulin-producing pancreatic beta cells in type 1 diabetes). In autoreactive T-cell responses, host peptides (autoantigens) are presented by HLA proteins that may differ in their ability to bind the autoantigen. As a result, specific HLA antigens are associated with disease.

At the classical HLA loci, large numbers of alleles can be typed (previously, as serological polymorphisms by using panels of antisera; more recently as DNA variants). HLA–disease association studies involve typing HLA gene variants in affected individuals and controls and calculating the frequencies of a specific antigen or DNA variant in the two groups. This allows calculation of the odds of a disease occurring in individuals with or without a particular genetic variant, and calculation of odds ratios (**Table 1**).

From Table 1 it is clear that possession of certain HLA antigens confers a substantially increased risk for certain disorders, and the odds ratios can be very impressive. If you carry an HLA-B27 antigen, for example, you have a much increased risk of developing ankylosing spondylitis (a form of inflammatory arthritis that affects the joints of the lower back). But HLA-B27 is merely a *susceptibility* factor and although the odds ratio approaches 70, only 1–5% of individuals with HLA-B27 develop ankylosing spondylitis.

at neighboring HLA genes in populations originating in northern Europe. Thus, in Denmark the frequencies of *HLA-A1* and *HLA-B8* are 0.311 and 0.237, respectively, but the frequency of the *HLA-A1–HLA-B8* haplotype is 0.191, more than 2.5 times the expected value (which would be $0.311 \times 0.237 = 0.074$).

Linkage disequilibrium might occur if a particular combination of alleles at neighboring loci were positively selected because they worked together to confer some advantage. However, linkage disequilibrium may often simply reflect reduced recombination between loci. This can happen in areas of the genome where there are low recombination rates. We now know, for example, that the human major histocompatibility complex (MHC) is a region of low recombination (with 0.49 cM per Mb of DNA, compared to a genomewide average of 0.92 cM/Mb).

When a new DNA variant emerges by mutation it will show very tight linkage disequilibrium with alleles at very closely linked loci. The linkage disequilibrium will be gradually eroded by recombination, but that will take a very long time for any locus that is physically very close to the locus with the new mutation.

Sharing of ancestral chromosome segments

Association studies depend on linkage disequilibrium, which in turn reflects shared chromosome segments in large numbers of people because of a very distant common ancestor. Throughout this book we talk about families—groups of people who share large parts of their genomes because of common recent descent. We speak about people being *related* to each other, but we are all related if we go far enough back in history.

What we mean by 'related' is having a *known* common ancestor (usually one that can be identified within the previous four generations). And when we say that two persons are unrelated we generally mean that they do not have any great-grandparent in common, and that they are unaware of any more distant common ancestor. So-called unrelated people do, however, share small common chromosome segments that they have inherited from more distant common ancestors. If the common ancestor lived a long time ago, each shared segment will be quite small but will be shared by a large number of descendants (see Figure 8.12A).

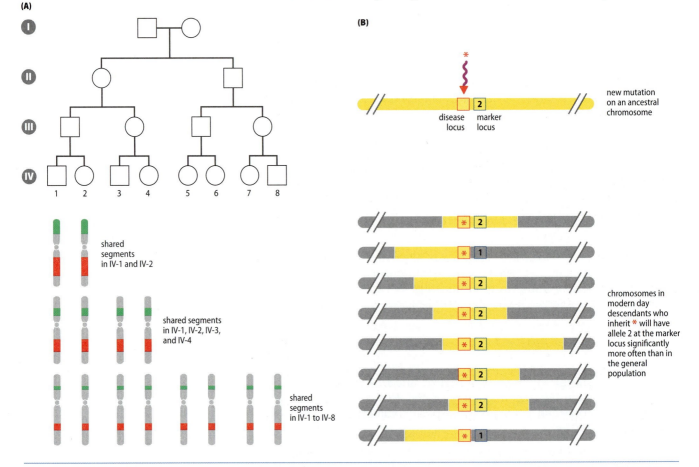

Figure 8.12 Shared ancestral chromosome segments and linkage disequilibrium in the immediate vicinity of an ancestral mutation.
(A) The more distant a common ancestor is, the smaller each shared chromosome segment will be, but the larger will be the number of people who share it. Here we imagine a highly idealized situation where sharing of two chromosome segments on a single chromosome extends to all eight individuals in generation IV (who have common great-grandparents). The extent of the shared region is greatest in sibs, but progressively decreases the further the individuals are separated from a common ancestor.
(B) Linkage disequilibrium around an ancestral mutation that confers disease susceptibility. Upper panel: imagine that the newly emergent mutation (red asterisk) appeared on a chromosome that had a minor (infrequent) allele 2 at a very closely linked SNP locus where allele 1 is the major allele. Lower panel: after passing down through multiple generations, meiotic homologous recombination will ensure that most of the original chromosome (yellow) will have been replaced by segments from other copies of the chromosome (gray). Descendants who inherited the part of the ancestral chromosome with the disease-susceptibility variant have an increased chance of having allele 2 at the very closely linked marker locus. Affected individuals will therefore have a significantly higher frequency of allele 2 at the marker locus than the general population, or than control unaffected individuals. (Adapted from Ardlie KG et al. [2002] *Nat Rev Genet* 3:299–309; PMID 11967554. With permission from Macmillan Publishers Ltd.)

Shared ancestral chromosome segments can explain linkage disequilibrium. Shared segments contain loci that have not been separated by recombination, and so there is a nonrandom association of alleles at linked loci within such segments. By chance, an ancestral chromosome segment might contain an allele that confers susceptibility to a complex disease. In that case, people living now who suffer from the same disease would tend to share that chromosome segment (**Figure 8.12B**).

Usually, the susceptibility allele is neither necessary nor sufficient to cause the complex disease; not everyone with the disease will have that allele, and not everyone with the susceptibility allele will have that disease. But, overall, people with the disease are more likely than unaffected people to have that ancestral chromosomal segment. This is the underlying principle that makes disease association studies possible.

Linkage disequilibrium decreases very rapidly with distance between alleles. If genomewide association studies are to be carried out successfully, a marker map with a very high density is therefore needed. Initial calculations suggested that 500,000 single nucleotide polymorphism (SNP) loci would be needed across the genome. The International HapMap Consortium has gone further by mapping and genotyping millions of SNP loci in different human populations, providing an excellent resource for genomewide association studies. The data from the HapMap project show that our nuclear genome is a mosaic of small blocks of sequence, **haplotype blocks**, in each of which there is very limited genetic diversity (see **Box 8.4** for the details).

How genomewide association studies are carried out

Genomewide association (GWA) studies began to really take off in the mid-2000s because of two technological developments. First, the International HapMap Project delivered hundreds of thousands and then millions of mapped SNP loci. Secondly, by the mid-2000s the extension of microarray technology allowed the automated genotyping of huge numbers of SNPs across the genome. We described the principles of microarray technology previously at the end of Section 3.2 (see Figure 3.9 on page 71). In the case of whole-genome SNP microarrays, the microarrays carry oligonucleotides that are specific for each allele at many hundreds of thousands of SNP loci across the genome, plus controls.

GWA projects have been designed to identify *common* variants (on the assumption that common complex diseases are often caused by common variants—we consider the rationale below). The bulk of GWA studies have therefore focused on case-control studies in which panels of affected individuals and matched controls are genotyped at hundreds of thousands of common SNPs (where the minor allele usually has a frequency of at least 0.05). SNPs are then identified in which allele frequencies are significantly different in cases than in controls (**Figure 8.13**).

SNP microarray hybridization typically involves many hundreds of thousands of parallel DNA hybridizations, one for each of the fixed oligonucleotides, and because such huge numbers of hybridization tests are being carried out, stringent statistical significance thresholds are required to assess the significance of individual hybridization results. One way of setting a more stringent genomewide significance threshold is to divide the standard P value of 0.05 by the number of tests. If the microarray hybridization involves one million different hybridization assays, for example, a stringent P value would then be $0.05/1,000,000 = 5 \times 10^{-8}$.

The genotype test statistics are calculated for each variant and referenced against statistics expected under the null hypothesis of no disease association. The data can be visualized in different types of plot as shown in **Figure 8.14**.

BOX 8.4 Haplotype blocks and the International HapMap Project.

Initial attempts to define ancestral chromosome segments began with high-resolution mapping of haplotype structure in defined small genome regions in populations of European ancestry. The results suggested that our nuclear DNA might be composed of defined blocks of limited haplotype diversity (**haplotype blocks**). **Figure 1A** illustrates an example—a haplotype block 84 kb long that spans most of the *RAD50* gene at 5q31. Eight common SNP loci were genotyped in this block, and two alleles at each of eight SNP loci means the potential for $2^8 = 256$ different haplotypes. Yet, within this block, almost every chromosome 5 that is tested has 1 of only 2 out of the 256 possible haplotypes—either the orange one in Figure 1A (which we can abbreviate by listing the nucleotides at the eight consecutive SNP loci as GGACAACC) or the green haplotype (AATTCGTG).

The low haplotype diversity is apparent in adjacent haplotype blocks. In Figure 1B, block 1 (the same block that is shown in Figure 1A) and the neighboring block 2 are dominated by two haplotypes, and the next two blocks by three and four haplotypes, respectively. It suggests that the DNA in block 1 was contributed mostly by two ancestors, and that in blocks 2, 3, and 4 by a different set of two, three, and four ancestors, respectively.

The International HapMap project set out to make comprehensive maps of linkage disequilibrium in the human genome. The project began by genotyping common single nucleotide polymorphisms (SNPs) in samples from four populations: the Yoruba from Nigeria (YRI); a white population from Utah, USA, descended from northern and western Europe (CEU); Han Chinese from Beijing (CHB); and Japanese from Tokyo (JPT). Hapolotype maps were constructed by genotyping 3.1 million SNPs (or about one every kilobase).

The HapMap project confirmed that humans show rather limited genetic variation (by comparison, chimpanzees show very much more genetic diversity). At a fairly recent stage in population history, the human population was reduced to a very small number—perhaps 10,000 or so individuals—that remained quite constant until fairly recently. First agriculture and then urbanization led to a very rapid massive expansion in population size to the current 7 billion individuals. As a result, about 90% of the genetic variation in humans is found in all human populations.

Overall, about 85% of our nuclear genome is a mosaic structure, composed of haplotype blocks. The average size of the haplotype blocks in the populations of European and Asiatic ancestry was 5.9 kb with an average of about 3.6 different haplotypes per block. In the Yoruban population there were an average of 5.1 different haplotypes per haplotype block and the blocks averaged 4.8 kb in size (all human populations originated in Africa, and African populations have greater genetic diversity).

Figure 1 Haplotype blocks. (A) Genotyping at eight SNP loci (vertical blue boxes) spanning most of the *RAD50* gene at 5q31 reveals an 84 kb haplotype block. Just two haplotypes account for the vast majority (76% and 18%, respectively) of the chromosomes 5 from a sampled European population. (B) Adjacent haplotype blocks at 5q31 including the block from (A), called block 1 here. Blocks 2, 3, and 4 were genotyped at respectively five, nine, and eleven SNP loci and had between two and four haplotypes shown in different colors at population frequencies given at the bottom. The dashed black lines signify locations where more than 2% of all chromosomes 5 are seen to switch from one common haplotype to another. (Adapted from Daly MJ et al. [2001] *Nat Genet* 29:229–232. With permission from Macmillan Publishers Ltd.)

(A)

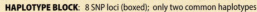

HAPLOTYPE BLOCK: 8 SNP loci (boxed); only two common haplotypes

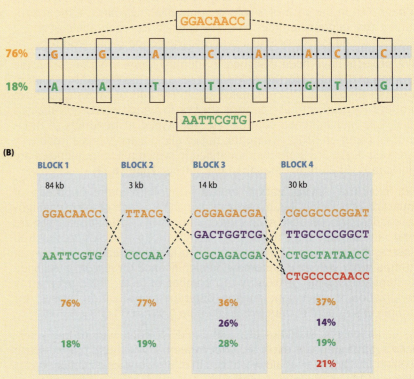

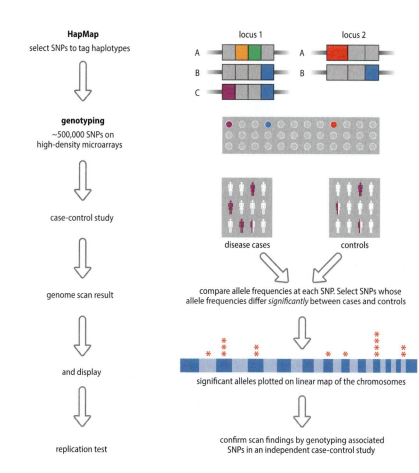

HapMap
select SNPs to tag haplotypes

genotyping
~500,000 SNPs on
high-density microarrays

case-control study

genome scan result

and display

replication test

locus 1   locus 2

disease cases   controls

compare allele frequencies at each SNP. Select SNPs whose
allele frequencies differ *significantly* between cases and controls

significant alleles plotted on linear map of the chromosomes

confirm scan findings by genotyping associated
SNPs in an independent case-control study

Figure 8.13 Carrying out a genomewide association (GWA) scan. Using HapMap data (which map linkage disequilibrium across the human genome), representative SNPs are selected that will differentiate (or *tag*) the common haplotypes at each locus. In this example, three common haplotypes (A, B, and C) at locus 1 are tagged by four SNPs (with strong color if present, or gray if absent). But just two SNPs (purple and blue) are sufficient to discriminate between the three haplotypes. Similarly, the two haplotypes at locus 2 can be distinguished by either the red (chosen here) or the blue SNP. The *tag SNPs* are then genotyped in disease cases and controls using microarrays, and the allele frequencies for each SNP are compared in the two groups. SNPs associated with disease at an appropriate statistical threshold are genotyped in a second independent sample of cases and controls to establish which of the associations from the primary scan are robust. (Adapted from Mathew CG [2008] *Nat Rev Genet* 9:9–14; PMID 17968351. With permission from Macmillan Publishers Ltd.)

Initial promising GWA subchromosomal locations need to be confirmed. To do this, candidate SNPs of high statistical significance are genotyped in an independent replication panel. In addition to low *P* values, extra confidence is obtained when the same location is replicated in independent studies on different populations. Confirmation can be achieved using a linkage disequilibrium test, notably the transmission disequilibrium test, described below. Then comes the more difficult part: identifying causal variants that are directly responsible for the susceptibility to disease.

Initial successes were obtained in identifying susceptibility factors of quite large effect; however, to map those with more modest effect, large numbers of cases are needed. A landmark paper by the Wellcome Trust Case Control Consortium in 2007 reported considerable success in using GWA studies to map susceptibility to seven complex diseases using 2000 cases for each disease and a common set of 3000 controls. There has since been quite an explosion in the number of GWA studies and a very considerable amount of success, as many initial findings have been replicated and confirmed. For an up-to-date list see the Hindorff et al. website under Further Reading.

The transmission disequilibrium test

The transmission disequilibrium test (TDT) looks for association in the presence of linkage (that is, it is a direct test for linkage disequilibrium and can confirm allelic association). The basis of the test is to collect samples from affected individuals and both parents and identify those SNP loci in which a parent is heterozygous and in which there is an informative meiosis that allows researchers to identify which allele was transmitted from the heterozygous parent to his/her affected child. The

Figure 8.14 Visualizing genomewide association (GWA) data. (A, B) Quantile–quantile (Q–Q) plots showing two types of distribution of observed test statistics generated in a GWA study. In case-control studies a chi-squared (χ^2) comparison of absolute genotype counts is calculated for each variant—red dots indicate idealized test results, and blue dots represent expected values under the null hypothesis of no association. In (A) the test results are consistently inflated across the distribution (which could signify population stratification effects). In (B) there is little evidence of association arising from population substructure effects (most of the red and blue dots coincide), but the deviations at the highly significant end of the scale in this example plot are compelling evidence for susceptibility loci with large effects. (C) A genomewide Manhattan plot (think of skyscrapers) displaying GWA findings according to their genomic positions (horizontal scale) and statistical significance (vertical scale; the negative $\log_{10}P$ scale helps reveal signals of particular interest). This plot is from a large study of coronary artery disease that shows newly discovered disease-susceptibility loci in blue, and previously discovered ones in red. The dashed horizontal line at position 7.30 on the vertical scale indicates the threshold of statistical significance (corresponding to $P = 5 \times 10^{-8}$ in this case). The most significant associations were with previously recorded SNPs in the immediate vicinity of the closely neighboring *CDKN2A* and *CDKN2B* genes at 9p21. (From Schunkert H et al. [2011] *Nat Genet* 43:333–338; PMID 21378890. With permission from Macmillan Publishers Ltd.)

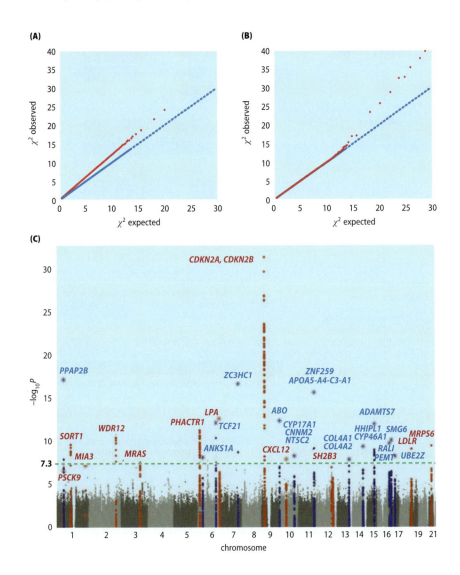

underlying idea is that if the disease is caused by a common susceptibility factor, a SNP locus is tightly linked to the disease locus. See **Figure 8.15** for the principle.

Moving from candidate subchromosomal region to identify causal genetic variants in complex disease

The short candidate subchromosomal regions identified by GWA studies are easily sequenced in affected individuals and controls. A SNP that shows a significant disease association is expected to be very closely linked to a genetic variant that predisposes to disease (by altering how a gene is expressed). However, moving from an associated SNP to the causal variant may be extremely difficult. In addition to associated SNPs that were part of the marker panel, there will be many other genetic variants in the candidate region that were not tested. Because they, too, are on a shared small chromosome segment with the disease allele, all of them will also be associated with the disease. How, then, do we identify the causal variant?

And then there is the problem that, unlike in monogenic disorders, the causal variant is often not causative! It will be absent in a proportion of people with the disease and will be found in many normal people—the 'causal' variant is merely a susceptibility factor. Take, for example, the hypothetical association of variant *a* of gene *A* with disease X. Imagine

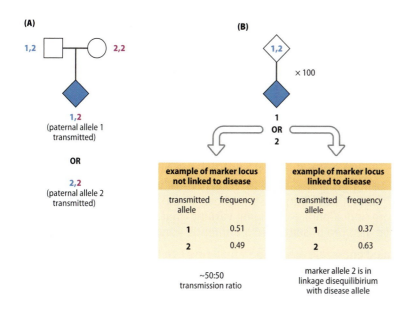

Figure 8.15 Principle of the transmission disequilibrium test (TDT). The TDT relies on being able to identify which of the two alleles at a marker locus in a heterozygous parent has been transmitted to an affected child. (A) A suitably informative pedigree. The father is heterozygous and the mother must transmit allele 2. If the affected child is a heterozygote with a 1,2 genotype, the father has transmitted allele 1; if, instead, the child is a 2,2 homozygote, the father has transmitted allele 2. (B) Imagine that we have collected samples from individuals affected with the same complex disease plus their parents and have identified 100 informative pedigrees in which we can track which of two marker alleles is transmitted from a heterozygous parent, just as in (A). If marker allele 1 is transmitted by a heterozygous parent in 51 out of the 100 examples (frequency 0.51), and marker 2 is transmitted in the other 49 cases, that is close to the predicted 50:50. If, however, allele 2 were to be transmitted in a disproportionately large number of families, say 63 out of 100, marker allele 2 and the disease allele can be inferred to be in linkage disequilibrium and the marker locus must be tightly linked to the disease locus.

that variant *a* has a frequency of 0.457 in 2000 affected cases and a frequency of 0.361 in 2000 controls. That might seem a small difference, but because of the large number of samples genotyped it would be highly significant.

Finally, the causal variant can often be expected to have a fairly subtle effect on gene expression and may often not be easily identified. For a causal variant to be common in the population it must be a comparatively ancient mutation (that has been transmitted from a very distant common ancestor to many people living now). An ancient mutation cannot have been exposed to strong natural selection (otherwise it would have been eliminated by now) and so it is unlikely to be the type of severe mutation (notably frameshifting, nonsense, and splice-site mutations) often seen in Mendelian disease. Instead, it would be more likely to be a subtle missense mutation that could confer partial loss of function, or a mutation that causes a moderate change in gene expression levels, such as a polymorphism in a noncoding regulatory DNA sequence that controls transcription (such as promoter) or translation (an miRNA-binding site within an untranslated region, and so on). If it were to be a frameshifting mutation, the effect on gene expression would need to be mild, as in the 3020insC variant at the *NOD2* locus that acts as a Crohn's disease susceptibility factor (as described in Figure 8.11A and the associated text).

Identifying causal variants

After sequencing the DNA of short candidate chromosome regions thrown up by GWA studies, many genetic variants will be identified in addition to SNPs from the GWA tests that map in this region. They will include other SNPs (and also occasional variants that are normally too rare to be common polymorphisms, including copy number variants, and single nucleotide variants that include missense mutations). Depending on the local crossover frequency, the candidate region can be small (enhanced recombination) or large (a region of low recombination).

The assorted variants that map to the same haplotype block as a SNP showing disease association will also be associated with disease. But the borders of haplotype blocks are not precisely defined (the linkage disequilibrium is always significantly less than 100%), and so there can be differences in the degree of association of the different variants within a block. Within each block are a limited number of haplotypes (see Box 8.4); variants that happen to be present in multiple haplotypes will be much more strongly associated than those on a single haplotype. Ideally, what we would hope to find is a peak area where a few variants show especially strong association, against a general background of association for the variants within a critical region (which can encompass adjacent haplotype blocks).

Variants showing the strongest association are prioritized for further investigation, including functional assays and bioinformatics analyses. For missense mutations, protein function tests can often be used. Additionally, *in silico* analyses can be used to assess the likely effect of amino acid substitution on the structure and function of the predicted protein, using programs such as PolyPhen-2 (http://genetics.bwh.harvard.edu/pph2/), SIFT (http://sift.jcvi.org/), and PROVEAN (http://provean.jcvi.org/index.php). Noncoding variants that subtly alter gene expression can be assayed in suitable experimental gene expression assays.

Detailed fine-scale association mapping can be carried out to home in on causal variants, and it is helped by a knowledge of how the protein works normally. In HLA associations with autoimmune disorders, for example, the genetic determinants might be expected to be variant amino acids involved in presenting autoantigens and that are important in peptide-binding. In rheumatoid arthritis, about 70% of affected individuals show HLA associations and have autoantibodies against natural cyclic citrullinated peptides (in which selected arginine residues have undergone modification to give a rare amino acid, citrulline). Fine-scale association mapping has shown that most of the HLA association with rheumatoid arthritis is due to five amino acids, all located in the peptide-binding grooves. Three occur in the HLA-DRβ1 chain, one in HLA-B, and one in HLA-DPβ1—see **Figure 8.16**, and see Raychaudhuri et al. (2012) under Further Reading for details of the fine association mapping.

The example in Figure 8.16 is atypical—the causal variants have not been identified at many GWA risk loci. Nevertheless, it has very often been possible to implicate a neighboring gene in the pathogenesis, based on the known properties of the gene (see examples below). In rare situations it might not be clear which gene is involved at a GWA risk locus, but sequencing of genes at that locus in very large numbers of affected individuals might reveal disease-associated rare variants (as described below), and that can confirm the identity of the susceptibility gene.

The limitations of GWA studies and the issue of missing heritability

Although early linkage studies were important in identifying subchromosomal locations for some important genetic variants underlying complex disease, the returns from later studies were low. GWA studies have had greater success: the electronic catalog of published GWA studies (see the Hindorff et al. website under Further Reading) lists more than 1000 significant associations, many of which have been replicated. However, as we describe in Section 8.3, the returns have been few for some disorders.

Despite initial high hopes, the common disease variants identified by GWA studies have very largely been of very weak effect—often with an odds ratio of 1.2 or less. Exceptions include some novel factors that strongly predispose to age-related macular degeneration (a leading cause of vision loss in older adults that results from a progressive deterioration of the macula, a central region of the retina). But many of the variants with high odds ratios were identified in the pre-GWA era (such as apoE4 in Alzheimer disease, the common *NOD2* alleles in Crohn's disease, and especially HLA alleles that remain, by some distance, the strongest known genetic variants in autoimmune disorders).

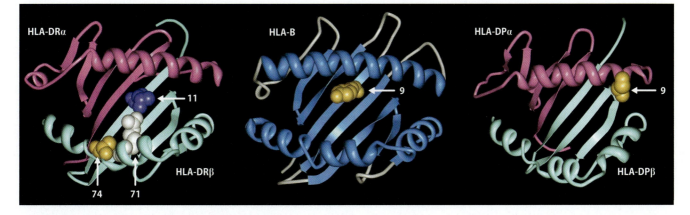

The first round of large GWA studies identified many variants that are significantly associated with individual complex diseases, but almost all of the variants were of weak effect: even the cumulative contributions of the identified variants were usually quite small. That is, the available GWA data could explain only a small proportion of the genetic variance of complex diseases. That raised the issue of the *missing heritability* that was not being detected.

More recent studies have also looked at copy number variation; however, it, too, is not a major contributor to complex disease. To explain the missing heritability, additional alternative explanations have been put forward, including those listed below.

Large numbers of common variants with very weak effect. GWA studies with a few thousand cases and controls are well suited to detecting susceptibility factors with odds ratios of 1.5 or more, but many genuine susceptibility factors might be missed if they have weaker effects (odds ratios of less than 1.2). To have a high chance of detecting these variants, GWA studies need to use much larger numbers of cases and controls, either directly or in meta-analyses which aggregate data from multiple individual studies.

Rare variants of large effect. As discussed below, a major limitation of GWA studies is that they are restricted to identifying associations with common (frequent) variants. Much of the disease susceptibility might conceivably be due to a heterogeneous set of rare variants with individually strong effects (high odds ratios)—see the next section.

Gene–gene and gene–environment interactions. The concept of heritability is flawed. It assumes additive effects by different loci, and the proportion of heritability explained by known GWA variants does not take into account genetic interactions between loci. Heritability is also traditionally separated into genetic and environmental components, but this is simplistic: genes interact with the environment. We will return to consider gene–gene and gene–environment interactions in Section 8.3.

The relative contributions of common and rare variants to complex disease susceptibility

Association studies identify factors that are present on chromosome segments shared by many individuals in the study group. Thus, they are limited to studying *common* nucleotide variants of ancient origin (the minor allele frequencies for selected SNPs are usually greater than 0.05, but at least greater than 0.01). But the missing heritability issue focused attention on the alternative that rare variants are important in complex disease. There remains some debate about two competing hypotheses, as detailed below.

Figure 8.16 HLA association with rheumatoid arthritis is mostly due to five amino acids located in the peptide-binding grooves of three HLA proteins. The three-dimensional ribbon models give direct views of the peptide-binding grooves of the HLA-DR, HLA-B, and HLA-DP proteins (the α and β chains of HLA-DR and HLA-DP are shown in pink and pale green, respectively). The five key amino acids that are important in HLA association with rheumatoid arthritis are shown—at positions 11, 71, and 74 in HLA-DRβ, and at position 9 in HLA-B and HLA-DPβ.

The common disease–common variant hypothesis

In this hypothesis, different combinations of common variants at multiple loci are believed to aggregate in specific individuals to increase disease risk. It offers an explanation for why there is such a steep falling away of disease risk in relatives of probands with a common disease (common diseases often appear as sporadic cases). And it was supported by known associations of common variants with complex diseases. For example, in the early 1990s the *APOE*ε4* (epsilon 4) allele (frequencies of 0.05–0.41 in different populations) was found to be a susceptibility factor for late-onset Alzheimer disease; the common *APOE*ε2* allele, by contrast, was found to be a protective factor, reducing disease risk (**Figure 8.17A**).

Common variants are expected to be of ancient origin (it takes a long time for a mutation to become established in the population). They are merely *susceptibility* factors, and so typically they have very weak deleterious effects (causing mild missense mutations or small changes in gene expression, for example).

How common deleterious alleles are maintained

Although they have weak effects, common disease-susceptibility alleles are deleterious. Why, then, have they not been eliminated by natural selection over the very many generations since they formed by mutation? Many complex diseases are of late onset, and because short human lifespans used to be common until quite recently, susceptibility alleles for aging-related disorders might have had very little effect on reproductive rates over large numbers of generations. Alleles causing diseases that manifest only later in life might therefore have been protected to a very considerable degree from natural selection.

Many common disease alleles also seem to have some advantages. Certain HLA alleles are susceptibility factors for specific autoimmune disorders but are also very important in allowing immune responses against certain viruses and other intracellular pathogens. The common *NOD2* alleles that confer susceptibility to Crohn's disease in many populations of European origin (see Figure 8.11) also seem to have been maintained by natural selection. Balancing selection is most probably involved—the deleterious haplotypes might confer some *heterozygote advantage* (such as the protection against malaria by the sickle-cell allele described in Section 5.4). Several different susceptibility factors conferring increased risk for type 1 diabetes are also known to be simultaneously protective factors for Crohn's disease.

Common disease variants may also have conferred some advantages in the recent past. The *'thrifty gene' hypothesis* proposes that certain genetic variants confer a selective advantage in populations exposed to

Figure 8.17 Risk of late-onset Alzheimer disease for the three human APOE alleles and amino acid differences between human alleles and the apoE of great apes. (A) Associated Alzheimer disease risk and amino acid differences between the common human alleles. Compared with those people who are homozygous for *APOE*ε3*, the most common allele, people with one copy of the *APOE*ε4* allele have a roughly threefold greater risk of Alzheimer disease, and people who have two *APOE*ε4* copies have a roughly fifteenfold increased risk. The mature protein made by *APOE* is called apoE with three common alleles apoE2, apoE3, and apoE4, each with 299 amino acids. They differ at positions 112 and 158 only, as shown. (B) Chimps and gorillas have a non-polymorphic apoE protein that resembles human apoE4. Dots indicate amino acid substitutions between human apoE4 and the chimp or gorilla apoE. Black dots indicate six amino acid differences that are common to the chimp and gorilla; orange dots indicate two extra differences in chimp and one in gorilla. The human *APOE*ε4* allele is regarded as our ancestral allele. Subsequent mutations gave rise to arginine/cysteine substitutions at positions 112 and 158 that increased in frequency because of a presumed selective advantage conferred by these variants.

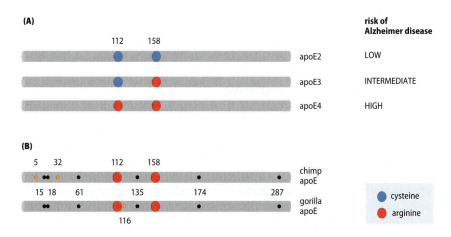

famine, and that they were advantageous in the past when food supplies were limited. But in modern societies in which food is plentiful, the same variants have become susceptibility factors for type 2 diabetes. The *APOE*ε4* allele, which predisposes to cardiovascular disease as well as to Alzheimer disease (and so will have an effect on reproductive rates), seems to have been the ancestral *APOE* allele (because of comparisons with the great apes—see **Figure 8.17B**). It has been imagined to have been selectively advantageous to early humans who had a low-calorie, low-fat diet. Over time, however, it has increasingly been replaced by the *APOE*ε3* allele, which offers the advantage of decreased cholesterol metabolism (reducing the risk of cardiovascular disease).

The common disease–rare variant hypothesis

The odds ratios of common risk alleles identified by GWA studies are generally very low. Doubts about how much common variants could contribute to the total disease susceptibility have propelled strong interest in an alternative explanation—that rare variants originating by comparatively recent mutations account for much, and perhaps most, of the susceptibility to complex diseases.

Moderately rare variants may have moderate effects (that are generally stronger than those of common susceptibility factors). Very rare variants might be expected to often have rather strong effects, and so be highly penetrant. Natural selection would ensure that variants with a strong effect would not persist for many generations in the population. They might often be confined to groups of related families, or to single families, or sometimes even to single cases (*de novo* variants). They would not appear on common haplotype blocks that have ancient origins. To balance their loss from the population, new deleterious mutations are created by random recurrent mutation.

One rationale for the common disease–rare variant hypothesis is that, given the great mutational heterogeneity of individual single-gene disorders, it is not clear why complex diseases should be any different. At one extreme, many complex diseases are known to have Mendelian subsets in which the pathogenesis is due to very rare mutations of extremely strong effect (so that the phenotype is highly penetrant). As long ago as 2002, reported mutational studies of the *NOD2* gene had indicated that there may be a heterogeneous set of rare *NOD2* variants in addition to the three common alleles shown in Figure 8.11A. And so it is not too much of a leap to propose that the DNA variants associated with a common disease might be heterogeneous—different frequency classes might be envisaged with effects that are, very roughly, inversely proportional to their frequency (**Figure 8.18**).

Different subsets of rare variants have been imagined, including moderately rare variants with frequencies that are just below the GWA study threshold (frequency less than 0.05), and very rare variants. The rarest variants, *private variants*, could be expected to arise *de novo* and would therefore explain a proportion of sporadic cases (but could not contribute to the heritability of the disease). They might be expected to be more common in disorders in which reproductive rates are notably reduced (where they might be expected to have strong effects). As described below, large-scale DNA sequencing studies began to be launched in the late 2000s to seek out rare variants associated with complex diseases.

Copy number variants associated with complex diseases

Because common SNP variants in GWA studies were thought to explain only a small proportion of the heritability of some complex diseases, intensive efforts have focused on two types of additional variation, each

Figure 8.18 Different classes of DNA variants contribute to complex diseases. Common and moderately common DNA variants have reached high to moderate frequencies in the population because they each originated by mutation a long time ago and have been passed down through very many generations to large numbers of descendants. The vast majority of the common or moderately common DNA variants that are associated with disease are of very weak effect; a very few have substantial effects, such as many HLA alleles in autoimmune disorders, and are represented by the cloud effect above the main boxes. Rare variants have originated fairly recently and so can have rather stronger effects than common variants, but they will eventually be removed by natural selection if the effects are strong enough. *De novo* variants can have large effects and are quickly removed from the population. Many complex diseases have Mendelian subsets in which the disease phenotype is caused by mutations that have a very strong effect, resulting in highly penetrant phenotypes, as shown in Figure 8.9.

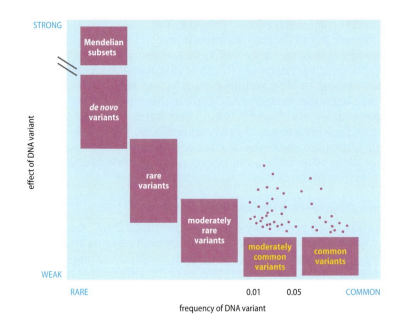

accounting for more variation in the general population than is due to common SNPs.

As described in Section 4.3, copy number variation in the human genome can range from changes to single nucleotides to change in copy number of a whole chromosome. The term **copy number variant** (**CNV**) is poorly defined, but it means changes in the copy number of sequences of intermediate length (from, say 100, nucleotides long upwards to a few megabases). Although the number of CNV loci is much less than the number of SNP loci, the larger sizes of CNV loci mean that if we compare two human genomes, the base pair differences due to CNVs significantly outnumber those due to SNPs.

The shorter CNV loci may sometimes have multiple length variants per locus, but for most CNV loci, especially the larger ones, the variation is due to simple deletions or duplications. Shorter CNVs can be studied by sequencing, but larger ones are normally studied by a hybridization method known as *comparative genome hybridization* that is carried out on microarrays. We describe the method in the context of genetic testing (Section 11.2, pages 438–439).

Some copy number variants are quite common (with a frequency of more than 1%) and are known as *copy number polymorphisms* (*CNPs*). CNPs have been found to be associated with a variety of different disorders, often by changing the copy number of genes in a clustered multigene family (see **Table 8.7** for examples).

Table 8.7 Examples of copy number polymorphisms (CNPs) associated with complex diseases. (Data from Girirajan S et al. [2011] *Annu Rev Genet* 45:203–226; PMID 21854229.)

CNP ALLELE	DISEASE
Deletion upstream of the *IRGM* gene (involved in the innate immune response)	Crohn's disease
Low-copy-number allele for a multiple-copy-number polymorphism with the gene encoding lipoprotein A (see Figure 2.12B)	coronary artery disease
High-copy-number allele that spans the β-defensin multigene family and so provides extra β-defensin genes (which make antimicrobial peptides that provide resistance to microbial colonization of epithelial cells)	psoriasis
Single copy of the complement C4 gene (instead of the normal two complement C4 gene copies)	systemic lupus erythematosus
Low-copy-number *FCGR3A* alleles (notably deletions). The *FCGR3A* gene makes the Fc portion of immunoglobulin G and is involved in removing antigen–antibody complexes from the circulation and in other antibody-dependent responses	

CNPs and CNVs in neuropsychiatric disorders

Neuropsychiatric disorders, which are associated with considerable phenotype heterogeneity and also reduced reproductive rates, show frequent changes in copy number for large DNA segments. Two intensively investigated disorders are schizophrenia (principally defined by long-standing delusions and hallucinations) and autism spectrum disorder (a heterogeneous spectrum of phenotypes defined by markedly abnormal social interaction and communication, beginning before 3 years of age; some affected individuals may have severe intellectual disability, but others lead relatively normal lives, with sometimes exceptional occupational achievement).

High-resolution karyotyping has shown that about 5% of individuals with autism spectrum disorder have cytogenetically visible chromosome rearrangements. Global screens also suggest that there is a greater load of subcytogenetic common CNPs and rare CNVs in individuals with autism than in controls. Duplications, as well as deletions, contribute to disease. Many of the CNVs are found to occur *de novo*; others are transmitted, sometimes from an unaffected parent.

Inherited CNVs are scarcely more frequent in autism spectrum disorder than in controls, but *de novo* CNVs in affected individuals are generally larger than in controls, and typically about three to six times more frequent. Many CNVs associated with autism spectrum disorder contain multiple genes (**Table 8.8**), even if the pathogenesis might be due to altered expression of a single gene in many cases. Schizophrenia-associated CNVs cover some of the same regions found to be associated with autism, such as 1q21.2 (deletion), 22q11.2 (deletion), and 16p11.2 (duplication), plus large deletions of variable size at the *NRXN1* locus at 2p16.3.

The *de novo* CNVs cannot contribute to heritability, and although the CNVs in autism spectrum disorder and schizophrenia are quite often of large effect, they account for only a small proportion of the observed genetic variance.

Recent explosive growth in human population has meant that most coding sequence variants are rare variants

Human populations have undergone something of a roller-coaster ride over the past 70,000 years or so. In this time there has been at least one recent *population bottleneck* when the global population was reduced to about just 10,000 individuals. (This explains why humans across the globe are much more genetically homogeneous than chimpanzees and gorillas, our closest evolutionary cousins.)

LOCUS SIZE	GENES	LOCATION	PATHOGENIC ALLELE	FREQUENCY IN ASD (%)
0.7 Mb	30 genes	16p11.2	deletion and duplication	0.8
~1 Mb	(*PTCHD1* and *PTCHD1AS*)	Xp22.1	deletion; mostly affecting upstream *PTCHD1AS* antisense noncoding RNA	0.5
Variable	*NRXN1*	2p16.3	mostly deletion	0.4
1.4 Mb[a]	22 genes	7q11.2	duplication	0.2
2.5 Mb	56 genes	22q11.2	deletion and duplication	0.2
1.5 Mb	14 genes	1q21.1	duplication	0.2
Variable	*SHANK2*	11q13.3	deletion	0.1

Table 8.8 Examples of loci that frequently undergo copy number variation in autism spectrum disorder (ASD). [a]The same region is deleted in Williams–Beuren syndrome. (Data from Devlin B & Scherer SW [2012] *Curr Opin Genet Dev* 22:229–237; PMID 22463983.)

Figure 8.19 After a long period of slow growth, the human population size has recently exploded. Census population size is presented on a logarithmic scale over the past 10,000 years, from about 5 million at 10,000 years ago (10 kya) to about 7 billion today. The depicted linear increase (on the logarithmic scale) through most of the time shown here denotes exponential growth of a relatively constant percentage increase in population size per year. An acceleration of that increase started in the past 1000–2000 years.

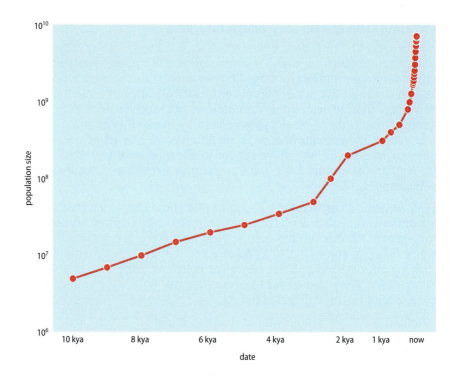

The global human population size of 10,000 or so was maintained without much increase over a rather long period until the development of agriculture provided a stimulus to population growth. Subsequently, urbanization took population growth to a new level, and particularly in the past 1000 years there has been explosive growth (**Figure 8.19**). With the recent population growth has come a burst of new variants created by mutation and distributed by meiosis. And with increased population size came increased competition between individuals and greater selection.

Because of the explosive population growth in the past few centuries, much of our genetic variation is very recent: the great majority of individual DNA sequence variants that affect how our genes are expressed occur at low frequency in the population. That became evident after exomes were sequenced in many thousands of people ('deep sequencing'). Two of the first such studies—published in mid-2012 and early 2013—examined more than 15,000 protein-coding genes in, respectively, 2,440 and 6,515 individuals of European or African descent. The outcome was that the vast majority of the variation in coding sequences is due to rare (low-frequency) variants that had previously been mostly unknown. In the study by Fu et al. (2013), listed under Further Reading, 73% of all protein-coding single nucleotide variants (SNVs) and 83% of the deleterious SNVs were estimated to have arisen in the past 5000 to 10,000 years. Because of their very recent origin, these variants are individually rare in the population, but because they account for so much variation they can be expected to make an important contribution to disease susceptibility.

Massively parallel DNA sequencing to identify rare sequence variants associated with complex disease

Much of the effort to uncover rare variants in complex disease involves candidate gene sequencing and exome sequencing in large-scale case-control studies. In the former case, much of the effort has focused on screening genes that have been implicated as likely (or possible) common susceptibility factors to see whether they also show rare disease-associated variants.

Using exome sequencing to identify risk factors in complex disease can be less straightforward than in identifying disease genes in monogenic disorders. Moderately rare sequence variants, with frequencies just below the 0.05 frequency cutoff that usually applies in GWA studies, may often have only mildly deleterious effects and be present at significant frequencies in the normal population, making them less easy to identify.

At least, moderately rare variants can be screened comparatively easily (by sequencing exomes from a panel of a few hundred cases, identifying variants found in multiple cases, and then genotyping these variants in very large panels of cases). Detecting very rare variants can also be done easily enough for candidate genes by using extremely large panels of affected individuals. And because they are very likely to have strong effects they can be expected to include deletions, frameshifting, and splice-site mutations that are comparatively easy to identify. However, carrying out blind genomewide screening to detect very rare variants would involve significant effort: at a minimum, exomes would need to be sequenced from very large numbers of individuals, and preferably whole-genome screening should be undertaken.

The rarest variants are *private mutations* that have arisen *de novo*; they can be spotted relatively easily if samples are also available from both parents of an affected individual. They are common in neuropsychiatric disorders, which are associated with considerable phenotype heterogeneity and also reduced reproductive rates.

De novo sequence variants

The first strong evidence for frequent rare variants associated with complex disease came from candidate gene studies of autism spectrum disorders. Genes lying within disease-associated CNVs were screened by DNA sequencing; highly penetrant disease-associated point mutations and deletions were found in various genes.

Although the frequency of *de novo* mutations in individuals with autism spectrum disorders does not significantly differ from that in controls, the affected individuals have a significantly greater proportion of nonsynonymous and nonsense mutations than in controls or unaffected sibs. That is, they have a greater *burden* (load) of deleterious mutations. In many cases, the implicated genes are involved in synapse function, including genes encoding neurexins and their ligands, the neuroligins, and genes that make synaptic scaffold proteins. Many others are expressed in brain and implicated in signaling pathways.

The overall contribution made by rare variants

Since about 2009 there has been intense debate as to the relative importance of common and rare variants to complex disease susceptibility. Large-scale sequencing of genes identified at some common risk loci in GWA studies has uncovered disease-associated rare variants, some of large effect. In age-related macular degeneration, for example, GWA studies had established that about 50% of the genetic susceptibility is attributable to common variants in certain complement factor genes (*CFH*, *C3*, *CFI*, and *C2-CFB*), and recent sequencing of these genes has also identified several rare variants of strong effect (Table 8.9).

Similar successes to those in Table 8.9 have been found at some other common susceptibility loci in other complex diseases, but there are also many cases in which they have not been identified. In age-related macular degeneration, for example, sequencing of many common susceptibility loci has failed to uncover disease-associated rare variants other than those in the complement pathway. And large-scale sequencing of 25 genes identified in GWA studies as susceptibility factors for different

GENE	VARIANT	ODDS RATIO	COMMENTS	SOURCE PMID
CFH	p.Arg1210Cys	23.11	defective binding to C3b. Also associated with a rare renal glomerular disease, atypical hemolytic uremic syndrome	22019782, 24036949
CFI	p.Gly119Arg	22.20	reduced expression	23685748
C3	p.Lys155Gln	2.2–3.8	reduces C3b binding to complement factor H, resulting in excessive alternative complement activation	24036949, 23046950, 24036952
C9	p.Pro167Ser	2.2		24036952

autoimmune disorders failed to find rare variants that were significantly associated with disease.

The converse can also be true. Genes that have highly penetrant rare disease-associated variants may not show common disease-susceptibility alleles, as seems to be the case for genes underlying Mendelian subsets of Alzheimer disease. We consider this in detail in Section 8.3.

8.3 Our Developing Knowledge of the Genetic Architecture of Complex Disease and the Contributions of Environmental and Epigenetic Factors

In this section we review the progress made in identifying genetic susceptibility to complex disease, especially since the development of large-scale GWA studies in 2007. We show how the data obtained are illuminating the molecular basis of complex diseases, and we take a look at possible clinical applications. Finally, we turn to the question of other factors that contribute to complex diseases, how environmental factors might have a role in disease, and how epigenetic chromatin modifications might be involved.

The success and the utility of genomewide association studies

GWA studies have exceeded initial expectations: thousands of robust associations with complex diseases have been recorded since the mid-2000s. In terms of numbers, the outstanding success has been in inflammatory bowel diseases, including Crohn's disease and ulcerative colitis (**Table 8.10**). Disorders in which the phenotype is highly homogeneous and/or where disease onset occurs late in life might be expected to have the best prospects of high returns from GWA studies. For highly heterogeneous disorders, GWA studies can be expected to have rather less success, as in autism spectrum disorder and schizophrenia (both of which have high heritability).

The hope that GWA studies would find many variants, with at least modest effects, have mostly not been realized. Variants identified in the pre-GWA study era include those with the strongest effects of any common variants, including HLA variants in autoimmune disorders, apoE4 in Alzheimer disease, and so on. Significantly, the great majority of variants identified by GWA studies have odds ratios of less than 1.3, and the overall contribution of the identified variants to the genetic variance is often quite small.

Increasingly, the trend has been to carry out meta-analyses that combine data from different studies. Extending the sample sizes allows greater

DISEASE	NUMBER OF ASSOCIATED DNA VARIANTS	
	Identified before 2007[a]	Additional variants identified from 2007[a] to mid-2013
Ankylosing spondylitis	0	13
Crohn's disease	3 (NOD2, IBD5, IL23R)	140
Multiple sclerosis	0	52
Primary biliary cirrhosis	0	28
Rheumatoid arthritis	2 (PAD14, CTLA4)	30
Systemic lupus erythematosus	2 (PTPN22, IRF5)	31
Type 1 diabetes	3 (INS, PTPN22, IRF5)	40
Ulcerative colitis	1 (IL23R)	133

Table 8.10 The impact of GWA studies in identifying non-HLA loci risk factors for autoimmune and inflammatory diseases. [a]The year 2007 marked the publication of the first large-scale GWA studies. (Updated from original assessment by Visscher PM et al. [2012] *Am J Hum Genet* 90:7–24; PMID 22243964.)

power to detect common risk variants, but any new variants identified in scale-up studies can be expected to have very weak effect. The study on coronary artery disease illustrated in Figure 8.14C illustrates the difficulties. In this study an initial GWA scan was carried out on a total of 85,000 individuals, and then the most significant SNPs were genotyped in an additional replication group of more than 56,000 individuals, but the odds ratios of the most significant hits were all below 1.2. A recent large-scale GWA study of schizophrenia (see Ripke et al. (2013) under Further Reading) was useful in identifying 13 novel risk loci, more than doubling the total number of risk loci. But the highest odds ratio was just 1.238; the same authors estimated that there might be more than 8000 SNP risk loci for schizophrenia, each of very weak effect.

The utility of GWA studies

How useful have GWA studies been? Early hopes for applications in predicting disease risk have been dampened; as outlined in the next subsection, this area is going to be a difficult one. GWA studies have nevertheless been extremely important, and there are several promising clinical applications—see Manolio (2013) under Further Reading. They include applications in drug development and drug toxicity (genetic inputs to these are considered in Chapter 9), identifying protective factors that confer resistance to infectious disease (with prospects for new disease treatments), and classifying diseases into subtypes. But the major immediate contribution of GWA studies has been to provide new insights into biological pathways and processes in complex disease. Those insights do not just illuminate molecular pathogenesis; they also permit new or alternative therapeutic approaches. We amplify these points in the next section, and further below.

Assessment and prediction of disease risk

We describe genetic testing in detail in Chapter 11. For now, note that two important parameters of any genetic test are its **sensitivity** (the proportion of all people who have the condition who are identified by the test) and its **specificity** (the proportion of all people who do not have the condition in whom the test result correctly predicts absence of the condition).

Identified genetic variants for complex disease susceptibility generally show rather low odds ratios. If genetic testing is ever to have high predictive accuracy in complex disease, a battery of tests would be needed. To measure the prediction accuracy of such testing, receiver-operating characteristic (ROC) curves are used. Here, the test sensitivity is plotted against 1 – specificity (the value of the specificity subtracted from 1.0). The area under the curve (AUC) is a measure of how well the test can distinguish between the tested people who have the condition and those who do not. AUC values range from 0.5 (providing no discrimination between those with the condition and those without it) to 1.0 (perfect discrimination). As shown in **Figure 8.20**, simulations show that AUC values can increase as more genetic susceptibility factors are included.

Note that a very high AUC predictor may be of little practical use when the disease is quite rare. HLA-B27 is very strongly associated with ankylosing spondylitis, a rare type of chronic arthritis that affects parts of the spine. Despite the very impressive odds ratio of close to 70 (see Box 8.3), and a test sensitivity and specificity each of 99%, the disease risk conferred by typing positive for HLA-B27 is low (in different populations only about 1–5% of individuals with HLA-B27 will develop the disease).

For many common complex diseases, even multiple variants identified by GWA studies fail to endow the genetic tests with any great predictive value (most SNP variants have odds ratios of less than 1.3). Current prediction of individual disease risk is not accurate because, for most diseases, only a small proportion of genetic variation in risk between people can be explained by known genetic variants. Type 1 diabetes is at the upper end of the scale: about 70% or more of familial (genetic) risk can be accounted for by a combination of the major histocompatibility complex (the dominant contributor) and more than 50 additional GWA risk loci. The predictive model has an AUC of close to 0.9, but that is still some distance from what would be desired.

Even if we were to know—and be able to test for—every single genetic risk factor for a disease, the resulting whole-genome genetic test would have only partial predictive success, because complex disease is caused by a combination of genetic and environmental factors. Depending on the heritability of a complex disease, the accuracy of genomewide genetic prediction would have an upper limit of 60–90% (assuming that we could identify every single genetic variant that affects risk and were able to

Figure 8.20 Predictive accuracy of testing for multiple genetic susceptibility factors in complex disease. A receiver-operating characteristic (ROC) curve plots sensitivity against (1 – specificity) for a test. The figure shows ROC curve simulations for testing with two, three, four, or five independent disease susceptibility factors; in this case susceptibility factors 1 to 5 are imagined to have relative disease risks of, respectively, 1.5, 2.0, 2.5, 3.0, and 3.5. (A relative risk of 2.0, for example, would mean that a person with the susceptibility factor has twice the risk of developing the disease compared with a person without it.) Testing for multiple susceptibility factors can lead to an increase in the area under the curve (AUC); the greater the AUC value, the more discriminating the test. (From Janssens AC et al. [2004] *Am J Hum Genet* 74:585–588; PMID 14973786. With permission from Elsevier.)

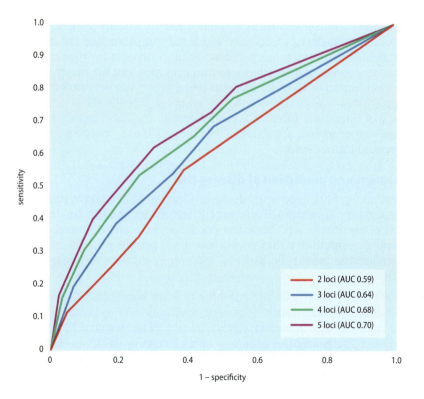

estimate their effects without error). To obtain truly accurate testing in complex disease, environmental factors need to be taken into account.

New insights into biological pathways in complex diseases may offer new approaches in classifying and treating disease

Genetic approaches and follow-up functional analyses to identify disease risk factors are crucially important in mapping the biological pathways of complex diseases at the molecular level. And with that knowledge comes the prospect of developing novel drug targets and treatments, and of defining novel **biomarkers** (in this case, biological molecules that can be objectively measured and evaluated as indicators of different stages of the disease process; they can be of help in assessing the efficacy of drugs or other new treatments).

After elucidating all the different genetic components, what was a single complex disease might often be revealed to be a collection of different diseases. That knowledge will allow better focusing of specific treatments according to the precise type of pathogenesis. Cancer genetics has led the way here, and we consider in Chapter 10 how genetic investigations are partitioning cancers, such as breast cancer, into multiple different subtypes.

In some cases, barely suspected or even entirely new pathways have been implicated in the pathogenesis of complex diseases. For example, genomewide genetic analyses have identified multiple complement genes as susceptibility loci in age-related macular degeneration, each of which works in the innate immune system. That was initially surprising, but it clearly indicates that variation in innate immune responses is important in risk for this disease. Genes that regulate lipid and extracellular matrix pathways were also implicated in pathogenesis, as were genes encoding regulators of angiogenesis (the formation of new blood vessels), notably vascular endothelial growth factor. As described in Chapter 9, targeted inhibition of this growth factor may be a promising novel treatment for a major subtype of this disease.

Alzheimer disease is another disease in which GWA studies have implicated novel genes in various biological pathways. Extracellular plaques are a central feature of the brain pathology and are largely composed of aggregated amyloid-β (see Box 7.7). Amyloid-β is now known to be a central focus of the disease pathways that link the genes underlying the early-onset autosomal dominant subsets of the disease (which regulate the production or maturation of amyloid-β) to those conferring susceptibility to late-onset complex Alzheimer disease, which are involved in pathways downstream of amyloid-β (Box 8.5).

The pathogenesis of inflammatory bowel disease

GWA studies have been very successful in Crohn's disease and ulcerative colitis. These inflammatory bowel diseases are distinguished by location (Crohn's disease can affect any part of the gastrointestinal tract; ulcerative colitis is restricted to the colon and rectum) and by the extent of affected tissue (the entire bowel wall in Crohn's disease; just the epithelial lining of the gut in ulcerative colitis). In each case, disease results from abnormal immune responses to the intestinal microbiota (Figure 8.21), but little detail was known of the molecular pathology before GWA studies.

Thanks to GWA studies we now know of a total of 163 risk loci for inflammatory bowel disease. Of these, 30 (including the common *NOD2* variants) are specific for Crohn's disease, 23 are specific for ulcerative colitis, and 110 are risk factors for both diseases. The associated genes

BOX 8.5 Common biological pathways in autosomal dominant and complex Alzheimer disease.

In rare dominantly inherited early-onset Alzheimer disease, the disease usually presents between ages 30 and 60 years; disease onset in the common non-Mendelian form normally occurs after age 65 years. The early-onset and late-onset forms have the same post-mortem brain pathology—abundant extracellular plaques, largely composed of amyloid-β (Aβ) peptides of slightly different sizes, and intracellular neurofibrillary tangles mostly made of tau protein.

The Aβ peptides are formed by cleavage of the 770-residue transmembrane amyloid-β precursor protein (APP) that works as a receptor on the surface of neurons. APP is involved in different neuronal functions, such as neuronal adhesion and the formation and growth of axons. Aβ peptides are known to be metal chelators, binding to metal ions such as copper, zinc, and iron, and reducing them; they also seem to have antimicrobial function. The Aβ peptides are thought to be the causative agent in Alzheimer disease, partly on the basis of the pathology and on the observation that Aβ is prone to aggregation in the same way as prions (see Box 7.7 on page 234), and partly on genetic analyses (described in the next section).

Genetic studies of early-onset Alzheimer disease implicate APP and enzymes cleaving it

Genomewide linkage studies of autosomal dominant early-onset Alzheimer disease have identified three causative genes: the *APP* gene, which produces APP, and *PSEN1* and *PSEN2*, which are both involved in processing APP to make Aβ. The APP processing reaction requires sequential cleavage by two endoproteinases: first, a β-secretase (also called BACE1) cuts off most of the large N-terminal extracellular portion of APP; then a multisubunit γ-secretase cleaves the transmembrane segment. The catalytic subunit of γ-secretase is a presenilin protein, either presenilin-1 or presenilin-2 (encoded by *PSEN1* and *PSEN2*, respectively).

γ-Secretase cleaves at alternative single locations to generate a series of Aβ isoforms of different lengths (**Figure 1**). The $A\beta_{42}$ isoform (42 residues long) is thought to be the greatest contributor to pathogenesis (it is more prone to forming amyloid aggregates) but is not normally produced in large quantities, unlike the predominant $A\beta_{40}$ isoform.

Excessive production of Aβ promotes pathogenesis and can be triggered by the excess production of upstream APP. Thus, mutations causing local *APP* gene duplication result in Alzheimer disease (plus cerebral amyloid angiopathy, a buildup of Aβ in the walls of the arteries in the brain); Alzheimer disease neuropathology almost invariably develops in individuals with Down syndrome with trisomy 21 (who have three *APP* gene copies—*APP* is located on chromosome 21).

Most causative *APP* mutations, however, are heterozygous missense mutations distributed close to APP cleavage sites (especially γ-secretase cleavage sites, suggesting that γ-cleavage events or their regulation are critical for development of disease). The *APP* missense mutations are thought to cause disease by increasing the $A\beta_{42}/A\beta_{40}$ ratio. The causative mutations in *PSEN1* and *PSEN2* are also heterozygous missense mutations; they, too, seem to have the effect of increasing the $A\beta_{42}/A\beta_{40}$ ratio.

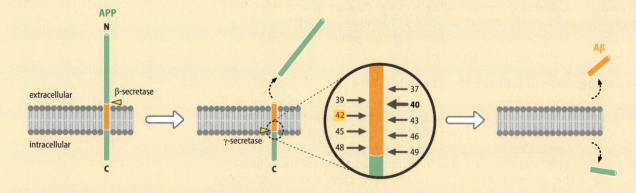

Figure 1 Production of amyloid-β from the APP amyloid-β precursor protein. The 770-residue APP is first cleaved by β-secretase, releasing most of the large extracellular region of APP. Subsequently, the membrane-bound γ-secretase cleaves at position 714 or 715, initially generating an amyloid-β (Aβ) peptide that is 48 or 49 bp long (orange rectangle). It can then go on to trim three nucleotides at a time, generating a set of isoforms of different lengths, from 37 to 49 amino acids long. Of these, $A\beta_{40}$ is the most frequent isoform, but $A\beta_{42}$ is especially prone to aggregation.

BOX 8.5 (continued)

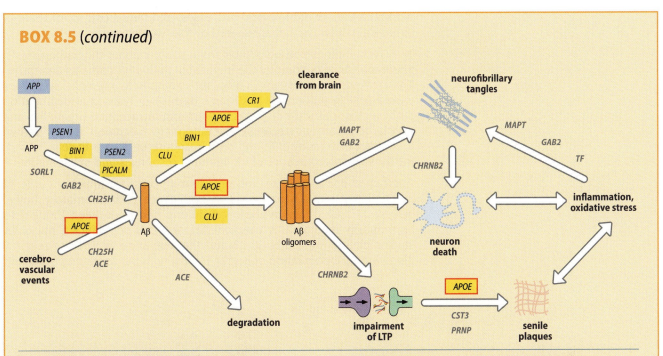

Figure 2 Early-onset Alzheimer disease genes and susceptibility factors for common late-onset Alzheimer belong to common pathways. Gene symbols highlighted in blue at the top left are early-onset disease genes identified by parametric linkage analyses in which causal variants are highly penetrant. Loci implicated from GWA studies are highlighted in yellow as common disease susceptibility factors with generally modest or weak effects. They include the previously implicated *APOE* gene (which has moderately strong effects and is shown with a distinguishing red border) plus some other well-replicated genes, including: *BIN1*, bridging integrator 1; *CLU*, clusterin; *CR1* (complement component 3b/4b receptor 1); *PICALM*, phosphatidylinositol binding clathrin assembly protein; and *CD33*. Gene symbols given in pale gray were implicated by functional studies. *SORL1* has also been implicated by DNA sequencing in cases of early-onset Alzheimer. LTP, long term potentiation. (Adapted from Bertram L & Tanzi RE [2008] *Nat Rev Neurosci* 9:768–778; PMID 22482448. With permission from Macmillan Publishers Ltd.)

The apolipoprotein E breakthrough and common pathways in early-onset and late-onset Alzheimer disease

Early linkage studies suggested a gene for late-onset Alzheimer disease at proximal 19q. Apolipoprotein E (ApoE), known to be encoded by a gene at 19q13, was identified as a component of senile plaques; differences were found between ApoE isoforms in binding to Aβ in *vitro*, and then the *APOE**ε4 allele was found to be quite strongly associated with Alzheimer disease (odds ratio close to 4 in many populations).

Apolipoprotein E is a key component of lipoprotein complexes that regulate lipid metabolism (by directing the transport and delivery of lipids from one tissue cell type to another). It is produced primarily in the liver, and then in the brain where it works in pathways involving Aβ (**Figure 2**). *APOE* has also been shown to be a modifier gene that affects the phenotype in autosomal dominant early-onset forms, for example in families with presenilin mutations.

Aβ metabolism involves a balance between the production from APP and its removal, either by enzymatic degradation (proteolysis) or receptor-mediated transport out of the brain via the blood–brain barrier (clearance). Pathogenesis results from an increase in the amount of Aβ or the amount of $A\beta_{42}$ relative to $A\beta_{40}$, and soluble Aβ oligomers may have a primary contribution (in addition to affecting synaptic transmission—by impairing long-term potentiation). They may exert some of their effects by regulating the production and phosphorylation of tau protein to induce the formation of neurofibrillary tangles that can cause neurons to die (see Figure 2). Aβ oligomers can further aggregate into fibrils that may ultimately be deposited in extracellular senile plaques. The plaques can provoke inflammation responses that can further contribute to pathogenesis.

GWA studies have identified additional variants, some of which have been well replicated and are considered established susceptibility factors. Like *APOE*, they have been implicated in pathways involving Aβ, but principally in the production of Aβ and its clearance from the brain (see Figure 2). However, several of the genes have a role in inflammation (*CR1* and *CLU*) or the innate immune response (*CD33;* not shown). None of the newly implicated variants in late-onset susceptibility have strong effects (typically odds ratios of 1.15 or 1.10). But as the biological pathways in disease are mapped, new targets become available for drug therapy.

work in a variety of biological processes, and they mostly participate in biological pathways that are shared by the two diseases.

The GWA findings caused a substantial rethink about the pathogenesis. The importance of some of the implicated pathways came as a surprise. The GWA risk variants implicated, for example, as many as five genes with a role in autophagy in Crohn's disease (a lysosomal degradation pathway that naturally disposes of worn-out intracellular organelles and very large protein aggregates). The autophagy machinery is now known to interact with many different stress response pathways in cells, including those involved in controlling immune responses and inflammation.

Another striking—and unexpected—finding was the important role of interleukin-23 (IL-23) pathways in both types of inflammatory bowel disease. Tissue injury in these conditions had once been thought to be primarily mediated by classical helper T-cell populations, but GWA studies clearly implicated IL-23 and the activation of Th17 (a recently discovered subpopulation of helper T cells) with the resulting production

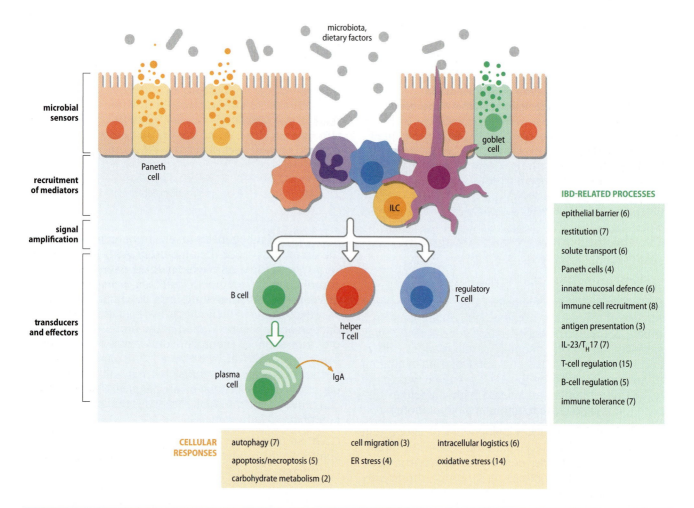

Figure 8.21 Genomewide association (GWA) studies provide valuable insights into the pathogenesis of inflammatory bowel disease (IBD). Maintaining intestinal homeostasis involves coordinated actions by epithelial cells, which include goblet and Paneth cells, and cells from both the innate and the adaptive immune systems. Barrier permeability permits microbial incursion (centre top) that is detected by cells of the innate immune system. Innate lymphoid cells (ILC) orchestrate appropriate responses, in part by releasing extracellular mediators that recruit cells of the adaptive immune system, including B cells, helper T cells, and regulatory T cells. Genetic variants, the microbiota, and immune factors affect the balance of these signals. Genes in linkage disequilibrium with IBD-associated SNPs are classified according to their function(s) in the context of intestinal homeostasis and immunity. Numbers of genes in each biological category are shown in brackets. Individual genes can be identified from the original figure. (From Khor B et al. [2011] *Nature* 474:307–317; PMID 21677747. With permission from Macmillan Publishers Ltd.)

of IL-17 and chronic inflammation. These findings prompted clinical trials using monoclonal antibodies against IL-23 to treat Crohn's disease.

Connections between different disease pathways

As the molecular pathology of complex diseases begins to be unraveled, connections are going to be made between molecular components and biological pathways in different diseases. Unsurprisingly, common sets of susceptibility factors and sometimes biological pathways are being found in autoimmune disorders. For example, the common R620W allele of the PTPN22 protein is known to modify disease risk in several autoimmune disorders. But what is now emerging are unexpected links between rather different diseases. According to the extent to which they share GWA variant profiles, heatmaps can be generated to compare the genetic profiles of different diseases (**Figure 8.22**).

Protective factors and the basis of genetic resistance to infectious disease

As well as susceptibility factors, which confer increased risk of disease, genetic investigations are identifying a series of **protective factors** that reduce disease risk. In **Table 8.11** we list some examples of protective factors identified in various common diseases. That information can suggest novel treatments; we give examples in Chapter 9.

Protective factors for one disease can, however, be susceptibility factors for another. Take, for example, the common *FUT2* non-secretor allele, a nonsense mutation in the gene that makes α(1,2)-fucosyltransferase. This enzyme completes the synthesis of H antigens, precursors of the ABO histo-blood group antigens that are found on cells in body fluids and on the surface of the intestinal mucosa. Homozygotes for the non-secretor allele fail to present ABO antigens in secretions and in the intestinal mucosa, and are strongly resistant to some strains of norovirus, the most common cause of nonbacterial gastroenteritis. But the same individuals

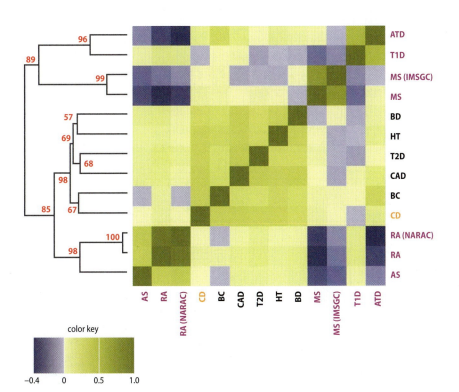

Figure 8.22 Heatmap showing correlations between different complex diseases according to GWA genetic variation profiles. Positive relationships between pairs of diseases are shown in the olive green colors; negative relationships are shown in purple colors. The abbreviations on the vertical and horizontal axes denote diseases with an autoimmune component (magenta), inflammatory disorders (orange), and other diseases (in black). Hierarchical clustering using these correlations as a distance metric is shown on the left; figures in red at the left are Approximately Unbiased (AU) probability values (as percentages) for each cluster, indicating how strongly the cluster is supported by data (clusters with AU probability values larger than 95% are strongly supported). Disease abbreviations: AS, ankylosing spondylitis; ATD, autoimmune thyroid disease; BC, breast cancer; BD, bipolar disorder; CAD, coronary artery disease; CD, Crohn's disease; HT, hypertension; MS, multiple sclerosis; T1D, type 1 diabetes; T2D, type 2 diabetes. IMSGC, International Multiple Sclerosis Genetics Consortium; NARAC, North American Rheumatoid Arthritis Consortium. (Adapted from Sirota M et al. [2009] *PLoS Genet* 5:e1000792.)

DISEASE	PROTECTIVE FACTOR	COMMENTS
Common non-infectious diseases		
Alzheimer disease	*APOE*ε2*	common allele
	APP*A673T	inhibits cleavage of APP, reducing production of amyloid-β
Coronary artery disease	*PCSK9*C679X*	*PCSK9* is important in cholesterol homeostasis, and inactivation reduces lipid levels; the C679X allele reaches a frequency of 1.8% in US black population
Coronary heart disease	blood group O	AB blood group is a significant risk factor
Crohn's disease	*PTPN22*R620W*	a common allele that is also a strong risk factor for type 1 diabetes and rheumatoid arthritis
	CARD9*IVS11+1 G>C	a rare splice variant that is highly protective (odds ratio 0.29)
Rheumatoid arthritis	*HLA-DRB1*1301*	affords protection in the 70% of affected cases who have anti-citrullinated protein antibodies
Infectious diseases		
HIV-AIDS	*CCR5Δ32*	rare 32 bp deletion reducing amount of a common helper T-cell receptor bound by HIV before infection. Homozygotes are essentially resistant to AIDS
Malaria (*Plasmodium falciparum*)	hemoglobin S	common in populations with high infection rates—see end of Section 5.4
Malaria (*P. vivax*)	inactivation of *DARC* (Duffy blood group)	*DARC* encodes a chemokine receptor that *Plasmodium vivax* binds to gain access to red blood cells. Homozygotes for inactivating mutations are protected
Norovirus-induced gastroenteritis	*FUT2*W143X*	homozygotes for this common nonsense mutation do not make α(1,2)-fucosyltransferase (see Figure 8.23 and referring text)

Table 8.11 Examples of protective variants or alleles that reduce the risk for a common disease. Non-HLA genes: *APOE*, apolipoprotein E; *APP*, amyloid protein precursor; *CARD9*, caspase recruitment domain family, member 9; *CCR5*, chemokine (C–C motif) receptor 5; *FUT2*, fucosyltransferase 2; *PCSK9*, proprotein convertase subtilisin/kexin type 9; *PTPN22*, protein tyrosine phosphatase, non-receptor type 22.

have an increased risk of Crohn's disease and type 1 diabetes, most probably because of alterations to the diverse microorganisms resident in the gut (**Figure 8.23**).

Protective factors are especially of interest in infectious disease. We may not think of infectious diseases as genetic diseases, but host genetic factors are important in susceptibility to infectious disease. Genetic variation is at its most extreme in components of our immune systems that

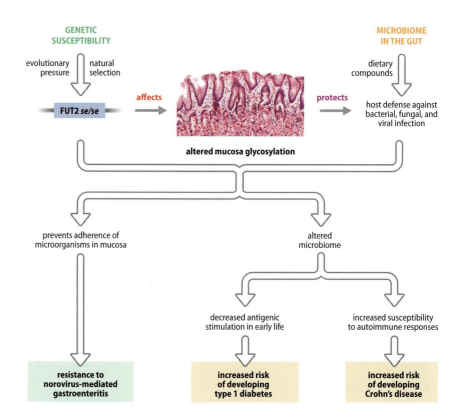

Figure 8.23 Possible implications of interactions between the FUT2 nonsecretor status and the gut microbiome in the pathogenesis of type 1 diabetes. The common *FUT2* nonsecretor (*se*) allele alters the profile of mucosa glycosylation and so prevents microorganisms from adhering to the mucosal epithelial cells and to the mucus layer lining the gastric epithelium. This null allele was naturally selected under evolutionary pressure to protect hosts against bacterial, fungal, and viral infections, but it also imbalances the microbiota in the gut. A possible consequence is decreased antigenic stimulation to the immune system in early life of subjects in modern societies that would predispose individuals homozygous for the nonsecretor allele to an increased risk of developing type 1 diabetes.

recognize and challenge foreign antigens. As a result, we differ widely in how we recognize and respond to pathogens, and some individuals may be resistant to disease caused by specific pathogen strains.

Lack of success with linkage analyses has suggested that major risk factors for infectious disease are comparatively rare, but GWA studies have had some success. The known loci are dominated by protective factors (but several HLA loci have alleles encoding predisposing or protective factors).

Large GWA studies in HIV/AIDS managed to re-confirm known class I HLA associations without detecting convincing new associations, but a GWA study has recently identified two novel resistance loci for severe malaria, and a large Chinese GWA study of leprosy has revealed five new risk loci. In the former study, one malaria resistance locus was identified within a gene (*ATP2B4*) that makes the main calcium pump of red blood cells, the host cells for the pathogenic stage of the *Plasmodium falciparum* parasite.

Given that common factors in infectious disease seem to have very weak effect and almost all of the known infectious disease risk factors map within exons, the momentum is swinging to exome sequencing studies.

Gene–gene interaction (epistasis) in complex disease

Human genetic association studies have not systematically taken into account **epistasis**, a form of gene interaction that was initially used to explain how alleles at one gene locus in experimental model organisms can mask (suppress) the phenotypic effect of alleles of another gene. Of course, stimulatory gene–gene interactions can also occur, and epistasis is now used to include any statistical deviation from the additive combination of different gene loci.

In the case of disease susceptibility, the risk conferred by a combination of two risk factors at different loci might be expected to be the multiplicative product of the individual risks if they work independently. But if the two risk factor loci were to interact (epistasis), the combined risk might deviate significantly from the multiplicative product of the individual risks.

How frequent are gene–gene interactions? Genes that operate in the same biological pathway are more likely to interact with each other, and many examples are well recognized in model systems. Unrecognized gene–gene interactions in humans may have caused estimates of heritability to be inflated (explaining why GWA variants usually account for so little of the assumed heritability), but few definitive examples of gene–gene interactions are known in humans.

Occasionally, association studies have identified risk factors that interact with each other. Thus, a recent GWA study of psoriasis identified variants at the *ERAP1* (endoplasmic reticulum aminopeptidase 1) locus that only influenced psoriasis susceptibility in individuals carrying the HLA-Cw6 risk allele. What was seen to be a compelling interaction between the *ERAP1* and *HLA-C* loci makes biological sense because *ERAP1* has an important role in the MHC class I antigen processing and presentation pathway (see Figure 1 in Box 8.3). Variants in *ERAP1* also affect risk for ankylosing spondylitis in those individuals carrying the HLA-B27 allele.

Some GWA scans have begun to include multilocus analyses, notably two-locus associations, but initial findings suggest that single-locus association can underlie and obscure apparent multilocus associations. For more background on the prospects and challenges in this area, see the review by Cordell (2009) under Further Reading.

Table 8.12 Examples of different types of environmental factors that contribute to common non-infectious and noncancerous disease. [a]But not ulcerative colitis—if anything, smoking protects against ulcerative colitis.

ENVIRONMENTAL	EXAMPLES
Teratogens and abnormal metabolite levels in uterine environment	low folic acid increases the risk of neural tube defects such as spina bifida
Unbalanced diets	over-consumption and excess of fatty foods in type 2 diabetes
Smoking	increased risk for many disorders, such as coronary artery disease, Crohn's disease[a], and aging-related macular degeneration
Exposure to infectious microorganisms	*Helicobacter pylori* infections in development of peptic ulcers
Commensal microorganisms	gut microbiota in inflammatory bowel disease, type 2 diabetes

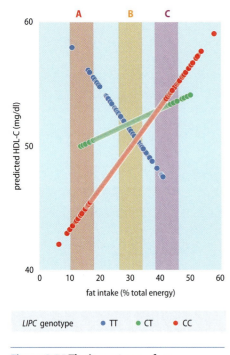

LIPC genotype ● TT ● CT ● CC

Figure 8.24 The importance of gene–environment interactions—an example. According to different total levels of dietary fat intake, variations in predicted values of high-density lipoprotein cholesterol (HDL-C) are shown for three genotypes at the −514(C/T) polymorphism (rs1800588) in the *LIPC* hepatic lipase locus. Low fat intake (band A) combined with the TT genotype (homozygous for the T allele) results in the highest HDL-C level. For a moderate fat intake (band B), there is no relationship between genotype and HDL-C level. For a high fat intake (band C), the TT genotype has the lowest HDL-C level. Gene–environment interactions are therefore important in identifying genetic and environmental determinants of medically relevant phenotypes such as HDL-C levels; depending on the dietary fat intake, one could variously conclude that the TT genotype produces high HDL-C levels (band A) or low HDL-C levels (band C), or that it is not associated with HDL-C levels at all (band B). (Adapted from Manolio TA et al. [2006] *Nat Rev Genet* 7:812–820; PMID 16983377. With permission from Macmillan Publishers Ltd.)

Gene–environment interactions in complex disease

Environmental factors are clearly important in infectious disease and in cancers (which we consider in Chapter 10). But they are increasingly being recognized to be important in complex diseases outside those two categories (as well as in some monogenic disorders); see **Table 8.12**. That should not be surprising because monozygotic (identical) twins are often seen to be discordant for complex diseases (see Table 8.4). Identical twins arise from the same zygote and might be expected to have identical DNA profiles. If one twin develops a complex disease, such as Crohn's disease, but the other lives a long healthy life, factors other than DNA can be expected to be important (although post-zygotic mutations could also have a role in some cases).

Striking evidence for the importance of environmental factors comes from increased risk for a specific disease that often befalls migrants who have moved from a community with a low general risk of that disease to join a society in which the disease is much more prevalent. In addition, as populations across the globe change their eating habits and lifestyles, there is a relentless rise in the frequency of conditions such as obesity and type 2 diabetes.

Gene–environment interactions are also important in the sense that they can make it difficult to detect a genetic (or environmental) effect if they are not identified and controlled for. That can lead to inconsistent disease associations when populations are variably exposed to certain environmental factors that modify the effect of a given genetic variant (**Figure 8.24**). Understanding gene–environment interactions can therefore allow us to develop protective strategies for complex diseases: by seeking to minimize exposure to an environmental factor, the harmful effect of a genetic susceptibility factor can be minimized.

A plethora of 'environmental factors'

The term *environment* has a multilayered meaning in this context—it effectively includes any component that alters disease risk without originating from DNA variation in our cells. There are external physical environments. Right from the earliest stages and throughout life, we are exposed to a range of diverse radiation sources, and also to infectious agents that can influence susceptibility to non-infectious diseases. In the womb, we are exposed to a uterine environment and will be variously affected by what our mother consumes during pregnancy. After birth and throughout life, we ingest, or have surface contact with, a huge range of additional foreign molecules. The molecules that we ingest intentionally (in food and drink, stimulants, and so on) may be considered, in part, lifestyle choices. They, too, along with the amount of physical and mental exercise that we experience and the degree of stress to which we are subjected, are important in disease susceptibility.

Then there is our internal **microbiome**, the diverse range of micro-organisms (*microbiota*) that constitute part of us. Mostly composed of bacteria, our personal microbiomes have 10 times more cells than we have and are mostly located within the gut (the gut microbiome in an average person has possibly 5000 different bacterial species. In addition to being beneficial to us (see above), our microbiomes have a major influence on susceptibility to disease—see above and the review by Virgin & Todd (2011) under Further Reading. Finally, there is the environment inside our body cells and how it links with the extracellular environment. Two important, and interlinked, components here are mitochondria (**Box 8.6**) and chromatin modifications, including DNA methylation (which we consider in the following section).

The study of gene–environment (G×E) interactions has traditionally involved case-control studies of candidate genes, but the advent of GWA studies has prompted hypothesis-free genomewide studies. They require large sample sizes, however—a G×E GWA study needs about four times as many samples as a standard GWA study to detect a main effect of the same magnitude. Various G×E GWA studies have been launched, such as a scan to identify genes that confer susceptibility to air pollution in childhood asthma.

Prospective cohort studies

Case-control studies are the most widely used method of investigating the genetic and environmental basis of complex disease. Cases and controls are typically investigated *retrospectively* (that is, the disease cases have already occurred, and subjects need to be quizzed about previous events such as exposure to environmental factors). As a result, the studies are open to all kinds of bias, for example in the selection of subjects to be studied.

Prospective cohort studies have the big advantage of removing much of the bias by studying individuals over a long time frame that commences *before* the onset of disease and involves the periodic assessment of subjects, including recording detailed information on the subjects and collecting samples for future laboratory tests. Because studies such as these do not select affected individuals, they need to be very large to ensure that eventually there will be statistically significant numbers of affected individuals.

A leading example of a prospective cohort study is the UK Biobank project. From 2007 to 2010 it recruited 503,000 British people aged between 40 and 69 years and will go on to follow them with periodic testing over a period of 30 years (**Table 8.13**). By comparing those who remain healthy with those who develop disease within the 30-year time frame of the study, researchers hope to gain important information on the genetic determinants of a range of common late-onset diseases (including cancers, heart diseases, stroke, diabetes, arthritis, osteoporosis, eye disorders, depression, and forms of dementia). The study will also help measure the extent to which individual diseases have genetic and environmental causes.

Epigenetics in complex disease and aging: significance and experimental approaches

How do environmental factors work to have an impact on complex disease? Somehow they must affect how our genes are expressed. They can do that at the DNA level by changing the DNA sequence in our cells (recall the environmental mutagens that we considered in Section 4.1).

BOX 8.6 Mitochondrial DNA haplogroups and mitochondrial DNA variation in common disease.

Because mitochondria are the power sources of our cells, the performance of cells (notably brain and muscle cells, which have high energy requirements) is hugely dependent on mitochondrial efficiency. In an environment where food (and therefore calories) is plentiful, mitochondria efficiently generate energy to keep cells in optimal condition; severely restricting calorie intake impairs mitochondrial and cell efficiency.

Mitochondria have an important influence on how nuclear genes are expressed, because they make the ATP and acetyl coenzyme A that is needed for diverse cell signaling pathways and for phosphorylating and acetylating histones in chromatin. They are also the principal generators of reactive oxygen species that damage our cells. Aging is a major risk factor for most common diseases, and accumulating oxidative damage through a lifetime is a principal contributor to increasing cellular inefficiency. The genetic control of mitochondrial function is mostly specified by nuclear genes, but mitochondrial DNA (mtDNA) is much more susceptible to mutation than is nuclear DNA (see Section 7.2).

Evolution of mtDNA haplogroups

Because mtDNA is strictly maternally inherited, mtDNA undergoes negligible recombination at the population level, and so SNPs in mtDNA form branches of an evolving phylogenetic tree. The major subdivisions of the world mtDNA phylogeny occurred more than 10,000 years ago and are called mtDNA **haplogroups**, which developed as humans migrated into new geographic regions, leading to region-specific haplogroup variation (**Figure 1**).

More than 95% of Europeans belong to one of ten major haplogroups, namely H, J, T, U, K (a subgroup of U), M, I, V, W, and X; several of these are associated with complex human traits (see below for examples). Each haplogroup defines a group or 'clade' of related mtDNAs containing specific sequence variants within the population. Mitochondrial DNA haplogroups influence the assembly and stability of the mitochondrial respiratory chain, the synthesis of respiratory chain proteins, and the propensity to develop intracellular oxygen free radicals, which are implicated in the pathophysiology of several common human diseases.

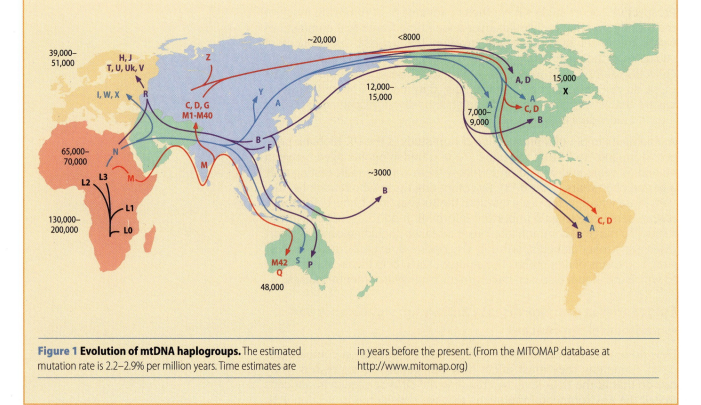

Figure 1 Evolution of mtDNA haplogroups. The estimated mutation rate is 2.2–2.9% per million years. Time estimates are in years before the present. (From the MITOMAP database at http://www.mitomap.org)

BOX 8.6 *(continued)*

Although contentious, there is some evidence to suggest that the distribution of human mtDNA haplogroups has been influenced by environmental pressures, including climate. Nuclear–mitochondrial DNA coevolution has been implicated in climatic evolution in other species.

Mitochondrial DNA and common disease

A variety of rare mitochondrial disorders are known to be due to variants of large effect in single genes in mtDNA, or rare multigenic mtDNA deletions and duplications. Common mtDNA polymorphisms with weak effects, notably SNPs, are known to alter the penetrance of these rare disorders. More recently, diverse association studies have shown that common mtDNA polymorphisms also influence susceptibility to a wide range of complex diseases, including neurodegenerative, psychiatric, cardiovascular, and many other diseases—see the supplementary table in Gomez-Duran A *et al.* (2010) *Hum Mol Genet* 19:3343–3353 (PMID 20566709) for a list of examples.

Certain mtDNA haplogroups have been shown to be associated with disease. The most common mtDNA haplogroup H, found in about 40% of Europeans, is associated with a more than twofold increased

change in surviving severe infection (sepsis), but subgroups of this haplogroup are emerging as a risk factor for late-onset degenerative diseases, including those affecting the nervous system. This raises the possibility that infectious disease has shaped mtDNA evolution in Europe over a relatively short period, increasing the frequency of mtDNA haplogroup H and thereby predisposing modern humans to late-onset common disease.

Despite the frequent implication of mtDNA in different common complex diseases, the precise haplogroup associations are not always consistent. This is partly due to the limited cohort size in some studies, restricting the statistical power. Another issue is the different frequency of mtDNA haplogroups in different ethnic groups. A further confounder is occurrence of the same base substitution on different branches of the phylogenetic tree as a result of recurrent mutation (called mtDNA *homoplasy*), which accounts for up to 20% of genetic variation in Europeans. The frequency of a sub-haplotype containing a functional homoplasy can vary in different populations, and the distribution of sub-haplogroups also varies in different populations. As a result, the major haplogroups can be associated with the disease in some populations but not in others.

Alternatively, if they are infectious agents they can introduce some novel genes or proteins that change how our cells work. Yet another way—and one that is now seen to be very common—is to change the epigenetic settings (the *epigenome*) in our cells. We described in Chapter 6 how epigenetic effects regulate gene expression (Section 6.2) and how they are important in some monogenic disorders (Section 6.3). And in Chapter 10 we illustrate how epigenetic effects are very important in cancer. Here, we focus on epigenetic effects in other complex diseases.

BASELINE QUESTIONNAIRE	BASELINE PHYSICAL MEASUREMENTS	FOLLOW-UP AND FUTURE MEASURES
Sociodemographic	blood pressure	stored blood, urine, saliva
Family history	weight, body impedance	repeat baseline assessment (20,000 participants)
Environmental	waist and hip circumference	access national health records: death, cancer, hospitalizations, primary care
Lifestyle	seated and standing heights	
Cognitive function	grip strength	
Food frequency	bone density	
Internet-administered 24-hour dietary questionnaire	mailed triaxial accelerometers	
	enhanced phenotyping (last 100,000–150,000 participants recruited): hearing, vascular reactivity, visual acuity, refractive error, intraocular pressure, corneal biomechanics, optical coherence tomography, fitness assessment	

Table 8.13 Components of the UK Biobank prospective cohort study. (From Manolio TA et al. [2012] *Am J Epidemiol* 175:859–866; PMID 22411865. With permission from Oxford University Press.)

Unlike the genome, which is very stable, the **epigenome**—effectively the chromatin states across all chromosomes (determined primarily by patterns of cytosine methylation, histone modification, and the positions of nucleosomes; see Section 6.2)—is comparatively fluid. In response to certain environmental cues (signals), the epigenome can be significantly altered, and that can result in important changes in gene expression.

As described in the second and third subsections below, epigenetic changes occur throughout the life of an individual. They are thought to be important in aging—a frequent risk factor in complex disease—and they can explain, at least in part, why identical twins develop to become different (different post-zygotic mutations arising in the identical twins can also be expected to play a part).

Early development is a period of rapid changes in the epigenomes of cells (including the global resetting of methylation marks; see Figure 6.14 on page 168). And at this stage epigenomes can be particularly sensitive to environmental factors. As described below, a popular theory holds that chronic adult diseases originate in early life, and perturbation of epigenetic settings by environmental factors is an attractive explanation. We consider ways in which that can happen below.

Experimental investigations

Because epigenomes are highly variable between different types of cell, analyzing epigenomes is potentially more complicated than genome analysis. As a result, the investigation of epigenetic factors in complex disease has lagged behind genome analysis. In the past few years, however, great strides have been taken in defining epigenetic settings in cells.

Analysis of global patterns of DNA methylation (the 'methylome') is comparatively advanced—the positions of 5-methylcytosines have been mapped across the genome to single nucleotide resolution in some cell types. Investigators can now carry out large-scale DNA methylation scans across the genome by using microarrays such as Illumina's Infinium Human Methylation450 bead chip (it scans 485,000 cytosine methylation sites distributed across virtually all protein-coding genes with an average of 17 CpG sites per gene region, including CpG sites in the promoter, untranslated sequences, first exon, and elsewhere in the gene body).

Genomewide investigations to identify the extent of epigenetic contributions to complex disease have been launched for different disorders, including common neurological and autoimmune disorders.

Epigenetic changes during aging

Aging, a very important risk factor for complex disease, is marked by progressive inefficiency in cell, tissue, and organ function. There are inherent limits on the efficiency of cellular processes, including endogenous errors in DNA replication, in DNA repair, and in the regulation of gene expression. As we age, therefore, both genetic and epigenetic changes accumulate progressively, and changes in the genomes and epigenomes of somatic stem cells (that are involved in maintaining tissue homeostasis) may be fundamental in the aging process.

The epigenetic changes that accumulate in our cells can be secondary to genetic changes (mutations in DNA sequences that regulate epigenetic mechanisms) or to inherent errors in the epigenetic regulation machinery. However, they can quite often be induced by environmental factors or as a result of stochastic (chance) factors. Both cytosine methylation and histone modification patterns change with aging. In the former case, for example, there is a progressive loss of cytosine methylation across the genome during aging, but against this general pattern of global hypomethylation (which includes very many methylation sites outside gene regions), hypermethylation occurs at promoters of certain genes.

Epigenetic changes in monozygotic twins

Epigenetic changes may constitute a major reason why monozygotic (identical) twins, who are initially extremely similar in appearance and behavior, go on to develop significant differences in various aspects of the healthy phenotype. And there is quite frequent discordance between identical twins for a variety of complex genetic disorders.

Identical twins derive from a single zygote (the embryo splits at a very early stage in embryonic development), and so they initially have identical genetic profiles (but may accumulate different post-zygotic mutations). Epigenetic differences between identical twins are initially minimal but can begin to occur even in prenatal development. Stochastic factors may be involved, such as different X-chromosome methylation patterns in female identical twins arising from the random choice of which parental X chromosome to inactivate (Section 5.2).

Environmental factors can also play a part in prenatal development, because the *in utero* environments can be different. The vast majority of identical twins are located in separate amniotic membranes, and in diamniotic twins there is an increased risk of congenital heart disease that usually affects just one twin. Differences in exposure to postnatal environmental factors most probably contribute to the significant epigenetic differences observed in older monozygotic twins (in both cytosine methylation and histone modification patterns).

The developmental origins of adult health and disease

Pioneering epidemiological studies have established that low birth weight confers an increased risk of developing different common adult diseases, including various cardiovascular diseases, hypertension, and stroke. When significantly fewer nutrients are provided to the fetus during pregnancy, reprogramming in early development seems to cause the fetus to develop a 'thrifty phenotype' with a low metabolic rate and reduced pancreatic beta cell mass and islet function. The *thrifty phenotype* is thought to be an adaptation that maximizes the chance of surviving in an adverse environment where calorie intake is restricted, but the altered metabolism is not well adapted to a later life where food is plentiful, increasing the risk of metabolic syndrome (with strong risk determinants for type 2 diabetes, obesity, and hypertension).

The effects of the Dutch *Hongerwinter*, a wartime famine that took place in western parts of the Netherlands for six months in 1944/1945, provide support for the 'thrifty phenotype' hypothesis. Women who endured semi-starvation conditions during mid- to late gestation gave birth to underweight babies who were then exposed in later life to normal levels of calorie intake, with an increased incidence of common metabolic and cardiovascular diseases. Individuals born to mothers who experienced starvation conditions during early gestation only, so that they were of average weight at birth, had even higher rates of obesity than those who suffered sharply reduced nutrition in mid to late gestation, and they also had an increased risk of schizophrenia. By implication, early gestation appears to be a particularly critical time in which environmental factors can have an influence.

Because other nutritional cues during infancy and childhood were also found to be associated with adverse effects in later life, the 'thrifty phenotype' hypothesis has broadened into a more general theory that proposes that a wide range of environmental conditions during embryonic development and early life determine susceptibility to different adult diseases.

Environmental factors seem to have an impact on development so as to increase the risk of disease in later life, but how do they work? The comparative plasticity of epigenomes makes them likely targets of environmental factors, and this is supported by data from experimental models and human studies (notably environmentally induced changes in DNA methylation patterns). Because epigenetic processes, such as DNA methylation and histone modifications, rely on metabolic factors, a differential availability of dietary components can be expected to influence epigenetic mechanisms. For example, methylation of cytosines and histones uses *S*-adenosylmethionine as a methyl donor, and dietary factors, notably folate (vitamin B_9), are known to have a key role in the pathway that produces *S*-adenosylmethionine. As described in Section 10.3, cancer studies have also shown a direct link between inflammation (which is often triggered by environmental factors) and epigenetic modification causing altered gene expression.

Transgenerational epigenetic effects

Epigenetic effects are clearly transmitted through mitosis so that chromatin states are heritable through cell generations. For example, when a liver cell divides it gives rise to two liver cells with the same type of epigenome (genomewide pattern of chromatin states) as the parent cell. But can epigenetic effects be transmitted through meiosis? Might a pattern of increased disease risk deriving from environmentally induced epigenetic modifications be passed on to children so that they, too, have increased disease risk?

Transgenerational epigenetic inheritance is common in plants but rare in animals. In the nematode worm *Caenorhabditis elegans*, experimental manipulation of specific chromatin modifiers in parents can result in an extended lifespan up to the third generation. Suggestive evidence for human transgenerational epigenetic effects comes from certain studies in northern Europe, such as a Swedish study that seems to link the availability of food supply during the early life of the *paternal* grandparents and longevity of the grandchildren, including associations with cardiovascular disease and diabetes. However, a defined mechanism for transgenerational inheritance in humans and animal models is currently missing. Interested readers should consult recent reviews such as the one by Grossnicklaus et al. (2013) listed under Further Reading.

Summary

- Identifying genes for monogenic disorders often relies on first finding a subchromosomal location for the disease gene. Genes are sought in that region, and promising candidates are tested for evidence of disease-associated mutations.

- Linkage analysis investigates whether alleles at two or more loci co-segregate in families. Two loci located on different chromosomes, or far apart on the same chromosome, are unlinked; alleles at the loci will have, just by chance, a 50% chance of being inherited together at meiosis. Alleles at linked loci (that are close together on a chromosome) will often be co-inherited as a haplotype.

- Genomewide linkage analyses can map genes for monogenic disorders. Several hundred polymorphic markers from defined genome locations are scanned. If a marker locus is physically close to the disease locus, a marker allele will tend to co-segregate with disease during meiosis.

- A structural chromosome abnormality may indicate a subchromosomal location for a disease gene, especially if it has arisen *de novo* in a sporadic affected case.

- In practice, whole-exome sequencing means sequencing the exons (plus the immediate flanking intron sequence) of protein-coding genes.

- A polygenic trait, such as adult height or blood pressure, shows a continuous range of values and a normal (bell-shaped) distribution within a population. The genetic susceptibility is due to alleles at many loci, each of weak effect.

- In complex (multifactorial) diseases no single gene locus dominates to the same extent as in monogenic conditions. As well as being polygenic, various nongenetic (environmental) factors have important roles in disease.

- In complex disease, the disease susceptibility needs to cross some high threshold value for a person to be affected. Affected people will have high-risk alleles at multiple susceptibility loci, and so their first-degree relatives will be at higher risk than the general population. Depending on environmental factors, the liability threshold value can change and often shows sex differences.

- The variance of a phenotype is the square of the standard deviation; the heritability is that proportion of the variance that is due to genetic factors.

- In diseases with a strong genetic component, siblings of an affected person have a much higher risk of disease than the general population, and monozygotic twins are much more likely to show concordance in disease status than are dizygotic twins.

- Heritability is not a fixed property—for any disease it varies between populations, and can vary within the same population when environmental factors change.

- Linkage analyses in complex disease are nonparametric. They test affected relatives only, usually affected sibs, and look for chromosomal segments that they share more often than expected by chance.

- Association studies offer a better way of finding susceptibility factors for complex diseases. They test affected individuals (cases) and unrelated controls from the same population to seek *statistical* associations between individual variants and the disease. The associations may arise because many people share a short chromosome segment inherited from a distant common ancestor who carried a susceptibility factor.

- The international HapMap project has defined ancestral chromosome segments in various human populations. The shared ancestral chromosome segments are usually very small (often just a few kilobases).

- Candidate gene association studies test for association between a disease and alleles of a specified gene of interest (often because of a suspected role in the disease).

- In genomewide association (GWA), densely spaced markers are needed (unlike linkage, association works over very short ranges only on the DNA). Thousands of SNP markers on each chromosome are scanned for alleles associated with disease.

- DNA variants may confer increased disease risk (susceptibility factors) or reduced risk (protective factors). Protective factors are especially important in infectious disease.

- Association studies identify haplotype blocks that harbor a disease susceptibility variant, but identifying the causal variant is difficult because of linkage disequilibrium, the nonrandom association between all the variants in the block.

- Association studies use case-control studies to study affected individuals (cases) and matched unaffected individuals (controls). The aim is to identify common alleles (on short chromosome segments) that have significantly different frequencies in cases and controls.

- Common disease susceptibility alleles are of ancient origin and are mildly deleterious (such as weak missense mutations). They avoid elimination by natural selection by having little effect on reproductive rates, or by simultaneously conferring some selective advantage now, or in the past.

- GWA studies have successfully identified thousands of disease-associated SNP markers; because almost all are of weak effect, they have limited use in predicting disease risk.

- Copy number variants (CNVs) occur at increased frequencies in some disorders. But generally CNVs do not make a large contribution to genetic susceptibility to disease.

- Because of recent explosive population growth, much of the variation that affects human gene expression is due to rare (low-frequency) single nucleotide variants (SNVs).

- Rare disease-associated SNVs and CNVs generally have strong effects.

- GWA studies have been of immense value in elucidating the biological pathways in complex diseases, with prospects for identifying new drug targets and treatments.

- Nongenetic factors are clearly very important in complex diseases, but standard case-control studies are limited in their ability to detect gene–environment interactions. Prospective cohort studies are more suited to that task; they study individuals over a long time frame that commences before the onset of disease.

- Environmental factors work at different levels to influence disease susceptibility. One important way is to alter the epigenetic settings of cells, resulting in altered gene expression. Altered epigenetic settings in early life are thought to alter the risk of various adult diseases such as diabetes and cardiovascular diseases.

Questions

Help on answering these questions and a multiple-choice quiz can be found at www.garlandscience.com/ggm-students

1. Candidate gene approaches have allowed the identification of human disease genes on the basis of prior information about the cause of a related phenotype. Give an example of successful disease gene identification based on prior knowledge of (a) a related human phenotype and (b) a related mouse phenotype.

2. What is a lod score? In standard genomewide linkage analyses the cutoff for statistical significance of linkage is a lod score of +3. How was this limit set?

3. The table below shows the percentage phenotype concordance in monozygotic (MZ) and dizygotic (DZ) twins in four hypothetical genetic diseases A to D. Which disease would you estimate to have the highest heritability and which the lowest heritability, and why?

DISEASE	CONCORDANCE IN MZ TWINS (%)	CONCORDANCE IN DZ TWINS (%)
A	37.5	16.2
B	15.2	11.7
C	19.2	7.9
D	17.2	1.6

4. The risk of developing a disease is sometimes expressed as a risk ratio, λ. What is meant by this ratio? Disorder A has a λ_s of 600 and disorder B has a λ_s of 6. What types of disease are A and B?

5. Genomewide linkage studies can often be carried out with just a few hundred DNA markers, but genomewide association studies often use hundreds of thousands of markers. Explain why this difference exists by explaining the very different designs of these two approaches.

6. What is meant by the odds ratio in case-control studies? Calculate the odds ratio from the table below.

	NUMBER OF CASES WITH DISEASE X	NUMBER OF UNAFFECTED CONTROLS
Possessing genetic variant X	850	297
Lacking genetic variant X	150	703

7. What is linkage disequilibrium? Which of the following haplotypes shows evidence of linkage disequilibrium, given the individual allele frequencies?
   (a) haplotype *A*1-B*3* with a population frequency of 0.101 (frequencies of 0.231 for *A*1* and 0.431 for *B*3*)
   (b) haplotype *C*2-D*1* with a population frequency of 0.071 (frequencies of 0.311 for *C*2* and 0.225 for *D*1*)
   (c) haplotype *E*1-F*1* with a population frequency of 0.205 (frequencies of 0.236 for *E*1* and 0.289 for *F*1*)
   (d) haplotype *X*2-Y*3* with a population frequency of 0.101 (frequencies of 0.532 for *X*2* and 0.434 for *Y*3*).

8. In genomewide association studies, the threshold for statistical significance, P, is often set at a very high value, often about 5×10^{-8}. Why so high?

Further Reading

General Genetic Mapping and Meiotic Recombination Frequency

Altshuler D, Daly MJ & Lander ES (2008) Genetic mapping in human disease. *Science* 322:881–888; PMID 18988837.

Cheung VG et al. (2007) Polymorphic variation in human meiotic recombination. *Am J Hum Genet* 80:526–530; PMID 17273974.

Coop G et al. (2008) High-resolution mapping of crossovers reveals extensive variation in fine-scale recombination patterns among humans. *Science* 319:1395–1398; PMID 18239090.

Ott J (1999) Analysis of Human Genetic Linkage, 3rd ed. Johns Hopkins University Press. [Authoritative, detailed account.]

Gene Identification and Exome Sequencing in Monogenic Disorders

Bamshad MJ et al. (2011) Exome sequencing as a tool for Mendelian disease gene discovery. *Nat Rev Genet* 12:745–755; PMID 21946919.

Collins FS (1995) Positional cloning moves from perditional to traditional. *Nat Genet* 9:347–350; PMID 7795639.

Gilissen C et al. (2012) Disease gene identification strategies for exome sequencing. *Eur J Hum Genet* 20:490–497; PMID 22258526.

Goodship J et al. (2000) Autozygosity mapping of a Seckel syndrome locus to chromosome 3q22.1-q24. *Am J Hum Genet* 67:498–530; PMID 10889046.

Najmabadi H et al. (2011) Deep sequencing reveals 50 novel genes for recessive cognitive disorders. *Nature* 478:57–63; PMID 21937992.

Piro RM & Di Cunto F (2012) Computational approaches to disease-gene prediction: rationale, classification and successes. *FEBS J* 279:678–696; PMID 22221742.

Heritability and Heritability Studies

Lichtenstein P et al. (2009) Common genetic determinants of schizophrenia and bipolar disorder in Swedish families: a population-based study. *Lancet* 373:234–239; PMID 19150704.

Visscher PM et al. (2008) Heritability in the genomics era—concepts and misconceptions. *Nat Rev Genet* 9:255–266; PMID 18319743.

Wells JCK & Stock JT (2011) Re-examining heritability: genetics, life-history and plasticity. *Trends Genet* 10:421–428; PMID 21757369.

Quantitative Traits and The Liability/Threshold Model

Falconer DS (1965) The inheritance of liability to certain diseases estimated from the incidence among relatives. *Ann Hum Genet* 29:51–76; doi 10.1111/j.1469-1809.1965.tb00500.x. [The original formulation of the liability threshold model to explain dichotomous traits.]

Lango Allen H et al. (2010) Hundreds of variants clustered at genomic loci and biological pathways affect human height. *Nature* 467:832–838; PMID 20881960.

Linkage Analysis in Complex Disease

Altmüller J et al. (2001) Genomewide scans of complex human diseases: true linkage is hard to find. *Am J Hum Genet* 69:936–950; PMID 11565063.

Hugot JP et al. (1996) Mapping of a susceptibility locus for Crohn's disease on chromosome 16. *Nature* 379:821–823; PMID 8587604.

Lander E & Kruglyak L (1995) Genetic dissection of complex traits: guidelines for interpreting and reporting linkage results. *Nat Genet* 11:241–247; PMID 7581446.

Risch N & Merikangas K (1996) The future of genetic studies of complex human diseases. *Science* 273:1516–1517; PMID 8801636.

Weeks DE et al. (2004) Age-related maculopathy: a genomewide scan with continued evidence of susceptibility loci within the 1q31, 10q26, and 17q25 regions. *Am J Hum Genet* 75:174–189; PMID 15168325.

Linkage Disequilibrium, Haplotype Blocks, and the HapMap Project

Ardlie KG et al. (2002) Patterns of linkage disequilibrium in the human genome. *Nat Rev Genet* 3:299–309; PMID 11967554.

Daly MJ et al. (2001) High-resolution haplotype structure in the human genome. *Nat Genet* 29:229–232; PMID 11586305.

The International HapMap Consortium (2005) A haplotype map of the human genome. *Nature* 437:1299–1320; PMID 16255080.

The International HapMap Consortium (2007) A second generation human haplotype map of over 3.1 million SNPs. *Nature* 449:851–862; PMID 17943122.

Slatkin M (2008) Linkage disequilibrium—understanding the evolutionary past and mapping the medical future. *Nat Rev Genet* 9:477–485; PMID 18427557.

Association Analysis and GWA Studies: General Reviews

Christensen K & Murray JC (2007) What genome-wide association studies can do for medicine. *N Engl J Med* 365:1094–1097; PMID 17360987.

Hindorff LA, MacArthur J, Morales J et al. A catalog of published genome-wide association studies. Available at http://www.genome.gov/gwastudies

Lewis CM & Knight J (2012) Introduction to genetic association studies. *Cold Spring Harbor Protoc* 3:297–306; PMID 22383645.

Manolio TA (2013) Bringing genome-wide association findings into clinical use. *Nat Rev Genet* 14:549–558; PMID 23835440.

Nature Reviews Genetics series on Genome-wide association studies. Available at http://www.nature.com/nrg/series/gwas/index.html [Multiple review articles since 2008.]

Visscher PM et al. (2012) Five years of GWAS discovery. *Am J Hum Genet* 90:7–24; PMID 22243964.

Assocation Studies in Specific Diseases

Jostins L et al. (2012) Host–microbe interactions have shaped the genetic architecture of inflammatory bowel disease. *Nature* 491:119–124; PMID 23128233.

Raychaudhuri S et al. (2012) Five amino acids in three HLA proteins explain most of the association between MHC and seropositive rheumatoid arthritis. *Nat Genet* 44:291–296; PMID: 22286218.

Ripke S et al. (2013) Genome-wide association analysis identifies 13 new risk loci for schizophrenia. *Nat Genet* 45:1150–1159; PMID: 23974872.

Timmann C et al. (2012) Genome-wide association study indicates two novel resistance loci for severe malaria. *Nature* 489:443–446; PMID 22895189.

Wellcome Trust Case Control Consortium (2007) Genome-wide association study of 14,000 cases of seven common diseases and 3,000 shared controls. *Nature* 447:661–678; PMID 17554300.

Missing Heritability, Mutational Heterogeneity, and Evolution of Common Alleles in Complex Disease

Cooper GM & Shendure J (2011) Needles in stacks of needles: finding disease-causal variants in a wealth of genomic data. *Nat Rev Genet* 12:628–640; PMID 21850043.

Manolio TA et al. (2009) Finding the missing heritability of complex diseases. *Nature* 461:747–753; PMID 19812666.

McClellan J & King MC (2010) Genetic heterogeneity in human disease. *Cell* 141:210–217; PMID 20403315. [Argues that rare mutations are responsible for a substantial portion of complex disease.]

Nakagome S et al. (2012) Crohn's disease risk alleles on the NOD2 locus have been maintained by natural selection on standing variation. *Mol Biol Evol* 29:1569–1585; PMID 22319155.

Raychaudhuri S (2011) Mapping rare and common causal alleles for complex human diseases. *Cell* 147:57–69; PMID 21962507.

Copy Number Variants in Complex Disease

Girirajan S et al. (2011) Human copy number variation and complex disease. *Annu Rev Genet* 45:203–226; PMID 21854229.

Girirajan S et al. (2013) Global increases in both common and rare copy number load associated with autism. *Hum Mol Genet* 22:2870–2880; PMID 23535821.

Malhotra D & Sebat J (2012) CNVs: harbingers of a rare variant revolution in psychiatric genetics. *Cell* 148:1223–1241; PMID 22424231.

Rare Variants: Disease Associations and Recent Evolution

Cirulli ET & Goldstein DB (2010) Uncovering the roles of rare variants in common disease through whole-genome sequencing. *Nat Rev Genet* 11:415–425; PMID 20479773.

Fu W et al. (2013) Analysis of 6,515 exomes reveals the recent origin of most human protein-coding variants. *Nature* 493:216–220; PMID 23201682.

Hunt KA et al. (2013) Negligible impact of rare autoimmune locus-coding region variants on missing heritability. *Nature* 498:232–235; PMID 23698362.

Lim ET et al. (2013). Rare complete knockouts in humans: population distribution and significant role in autism spectrum disorders. *Neuron* 77:235–242; PMID 23352160.

Keinan A & Clark AG (2012) Recent explosive human population growth has resulted in an excess of rare genetic variants. *Science* 336:740–743; PMID 22582263.

Kiezun A et al. (2012) Exome sequencing and the genetic basis of complex traits. *Nat Genet* 44:623–630; PMID 22641211.

Rivas MA et al. (2011) Deep resequencing of GWAS loci identifies independent rare variants associated with inflammatory bowel disease. *Nat Genet* 43:1066–1073; PMID 21983784.

Tennessen JA et al. (2012) Evolution and functional impact of rare coding variation from deep sequencing of human exomes. *Science* 337:64–69; PMID 22604720.

Genetic Risk Prediction in Complex Disease

Janssens AC et al. (2006) Predictive testing for complex disease using multiple genes: fact or fiction? *Genet Med* 8:395–400; PMID 16845271.

Jostin L & Barrett JC (2011) Genetic risk prediction in complex disease. *Hum Mol Genet* 20:R182–R188; PMID 21873261.

Visscher PM & Gibson G (2013) What if we had whole-genome sequence data for millions of individuals? *Genome Med* 5:80; PMID 24050736.

Wray NR et al. (2010) The genetic interpretation of area under the ROC curve in genomic profiling. *PLOS Genet* 6:e1000864; PMID 20195508.

Genetic Architecture and Biological Pathways in Complex Disease

Bertram L & Tanzi RE (2012) The genetics of Alzheimer's disease. *Prog Mol Biol Transl Sci* 107:79–100; PMID 22482448.

Hill AVS (2012) Evolution, revolution and heresy in the genetics of infectious disease susceptibility. *Phil Trans R Soc B* 367:840–849; PMID 22312051.

Khor B, Gardet A & Xavier RJ (2011) Genetics and pathogenesis of inflammatory bowel disease. *Nature* 474:307–317; PMID 21677747.

King RA, Rotter JI & Motulsky AG (eds) (2002) The Genetic Basis of Common Disease, 2nd ed. Oxford University Press.

Sullivan PF et al. (2012) Genetic architectures of psychiatric disorders: the emerging picture and its implications. *Nat Rev Genet* 13:537–551; PMID 22777127 .

Gene–Gene Interactions in Complex Disease

Cordell HJ (2009) Detecting gene–gene interactions that underlie human diseases. *Nat Rev Genet* 10:392–404; PMID 19434077.

Strange A et al. (2010) A genome-wide association study identifies new psoriasis susceptibility loci and an interaction between HLA-C and ERAP1. *Nat Genet* 42:985–990; PMID 20953190.

Zuk O et al. (2012) The mystery of missing heritability: genetic interactions create phantom heritability. *Proc Natl Acad Sci USA* 109:1193–11988; PMID 22223662.

Gene–Environment Interactions in Complex Disease

Manolio TA et al. (2006) Genes, environment and the value of prospective cohort studies. *Nat Rev Genet* 7:812–820; PMID 16983377.

Thomas D (2010) Gene–environment wide association studies: emerging approaches. *Nat Rev Genet* 11:259–272; PMID 20212493.

Virgin HW & Todd JA (2011) Metagenomics and personalized medicine. *Cell* 147:44–56; PMID 21962506. [Considers how the human microbiome influences disease risk.]

Epigenetic Effects in Complex Disease

Czyz W et al. (2012) Genetic, environmental and stochastic factors in monozygotic twin discordance with a focus on epigenetic differences. *BMC Med* 10:93; PMID 22898292.

Feil R & Fraga MF (2012) Epigenetics and the environment: emerging patterns and implications. *Nat Rev Genet* 13:97–109; PMID 22215131.

Grossniklaus U et al. (2013) Transgenerational epigenetic inheritance: how important is it? *Nature Rev Genet* 14:228–235; PMID 23416892.

Kaelin WG Jr & McKnight SL (2013). Influence of metabolism on epigenetics and disease. *Cell* 153:56–69; PMID 23540690.

Petronis A (2010) Epigenetics as a unifying principle in the aetiology of complex traits and diseases. *Nature* 465:721–727; PMID 20535201.

Portela A & Esteller M (2010) Epigenetic modifications and human disease. *Nat Biotechnol* 28:1057–1068; PMID 20944598.

Rakyan VK et al. (2011) Epigenome-wide association studies for common human diseases. *Nat Rev Genet* 12:529–541; PMID 21747404.

Relton CL & Smith GD (2010) Epigenetic epidemiology of common complex disease: prospects for prediction, prevention and treatment. *PLoS Med* 7:e1000356; PMID 21048988.

Genetic Approaches to Treating Disease

Treatment of genetic disease and genetic treatment of disease are two separate matters. The cause of a disease (whether mostly genetic or mostly environmental) and its treatability are quite unconnected. Standard medical treatments that are intended to alleviate disease symptoms—hearing aids or cochlear implants for treating profound deafness, for example—are just as applicable if the disease is mostly genetic or mostly environmental. In this chapter the primary focus is on how genetic technologies are being applied to treat disease, but we begin by taking a broader look at different treatment strategies for genetic disorders.

For the great majority of genetic conditions, even for single-gene disorders, existing treatments are lacking or unsatisfactory. **Figure 9.1** represents a snapshot taken in 1999 when the treatability of 372 genetic diseases was assessed by Charles Scriver and Eileen Treacy. The situation has improved since then but we still have a long way to go.

Causative genes for many single-gene disorders have been identified only quite recently, and it may take many years of research to identify how the underlying genes function normally in cells and tissues. Armed with that knowledge, we might hope to devise better treatments in the future. Gene therapy offers the opportunity for highly effective treatment for some monogenic disorders and, as we describe below, there have been exciting breakthroughs in recent years. For some other monogenic disorders there are extremely difficult obstacles to devising effective therapies—we provide examples below.

Reasonably satisfactory treatments exist for some complex diseases, such as diabetes; for many others the treatments are less than satisfactory, or ineffective. By definition, complex diseases are complex at the genetic level: until very recently we knew very few of the underlying genetic factors, but the principal ones will be revealed by genetic studies. In some cases, genetic studies will be able to divide individual complex diseases into subtypes, allowing different treatments to be tailored to suit different disease subtypes. The emerging information will place us in a better position to develop novel, more effective treatments.

Environmental factors are clearly very important in complex diseases and have been notably well documented in many cancers. Some environmental factors are also well recognized in some noncancer conditions. Cigarette smoking is a powerful factor in age-related macular degeneration and emphysema, for example, and the importance of a healthy diet and regular exercise is well recognized in conditions such as type 2 diabetes. Considerable work needs to be done to extend our knowledge of contributory environmental factors. That will provide opportunities for effective interventions, because exposure to an environmental factor can often be modified.

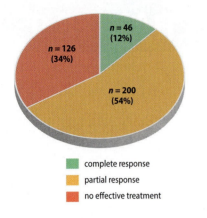

Figure 9.1 Treatment of genetic disorders has often not been very effective. The data here record the response to treatment for 372 single-gene disorders in which the underlying gene or its biochemical function were known and representative treatment data were available by 1999. (Data from Scriver CR & Treacy EP [1999] *Mol Genet Metab* 68:93–102; PMID 10527662.)

In this chapter we are primarily concerned with molecular approaches to treating disease. We deal principally with genetic approaches, but we also take a brief look at the application of small molecule drugs. We explore applications in treating both single-gene disorders and most classes of complex disease, but we cover cancer therapies separately in Chapter 10.

In Section 9.1 we give an overview. First, we look at how treatments can be classified into different categories. We take a broad view of the different levels at which disease can be treated, and explore the different genetic technology inputs that can be applied. In Section 9.2 we cover genetic inputs to treating disease with therapeutic chemical drugs (small molecule drugs) and biological drugs (therapeutic proteins), and we explore aspects of pharmacogenetics that deal with the different responses of patients to small molecule drugs and aspects of drug metabolism. Variation in how we respond to chemical drugs is very important: it leads to hundreds of thousands of fatalities per year.

In Section 9.3 we cover the principles and general methodology of different therapeutic methods involving the genetic modification of a patient's cells (gene therapy). In this section we also describe related stem cell therapy methods. All the methods described above need to be tested in animal disease models before clinical trials are carried out, and we deal with different approaches to disease modeling in the closing section of Section 9.3.

Finally, we describe in Section 9.4 how gene therapy has been applied in clinical trials, assess the progress that has been made, and consider future prospects. Ethical issues are considered later, in Chapter 11.

9.1 An Overview of Treating Genetic Disease and of Genetic Treatment of Disease

In this introductory section we first look at broad categories of treating genetic disease and illustrate the diversity of treatments with examples from inborn errors of metabolism. We then consider the different levels at which molecular-based disease treatments can be applied.

Three different broad approaches to treating genetic disorders

Two types of treatment can be used, according to whether pathogenesis is due to some defined genetic deficiency or whether a disease phenotype is due to some positively harmful effect (rather than a lack of some important gene product or metabolite). A third type of treatment seeks to reduce susceptibility to disease by understanding the pathway involved (**Figure 9.2**). We expand on these themes in the sections below, taking into account both current practice and experimental therapies.

Augmentation therapy for genetic deficiencies

In some genetic disorders the problem is the loss of some normal function. In principle, these disorders might be treated by **augmentation therapy**: something is provided to the patient that *supplements* a severely depleted, or missing, factor, thereby overcoming the deficiency and restoring function. Different types of supplement can be provided to restore function at different levels. At the level of the somatic phenotype, treatment can be conventional—providing cochlear implants or hearing aids to treat hereditary deafness, for example.

At the molecular level, the phenotype can be restored by providing a purified gene product that is lacking—a missing enzyme, say, in many inborn

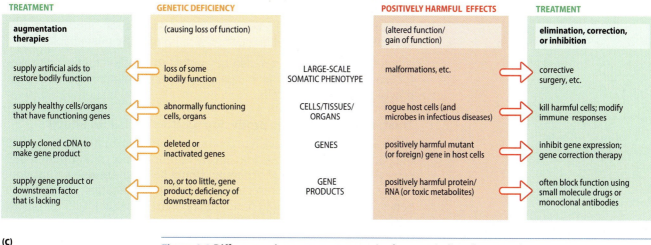

Figure 9.2 Different major treatment strategies for genetic disorders. Note that some strategies are experimental. Brackets indicate applicability of a strategy to infectious disease. (A) Augmentation therapies for phenotypes that result from deficiency of some gene product. The idea is to compensate for the genetic deficiency by supplying purified functional gene product directly (protein augmentation) or a purified downstream factor that is required but lacking, or by indirectly supplying cloned DNA or healthy cells (either from a donor, or genetically modified cells from the patient) to make the missing gene product. (B) Therapies for phenotypes that result from positively harmful cells or molecules. Some therapies work at the cell or tissue level to deal with rogue cells that behave abnormally to cause disease (cancer cells or immune system cells that attack host cells in autoimmune and inflammatory diseases); others work at the gene or gene product level to prevent the harmful effects of a gain-of-function mutation (or a gene from a pathogenic microorganism), or seek to eliminate or reduce the production of elevated toxic metabolites in inborn errors of metabolism. (C) Disease prevention strategies include altering exposure to environmental triggers, such as through extreme dietary modifications in some inborn errors of metabolism, and the use of drugs such as statins to alter disease susceptibility.

errors of metabolism. Alternatively, when the gene product works in a biological pathway required to synthesize some important downstream factor, such as a lipid hormone, it might be a lack of the downstream factor that is treated (by providing purified lipid hormone, in this case). At a higher molecular level, the aim of some types of gene augmentation therapy has been to transfer a cloned cDNA into the tissues of a patient where it can be expressed to make a missing protein.

At the cellular and organ levels, healthy cells and organs can be transplanted into a patient to make a product that the patient lacks. That can involve transplanting cells from a donor, as in bone marrow transplantation or organ transplantation. More recently, some cellular gene therapies have been used very successfully; here, the cells of the patient are genetically modified so that they can now express the desired gene product.

Applicability of molecular augmentation therapy

Recessive disorders (where both alleles have lost their function) are more suited to molecular augmentation therapy than are dominant disorders. Affected individuals often cannot make any functional copies of some normal gene product. Even a modest efficiency in delivery (of healthy cells, genes, or proteins) to an affected individual can often allow effective treatment, and recently there have been substantial breakthroughs. As illustrated below, however, augmentation therapy is currently not practical for some recessive disorders—it can often be difficult to get efficient delivery and production of the desired molecules.

In dominant disorders due to haploinsufficiency, the disease occurs even when one allele is normal and present in all diploid cells; efficient delivery and production of the missing gene product would be essential. But there is the added problem that the underlying gene is also dosage-sensitive (see Box 7.6). Very precise augmentation therapy—a very difficult prospect—would be needed, and is currently unavailable.

Augmentation therapy can also be applied to certain complex diseases, for example by treating diabetes using purified insulin, or by transplantation of pancreatic islet cells.

Treatment for disorders producing positively harmful effects

A second, different approach to treatment is needed for diseases in which the pathogenesis involves a positively harmful effect, rather than a deficiency. Here, augmentation therapy cannot be used: something has gone wrong that cannot be corrected by simply administering some normal gene, normal gene product, or normal cells to the patient. Different methods are needed (see Figure 9.2B).

The harmful effect might be treatable at the somatic phenotype level, as in the case of some developmental malformations: corrective surgery is highly effective, for example, in treating various complex disorders such as congenital heart defects, cleft lip and palate, and pyloric stenosis.

At the molecular level, treatments can be conducted at different stages. In many inborn errors of metabolism the problem is elevated levels of harmful metabolites that can be tackled in different ways, as described in the next major section. A more general problem is presented by actively harmful gene products from a mutant gene. Examples include mutant prion proteins and β-amyloid, which are liable to form protein aggregates that can kill cells (see Box 7.7), and also harmful proteins or RNAs formed after the unstable expansion of short oligonucleotide repeats (see Section 7.2). Dangerous mutant gene products may be combated by using a small molecule drug or therapeutic monoclonal antibody to bind selectively to the mutant molecule and inhibit its activity.

In some cases the therapy seeks to inhibit the expression of a harmful gene selectively at the mRNA level by using novel RNA interference strategies, as described in detail in Section 9.3. (As well as targeting harmful mutant RNAs, the same approach can also be targeted to silence the genes of intracellular pathogens in infectious diseases.) In addition, at the gene level, experimental corrective gene therapy has the potential to reverse a mutation and restore the original sequence, as described below.

At the cellular level, the problem may manifest itself as harmful cells. Some mutations can induce cells to behave abnormally, proliferating excessively to cause cancers that can be treated by long-standing methods (surgical excision, radiation, and chemotherapy) and by targeted chemical and biological drugs and cancer gene therapies (as described in Chapter 10). In some genetic disorders, the problem is excessive immune responses in which certain immune system cells inappropriately attack host cells (in autoimmune disorders such as rheumatoid arthritis, and in inflammatory diseases such as Crohn's disease). Here, there is the potential to employ therapies that down-regulate immune responses, but in some cancer gene therapies the exact opposite approach has been taken (up-regulating immune responses in an attempt to kill cancer cells).

Treatment by altering disease susceptibility

A third way of treating disease seeks to reduce susceptibility to disease in some way, and offers methods of treating certain monogenic disorders and also some complex diseases (see Figure 9.2C). In some inborn errors of metabolism, the blockage at one step in a metabolic pathway can drive alternative pathways that cause a buildup of toxic metabolites. However, that can sometimes be overcome by reducing disease susceptibility, such as by removing an environmental trigger. And in some diseases, key susceptibility factors can be manipulated to reduce the chances that a disease will recur, or to reduce the effects of a progressive disease.

Very different treatment options for different inborn errors of metabolism

The concept of inborn errors of metabolism was founded on the work of Archibald Garrod on alkaptonuria in the early 1900s, and they were the first genetic disorders to be understood at the biochemical level. Since then we have developed a quite detailed understanding of the molecular pathology for many of the disorders.

Early optimism that disease treatments would follow once we knew all the major salient details of the pathology has suffered quite a setback: effective treatment is available for a rather small percentage of these disorders. Nevertheless, there has been a steady improvement. In a 25-year longitudinal study of 65 inborn errors of metabolism published in 2008, significant improvement was recorded: 31 disorders showed no significant response to treatment in 1983 but only 17 in 2008, and the number of conditions fully responding to treatment jumped from 8 in 1993 to 20 in 2008. As we describe later, there have been some very important and encouraging successes using drug and gene therapy since 2008.

In this section, we use inborn errors of metabolism to illustrate the different ways in which treatment can be offered (each of the three general strategies shown in Figure 9.2 has been successfully applied), and why effective treatment will be difficult for some disorders.

Two broad phenotype classes

For the simplest cases, picture a metabolic pathway composed of sequential steps, each catalyzed by an individual enzyme, and imagine that loss-of-function mutations have inactivated the gene encoding one of the enzymes. The resulting absence of that enzyme will lead to a lack of downstream product plus a buildup of substrate proximal to (before) the blocked step.

Sometimes, and often in biosynthetic pathways, the most noticeable effect is the lack of end product. In these cases, augmentation therapy can compensate for a lack of a gene product, or of some other downstream molecule whose production depends on the gene product.

In other cases, as previously described for phenylketonuria (see Box 7.8), the buildup of precursors proximal to the blocked step drives alternative pathways, producing abnormal concentrations of some metabolites that have harmful effects. A disorder such as this is caused by a recessive loss of function at the disease locus, but the disease phenotype results from a buildup of positively harmful metabolites and requires different treatment strategies. As we show below, the disease phenotype for some disorders may have components that are treatable by augmentation therapy, and others that are due to positively harmful effects.

Augmentation therapy

Here the missing product is provided to overcome the deficiency. It might be the gene product itself (protein augmentation) or a critically important downstream factor that it regulates. For example, recessive congenital hypothyroidism (OMIM 275200) is due to a deficiency in thyroid hormone (a lipid hormone whose production is largely dependent on thyroid-stimulating hormone). Affected individuals have a mutant thyroid-stimulating hormone receptor that fails to respond to thyroid-stimulating hormone, but they are effectively treated with purified thyroid hormone. Mutations in *CYP21A2* cause 21-hydroxylase deficiency by disturbing the production of steroid hormones. The effects include greatly reduced levels of two classes of steroid hormones: mineralocorticoids (such as aldosterone) and glucocorticoids (such as cortisol); the deficiencies in these hormones can be treated by augmentation therapies (**Figure 9.3A**).

Augmentation therapy can also involve transplanting cells or organs that can make the required functional product. Bone marrow transplantation, effectively a way of transplanting hematopoietic stem cells, has been frequently used to treat disorders of blood cells (or other cells originating from hematopoietic stem cells). Liver transplantation has been used for many serious inborn errors of metabolism (many metabolic enzymes are synthesized by the liver). Because of the possibility of graft rejection, organ transplantation is a serious matter; it is indicated when the disorder is expected to progress to organ failure—the treatment is intended to save lives, not just cure patients.

A more recent, and quite different, treatment uses a form of gene therapy: the gene product is obtained after the cells of a patient have been genetically modified to contain and express a functional copy of the relevant gene (as detailed in Sections 9.3 and 9.4).

Treating or preventing harmful effects of elevated metabolites

When abnormally elevated levels of metabolites cause disease, different treatments can be devised. One way is use drugs to cause a compensatory change in metabolite levels. Take type 1 tyrosinemia (PMID 20301688). This disorder results from a deficiency of fumarylacetoacetate hydrolase, which catalyzes the terminal step in the tyrosine degradative pathway (**Figure 9.3B**). The resulting buildup of precursors leads to liver and renal tubule dysfunction, and untreated children may have repeated neurologic crises. Oral administration of nitisinone (also called NTBC) provides effective treatment—it inhibits a proximal (upstream) enzyme step, causing a compensatory reduction in fumarylacetoacetate hydrolase levels (see Figure 9.3B).

Normal levels of elevated metabolites may be able to be restored by the removal of excess amounts from the body. Phlebotomy is a possibility if the excess metabolite is present in blood (**Table 9.1**). Alternatively, some indirect means can be used to force the metabolite into another metabolic pathway to decrease the levels to normal values—see the example of dealing with excess ammonia in urea cycle disorders in **Figure 9.4**.

A different approach seeks to *prevent* the buildup of toxic levels of metabolite. In some cases, *substrate restriction* is used: the diet is modified to severely reduce or eliminate the intake of a substrate of the deficient enzyme. That can work very successfully when the blocked enzyme is at the start of a pathway that metabolizes a dietary component. This approach prevents the harmful effects of toxic metabolites that build up in phenylketonuria (as described in Box 7.8). Even in this case, the treatment requires lifelong compliance with a rather restricted, and difficult, diet. Prevention is sometimes possible at the prenatal level, as in the case of 21-hydroxylase deficiency (see Figure 9.3A).

Mixed success in treatment

Treatment of some inborn errors of metabolism is very effective, as with phenylketonuria. However, the treatment can be sub-optimal in some cases, and very difficult or essentially nonexistent in others for various reasons. If a disorder is congenital and harmful effects have occurred during development, treatment options may be limited. In some cases, potential therapy can be frustrated by delivery problems. In Tay–Sachs

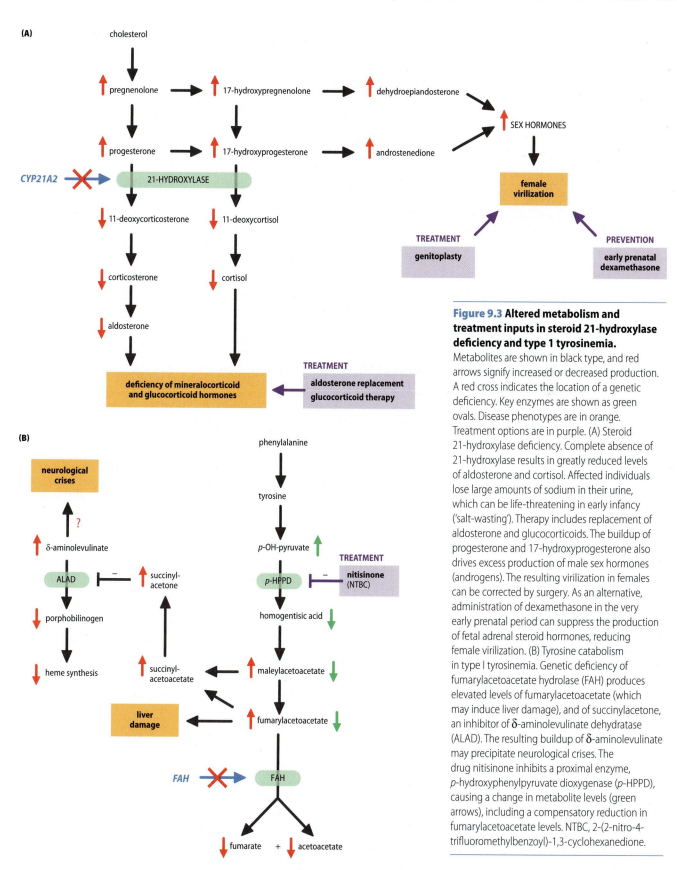

(A)

(B)

Figure 9.3 Altered metabolism and treatment inputs in steroid 21-hydroxylase deficiency and type 1 tyrosinemia.
Metabolites are shown in black type, and red arrows signify increased or decreased production. A red cross indicates the location of a genetic deficiency. Key enzymes are shown as green ovals. Disease phenotypes are in orange. Treatment options are in purple. (A) Steroid 21-hydroxylase deficiency. Complete absence of 21-hydroxylase results in greatly reduced levels of aldosterone and cortisol. Affected individuals lose large amounts of sodium in their urine, which can be life-threatening in early infancy ('salt-wasting'). Therapy includes replacement of aldosterone and glucocorticoids. The buildup of progesterone and 17-hydroxyprogesterone also drives excess production of male sex hormones (androgens). The resulting virilization in females can be corrected by surgery. As an alternative, administration of dexamethasone in the very early prenatal period can suppress the production of fetal adrenal steroid hormones, reducing female virilization. (B) Tyrosine catabolism in type I tyrosinemia. Genetic deficiency of fumarylacetoacetate hydrolase (FAH) produces elevated levels of fumarylacetoacetate (which may induce liver damage), and of succinylacetone, an inhibitor of δ-aminolevulinate dehydratase (ALAD). The resulting buildup of δ-aminolevulinate may precipitate neurological crises. The drug nitisinone inhibits a proximal enzyme, *p*-hydroxyphenylpyruvate dioxygenase (*p*-HPPD), causing a change in metabolite levels (green arrows), including a compensatory reduction in fumarylacetoacetate levels. NTBC, 2-(2-nitro-4-trifluoromethylbenzoyl)-1,3-cyclohexanedione.

syndrome (PMID 20301397), for example, deficiency in hexosaminidase A leads to an inexorable buildup of a sphingolipid GM2 ganglioside to toxic levels, causing damage to brain cells.

TYPE OF TREATMENT	TYPE OF ACTION	EXAMPLES AND COMMENTS
Augmentation therapy	protein augmentation therapy	enzyme replacement therapies for many inborn errors of metabolism; provision of blood clotting factors in hemophilias (see Section 9.2 for details)
	hormone replacement therapy	thyroid hormone for infants with congenital hypothyroidism; growth hormone for growth hormone deficiency (see also Figure 9.3A)
	bone marrow transplantation	useful for disorders affecting blood cells and some other immune system cells (as illustrated in Figure 9.20), such as mucopolysaccharidosis type 1 (Hurler syndrome, OMIM 607014)
	organ transplantation	liver transplantation has been used successfully for various inborn errors of metabolism, including α_1-antitrypsin deficiency and urea cycle disorders
	gene augmentation therapy	successfully used for different types of severe combined immunodeficiency, hemophilia (see Sections 9.3 and 9.4)
Counteracting harmful effects of abnormally elevated metabolites	artificially manipulated excretion of metabolite	periodic blood removal is a very effective treatment for directly removing excess iron in the iron overload condition hemochromatosis (OMIM 235200)
	shunting of elevated metabolite into a side metabolic pathway	for urea cycle disorders the buildup of toxic ammonia can be alleviated by sodium benzoate treatment, driving excess ammonia to be metabolized in a side pathway (see Figure 9.4)
	inhibition of a proximal step in a pathway leading to harmful metabolites	babies with type 1 tyrosinemia (OMIM 276700) are unable to metabolize tyrosine effectively and suffer liver damage from toxic intermediates. The drug NTBC prevents buildup of the toxic intermediates by inhibiting a proximal enzyme (Figure 9.3B)
Prevention (avoiding or reducing susceptibility)	substrate restriction (the diet is modified to severely reduce or eliminate intake of a substrate for a deficient enzyme)	reduced intake of phenylanine in phenylketonuria (see Box 7.8). Elimination of galactose in galactosemia (OMIM 230400); affected individuals completely lack galactose-1-phosphate uridyltransferase. Galactose is a component of the lactose in milk but is inessential, and milk is completely withdrawn from the diet
	reduction of a susceptibility factor	in familial hypercholesterolemia, *LDLR* gene mutations result in low levels of the low-density lipoprotein receptor; the resulting elevated plasma low-density lipoprotein cholesterol levels predispose to cardiovascular disease but can be effectively lowered by statins, drugs that inhibit a proximal enzyme, HMCoA reductase, in the cholesterol biosynthesis pathway

Table 9.1 Examples of different types of treatment of inborn errors of metabolism.

Genetic treatment of disease may be conducted at many different levels

Any disease, whether it has a genetic cause or not, is potentially treatable using a range of different procedures that apply genetic manipulations or genetic knowledge in some way (**Figure 9.5**).

Sometimes genetic techniques form part of a treatment regime that also involves conventional small molecule drugs or vaccines.

Figure 9.4 Reducing elevated metabolite levels by shunting the metabolite into an alternative metabolic pathway. The urea cycle normally serves to convert ammonia (NH_3), which is neurotoxic, to nontoxic urea. But in urea cycle disorders, ammonia cannot be converted to urea and builds up. In ornithine transcarbamylase (OTC) deficiency (OMIM 311250), the metabolic block causes an increase in levels of the proximal metabolites, carbamoyl phosphate and ammonia (vertical red arrows). The therapy here involves treating a patient with large amounts of sodium benzoate and takes advantage of a normally minor pathway in which some ammonia is naturally converted into small amounts of glycine. Benzoate ions conjugate with glycine to form hippurate, which is excreted in urine. By removing glycine, the treatment drives the production of replacement glycine from ammonia (thick purple horizontal arrow), thereby reducing ammonia levels (vertical purple arrow).

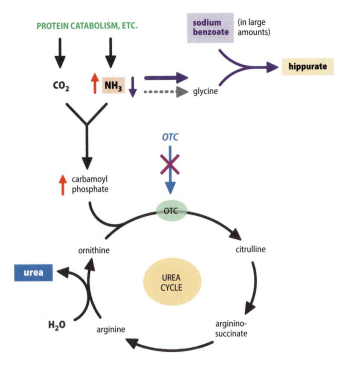

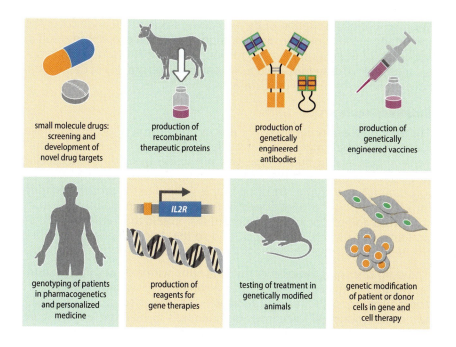

Pharmacogenetics is concerned with how the actions of drugs and the reactions to them vary according to variation in the patient's genes. Genotyping of individuals might then be used to predict patterns of favorable and adverse responses to specific drug treatments. Such genotyping may become routine as massively parallel DNA sequencing permits extensive screening of genes in vast numbers of people.

New targets for drug development are being identified using a knowledge of genetics and cell biology. Genetic techniques can also be used directly in producing drugs and vaccines for treating disease. Another active area concerns treating disease with therapeutic proteins that are produced or modified by genetic engineering. Genes are cloned and expressed in suitable cultured cells or organisms to make large amounts of a specific protein that is then purified (*recombinant proteins*), including hormones, blood factors, and enzymes, and especially genetically engineered antibodies.

Gene therapies are the ultimate genetic application in treating disease; they rely on genetically modifying the cells of a patient. Delivering therapeutic constructs into the stem cells of a patient is particularly valuable when the disease primarily affects short-lived cells, such as blood cells. Animal models are especially important resources for testing new therapies before they are used in clinical trials. As described in Section 9.3, the vast majority of animal models of disease have been generated by the genetic manipulation of rodents, notably mice.

9.2 Genetic Inputs into Treating Disease with Small Molecule Drugs and Therapeutic Proteins

Chemical treatments for disease are developed by the pharmaceutical industry. Previously they relied almost exclusively on hydrocarbon-based small molecule drugs synthesized by standard chemical reactions (see **Figure 9.6** for some examples). More recently they have been joined by a new class of biological drugs based on therapeutic proteins.

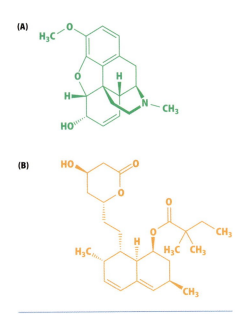

Figure 9.6 Structures of two widely prescribed small molecule drugs. (A) Codeine (= methylmorphine). (B) Simvastatin, a popular statin used to reduce cholesterol levels.

We give a brief description of the two classes below. Then we cover the important question of how genetic variation between people can result in very wide differences in how we respond to therapeutic drugs, and consider aspects of drug metabolism. We finish with examples of therapeutic uses of the two principal drug classes.

Small molecule drugs

The conventional drug discovery process has involved screening huge numbers of small molecules for evidence that they can reduce pathogenic effects. A drug such as this typically works by binding to a specific target protein that has a key role in the pathogenesis—often a receptor, ion channel, or enzyme. The drug is able to bind to the protein by fitting into some cleft, groove, or pocket at a key position in the protein structure. In so doing, it blocks the ability of the protein to interact with other proteins, or with other molecules, and so the drug can effectively block the function of a protein that is key to pathogenesis.

The drug screening process normally begins with assays in cell culture and in animal models to see whether a candidate drug has some encouraging properties. Promising candidates may be used in clinical trials, when their potential usefulness is monitored in different ways (**Figure 9.7**). To bring a drug to market is both costly (about US $1 billion) and time-consuming (typically 12–15 years). Sometimes, however, a drug previously developed to treat one type of disease can be used to treat other diseases. Drug 'repurposing' is valuable because the drug has already been through lengthy and expensive clinical trials to assess its safety profile.

Small molecule drugs have been with us for some time. Two important questions are: how effective are they, and how safe are they? Although the therapeutic value of many small molecule drugs on the market is questionable, many others have undoubtedly been of great service. But individual drugs affect different people in different ways. As explained below, many of our drug-handling enzymes are polymorphic; genetic variation between individuals has an important influence on both the efficacy and the safety of drugs.

All drugs currently on the market act through only a few hundred target molecules (they were first developed when information about possible targets was scarce), and the declining number of new drug applications and approvals over the past few years has reflected a crisis in drug target identification and validation.

New approaches

Recently, both genomics and genetic engineering have been making an impact on drug development. Genomic advances offer a broader perspective on how genetic variation affects drug metabolism, and the ways

Figure 9.7 Major stages in drug development. The bullet lists indicate the principal parameters that are tested at each stage. Pharmacokinetic testing assesses the absorption, activation, metabolism, and excretion of drugs. Pharmacodynamics monitors what the drug does to the body. Successive stages toward regulatory approval are increasingly expensive. Any effects of genetic variation among patients that might influence marketability need to be identified as early as possible in the process, to avoid unnecessary expenditure.

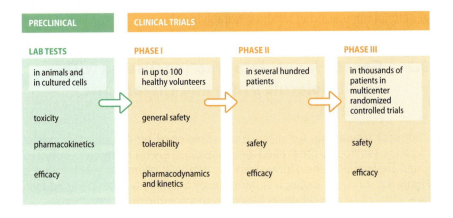

PRECLINICAL	CLINICAL TRIALS		
LAB TESTS	**PHASE I**	**PHASE II**	**PHASE III**
in animals and in cultured cells	in up to 100 healthy volunteers	in several hundred patients	in thousands of patients in multicenter randomized controlled trials
toxicity	general safety		
pharmacokinetics	tolerability	safety	safety
efficacy	pharmacodynamics and kinetics	efficacy	efficacy

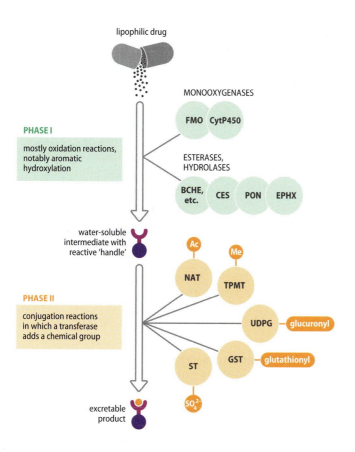

lipophilic drug

PHASE I

mostly oxidation reactions, notably aromatic hydroxylation

MONOOXYGENASES

FMO CytP450

ESTERASES, HYDROLASES

BCHE, etc. CES PON EPHX

water-soluble intermediate with reactive 'handle'

PHASE II

conjugation reactions in which a transferase adds a chemical group

Ac Me

NAT TPMT

UDPG — glucuronyl

ST GST — glutathionyl

SO₄²⁻

excretable product

Figure 9.8 Two major stages in drug metabolism. Phase I drug reactions typically result in a more polar drug derivative with a reactive group, a molecular 'handle' that makes it easier to accept a chemical group donated by a phase II enzyme. Phase I enzymes are often monooxygenases, notably cytochrome P450 enzymes (CytP450), but also include various other enzymes, notably esterases and other hydrolases. In phase II drug metabolism one of a variety of different transferase enzymes adds a chemical group that facilitates excretion. Note that the sequence shown here occurs commonly, but phase II reactions can sometimes occur without a previous phase I reaction. BCHE, butyrylcholinesterases; CES, carboxyesterases; EPHX, epoxide hydrolases; FMO, flavin-containing monooxygenases; GST, glutathione S-transferase; NAT, N-acetyltransferase; PON, paraoxonases; TPMT, thiopurine methyltransferase; UDPG, UDP glucuronosyltransferases; ST, sulfotransferases.

posing any great risks to health. If the concentration is below this range, the therapeutic benefit might be insufficient (drug underdose); if above this range, there is an increasing risk of toxicity (drug overdose).

When drugs are detoxified by metabolism, the concentration of active drug falls; repeated drug doses are required to maintain drug concentrations within the safe therapeutic range. The speed at which a drug is metabolized has consequences for both *drug efficacy*—the degree to which the drug gives therapeutic benefit—and safety. Individuals who eliminate or inactivate drugs comparatively slowly ('slow metabolizers') will have a longer or stronger response to a given concentration of the drug than fast metabolizers will. They can be at risk of a drug overdose if given the usual dose (**Figure 9.9**). They may also be more at risk of adverse reactions if breakdown products from the drug are toxic. Ultrafast metabolizers might gain little therapeutic benefit from a drug (see Figure 9.9).

Individual drugs can be metabolized by multiple different enzymes; in addition to genetic variation in different genes, environmental factors make significant contributions to how a drug is metabolized. As a result, drug response phenotypes are often multifactorial. Sometimes, however, a specific drug might be metabolized by just one enzyme; genetic variation in the gene that makes that enzyme can have a predominant contribution to the phenotype—we provide examples in the next sections.

Genetic variation in cytochrome P450 enzymes in phase I drug metabolism

The great majority of phase I drug reactions are carried out by monooxygenases, notably heme-containing enzymes belonging to the cytochrome P450 superfamily (the name reflects a common spectral absorption peak at 450 nm). Cytochrome P450 enzymes are often regarded as mixed function oxidases because they catalyze different types of reaction. In addition

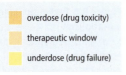

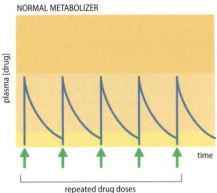

NORMAL METABOLIZER

plasma [drug]

time

repeated drug doses

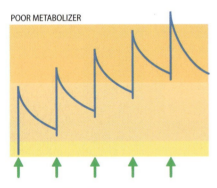

POOR METABOLIZER

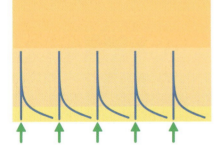

ULTRAFAST METABOLIZER

Figure 9.9 The effect of different drug metabolizing rates on plasma drug concentration. The *therapeutic window* is the range of plasma drug concentrations that are of therapeutic benefit without causing extra safety risks due to drug toxicity. Normal metabolizers are expected to benefit from having drug concentrations in the therapeutic window for long periods. If given the normal drug concentration, poor metabolizers can suffer from an overdose: the failure to metabolize the drug quickly means that the drug concentration progressively increases to very high, unsafe levels after repeated doses. Ultrafast metabolizers might gain little therapeutic benefit, because the drug is rapidly cleared from the plasma after each drug dose. See Figure 9.10 for specific examples of different classes of drug metabolism.

to specific drugs, their substrates can be endogenous chemicals, notably certain steroids, and xenobiotics in our food and in the environment.

We have close to 60 different cytochrome P450 genes. According to evolutionary relationships based on amino acid sequence homology, they are classified into several families and subfamilies. Thus, for example, the *CYP2C19* gene gets its name from cytochrome P450 family 2, subfamily C, polypeptide 19.

Six cytochrome P450 enzymes catalyze 90% of the phase I reactions on commonly used drugs. CYP3A4 is the most versatile, in that it is involved in metabolizing about 40% of all drugs; CYP2D6 is another prolific drug-handler. Individual drugs are also often substrates for more than one P450 enzyme. The antidepressant amitriptyline, for example, can be metabolized by each of CYP1A2, CYP2C19, and CYP2D6.

Specific cytochrome P450 enzymes can be induced or inhibited by certain drugs. That can result in unexpected interactions between these drugs and those that are substrates of the enzyme. Comprehensive drug interaction lists are maintained in the Cytochrome P450 Drug Interaction Table (see Flockhart [2007] in Further Reading).

The wide and overlapping specificities of cytochrome P450 enzymes mean that it can often be difficult to correlate how a person metabolizes a specific drug with the activity of any one P450 enzyme. Nevertheless, some drugs are metabolized by just one P450 enzyme, and DNA variation in just a single gene can result in wide differences between people in how they metabolize the drug. Often the variation is due to simple mutations that change key amino acids or that inactivate gene expression, but occasionally excess gene activity occurs when there are multiple copies of the same cytochrome P450 gene.

Genetic variation in CYP2D6 and its consequences

The CYP2D6 enzyme illustrates how different types of genetic variation in a single enzyme can have a marked effect on the metabolism of certain drugs. As a result of severe inactivating mutations or deletions in both *CYP2D6* alleles, rare individuals have a very low activity of the enzyme. When treated with certain drugs that are normally metabolized by this enzyme alone, they fail to metabolize and excrete the drug (with high plasma levels for the drugs and low levels of expected catabolic products in urine samples).

Individuals with very low CYP2D6 activity are classified as poor metabolizers (of drugs for which this enzyme has the predominant role) and are comparatively frequent in Caucasian populations. As well as poor metabolizers, people can be classified into three additional groups: intermediate metabolizers; extensive metabolizers (with one or two active *CYP2D6* alleles); and ultrafast metabolizers (**Figure 9.10**).

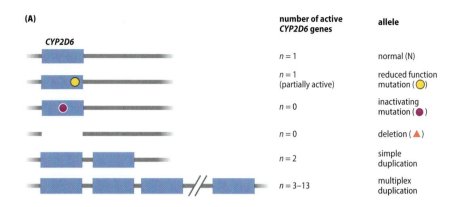

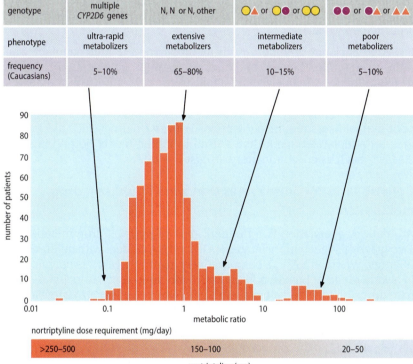

Figure 9.10 *CYP2D6* **allele classes and correlation between genotypes and drug-metabolizing abilities.** (A) *CYP2D6* allele classes. Variation in gene copy number occurs by unequal crossover; multiple *CYP2D6* genes are quite common in some populations, most probably as a result of natural selection (much like the occurrence of multiple α-amylase genes in some populations, as described in Figure 4.11). (B) *CYP2D6* genotypes and drug-metabolizing ability. The large middle panel shows a range of CYP2D6's drug-metabolizing ability in a study group and how it relates to different classes of genotype (top panel). To assay CYP2D6 activity, a urinary *metabolic ratio* is used: after a standard drug dose, the urinary concentration of the substrate drug (debrisoquine in this case) is measured and divided by that of its metabolic product. High ratios show poor conversion as a result of low enzyme activity. Low metabolizers are at risk of drug overdose and should be given lower drug doses. As an example, the lower panel shows recommended doses of the antidepressant nortriptyline (also metabolized by CYP2D6) that are arranged on a sliding scale according to the metabolic ratio shown above. (Adapted from Meyer UA [2004] *Nature Rev Genet* 5:669–676; PMID 15372089. With permission from Macmillan Publishers Ltd.)

People with very low CYP2D6 activity can show unusually marked sensitivity to certain drugs; they are also at risk of drug overdose if prescribed with normal doses of certain beta-blockers and tricyclic antidepressants. Because CYP2D6 is also the enzyme that converts codeine to morphine, people with very low CYP2D6 activity also get very little painkilling benefit from codeine. Ultrafast metabolizers may also get little benefit from drugs that are principally metabolized by CYP2D6 (because the drug is metabolized and detoxified so quickly).

Genetic variation in other cytochrome P450 enzymes

CYP3A4 activity in the liver shows extensive variability between individuals, but unlike for CYP2D6 there are very few coding sequence variants. And unlike CYP2D6, CYP3A4 is inducible; regulatory mutations are thought to be significant contributors to the variability.

CYP3A4 is highly related to CYP3A5 (and to CYP3A7, which is normally expressed only at fetal stages); the drug-metabolizing activities of CYP3A4 and CYP3A5 strongly overlap, complicating matters. However,

Table 9.2 Significant population differences in the frequency of CYP2D6 and CYP2C19 poor metabolizers. [a]Excluding the Indian subcontinent. (Data from Burroughs VJ, Maxey RW & Levy RA [2002] *J Natl Med Assoc* 94(10 Suppl): 1–26; PMID 12401060.)

POPULATION ORIGIN	FREQUENCY (%) OF CYP2D6 POOR METABOLIZERS	FREQUENCY (%) OF CYP2C19 POOR METABOLIZERS
Amerindian	0	2.0
Caucasian	7.2	2.9
East and South East Asian[a]	0.5	15.7
Middle East and North African	1.5	2.0
Polynesian	1.0	13.6
Indian subcontinent (Sri Lankan)	0	17.6
Sub-Saharan African	3.4	4.0

CYP3A5 is less biologically active than CYP3A4: in Caucasian populations *CYP3A5* null alleles predominate and only 10% of the alleles make an active enzyme.

CYP2C9 deficiency is important in metabolizing the anticoagulant warfarin (as described below), and also results in an exaggerated response to tolbutamide, a hypoglycemic agent used in treating type 2 diabetes. CYP2C19 deficiency shows marked differences between different ethnic groups, as revealed by the frequency of poor metabolizers (**Table 9.2**); poor metabolizers require a lower dose of certain drugs such as clopidogrel, an antiplatelet agent that is used to inhibit blood clotting in some diseases such as coronary artery disease.

Genetic variation in enzymes that work in phase II drug metabolism

Aromatic *N*-acetylation is a frequent type of phase II drug metabolism. Two types of *N*-acetyltransferase, NAT1 and NAT2, deal with different sets of drugs. Whereas NAT1 is comparatively invariant, NAT2 is polymorphic in a wide range of human populations, with rapid acetylators (who eliminate drugs rapidly) and slow acetylators (who have low NAT2 levels).

Variation in acetylating abilities of NAT2 can be clearly seen as a bimodal distribution using the drug isoniazid, which is used to treat tuberculosis (**Figure 9.11**). The proportion of slow acetylators is highly variable between different ethnic groups but can be very high in some populations, notably in Caucasian populations (**Table 9.3**).

Figure 9.11 A bimodal distribution of plasma levels of isoniazid as a result of genetic polymorphism in the *NAT2* (*N*-acetyltransferase 2) gene. Plasma concentrations were measured in 267 normal subjects 6 hours after an oral dose of isoniazid. Fast acetylators removed the drug rapidly. The number of slow acetylators (presumed to be homozygotes or compound heterozygotes for severe inactivating mutations) was almost the same as the number of fast acetylators. This suggests that only about 30% of *NAT2* alleles in the study group were producing active enzyme. (Adapted from Price Evans DA, Manley KA & McKusick VA [1960] *Br Med J* ii:485–491. With permission from BMJ Publishing Group Ltd.)

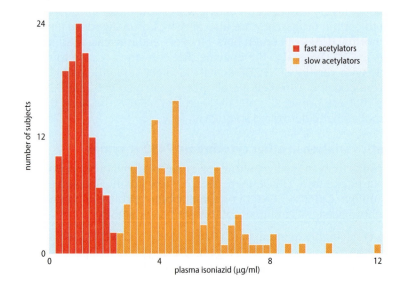

The slow acetylators take longer to eliminate drugs (and other xenobiotics) and so often show enhanced sensitivity to drugs metabolized by NAT2; they also appear to be more susceptible to certain cancers, notably bladder cancer. Nevertheless, natural selection appears to have driven an increase in frequency of the slow-acetylator phenotype in some populations. A possible explanation is that certain chemicals in well-cooked meat are converted by NAT2 into carcinogens; individuals with slow-acetylator phenotypes would be comparatively protected in populations that have had a long tradition of eating well-cooked meat.

Variation in other phase II enzymes is also significant. Polymorphism in thiopurine methyltransferase is a particular clinical concern when using certain immunosuppressant drugs such as 6-mercaptopurine, which is commonly used to treat childhood leukemia. (About 1 in 300 children do not express thiopurine methyltransferase; for them 6-mercaptopurine is toxic.) The glutathione S-transferase (GST) superfamily includes some enzymes such as GSTM1 and GSTT1 that are encoded by genes which are susceptible to gene deletion as a result of unequal crossover. As a result, inactive alleles are very common (about 50% of the GSTM1 alleles in people of northern European ancestry are gene deletions, for example). People with consequently low levels of these enzymes find it difficult to cope with high doses of drugs that are processed by them.

The UDP glucuronosyltransferase superfamily includes polymorphic enzymes that are involved in metabolizing different substrates, including drugs used in cancer chemotherapy. For example, the prodrug irinotecan is converted into an active anti-tumor form in the liver (it inhibits DNA topoisomerase, an enzyme needed for DNA replication) and is normally processed by the enzyme UGT1A1. The common UGT1A1*28 promoter polymorphism results in reduced production of this enzyme, and is frequent in many populations (1 in 3 people are homozygotes in sub-Saharan Africa, and 1 in 5 in the Indian subcontinent). The *28/*28 homozygotes consequently have a much higher risk of serious bone marrow and gastrointestinal toxicity.

Altered drug responses resulting from genetic variation in drug targets

In the sections above we have considered genetic variation in enzymes that perform phase I and phase II drug metabolism, and how that variation affects drug pharmacokinetics (including how fast a drug is metabolized and excreted). In this section we consider how genetic variation affects the pharmacodynamics of drugs.

The efficacy of a drug is partly determined by genetic variation in *drug targets*, the molecules that the drug must interact with in the cells of the target tissue. Drug targets will typically include receptors, signaling molecules and other molecular components of biological pathways that the drug interacts with to have its pharmacogenetic effect.

Genes encoding drug receptors quite often show polymorphisms or variants that lead to clinically significant altered responses to drugs. Examples of clinically significant genetic variation in drug receptors include variants of beta-adrenergic receptors, cell surface receptors that have central roles in the sympathetic nervous system. Two of these receptors, ADRB1 and ADRB2, are widely used as drug targets in therapeutic approaches for various common and important diseases including asthma, hypertension, and heart failure. Genetic variation in both ADRB1 and ADRB2 has been linked to altered responses to drugs. Other examples include variation in the H2RA serotonin receptor and the RYR1 ryanodine receptor (**Table 9.4**).

POPULATION ORIGIN	FREQUENCY (%) OF NAT2 SLOW ACETYLATORS
Caucasian	58
Chinese	22
Eskimo	6
Japanese	10
Sub-Saharan African	51

Table 9.3 Significant population differences in the frequency of NAT2 slow acetylators. (Data from Wood AJ & Zhou HH [1991] *Clin Pharmacokinet* 20:350–373; PMID 1879095.)

Table 9.4 Examples of how genetic variation in drug targets cause altered responses to therapeutic drugs. [a]Note that the R16G polymorphism has significant effects on short-acting agonists such as albuterol but has little effect on long-acting agonists, which now constitute the more widely used treatment.

DRUG TARGET	FUNCTION	POLYMORPHISM OR VARIANT	EXAMPLE OF DRUG TREATMENT	EFFECT OF POLYMORPHISM OR VARIANT
ACE	angiotensin-converting enzyme	Alu repeat insertion (Alu+)/deletion (Alu−) polymorphism in intron of *ACE* gene	use of ACE inhibitors such as captopril and enalapril in treating heart failure	drugs are more effective in Alu−/Alu− homozygotes
ADRB1	β_1 adrenergic receptor	common R389G polymorphism	beta-blockers, such as bucindol, for reducing risk of heart disease	reduced cardiovascular response to drugs
ADRB2	β_2 adrenergic receptor	common R16G polymorphism	albuterol for treating asthma	homozygotes are much less likely to respond to treatment[a]
RYR1	ryanodine receptor	different mutations	inhalation anesthetics	potentially fatal (see Box 9.1)

Some therapeutic drugs are designed to specifically inhibit key enzymes that have pivotal roles in biological pathways underlying common diseases. Examples include statins, which were designed to inhibit HMG CoA reductase (lowering cholesterol levels, and so reducing blood pressure and the risk of cardiovascular disease); warfarin, an inhibitor of a key enzyme in the maturation of several blood-clotting factors (see the next section); and inhibitors of angiotensin-converting enzyme (ACE). The last enzyme is also an important regulator of blood pressure (and several other functions), and an insertion/deletion polymorphism due to the variable presence of an Alu repeat in intron 15 of the *ACE* gene is associated with variation in ACE activity. People who are homozygous for the Alu deletion allele have about twice the enzyme activity of those who have two Alu insertion alleles; this difference is thought to be an important contributor to variable responses to ACE inhibitors (see Table 9.4).

When genotypes at multiple loci in patients are important in drug treatment: the example of warfarin

We give a summary of serious adverse drug reactions in **Box 9.1**. In most cases, the genetic variation associated with the effect is primarily confined to a single locus. However, for many drugs genetic variation at multiple loci might be important. One common example where we know some of the details concerns treatment with the anticoagulant warfarin.

Warfarin is prescribed for patients at risk of developing clots within blood vessels (thrombosis), including clotting that can block arteries (embolism). Delivering the optimal warfarin dosage is clinically very important because there is a narrow therapeutic window: if the administered warfarin level is too low, the patient remains at risk of thrombosis and embolism; if it is too high, there is a risk of life-threatening hemorrhage. The final warfarin dose is critical, but because of genetic variation the optimal dose varies enormously between individuals.

Chemically, warfarin is a mixture of two isomers. The (*S*)-warfarin isomer is three to five times as potent as the (*R*)-isomer, has a shorter half-life, and is metabolized predominantly by CYP2C9. Two common polymorphisms in CYP2C9 result in a large decrease in enzyme activity, down to 12% of wild-type activity for CYP2C9*1, and just 5% of normal enzyme activity for CYP2C9*3. Patients with one or two of these common alleles are at increased risk of hemorrhage (presumably because drug metabolism takes longer). It soon became clear, however, that these polymorphisms could explain only a part of the genetic variation in the final warfarin dose.

In 2004 the drug target of warfarin was found to be vitamin K epoxide reductase complex subunit 1 (VKORC1), an enzyme that converts oxidized vitamin K to its reduced form; afterward, association studies showed that genetic variation in VKORC1 was associated with variation in the final warfarin dose. Vitamin K is an indispensable cofactor for the enzyme that converts inactive clotting proteins to give four of our blood clotting factors (factors II, VII, IX, and X). By inhibiting vitamin K epoxide reductase, warfarin inhibits the recycling of vitamin K; the consequent decreased supply of vitamin K inhibits the formation of these four clotting factors.

Subsequently a genomewide association study also implicated a common V433M polymorphism in CYP4F2, a cytochrome P450 enzyme that works as a vitamin K oxidase (**Figure 9.12**). Variations in VKORC1 and CYP2C9 remain the largest genetic determinants, accounting for about 40% of the variation in the final warfarin dose. However, other factors, such as aging and the simultaneous administration of other medicines, also have a very significant influence on the dose of warfarin.

Translating genetic advances: from identifying novel disease genes to therapeutic small molecule drugs

Many previously unknown genes have been identified in the long-standing quest to identify the molecular basis of genetic disorders. Labor-intensive studies have often been required to work out what the gene normally does, and how it goes wrong in disease. As a result, it can sometimes take some time to identify suitable drug targets before proceeding to drug development. We provide four examples to illustrate differential progress in moving from identifying novel genes for monogenic disorders to developing small molecule drugs.

The first example illustrates enduring difficulties but a recent small step forward; the second is a success story in which genetic advances led to a promising drug target that was screened to identify a new class of drugs that are of great value; and in the third and fourth examples, working out what the disease normally does and how it was part of a molecular pathway allowed the application of previously identified drugs that work in the same pathway and that look highly promising.

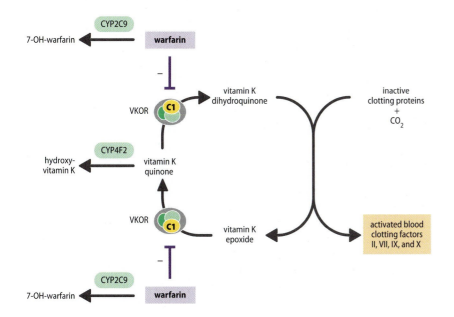

Figure 9.12 Roles of warfarin in anticoagulation and genetic variants affecting final warfarin drug dose. Warfarin is a therapeutic anticoagulant that is prescribed for people at risk of thrombosis and embolism. It works by inhibiting VKORC1, the C1 subunit of the vitamin K epoxide reductase complex (VKOR) and so decreases the supply of vitamin K that is essential for activating four blood clotting factors (factors II, VII, IX, and X; they are activated by adding a carboxyl group to a side chain of a specific glutamic acid residue). In addition to VKORC1, genetic variation in at least two cytochrome P450 enzymes is known to be associated with variation in the final warfarin dose needed: CYP2C9 converts warfarin to an inactive form, 7-hydroxywarfarin, and CYP4F2 metabolizes vitamin K quinone.

BOX 9.1 When prescribed drugs can be dangerous and sometimes deadly, depending on a patient's genotype.

Therapeutic drugs and other drugs administered in medical procedures (such as anesthetics) can produce extreme responses in some people. Adverse drug reactions are very common, being responsible for a significant proportion of all hospital admissions (nearly 7% in a UK study) and can result in disability or permanent damage, birth defects, and an extraordinary number of fatalities (about 100,000 deaths each year in the USA alone).

Adverse drug reactions have various causes. Type A reactions are relatively common and dose-dependent; they are predictable from the drug pharmacology and are usually mild. Type B reactions are idiosyncratic reactions that are not related simply to drug dose; they are rare but can often be severe (**Table 1**). Genetic variants are important in both types of reaction.

DRUG-INDUCED INJURY OR TOXICITY	EXAMPLES OF ASSOCIATED DRUGS	COMMENTS AND EXAMPLES
Apnea (respiratory paralysis)	suxamethonium (succinyl choline)	this drug works as a fast-acting muscle relaxant and is used before surgery. Normally the effects of the drug wear off quite quickly when the drug is metabolized by the enzyme butylcholinesterase. Low metabolizers are at risk of apnea—they remain paralyzed and unable to breathe after surgery because they cannot regain their muscle function quickly enough and may require extended ventilation
Prolongation of cardiac QT interval	thioridazine, clarithromycin, terfenadine	induced by many different drugs. Associated with polymorphic ventricular tachycardia, or torsades de pointes (see **Figure 1**), which can be fatal
Hematologic toxicity	6-mercaptopurine azathioprine	thiopurine *S*-methyltransferase (TPMT) inactivates these immunosuppressant drugs by adding a methyl group. In people with two low-activity TPMT alleles, the drugs are metabolized slowly; if normal doses are given, the drugs accumulate and can result in life-threatening bone marrow toxicity
Hemorrhage	warfarin	see above
Hypersensitivity reactions	abacavir, carbamazepine, allopurinol	inappropriate immune reactions to otherwise nontoxic drugs can have broad manifestations. When treated with the anti-HIV drug abacavir, about 5% of patients demonstrate skin, gastrointestinal, and respiratory hypersensitivity reactions that can sometimes be fatal. Treatment with the anticonvulsant carbamazepine or allopurinol (used in treating gout) can induce cutaneous adverse drug reactions, including toxic epidermal necrolysis
Liver injury	flucloxacillin, isoniazid, allopurinol	individuals with certain HLA antigens are at increased risk of induced liver disease for some drugs, such as *HLA-B*5701* in the case of flucloxacillin
Muscle toxicity	halothane, isoflurane, various statins	statins and several other drugs are associated with usually mild myopathies. Sometimes, however, more severe cases show rhabdomyolysis (breakdown of muscle tissue) that can result in death. In response to inhalation anesthetics (halothane, isoflurane), individuals who have inactivating mutations in the ryanodine receptor gene develop life-threatening rhabdomyolysis and an extreme rise in temperature, a form of malignant hyperthermia (OMIM 145600)

Table 1 Some classes of severe type B adverse drug reactions.

Cystic fibrosis: not an easy prospect

In 1989 cystic fibrosis was found to be caused by mutations in a previously unstudied gene. The novel cystic fibrosis transmembrane regulator (*CFTR*) gene was predicted to regulate transmembrane conductance, and was subsequently shown to function as a channel that allows chloride ions to pass through the cell membrane. A great deal was discovered about aspects of CFTR biology that would inform diverse fields such as protein trafficking and membrane transport. However, a retrospective profile in the journal *Nature* in 2009 (20 years since discovery of the gene) reported the uncertainty in understanding how *CFTR* mutations cause the disease, and the lack of cystic fibrosis therapies emanating from the discovery of the *CFTR* gene. Jack Riordan, one of the major contributors to *CFTR* gene cloning, was quoted as saying "the disease has contributed much more to science than science has contributed to the disease."

BOX 9.1 (continued)

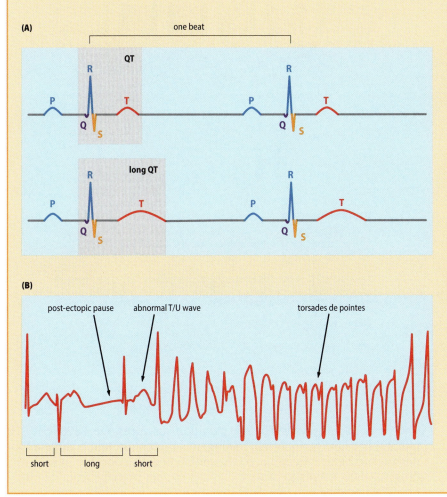

(A)

one beat

QT

long QT

(B)

post-ectopic pause   abnormal T/U wave   torsades de pointes

short   long   short

Figure 1 Drug-induced prolongation of the cardiac QT interval and torsades de pointes. (A) Cardiac depolarization–repolarization cycle. Specific repeated features are labeled from P to T. The *QT interval*, the shaded interval that spans the onset of the QRS complex until the end of the T wave, represents the time taken for one complete cycle of ventricular depolarization and repolarization. Certain drugs can prolong the QT interval (*long QT*), and this can sometimes induce a rapid beating of the heart which often manifests itself as torsades de pointes (TdP). (B) Cardiac rhythm profile in a patient with drug-induced TdP. Notice the short–long–short initiating ventricular cycle, pause-dependent long QT interval, and abnormal TU wave preceding the development of TdP. This type of ventricular arrhythmia can self-terminate, but it can also degenerate into potentially fatal arrythmias such as ventricular fibrillation. (B, Adapted from Yap YG & Camm AJ [2003] *Heart* 89:1363–1372; PMID 14594906. With permission from BMJ Publishing Group Ltd.)

One of the complications is that six different *CFTR* mutation classes can be recognized, according to their effect on how the gene normally works in cells. Some are inactivating mutations that result in failure to make any protein and some result in reduced protein synthesis. In other cases a protein is made but doesn't get to the membrane, or the stability of the protein is reduced. And finally in some cases the CFTR protein does get incorporated into the membrane but the ion channel regulation is defective, or there is decreased channel conductance. According to the mutation class, different types of treatment approach are needed, but for a long period there were no effective drug treatments.

Recent developments, however, have been more encouraging. In 2012 ivacaftor (marketed as Kalydeco by Vertex Pharmaceuticals) became the first drug approved by the US Food and Drug Administration (FDA) to target a cause of cystic fibrosis rather than the condition's symptoms. It has been targeted to treat patients with the G551D mutation, which causes the chloride channel to fail to open (**Figure 9.13**). Ivacaftor works by helping to reopen the chloride channel. While ivacaftor may well result in marked improvement in longevity, quality of life, treatment burden, and so on, it is directed at just this one mutation, and is applicable to just 4% of cystic fibrosis patients (those who have at least one G551D mutation).

Figure 9.13 Therapeutic targeting of aberrant chloride ion channels in one form of cystic fibrosis. The cystic fibrosis transmembrane regulator (CFTR) comprises two membrane-spanning domains, which act to form a channel that regulates the passage of chloride ions between the intracellular and extracellular environments. The *CFTR*G551D mutation results in a mutant CFTR in which glycine is replaced by aspartate at position 551. It can migrate to take up its position in the membrane but, as shown, it fails to function as an ion channel, with the result that the normally thin protective layer of watery mucus becomes thick and sticky and prone to bacterial infection. The drug ivacaftor (Kalydeco™) is able to bind to the mutant CFTR protein so as to restore some of the original function, but the common F508del mutant CFTR is not able to get to the membrane at all because of a protein folding problem (as described in Section 7.6).

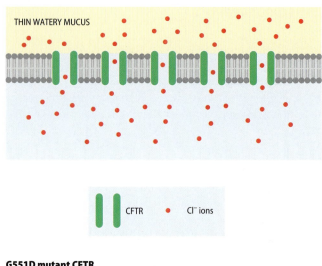

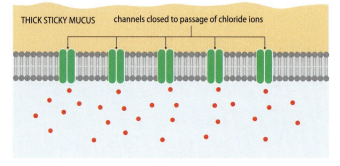

The predominant cystic fibrosis mutation produces the F508del CFTR mutant; because of abnormal folding, this mutant protein is targeted for destruction before it can get to the membrane. But another new Vertex drug, VX-809, seems to protect F508del mutant proteins from degradation. If it could help the F508del mutant protein to get to the membrane, ivacaftor might be able to assist in ensuring that the chloride channel remains open.

Familial hypercholesterolemia: new and valuable drugs

With a frequency of more than 0.2% in most populations, familial hypercholesterolemia (OMIM 143890), an autosomal dominant disorder, is the most common single-gene disorder. Affected persons have extremely high cholesterol levels, irrespective of diet. Most cases are due to mutations in the low-density lipoprotein receptor gene, *LDLR*. Heterozygotes typically develop coronary artery disease in the fourth or fifth decade; rare homozygotes are much more severely affected, and most suffer a heart attack before the age of 20 years.

LDLR imports cholesterol-containing low-density lipoprotein into liver cells, where it represses cholesterol synthesis as part of a homeostatic mechanism. An *LDLR* loss-of-function mutation results in less LDLR being made, resulting in an increase in endogenous cholesterol synthesis. Hydroxymethylglutaryl (HMG) CoA reductase was known to be the rate-limiting enzyme in the endogenous cholesterol biosynthesis pathway, and so represented a very promising drug target.

Screening for small molecules that inhibit HMG CoA reductase led to a new class of drugs called statins that are effective in lowering cholesterol, such as the simvastatin drug whose structure is shown in Figure 9.6B. Drugs such as these have been widely prescribed to reduce the general risk of heart disease.

Marfan syndrome: advantages of a mouse model

Treatment for Marfan syndrome (PMID 20301510), a systemic autosomal dominant disorder caused by mutations in the fibrillin locus *FBN1*, illustrates the value of mouse models. Because fibrillin is essential for the formation of elastic fibers in connective tissue, the disorder might have been expected to be due to loss of tissue integrity. However, certain clinical observations had suggested a more complicated pathogenesis. For example, various features—including bone overgrowth, craniofacial features, and valve and lung abnormalities—argued for altered cell behavior during morphogenesis. To model Marfan syndrome, mice mutants were constructed to be deficient in fibrillin 1, and comprehensive analyses revealed that different aspects of the phenotype were due to excessive signaling by transforming growth factor β (TGFβ); that is, fibrillin 1 has a role in repressing TGFβ signaling. Angiotensin II-receptor blockers, such as losartan, were known to work as TGFβ antagonists, and treatment with losartan was found to mitigate aspects of the phenotype in the mouse models.

Premature enlargement of the aortic root, leading to separation of the layers within the aortic wall (aortic dissection) is the main cause of premature death in Marfan syndrome, and data from the mouse models suggested that it is caused by excessive TGFβ signaling. In a small clinical trial published in 2008, administration of losartan or irbesartan achieved a remarkable reduction in the rate of dilation of the aortic root in 18 patients, from a mean of 3.54 mm in diameter per year during previous (not very effective) medical therapy to a mean of just 0.46 mm per year during therapy with the angiotensin II-receptor blockers.

Tuberous sclerosis: from a biological pathway to a promising drug

Tuberous sclerosis complex (PMID 20301399) is an autosomal dominant disorder in which benign (noncancerous) tumors develop in many organs and can disrupt how they function (tumors of the central nervous system and kidneys are the leading causes of the morbidity and mortality). Additional abnormalities of cell migration and function in the brain lead to seizures, autism, and learning difficulties. The disorder can be caused by mutations in the *TSC1* or *TSC2* genes that encode components of the TSC1–TSC2 protein complex. Until these novel genes were identified in the 1990s, nothing was known about the molecular pathogenesis of the disorder, and the only treatment options were the surgical removal of tumors, which was often problematic.

The TSC1–TSC2 complex was found to be part of the mTORC1 growth signaling pathway. When the TSC1–TSC2 complex is disrupted by mutations in either *TSC1* or *TSC2*, mTORC1 signaling is constitutively active; downstream targets become activated by phosphorylation, driving protein synthesis and cell growth (**Figure 9.14**). A principal subunit of the mTORC1 complex is the mTOR protein, whose name originates from mammalian target of rapamycin. Rapamycin, also called sirolimus, is an antifungal antibiotic that was first isolated from a strain of *Streptomyces* in the 1970s. Both it and a closely related drug, everolimus, are powerful mTOR inhibitors. They had initial applications as immunosuppressants to prevent organ transplant rejection and then as anti-proliferative agents for treating certain cancers.

The discovery that tuberous sclerosis complex was due to unregulated growth because of abnormal regulation of the mTORC1 pathway

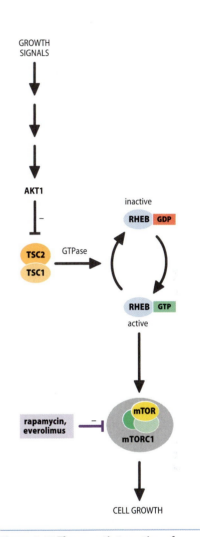

Figure 9.14 Therapeutic targeting of mTORC1 signaling in tuberous sclerosis complex. In tuberous sclerosis complex, the problem is that mutations in either *TSC1* or *TSC2* disrupt the TSC1–TSC2 protein complex. Normally, TSC2 acts as a GTPase and stimulates the formation of an inactive form of the RHEB regulator of the mTORC1 complex (but can be countermanded when growth signals repress TSC1–TSC2). But in tuberous sclerosis complex, the disruption of the TSC1–TSC2 complex causes the RHEB regulator to be activated so that mTORC1 signaling is constitutively active, and growth is no longer regulated as normal. Rapamycin (also called sirolimus) and everolimus, an *O*-(2-hydroxyethyl) derivative of sirolimus, are effective inhibitors of mTOR, a major subunit of the mTORC1 complex, and work to suppress cell growth. AKT1 is also known as protein kinase B.

prompted very recent clinical trials with mTOR inhibitors. It now seems that everolimus is an effective and safe drug for treating various aspects of tuberous sclerosis, including angiomyolipomas and subependymal giant cell astrocytomas (SEGAs; see Further Reading). Because renal angiomyolipomas are very common in adult patients and can lead to renal failure (and a need for dialysis), and because the surgical removal of SEGAs can be particularly difficult, treatment with everolimus appears to be a major advance.

Translating genomic advances and developing generic drugs as a way of overcoming the problem of too few drug targets

Most small molecule drugs work by binding to specific protein targets, blocking their interactions with other molecules. However, only a rather small percentage of protein targets are susceptible to drugs. In many cases, small molecule drugs cannot block interactions between two types of protein because the interacting surfaces of the proteins are too smooth (small molecule drugs are most effective when they can sneak into clefts and pockets within proteins). Some types of protein are easier targets: more than 50% of drug targets belong to one of four types of protein (class I G-protein-coupled receptors, nuclear receptors, ligand-gated ion channels, and voltage-gated ion channels); protein kinases are another favorite target.

Translating genomic advances

A survey published in 2006 estimated that there were just 324 targets for all approved drugs; that figure will have increased since then, but novel drug targets are badly needed. A potential source of help could be the massive data sets generated by the combination of genomics, high-throughput transcriptomics, proteomics, and bioinformatics.

The recent increase in data has been quite extraordinary. For example, by mid-2012 one million sets of gene expression data had been entered into the Gene Expression Omnibus (GEO; http://www.ncbi.nlm.nih.gov/geo/) or the ArrayExpress database of functional genomics data (http://www.ebi.ac.uk/arrayexpress/).

New algorithms are being trained to recognize gene signatures for individual diseases by logging exceptional gene expression changes associated with a specific disease. That might lead to the identification of previously unstudied genes that can then be studied further as drug targets (or disease biomarkers). Large-scale functional genetic screens are also being carried out in a variety of model organisms to identify new drug targets.

The treatment of infectious diseases will surely benefit from the recent investment in genome projects for a wide range of pathogenic microbes and follow-up functional and proteomic studies. The aim will be to identify pathogen-specific gene products that might make good drug targets, and also proteins on pathogen cell surfaces that might be targets for vaccines.

Developing generic drugs

Another potential solution to obstacles in identifying novel drug targets is to identify generic drugs that are not focused on a specific gene product. Because they might have quite widespread applicability, generic drugs might also have the merit of reducing costs (which can be extraordinary for 'orphan drugs' used to treat rare diseases: the cost of treatment with Kalydeco for the G551D type of cystic fibrosis has been estimated to be approximately US $294,000 per patient per year).

A prominent class of generic drugs is made up of those drugs that can suppress stop codons so that translation continues (translational 'readthrough'). The potential applications are huge: nonsense mutations are responsible for causing anywhere from 5–70% of individual cases for most inherited diseases. Aminoglycoside antibiotics such as gentamycin were known to cause readthrough of premature stop codons in mammalian cells but were not clinically useful, partly because of a lack of potency and partly because of toxicity problems.

A recently identified drug, Ataluren (or PTC124), seemed more promising: it appeared to cause readthrough of premature nonsense mutations (notably UGA), without affecting the recognition of normal stop codons, and with little evidence of toxicity. However, there have been concerns that PTC124 activity is due to off-target effects (see PMID 23824517). Recently completed phase III trials indicate that, when treated with Ataluren, patients with cystic fibrosis due to a *CFTR* nonsense mutation showed a modest improvement in lung function only.

Developing different drugs: therapeutic recombinant proteins produced by genetic engineering

Certain genetic disorders that result from deficiency of a specific protein hormone or blood protein can be treated by administering an external supply of the missing protein. To ensure greater stability and activity, the proteins are often conjugated with poly(ethylene glycol) (PEG). The increased size of the protein–PEG complex means reduced renal clearance, so that the protein spends more time in the circulation. Adding PEG can also make the protein less immunogenic.

Therapeutic proteins were often previously extracted from animal or human sources, but there have been safety issues. Many hemophiliacs contracted AIDS and/or hepatitis C after being treated with factor VIII prepared from unscreened donated blood. And some children succumbed to Creutzfeld–Jakob disease, after having injections of growth hormone extracted from unscreened pituitary glands from cadavers.

A safer, but rather expensive, alternative is to use therapeutic 'recombinant' proteins. They are produced by cloning human genes and expressing them to make protein, usually within mammalian cells, such as human fibroblasts or the Chinese hamster ovary cell line. (Mammalian cells are often needed because many proteins undergo post-translational modification such as glycosylation, and the pattern of modification shows species differences.) Recombinant human insulin was first marketed in 1982; **Table 9.5** also gives several subsequent examples.

Some human proteins are required in very high therapeutic doses, beyond the production capabilities of cultured cell lines. Transgenic animals such as transgenic sheep or goats are an alternative source; in these the desired protein is secreted in the animal's milk, aiding purification. In 2009, Atryn became the first therapeutic protein produced by a transgenic animal to be approved by the FDA. Atryn is an antithrombin expressed in the milk of goats, and was designed to be used in therapy to prevent blood clotting.

Genetically engineered therapeutic antibodies with improved therapeutic potential

One class of recombinant protein has notably been put to therapeutic use: genetically engineered antibodies. As detailed in Section 4.5, each of us has a huge repertoire of different antibodies that act as a defense system against innumerable foreign antigens. Antibody molecules function

Table 9.5 Examples of therapeutic recombinant proteins. For genetically engineered therapeutic antibodies, see Table 9.6.

RECOMBINANT PROTEIN	FOR TREATMENT OF
Insulin	diabetes
Growth hormone	growth hormone deficiency
Blood clotting factor VIII	hemophilia A
Blood clotting factor IX	hemophilia B
Interferon α	hairy cell leukemia; chronic hepatitis
Interferon β	multiple sclerosis
Interferon γ	infections in patients with chronic granulomatous disease
Tissue plasminogen activator	thrombotic disorders
Leptin	obesity
Erythropoietin	anemia

as adaptors: they have binding sites for foreign antigens at the variable end, and binding sites for effector molecules at the constant end. Binding of an antibody may be sufficient to neutralize some toxins and viruses; more usually, the bound antibody triggers the complement system and cell-mediated killing.

Artificially produced therapeutic antibodies are designed to be mono-specific (specific for a single antigen). Traditional monoclonal antibodies (mAbs) are secreted by *hybridomas*, immortalized cells produced by fusing antibody-producing B lymphocytes from an immunized mouse or rat with cells from an immortal mouse B-lymphocyte tumor. Hybridomas are propagated as individual clones, each of which can provide a permanent and stable source of a single mAb.

The therapeutic potential of mAbs produced like this is, unfortunately, limited. Rodent mAbs, raised against human pathogens for example, have a short half-life in human serum, often causing the recipient to make anti-rodent antibodies. And only some of the different classes can trigger human effector functions.

Genetically engineered antibodies

Genetic engineering can modify rodent monoclonal antibodies so that they become more stable in humans: some or all of the rodent protein sequence is replaced by the human equivalent. (At the DNA level, sequences encoding parts of the rodent antibody are replaced by equivalent human sequences.) The resulting hybrid DNA is expressed to generate an antibody that is part human and part rodent, such as an antibody with the original rodent variable chains but human constant regions (a chimeric antibody; see **Figure 9.15**).

Subsequently, *humanized antibodies* were constructed in which all the rodent sequence was replaced by human sequence, except for the complementarity-determining regions (CDRs), the hypervariable sequences of the antigen-binding site (see Figure 9.15). More recently, it has been possible to prepare *fully human antibodies* by different routes. For example, mice have been genetically manipulated to delete their immunoglobulin loci and replace them with an artificial chromosome containing the entire human heavy-chain and γ light-chain loci so that they can make fully human antibodies only.

From inauspicious beginnings in the 1980s, mAbs have become the most successful biotech drugs ever, and the market for mAbs is the fastest-growing component of the pharmaceutical industry. Of the therapeutic mAbs currently in use, the eight bestsellers together generate an annual

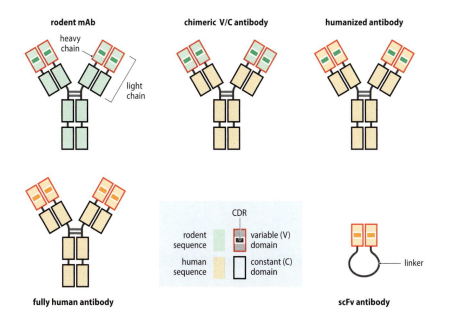

rodent mAb

heavy chain

light chain

chimeric V/C antibody

humanized antibody

CDR

rodent sequence — variable (V) domain

human sequence — constant (C) domain

fully human antibody

scFv antibody

linker

Figure 9.15 Using genetic engineering to make improved therapeutic antibodies. Classical antibodies consist of heavy and light chains with variable (V) and constant (C) domains. Rodent monoclonal antibodies (mAbs) are monospecific antibodies synthesized by hybridomas. Chimeric V/C antibodies are genetically engineered to have human constant domains joined to rodent variable domain sequences (containing the critically important hypervariable complementarity-determining region, CDR). Humanized antibodies can be engineered so that all the sequence is human except the hypervariable CDR. More recently, it has been possible to obtain fully human antibodies by different routes. Genetic engineering has also been used to make single-chain antibodies composed of two variable domains only, connected by a linker peptide. These single-chain variable fragment (scFv) antibodies are particularly well suited to working within the reducing environment of cells, and can serve as *intrabodies* (intracellular antibodies) by binding to specific antigens within cells. Depending on the length of the linker, they bind their target as monomers, dimers, or trimers. Multimers bind their target more strongly than monomers.

income of more than US $30 billion. Several hundred additional mAb products are in the pipeline. Of the FDA-approved mAbs, most are partly or fully human and the great majority are aimed at treating autoimmune or immunological disease or cancers (**Table 9.6**). In the latter case, the latest antibodies are being developed as antibody–drug conjugates so that the antibodies deliver powerful toxins to kill cancer cells.

Intrabodies

A more recently developed—and potentially promising—class of therapeutic antibody has been engineered to have a single polypeptide chain. Single-chain variable fragment (scFv) antibodies have almost all the binding specificity of a mAb but are restricted to a single non-glycosylated variable chain (see Figure 9.15). They can be made on a large scale in bacterial, yeast, or even plant cells.

Unlike standard antibodies (which have four polypeptide chains linked by disulfide bridges), scFv antibodies are stable in the reducing environment within cells. They are therefore well suited to acting as intracellular antibodies (**intrabodies**). Instead of being secreted like normal antibodies,

DISEASE CATEGORY	TARGET	mAb GENERIC NAME (TRADE NAME)	DISEASE TREATED
Autoimmune disease or immunological	CD11a	efalizumab (Raptiva)	psoriasis
	IgE	omalizumab (Xolair)	asthma
	Integrin α₄	natalizumab (Tysabri)	multiple sclerosis
	TNFα	certolizumab pegol	Crohn's disease, rheumatoid arthritis
		adalimumab (Humira)[a]	
Cancer	EGFR	panitumumab (Vectibix)[a]	colorectal cancer
	HER2	trastuzumab (Herceptin)	metastatic breast cancer
	VEGF	bevacizumab (Avastin)	colorectal, breast, renal, NSCL cancer
Other diseases	RSV	palivizumab (Synagis)	respiratory syncytial virus prophylaxis
	VEGF	ranibizumab (Lucentis)	age-related macular degeneration

Table 9.6 Examples of licensed therapeutic monoclonal antibodies (mAbs). CD11a, white blood cell antigen; IL2R, interleukin type 2 receptor; IgE, immunoglobulin E; TNFα, tumor necrosis factor α; EGFR, epidermal growth factor receptor; HER2, human epidermal growth factor receptor 2; NSCL cancer, non-small-cell lung cancer; VEGF, vascular endothelial growth factor; RSV, respiratory syncytial virus. [a]Fully human antibodies; all others are humanized antibodies (see Figure 9.15 for an explanation of different monoclonal antibody classes).

they are designed to bind specific target molecules within cells and can be directed as required to specific subcellular compartments.

Intrabodies can carry effector molecules that perform specific functions when antigen binding occurs. However, for many therapeutic purposes they are designed simply to block specific protein–protein associations within cells. As such, they complement conventional drugs. Protein–protein interactions usually occur across large, flat surfaces and are often unsuitable targets for small molecule drugs (that normally operate by fitting snugly into clefts on the surface of macromolecules). Promising therapeutic target proteins for intrabodies include mutant proteins that tend to misfold in a way that causes neurons to die, as in various neurodegenerative diseases including Alzheimer, Huntington, and prion diseases.

9.3 Principles of Gene and Cell Therapy

Gene therapy involves the direct genetic modification of cells to achieve a therapeutic goal. The genetic modification can involve the insertion of DNA, RNA, or oligonucleotides. Gene therapy can be classified into two types, according to whether germ-line cells or somatic cells are genetically modified. Germ-line gene therapy would produce a permanent modification that can be transmitted to descendants; this could be achieved by modifying the DNA of a gamete, zygote, or early embryo. Germ-line gene therapy that involves modifying nuclear DNA is widely banned in humans for ethical reasons (as detailed in Section 11.5), but, as described below, modification of mtDNA of germ-line cells is being actively considered for treating certain mitochondrial DNA disorders. Somatic cell gene therapy seeks to modify specific cells or tissues of the patient in a way that is confined to that patient.

All current gene therapy trials and protocols are for somatic cell therapy. However, as gene therapy successes accumulate and the technologies become increasingly refined and safe, the idea of extending the technology to produce 'designer babies' can be expected to come more to the forefront. That, too, would raise ethical concerns, as discussed in Chapter 11.

Gene therapy has had a checkered history. Tremendous initial excitement—and quite a bit of hype—was followed by a fallow period of disappointing results and safety concerns (with unexpected deaths of patients arising from unforeseen deficiencies in the treatment methods). More recently, there have been significant successes, and a greater appreciation of safety risks.

In this section and Section 9.4, we are mostly concerned with gene therapy for inherited disorders, which has very largely been focused on recessive Mendelian disorders. We also describe some approaches to treating infectious disease by using gene therapy. Cancer gene therapy has largely been of limited practical use; we describe approaches to treating cancer in Chapter 10.

The first real successes of gene therapy were not achieved until the early 2000s and involved very rare cases of severe combined immunodeficiencies. They took advantage of previous experience of bone marrow transplantation, which is effectively a type of stem cell therapy.

As we describe later, *stem cells* are cells that have the property of being able to renew themselves and also being able to give rise to more specialized cells. For many types of gene therapy, it is important to maximize gene transfer into stem cells. Cell therapies based on the genetic modification of stem cells are also fundamental in *regenerative medicine*, in which the object is to treat disease by replacing cells or tissues that have been lost through disease or injury.

In this section we consider the principles underlying gene and cell therapy. In Section 9.4 we deal with the progress that has been made, and discuss future prospects.

Two broad strategies in somatic gene therapy

In somatic cell gene therapy, the cells that are targeted are often those directly involved in the pathogenic process (but in Chapter 10 we will consider certain types of cancer gene therapy in which the object is to genetically modify *normal* cells in a patient).

Using molecular genetic approaches to treat disease might involve many different strategies. But at the level of the diseased cells there are two basic strategies: disease cells are simply modified in some way so as to alleviate disease, or they are selectively killed. Within each of the two main strategies are different substrategies, as described below.

Modifying disease cells (**Figure 9.16A**). According to the molecular pathology, different strategies are used. If the problem is loss of function,

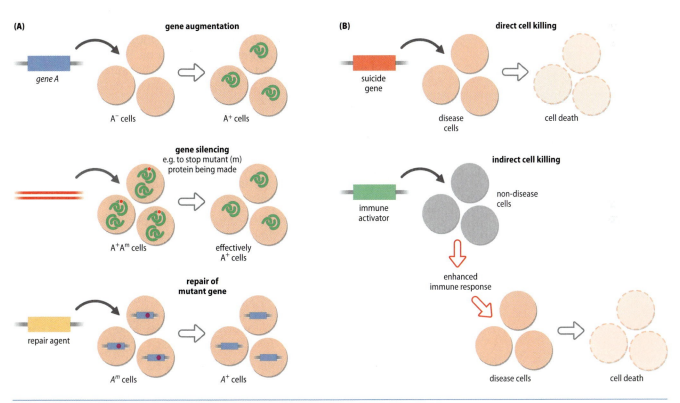

Figure 9.16 Different general types of gene therapy strategy.
(A) Therapies aimed at modifying disease cells. Gene augmentation therapy can be applied to loss-of-function disorders but is currently limited to treating recessive disorders (in which the disease results from a lack, or an almost complete lack, of some gene product). The object is simply to transfer a cloned working gene copy into the cells of the patient in order to make some gene product that is lacking. *Gene-silencing therapy* can be applied to disorders that result from positively harmful gene products. If the disease is caused by a gain-of-function mutation that produces a harmful mutant gene product, A^m, in addition to the normal gene product, A^+, one might try to specifically inhibit the expression of the mutant allele

without inhibiting expression of the normal allele. The same approach can be applied to treating autoimmune and infectious diseases. Additional approaches seek to repair the DNA lesion (by converting the mutant allele sequence, *Am*, to the normal allele sequence, *A*), or to minimize the effect of a pathogenic mutation (not shown). (B) Therapies aimed at killing cells. Cancer gene therapies often rely on killing cancer cells. The aim is to kill cells directly (by inserting and expressing cloned genes that will give rise to some cytotoxic product and cell death) or indirectly (by transferring genes into non-disease cells, such as immune system cells, to provoke an immune response directed at tumors).

a simple solution (in theory) is to add functioning copies of the relevant gene. In infectious diseases and in genetic disorders in which the pathogenesis results from a gain of function, there is some harmful or toxic gene product within cells. The approach then might be to selectively inhibit the expression of the harmful gene product without affecting the expression of any normal genes. This can often be done by selectively blocking transcription or by targeting transcripts of a specific gene to be destroyed (*gene silencing*). Yet other approaches seek to repair a genetic lesion (bottom panel of Figure 9.16A) or find a way of minimizing its effect. We detail the approaches below.

Killing disease cells (**Figure 9.16B**). This approach is particularly appropriate for cancer gene therapy. Traditional cancer treatments have often relied on killing disease cells by using blunt instruments, such as high-energy radiation and harmful chemicals that selectively kill dividing cells. Gene therapy approaches can kill harmful cells either directly or by modifying immune system cells to enhance immune responses that can kill the harmful cells.

The delivery problem: designing optimal and safe strategies for getting genetic constructs into the cells of patients

In gene therapy, a therapeutic *genetic construct* of some type—often a cloned gene, but sometimes RNA or oligonucleotides—is transferred into the cells of a patient. (A nucleic acid molecule introduced in this way is often referred to as a **transgene**.) Depending on the disease, the cells that need to be targeted can be very different. This means that different strategies are needed, and depending on the target cells some disorders are easier to treat in principle than others.

Consider access to the desired target cells. Some cells and tissues—notably blood, skin, muscle, and eyes—are very accessible; others, such as brain cells, are not easily accessed. Then there is the question of overcoming various barriers that impede the transfer and expression of genetic constructs. Strong immune responses constitute important barriers; as we will see, mechanical barriers can also be important.

Another significant difference occurs between cells that are short-lived and need to divide from time to time to replenish the lost cells (such as blood and skin cells), and those that are long-lived, such as terminally differentiated muscle cells. That is an important distinction because for nondividing cells the key parameters would simply be the efficiency of transfer of the therapeutic construct into the cells of the patient and the degree to which the introduced construct was able to function in the expected way. However, for dividing cells we also need to take into account what happens to the descendant cells.

Even if we were to achieve significant success in getting the desired genetic construct into short-lived cells, the cells that have taken up the genetic construct are going to die and will be replaced by new cells. Certain **stem cells** divide to continuously replenish cells lost through aging, illness, or injury (see **Box 9.2** for a brief overview of stem cells). To ensure that copies of the therapeutic construct keep getting into newly dividing cells, therefore, it would be best to target the relevant stem cells if possible, and get the therapeutic construct integrated into chromosomes (so that it gets replicated, allowing copies to be passed to both daughter cells at cell division).

Efficiency and safety aspects

In any gene delivery system used in gene therapy, two key parameters are fundamental: efficiency and safety. Most gene therapy methods rely

BOX 9.2 An overview of stem cells and artificial epigenetic reprogramming of cells.

Stem cells have two essential properties: they can self-renew and they can also give rise to more **differentiated** (more specialized) cells. As well as undergoing normal (symmetric) cell divisions, stem cells can undergo asymmetric cell division to give two *different* daughter cells. One daughter cell is identical to the parent stem cell, allowing self-renewal; the other daughter cell is more specialized and can undergo further rounds of differentiation to give terminally differentiated cells (**Figure 1**).

Two fundamental classes of stem cell are used in experimental investigations. Somatic stem cells occur naturally in the body, and give rise to a limited number of differentiated cell types. Some of them have been cultured and used experimentally. Other stem cell lines are **pluripotent**—they have the capacity to give rise to all of the different cell types in the body.

The first pluripotent stem cell lines to be developed were **embryonic stem (ES) cell lines**. They are derived from naturally pluripotent cells of the very early embryo. Another approach relied on artificially changing the normal epigenetic settings of cells (epigenetic **reprogramming**). Different types of reprogramming are possible. A skin cell, for example, can be reprogrammed to make another type of differentiated cell, such as a neuron (**transdifferentiation**). Or a differentiated cell may be reprogrammed so that it reverts to an unspecialized pluripotent state (**dedifferentiation**), and is then subsequently induced to form a desired differentiated cell.

Somatic stem cells

These cells, sometimes also called adult stem cells or tissue stem cells, occur naturally at low frequencies in the body, and are used to replace cells that have naturally short life spans (notably in the blood, skin, intestines, and testis) or to replenish cells that are lost in disease or injury. (Our powers of tissue regeneration are, however, rather limited.)

Most somatic stem cells give rise to a limited set of differentiated cells. Some, such as spermatogonial stem cells, are unipotent—they give rise to only one type of differentiated cell. Others, such as hematopoietic stem cells (see Figure 9.20), give rise to several different classes of differentiated cell and are multipotent. Cultured somatic stem cell lines have been used for studying differentiation, and purified populations of genetically modified somatic stem cells, notably hematopoietic stem cells, have been used in gene therapy.

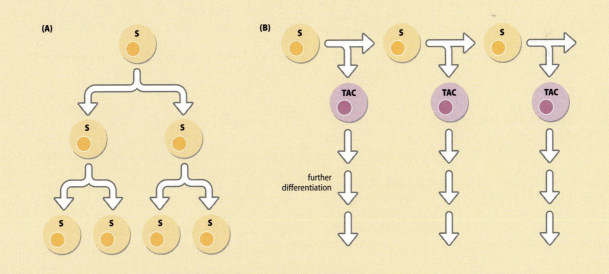

Figure 1 Symmetric and asymmetric stem cell divisions.
(A) Populations of stem cells (S) can expand quickly by symmetrical cell division during growth and when there is a rapid need for new stem cells (to replace cells lost through disease or injury). (B) Stem cells give rise to more differentiated cells by asymmetric cell divisions—the stem cell produces two different daughter cells. One daughter cell is a stem cell identical to the parent cell. The other is a stem cell derivative, sometimes called a **transit amplifying cell** (TAC), that is more differentiated than its sister or parent cell and can subsequently undergo additional differentiation steps to form a terminally differentiated cell.

BOX 9.2 (*continued*)

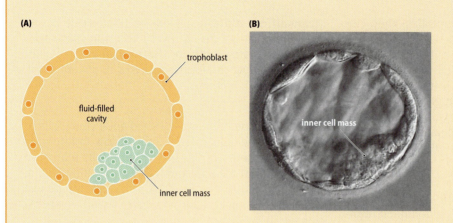

(A)

trophoblast

fluid-filled
cavity

inner cell mass

(B)

inner cell mass

Figure 2 Blastocyst structure and the inner cell mass. (A) The blastocyst is a hollow, fluid-filled ball of cells with two distinct cell populations: an outer cell layer (the trophoblast) and an inner group of cells located at one end (the inner cell mass). (B) A 6-day-old human blastocyst containing about 100 cells, showing the location of the inner cell mass. (B, Courtesy of M. Herbert, Newcastle University, UK.)

Embryonic stem cell lines

Cells from the very early mammalian embryo (including the zygote and the cells that descend from it through the first few cleavage divisions) are entirely unspecialized and are said to be *totipotent*—they can give rise to every type of cell in both the embryo and in the extra-embryonic membranes. Subsequently, the **blastocyst** forms as a hollow ball of cells with two quite distinct layers: an outer layer of cells known as the *trophoblast* (which will give rise to the extra-embryonic membranes such as the chorion and the amnion), and a group of inner cells, the **inner cell mass**, located at one end of the blastocyst (**Figure 2**).

The cells from the inner cell mass are pluripotent and can be cultured to make a pluripotent embryonic stem (ES) cell line. ES cell lines can be experimentally induced to make desired types of differentiated cell (including derivatives of the three germ cell layers—ectoderm, mesoderm, and endoderm—and also germ cells), and have been vitally important for making animal (mostly, mouse) models of disease, as described in Box 9.3. Human ES cell lines are produced from cells from surplus embryos in assisted reproduction (*in vitro* fertilization; IVF) clinics. They may have promise in cell therapy if immune responses in recipients are minimized in some way, but because human ES cell lines are derived from human embryos, the creation of new ES lines remains controversial.

Cell reprogramming

For decades, cell differentiation in mammals was thought to be irreversible. Then a cloned sheep called Dolly proved that terminally differentiated mammalian cells could be reprogrammed to become unspecialized cells (resembling the pluripotent cells of the early embryo). However, cloning mammals is extremely arduous and technically difficult.

Comparatively simple methods can now be used to re-set the epigenetic marks of a cell. For example, terminally differentiated mammalian cells can often be induced to dedifferentiate or transdifferentiate by providing certain key transcription factors, or by inducing the cells to make them.

The first approach was to use genetic methods. By transferring genes encoding key pluripotency transcription factors into differentiated cells it is possible to reverse differentiation to make **induced pluripotent stem (iPS) cells**. Like ES cells, iPS cells can be directed to differentiate into more specialized cells (**Figure 3**). Because iPS cells may retain some characteristics of their progenitor cells, they are less robust than ES cells.

Human cells have also been reprogrammed by simply adding protein transcription factors or certain low molecular weight chemicals. A simple environmental shock (exposing the cells briefly to a low pH environment) has recently been claimed to induce dedifferentiation of mouse cells. The treated cells appear to become quite unspecialized and to contribute to both the embryo and the placenta in development, but these findings have been controversial and need confirmation.

From a medical perspective, iPS cells have two potentially exciting applications. First, they can be induced to make human cellular models of disease. Animal disease models have been very valuable because they allow the use of invasive studies to understand the molecular basis of human disease. But they are only *models*; they quite often show important differences from humans. Accessible skin cells from a

BOX 9.2 (*continued*)

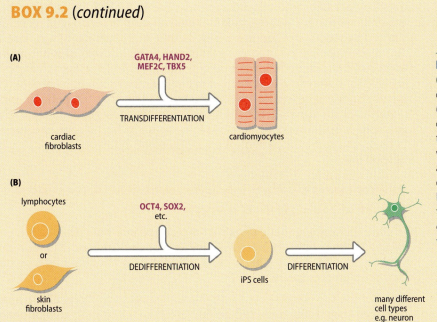

(A)

GATA4, HAND2,
MEF2C, TBX5

TRANSDIFFERENTIATION

cardiac
fibroblasts

cardiomyocytes

(B)

lymphocytes

or

skin
fibroblasts

OCT4, SOX2,
etc.

DEDIFFERENTIATION

iPS cells

DIFFERENTIATION

many different
cell types
e.g. neuron

Figure 3 Cell reprogramming.
(A) An example of direct reprogramming. Here, certain cardiac transcription factors (in pink) induce the transdifferentiation of fibroblasts to myocytes. (B) Reprogramming to give induced pluripotent stem (iPS) cells with the use of transcription factors, such as OCT4 and SOX2, that are important in embryonic development and pluripotency. By providing suitable transcription factors, the resulting iPS cells can be induced to differentiate to give a desired cell type.

patient can now be reprogrammed to become iPS cells that can then be directed to differentiate into cells relevant to the disease process (such as normally inaccessible neurons for a neurodegenerative disorder). The genetically impaired disease cell lines will be useful for drug screening (testing for toxicity, efficacy, and so on) and for studying the molecular basis of disease in human cells.

The second potential application is to provide genetically modified cells for therapeutic purposes. The advantage here is that iPS cells can be made from the cells of a patient, genetically modified, and then returned to the patient without provoking an immune response. Successful environmentally induced reprogramming of human cells may transform the prospects of using dedifferentiated human cells therapeutically. We describe this aspect in more detail in section 9.4.

on transferring genes into the cells of a patient and expressing them to make some product. For the gene delivery method to be effective, it is important to maximize transfection efficiencies for optimal target cells and to get long-lasting high-level expression of the therapeutic genes.

For disorders in which target cells are short-lived, the relevant stem cells should be targeted; however, the problem is that the stem cells might occur at very low frequencies. For blood disorders, happily, it is possible to obtain preparations of bone marrow cells or peripheral blood lymphocytes from patients, grow the cells in culture, and enrich for hematopoietic stem cells. The purified cells can be genetically modified in culture to overcome a genetic defect and then returned to the patient, a type of *ex vivo* gene therapy, as described in the next section.

As we describe below, viral vectors are commonly used to get therapeutic gene constructs into cells at high efficiency, and they often allow high-level expression of the therapeutic transgenes. Some viral vectors are deliberately used because they are adept at getting DNA inserted into chromosomes, which is important when targeting tissues in which cells are short-lived. But the features that make the gene therapy process efficient come with significant safety risks.

One important risk concerns the integration of some therapeutic recombinant viruses into chromosomes—there has been little control over where they will insert into the genomic DNA of patient cells. They might insert by accident into an endogenous gene and block its function, but the greatest danger is the accidental activation of an oncogene, causing tumor formation. An additional risk is that the patient might mount a strong immune or inflammatory response to high levels of what might appear to be foreign molecules. Components of viral vectors might pose such risks, but even if a perfectly normal therapeutic human gene were inserted and expressed to give a desired protein that the patient completely lacked (through constitutional homozygous gene deletion, for example), an immune response might occur if the protein had never been produced by the patient. We provide some detail on these issues in Section 9.4.

Different ways of delivering therapeutic genetic constructs, and the advantages of *ex vivo* gene therapy

In gene therapy, a genetic construct is inserted into the cells of a patient by using either viral delivery systems or nonviral methods. In general, viral vector systems are much more efficient than nonviral methods but they pose greater safety risks.

Using viruses to transfer DNA into human (or other animal) cells (**transduction**) might be expected to be efficient: over long evolutionary timescales, viruses have mastered the process of infecting cells, inserting their genomes and getting their genes to be expressed. Depending on the type of virus, a virus may have a DNA or RNA genome that may be single-stranded or double-stranded, but to be useful for ferrying genes into cells, a virus vector is used that is a modified double-stranded DNA copy of the viral genome (making it easy for a therapeutic DNA to be joined to it to form a recombinant DNA).

Viral vectors for use in gene therapy have been engineered to lack most, and quite often all, of the coding capacity of the original viral genome. The idea is that the recombinant DNA (virus vector plus therapeutic DNA) can nevertheless get packaged into a viral protein coat to make a recombinant virus that is still efficient at infecting cells. In some cases the recombinant viral DNA can integrate into the nuclear genome of a cell and so provide the means for long-lasting therapeutic gene expression, but the integration of vectors poses safety risks. When using other types of virus vector, the recombinant viral DNA does not integrate into the host cell genome. Instead, it remains as an extrachromosomal **episome** in cells.

The nonviral transfer of DNA, RNA, or oligonucleotides into human or other animal cells (**transfection**) is much less efficient than viral transduction; the overall amount of transgene expression is therefore more limited. The transfection procedures also do not result in an appreciable integration of DNA into the genome of the cell. As a result, transfection has the advantage of greater safety in therapeutic applications, but with reduced efficiency. In addition, transfection methods do not have the same size constraints for the packaged nucleic acid that applies to virus vectors; they can be used to ferry very large nucleic acids.

In vivo and *ex vivo* gene therapy

Some types of gene therapy procedure occur *in vivo*: the transfer of the therapeutic constructs is carried out *in situ* within the patient. Often the therapeutic construct is injected directly into an organ (such as muscle, eye, or brain). It may in some cases be introduced indirectly to target cells. For example, genes that are important in vision have been successfully delivered into the eyes of patients with hereditary loss of vision with quite good outcomes, as detailed in Section 9.4. We describe below how

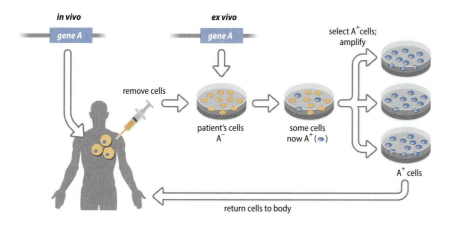

Figure 9.17 *Ex vivo* and *in vivo* gene therapy. In *ex vivo* gene therapy, cells are removed from the patient and genetically modified in some way in the laboratory (in this case we illustrate a gene augmentation procedure in which a therapeutic transgene, *A*, is expressed to make a gene product, A, that is lacking in the cells of the patient). The modified cells are selected, amplified in culture and returned to the patient. The procedure allows detailed checking of genetically modified cells to ensure that they have the correct genetic modification before they are returned to the patient. For many tissues, this is not possible and the cells must be modified within the patient's body (*in vivo* gene therapy).

certain viruses are known to infect human cells of a particular type and how that property has also been exploited to increase the efficiency of delivering therapeutic genes to the desired target cells.

Because there is no way of selecting and amplifying cells that have taken up and (sometimes) expressed the genetic construct, the success of *in vivo* gene therapy is crucially dependent on the general efficiency of gene transfer and, where appropriate, expression in the correct tissue.

Ex vivo gene therapy means removing cells from a patient, culturing and genetically modifying them *in vitro*, and then returning suitably modified cells back to the patient (**Figure 9.17**). Because the cells of the patient are genetically modified in the laboratory, they have the enormous advantage that the cells can be analyzed at length to identify those in which the intended genetic modification has been successful. The correctly modified cells can then be amplified in culture and injected back into the patient.

In practice, *ex vivo* gene therapy has been directed at certain disorders— mostly blood disorders but also some storage disorders—in which the genetically modified cells are bone marrow cells that have been taken from the patient and then treated in such a way so as to enrich for hematopoietic stem cells. As described in Section 9.4, this procedure has been at the core of a series of successful gene therapies.

Nonviral systems for delivering therapeutic genetic constructs: safety at the expense of efficiency

Interest in nonviral vector delivery systems has mostly been propelled by safety concerns over the use of viral vectors. The nonviral vector systems are certainly safer—they do not integrate into chromosomes and they are not very immunogenic; but they typically exhibit poor transfer rates and low-level transgene expression.

The therapeutic gene is typically carried in a plasmid vector, but transport of plasmid DNAs into the nucleus of nondividing cells is normally very inefficient (the plasmid DNA often cannot enter nuclear membrane pores). Various tricks can be used to help get the plasmids into the nucleus (such as conjugating specific DNA or protein sequences that are known to facilitate nuclear entry, or compacting the DNA to a small enough size to pass through the nuclear pores). Because the transfected DNA cannot be stably integrated into the chromosomes of the host cell, nonviral methods of therapeutic gene delivery are more suited to delivery into tissues such as muscle which do not regularly proliferate, and in which the injected DNA may continue to be expressed for several months.

Different delivery systems can be used. In some cases, naked DNA has been injected directly with a syringe and needle into a target tissue such as skeletal muscle. More efficient transfer is often obtained by using **liposomes**, synthetic vesicles that form spontaneously when certain lipids are mixed in aqueous solution. Phospholipids, for example, can form bilayered vesicles that mimic the structure of biological membranes, with the hydrophilic phosphate groups on the outside and the hydrophobic lipid tails in the inside.

Cationic liposomes are the most common type of liposome used to deliver therapeutic genes into cells. The lipid coating allows the DNA to survive *in vivo*, bind to cells and be endocytosed into the cells (**Figure 9.18**).

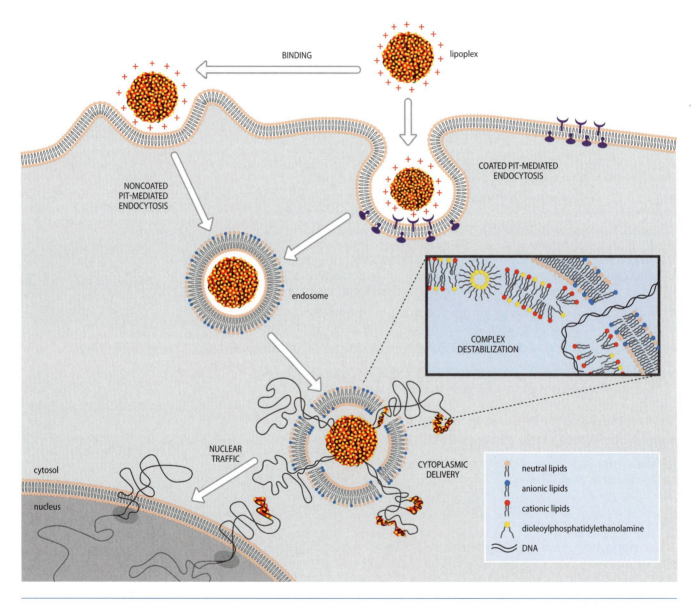

Figure 9.18 Cationic liposomes as vectors for gene delivery into mammalian cells. The gene to be transferred is complexed with cationic liposomes to form *lipoplexes* that can interact with cell membranes. The lipoplexes are taken up by cells through different endocytosis pathways in which the cell membrane invaginates to form a pit. Large lipoplexes are taken up by pits coated with clathrin complexes (top center); small lipoplexes are taken up by noncoated pits (top left). In either case the lipoplexes become trapped in endosomes that are targeted for destruction by lysosomes, where the DNA would be degraded if it were unable to escape. The inclusion within the liposomes of certain helper lipids—often electrically neutral lipids—helps to destabilize the endosomal membranes, causing the passenger DNA to escape to the cytoplasm. For the DNA to be transcribed it must pass to the nucleus. In dividing cells, the breakdown of the nuclear envelope during mitosis allows the DNA to gain access to the nucleus, but in nondividing cells the precise mechanism of entry into the nucleus is unclear. (From Simões S et al. [2005] *Expert Opin Drug Deliv* 2:237–254; PMID 16296751. With permission from Informa Healthcare.)

Unlike viral vectors, the DNA–lipid complexes are easy to prepare and there is no limit to the size of DNA that can be transferred. However, the efficiency of gene transfer is low, with comparatively weak transgene expression. Because the introduced DNA is not designed to integrate into chromosomal DNA, transgene expression may not be long-lasting.

Another method uses compacted DNA nanoparticles. Because of its phosphate groups, DNA is a polyanion. Polycations bind strongly to DNA and so cause the DNA to be significantly compacted. To form DNA nanoparticles, DNA is complexed with a PEG-substituted poly-L-lysine known as PEG-CK30 (because it contains 30 lysine residues and an N-terminal cysteine to which the PEG is covalently bound). Within this complex, the DNA forms a very condensed structure. Because of their much reduced size, compacted DNA nanoparticles are comparatively efficient at transferring genes to dividing and nondividing cells and have a plasmid capacity of at least 20 kb.

Viral delivery of therapeutic gene constructs: relatively high efficiency but safety concerns

Viruses have a DNA or RNA genome packaged within an outer protein coat (capsid). They normally attach to suitable host cells by recognizing and binding to specific receptor proteins on the host cell surface. Some viruses infect a broad range of human cell types and are said to have a broad **tropism**. Other viruses have a narrow tropism: they bind to receptors expressed by only a few cell types. For example, herpes viruses are tropic for cells of the central nervous system. The natural tropism of viruses may be retained in vectors or genetically modified in some way, so as to target a particular tissue.

Enveloped viruses have the capsid enclosed by a lipid bilayer containing viral proteins. Some of them enter cells by fusing with the host plasma membrane to release their genome and capsid proteins into the cytosol. Other enveloped viruses first bind to cell surface receptors and trigger receptor-mediated endocytosis, fusion-based transfer, or endocytosis-based transfer.

For an introduced transgene to be expressed, it needs to be ferried to the nucleus. Some viruses can gain access to the nucleus only after the nuclear envelope has dissolved during mitosis. They are limited to infecting dividing cells. Other viruses have devised ways to transfer their genomes efficiently through nuclear membrane pores, so that both dividing and nondividing cells can be infected.

To allow the easy insertion of a therapeutic gene construct into a viral vector, the vector is in a double-stranded DNA form. Some vectors used in gene therapy are based on **retroviruses**, single-stranded RNA viruses that can be converted into complementary DNA. After infecting cells, retroviruses encode a reverse transcriptase to make a cDNA copy of their RNA genome. The resulting single-stranded DNA is used to make a double-stranded DNA that gets inserted into a host cell chromosome.

Integrating and non-integrating viral vectors

Integrating vectors allow therapeutic genes to be inserted into chromosomes of cells and to be passed on to any descendant cells (an important advantage if the target cells are blood cells or other cell types that have a high rate of cell turnover). They are typically based on retroviruses, which are adept at getting genes into chromosomes. The vector is made by isolating viral replicative forms that consist of double-stranded DNA, and genetically modifying them in various ways.

(A)

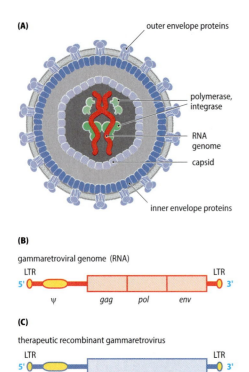

outer envelope proteins

polymerase, integrase

RNA genome

capsid

inner envelope proteins

(B)

gammaretroviral genome (RNA)

LTR   LTR
5'   3'

ψ   gag   pol   env

(C)

therapeutic recombinant gammaretrovirus

LTR   LTR
5'   3'

ψ   therapeutic gene

Figure 9.19 Structure of a gammaretrovirus, its genome, and a therapeutic recombinant gammaretrovirus. (A) Structure of a gammaretrovirus. These are enveloped viruses with an outer lipid bilayer to which are attached inner (matrix) proteins plus outer proteins that serve to bind to receptors on host cell surfaces. The genome, consisting of two identical RNA molecules, is enclosed within the capsid along with key proteins (reverse transcriptase, integrase). (B) The gammaretroviral genome is a single-stranded RNA that contains three transcription units: *gag* (makes internal proteins), *pol* (makes reverse transcriptase and some other proteins), and *env* (makes viral envelope proteins). Additional regulatory sequences (shown in yellow) include a ψ (psi) sequence that needs to be recognized by viral proteins for assembly of the RNA into a virus particle, and promoter/enhancer elements in the long terminal repeats (LTR). (C) Vectors based on gammaretroviruses start with a double-stranded DNA form of the retrovirus (replicative form) that is then genetically engineered to delete the *gag*, *pol*, and *env* transcription units and replace them by the therapeutic gene. The resulting *vector construct*, a therapeutic recombinant retrovirus, is transfected into special vector producing cells that have been genetically modified to contain *gag*, *pol*, and *env* sequences in their chromosomal DNA. Vector construct RNA transcripts are packaged into viral protein coats using Gag, Pol, and Env proteins supplied by this cell. Recombinant viral genomes are packaged into infective (but replication-deficient) virus particles that can be recovered and purified. (A, Adapted from ViralZone: http://viralzone.expasy.org/. With permission from the Swiss Institute of Bioinformatics.)

Initially, integrating vectors were based on gammaretroviruses (formerly called oncoretroviruses), a class of retroviruses with simple genomes (**Figure 9.19**). As described below, however, there have been important safety issues with gammaretrovirus vectors. As a result, modern clinical trials use safer integrating vectors, often based on a class of retroviruses known as lentiviruses (notably HIV), which have more complex genomes. Lentiviral vectors also have the ability to target both nondividing and dividing cells.

Non-integrating vectors are traditionally based on DNA viruses, and they can be especially useful when the object is to get high-level expression in nondividing target cells, such as muscle. Vectors based on adenovirus have been popular because they can permit very high levels of gene expression, but here, too, there have been safety issues (which relate to their immunogenicity). Safer vectors based on adeno-associated virus (AAV) have subsequently been widely used. See **Table 9.7** for properties of six major classes of viral vectors used in gene therapy.

The importance of disease models for testing potential therapies in humans

Cellular disease models can be very helpful in understanding the molecular basis of disease, and they can be used in drug screening and drug toxicity assays. Recent advances in stem cell technology have allowed the production of a wide range of human cellular disease models. Readily accessible blood or skin cells from a patient can be genetically reprogrammed so that they are converted to some desired cell type that is principally involved in the pathology, such as normally inaccessible neurons. We cover the relevant technology—induced pluripotent stem cells—in Section 9.4.

To test novel therapeutic approaches, a robust whole-animal model of disease is necessary. Some animal models of disease, such as the *mdx* mouse model of muscular dystrophy, originated by spontaneous mutation, but the vast majority are artificially generated by genetic manipulation. Primate models might be expected to be the most faithful disease models,

VIRUS CLASS	VIRAL GENOME	CLONING CAPACITY	INTERACTION WITH HOST GENOME	TARGET CELLS	TRANSGENE EXPRESSION	VECTOR YIELD[b]; OTHER COMMENTS
Gammaretroviruses (=oncoretroviruses)	ssRNA; ~8–10 kb	7–8 kb	integrating	dividing cells only	long-lasting	moderate vector yield; risk of activation of cellular oncogene
Lentiviruses (notably HIV)	ssRNA; ~9 kb	Up to 8 kb	integrating	dividing and nondividing cells; tropism varies	long-lasting and high-level expression	high vector yield; risk of oncogene activation
Adenoviruses	dsDNA; 38–39 kb	often 7.5 kb but up to 34 kb	non-integrating	dividing and nondividing cells	transient but high-level expression	high vector yield; imunogenicity can be a major problem
Adeno-associated viruses (AAVs)	ssDNA; 5 kb	<4.5 kb	mostly non-integrating[a]	dividing and nondividing cells; individual strains can be selectively tropic	high-level expression in medium to long-term (year)	high vector yield; small cloning capacity but immunogenicity is less significant than for adenovirus
Herpes simplex virus	dsDNA; 120–200 kb	>30 kb	non-integrating	central nervous system	potential for long-lasting expression	able to establish lifelong latent infections

Table 9.7 Six major classes of viral vectors used in gene therapy protocols. Abbreviations: ss, single-stranded; ds, double-stranded. [a]Recombinant AAVs very occasionally integrate but are mostly episomal. [b]High vector yield, 10^{12} transducing units/ml; moderate vector yield, 10^{10} transducing units/ml.

but for decades the preferred disease models have been rodent models, notably mice. There are several reasons: rodents breed quickly and prolifically; they are reasonably closely related to humans, sharing 99% of our genes; maintaining rodent colonies is not too expensive; and there are fewer ethical concerns than with primate models. An additional compelling reason is that for decades certain important genetic manipulation technologies have effectively been available in mice only.

The vast majority of rodent disease models have been created by genetically modifying the germ line, in which foreign DNA is typically engineered into the chromosomal DNA of germ-line cells. One way is to make a **transgenic animal** disease model by inserting a transgene (= any foreign DNA) into the zygote. This approach can be used in a wide range of different animals.

A second, powerful, technology called **gene targeting** relies on first genetically modifying embryonic stem (ES) cells in culture. The modified ES cells are then transferred into the early embryo in a way that can give rise to an animal with genetically modified cells, including modified germ-line cells. Only a few ES cell strains are suitable for the gene targeting procedure, and certain mouse ES cell lines have been particularly amenable (which explains why mouse disease models are so prevalent). The technology is so sophisticated that we can, in principle, make any desired change to the genome sequence of a mouse—even substituting a single nucleotide—at virtually any position we choose. See Box 9.3 for the salient details.

Inevitably (because it is simpler to do so), most of the artificially created disease models are intended to replicate monogenic disorders. Some good disease models have been produced, and they have been very helpful in allowing us to gain insights into the molecular basis of human diseases, and in testing gene therapies and other new treatments.

When rodent disease models can be inadequate

Rodent models are generally extremely valuable, but are limited in some ways. Mice are small and are less well-suited than larger mammals to physiological analyses. Larger animal disease models including dog, pig, and sheep models have been constructed for some disorders.

Because of species differences, rodent models may quite often fail to replicate some aspects of the human phenotype that they were intended to

BOX 9.3 Two popular ways of making mouse disease models.

Transgenesis through pronuclear microinjection

One important route for making transgenic mice (or other transgenic animals) is to inject a transgene into the zygote so that the exogenous DNA gets into the genome of the zygote. There is usually no control over where the transgene integrates. The resulting animal will have the transgene in all cells and can transmit it to future generations (**Figure 1**).

This method is often used for modeling dominantly inherited disease due to gain of function or over-expression. In the former case, for example, the transgene might often be a mutant human cDNA with an attached promoter sequence to drive expression of the mutant protein in the same cells as those in which it is expressed in humans. Larger transgenes are possible, too, and have sometimes included artificial human chromosomes.

Gene targeting in embryonic stem cells

Another popular way of getting foreign DNA into the germ line begins with cultured mouse **embryonic stem (ES) cells** (which are derived from pluripotent cells of the early mouse embryo). ES cells are immortal and can be used to give rise to all cells of the organism, including gametes. A selected mouse ES cell line is genetically modified in culture and then transferred into an isolated mouse blastocyst. The genetically modified blastocyst is implanted into a mouse, which is bred to obtain genetically modified mice.

Genetic modification of an ES cell line in culture has the big advantage that a very precise change—sometimes just a specific single nucleotide change—can be made to order within any individual gene or locus of interest in intact ES cells (**gene targeting**). First, a mouse plasmid DNA clone that contains a sequence from the desired target gene is genetically modified in the laboratory so that it contains a desired sequence change. The mutated plasmid is then linearized and transfected into cultured mouse ES cells. In a very small percentage of cells, homologous recombination between the introduced plasmid and the corresponding endogenous mouse gene sequence allows swapping of the mutated sequence for the normal sequence. The few genetically modified ES cells can be selected by using a metabolic and drug selection scheme. They are then placed in an isolated mouse blastocyst, which is then implanted into a foster mother to introduce the mutation into the germ line (**Figure 2**).

The gene targeting approach to germ-line modification is extremely powerful and is widely used to make loss-of-function mutations to inactivate a gene (these mutations are known as **gene knockouts**, and mice containing them are called knockout mice). Homozygous loss-of-function

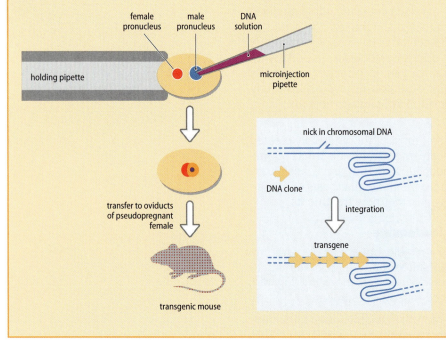

female pronucleus
male pronucleus
DNA solution

holding pipette

microinjection pipette

transfer to oviducts of pseudopregnant female

transgenic mouse

nick in chromosomal DNA

DNA clone

integration

transgene

Figure 1 Construction of transgenic mice by pronuclear microinjection.
A fine-pointed microinjection pipette is used to pierce first the oocyte and then the male pronucleus (which is bigger than the female pronucleus), delivering an aqueous solution of a desired DNA clone. The introduced DNA integrates at a *nick* (single-stranded DNA break) that has occurred randomly in the chromosomal DNA. The integrated transgene usually consists of multiple copies of the DNA clone. Surviving oocytes are reimplanted into the oviducts of foster females. DNA analysis of tail biopsies from resulting newborn mice checks for the presence of the desired DNA sequence.

BOX 9.3 (continued)

mouse mutants are often used to model human recessive disorders, but the method also delivers heterozygous mutants that might show phenotypes too. If the homozygous condition is lethal, the mutation can be maintained in heterozygotes, and the mutant strain can be stored for decades by freezing cells in liquid nitrogen. Variant gene targeting methods can be used to make subchromosomal duplications and deletions, translocations, and so on (*chromosome engineering*).

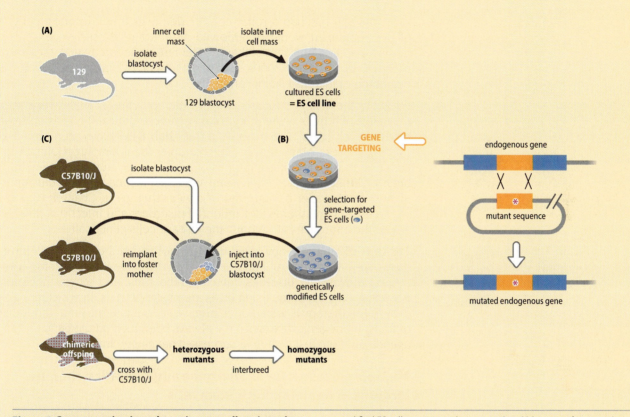

Figure 2 Gene targeting in embryonic stem cells to introduce mutations into the mouse germ line. (A) Embryonic stem (ES) cell lines are made by excising blastocysts from the oviducts of a suitable mouse strain. Cells from the inner cell mass are cultured to eventually give an ES cell line. (B) An ES cell line can be genetically modified in culture by transfecting a linearized plasmid containing a DNA sequence (orange box) that is identical to part of the endogenous target gene, except for a genetically engineered desired mutation (magenta asterisk). Double recombination (X) allows the desired mutation to be introduced into the endogenous gene. (C) The modified ES cells are injected into an isolated blastocyst from another mouse strain with a different coat color, and the blastocyst containing modified ES cells is implanted into a foster mother of the same strain. Subsequent development of the introduced blastocyst can generate chimeric offspring that can be readily identified because they have differently colored coat patches. Backcrossing of chimeras can produce heterozygous mutants (if the genetically modified ES cells have contributed to the germ line); subsequent interbreeding generates homozygous mutants.

mimic. In some cases, they are simply inadequate to the task. Disorders such as autism, schizophrenia, and Alzheimer disease cannot be fully replicated in mice (which lack the complex cognitive and social abilities of primates). Many neuroactive drugs have shown early promise in mice but failed in human trials.

As a result of these difficulties, and because of the recent emergence of what promises to be a transformative technology, there has been renewed interest in making primate disease models. This transformative technology is a new form of *genome editing* using the CRISPR-Cas system, which should offer rapid ways of carrying out gene targeting in a wide range of animals. Because it also offers interesting therapeutic potential we consider this method in the next section.

9.4 Gene Therapy for Inherited Disorders and Infectious Disease: Practice and Future Directions

Gene therapy has had a roller-coaster ride over three decades; periods of over-optimism would be followed by bouts of excessive pessimism in response to significant setbacks. The first undoubted successes were reported in the early 2000s, and the number of successful reports is beginning to increase significantly.

By 2012, the Wiley database of gene therapy clinical trials worldwide (available at http:// www.wiley.co.uk/genmed/clinical/) had listed close to 2000 such trials. The majority (more than 64%) have been aimed at treating cancers. Many cancer gene therapies focus simply on killing cancer cells and so use different types of gene therapy approach; they have been of limited clinical value, and we consider the general difficulties against the broader background of cancer therapy in Chapter 10. Here we focus on gene therapy for inherited disorders and infectious disease, where the approach is to modify the disease cells genetically.

The 36% of gene therapy trials that are not focused on treating cancer have been split almost equally between monogenic disorders, complex cardiovascular diseases, infectious diseases, and other categories. However, only 3% of the listed trials are phase III trials, in which the efficacy of the therapy is tested on a large scale. Despite the limited number of trials, monogenic diseases have always been high on the gene therapy agenda, and the first definitive successes have been in that area.

Multiple successes for *ex vivo* gene augmentation therapy targeted at hematopoietic stem cells

Successful *ex vivo* gene therapy trials have been carried out for various blood disorders and some storage disorders by targeting bone marrow cells or peripheral blood lymphocytes that have been enriched for hematopoietic stem cells. Our blood cells are short-lived and need to be replaced by new cells that are derived from self-renewing hematopoietic stem cells. These cells, which are found mostly in the bone marrow (and to a smaller extent in peripheral blood), give rise to all of the many different types of blood cell and also some tissue cells that have immune functions (**Figure 9.20**).

For some of the disorders treated in this way, alternative treatments have sometimes been used, but they are either very expensive or very risky (see below). For some blood disorders, treatment with purified gene product (such as recombinant proteins) is an option, but it is extremely expensive. Bone marrow transplantation has occasionally been used.

In **allogeneic** bone marrow transplantation the donor is often a family relative, such as a sibling, but complete HLA matching of donor and patient is rare (even for siblings there is only a 1 in 4 chance) and sometimes transplantation is attempted using partial HLA matching between donor and recipient. That may result in a severe *graft-versus-host disease*, in which immune system cells originating from the donor bone marrow interpret the cells of the patient as being foreign and mount a strong immune response against them. As a result, the procedure carries a 10–15% mortality risk that increases to 35% if the recipient has previously received irradiation treatment (in an attempt to kill many of the original hematopoietic stem cells, allowing the transplanted stem cells to expand and become the dominant type).

The advantage of *ex vivo* gene therapy is that it is much less expensive than using purified proteins and is much less risky than bone marrow transplantation because the genetically modified transplanted cells derive originally from the patient (**autologous** transplantation).

Safety issues in gammaretroviral integration

The first gene therapy successes came in treating severe immunodeficiencies. In severe combined immunodeficiency (SCID) the functions of both B and T lymphocytes are defective. Affected individuals have virtually no functioning immune system and are extremely vulnerable to infectious disease.

The most common form of SCID is X-linked; inactivating mutations in the *IL2RG* gene means a lack of the γ_C (common gamma) subunit for multiple

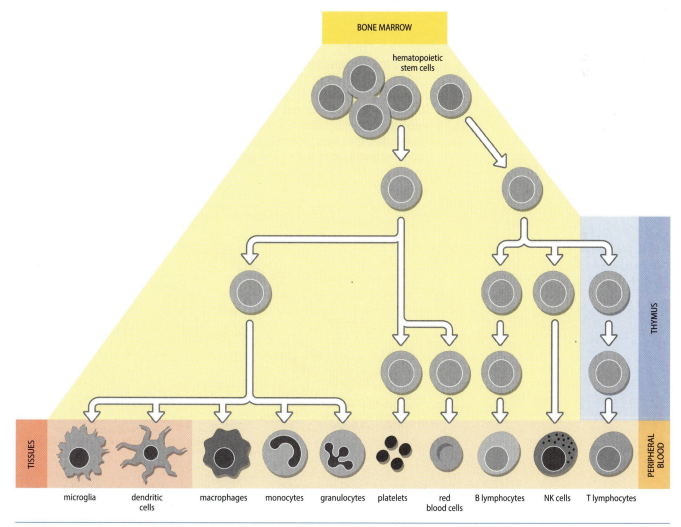

Figure 9.20 All blood cells and some tissue immune system cells originate from hematopoietic stem cells. All differentiated blood cells have limited life spans, and there is a continuous cycle of cell death and cell replacement. The replacement blood cells are derived from hematopoietic stem cells that are particularly concentrated in the bone marrow.

Hematopoietic stem cells also give rise to some tissue cells, including tissue macrophages (such as microglia, the resident macrophages of the brain and spinal cord) and dendritic cells (a class of immune system cells that work in antigen presentation in varied tissues). NK cells, natural killer cells.

interleukin receptors, including interleukin receptor 2. (Lymphocytes use interleukins as **cytokines** or chemical messengers that help in intercellular signaling, in this case between different types of lymphocyte and other immune system cells; lack of the γ_c cytokine receptor subunit has devastating effects on lymphocyte and immune system function.) Another common form of SCID is due to adenosine deaminase (ADA) deficiency; the resulting buildup of toxic purine metabolites kills T cells. B-cell function is also impaired because B cells are normally regulated by certain types of regulatory T cell.

The first SCID gene therapy trials involved *ex vivo* gammaretroviral transfer of *IL2RG* or *ADA* coding sequences into autologous patient cells. To aid the chances of success, bone marrow cells from the patient were further enriched for hematopoietic stem cells by selecting for cells expressing the CD34 surface antigen, a marker of hematopoietic stem cells (**Figure 9.21**). By 2008, 17 out of 20 patients with X-linked SCID and 11 out of 11 patients with ADA-deficient SCID had been successfully treated and retained a functional immune system (for more than 9 years after treatment in the earliest patients).

Although the use of integrating retrovirus vectors was beneficial in terms of efficiency, the chromosomal insertion of the transgenes was unsafe and led to the development of leukemia in several patients (**Box 9.4**). The same kind of approach has been successfully applied to some other blood disorders. However, oncogene activation in some other cases also led to leukemia in patients or silencing of the inserted transgenes. In the first major breakthrough in gene therapy for β-thalassemia, when a patient benefited from treatment, the transgene was inserted into the *HMGA2* gene (which encodes a protein that works in gene regulation). That resulted in the over-expression of a truncated HMGA2 protein; long-term follow-up will be required to confirm safety and efficacy.

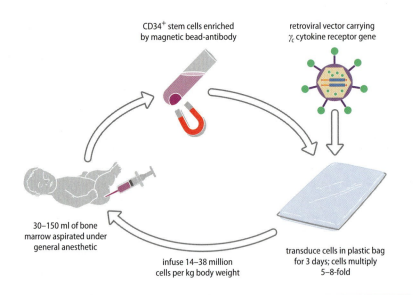

CD34$^+$ stem cells enriched by magnetic bead-antibody

retroviral vector carrying γ_c cytokine receptor gene

30–150 ml of bone marrow aspirated under general anesthetic

infuse 14–38 million cells per kg body weight

transduce cells in plastic bag for 3 days; cells multiply 5–8-fold

Figure 9.21 The first successful gene therapy: *ex vivo* gene therapy for X-linked severe combined immunodeficiency disease. Bone marrow cells were removed from the patient and antibody affinity was used to enrich for cells expressing the CD34 antigen, a marker of hematopoietic stem cells. To do this, bone marrow cells were mixed with paramagnetic beads coated with a CD34-specific monoclonal antibody; beads containing bound cells were removed with a magnet. The transduced stem cells were expanded in culture before being returned to the patient. For details, see articles with PMID numbers 10784449 and 11961146.

BOX 9.4 Two major safety issues in gene therapy trials.

As described below, the practice of gene therapy has been marred by serious illness and even fatalities in a few participating subjects. (Although tragic, the partial successes provided some hope and benefit to seriously affected patients, and fundamental information for how to improve the gene therapy design.) Two serious safety issues have been highlighted. One applies to integrating vectors—the lack of control over the integration event can lead to altered gene expression and tumor formation. The second safety issue is a general one and concerns the immunogenicity of the therapeutic genetic construct.

Problems with integrating vectors

From an efficiency point of view, integrating retroviral vectors are highly desirable—by allowing therapeutic transgenes to be stably transmitted after cell division they can offer long-lasting transgene expression in populations of cells that need to undergo cell division, such as white blood cells. However, early gene therapy trials for severe combined immunodeficiency (SCID) showed that the use of conventional gammaretroviral vectors (based on the murine Moloney leukemia virus) was not safe. Five of the 19 individuals who had been treated for X-linked SCID as shown in Figure 9.21 went on to develop T-acute lymphoblastoid leukemia.

It has become clear that gammaretroviral vectors have a pronounced tendency to integrate close to transcriptional start sites, and the long terminal repeats carry very powerful promoter and enhancer sequences that can readily activate the expression of neighboring host cell genes. Although the locations of the integration site might have been expected to vary considerably, surprisingly in four of the five patients who developed leukemia there was inappropriate activation of the same gene, the proto-oncogene *LMO2*. (Activation of *LMO2* is now known to promote the self-renewal of thymocytes so that committed T cells accumulate the additional genetic mutations required for leukemia transformation.)

Subsequent clinical trials have used safer, *self-inactivating* retrovirus vectors in which powerful promoter/enhancers in the long terminal repeats are deleted and replaced by more moderate mammalian control sequences without significant effects on recombinant vector performance. Lentivirus vectors have been preferred because they are less disposed to integrate next to transcriptional start sites. But even the modern vectors are not entirely safe, and there is little control over where they integrate. Efforts are currently being made to further reduce safety risks by developing technologies that direct integration to occur only at certain safe sites within the genome (called 'safe harbors'; see the review by Sadelain et al. [2012] under Further Reading).

Problems with immunogenicity

Any introduced therapeutic genetic construct is potentially immunogenic. Nonviral vectors are less of a problem here, and constructs such as synthetic liposomes can be designed to be weakly immunogenic. By contrast, viral vectors can be a very significant problem. Humans often have antibodies against strains of viruses from which gene therapy vectors have been developed. And in the case of non-integrating vectors, gene therapy protocols often involve giving repeated doses (the expression of therapeutic transgenes often declines with time), which can exacerbate immune responses.

This immunogenicity problem was highlighted by the tragic death of Jesse Gelsinger just a few days after receiving recombinant adenoviral particles by intrahepatic injection in a clinical gene therapy trial for ornithine transcarbamylase deficiency. The procedure provoked a massive inflammatory reaction that resulted in multiple organ failure. Since then, adeno-associated virus vectors have often been used in preference to adenovirus vectors because the immunogenicity concerns are less for these vectors.

Increased safety profiles using lentiviral vectors

More recently, *ex vivo* gene therapy trials have largely used self-inactivating lentivirus vectors. They have the advantage of long-lasting high-level expression but are much safer. Abnormal activation of an endogenous gene is very rarely triggered when lentivirus vectors integrate into chromosomes (they do not have quite the same tendency as gammaretroviruses to insert close to transcriptional start sites; and in the self-inactivating lentiviral vectors the very strong viral promoter and enhancer sequences in the long terminal repeats have been deleted and replaced by more appropriate mammalian promoter sequences).

The first successful gene therapy using a lentivirus vector was for a lipid storage disorder that primarily affects the brain. X-linked adrenoleukodystrophy (OMIM 300100) is a progressive neurodegenerative disorder that is accompanied by adrenal insufficiency. Although the adrenal insufficiency is treatable, there is no effective treatment for the neurodegeneration. Affected boys have inactivating mutations in the *ABCD1* gene and usually die in adolescence.

ABCD1 normally makes a peroxisomal membrane protein, ALDP, that is important for the natural degradation of very-long-chain fatty acids in the peroxisomes; a deficiency of this protein results in the harmful buildup of very-long-chain fatty acids. The result is a progressive loss of the lipid-rich myelin sheath of nerve cells (and axon degeneration) and an impaired ability to convert cholesterol into steroids, so that the adrenal glands fail to make steroid hormones.

Ex vivo gene therapy was designed to halt the progression of adrenoleukodystrophy by transducing an autologous cell population enriched in hematopoietic stem cells with a recombinant HIV vector containing an *ABCD1* coding sequence. The transduced stem cells gave rise to myelomonocytic cells (with characteristics of both granulocytes and monocytes) that migrated into the central nervous system to replace diseased microglial cells and relieve the lipid storage problem.

In vivo gene therapy: approaches, barriers, and recent successes

In vivo gene therapy involves the transfer (usually direct) of a genetic construct into post-mitotic disease cells at specific sites in the body (such as muscles, eyes, brain, liver, lung, heart, and joints). Because the intended target cells are nondividing cells, there is no need to insert genes into chromosomes, and so the viral vectors that are used are typically based on non-integrating DNA viruses.

Delivery using adenovirus and adeno-associated virus vectors

Early *in vivo* gene therapy trials often used adenovirus vectors to transfer therapeutic transgenes. These allow high-level expression, and some adenovirus vectors can accept inserts as large as 35 kb (much larger than the vast majority of full-length human cDNA sequences). However, harmful immune and inflammatory responses have sometimes resulted (see Box 9.4). The vectors are non-integrating, and the expression of introduced transgenes is often somewhat transient; short-term and repeated administration would be necessary for sustained expression, but that only exacerbates the immune response.

Adeno-associated viruses (AAVs) are nonpathogenic and are quite unrelated to adenoviruses (their name comes from their natural reliance on simultaneous infection by a helper virus, often an adenovirus). Their most important advantage is that they can permit the robust *in vivo* expression of transgenes in various tissues over several years while exhibiting little immunogenicity and little or no toxicity or inflammation. Multiple different serotypes of AAV have been isolated, and some have a usefully narrow tropism, such as AAV8 (strongly tropic for the liver). A downside is that a maximum of just 4.5 kb of foreign DNA can be inserted into an AAV vector.

Amenability of disorders to *in vivo* gene therapy

Different disorders may be more or less amenable to *in vivo* gene therapy, largely depending on the efficiency of transgene transfer and expression.

That, in turn, partly depends on different types of barrier. Immunological barriers are particularly important when using recombinant virus vectors: as well as posing safety risks, immunological responses can result in transgene silencing (increased host cell cytokine signaling often attenuates the influence of viral promoters).

In addition to immunological barriers, mechanical barriers can also be a major obstacle. Take disorders that primarily affect the lungs, such as cystic fibrosis. Gene delivery to the airways might seem a very attractive option, given that lung epithelial cells interface directly with the environment. But a combination of immunological and mechanical barriers is a formidable obstacle to gene therapy. Lung epithelial cells are locked together by intercellular *tight junctions*, and large numbers of macrophages are on patrol, readily intercepting and destroying viral vectors. And finally, there is a natural layer of mucus on the epithelial surface that becomes thicker in individuals with cystic fibrosis.

Some parts of the body are *immunologically privileged sites* in which immune responses to foreign antigen are much weaker than in most other parts of the body (as a result of blood–tissue barriers or a lack of lymphatics, for example). They include the brain and much of the eyes. Additional advantages of the eyes are their accessibility and also their compactness (compare the need for multiple injections at diverse skeletal muscle sites in disorders such as Duchenne muscular dystrophy).

The liver, too, is a quite accessible organ (via direct injection, injection into the hepatic portal vein, or even injection into a peripheral vein); because it has a primary role in biosynthesis, the liver has become a popular target for gene delivery. A wide range of metabolic disorders are caused by defective synthesis of proteins manufactured in the liver (such as blood clotting factors VIII and IX, which are deficient in hemophilia, and many enzymes in inborn errors of metabolism).

Two recent examples of successful *in vivo* gene therapy

Hemophilia B (OMIM 306900) is an X-linked recessive disorder caused by a deficiency of blood clotting factor IX. The disorder can be treated by protein therapy (using clotting factor concentrates), but at huge cost. Remarkably, a single intravenous injection of a recombinant AAV construct with a factor IX coding sequence could successfully treat patients with hemophilia for more than a year, even though factor IX expression levels were about 10% or less of the normal values, see Nathwani et al. (2011) under Further Reading.

In type 2 Leber congenital amaurosis (OMIM 204100), the principal clinical feature—profound loss of vision—usually presents at birth. In the type 2 form, the blindness results from inactivating mutations in both copies of the *RPE65* gene, causing severe retinal degeneration (*RPE65* encodes a retinal pigment epithelium enzyme). Different *in vivo* gene therapy trials have involved injecting a recombinant AAV construct containing a transgene with the *RPE65* coding sequence into the subretinal space, allowing the transduction of retinal pigment epithelial cells (**Figure 9.22A**). The trials showed the procedure to be both safe and of considerable clinical benefit. In the largest clinical trial, all patients demonstrated increased pupillary response (**Figure 9.22B**) and increased visual field, and a majority of patients demonstrated improved visual acuity.

Complex disease applications: the example of Parkinson disease

In vivo gene therapy has often been applied in gene therapy for cancer, and clinical trials are being carried out for some other complex diseases, such as for Parkinson disease, a common neurodegenerative condition

Figure 9.22 *In vivo* recombinant AAV gene therapy for type 2 Leber congenital amaurosis restores pupillary light response.
(A) Surgical injection into the subretinal space of an AAV2-*RPE65* construct with vector doses ranging from 1.5×10^{10} to 2×10^{11} viral genomes. (B) Results of a 'swinging flashlight' test, in which first the left eye and then the right is exposed to a light stimulus. Normally, both pupils constrict when one eye is exposed to light. Untreated (pre-injection) subjects with type 2 Leber congenital amaurosis show no response to light stimulus in either eye. Three months after treatment by the injection of vector into the right eye, both pupils constrict after light stimulus to the right eye, but neither constricts after light stimulus to the left eye (both pupils continue to dilate as though no light exposure had occurred). (A, From Mingozzi F & High KA [2011] *Nature Rev Genet* 12:316–328; PMID 21499295. With permission from Macmillan Publishers Ltd. B, Courtesy of Jean Bennett, University of Pennsylvania, USA.)

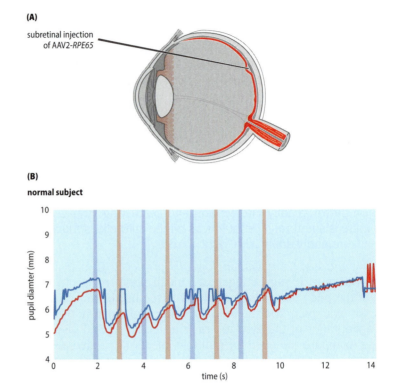

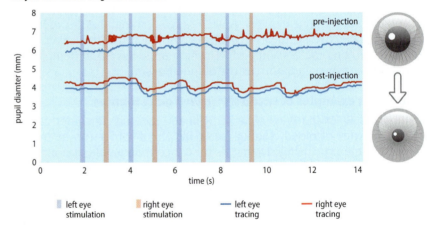

whose hallmark is the loss of neurons of the substantia nigra (the region of the midbrain where dopamine is produced), with consequent muscular rigidity, tremor, and difficulty in initiating movements. Drug treatment (to facilitate dopaminergic neurotransmission) benefits most patients initially, but those with advanced Parkinson disease often develop serious drug-related complications, notably dyskinesia (a difficulty or distortion in making voluntary movements) and motor fluctuations.

Recently, *in vivo* gene therapy trials have used recombinant AAV vectors in bilateral infusion of the glutamic acid decarboxylase gene, *GAD*, directly into neurons of the subthalamic nucleus (which sends excitatory efferent axons to the substantia nigra). The GAD enzyme catalyzes the synthesis of γ-aminobutyric acid, the major inhibitory neurotransmitter in the brain; the rationale is to dampen down the activity of the subthalamic nucleus. A phase II trial indicates that the gene therapy procedure is safe and that it seems to afford some clinical improvement.

RNA targeting therapies: gene silencing by RNA interference and modulation of RNA splicing

Gene-silencing therapy using RNA interference

Different diseases are potentially amenable to treatment based on **gene silencing** (the expression of the gene is severely repressed in some way). In some cases, the disease is due to a gain-of-function mutation or to a dominant-negative effect: the problem is a mutant resident gene that is doing something positively harmful. Here, the strategy must be to selectively inhibit the expression of the mutant gene, without affecting the normal allele too much. There is a similar opportunity for treating infectious diseases, which might be treated by targeting a pathogen-specific gene (or gene product).

Different technologies can be used to achieve gene silencing. Initial attempts used gene-specific antisense RNAs to bind to transcripts of a mutant gene and then selectively block expression; alternatively, modified ribozymes (RNA enzymes) would be designed to cleave a specific RNA transcript. However, large single-stranded RNA is very prone to degradation. The most popular current approach takes advantage of a natural gene-silencing phenomenon, RNA interference (RNAi; see **Box 9.5**).

RNAi therapy is not straightforward: complete gene silencing is difficult to obtain. There can also be the risk of off-target effects, in which very closely related sequences that occur by chance in other genes become collateral targets. A variety of clinical gene trials have been or are being carried out—see **Table 9.8** for some examples. Although the therapeutic potential of RNAi therapies might be high, the technology needs to be refined. To translate therapeutic potential into real clinical benefits, some residual problems need to be overcome, notably the inability to deliver RNA constructs to cells with adequate pharmacokinetics, while avoiding side effects.

Modulation of splicing

One unusual approach to the treatment of disease is to force a disease gene to undergo a specific altered splicing pattern that somehow reduces the normally harmful effect of the pathogenic mutation. That can be done, for example, by designing an antisense oligonucleotide that will bind to a specific splice junction in a target gene. By blocking the splice junction from interacting with the spliceosomal machinery, the antisense oligonucleotide can induce skipping of the exon.

CLINICAL INDICATION	TARGET GENE	RNAi THERAPY TRIALS
Age-related macular degeneration	*VEGFR1* (vascular endothelial growth factor receptor 1)	phases I and II completed
Epidermolytic palmoplantar keratoderma	mouse keratin 9 gene	preclinical on mouse model (PMID 22402445)
Glaucoma	*ADRB2* (β2-adrenergic receptor)	phases I and II completed
Huntington disease	mouse huntingtin gene	preclinical on mouse model (PMID 22939619)
Kidney injury acute renal failure	*TP53* (p53)	phase I completed
Pachyonychia congenita	*KRT6A* (keratin 6A) (N171K mutant)	phase I completed
Hepatitis C	*MIR122* (miR-122 miRNA)	phase II completed
RSV in lung transplant patients	nucleo-capsid gene	phases II and IIb completed

Table 9.8 Examples of noncancer RNAi therapy clinical and preclinical trials. Data on clinical trials abstracted from Davidson & McCray (2011); see under Further Reading. Additional preclinical trials on mouse models are referenced by giving the PMID numbers. RSV, respiratory syncytial virus.

BOX 9.5 Gene silencing by RNA interference.

RNA interference (**RNAi**) is an innate defense mechanism that protects cells against invading viruses and also from excess activity by transposable elements. A small percentage of our resident transposable elements are actively transposing (an evolutionary advantage in providing a source of novel exons and novel regulatory sequences that can be incorporated into genes); if that percentage were allowed to become too great, the genome could be overwhelmed by transposons inserting into essential genes.

The trigger for RNA interference is the presence of double-stranded RNA (which is uncommon in our cells but can be formed by invading viruses and by the association of sense and antisense transcripts from highly repeated transposable elements). Double-stranded RNA is detected and cleaved by a special class of ribonuclease called dicer to give fragments that are 21 bp long but with recessed 5′ ends, known as **short interfering RNA** (**siRNA**). They are recognized by special protein complexes called RNA-induced silencing complexes (RISC) that initiate a pathway leading to the destruction of any RNA transcripts that contain the same nucleotide sequence as the siRNA (**Figure 1**).

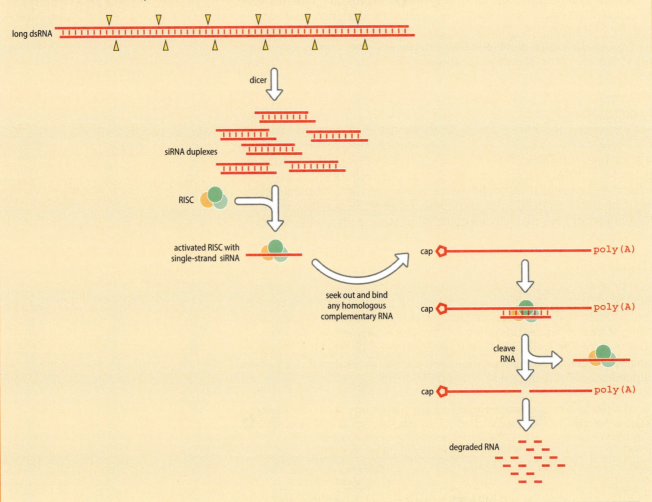

Figure 1 RNA interference. Long double-stranded (ds) RNA is an unusual structure in our cells. It is cleaved by the cytoplasmic ribonuclease dicer at asymmetrically positioned sites on the two strands at intervals of 21 nucleotides (yellow triangles). The resulting short interfering RNA (siRNA) consists of a duplex of two 21-nucleotide sequences with overhangs of two nucleotides at the 3′ ends. The siRNA duplexes are bound by RNA-induced silencing complexes (RISC) that degrade one of the siRNA strands to give an activated RISC complex with a single siRNA strand (the *guide* RNA). The RISC–siRNA complex is now activated and will bind (by RNA–RNA base pairing) to any RNA sequence that is complementary in sequence to the guide RNA, such as a specific viral mRNA sequence as shown. The large cleaved RNA fragments, lacking a protective cap or poly(A) sequence, are vulnerable to attack by cellular exonucleases and are rapidly degraded. Note that long dsRNA in mammalian cell culture results in the indiscriminate destruction of mRNA, so siRNAs need to be 21 nucleotides long.

BOX 9.5 (continued)

(A)

(B)

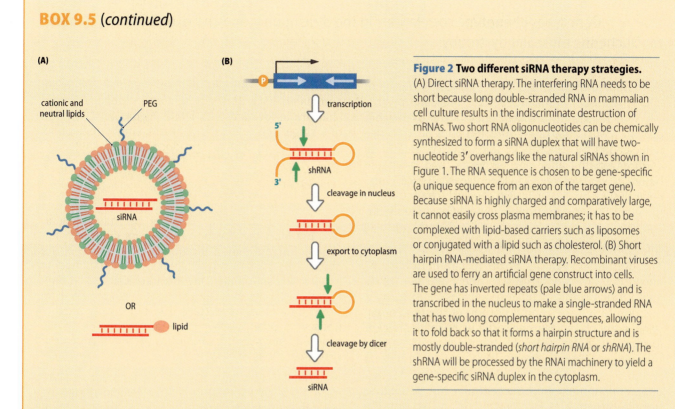

Figure 2 Two different siRNA therapy strategies.
(A) Direct siRNA therapy. The interfering RNA needs to be short because long double-stranded RNA in mammalian cell culture results in the indiscriminate destruction of mRNAs. Two short RNA oligonucleotides can be chemically synthesized to form a siRNA duplex that will have two-nucleotide 3′ overhangs like the natural siRNAs shown in Figure 1. The RNA sequence is chosen to be gene-specific (a unique sequence from an exon of the target gene). Because siRNA is highly charged and comparatively large, it cannot easily cross plasma membranes; it has to be complexed with lipid-based carriers such as liposomes or conjugated with a lipid such as cholesterol. (B) Short hairpin RNA-mediated siRNA therapy. Recombinant viruses are used to ferry an artificial gene construct into cells. The gene has inverted repeats (pale blue arrows) and is transcribed in the nucleus to make a single-stranded RNA that has two long complementary sequences, allowing it to fold back so that it forms a hairpin structure and is mostly double-stranded (*short hairpin RNA* or *shRNA*). The shRNA will be processed by the RNAi machinery to yield a gene-specific siRNA duplex in the cytoplasm.

Using RNAi to selectively silence any cellular gene

The pathway shown in Figure 1 is concerned with natural *gene silencing*; it destroys transcripts from the genes of invading viruses or from transposable elements. It can be exploited experimentally to selectively silence a single predetermined gene of interest within cells. In that case, complementary oligonucleotides are designed to represent a sequence from an exon of the gene to be silenced. The oligonucleotides are used to form a siRNA duplex which is transfected into cells. An alternative is to make a gene construct that can be expressed within cells to produce siRNA.

In either case, RISC complexes can be activated by the gene-specific siRNA and will then induce silencing of the chosen gene.

Gene silencing induced by RNAi is a favorite way of getting information about what a gene does in cultured cells. The silencing is never totally effective, and the procedure is referred to as a **gene knockdown** (rather than a *gene knockout*). Much the same procedure is used in RNAi therapy. Here the object is to silence a positively harmful gene in the cells of patient, and two different strategies are used to get the siRNA into cells of patients (**Figure 2**).

Induced exon skipping might be applicable in cases where the exon contains a harmful mutation and the number of nucleotides in the exon is exactly divisible by three (to maintain the reading frame). Or it might be used to correct a frameshifting deletion. But that would only be appropriate in rare cases when loss of an exon does not have a catastrophic effect. Duchenne muscular dystrophy provides some examples in which some success has been achieved in clinical trials (**Box 9.6**).

Future prospects and new approaches: therapeutic stem cells, cell reprogramming, and genome editing

Despite recent successes, gene therapy is still very much at an experimental stage, and its applicability is going to be limited. The first approval for a clinical gene therapy by an official regulatory authority was given

BOX 9.6 Genotype–phenotype correlations in the dystrophin gene and exon skipping therapy for Duchenne muscular dystrophy.

Duchenne muscular dystrophy (DMD) is a severe and progressive X-linked recessive muscular dystrophy that results from a deficiency of the dystrophin protein (PMID 20301298). Affected boys need to use wheelchairs by 12 years of age, develop additional cardiomyopathy after age 18 years, and often die before 30 years of age. A milder disease, Becker muscular dystrophy (BMD), is also caused by mutations in the giant (2.4 Mb) dystrophin gene; affected individuals have late-onset weakness of the skeletal muscles and cardiomyopathy.

The dystrophin gene is prone to internal deletions because of its very large size (there are numerous long introns with repetitive sequences). Surprisingly, deletion of a large central portion of the dystrophin gene, as much as 1 Mb, can result in the milder Becker form of muscular dystrophy, but deletion of a single nucleotide within an exon in that same 1 Mb region results in severe DMD. Two observations explain that apparent anomaly. First, the central region of the dystrophin protein acts as a flexible linker between the functionally important regions of the protein, which are domains at the N-terminal and C-terminal ends (**Figure 1A**); deletion of quite a large part of the central region reduces the performance of the protein but some functional protein remains. Secondly, internal deletions that maintain the same reading frame are associated with BMD; those that produce a frameshift are associated with DMD (Figure 1B). Deletion of a single central exon, or even a single nucleotide, would result in DMD if it produced a frameshift.

Exon skipping to restore the translational reading frame in DMD therapy

Gene therapy for DMD is beset by many difficulties, not least the question of efficient gene transfer. In gene augmentation therapy, the huge size of the dystrophin coding sequence poses a challenge (a full-length dystrophin cDNA is more than 13 kb long). Certain adenovirus vectors could accommodate an insert of that size (see Table 9.7), but repeated injections with adenovirus vectors are not an attractive prospect because of safety (immunogenicity) concerns (see Box 9.4).

As a result, alternative approaches for treating DMD have been developed. One involves up-regulating the utrophin gene. The utrophin protein is evolutionarily and functionally related to dystrophin and can substitute in part for its function; extra production of utrophin in cells of patients might therefore partly compensate for a loss of dystrophin. A second approach is to induce exon skipping.

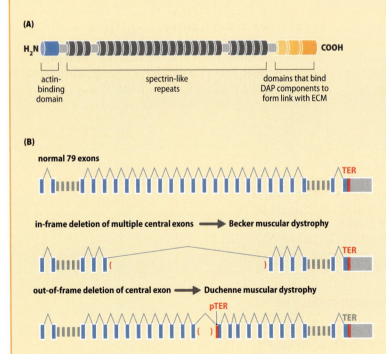

(A)

H₂N ▬▬▬▬▬▬▬▬▬▬▬▬ COOH

actin-binding domain | spectrin-like repeats | domains that bind DAP components to form link with ECM

(B)

normal 79 exons

in-frame deletion of multiple central exons ➝ Becker muscular dystrophy

out-of-frame deletion of central exon ➝ Duchenne muscular dystrophy

Figure 1 Dystrophin structure and different consequences of deletions in the dystrophin gene. (A) Dystrophin structure. The functionally important parts are domains in the N-terminal and C-terminal regions. Dystrophin provides a structural link between actin filaments of the cytoskeleton and, through a membrane-bound dystrophin-associated protein (DAP) complex, the extracellular matrix (ECM). (B) The effect of in-frame and out-of-frame deletions of central exons in the dystrophin gene. Large deletions in the dystrophin gene can remove many of the central exons but are associated with Becker muscular dystrophy (BMD) if the reading frame is not disrupted (in-frame deletion). If a single central exon with a total number of nucleotides not exactly divisible by three is deleted, a frameshift results. The resulting premature termination codon (pTER)—that often occurs within the next exon, as shown here—will trigger *nonsense-mediated decay* of dystrophin mRNA. The ensuing failure to make any protein will result in Duchenne muscular dystrophy (DMD).

BOX 9.6 (*continued*)

Internal deletions are a common cause of pathogenesis in the dystrophin gene, and many deletions that affect central exons result in a frameshift, resulting in severe DMD. Exon skipping therapy seeks to restore the translational reading frame, converting the effect of a central deletion to resemble that of the in-frame deletions associated with milder BMD. **Figure 2** shows the example of how induced skipping of exon 51 could restore the translation reading frame in patients with deletion of exon 50 of the dystrophin gene; skipping of exon 51 can also restore the translational reading frame for several other common deletions, including deletions of exons 52, exons 45–50, exons 48–50, and exons 49–50. In total about 25% of DMD-associated deletions might have the translational reading frame restored by the skipping of exon 51, and that would make treatment available to 15% of DMD patients. Skipping of other exons can further extend this type of treatment.

Local intramuscular injections of an antisense oligonucleotide to induce skipping of exon 51 can restore dystophin production in muscle fibers of patients with the appropriate types of dystrophin exon deletion. Follow-up systemic administration of the oligonucleotide (with access to the circulation via abdominal subcutaneous injections) did not elicit any serious adverse reactions and the procedure seems to work quite well. In 10 of the 12 treated boys (aged 7 to 13 years), new dystrophin expression was observed in between about 60% and 100% of muscle fibers, and clinical benefit was significant as measured by improved walking statistics when compared with controls.

The oligonucleotides that are used in therapeutic exon skipping need to be more stable than conventional oligonucleotides. Serepta Therapeutics uses phosphorodiamidate morpholino oligonucleotides; Prosensa uses 2′-*O*-methyl-modified ribose molecules with a full-length phosphorothioate backbone. Even then, the oligonucleotides have a limited half-life after systemic administration (for example, 29 days in the latter case).

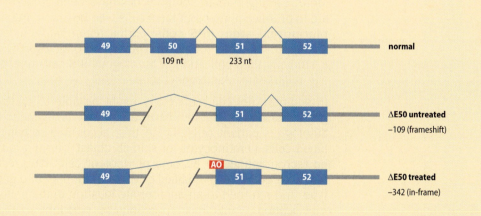

Figure 2 Example of exon skipping therapy for Duchenne muscular dystrophy. Illustration of how skipping exon 51 could restore the translational reading frame in patients who have a deletion of exon 50 (ΔE50) in the dystrophin gene. Deletion of the 109-nucleotide (nt) exon 50 results in the splicing of exon 49 to exon 51. The loss of the 109 nucleotides (not a multiple of three) produces a frameshift and severe DMD. Therapy that causes skipping of exon 51 in ΔE50 patients will result in the splicing of exon 49 to exon 52. That means a loss of 109 + 233 = 342 nucleotides, which is exactly divisible by three; the translational reading frame is maintained even although the coding capacity is reduced (with a loss of 114 amino acids). The exon skipping occurs because a specific antisense oligonucleotide (AO) is used to bind to and blockade the splice junction at the start of exon 51.

in November 2012: clinical treatment of lipoprotein lipase deficiency using Glybera, a recombinant adeno-associated virus, was approved for use within the European Union. Others will follow soon. Although very important for some rare recessive disorders, gene augmentation therapy is unlikely to make a broad impact in disease treatment any time soon, mostly because of the obstacles to gene delivery and expression for many disorders. In addition, as discussed above, therapies based on gene silencing need to meet high requirements—virtually complete gene silencing without off-target effects—before they can make major inroads into clinical application.

Therapeutic embryonic stem cells

Because they have the capacity to make any type of body cell, human pluripotent stem cells offer the prospect of **regenerative medicine**. If efficiently directed down the correct differentiation pathway, they might allow cell augmentation therapy, providing replacement cells to supplement a deficiency of functioning cells in a patient. In principle, a range of complex diseases might be treated, such as diabetes (deficiency in pancreatic beta cells), Parkinson disease (deficiency in dopaminergic neurons), or stroke. Injuries that lead to cell damage might also be treatable, such as spinal cord injuries.

Human ES cell lines were first reported in 1998. Various clinical trials using human ES cells have recently been launched, including the treatment of certain degenerative blindness diseases (such as Stargardt macular dystrophy, and age-related macular degeneration), by directing ES cells to make retinal pigment epithelial cells. The retinal pigment epithelium (RPE) is a single layer of cells beneath the retina that nourishes and protects retinal cells, and the ES-derived RPE cells are injected into the subretinal space.

Therapeutic induced pluripotent stem cells

More recently, human somatic cells have been induced to dedifferentiate to produce pluripotent stem cells (see Box 9.2). Autologous **induced pluripotent stem** (**iPS cells**) permit regenerative medicine with little risk of provoking an immune response. An ongoing clinical trial in Japan is studying the application of autologous iPS cells to regenerate RPE in age-related macular degeneration.

In addition, iPS cell technology might be able to extend the range of genetic disorders that can be treated by *ex vivo* genetic modification of autologous patient cells. To do this, accessible somatic cells (such as blood cells, keratinocytes, or skin fibroblasts) from a patient would first be reprogrammed to become pluripotent cells. The resulting iPS cells would be genetically modified *ex vivo*, differentiated to give a desired cell type, and returned to an appropriate location in the patient. The genetic modification would most probably be performed by *genome editing* (see below).

iPS cells are more acceptable ethically than ES cells. Differentiated cells formed from iPS cells might also be expected to have negligible immunogenicity when transplanted back into a patient, unlike differentiated cells produced from ES cells, which would be expected to induce an immune response (the former are *syngeneic*, genetically identical to the cells of the patient; ES cell lines are necessarily *allogeneic*—they cannot be made from the cells of a patient).

The cost of making iPS cells from individual patients is likely to be prohibitive. Banks of different iPS cell lines are being set up in some countries from donors that are homozygous for frequently occurring HLA haplotypes. According to the degree of HLA matching, an optimal iPS cell line would be selected for use with a patient.

A possible alternative to iPS cell technology is to dedifferentiate cells simply by using environmental shocks. As described in Box 9.2, that has been successfully done for mouse cells by briefly exposing them to a low pH environment, and the method is extremely simple and affordable. If the technology can be easily applied to human cells, it could be transformative.

There remain some formidable challenges. There are practical difficulties in accurately and efficiently directing iPS or ES cells to undergo several successive differentiation steps to get ultimately to a desired differentiated cell type (incomplete reprogramming can occur in iPS cells). And there are safety considerations: as well as immunogenicity concerns, human ES cell lines are genetically unstable, and for both ES and iPS cells there is also the possibility that incomplete differentiation might lead to some pluripotent cells being transmitted to the patient that could form teratomas, which are tumors composed of heterogeneous cell types derived from the different embryonic germ layers.

Therapeutic cell reprogramming by transdifferentation

An interesting alternative is to induce direct reprogramming of cells *in vivo*. For example, consider heart attacks (myocardial infarction). Here, the injury heals by scar formation rather than regeneration of the heart muscle. The heart then pumps less efficiently, leading to high chances of subsequent heart failure. But what if we could regenerate heart muscle by reprogramming resident cells in the heart?

Only about 30% of heart cells are muscle cells (cardiomyocytes); the majority (about 65%) are connective tissue cells (cardiac fibroblasts). To regenerate heart muscle we might seek to reprogram cardiac fibroblasts so that some of them become additional cardiomyocytes. That might be done by injecting recombinant viruses with genes encoding key cardiac transcription factors that naturally work in controlling heart cell identity and/or heart development (see Figure 3A in Box 9.2). When that is done in live mice in which myocardial infarction has previously been induced (by occluding a coronary artery), the genes encoding transcription factors are expressed in dividing cells, notably the cardiac fibroblasts. The ensuing reprogramming makes these cells behave as additional cardiomyocytes that beat and function in other ways like normal cardiomyocytes. And the effect is to improve cardiac function. The method needs refining but might hold considerable promise.

Therapeutic genome editing

Ex vivo therapeutic **genome editing** is another interesting approach that involves changing the sequence at one predetermined gene in cells taken from a patient, selecting cells that have the correct sequence, and re-inserting them into the patient. Essentially, genome editing is used as a form of gene targeting. It depends on the ability to make a single double-stranded DNA break at a *unique predetermined location* in the genome of an intact cell.

A double-strand break will activate cellular DNA repair mechanisms. Recall from Section 4.2 that two DNA repair pathways are dedicated to repairing double-strand DNA breaks. DNA repair mediated by homologous recombination (HR) allows highly accurate repair, but often the alternative nonhomologous end joining (NHEJ) is used and that method is inaccurate—it quickly splices the broken ends and very small mistakes are made (sometimes a nucleotide is deleted or inserted, for example).

The HR pathway can be used to repair a pathogenic mutation; the NHEJ pathway can be used to selectively inactivate just one predetermined gene.

An early genome editing approach uses genetically engineered endonucleases known as **zinc finger nucleases** to make a unique double-strand break. These enzymes are designed to work in pairs, each with two functional components. One is a protein domain that cleaves DNA (it derives from a rare type of restriction endonuclease in which the DNA-recognizing and DNA-cleaving activities are on physically separate domains). The second component is a series of different zinc finger modules. Each zinc finger module naturally recognizes a specific trinucleotide sequence, and a combination of zinc finger modules can be assembled to recognize a specific DNA sequence (**Figure 9.23**).

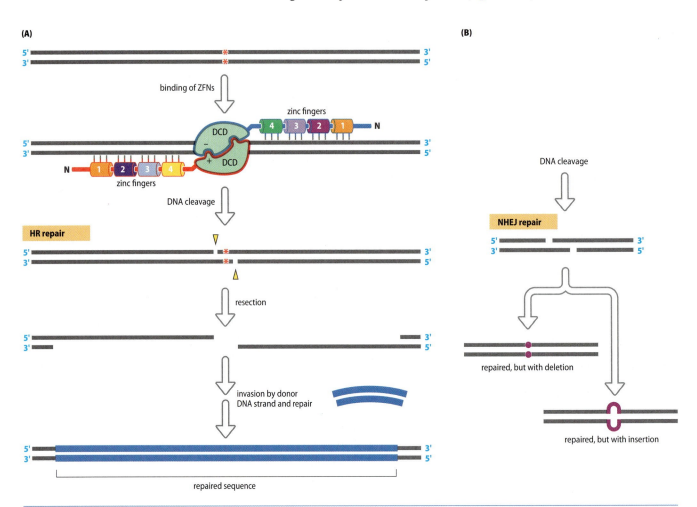

Figure 9.23 Genome editing within intact cells, using artificial zinc finger nucleases. (A) *Therapeutic gene repair.* Zinc finger nucleases consist of a DNA-cleaving domain (DCD) plus a series of *zinc fingers*, protein modules that are often found in transcription factors and that individually recognize a specific trinucleotide sequence in DNA. To cut at a unique location so as to repair a pathogenic mutation (red asterisks), a pair of zinc finger nucleases is used, each with a DNA-cleaving domain plus a series of modular zinc fingers. In this example, the two zinc finger nucleases are shown by red or blue outlines and each has four zinc fingers that are designed to recognize a specific 12-nucleotide sequence close to the mutation site. The two complementary DNA-cleaving domains work together to make a double-strand DNA break that activates the cellular pathways that naturally repair such breaks. In the homologous recombination (HR) DNA repair pathway, the 5′ ends of the double-strand break are first trimmed back (resection), allowing strand invasion by exogenous donor DNA strands from a provided DNA clone that has the correct (normal) DNA sequence for the region spanning the mutation site. These strands are used as templates for synthesizing a new normal DNA sequence to replace the mutant sequence. (B) *Gene inactivation.* Imagine that zinc finger nucleases have been targeted to cleave a *normal* gene, and that this time the inaccurate nonhomologous end joining (NHEJ) repair pathway has been used that can by mistake lead to the deletion or insertion of one or two nucleotides. The result would often be to introduce an inactivating mutation.

BOX 9.7 Curing infectious diseases by rendering cells resistant to infection by viruses.

As previously described in Section 8.3, genetic variation between people causes differences in susceptibility to infectious diseases. And in some extreme cases people seem to be highly resistant to infection by a disease-causing virus. That can happen when a person naturally does not make a host cell receptor that the virus interacts with to infect cells.

Take the human immunodeficiency virus HIV. Its point of initial attack is to infect CD4 helper T cells. By attacking and killing helper T cells (regulatory immune system cells with a major role in helping to protect us against viruses), HIV compromises the immune system; people with AIDS are unable to fight off common infections and they develop various virus-induced cancers.

To latch onto a helper T cell, HIV first binds to a CD4 receptor on a T cell and then interacts with a co-receptor—often the chemokine (C–C motif) receptor 5 (CCR5). Unlike CD4, the CCR5 receptor is not so important in T-cell function, and some normal people have defective CCR5 receptors—a *CCR5* allele with an inactivating 32 bp deletion (*CCR5-Δ32*) is carried by 5–14% of individuals of European descent. Heterozygotes with one *CCR5-Δ32* allele are more

resistant to HIV infection than the normal population, and homozygotes are highly resistant to HIV infection.

CCR5 inactivation as a cure for HIV infection

The idea that HIV-AIDS could be cured by artificially inactivating *CCR5* was promoted by the famous 'Berlin patient' study. First reported in 2009, an HIV patient with acute myeloid leukemia received allogeneic CD34+ peripheral blood stem cells from an HLA-identical donor who had been screened for homozygosity for the *CCR5-Δ32* allele. Four years later, and after discontinuation of anti-retroviral therapy, the patient appears to be free from HIV, indicating that this could be the first complete cure for HIV infection.

The Berlin patient study was clearly an exceptional situation, and various follow-up studies have sought to extend resistance to HIV by inactivating *CCR5* in autologous T cells. Among them are phase I and phase 2 gene therapy trials carried out by Sangamo Biosciences in which zinc finger nucleases are used to target and inactivate the *CCR5* gene (see Figure 9.23 for the general approach).

There are multiple potential applications in therapeutic DNA repair, but there has also been a significant focus on using this method to try to cure infectious diseases. The aim is to inactivate cell receptors needed by viruses to attach to and infect host cells, and clinical trials are underway, notably for treating HIV-AIDS (see **Box 9.7** for the background).

Genome editing with zinc finger nucleases is technically arduous. Instead of using zinc finger nucleases to make the unique double-strand break, a newer genome editing method uses RNA-guided cleavage based on the bacterial CRISPR-Cas antivirus defense system (**Figure 9.24**). The method, which is simpler and seems to allow precise and specific cleavage, is likely to be transformative. Future therapeutic applications can be expected, but an immediate use will be in inactivating genes to make novel animal disease models and to study gene function.

Figure 9.24 Principle of target recognition in genome editing using CRISPR-Cas9 to create a unique double-strand break. The CRISPR-Cas9 prokaryotic antiviral defense system uses guide RNA sequences that have a unique 5′ guide sequence (shown here as thick red or thick mauve lines) that is designed to base pair to a specific target sequence and a common 3′ part (thin orange lines with hairpins). Hybridization to the target allows an associated nuclease, Cas9, to cleave the DNA. The system has been adapted to allow genome editing in a complex genome by making a double-strand break at a unique location within the genome. For the desired region where the double-strand break is to be made, two closely neighboring target sequences are selected (here labeled 'left' and 'right' targets), and guide sequence RNAs are designed to base pair to the two target sequences to allow cleavage by the associated Cas9 nucleases. The resulting double-strand break (yellow darts indicate points of cleavage) is then subject to repair and can be used in principle in the repair or inactivation of a target gene just as in genome editing with zinc finger nucleases (see Figure 9.23).

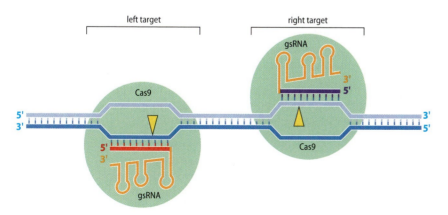

The prospect of germ-line gene therapy to prevent mitochondrial DNA disorders

Mutations in mitochondrial DNA (mtDNA) are a significant cause of human disease: pathogenic mutations are found in at least 1 in 200 of the population, and cause severe multisystem disease in approximately 1 in 10,000 of the population. Pathogenic mtDNA can be maternally inherited, but there are no effective treatments for mitochondrial DNA disorders.

In the clinical management of mtDNA disorders, the emphasis has therefore been on prevention. Preimplantation and prenatal diagnosis (as described in Chapter 11) are well established in clinical genetic practice as a way of selecting unaffected embryos. However, the results can be difficult to interpret for patients with heteroplasmic mtDNA mutations (when there may be variable numbers of mutant and normal mtDNAs in each cell). In addition, an increasingly large group of diseases are recognized to be caused by homoplasmic mtDNA mutations (all the mtDNA molecules are mutant). Here, prevention is not an option—all the offspring would inherit the pathogenic mutation in the maternal egg, and this type of genetic defect can be associated with a very high disease recurrence risk.

An entirely different way of trying to prevent the transmission of homoplasmic mutations is to replace maternal mtDNAs by mtDNAs from an asymptomatic donor. This type of approach has been used in mouse and primate models, with encouraging results. Two recent studies that used slightly different approaches have been carried out in human embryos *in vitro* (**Figure 9.25**). The resulting human embryos appear to be viable *in vitro*, and the degree of mutant DNA carryover is low or undetectable.

The use of these techniques in a clinical context is currently illegal, and the ethics of this type of disease prevention is currently being debated. If the procedure were to be legalized, it would probably be the first example of germ-line gene transfer in humans. We will return to consider the associated ethical considerations in Chapter 11.

Summary

- Treatment for inborn errors of metabolism sometimes involves supplementing a genetic deficiency, but often the treatment is directed at reducing the harmful effects of abnormally elevated metabolites.

- Drug development typically involves screening hydrocarbon-based small molecules for compounds that will bind to medically important protein targets. By binding to a protein, the drug affects its function in some way.

- Genetic variation means that different individuals can respond very differently to drugs; adverse reactions to drugs are very common and cause very many fatalities.

- The pharmacokinetics of a drug describes how it is absorbed, activated (in the case of a prodrug), metabolized, and excreted; pharmacodynamics describes the effect it has on the body.

- Phase I drug metabolism reactions are typically oxidative reactions carried out by monooxygenases; phase II reactions are conjugative reactions in which a transferase enzyme adds a chemical group. The overall effect is to convert lipophilic hydrocarbon drugs into more polar forms that can be excreted more easily.

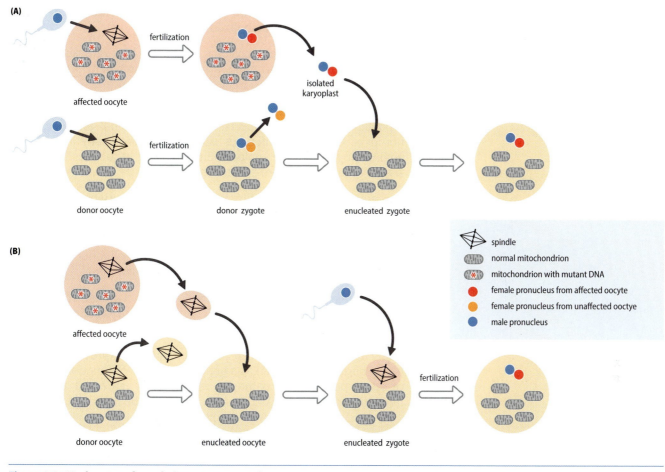

Figure 9.25 Nuclear transfer techniques to prevent the transmission of mtDNA disease. (A) Pronuclear transfer technique. An oocyte with mutant mtDNA is fertilized; the normal *karyoplast* (combined male and female pronuclei) is isolated and then transferred into an enucleated donor zygote with normal mitochondria. (B) Metaphase II spindle transfer technique. The spindle is transferred from an oocyte that has mutant mtDNA into a mitochondrial donor oocyte followed by intracytoplasmic sperm injection fertilization. Both techniques involve monitoring embryo development following transfer. (Adapted from Craven L et al. [2011] *Hum Mol Genet* 20:R168–R174; PMID 21852248. With permission from Oxford University Press.)

- In addition to dealing with artificial drugs, drug-metabolizing enzymes handle unusual exogenous chemicals (xenobiotics) in our diet and environment. They are often highly polymorphic because xenobiotics originating from other organisms are under genetic control and potentially harmful to us.

- The therapeutic window is the range of drug concentrations in which pharmaceutical benefit is achieved without safety risks.

- Poor drug metabolizers are at risk of a drug overdose (the drug does not get cleared quickly; repeated drug doses drive up the concentration). Others are ultrafast metabolizers and may get little therapeutic benefit (the drug is cleared too rapidly).

- Six cytochrome P450 enzymes carry out 90% of phase I drug metabolism. Each handles the metabolism of multiple drugs; conversely, some individual drugs may be metabolized by two or more cytochrome P450 enzymes.

- When a drug is metabolized principally by one enzyme, genetic variation in that enzyme can be mostly responsible for large differences between individuals in the ability to metabolize that drug. For some other drugs, such as warfarin, several different genetic factors determine how the drug is metabolized.

- Therapeutic recombinant proteins are made by expressing cloned human genes in cells to make a human 'recombinant protein' that can be purified and used to treat a genetic deficiency of that protein.

- Therapeutic antibodies are usually designed to bind to harmful gene products to block their effects. Rodent monoclonal antibodies are not ideal (with limited lifetimes after injection into patients); genetic engineering allows the replacement of rodent sequences by human sequences to make more effective antibodies.

- Genetically engineered antibodies with a single variable polypeptide chain can work as intracellular antibodies by binding harmful proteins within cells.

- Gene therapy means inserting a nucleic acid or oligonucleotide into the cells of a patient to counteract or alleviate disease.

- In gene augmentation therapy, diseased cells that are genetically deficient for some product are supplemented by transfecting a cloned gene to make the missing product inside the cells.

- Some therapies target RNA. In gene silencing, the expression of a positively harmful gene (such as a gene with a gain-of-function mutation or one expressed by a pathogen) is selectively repressed, usually by inhibiting the RNA. RNAs can sometimes also be induced to undergo alterative splicing to counteract disease.

- Stem cells are cells that can both renew themselves and give rise to more differentiated (more specialized) cells. Pluripotent embryonic stem cells are artificially cultured cells derived from the very early embryo that can be induced to give rise to virtually any differentiated cell. Somatic stem cells help to replace a limited set of short-lived cells.

- In cell reprogramming, the epigenetic settings of cells are artificially altered to induce changes in gene expression so that the cells acquire the characteristics of a different cell type. Differentiated cells can be induced to dedifferentiate to become unspecialized pluripotent stem cells or to form a different type of somatic cell (transdifferentiation).

- Virus vectors are more efficient but less safe than nonviral vectors in transporting therapeutic genetic constructs into cells.

- Retrovirus vectors can allow a construct to be inserted into the chromosomes of a cell. That is highly desirable when targeting short-lived cells that are replenished by stem cells; if a therapeutic transgene integrates into the stem cell, it will be transmitted by cell division.

- *Ex vivo* gene therapy involves removing cells from a patient, genetically modifying them in culture and returning the genetically modified autologous cells to the patient. It has been used to treat disorders by genetic modification of impure populations of hematopoietic stem cells that give rise to blood cells or some types of tissue immune system cell.

- Animal disease models are usually created by genetically modifying the germ line to mimic a human phenotype. Primate models might be expected to be the best disease models; however, for practical and ethical reasons, rodent models have been popularly used.

- RNA interference is a natural gene-silencing mechanism that has evolved for cellular defense against virus attack or excessive transposon activity. It can be used experimentally to silence any gene of interest within the genome, and is the most actively used method for treating disease by gene-silencing therapy.

- Genome editing involves designing a double-strand break at a unique position in the genomes of intact cells. Applications include gene targeting to make animal models of disease and therapeutic gene inactivation to render cells resistant to infection by viruses.

Questions

Help on answering these questions and a multiple-choice quiz can be found at www.garlandscience.com/ggm-students

1. Treatment of genetic disorders sometimes involves augmentation therapies. What does this mean? Give examples of how augmentation therapy can be deployed at the following levels: (a) somatic phenotype; (b) cellular level; (c) gene level; (d) gene product/metabolite level.

2. In the metabolic pathway below, a major pathway that converts metabolite A to metabolite F in five steps is controlled by enzymes 1 to 5. In addition, trace amounts of metabolite X are normally produced in a minor side pathway.

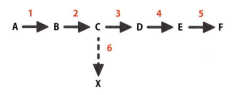

If the gene that produces enzyme 4 were to be homozygously inactivated, how would you expect the concentrations of the different metabolites to be affected? The genetic deficiency in enzyme 4 might produce a disease phenotype that could be treated by augmentation therapy, and it could simultaneously result in a disease phenotype that is not treatable by augmentation therapy. Explain how the two different disease phenotypes could arise.

3. Genes that encode drug-handling enzymes often show quite high levels of polymorphism. Why should that be?

4. For some drugs there is a narrow therapeutic window. With reference to the anticoagulant warfarin, explain what this means.

5. Genetically engineered antibodies have become an important class of therapeutic drug. Describe briefly the different classes of genetically engineered antibody and their applicability to treating genetic disorders.

6. What are the main advantages and disadvantages of using retrovirus and adenovirus vectors in gene therapy?

7. Genetically modified mice are frequently used to model human disorders by introducing mutations into the mouse germ line. This is most often achieved by two methods that involve either gene targeting of mouse embryonic stem cells or inserting a transgene into the zygote. What classes of human genetic disorder are most readily modeled by each of these methods?

Further Reading

General Overviews

Dietz H (2010) New therapeutic approaches to mendelian disorders. *N Engl J Med* 363:852–863; PMID 20818846.

Treacy EP, Valle D & Scriver CR (2000) Treatment of genetic disease. In The Metabolic and Molecular Bases of Inherited Disease, 8th ed. (Scriver CR, Beaudet AL, Valle D, Sly WS et al. eds), pp. 175–191. McGraw-Hill.

Pharmacogenetics and Pharmacogenomics

Evans WE & McLeod HL (2003) Pharmacogenomics—drug disposition, drug targets and side effects. *N Engl J Med* 348:538–549; PMID 12571262. [A general review.]

Flockhart DA (2007) Drug Interactions: Cytochrome P450 Drug Interaction Table. Indiana University School of Medicine, Division of Clinical Pharmacology. http://medicine.iupui.edu/clinpharm/ddis/clinical-table/

Meyer UA, Zanger UM & Schwab M (2013) Omics and drug response. *Annu Rev Pharmacol Toxicol* 53:475–502; PMID 23140244.

Pharmacogenomics Knowledge Base (PharmGKB). http://www.pharmgkb.org [A knowledge resource that encompasses clinical information including dosing guidelines and drug labels, potentially clinically actionable gene–drug associations and genotype–phenotype relationships.]

Roden DM (2004) Drug-induced prolongation of the QT interval. *N Engl J Med* 350:1013–1022; PMID 14999113.

Wang L, McLeod HL & Weinshilboum RM (2011) Genomics and drug response. *N Engl J Med* 364:1144–1153; PMID 21428770.

Wei CY et al. (2012) Pharmacogenomics of adverse drug reactions: implementing personalized medicine. *Hum Mol Genet* 21:R58–R65; PMID 22907657.

Weinshilboum R (2003) Inheritance and drug response. *N Engl J Med* 348:529–537; PMID 12571261. [A general review.]

Small Molecule Drug Therapy for Genetic Disorders

Baker M (2012) Gene data to hit milestone. *Nature* 487:282–283; PMID 22810669.

Barrett PM et al. (2012) Cystic fibrosis in an era of genomically-guided therapy. *Hum Mol Genet* 21:R66–R71; PMID 22914736.

Brooke BS et al. (2008) Angiotensin II blockade and aortic-root dilation in Marfan's syndrome. *N Engl J Med* 358:2787–2795; PMID 18579813.

Curatolo P & Moavero R (2012) mTOR inhibitors in tuberous sclerosis complex. *Curr Neuropharmacol* 10:404–415; PMID 23730262.

Franz DN et al. (2013) Efficacy and safety of everolimus for subependymal giant cell astrocytomas associated with tuberous sclerosis complex (EXIST-1): a multicenter, randomized, placebo-controlled phase 3 trial. *Lancet* 381:125–132; PMID 23158522.

Pearson H (2009) Human genetics: one gene, twenty years. *Nature* 460:164–169 (PMID 19587741). [A perspective on the failure to develop novel cystic fibrosis therapies following the discovery of the underlying CFTR gene 20 years before.]

Peltz SW et al. (2013) Ataluren as an agent for therapeutic nonsense suppression. *Annu Rev Med* 64:407–425; PMID 23215857.

Therapeutic Antibodies and Proteins

Cardinale A & Biocca S (2008) The potential of intracellular antibodies

for therapeutic targeting of protein-misfolding diseases. *Trends Mol Med* 14:373–380; PMID 18693139.

Dimitrov DS (2012) Therapeutic proteins. *Methods Mol Biol* 899:1–26; PMID 22735943.

Kim SJ et al. (2005) Antibody engineering for the development of therapeutic antibodies. *Mol Cells* 20:17–29; PMID 16258237.

Reichert JM (2008) Monoclonal antibodies as innovative therapeutics. *Curr Pharm Biotechnol* 9:423–430; PMID 19075682.

Animal Disease Models for Testing Therapies

Shen H (2013) Precision gene editing paves way for transgenic monkeys. *Nature* 503:14–15; PMID 24201259.

Strachan T & Read AP (2010) Human Molecular Genetics, 4th ed. Garland Science. [Chapter 20 gives a detailed account of the technologies involved in making different animal models and the extent to which the phenotypes in animal models faithfully replicate that of human disorders that they were intended to mimic.]

Gene Therapy: General

Cao H et al. (2011) Gene therapy: light is finally in the tunnel. *Prot Cell* 2:973–989; PMID 22231356.

Fischer A & Cavazzana-Calvo M (2008) Gene therapy of inherited diseases. *Lancet* 371:2044–2047; PMID 18555917.

Ginn SL et al. (2013) Gene therapy clinical trials worldwide to 2012—an update. *J Gene Med* 15:65–77; PMID 23355455.

Kaufmann KB et al. (2013) Gene therapy on the move. *EMBO Mol Med* 5:1–20; PMID 24106209.

Kay MA (2011) State-of-the-art gene-based therapies: the road ahead. *Nature Rev Genet* 12:316–328; PMID 21468099.

Mingozzi F & High KA (2011) Therapeutic *in vivo* gene transfer for genetic disease using AAV: progress and challenges. *Nature Rev Genet* 12:316–328; PMID 21499295.

Naldini L (2011) *Ex vivo* gene transfer and correction for cell-based therapies. *Nature Rev Genet* 12:301–315; PMID 21445084.

Sadelain M et al. (2012) Safe harbours for the integration of new DNA in the human genome. *Nature Rev Cancer* 12:51–58; PMID 22129804.

Somatic Gene Therapy for Specific Diseases

Aiuti A et al. (2009) Gene therapy for immunodeficiency due to adenosine deaminase deficiency. *N Engl J Med* 360:447–458; PMID 19179314.

Cartier N et al. (2009) Hematopoietic stem cell gene therapy with a lentiviral vector in X-linked adrenoleukodystrophy. *Science* 326:818–823; PMID 19892975.

Cavazzana-Calvo M et al. (2010) Transfusion independence and *HGMA2* activation after gene therapy of human β-thalassaemia. *Nature* 467:318–312; PMID 20844535.

Hütter G et al. (2009) Long-term control of HIV by *CCR5* Delta32/Delta 32 stem cell transplantation. *N Engl J Med* 360:692–698; PMID 19213682.

LeWitt PA et al. (2011) AAV2-GAD gene therapy for advanced Parkinson's disease: a double-blind, sham-surgery controlled, randomised trial. *Lancet Neurol* 10:309–319; PMID 21419704.

Maguire AM et al. (2008) Safety and efficacy of gene transfer for Leber's congenital amaurosis. *N Engl J Med* 358:2240–2248; PMID 18441370.

Nathwani AC et al. (2011) Adenovirus-associated virus vector-mediated gene transfer in hemophilia B. *N Engl J Med* 365:2357–2365; PMID 22149959.

Clinical Trials Databases

ClinicalTrials.gov. www.clinicaltrials.gov [A comprehensive US government site.]

Gene Therapy Clinical Trials Worldwide. http://www.wiley.co.uk/genmed/clinical/ [Provided by the *Journal of Gene Medicine* and published by Wiley.]

Therapeutic Gene Silencing and Modulated Splicing

Davidson BL & McCray PB Jr (2011) Current prospects for RNA interference-based therapies. *Nature Rev Genet* 12:320–340; PMID 21499294.

Spitali P & Aartsma-Rus A (2012) Slice modulating therapies for human disease. *Cell* 148:1085–1088; PMID 22424220.

Van Deutekom JC et al. (2007) Local dystrophin restoration with antisense oligonucleotide PRO051. *N Engl J Med* 357:2677–2686; PMID 18160687.

Yu D et al. (2012) Single-stranded RNAs use RNAi to potently and allele-selectively inhibit mutant huntingtin expression. *Cell* 150:895–908; PMID 22939619.

Stem Cells and Cell Therapy

Bellin M et al. (2012) Induced pluripotent stem cells: the new patient? *Nature Rev Mol Cell Biol* 13:713–726; PMID 23034453. [Review on the uses of iPS cells as cellular models of disease.]

Borooah S et al. (2013) Using human induced pluripotent stem cells to treat retinal disease. *Prog Retin Eye Res* 37:163–181; PMID 24104210.

Cherry ABC & Daley GQ (2013) Reprogrammed cells for disease modeling and regenerative medicine. *Annu Rev Med* 64:277–290; PMID 23327523.

Song K et al. (2012) Heart repair by reprogramming non-myocytes with cardiac transcription factors. *Nature* 485:599–604; PMID 22660318.

Takahashi K & Yamanaka S (2013) Induced pluripotent stem cells in medicine and biology. *Development* 140:2457–2461; PMID 23715538.

Therapeutic Genome Editing

Ran FA et al. (2013) Double nicking by RNA-guided CRISPR Cas9 for enhanced genome editing specificity. *Cell* 154:1380–1389; PMID 23992846.

Urnov FD et al. (2010) Genome editing with engineered zinc finger nucleases. *Nature Rev Genet* 11:636–646; PMID 20717154.

Yusa K et al. (2011) Targeted gene correction of α_1-antitrypsin deficiency in induced pluripotent stem cells. *Nature* 478:391–394; PMID 21993621.

Potential Germ-Line Gene Therapy to Prevent mtDNA Disorders

Craven L et al. (2011) Mitochondrial DNA disease: new options for prevention. *Hum Mol Genet* 20:R168–R174; PMID 21852248.

Tachibana M et al. (2013) Towards germline gene therapy of inherited mitochondrial diseases. *Nature* 493:627–631; PMID 23103867.

Cancer Genetics and Genomics

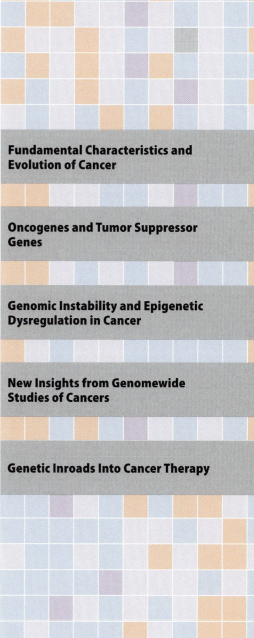

CHAPTER 10

Fundamental Characteristics and Evolution of Cancer

Oncogenes and Tumor Suppressor Genes

Genomic Instability and Epigenetic Dysregulation in Cancer

New Insights from Genomewide Studies of Cancers

Genetic Inroads Into Cancer Therapy

You may wonder why cancer merits a separate chapter in this book and not, say, neurology or cardiology. The answer is that, for the most part, the genetic basis of cancers is quite different from that of other diseases. And, as we will see, cancer genetics has a special complexity.

In Section 10.1 we give an overview of the primary distinguishing biological capabilities of cancer cells, and we outline the broad multi-stage evolution of cancers and describe how intratumor heterogeneity evolves. Section 10.2 is mostly devoted to considering the principles underlying two fundamental classes of genes that are important in cancer development: oncogenes and tumor suppressor genes. As cancers evolve, genomic instability and epigenetic dysregulation become increasingly prominent; we consider selected aspects in Section 10.3.

Genomewide molecular profiling studies—notably genomewide sequencing—are transforming our understanding of cancer, and in Section 10.4 we take a look at new insights emerging from the burgeoning cancer genome studies. Finally, in Section 10.5 we consider the challenges and prospects in deriving clinical benefit from all the extraordinary information coming out of cancer genetics and cancer genome studies. But we defer describing applications in cancer diagnosis, screening and prevention until Chapter 11, when we consider them within a broad general context of genetic diagnosis and screening.

10.1 Fundamental Characteristics and Evolution of Cancer

The defining features of unregulated cell growth and cancer

The term **cancer** is applied to a heterogeneous group of disorders whose common features are uncontrolled cell growth and cell spreading; abnormal cells are formed that can invade adjacent tissues and spread to other parts of the body through the blood and lymph systems. The general process of cancer formation is known as *carcinogenesis*.

Aberrant regulation of cell growth results in an abnormal increase in cell numbers; growths can result that appear normal or abnormal. A growth containing excessive numbers of cells that appear to be virtually the same as those in the normal tissue is said to be *hyperplastic*; a growth that has cytologically abnormal cells is said to be *dysplastic*.

Sometimes a growth formed by excessive cell proliferation is localized. That is, it shows no signs of invading neighboring tissue, and is described as a **benign tumor**. Benign tumors are self-limiting: they grow slowly and can often be surgically removed with low risk of recurrence. They often do not present much danger. Sometimes, however, they grow quite

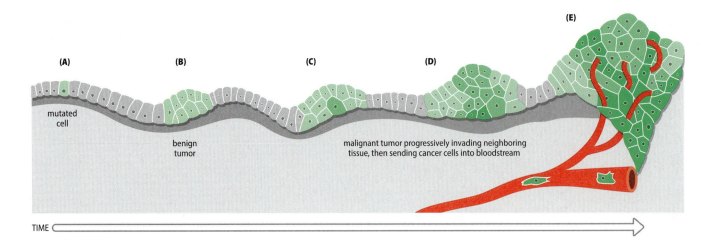

Figure 10.1 Progressive changes in the formation of malignant tumors. The initial mutated cell (A) can develop into a benign tumor (B) through the loss of some normal controls on cell division. Subsequent DNA and epigenetic changes can cause tumor cells to lose further normal controls to become a malignant tumor (C to E) that aggressively invades neighboring tissue. Cells from the malignant tumor can detach themselves and enter the bloodstream (as shown here) or the lymphatic system. In this way they are carried to remote sites in the body where they can exit the circulation and invade neighboring tissues to establish secondary tumors (*metastasis*—for detail of the mechanism see Figure 10.2). (From the Website of the National Cancer Institute [http://www.cancer.gov].)

large over time, and simply by expanding, they can press on neighboring structures in a way that can cause disease. For example, in tuberous sclerosis complex (caused by mutations in *TSC1* or *TSC2*, genes that work in the mTOR growth signaling pathway), benign tumors usually form in multiple different organs. By growing to a large size, they can sometimes disrupt organ function.

In the more than 100 different diseases that we call cancers, the abnormal cells resulting from uncontrolled cell growth have an additional defining property: they can spread. In these diseases the tumors may initially be benign, but they often progress to become **malignant tumors** (which are also commonly called cancers).

Malignant tumors have two distinguishing features: they can invade neighboring tissues, and the cells can break away and enter the lymphatic system or bloodstream to be carried to another location where they cross back into tissues to form secondary tumors (**Figure 10.1**). Spreading to more distant sites in the body is known as **metastasis**; **Figure 10.2** shows dissemination via the bloodstream, with cancer cells crossing capillary walls and migrating through the extracellular matrix.

The tumors (also called *neoplasms*) in cancer can be broadly classified as solid or liquid. Solid tumors form discrete masses composed of epithelial or mesenchymal (stromal) cells. 'Liquid tumors' are made up of neoplastic cells whose precursors are normally mobile blood cells; they include leukemias and also lymphomas (which, although generally forming solid masses in lymph nodes, are able to travel through the lymphatic system). According to the type of tissues or cells in which they arise, the tumors are classified into different categories (**Table 10.1**).

Cancers form when cell division is somehow affected to cause uncontrolled cell proliferation. Changes at the DNA and chromatin levels are the primary contributors. A small minority of human cancers are associated with a specific virus, as described below. Mostly, however, human cancers form because of a series of mutations and epigenetic dysregulation in certain cancer-susceptibility genes.

Recall from Section 4.1 that mutations frequently arise through errors in DNA replication and DNA repair; tissues that have actively dividing cells are therefore prone to forming tumors. Cells divide in development as an organism grows, and some childhood tumors can arise from mutations in cells of the embryo. Although the great majority of our cells are not actively dividing, an adult human has roughly one trillion (10^{12}) rapidly multiplying cells. There is a need to replace certain types of cell that have a high turnover, notably cells in the blood, skin, and the gastrointestinal

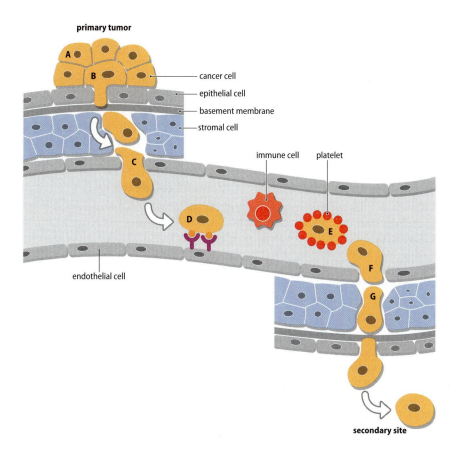

Figure 10.2 The multiple steps taken by metastatic cells to seed secondary tumors. Metastatic cells must first break free from the primary tumor. To accomplish this, cancer cells (A) reduce adhesion to neighboring cells and (B) clear a path for migration into the vasculature-rich **stroma**. Once at the vasculature, cells can freely enter the bloodstream if the vasculature is discontinuous, as in certain regions of the liver, bone marrow, and kidneys. *Intravasation* (C) is required if the vasculature is continuous; metastatic cells either cause endothelial cells to retract (by releasing compounds such as vascular endothelial growth factor) or induce endothelial cell death (by releasing reactive oxygen species and factors including matrix metalloproteinases). In the bloodstream, cancer cell distribution is determined by blood flow and interactions between cancer cells and the secondary organs that they colonize: cells can get trapped in narrow capillary beds, such as those of the lung and liver, and can also express receptors that bind to metastasis-supporting sites (D) or to platelets (E), which protect the cancer cells from the immune system. Cancer cells can circulate for more than 2 hours, suggesting that they do not always become lodged in the first capillary beds that they reach. After reaching the secondary site, cancer cells can leave the bloodstream (F) by *extravasation* (inducing endothelial cell retraction or death). To proliferate in the secondary site, cancer cells co-opt the local environment by releasing pro-inflammatory compounds and proteinases that induce their neighbors to release growth factors (G). (Adapted from Schroeder A et al. [2012] *Nature Rev Cancer* 12:39–50; PMID 22193407. With permission from Macmillan Publishers Ltd.)

tract. Thus, for example, each day about 4% of the keratinocytes in our skin and a remarkable 15–20% of the epithelial cells of the colon die and are replaced.

The short-lived cells that need to be replaced regularly are ones that interface, directly or indirectly, with the environment, and continuous turnover of these cells is a protective measure. Stem cells are the key cells responsible for manufacturing new body cells to replace the lost cells. As we explain below, considerable evidence suggests that cancers are often diseases of stem cells.

Cancer as a battle between natural selection operating at the level of the cell and of the organism

We are accustomed to thinking of how Darwinian natural selection works at the level of the organism: the key parameter is the reproductive success rates of individuals. *Selection pressure* is the effect of natural selection on allele frequencies. It ensures that a deleterious allele—one that reduces reproductive fitness—will normally be at a low frequency in the population (according to the penetrance, the frequency will be maintained by

TISSUE/CELLS OF ORIGIN	TUMORS
Epithelial tissue (single-layer or bilayer)	adenoma (benign); adenocarcinoma (malignant)
Epithelial tissue (multi-layer, as in skin and bladder)	papilloma (benign); squamous cell carcinoma (malignant, in skin); transitional cell carcinoma (malignant, in bladder)
Blood forming tissue (notably bone marrow)	lymphoma (of lymphocytes); leukemia (of leukocytes)
Stromal (mesenchymal) tissue	–oma (benign) or –sarcoma (malignant)[a]
Glial cells	Gliomas

Table 10.1 Major categories of tumors according to tissue or cells of origin. [a]For example, fibroma and fibrosarcoma (fibroblasts); osteoma and osteosarcoma (bone); chondroma and chondrosarcoma (cartilage); hemangioma and hemangiosarcoma (endothelial cells).

new mutation, and by transmission by unaffected carriers). Germ-line mutations normally make the key contribution to noncancerous genetic disorders; somatic mutations usually have minor roles.

Cancers are different. Yes, occasionally cancers can run in families, and germ-line mutations are clearly important in some cases; however, all cancers have multiple somatic mutations, and the genetic contribution to cancers is dominated by somatic (post-zygotic) mutation. That happens because natural selection also operates at the level of *cells*, and the principal defining feature of cancer is uncontrolled growth in cell number.

The balance between cell proliferation and cell death

Growth occurs when there is a positive net balance between cell proliferation and cell death. Cell proliferation is required for growth, but a complicated series of controls ensures that normally our cells do not divide in an uncontrolled fashion; sometimes there is a need for brakes to be applied, and cells are ordered to undergo cell cycle arrest. Cell death is a natural way of removing inefficient cells, or cells that are unwanted or potentially dangerous; like cell proliferation, cell death is also highly regulated.

The mechanisms that regulate cell proliferation and cell death involve sophisticated intercellular signaling. Some signaling pathways send instructions for certain cells to proliferate or undergo cell cycle arrest; other pathways induce the death of cells that are undesirable in some way (**apoptosis**). Classical cancer-susceptibility genes were identified as working to regulate cell division or having direct roles in growth-signaling pathways. Additional cancer-susceptibility genes were found to have roles in apoptosis, but as well as these types of gene, there are many types of non-classical cancer-susceptibility gene that do not function directly in these areas. They might instead have indirect roles. For example, they might be important in DNA repair or genome maintenance—when such genes are faulty, there can be consequential effects on the genes that directly regulate cell growth or apoptosis. See **Figure 10.3** for a summary.

Throughout development there is an overarching priority for increased numbers of cells to sustain rapid growth of the organism—not in an unconstrained way, of course, but executed according to detailed prescribed body plans and the requirements of intricate tissue architecture, and so on. But many cells are also lost during development. In addition to short-lived cells, and cells that are deleted as part of the natural process of sculpting our tissues and organs, many cells are deleted from our developing immune system (to protect the body from attack by cells with receptors that bind to self-antigens) and from the developing nervous system (to weed out neurons with unproductive interneuron connections).

Figure 10.3 Major classes of cancer gene as positive or negative regulators of cell proliferation (cell growth). Green arrows indicate stimulatory effects; red T-bars indicate inhibitory effects. Classical oncogenes are genes that are involved in promoting cell division and cell proliferation by participating in regulation of the cell cycle or in cell growth signaling pathways. However, there are many other genes that have similar effects, including genes that inhibit apoptosis, and also genes that make telomerase, which is involved in promoting continued cell division. Classical tumor suppressor genes function by directly suppressing or restricting cell division, but genes that regulate genome maintenance (and epigenome stability) can have indirect but similar effects. Apoptosis-promoting genes also affect cell growth and include some classical tumor suppressor genes such as *TP53*.

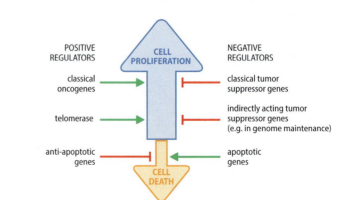

But when we reach adulthood, growth is restricted. By then, most of our cells are non-dividing cells, but a significant minority continue to divide to replace certain categories of cells that turn over rapidly. Apoptosis is also used in adults to ensure the destruction of damaged cells and potentially harmful cells, such as virus-infected cells.

Why we do not all succumb to cancer

Cancer develops progressively when a series of successive mutations disrupt the normal controls that limit cell proliferation, or that induce apoptosis. Cells become able to break free from suppressive controls exerted by their neighbors in a tissue microenvironment. This is where natural selection at the cellular level is important: each successive mutation that disrupts normal controls on cellular proliferation and apoptosis confers an additional selective growth advantage on its descendants. As a result, there is strong selection pressure on cells to evolve through a series of stages into tumor cells.

If there is such strong selection pressure on cells to evolve into tumor cells, why do we not all succumb to cancer? Certainly, if we were to live long enough, cancer would be an inevitable consequence of random mutations. However, an opposing force of natural selection works at the level of the organism (to keep us healthy and free from tumors—at least until we have produced and raised children). It involves different mechanisms, not least immunosurveillance to detect and kill cancer cells (individuals whose immune systems are suppressed are more susceptible to cancer).

Luckily for us, natural selection working at the organismal level also operates over a much longer timescale than does the selection pressure in favor of tumor cell formation. Cancer cells can successfully proliferate and form tumors within an individual person, but they do not leave progeny beyond the life of their human host; tumorigenesis processes must start afresh in a new individual. But individuals who have efficient cancer defense mechanisms are able to pass on good anti-cancer defense genes to their offspring, and the anti-cancer defense systems continue to evolve from generation to generation.

Cancer cells acquire several distinguishing biological characteristics during their evolution

As described in the next section, the development of tumors occurs as a series of stages during which genetic (and epigenetic) changes progressively accumulate in cells. During these stages the cells progressively acquire different biological capabilities that mark them out as cancer cells.

By definition, cancer cells show unregulated cell proliferation, and tumors develop by breaking away from normal control systems. They switch off various brakes that normally place limits on cell proliferation and genome instability, and counteract death (apoptosis) signals from neighboring cells. Cancer cells also lose the contact inhibition of normal cells that places limits on cell growth. Partly by overcoming these negative signals, they become masters of their own growth, and go on to acquire the characteristic ability to replicate indefinitely (Box 10.1).

To support continued cell proliferation, cancer cells re-adjust their metabolism. Thus, they show increased flux through the pentose phosphate pathway (PPP) and elevated rates of lipid biosynthesis, and they take up and use glucose at much higher rates than normal cells. (The last characteristic can be used by imaging systems to differentiate cancer cells from normal cells, so that the spread of cancer cells in the body can be visualized; Figure 10.27 provides an example.)

BOX 10.1 Telomere shortening and the pressure on cancer cells to become immortal by activating telomerase expression.

Normal human cells can be grown in culture for limited periods. Fetal cells, for example, can divide between 40 and 60 times in culture before reaching a state—called *senescence*—after which they cannot grow any further. Cancer cells, however, have unusual growth properties. In culture they do not exhibit contact inhibition, do not require adhesion to a solid substrate, and, notably, can replicate indefinitely, and so are immortal. (The most outstanding human example is the HeLa cell line; developed from a cervical cancer biopsy in the early 1950s, it has been extensively propagated to become the most intensively studied human cell line.)

The end-replication problem

The above observations on the replicative behavior of cells relate to the end-replication problem: how can the extreme ends of linear chromosomes be replicated when new DNA strands grow in the 5′ → 3′ direction only? During DNA replication, new DNA synthesis is catalyzed by DNA polymerases that use an existing DNA strand as a template. As the replication fork advances, one new DNA strand is made in the same direction as the direction of travel for the replication fork, and can be synthesized continuously in the 5′ → 3′ direction. However, the other strand can be made only by synthesizing successive short pieces of DNA (Okazaki fragments) in the opposite direction to that of the replication fork, and there is

a problem with completing the synthesis at the very end (**Figure 1**). Because of the end-replication problem, using just DNA-dependent DNA polymerases means that a small amount of DNA will be lost from each telomere after every cell division.

The telomeres of our chromosomes have tandem TTAGGG repeats extending over several kilobases of DNA (see Figure 1.10 on page 9), but because of the end-replication problem (plus oxidative damage and other end-processing events), the arrays of telomeric TTAGGG repeats normally shorten with each cell division (the number of telomere repeats lost varies between different cell types but is often in the range of 5–20 repeats). When a few telomeres become critically shortened, there is a growth arrest state, at which time DNA damage signaling and cellular senescence is triggered. In the absence of other changes, cells can remain in a senescent state for years.

Telomerase solves the end-replication problem

The end-replication problem can be solved—and telomeres restored to full-length—when cells express telomerase, an RNA-dependent DNA polymerase. Telomerase is a ribonucleoprotein consisting of a reverse transcriptase and a noncoding RNA (ncRNA). The ncRNA has a hexanucleotide sequence that is complementary in sequence to the telomere repeat;

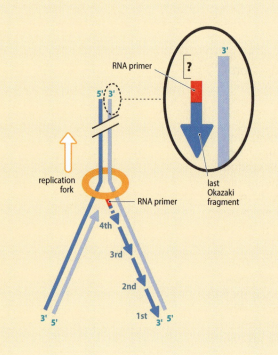

Figure 1 The problem with replicating the extreme ends of DNA in linear chromosomes. In normal DNA replication by DNA-dependent DNA polymerases, an existing DNA strand is used as a template for making a complementary new DNA strand. Here, as the replication fork advances in the upward direction it can synthesize a continuous DNA strand upward in the 5′ → 3′ direction from one original DNA strand (colored deep blue) but for the pale blue original strand the 5′ → 3′ direction for DNA synthesis is in the opposite direction to the upward direction of the replication fork. The DNA must be synthesized in short pieces, called Okazaki fragments, starting from a position beyond the last fragment and moving backward toward it. (DNA-dependent polymerases use short RNA primers to initiate the synthesis of DNA, but the RNA primers are degraded, DNA synthesis fills in, and adjacent Okazaki fragments are ligated.) The question mark indicates a problem that is reached at the very end: how is synthesis to be completed when there can be no DNA template beyond the 3′ terminus?

BOX 10.1 (continued)

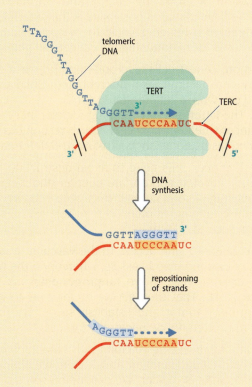

Figure 2 Telomerase uses a reverse transcriptase and a noncoding RNA template to make new telomere DNA repeats. The telomerase reverse transcriptase (TERT) is an RNA-dependent DNA polymerase: it uses an RNA template provided by the other subunit, TERC (telomerase RNA complex). Only a small part of the RNA is used as a template—the hexanucleotide that is shaded—and so the telomeric DNA is extended by one hexanucleotide repeat (blue shading). Repositioning of the telomeric DNA relative to the RNA template allows the synthesis of tandem complementary copies of the hexanucleotide sequence in the RNA template.

advantage, about another 20–40 cell divisions might be required to achieve a cell population size that is sufficient for another spontaneous mutation to occur in a cell with the previous mutation. Premalignant cells would therefore often be expected to come up against the barrier of replicative senescence before they have sustained sufficient mutations to form malignant tumors.

Tumor cells are able to bypass replicative senescence by suppressing tumor suppressors, such as p53 and the RB1 retinoblastoma protein. However, after a few additional cell divisions past the point at which senescence normally occurs, the cells enter a *crisis* state. Now the telomeres can be so short that DNA repair mechanisms do not recognize that they are legitimate ends of chromosomes; instead, they treat them as double-strand DNA breaks. As a result, chromosomes can undergo end-to-end fusions. The resulting chromosomes have two centromeres and may be pulled in opposite directions at mitosis. That causes further broken ends, new cycles of chromosome fusion and breakage, and an acceleration of genome instability.

Rare cells that escape this crisis stage are almost always able to do so by having reactivated expression of telomerase: the telomeres are stabilized and the cell becomes immortal. However, the telomerase produced is not present at excess (the telomeres in cancer cells with stem cell-like properties are generally of the same length or shorter than those in adjacent normal cells).

it serves as a template from which the reverse transcriptase can make tandem telomere repeats (**Figure 2**).

Telomerase is expressed during the early stages of human embryogenesis (and in embryonic stem cells). However, its expression is subsequently repressed in most somatic cells except the male germ line, activated lymphocytes, and stem cells found in certain regenerative tissues.

Selection pressure on cancer cells to activate telomerase

The repression of telomerase and the resulting erosion of telomeres in our cells is thought to be yet another defense system to stop cancer from developing during our long lifetimes. Cancer cells require multiple successive mutations to become malignant. After one mutation has led to some growth

Although apparently exposed to aerobic conditions, cancer cells nevertheless derive their energy from glycolysis, rather than from oxidative phosphorylation. Glycolysis is normally used by cells in anaerobic conditions; the process involves converting glucose to first pyruvate, and then lactate, and energy production is inefficient (2 molecules of ATP generated per molecule of glucose). Under aerobic conditions, normal cells convert glucose to pyruvate and then transport the pyruvate into the mitochondria to be catabolized in the tricarboxylic acid (Krebs) cycle, generating up to 36 molecules of ATP per molecule of glucose.

Why, under aerobic conditions, cancer cells normally use the much more inefficient glycolysis system of producing energy (the Warburg effect) remains poorly understood. One favored possibility is that intermediates in the glycolytic pathway serve as building blocks for macromolecule synthesis needed by cancer cells for maintaining the cancer cell phenotype.

During cancer progression, cancer cells also undergo epigenetic reprogramming so that they can become less differentiated. Solid cancers show a plastic phenotype, with a differentiated tumor mass and also undifferentiated areas. The latter, notably marking regions that form an invasive front as the cancer spreads, allow flexibility to respond to different environments, and metastases can show striking re-differentiation.

To ensure their survival, cancer cells need to avoid destruction by the immune system and they develop appropriate counter-attacking measures. Not only that, but they also maximize their ability to survive by invading host tissue and co-opting normal cells to help them, and by sending out cells to form secondary tumors.

It is also common for cancers to gain access to the vascular system by inducing the sprouting of existing blood vessels, whereupon the tumors become linked to the existing vasculature (**angiogenesis**), as in the tumor shown in Figure 10.1E. That then allows tumor cells to escape more readily from a primary tumor and establish secondary tumors, although angiogenesis may often not be necessary for metastasis.

Table 10.2 provides a summary by listing 10 biological characteristics that cancer cells acquire which Doug Hanahan and Robert Weinberg have recently proposed as hallmarks of cancer. We expand on some of these points in Section 10.2, but also provide details on some other of the points in later sections.

The initiation and multi-stage nature of cancer evolution and why most human cancers develop over many decades

Epidemiology studies have shown that age is a very large factor in cancer incidence (the rate at which it is diagnosed). For example, the age–incidence plots for epithelial cancers suggested that the risk of death from

Table 10.2 Ten acquired biological capabilities proposed as hallmarks of cancer by Douglas Hanahan and Robert Weinberg. IGF, insulin growth factor; TGFβ, transforming growth factor β; VEGF, vascular endothelial growth factor. (Adapted from Hanahan D & Weinberg R [2011] *Cell* 144:646–674; PMID 21376230. With permission from Elsevier.)

ACQUIRED BIOLOGICAL CAPABILITY	EXAMPLES OF HOW THE BIOLOGICAL CAPABILITY IS ACQUIRED
Self-sufficiency in growth signaling	activate cellular oncogene
Insensitivity to signals suppressing growth	inactivate *TP53* to avoid p53-mediated cell cycle arrest
Ability to avoid apoptosis	produce IGF survival factor
Replicative immortality	switch on telomerase (Box 10.1)
Genome instability	inactivate certain genes involved in DNA repair
Induction of angiogenesis	produce factor that induces VEGF
Tissue invasion and metastasis	inactivate E-cadherin
Ability to avoid immune destruction	paralyze infiltrating cytotoxic T lymphocytes and natural killer cells by secreting TGFβ or other immunosuppressive factor
Induction of tumor-promoting inflammation	redirect inflammation-causing immune system cells that infiltrate the tumor so that they help in various tumor functions (see Table 10.3)
Reprogramming energy metabolism	induce aerobic glycolysis

this cause increases roughly as the fifth or sixth power of elapsed lifetime. That observation suggested that perhaps six to seven independent events might be required for an epithelial cancer to develop (if the probability of an outcome is a function of some variable raised to the power n, a total of $n+1$ independent events, each occurring randomly, are required for the outcome to be achieved).

The epidemiology studies provided an early indication of the multi-stage nature of cancer, and a suggestion of the number of critical steps involved. Now we know that as normal cells evolve to become cancer cells, they pick up many somatic changes—both genetic and epigenetic. A small subset of the genetic changes, known as **driver mutations**, result in altered expression of genes that confer a growth advantage to their descendants. They are positively selected and causally implicated in cancer development. The remainder are *passenger mutations*.

A cell with a driver mutation that its neighbors lack usually possesses a small growth advantage, of the order of just a 0.4% increase in the difference between new cell formation and cell death. The growth advantage is small because we have multiple layers of defense against cancer. Many tumor cells succumb to our natural defenses, or are disadvantaged by certain karyotype changes. Despite the high attrition rate of tumor cells, the small growth advantage can ultimately lead to a large mass, containing billions of cells, but that usually takes many years.

Clonal expansion and successive driver mutations

Tumors are monoclonal: all the cells descend from a single starting cell. Strong evidence for that supposition came from studies of B-cell lymphomas. Recall from Section 4.5 that individual B cells in a person make different immunoglobulins, but the cells in individual B-cell lymphomas all make the same type of immunoglobulin.

Preferential clonal expansion of the mutant cells produces an expanded target (more cells) for a second driver mutation to occur in one of the mutant cells. As the process continues, cells progressively acquire more mutations (**Figure 10.4A**) that cause them to become ever more like a cancer cell.

Figure 10.4 Driver mutations in the multi-stage evolution of cancer. (A) *General process.* Each successive driver mutation gives the cell in which it occurs a growth advantage, so that it forms an expanded clone and thus presents a larger target for the next mutation. Orange cells carry driver mutation 1; red cells have sequential driver mutations 1 and 2; and purple cells have driver mutations 1, 2, and 3. (B) *Genetic alterations and the progression of colorectal cancer.* The major signaling pathways that drive tumorigenesis are shown at the transitions between each tumor stage. One of several driver genes that encode components of these pathways can be altered in any individual tumor. Small and large adenomas appear as intestinal polyps that are benign but can progress to become carcinomas, cancers that invade the underlying tissue. Patient age indicates the time intervals during which the driver genes are usually mutated. Note that this model may not apply to all tumor types. PI3K, phosphoinositide 3-kinase pathway; TGFβ, transforming growth factor β pathway. (B, from Vogelstein B et al. [2013] *Science* 339:1546–1558; PMID 23539594. With permission from the AAAS.)

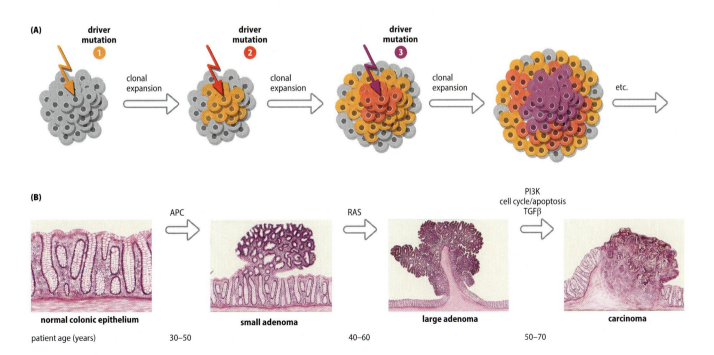

(A)

driver mutation 1 — clonal expansion — driver mutation 2 — clonal expansion — driver mutation 3 — clonal expansion — etc.

(B)

| | APC | | RAS | | PI3K cell cycle/apoptosis TGFβ | |
| normal colonic epithelium | | small adenoma | | large adenoma | | carcinoma |

| patient age (years) | 30–50 | | 40–60 | | 50–70 | |

In some cases, just a few driver mutations are required. Figure 10.4B illustrates a classic example: the gradual transformation from normal epithelium to carcinoma in the development of colon cancer. The initial driver mutation is almost always one that affects the Wnt signaling pathway usually through loss of function of the *APC* gene at 5q21, but there can be more flexibility in the order of the subsequent genetic changes.

Cancer development through accelerated mutation

The average rate of mutation in human cells is low (about 10^{-6} per gene per cell) and the majority of cancer-causing mutations are recessive at the cellular level, so that both alleles need to be mutated. Cancer might therefore be expected to be highly improbable: the chance that any cell would receive successive mutations, often in both alleles, at several cancer-susceptibility loci would normally be vanishingly small.

Cancer nevertheless is common, and altered expression at a few cancer-susceptibility loci can be sufficient. One major explanation is that early driver mutations greatly increase the probability of later mutations in two ways. The first way is by giving the cell a growth advantage. If cells with a driver mutation have an increased growth rate, they will produce more progeny than other cells and will produce an expanded target of mutant cells, thereby increasing the probability of a subsequent mutation (see Figure 10.4A).

The second way of increasing the probability of later mutations in cancer is by destabilizing the genome. Chromosome instability is a feature of most tumor cells, producing grossly abnormal karyotypes with abnormal numbers of chromosomes and frequent structural arrangements that can activate oncogenes or cause a loss of tumor suppressor genes. In some cancers, a form of global DNA instability occurs—it is caused by mutations in key DNA repair genes and can result in greatly elevated mutation rates. Mutations in genes that regulate epigenetic modifications result in additional epigenetic instability that can result in altered expression at cancer-susceptibility loci. Additionally, some types of epigenetic change cause genome instability. We will explore the detail of genome instability and epigenetic dysregulation in cancer in Section 10.3.

Mutation accumulation and age of cancer onset

Tumors gradually acquire mutations to evolve from benign to malignant lesions. Because that takes some time, cancer is primarily a disease of aging. In self-renewing tissues—such as epithelial cells lining the gastrointestinal tract and genitourinary epithelium—the cells contain DNA that has progressively accumulated mutations through multiple DNA replication cycles in progenitor cells (recall that errors in DNA replication and post-replicative DNA repair are frequent causes of mutations). Thus a colorectal tumor in a person in their eighties or nineties will have nearly twice as many somatic mutations as in a morphologically identical colorectal tumor in a person half their age. (The difference in ages when the tumor presented will reflect when crucially important driver mutations occurred.)

Cells in tissues associated with some other cancers do not replicate, and the tumors associated with these cells have fewer mutations, such as in glioblastomas (advanced brain tumors formed from nonreplicating glial cells) and pancreatic cancers (epithelial cells of the pancreatic duct also do not replicate). Initiating driver mutations in these cases must occur in cells that have had comparatively few mutations.

Some types of cancer commonly arise in childhood. Pediatric tumors often occur in tissues that are not self-renewing, and such tumors typically have fewer mutations than adult tumors. However, leukemias and lymphomas, which are diseases of self-renewing blood cells, can also often develop early in life. Here the precursor cells are already mobile and invasive and are thought to require fewer DNA changes than in solid tumors, in which the tumor cells require additional mutations to confer these biological capabilities.

Another possibility is that childhood cancers can arise from an initiating mutation that arises in embryonic cells. Progenitor cells in the embryo resemble cancer cells—they are poorly differentiated and rapidly dividing. If they receive a cancer-predisposing mutation, they are much more likely to develop into tumors at an early stage than more differentiated cells with the same mutation.

Intratumor heterogeneity arises through cell infiltration, clonal evolution, and differentiation of cancer stem cells

Although tumors are considered to be monoclonal (composed of cells derived from a single ancestral cell), that does not mean that the cells in a tumor are the same. Instead, tumors often have quite complicated tissue architectures and are composed of functionally different cells.

A first level of intratumor heterogeneity exists because tumors are usually made up of both tumor cells proper (that originate from a single cell, and so are monoclonal), and also various unrelated cells that infiltrate the tumor from the surrounding environment. Different types of stromal cells, including immune infiltrating cells, become part of the tumor microenvironment and can be redirected to support tumor activities (**Figure 10.5A** and **Table 10.3**).

A second level of intratumor heterogeneity exists because the tumor cells proper within a tumor can become functionally distinct from each other. That can happen as a result of differential genetic changes. In addition, differential epigenetic changes can occur and some tumors can clearly be seen to have cells at different stages of differentiation.

Cells that have descended from the originating cancer cell can acquire new mutations conferring some additional growth advantage, or other advantageous tumor-associated biological capability. A cell with an advantageous mutation such as this can form a subclone that competes with and outgrows the other cells. The process continues with new

STROMAL CELL CATEGORY	EXAMPLES OF CELL TYPES	FUNCTIONS IN SUPPORT OF:			
		Mitogenesis	Angiogenesis	Tissue invasion	Metastasis
Infiltrating immune cells	macrophages	+	+	+	+
	mast cells	+	+	+	
	neutrophils	+	+	+	+
	T cells (notably of the Th2-CD4 class and regulatory T cells); B cells	+			
Cancer-associated fibroblastic cells	activated tissue fibroblasts	+	+	+	+
Endothelial cells	endothelial tip, stalk, tube		+		
Pericytes	mature/immature pericytes		+		

Table 10.3 Different categories of stromal cell types can support the tumor microenvironment. The table gives a quite selective list, both of the different stromal cell types that support tumors and of their functions. For a fuller account, see Table 2 of Hanahan D & Coussens LM (2012) *Cancer Cell* 21:309–322; PMID 22439926.

Figure 10.5 Cell heterogeneity within tumors. (A) Tumor formation involves co-evolution of neoplastic and non-neoplastic cells in a supportive and dynamic microenvironment that includes different stromal cells—cancer-associated fibroblasts, vascular endothelial cells (including pericytes), and diverse infiltrating immune cells—and the extracellular matrix. The tumor microenvironment offers structural support, access to growth factors, vascular supply, and immune cell interactions. The immune cells include cell types normally associated with tumor-killing abilities as well as immune cells that can have tumor-promoting properties (see Table 10.3). (B) Different tumor subclones may show differential gene expression due to both genetic and epigenetic heterogeneity. Cells from some subclones may intermingle (subclones 1 and 2) or be spatially separated (subclone 3), sometimes by physical barriers such as blood vessels. Within a subclonal population of tumor cells there may be intercellular genetic and nongenetic variation. For example, in the expanded square (which represents a section taken from a spatially separated subclone), differences in chromosome copy number between cells are revealed by the hybridization signals obtained with fluorescent probes for the centromeres of chromosome 2 (red) and chromosome 18 (green), against a background stain of blue for DNA. (A, From Junttila MR & de Sauvage FJ [2013] *Nature* 501:346–354; PMID 24048067. With permission from Macmillan Publishers Ltd. B, From Burrell RA et al. [2013] *Nature* 501:338–345; PMID 24048066. With permission from Macmillan Publishers Ltd.)

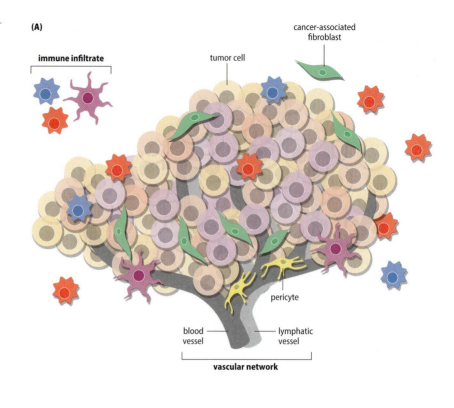

(A)

immune infiltrate

cancer-associated fibroblast

tumor cell

pericyte

blood vessel

lymphatic vessel

vascular network

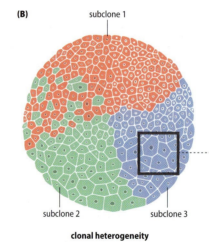

(B)

subclone 1

subclone 2

subclone 3

clonal heterogeneity

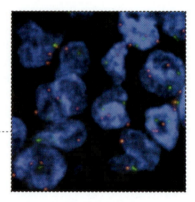

intercellular genetic and nongenetic heterogeneity

subclones competing against previous subclones. Subclones may intermingle or be spatially distinct (Figure 10.5B).

After the appearance of successive subclones, a tumor might be dominated by cells from a recent particularly successful subclone but still contain some cells from previously dominant subclones. Clonal evolution by acquisition of new mutations—both driver and passenger mutations—might therefore explain how functionally different types of tumor cell could arise within the same tumor. Despite being functionally divergent, the cells in the different subclones may or may not be recognizably different in appearance.

Subclones of a primary tumor can also undergo mutations that will drive genetic divergence leading to metastases—see the paper by Wu et al. (2012) under Further Reading for the example of clonal selection driving genetic divergence of metastases in medulloblastoma. We will return to this topic in Section 10.4 when we consider insights from cancer genome sequencing.

In addition to clonal evolution, the concept of cancer stem cells has been invoked to explain intratumor heterogeneity. That is, self-renewing tumor cells have been proposed to give rise to all the different types of tumor cell within a tumor by progressive differentiation (Box 10.2). Because cancer stem cells are very long-lived and can potentially regenerate tumors or seed metastases starting from a single cell, there are important implications for cancer therapy.

Although the concept of cancer stem cells and clonal evolution might seem to provide alternative explanations for intratumor cell heterogeneity, they are not mutually exclusive. There is some evidence that even the stem cells within some tumors are heterogeneous as a result of mutation-induced divergence.

10.2   Oncogenes and Tumor Suppressor Genes

By early 2014, more than 500 human genes had been identified as having a causal role in cancer. The relevant data were obtained by different approaches: analyzing tumor-associated chromosomal rearrangements (notably translocations), identifying tumor-specific changes in gene copy number, and by identifying tumor-specific mutations (by comparing tumor DNA sequences with the corresponding DNA sequence in normal cells from the same individual).

Two fundamental classes of cancer gene

The key cancer-susceptibility genes—those in which driver mutations occur—can be grouped into two fundamental classes, according to how they work in cells. Some are dominant at the cellular level: mutation of a single allele is sufficient to make a major, or significant, contribution to the development of cancer. Others are recessive: both alleles need to be inactivated to make a significant contribution to cancer.

Oncogenes are the exemplars of dominantly acting cancer-susceptibility genes. In our cells the normal copies of these genes (sometimes called proto-oncogenes) often function in growth signaling pathways to promote cell proliferation or inhibit apoptosis, but as we describe below they can also work in other cellular functions. After receiving a suitably activating mutation, a single allele of a proto-oncogene can make a significant contribution to the tumorigenesis process. (For some other genes, however, an inactivating mutation in a single allele can also make a significant contribution to the development of a cancer.)

Tumor suppressor genes are the exemplars of recessive cancer-susceptibility genes. Classical tumor suppressors work in the opposite direction to oncogenes—to suppress cell proliferation by inducing cell cycle arrest. Certain other tumor suppressor genes work to promote the apoptosis of deviant cells. When both alleles of a classical tumor suppressor gene are inactivated, that locus can make a significant contribution to cancer development.

A common analogy imagines an oncogene as the accelerator of a car and a tumor suppressor as the brake. The car will run out of control if the accelerator is jammed on (inappropriately activated) or if the brake fails. The cell is more complicated than this analogy allows: it has many different types of accelerator and brake to regulate cell growth, and usually several of the cell's accelerators and brakes need to be faulty to cause real damage.

BOX 10.2 Cancer as a disease of stem cells.

Somatic cells often have rather short lives—only about a week or so, on average, in the case of intestinal epithelial cells. Short-lived cells need to be replaced periodically by new cells produced ultimately from the relevant tissue stem cells. The stem cells are capable of two types of cell division: symmetrical cell division and asymmetrical cell division. If the numbers of stem cells get too low for any reason, they can quickly regenerate by multiple successive symmetrical cell divisions, each producing daughter cells that are identical to the parent stem cell.

Asymmetrical stem cell divisions are reserved for making differentiated cells. In this case, when the stem cell divides it gives rise to one daughter cell that is identical to the parent cell (step a in Figure 1), plus a daughter cell that is more differentiated, a **transit-amplifying cell** (step b in Figure 1).

Newly formed transit-amplifying cells go through multiple symmetrical cell divisions to produce very large numbers of cells that subsequently undergo differentiation to give rise to the highly differentiated, comparatively short-lived cells. Because the latter cells have short lives mutation in them could never lead to cancer. Instead, cancer must arise in a longer-lived progenitor cell.

Because the lineage of stem cells represents the only stable repository of genetic information within the tissue, the genomes of stem cells need to be protected as far as possible from mutation. First, the stem cell compartment is physically separated to reduce contact with potential mutagens. In the case of intestinal epithelial cells, for example, the stem cell compartment lies at the base of the intestinal crypts (see Figure 8.11B for the latter); here they are far removed from the epithelial cell lining that comes into contact with potentially hazardous mutagens in our diet. Secondly, the stem cells rarely need to divide: the transit amplifying cells can undergo exponential expansion from a single stem cell.

Despite efforts to maintain the genomes of tissue stems cells, mutations can arise in stem cells so that they form cancer stem cells. That can cause problems for treating cancers because cancer stem cells are relatively resistant to being killed by cytotoxic chemicals and radiation, and can keep on regenerating the more differentiated tumor cells. Because of their very long lifetimes, stem cells may be direct targets for mutagenesis, leading to the formation of cancer stem cells. Additionally, mutations can be conveyed indirectly into the stem cell pool after mutated transit amplifying cells are dedifferentiated by epigenetic changes to become cancer stem cells.

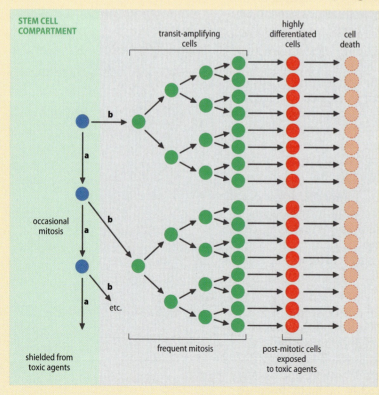

Figure 1. **Epithelial tissue as an example of cell differentiation from stem cells and protection of the stem cell genome**. Each stem cell (*blue*) divides only occasionally in an asymmetric fashion (steps a and b) to generate a new stem cell daughter and a more differentiated transit-amplifying daughter cell (*green*). Transit-amplifying cells undergo repeated rounds of growth and division, leading to exponential increase in cell numbers. Eventually, the products of these cell divisions undergo further differentiation into post-mitotic highly differentiated cells (*red*). The highly differentiated cells, which are often in direct contact with various toxic agents, are frequently shed (so that any harmful mutations that arise in these cells are quickly lost from the tissue). The stem cells are protected from the potentially mutagenic effects of toxic agents because they are shielded by an anatomical barrier. (Adapted from Weinberg RA [2014] Biology of Cancer, 2nd edn., Garland Science.)

BOX 10.2 (*continued*)

Supportive evidence for cancer stem cells comes from analyses of many types of leukemia. For example, in chronic myelogenous leukemia (CML) the Philadelphia chromosome (a specific type of chromosomal translocation that is believed to be the initiating genetic lesion; see Figure 10.7) may often be evident in different hematopoietic cell types within CML patients, being found in B and T lymphocytes, neutrophils, granulocytes, megakaryocytes, and so on. The cell in which the Philadelphia chromosome first arose was presumably a precursor of all those different blood cell types, a hematopoietic stem cell.

The idea of cancer stem cells is supported by hierarchical cell organizations for certain types of cancer. In some types of neuroblastoma and myeloid leukemia, for example, the cancer evolves so that some of the tumor cells differentiate and have limited capacity for proliferation, despite retaining the oncogenic mutations of their malignant precursors.

Further evidence comes from flow cytometry, which allows separation of phenotypically distinct subpopulations of live cancer cells. The tumorigenic properties of the subpopulations can then be studied by transplanting them into immunocompromised mice. Using this approach, it became clear that only a small proportion of cancer cells in leukemia and breast cancer proliferate extensively, and they express specific combinations of cell surface markers. For example, breast-cancer-initiating cells were found to express CD44, but showed very little or no expression of CD24; that is, the cancer-initiating cells were found to be $CD44^+CD24^-$ cells, a small minority of the population of tumor cells. Similarly, leukemia-initiating cells were found to be a minority population of $CD34^+CD38^-$ cells. Follow-up studies indicate that many other cancers might also follow the cancer stem cell model.

As well as standard oncogenes and tumor suppressor genes, various other types of cancer-susceptibility gene have been identified. As described below, some work in DNA repair and genome maintenance. Some others support certain biological capabilities of cancer cells; they include genes encoding telomerase, and genes involved in energy metabolism and angiogenesis, among others.

Viral oncogenes and the natural roles of cellular oncogenes

Oncogenes were discovered after it became clear that certain cancers in chickens and rodents were induced by viruses. (Note that most human cancers are not caused by viruses, but some viruses are known to be linked to specific human cancers: Epstein–Barr virus is linked to nasopharyngeal carcinoma and lymphomas; some papillomaviruses are linked to cervical and oropharyngeal squamous cell carcinoma; chronic hepatitis B virus infection is linked to hepatocellular carcinoma; acute transforming human T-cell lymphotropic virus is linked to acute T-cell leukemia; and human herpesvirus-8 is linked to Kaposi's sarcoma.)

Among the viruses found to cause cancers in chickens and rodents were types of acute transforming retrovirus (also called oncoretroviruses) that could cause cells in culture to change their normal growth pattern to resemble that of tumors (a process known as **transformation**). Whereas normal versions of these retroviruses had the three standard transcription units (*gag*, *pol*, and *env*; see Figure 9.19B), the oncoretroviruses had an altered genome in which part of the viral genome had been replaced by an altered copy of a cellular gene that, when expressed, promoted cellular proliferation. The normal cellular gene (sometimes called a *proto-oncogene*) is subject to many controls and normally promotes cell proliferation only when there is a natural need for cell proliferation, but it becomes an oncogene when inappropriately expressed.

The copy of the cellular proto-oncogene in an oncovirus is strongly expressed because it is subject to regulation by the powerful promoter/

Table 10.4 Major functional classes of cellular oncogenes that work in cell growth signaling or apoptosis.

FUNCTIONAL CLASS	EXAMPLES OF ONCOGENES	ONCOGENE PRODUCT
Secreted growth factor	FGF4	fibroblast growth factor type 4
	PDGFB	platelet-derived growth factor
Growth factor receptor; receptor tyrosine kinase	EGFR	epidermal growth factor receptor
Signal transduction component	ABL	cytoplasmic tyrosine kinase
	HRAS	homolog of Harvey rat sarcoma viral oncogene, a GTPase
	PIK3CA	catalytic subunit of a lipid kinase, phosphatidylinositol kinase
Transcription factor	GLI1	zinc finger protein that works as a transcriptional activator
	MYC	homolog of myelocytomatosis viral oncogene. Activates transcription of various growth-related genes
Cell cycle regulator	CCND1	cyclin D
Apoptosis inhibitor	BCL2	mitochondrial protein

enhancer sequences in the viral long terminal repeats. As a result, it is capable of driving abnormal cellular proliferation leading to cancer. The viral oncogenes often lack introns present in the cellular proto-oncogenes.

Many cellular oncogenes work naturally within growth signaling pathways at different levels—as growth factors, receptors, transcription factors, and different types of regulatory protein. Some other oncogenes normally serve to suppress apoptosis (see **Table 10.4** for examples).

Oncogene activation: chromosomal rearrangements resulting in increased gene expression or gain-of-function mutations

Proto-oncogenes are activated by a DNA change that is dominant at the cellular level (and normally affects just a single allele). In the subsections below, we describe the three ways in which this can occur. Two of the three types of DNA change result in increased gene expression, either as a result of a large increase in gene copy number (*gene amplification*) or via transcriptional activation (in which a translocation brings the proto-oncogene into close proximity to regulatory sequences of an actively transcribed gene). The third class is made up of activating point mutations that alter how the protein behaves.

Note that cells have multiple anti-cancer defense systems, and the activation of a single cellular proto-oncogene is usually not oncogenic by itself. If we experimentally activate a cellular proto-oncogene in cultured cells, the usual effect is to induce cell cycle arrest (the abnormal proliferative signals usually induce cellular defense mechanisms that shut down cell proliferation); multiple genetic (and epigenetic) changes are needed to induce cancer.

Activation by gene amplification

Tumor cells often contain abnormally large numbers—often hundreds of copies—of a structurally normal oncogene. The *MYCN* oncogene, for example, is frequently amplified in late-stage neuroblasts (**Figure 10.6A**) and in rhabdomyosarcomas; *ERBB2* (also called *HER-2*) is often amplified in breast cancers.

The amplification mechanism is not simple tandem amplification; instead, there seem to be complex rearrangements that bring together

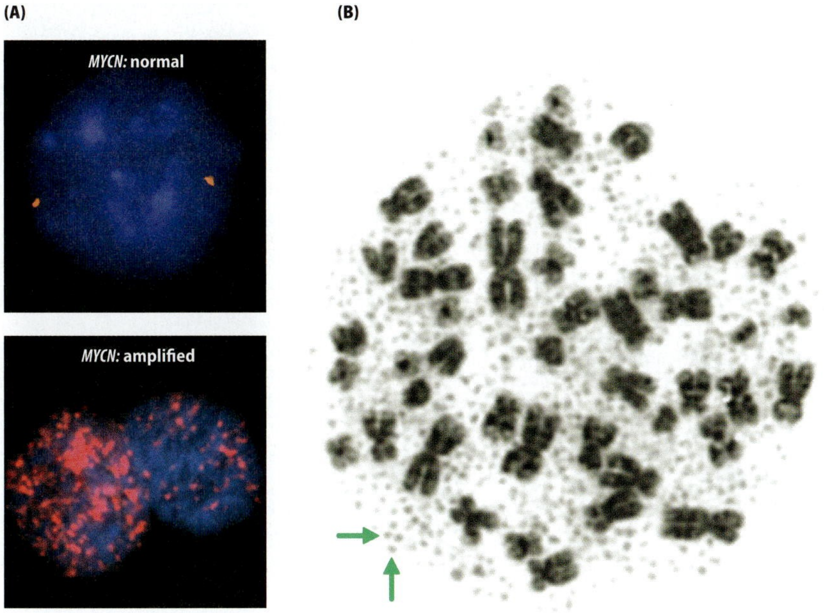

(A)

MYCN: normal

MYCN: amplified

(B)

Figure 10.6 Amplification of the *MYCN* gene and formation of double minutes in neuroblastoma cells. (A) Fluorescence *in situ* hybridization (FISH) images using a labeled *MYCN* gene probe, showing two copies of the gene (red signals) in normal cells against a background of DNA staining (shown in blue). In neuroblastoma cells, the *MYCN* gene can undergo extensive amplification to produce many dozens or even hundreds of *MYCN* gene copies (as shown at the bottom). (B) A metaphase chromosome preparation from a neuroblastoma tumor sample, showing *double minute* chromosomes (which appear as a cloud of very small dots; arrows indicate two of them). (A, Courtesy of Nick Bown, NHS Northern Genetics Service, Newcastle upon Tyne, UK; B, Courtesy of Paul Roberts, NHS Cytogenetics Service, Leeds, UK.)

sequences from several different chromosomes. The amplification may manifest itself in two forms. An extrachromosomal form is known as *double minutes*, which are so called because they are tiny, paired acentric chromatin bodies that are separated from chromosomes and contain multiple copies of just a small set of genes (**Figure 10.6B**). A corresponding intrachromosomal form (in which multiple repeated copies integrate into chromosomes) gives rise to *homogeneously staining regions*. For a list of genes amplified in cancer, see the Cancer Gene Census at http://cancer.sanger.ac.uk/cancergenome/projects/census/.

Translocation-induced gene activation

Chromosomal translocations occur when DNA molecules receive double-strand breaks and are then re-joined incorrectly so that pieces of different DNA molecules are joined together. When that happens, an oncogene is often inappropriately transcriptionally activated and so there can be a selective growth advantage.

Translocations that activate oncogenes are common in cancer (more than 300 cancer-associated translocations are listed within the Cancer Gene Census database—see http://cancer.sanger.ac.uk/cancergenome/projects/census/). In many cases, the translocations result in the formation of clearly chimeric genes that result in the constitutive expression of oncogene sequences. In other cases, the oncogene sequence is not interrupted by a breakpoint; instead it is simply brought into close proximity to regulatory sequences in another gene that is actively expressed (see **Table 10.5** for some examples).

The Philadelphia (Ph[1]) chromosome, occurring in 90% of individuals with chronic myeloid leukemia, illustrates how a translocation gives rise to cancer via a chimeric gene. It results from a balanced reciprocal translocation with breakpoints near the start of the *ABL1* oncogene at 9q34 and close to the end of *BCR* gene at 22q11 (**Figure 10.7**). The resulting *BCR–ABL1* fusion gene on the Philadelphia chromosome (with the *ABL1* coding sequence positioned downstream of the *BCR* gene sequence and *BCR* promoter) produces a large protein that carries the ABL1 polypeptide sequence at its C-terminal end. This fusion protein acts as a growth-stimulating tyrosine kinase that is *constitutively* active and so drives cell proliferation.

TUMOR TYPE	ONCOGENE (CHROMOSOME LOCATION)	INTERACTING GENE (CHROMOSOME LOCATION)
Acute lymphoblastoid leukemia (ALL)	*MLL* (11q23)	*AF4* (4q21)
		AF9 (9p22)
		AFX1 (Xq13)
		ENL (19p13)
Acute myeloid leukemia (AML)	*FUS* (16p11)	*ERG* (21q22)
Acute promyelocytic leukemia	*PML* (15q24)	*RARA* (17q21)
Burkitt's lymphoma	*MYC* (8q24)	*IGH* (14q32)
Chronic myeloid leukemia (CML)	*ABL* (9q34)[a]	*BCR* (22q11)[a]
Ewing sarcoma	*EWS* (22q12)	*FLI1* (11q24)
Follicular B-cell lymphoma	*BCL2* (18q21)	*IGH* (14q32)
T cell leukemia	*LMO1* (11p15)	*TRD* (14q11)
	LMO2 (11p13)	
	TAL1 (1p32)	

Tumors of B and T cells, including various lymphomas and leukemias, often result from translocations with breakpoints in an immunoglobulin heavy-chain or light-chain gene, notably *IGH*, or a T-cell receptor gene such as *TRA*, *TRB*, or *TRD*. Developing B and T cells are unusual in that programmed DNA rearrangements are required to rearrange, respectively, immunoglobulin and T-cell receptor genes so that they produce cell-specific immunoglobulin or T-cell receptor chains, as detailed in

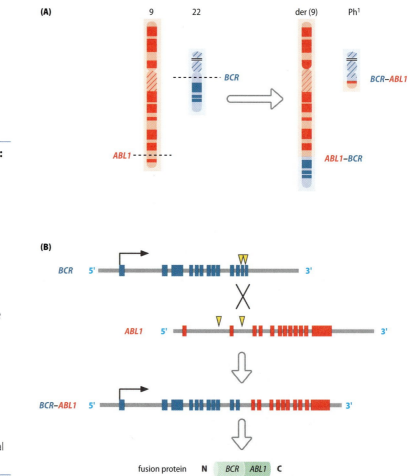

Figure 10.7 The Philadelphia chromosome: oncogene activation following a translocation that produces a chimeric gene. (A) A common t(9;22) translocation in chronic myeloid leukemia joins exons of the *BCR* (breakpoint cluster region) gene and the *ABL1* oncogene to form chimeric genes. The smaller of the two derivative chromosomes is the Philadelphia (Ph[1]) chromosome. (B) Formation of a chimeric *BCR–ABL1* gene on the Philadelphia chromosome. Colored vertical bars represent exons. Yellow arrowheads indicate observed breakpoints in different patients, and the large X indicates the position of re-joining of the broken chromosomes—in this case we show recombination at the first (more 5′) of the breakpoints in the *BCR* gene, and at the second (more 3′) of the breakpoints in the *ABL1* gene. The resulting chimeric *BCR–ABL1* gene produces a large protein with constitutively active tyrosine kinase activity, which does not respond to normal controls.

Section 4.5. Because of the natural need to produce double-strand breaks in the immunoglobulin or T-cell receptor genes, there is a higher chance that these genes will participate in translocations. And because these large genes contain many different enhancer sequences, translocations can often result in the transcriptional activation of an oncogene that lies close to the reciprocal breakpoint (see Table 10.5).

Gain-of-function mutations

Oncogenes can also be activated by certain point mutations that make a specific change at one of a few key codons. Activating mutations in some cellular oncogenes are particularly common, especially when the genes make a product that links different biological pathways connected to cell proliferation and growth.

Take as an example the human homologs of the Ras (**ra**t **s**arcoma) oncogene family. There are three human Ras genes—*HRAS*, *KRAS*, and *NRAS*—that make highly related 21 kDa Ras proteins with 188 or 189 amino acids. The Ras proteins work as GTPases and mediate growth signaling by receptor tyrosine kinases in mitogen-activated protein (MAP) kinase pathways. They are heavily implicated in cancer because they act as signaling hubs (a single Ras protein interacts with multiple intercellular signaling proteins and can transmit a signal from a receptor tyrosine kinase to a variety of different downstream signaling pathways, thereby affecting multiple processes). About one in six human cancers has activating mutations in one of the RAS genes, most commonly in *KRAS* (which is naturally expressed in almost every tissue). More than 99% of the activating Ras mutations are in one of only three key codons, codon 12 (Gly), codon 13 (Gly), and codon 61 (Gln).

The bias toward missense mutations, and the very pronounced narrow distribution of where the mutations occur, distinguishes oncogenes from tumor suppressor genes (see **Figure 10.8** for examples).

Tumor suppressor genes: normal functions, the two-hit paradigm, and loss of heterozygosity in linked markers

Tumor suppressor genes make products that keep cells under control by restraining cell proliferation. Some tumor suppressor genes, sometimes called *gatekeeper genes*, do this directly. Their products regulate cell division—by regulating the cell cycle and inducing cell cycle arrest, as required, or by working in upstream growth signaling pathways—or they promote apoptosis. Other types of tumor suppressor gene work more indirectly. They include *caretaker genes*, which help to maintain the integrity of the genome as detailed below, and *landscaper genes* that control the stromal environment in which the cells grow.

Unlike oncogenes, a tumor suppressor gene contributes to cancer when the gene is lost or inactivated in some way. Whereas mutated oncogenes act in a dominant manner at the cellular level, mutated tumor suppressor genes often act in a recessive manner. For classical tumor suppressor genes, inactivation of one copy of a tumor suppressor gene has little effect; the additional loss or inactivation of the second gene is required in the tumorigenesis process. For these genes, the tumor suppressor locus needs to sustain two 'hits' to make a significant contribution to tumorigenesis (**Figure 10.9A**).

Familial cancers and the two-hit paradigm

From what we have described so far, the idea of familial cancers might seem strange; nevertheless, they do account for a minority of cancers. Familial cancers nearly always involve inheritance of a loss-of-function

Figure 10.8 Oncogenes differ from tumor suppressor genes in the distribution and range of cancer-associated mutations. The distributions of cancer-associated missense mutations (red arrowheads) and mutations introducing a premature termination codon (PTC; blue arrowheads) are mapped to the corresponding regions of protein products for two representative oncogenes (*PIK3CA* and *IDH1*) and two tumor suppressor genes (*RB1* and *VHL*). Colored bars on the pale green background represent functional domains and motifs. The data were collected from genomewide studies annotated in the COSMIC database (release version 61). For *PIK3CA* and *IDH1*, mutations obtained from the COSMIC database were randomized, and the first 50 are shown. For *RB1* and *VHL*, all mutations recorded in COSMIC were plotted. Note the predominance of missense mutations in the oncogenes and how they are restricted to just a very few codons. Abbreviation: aa, amino acid residues. (From Vogelstein B et al. [2013] *Science* 339:1546–1558; PMID 23539594. With permission from the AAAS.)

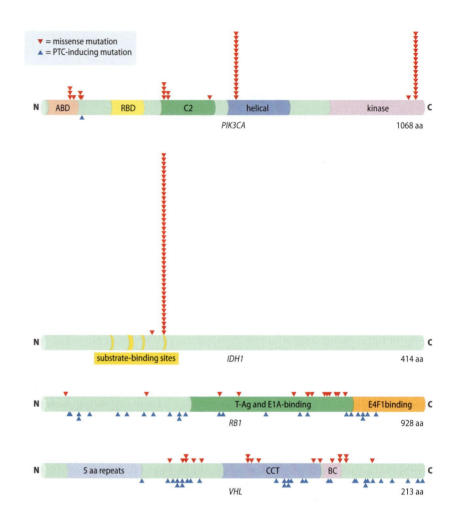

allele in a tumor suppressor gene (but see below for the example of inherited mutations in the *RET* oncogene).

The two-hit hypothesis proposed by Alfred Knudson explained why certain tumors can occur in hereditary or sporadic forms. In the hereditary form, one inactivating mutation (the first hit) in a tumor suppressor gene is inherited and the second hit occurs in the somatic cancer progenitor cell; in the sporadic form, two successive inactivating hits, one in each allele, occur in a somatic cell to initiate tumorigenesis.

Retinoblastoma, a cancer of the eye that represents 3% of childhood cancers, provided the first support for the two-hit hypothesis. In retinoblastoma, tumors can occur in both eyes or in one eye. People with bilateral tumors often transmit the disorder to their children, but the children of a person with a unilateral retinoblastoma usually do not have retinoblastoma.

Statistical modeling indicated that hereditary cases of retinoblastoma probably developed after only one somatic mutational event. People with bilateral retinoblastomas were postulated to have inherited an inactivating mutation in one copy of a retinoblastoma-susceptibility locus, now called *RB1*; in that case each nucleated cell in the body would have one inactive *RB1* allele. Retinoblastomas develop from many poorly differentiated retinoblast progenitor cells that proliferate rapidly. There is therefore a high chance that within a population of a million or so retinoblasts carrying an inactivated *RB1* allele, more than one cell sustains an additional inactivating mutation in the second *RB1* allele. If multiple tumors can form, bilateral tumors are likely to occur.

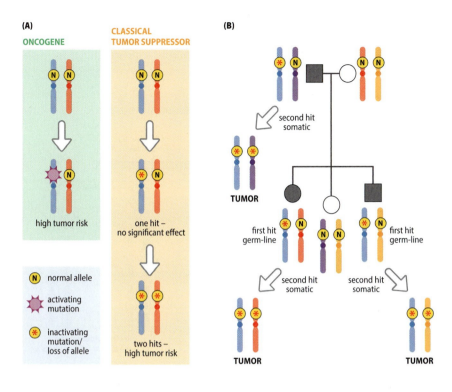

(A)

ONCOGENE

CLASSICAL TUMOR SUPPRESSOR

(B)

high tumor risk

one hit – no significant effect

two hits – high tumor risk

N normal allele

activating mutation

inactivating mutation/ loss of allele

second hit somatic

TUMOR

first hit germ-line

first hit germ-line

second hit somatic

second hit somatic

TUMOR

TUMOR

Figure 10.9 Classical tumor suppressor genes and the two-hit hypothesis.
(A) Activating mutations in a single allele of an oncogene are sufficient to confer a high risk of tumorigenesis. For a classical tumor suppressor locus to make a significant contribution to tumorigenesis, both alleles need to lose their function (the loss of function may occur through mutational inactivation or loss of the allele, or sometimes epigenetic silencing). Some tumor suppressors do not follow this simple model.
(B) Cancers due to mutations at a classical tumor suppressor locus are recessive at the cellular level (both alleles need to be inactivated) but cancer susceptibility can still be dominantly inherited. Inheritance of a single germ-line mutation (first hit, on the pale blue chromosome here) means that each cell of the body already has one defective allele and there is a very high chance of some cells receiving a second (somatic) hit. In sporadic forms of the disease, tumors are thought to arise by two sequential somatic mutations in the same cell.

If, however, two normal *RB1* alleles have been inherited, each tumor must occur by two successive hits at the *RB1* locus in one somatic cell. Unless the first somatic mutation just happened to occur very early in embryogenesis, the chances that two sequential somatic mutations would cause a loss of function of both *RB1* copies in more than one cell would be expected to be very rare. That makes unilateral retinoblastoma the expected outcome; the age of onset is generally later than in cases with inherited *RB1* mutations.

Note that while people with bilateral tumors can be confidently expected to have inherited a germ-line *RB1* mutation, a minority of people with unilateral tumors also have a germ-line mutation (by chance, tumor formation has only occurred in one eye). Inheritance of retinoblastoma susceptibility is dominant, but incompletely penetrant.

The two-hit paradigm explains why the cancer can be transmitted in a dominant fashion, even although the phenotype is recessive at the cellular level (both alleles of a tumor suppressor gene need to be inactivated or silenced); see **Figure 10.9B**. It also applies well to some other cancers that exist in both familial and sporadic forms but not, as we shall describe below, to some other cancers.

Loss of heterozygosity

For tumor suppressor genes, the initial hit is typically confined to the tumor suppressor locus—usually an inactivating point mutation. Inactivation of the second allele can also occur by a locus-specific DNA change—a point mutation, gene deletion or gene conversion—or sometimes by epigenetic silencing.

Often, however, the second allele is inactivated by large-scale DNA changes (loss of the whole chromosome, or a substantial part of it) or by mitotic recombination (**Figure 10.10**). In that case *loss of heterozygosity* will be evident: linked DNA markers that are constitutionally heterozygous in normal blood cells from an individual contain just a single allele in the tumor sample.

Figure 10.10 Different types of second hit at a tumor suppressor locus, some readily detectable by screening for loss of heterozygosity, and others not. Here the first hit at the tumor suppressor locus is shown as a small-scale inactivating mutation on the blue chromosome. A second hit that involves a large-scale (chromosomal) change (such as loss of the purple chromosome or loss of a part of that chromosome by mitotic recombination) can result in obvious *loss of heterozygosity* (readily detectable at the level of cytogenetic or DNA marker analysis). Sometimes, however, the second hit can be an inactivating mutation at the second allele or an epigenetic silencing event encompassing the tumor suppressor locus. In these cases, both alleles are unable to be expressed but loss of heterozygosity would not be evident by either cytogenetic analyses or DNA analyses using flanking markers.

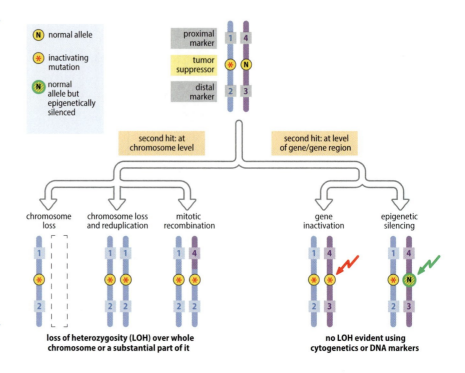

Loss of heterozygosity has been used as a way of mapping tumor suppressor genes. Paired samples of blood and the relevant tumor from individuals are screened with DNA markers from across the genome to identify chromosomes and, more profitably, chromosomal regions that show convincing loss of constitutional marker heterozygosity in the tumor samples. Analysis of multiple different tumors might lead to the identification of a quite small subchromosomal region defined by different mitotic crossovers or other breakpoints observed.

The key roles of gatekeeper tumor suppressor genes in suppressing G$_1$–S transition in the cell cycle

Understanding how cell division is regulated is of paramount importance in understanding cancer. Protein complexes made up of cyclins and cyclin-dependent kinases (CDKs) have key roles in regulating the cell cycle at certain cell cycle *checkpoints*.

The regulation of G$_1$, the phase of the cell cycle when cells make the decision whether or not to divide, is pivotal in tumorigenesis. A principal checkpoint occurs late in G$_1$, close to the G$_1$/S boundary, and is subject to intense regulation. A complex of cyclin E and the CDK2 protein works at this checkpoint to promote the transition from G$_1$ to S phase (which will usually commit the cell to cell division).

The CDK2–cyclin E complex is in turn regulated by interconnecting pathways. The control system has two arms in which the tumor suppressor proteins RB1 and p53 have commanding roles; another three tumor suppressor proteins, p14, p16, and p21 (the numbers refer to initially estimated molecular weights in kilodaltons) support p53 and RB1 in putting a brake on cell division (**Figure 10.11**).

Growth signaling pathways can induce a loss of RB1 function by stimulating CDK4–cyclin D or CDK6–cyclin D complexes that inactivate RB1 (by phosphorylating it) and the negative regulator MDM2 (which adds ubiquitin residues to target RB1 for destruction to keep RB1 levels low when growth is needed). Otherwise, p16 and p14 work to suppress the inhibition of RB1 and so put a brake on cell growth (see Figure 10.11).

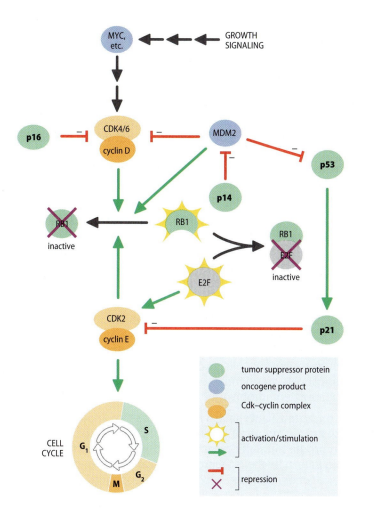

Figure 10.11 Major roles for p53, RB1, and accessory tumor suppressor proteins as brakes on cell growth. To permit cell growth, the CDK2–cyclin E complex promotes the G_1–S transition and is stimulated to do so by the E2F transcription factor. (E2F activates the transcription of multiple genes whose products are required for progression to S phase, notably cyclin E.) Five tumor suppressor proteins work in the opposite direction, as brakes on cell growth. RB1 inhibits the E2F transcription factor by binding to it to keep it in an inactive form. It, in turn, is repressed by CDK4/6–cyclin D and MDM2 but is assisted by proteins that repress its inhibitors: p16 (also called INK4a because it inhibits cyclin-dependent kinase 4) and p14 (also called ARF; it inhibits MDM2). p53 and p21 work in a pathway that bypasses E2F to inhibit the CDK2–cyclin E complex directly. Normally, p53 concentrations are kept low in cells but are increased in response to severe DNA damage. Elevated p53 both suppresses cell division (as shown here) and also stimulates apoptosis pathways (see Figure 10.12).

Elevated levels of p53 protein can stimulate p21 to inhibit the CDK2–cyclin E complex so as to induce cell cycle arrest. However, p53 is normally kept at relatively low concentrations in cells, mostly because MDM2 binds to p53 and adds ubiquitin groups to it, targeting it for destruction.

Signals from different sensors that detect cell stress—such as sensors of DNA damage—result in phosphorylation of p53. Phosphorylated p53 is not bound by MDM2 and so p53 levels increase. This can lead to cell cycle arrest (see Figure 10.11), which provides the opportunity to repair DNA, or to apoptosis when the DNA damage is too severe to repair. As detailed below, p53 has a pivotal role in cancer and is a rather unconventional tumor suppressor.

Note that although very different in sequence, the p14 and p16 tumor suppressors are both made from alternative splicing of a single gene, *CDKN2A* (see Figure 6.6B on page 156), and loss-of-function mutations in this one gene can inactivate both the RB1 and p53 arms of the cell cycle control system. Not surprisingly, *CDKN2A* mutations are important in tumorigenesis, and homozygous deletion or inactivation of this gene is quite common in cancers.

The additional role of p53 in activating different apoptosis pathways to ensure that rogue cells are destroyed

Cells that are unwanted, heavily damaged, or actively dangerous are normally induced to commit suicide through **apoptosis** (programmed cell death) pathways. Some apoptosis pathways work through a cell surface receptor that receives a 'death signal' from neighboring cells (examples include FAS receptors and other members of the tumor necrosis factor receptor superfamily). Other pathways, such as the mitochondrial apoptosis pathway, respond to certain types of internal damage such as that caused by harmful reactive oxygen species or exposure to dangerous levels of ionizing radiation.

In most cases the apoptosis pathway ends by triggering the cell to produce certain caspases, proteolytic enzymes that wreak havoc when they inactivate all kinds of important proteins in the cell; an endonuclease is also activated that cleaves DNA into small fragments. Because each of our normal cells has the potential to commit suicide, apoptotic pathways need to be very tightly regulated.

Various cancer-associated genes make products that regulate apoptosis. They include some tumor suppressors, notably *TP53*. When an unexpected double-strand break occurs in DNA, the DNA damage response activates high-level expression of p53. In response, p53 may activate transcription of various apoptosis-promoting genes in different apoptosis pathways (**Figure 10.12**).

From the above, we can see that p53 has dual central roles: it inhibits excessive cell proliferation, and it also acts as a 'guardian of the genome' by inducing apoptosis in response to double-strand DNA breaks (which are common in cancer cells). To promote cell proliferation and inhibit apoptosis, cancers frequently seek to inactivate both *TP53* alleles, and *TP53* is the most commonly mutated gene in cancer.

Note that oncogenes also have a role in inhibiting apoptosis. For example, the oncogene *BCL2* works in the mitochondrial apoptosis pathway, where its protein product inhibits cytochrome *c* release from mitochondria and is inhibited in turn by the BAX protein. In cancer cells, over-expression of certain oncogenes, such as *BCL2*, inhibits apoptosis.

Figure 10.12 Regulation of different apoptosis pathways by p53. When actively expressed at high concentrations, p53 stimulates the transcription of various genes to produce increased quantities of apoptosis-promoting proteins (indicated by vertical red arrows). They include cell surface receptors that are able to recognize death signals from neighboring cells, such as FAS receptors, and regulators of the mitochondrial apoptosis pathway, notably BAX and APAF1. FAS receptors are monomers, but when contact is made with the trimeric FAS ligand (FASLG) they form trimers. The FAS trimers recruit an adaptor (FADD), forming a platform for binding and activating procaspase 8. The BAX1 protein forms oligomers within the mitochondrial outer membrane that act as pores, allowing the release of cytochrome *c* into the cytosol. The released cytochrome *c* binds and activates APAF1, which in turn binds and activates procaspase 9. Activated procaspases 8 and 9 ultimately lead to mature effector caspases that destroy the cell.

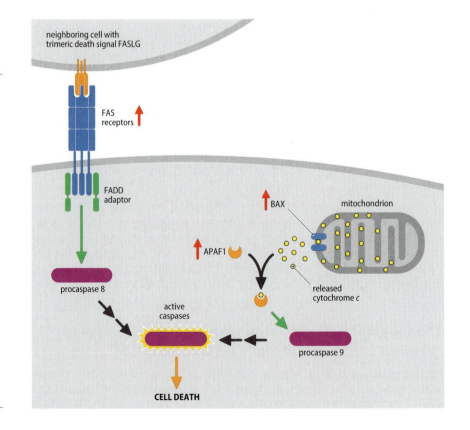

Rare familial cancers and the need for a revision of the classical two-hit tumor suppressor paradigm

Familial cancer is comparatively rare. Some rare examples are known of heritable oncogene mutations that cause cancer. For example, germ-line missense mutations in the *RET* proto-oncogene are found in familial thyroid cancer. But the great majority of familial cancers have germ-line mutations in tumor suppressor genes, including both gatekeeper genes such as *RB1* (with normal roles in restraining cell proliferation and/or promoting apoptosis) and caretaker genes (with genome maintenance roles, notably in DNA repair); see **Table 10.6** for some examples.

In retinoblastoma, few driver mutations are thought to be required for tumorigenesis (embryonic retinal progenitor cells are both poorly differentiated and rapidly proliferating, and so these cells already have two important tumor cell characteristics). The two-hit paradigm applies to additional types of cancer in which investigation of rare familial forms led to the identification of a tumor suppressor gene that was then found to be mutated in more common sporadic forms.

Some cancers that exist in both heritable and sporadic forms do not, however, readily fit the classical two-hit tumor suppressor paradigm. Major tumor suppressor genes implicated in the common sporadic tumors are often different from those involved in familial forms. This can be explained at least in part by disease heterogeneity. For example, *BRCA1*, the principal tumor suppressor gene implicated in familial breast cancer, is inactivated in only 10–15% of sporadic breast cancers. The latter form a recognizably distinct subset of sporadic breast cancers (and in these

FAMILIAL CANCER TYPE	GENE[a]	NORMAL FUNCTION OF GENE PRODUCT(s)
Defect in gatekeeper gene		
Familial adenomatous polyposis coli	*APC*	multiple functions, notably in signal transduction (Wnt pathway)
Familial melanoma	*CDKN2A*	two unrelated protein products, p14 and p16, facilitate p53-mediated cell cycle arrest (see Figure 10.11)
Gastric carcinoma	*CDH1*	regulator of cell–cell adhesion
Gorlin syndrome (basal cell carcinoma, medulloblastoma)	*PTCH*	sonic hedgehog receptor
Juvenile polyposis coli	*DPC4*	signal transduction (TGFβ pathway)
	SMAD4	
Li–Fraumeni syndrome (multiple different tumors)	*TP53*	the p53 transcription factor induces cells to undergo cell cycle arrest (see Figure 10.11) or apoptosis (see Figure 10.12)
Neurofibromatosis type 1 (NF1)	*NF1*	negative regulation of Ras oncogene
Neurofibromatosis type 2 (NF2)	*NF2*	cytoskeletal protein regulation
Retinoblastoma	*RB1*	acts as a brake on the cell cycle (see Figure 10.11)
Wilms tumor (childhood kidney tumor)	*WT1*	a transcriptional repressor protein with multiple functions including regulating the fetal mitogen insulin-like growth factor
Defect in caretaker gene		
Familial breast/ovarian cancer	*BRCA1*	makes product that interacts with double-strand DNA repair complex/components
	BRCA2	
Hereditary non-polyposis colorectal cancer (Lynch syndrome)	*MLH1*	DNA mismatch repair
	MSH2	

Table 10.6 Examples of familial cancers resulting from germ-line mutations in tumor suppressor genes. Gatekeeper genes include classical tumor suppressors that work in regulating cell division or upstream growth signaling pathways. Caretaker genes include other tumor suppressors that work in DNA repair or DNA damage responses. [a]Predisposing locus that shows germ-line mutations.

cases any second hit occurs by epigenetic silencing). Other data from many cancers have prompted the need for a radical overhaul of the classical two-hit suppressor hypothesis, as described in the next subsections.

Haploinsufficiency and gain-of-function mutations

Some cancer-susceptibility genes seem to lose the function of one allele but the second allele seems perfectly normal at the DNA level. Sometimes the second allele is epigenetically silenced. In other cases, however, inactivating a single allele seems to be sufficient to induce a tumorigenic change; that is, a significant contribution to tumorigenesis can be made by heterozygous loss of function, or *haploinsufficiency*.

Examples include some tumor suppressor genes involved in genome stability for which homozygous inactivation would be expected to lead to cell death (but can be averted by a third hit such as mutation in *TP53*). And mutation of just a single allele of certain tumor suppressor genes, such as *BRCA1*, has been shown to lead to genome instability in cultured cells and animal models.

Gain-of-function mutations can also occur in some tumor suppressor genes; in that case, a single mutated tumor suppressor allele can behave like an oncogene. For example, missense mutations are very common in *TP53*, which makes the p53 tumor suppressor, and the resulting mutant p53 proteins can behave in a dominant-negative fashion (**Box 10.3**).

A revised model for tumor suppression

Within the past decade, accumulating evidence from various sources suggests that even partial inactivation of tumor suppressors can make vital contributions to tumorigenesis. A revision of the classical two-hit suppressor hypothesis proposes that even subtle changes to the dosage of tumor suppressors can quite often make a substantial difference; the dosage effects can be highly tissue-specific and dependent on the context, such as the genetic background—see **Figure 10.13** for the strong example of dosage effects in the *PTEN* tumor suppressor and see the review by Berger et al. (2011) under Further Reading.

The significance of miRNAs and long noncoding RNAs in cancer

Irrespective of the class of cancer-susceptibility gene, the normal products are almost always proteins. However, hundreds of different noncoding RNAs are also known to be aberrantly expressed in cancer, including both microRNAs (miRNAs) and long noncoding RNAs (lncRNAs).

Figure 10.13 Tissue specificity and dependence on the genetic context of phenotypes resulting from reduced expression of the *PTEN* tumor suppressor gene. The effect of complete loss of *PTEN* expression can be very dependent on the tissue, as shown by the alternative phenotypes in prostate tissue and blood. In prostate tissue, cells are induced to senesce by wild-type p53; however, if the tumor has mutant p53, aggressive cancer results. In blood, bone marrow failure (resulting from hematopoietic stem cell exhaustion) occurs against a wild-type genetic background, but the co-occurrence of aneuploidy, additional mutations, and mutant p53 means that leukemia results. (Adapted from Berger AH et al. [2011] *Nature* 476:163–169; PMID 21833082. With permission from Macmillan Publishers Ltd.)

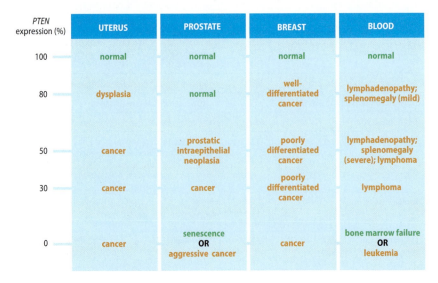

PTEN expression (%)	UTERUS	PROSTATE	BREAST	BLOOD
100	normal	normal	normal	normal
80	dysplasia	normal	well-differentiated cancer	lymphadenopathy; splenomegaly (mild)
50	cancer	prostatic intraepithelial neoplasia	poorly differentiated cancer	lymphadenopathy; splenomegaly (severe); lymphoma
30	cancer	cancer	poorly differentiated cancer	lymphoma
0	cancer	senescence OR aggressive cancer	cancer	bone marrow failure OR leukemia

Certain miRNAs are known to have important regulatory roles in processes relating to cancer, such as cell cycle control, cellular senescence, apoptosis, and DNA damage responses. Dysregulation of miRNA expression is frequent in cancer, and certain miRNA genes can be lost in cancer cells. On the basis of these observations alone, various miRNAs have been viewed as behaving as oncogenes or tumor suppressor genes. For example, the *MIR15A* and *MIR16-1* genes at 13q14 have been regarded as producing tumor suppressor miRNAs on the basis that they normally induce apoptosis by targeting BCL-2 but are frequently deleted or downregulated in chronic lymphocytic leukemia.

As we describe below, genomewide DNA sequencing is providing important insights into the genetic basis of cancer. Although modern whole-exome sequencing includes a wide coverage of many miRNAs in addition to protein-coding DNA, tumor-specific mutations in miRNA genes have not been evident. The dysregulated expression of miRNA and the loss of certain miRNA genes in cancers may be important events in cancer progression, but there is little direct evidence that miRNAs have unambiguous roles as oncogenes or tumor suppressor genes. Nevertheless, miRNA expression patterns may help us to dissect different disease subgroups, and there is interest in using miRNAs as therapeutic targets and as cancer biomarkers.

More recently, the possible roles of lncRNAs in cancer have begun to be investigated. They have been less well studied (and, for example, have not been covered in whole-exome sequencing studies), but there is strong evidence that some lncRNAs are important in cancer. In mice, for example, the *Xist* gene is not just involved in X-chromosome inactivation but also suppresses cancer *in vivo*; if *Xist* is deleted in blood cells, mutant females develop a highly aggressive myeloproliferative neoplasm and myelodysplastic syndrome with 100% penetrance (see also **Table 10.7**).

10.3 Genomic Instability and Epigenetic Dysregulation in Cancer

An overview of genome and epigenome instability in cancer

Genome instability is an almost universal characteristic of cancer cells, and frequently results from defects in chromosome segregation or DNA repair. By weakening the capacity to maintain the integrity of the genome, more DNA changes will be generated for natural selection to work on to drive tumor formation. Eventually, a cell can build up a sufficient number of DNA changes to become an invasive cancer cell.

Genomic instability can manifest itself at the chromosomal level or at the DNA level. Chromosomal instability (sometimes abbreviated as CIN)

LONG NONCODING RNA	INVOLVEMENT IN CANCER
ANRIL	represses expression of both p14 and p16 tumor suppressors; up-regulated in prostate cancer
H19	ectopic expression promotes cell proliferation; up-regulated in gastric cancer
HOTAIR	promotes cancer metastasis; up-regulated in breast, gastric, and colorectal cancers
PTENP1	regulator of the PTEN tumor suppressor gene; lost in many human cancers
XIST	involved in X-chromosome inactivation but down-regulated in various cancer cell lines and a powerful suppressor of hematological malignancies *in vivo* in mice

Table 10.7 Examples of the involvement of long noncoding RNAs in cancer. For more examples, see Cheetham SW et al. (2013) *Br J Cancer* 108:2419–2425; PMID 23660942.

BOX 10.3 A central role in cancer for the *TP53* gene that makes a non-classical tumor suppressor, p53.

The *TP53* gene at 17p13 has a central role in cancer, being mutated in nearly half of all tumors. The gene product, p53, has many roles and is involved in numerous different features of cancer. However, much of its importance comes from its role as a 'guardian of the genome'—it connects DNA damage, a common feature in cancer cells (which frequently undergo genome instability as described in Section 10.3), to decisions to induce cell cycle arrest (see Figure 10.11) or apoptosis (see Figure 10.12).

The p53 control mechanism that seeks to nip tumorigenesis in the bud can never be a failsafe mechanism; as a back-up, two p53-related proteins, p63 and p73, are produced with functions that partly overlap those of p53. Nevertheless, p53 has the dominant role.

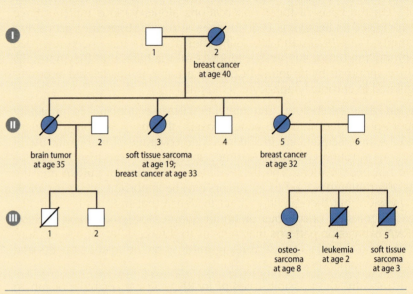

Figure 1 A typical pedigree of Li–Fraumeni syndrome. (Adapted from Malkin D [1994] *Annu Rev Genet* 28:443–465; PMID 7893135. With permission from Annual Reviews.)

As befits its crucial role, p53 is expressed in virtually all cells. Germ-line mutations in *TP53* underlie Li–Fraumeni syndrome (OMIM 151623), a dominantly inherited disorder in which affected individuals within a family can present with different early-onset tumors (**Figure 1**).

p53 as a nonclassical tumor suppressor

In several ways, p53 does not behave as a classical tumor suppressor. In most tumor suppressor genes (such as *RB1*, *APC*, *NF1*, *NF2*, and *VHL*) the primary mutations are mostly deletion or nonsense mutations that result in little or no expression of the respective proteins. *TP53* is different: the great majority of small-scale cancer-associated mutations are single nucleotide missense mutations that are very largely clustered within the central DNA-binding domain.

Six codons are predominantly mutated within the DNA-binding domain, and the missense mutations fall into two classes (**Figure 2**). In the DNA contact class, the missense mutation alters an amino acid that is normally used to make direct contact with the DNA of genes regulated by p53. The conformation class of mutations disrupt the structure of the p53 protein.

The mutated p53 proteins have multiple properties that distinguish them from wild-type p53. First,

unlike wild-type p53, mutant p53s do not participate in a self-limiting regulation. In normal cells, the amount of p53 is kept low because p53 is negatively

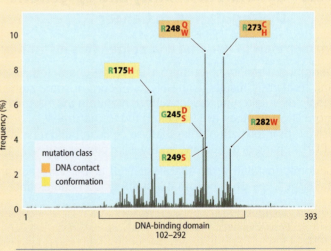

Figure 2 *TP53* **missense mutations are very largely confined to the DNA-binding domain, with six hotspots.** Vertical black lines indicate frequencies of missense mutations at each of the 393 codon positions. Two types of amino acid replacement are seen at codons 245, 248, and 273 (for example, at codon 273 arginine is replaced by cysteine or histidine). (Adapted from Freed-Pastor WA & Proves C [2012] *Genes Dev* 26:1268–1286; PMID 22713868. With permission from Cold Spring Harbor Laboratory Press.)

BOX 10.3 (continued)

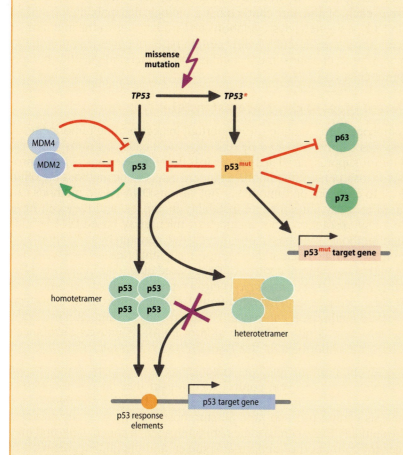

Figure 3 Missense p53 mutants have multiple novel properties and can show dominant-negative interactions with wild-type p53. Wild-type p53 works as a homotetramer to recognize and bind DNA sequences with specific motifs (p53 response elements) in the control regions of the p53 target genes. The p53 missense mutants suppress both wild-type 53 and the related p63 and p73 transcription factors. Mutant p53 is produced in very large amounts (unlike wild-type p53, it is not subject to self-regulation through stimulation of the MDM2 repressor) and interferes with normal p53-mediated transcription by interacting with wild-type 53 to form unproductive heterotetramers. Instead, mutant p53 stimulates the transcription of different genes.

regulated by MDM2 (and MDM4), and p53 positively regulates the production of its major antagonist MDM2; in cells with missense mutations in *TP53*, large amounts of mutant p53 are produced because mutant p53 fails to stimulate the production of MDM2. Mutant p53 can work in a dominant-negative fashion. It suppresses wild-type p53 and also the related p63 and p73 transcription factors (which show high sequence homology to p53 in some domains), and it antagonizes the interaction of wild-type p53 and the recognition sequences it must bind in its target genes (**Figure 3**). Instead, mutant p53 works as a rather different type of transcription factor by stimulating transcription of quite different target genes, including many genes that stimulate cellular proliferation or that inhibit apoptosis; see the review by Freed-Pastor & Prives (2012) under Further Reading.

is a particularly common form of genome instability. Tumor cells typically have grossly abnormal karyotypes (extra or missing chromosomes and many structural rearrangements), and they often show chromosomal instability in culture.

In addition, genome instability may be evident at the DNA level in cancer cells. As detailed below, a genomewide form of DNA instability is especially evident in some types of colon cancer. Sporadic colorectal tumors either show chromosome instability (in most cases) or global DNA instability (in about 15% of cases), but not both: the instability seems to be the result of natural selection.

A type of localized DNA instability is also occasionally seen: *kataegis* is a form of clustered hypermutation that was first reported in 2012, and we consider the details within the context of cancer genomics in Section 10.4.

Epigenetic dysregulation is a feature of all cancer cells, ranging from apparently normal precursor tissue to advanced metastatic disease. As well as being important in cancer progression (and helping cancer cells achieve each of the ten characteristic biological capabilities listed in Table 10.2), epigenetic dysregulation can be a key step in the initiation of cancer. And as we describe below certain types of epigenetic dysregulation also cause chromosome instability and accelerated genetic changes in tumor cells.

Different types of chromosomal instability in cancer

Chromosome abnormalities are important in accelerating tumorigenesis: oncogenes can be activated by rearrangements such as translocations, and tumor suppressor alleles can be lost through deletions, whole chromosome loss, or recombinations. Standard cytogenetic methods are often difficult to carry out on tumor cells, but various DNA-based methods can be used to study chromosome instability in cancer, and they can have quite a high resolution. They include two microarray-based DNA hybridization methods—one based on comparative genome hybridization, and one on SNP (single nucleotide polymorphism) analyses—which we introduce within the general context of DNA-based diagnosis in Section 11.1. Another method is *spectral karyotyping*, a type of multicolor chromosome FISH (fluorescence *in situ* hybridization). Unlike the microarray hybridization-based methods, it can reveal balanced chromosome abnormalities (in which there is no net loss or gain of DNA), as well as unbalanced chromosome abnormalities; see **Figure 10.14** for an application.

A major source of aneuploidies in cancer cells is defects in the spindle checkpoint, the cell cycle control mechanism that checks for correct chromosome segregation (it normally ensures that the anaphase stage of mitosis cannot proceed until all chromosomes are properly attached to the spindle). Extra centrosomes are often seen in cancer cells and may trigger the formation of abnormal spindles and unequal segregation of chromosomes into the daughter cells.

Structural chromosome abnormalities in cancer cells can arise in different ways. The most common source is an abnormal response to unrepaired DNA damage. As detailed in Section 4.2 we have complex DNA repair systems that can never be 100% efficient. Normally DNA damage responses act as a backup: they trigger apoptosis if the DNA damage is severe, or they arrest the cell cycle so that an unrepaired defect can be repaired. Defects in DNA repair of DNA damage responses allow unrepaired or damaged DNA to be passed on to daughter cells.

Failure to repair double-strand DNA breaks is an important source of structural chromosome abnormalities and can be precipitated by inactivation of key caretaker genes that function in this repair pathway, including the breast-cancer-associated *BRCA1* and *BRCA2* genes, which work in homologous recombination-mediated DNA repair. Proteins such as the ATM (ataxia–telangiectasia mutated) protein kinase work as sensors to detect unprogrammed double-strand DNA breaks. They then activate signaling mediators, which in turn recruit effectors to repair the damage. As well as DNA repair, the DNA damage response involves arresting the cell cycle, notably by activating p53 (**Figure 10.15**).

Chromothripsis

Sometimes chromosome breakage can involve an extensive localized rearrangement of chromosomes. In the process of *chromothripsis* large numbers of chromosomal rearrangements are generated in what appear to be single catastrophic events (**Figure 10.16**). The chromosome

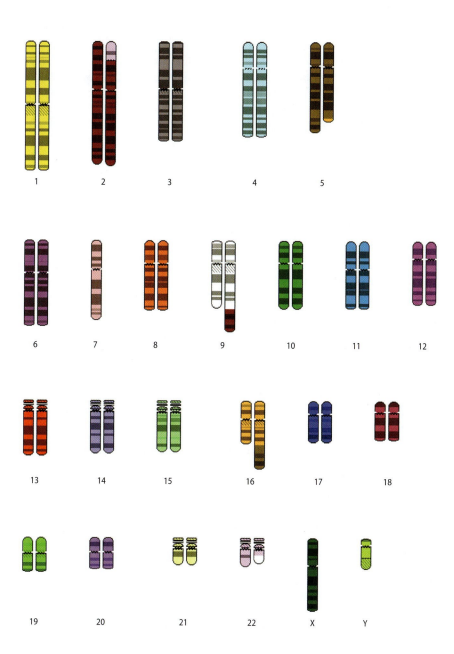

Figure 10.14 An example of using spectral karyotyping to analyze chromosomes in tumor cells. Spectral karyotyping (SKY) is a variant of chromosome fluorescence *in situ* hybridization (FISH) in which cocktails of many fluorescently labeled DNA probes from different regions of chromosomes are used to 'paint' chromosomes so that entire chromosomes are labeled with a specific fluorochrome and become fluorescent. Different chromosomes are painted with different combinations of multiple fluorescent labels. An image analyzer scans the fluorescent signals and can discriminate between the different fluorescence signals used for each of the 24 different chromosomes. To help us visualize the result it assigns artificial ('false') colors for each chromosome signal. In this example, there is a three-way variant of the standard 9;22 translocation (involving chromosome 2), plus an additional 5;16 translocation and the loss of one copy of chromosome 7. The karyotype is interpreted as 45,XY,t(2;9;22)(p21;q34;q11),t(5;16)(q31;q24),–7. (Case 6 of H Padilla-Nash, reproduced with permission from the NCBI SKY archive [http://www.ncbi.nlm.nih.gov/sky/].)

rearrangements may occur by chromosome shattering and aberrant rejoining of fragments by error-prone end-joining DNA repair pathways, or by aberrant DNA replication-based mechanisms. Chromothripsis may not be common in many cancers, but it is significantly more frequent in cells with mutated p53.

Telomeres and chromosome stability

In human cells, the telomeres shorten at cell division (usually by about 30–120 nucleotides at each cell division). By inactivating normal controls on cell growth, cancer cells can reach a stage where some telomeres become so short that the cell can misinterpret the ends of seriously short-ened telomeres as breaks in double-strand DNA. That alerts a DNA repair pathway that attempts a repair by fusing chromosomes at their ends. The resulting chromosomes with two centromeres may be pulled in oppo-site directions at mitosis, causing further broken ends and new cycles of chromosome fusion and breakage. Cancer cells seek to avoid this type of

Figure 10.15 Some cellular signaling responses to double-strand DNA breaks and different roles of BRCA1 and BRCA2 in homologous recombination-based DNA repair. Homologous recombination (HR) appears to be the major mechanism for repairing double-strand breaks in proliferating cells. Green arrows indicate stimulatory reactions; the red T-bar indicates inhibition. The ATM protein kinase is a prominent sensor of DNA damage. It is activated by phosphorylation (P), and in turn causes phosphorylation-mediated activation of CHEK2, which similarly activates p53 and BRCA1. Phosphorylated BRCA1 has multiple roles including activating protein complexes that are directly involved in HR-mediated double-strand (ds) DNA repair. These complexes—shown here by curly brackets—include the MRE11–RAD50–NIBRIN (MRN) complex, and also a complex in which activated BRCA1 recruits BRCA2 through an intermediary binding protein PALB2. As well as activating DNA repair, the DNA damage response may initiate cell cycle arrest, notably by activating p53 (which works at the G_1–S checkpoint); if DNA damage cannot be readily repaired, it can also activate apoptosis through enhanced p53 production.

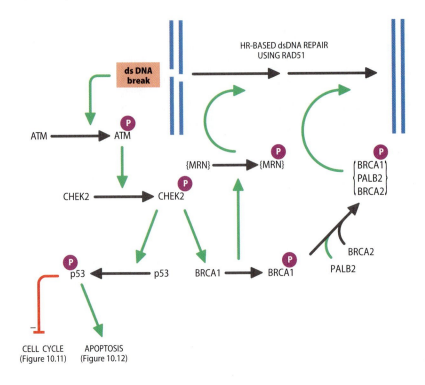

chromosome instability, and most frequently the solution involves ensuring that telomerase somehow becomes expressed (see Box 10.1).

Deficiency in mismatch repair results in unrepaired replication errors and global DNA instability

Mutation in genes involved in different types of DNA repair leads to cancer. Defective homologous recombination-based DNA repair is associated with various types of cancer, notably breast, ovarian, and pancreatic cancer. Genetic deficiency in components of nucleotide excision repair produces syndromes with increased cancer susceptibility,

Figure 10.16 Chromothripsis, a catastrophic shattering and rearrangement of the DNA sequence on selected chromosome regions in cancer cells. In what appears to be a single catastrophic event, multiple chromosome rearrangements appear to be simultaneously generated within a chromosome region. The chromosome rearrangements can occur by chromosome shattering and the rejoining of fragments by error-prone end-joining DNA repair pathways, or by aberrant DNA replication-based mechanisms. The rejoining typically leads to a rearranged order for component sequences, and the loss of some DNA fragments. Some fragments may persist extrachromosomally after amplification to form *double minute* chromosomes (a similar process to the oncogene amplification shown in Figure 10.6B). For a review, see Wyatt AW & Collins CC (2013) *J Pathol* 231:1–3; PMID 23744564. (Adapted from Tubio JMC & Estivill X [2011] *Nature* 460:476–477; PMID 21350479. With permission from Macmillan Publishers Ltd; and Stephens PJ et al. [2011] *Cell* 144:27–40; PMID 21215367. With permission from Elsevier.)

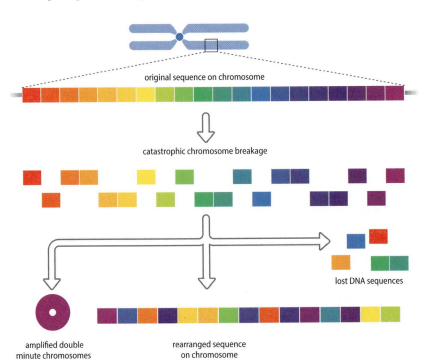

notably xeroderma pigmentosum (OMIM 278700). Genetic defects in base excision repair are associated with certain neurological disorders and occasionally cancer.

Germ-line mutations in both copies of the *MUTYH* gene (which is involved in the repair of adenines that are inappropriately base-paired with guanine, oxoguanine, or cytosine) result in an autosomal recessive form of familial adenomatous polyposis (FAP), a type of hereditary colon cancer in which multiple polyps (adenomas) develop. Deficiency in **mismatch repair**, which corrects errors in replication that for some reason have not been detected by the proofreading activity of a DNA polymerase, results in a global form of DNA instability and is most commonly associated with colon cancer.

The mechanism of mismatch repair

The mismatch repair (MMR) components work closely with the DNA replication machinery. In human cells, three types of protein dimer carry out most of the repairs (**Figure 10.17A**). Two of them—hMutSα and hMutSβ—are needed to identify base mismatches. The former identifies base–base mismatches but can also handle mismatching due to single-nucleotide insertions or deletions; hMutSβ can spot base mismatching for different sizes of very short insertions or deletions (which frequently occur at short tandem repeats as a result of replication slippage, the tendency for DNA polymerase to stutter or skip forward at tandem repeats; see Figure 4.8).

The MMR machinery cannot simply repair one of the two strands at random: there has to be a way of distinguishing the original (correct) strand from the newly replicated strand with the incorrect sequence that needs to be repaired. Before being repaired by DNA ligase, nicks (single-strand breaks) are common on a freshly replicated DNA strand, and in human (and eukaryotic) cells the strand distinction is achieved by identifying a nearby nick on the newly replicated DNA strand. Then hMutLα cleaves the newly replicated strand close to the mismatch and recruits an exonuclease to excise a short stretch of DNA containing the replication error so that the DNA can be resynthesized and repaired (**Figure 10.17B**).

Consequences of defective mismatch repair

Loss of function for both alleles of a mismatch repair gene can result in a form of global DNA instability (in which replication errors in newly synthesized DNA go uncorrected) and Lynch syndrome I (also called

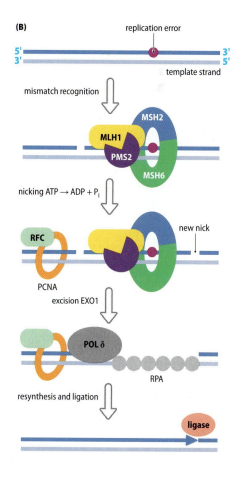

(A)

protein dimers	function	subunit
hMutSα	recognizes base–base mismatches and single nucleotide insertions/deletions	MSH2
		MSH6
hMutSβ	recognizes short insertions/deletions caused by replication slippage	MSH2
		MSH3
hMutLα	forms complex with hMutS and DNA; contributes PSM2 endonuclease to make nick	MLH1
		PMS2

(B)

Figure 10.17 Mismatch repair for correcting replication errors. (A) Major classes of MutS or MutL dimers in human mismatch repair. (B) Mechanism of 5′-directed mismatch repair in eukaryotic cells. Replication errors on a newly synthesized strand result in base mismatches that can be recognized by a MutS–MutL complex. The MutS component works as a clamp that can slide along the DNA, allowing it to scan for a base–base mismatch (MutSα) or unpaired insertion/deletion loop (often MutSβ). MutLα, which has an endonuclease function, can form a ternary complex with MutS and DNA. After the newly replicated DNA has been identified (by having a preexisting nick in the DNA), PCNA (proliferating cell nuclear antigen) and RFC (replication factor C) are loaded onto the newly replicated DNA, where they help trigger the endonuclease function of PMS2 to make a new nick close to the replication error. EXO1 exonuclease is recruited to excise the sequence containing the replication error, making a gapped DNA. The resulting stretch of single-stranded DNA (stabilized by binding the RPA protein) is used as a template for the resynthesis of the correct sequence using high-fidelity DNA polymerase δ, followed by sealing with DNA ligase I. (Adapted from Geng H & Hsieh P [2013] In DNA Alterations in Lynch Syndrome: Advances in Molecular Diagnosis and Genetic Counseling [M. Vogelsand, ed.]. With permission from Springer Science and Business Media.)

```
121                                        130
TGC ATT ATG AAG GAA AAA AAA AAG CCT GGT
 C   I   M   K   E   K   K   K   P   G
```

Figure 10.18 A long homopolymeric region in TGFRB2-coding DNA is a weak spot in the defense against excessive cell proliferation in colorectal cells. The nucleotide sequence of codons 121–130 within exon 3 of the *TGFRB2* (transforming growth factor receptor β-2) gene is shown with predicted amino acids below. The sequence contains a perfect run of 10 adenines that is vulnerable to insertion or deletion by replication slippage (Figure 4.8), especially when cells become defective at mismatch repair. Resultant frameshift mutations lead to a failure to make the 567-residue TGFBR2 protein, with adverse consequences for TGFβ signaling.

hereditary nonpolyposis colon cancer; OMIM 120435). In some tumors, this can readily be detected by analyzing a selection of standard microsatellite DNA markers, which will show higher frequencies of minor additional bands. Tumors with this *microsatellite instability* are often classified as MSI-positive (or MIN-positive) tumors.

In cells in which the MMR machinery is defective, the mutation rate increases about 1000-fold and so generates large numbers of mutations to drive carcinogenesis. In coding sequences, inefficient repair of base mismatches and replication slippage errors can result in gene inactivation or mutant proteins: long runs of a single nucleotide are particularly vulnerable to frameshifting insertions or deletions, and nucleotide substitutions can result in nonsense or missense mutations.

Defective mismatch repair can occur occasionally in other types of tumor, but it is particularly associated with colon cancer. Why should that be? One explanation is that MMR deficiency sabotages a key defense system that protects against colorectal cell proliferation. In the colorectum, transforming growth factor β (TGFβ) is a particularly strong inhibitor of cell proliferation, and it specifically binds to a receptor on the surface of the cells of which the TGFBR2 protein is a key component. However, the *TGFBR2* gene is readily inactivated as a result of mismatch repair deficiency because it has a long sequence of adenines that make it vulnerable to frameshifting insertions and deletions (**Figure 10.18**). Somatic mutations in *TGFBR2* are found in about 30% of sporadic colorectal cancer but are very frequent in MSI-positive colorectal cancer.

Epigenetic dysregulation in cancer and its effects on gene expression and genome stability

Recall that in somatic cells much of the genome (including the heterochromatic regions and a significant, but variable, fraction of the euchromatin) is transcriptionally silenced. This is achieved by epigenetic modifications—DNA methylation, histone modifications, and nucleosome repositioning—that attract specific proteins to compact the DNA and deny access to the transcription machinery.

The epigenetic modifications ensure that cells have distinctive chromatin patterns. They allow specific gene expression patterns to be established that determine the identity of a cell (so that it behaves as a T cell or a cardiomyocyte, for example). In addition they help to maintain genome stability (by maintaining the stability and function of centromeres and telomeres, and by suppressing excess activity by transposons).

One rationale for epigenetic dysregulation in tumors is that it allows cancer cells to revert to less differentiated states, permitting more flexibility for the cancer cell to adapt to changing environments, and to assist the transformation required for the progression to cancer. It was initially thought to result simply from genetic changes in genes controlling epigenetic regulation (for a comprehensive list, see the review by Timp and Feinberg [2013] under Further Reading).

More recently, it has become clear that epigenetic changes can also initiate cancer formation. Thus, for example, tissue inflammation is now thought to result in altered cell signalling that can induce dedifferentiation, converting more differentiated cells into cells with the properties of stem cells. The latter cells can initiate tumor growth—see the paper by Schwitalla *et al.* (2013) under Further Reading. A further role for epigenetic dysregulation is in promoting chromosome instability, notably by inducing hypomethylation of highly repetitive DNA sequences that are normally heavily methylated (see below for more details).

Aberrant DNA methylation

DNA methylation profiles are much easier to obtain from solid tumors than are histone modification profiles; for that reason, much of our information on the epigenetic profiles of tumors has come from studies on DNA methylation. In human cells, DNA methylation is almost exclusively restricted to certain cytosines that have a neighboring guanine within the dinucleotide CG (or CpG, as it is often called). Methylated cytosines can be distinguished from unmethylated cytosines by treating DNA with sodium bisulfite. (Sodium bisulfite changes all unmethylated cytosines to uracils, which become thymines in replicated DNA; methylated cytosines do not react and are unchanged. We describe the method in detail in the context of diagnostic applications; see Figure 11.13 on page 454.)

In somatic mammalian cells, about 70–80% of the cytosines present in CG dinucleotides are present as 5-methylcytosine. The 5-meCG sequences are recognized and bound by specific proteins that are important in helping to organize the chromatin into compact formations that lead to transcriptional silencing. In cancer, however, the DNA methylation patterns are changed in two ways: extensive hypomethylation and selective hypermethylation.

Across the genome as a whole, cancer cells typically show a significant reduction in DNA methylation (*hypo*methylation). That includes very many genes; long blocks of sequences, enriched in repetitive DNA but containing about one-third of transcriptional start sites, are hypomethylated.

Loss of methylation in constitutively heterochromatic regions may produce aberrant transcriptional expression of highly repetitive DNA sequences, resulting in widespread chromosomal instability. That seems to be a very common event in early adenomas, for example, occurring shortly after the disturbance to the Wnt signalling pathway (mutations in *APC* or equivalent) shown in Figure 10.4B. There is uncertainty about how this happens. One hypothesis is that the demethylation of highly repetitive DNA sequences allows normally silenced retrovirus-like elements and other related transposable elements in the genome to become active and jump to new locations in the genome, creating havoc. Constitutional DNA hypomethylation and chromosomal instability are also features of some human disorders, such as ICF1 (immunodeficiency with centromeric instability and facial anomalies; OMIM 242860), an autosomal recessive disorder that often results from mutations in the *DNMT3B* DNA methyltransferase gene.

DNA *hyper*methylation commonly occurs at the promoters of a few hundred genes in cancer cells, including tumor suppressor genes, DNA repair genes, and genes encoding certain transcription factors that are important in differentiation. Some tumor suppressor genes, such as *CDKN2A* and *MGMT*, are frequently silenced in a wide range of tumors; for others, silencing is limited to certain types of cancer: *VHL* in renal cancer, for example, and *BRCA1* in breast and ovarian cancer. A more extensive form of DNA hypermethylation occurs in some cases for certain cancers; we discuss this later in the chapter, in the context of links between metabolic and epigenetic dysregulation in cancer.

Genome–epigenome interactions

Genetic and epigenetic alterations in cancer used to be regarded as separate mechanisms. Now we know that they work closely together to promote cancer development. They may be viewed as complementary ways in which cancer cells seek to attain greater *plasticity* in an effort to accelerate the normal rate of genetic changes, and so escape normal cellular controls. See **Table 10.8** for examples of genome–epigenome interactions in cancer.

Table 10.8 Examples of genome–epigenome interactions in cancer. [a]See Section 10.4. [b]By one estimate, over 60% of the point mutations in the genomes of tumors of internal organs (in tissues which are shielded from UV radiation) arise in CpG sequences. [c]As a result of genetic changes (top two rows) or non-genetic changes, such as altered cell signalling in inflammation, and so on.

	GENOME		EPIGENOME	
Change	Mutation in genes encoding an epigenetic regulator	⇨	Epigenetic dysregulation	**Effect**
	Mutation in certain genes encoding a metabolic regulator[a]	⇨	Epigenetic dysregulation	
Effect	C → T substitutions	⇦	Deamination of 5-meC[b]	**Change**
	Chromosome instability	⇦	Hypomethylation of highly repetitive DNA	
	Silencing of tumor-suppressor genes	⇦	Induced epigenetic changes[c]	

10.4 New Insights from Genomewide Studies of Cancers

Until quite recently, molecular genetic studies of cancer cells had focused on individual genes of interest. Databases were established to store information on DNA changes associated with cancer-associated changes in important cancer-susceptibility genes, such as the International Agency for Research on Cancer's *TP53* database (http://p53.iarc.fr/) and the Breast Cancer Information Core of the US National Human Genome Research Institute (http://research.nhgri.nih.gov/projects/bic/).

Once the sequence of the human (euchromatic) genome had been obtained, the age of cancer genomics could begin. Different genome-wide screens were devised to get comprehensive data from cancer cells, beginning with microarray studies that reported the relative abundance of transcripts from thousands of different human genes. To seek out novel cancer-susceptibility genes, whole-genome association studies have been used (of the kind described in Section 8.2), but whole-exome and whole-genome sequencing have been especially fruitful. More recently, high-resolution genomewide DNA methylation screens have been carried out for some types of cancer.

After the launch of the Cancer Genome Project in the UK in 2000, and The Cancer Gene Atlas (TCGA) in the USA in 2006, the International Cancer Genome Consortium (ICGC) was created in 2007 to coordinate efforts on a global scale. The burgeoning data coming out of the cancer genome projects are stored in dedicated databases and can be navigated with dedicated Web browsers (**Table 10.9**). In addition to transforming our understanding of cancer—we describe below some examples of new insights that have emerged—the new data will also have important consequences for cancer diagnosis and treatment.

Genomewide gene expression screens to enable clinically useful gene expression signatures

Pathologists have long been able to grade tumors, classifying them by how abnormal they appear under a microscope and by how quickly the tumor is likely to grow and spread. A common classification uses four grades, from grade 1 tumors (well-differentiated cells that are similar in appearance to the corresponding normal cells and multiply slowly) to grade 4 (the cells appear undifferentiated and tumors tend to grow rapidly and spread quickly). Molecular genetics offers the hope of greater resolution in discriminating between different tumors that might then have important clinical applications. After cytogenetic analyses, the next genomewide study to produce a significant advance came from studying gene expression profiles.

Genomewide gene expression screens typically use microarrays containing cDNA clones, or more recently (multiple different) gene-specific

ELECTRONIC RESOURCE	DESCRIPTION	WEBSITE URL
COSMIC (Catalog of Somatic Mutations in Cancer) database	stores and displays somatic mutation information and related details and contains information relating to human cancers; includes a census of human cancer genes at http://cancer.sanger.ac.uk/cancergenome/projects/census/	http://cancer.sanger.ac.uk/ cancergenome/projects/cosmic/
International Cancer Genome Consortium (ICGC)	international consortium whose goal is to get a comprehensive description of genomic, transcriptomic, and epigenomic changes in 50 different tumor types and/or subtypes that are of clinical and societal importance across the globe	http://icgc.org
The Cancer Genome Atlas (TCGA)	cancer genomics network of research centers in the USA; data are available at http://www.intogen.org/tcga and can be browsed at http://www.gitools.org/tcga	http://cancergenome.nih.gov
UCSC Cancer Genomics Browser	a set of Web-based tools to display, investigate, and analyze cancer genomics data and its associated clinical information	https://genome-cancer.ucsc.edu

oligonucleotide probes for thousands of human genes. Collectively, the microarray offers a way of tracking the expression of the known human genes by hybridization to labeled RNA or cDNA from tissue or cell samples (Section 3.2 explains the principles of microarray hybridization).

The screens involve *comparative hybridization*: the individual gene expression profiles from test samples (tumors or cancer cells in this case) are referenced against an internal standard. For example, cDNA from tumor cells might be labeled with a Cy5 fluorophore (red) and used with an internal control of cDNA prepared from normal human cells. The internal control is labeled with a suitably different fluorophore (such as Cy3, which fluoresces in the green spectrum).

Large-scale cancer gene expression screens began to be carried out in the late 1990s. In an early study of breast cancer, for example, a microarray with cDNA clones for 8102 human genes was used to interrogate Cy5-labeled cDNA samples prepared from individual breast tumors; the internal standard was Cy3-labeled cDNA from a mix of cultured human cell lines. The initial screen revealed a total of 1753 genes that showed interesting variation in expression (at least fourfold variation in the red–green fluorescence ratio in at least three or more of the analyzed samples). A follow-up study identified a smaller subset of 456 cDNAs that could be used to distinguish tumor subclasses, as shown in **Figure 10.19**.

Microarray expression data are often displayed as a matrix in which each column shows the data for one sample and each row shows the data for one gene. The cells within the matrix are colored to signify different ratios of the test (tumor) sample cDNA fluorescence signal (red) to the internal standard fluorescence signal (green), giving the appearance of a heatmap. The output is organized by hierarchical clustering methods, which place samples with similar expression data close together (seen more clearly at the top of Figure 10.19B) and also place genes with similar expression profiles close together, such as the genes within each of the gene clusters (i) to (v) in Figure 10.19.

Clinical applications

The desire for a more comprehensive way of classifying tumors of a particular cancer, such as breast cancer, has been motivated by the hope of improved diagnosis and the possibility of targeting treatment more effectively (for any one type of cancer, tumors of the same grade are often heterogeneous in terms of both the prognosis and the effects of individual treatments). Gene expression studies can also be used to assess the chemosensitivity of cancer cell lines that can be good proxies for certain types of tumor.

Large-scale gene expression studies have been important in providing a novel classification of individual tumor types (such as the one shown in Figure 10.19 for breast cancer subtypes). The resolution is a step up

Table 10.9 Examples of databases, Web browsers, and networks in cancer genomics.

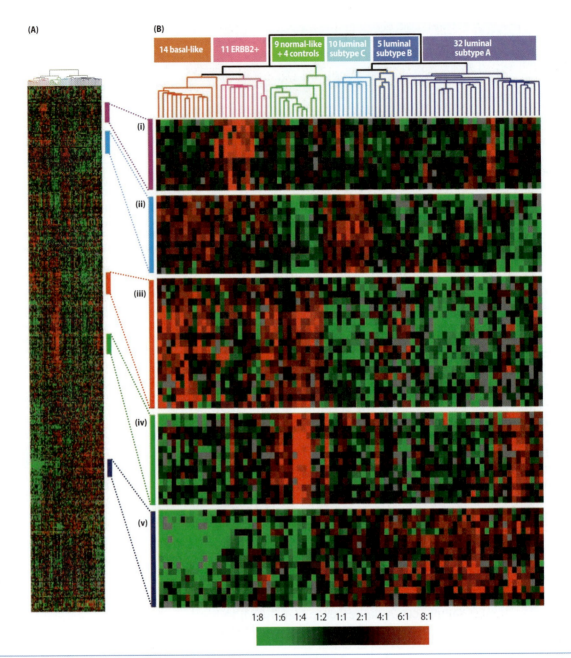

(A)

(B)

| 14 basal-like | 11 ERBB2+ | 9 normal-like + 4 controls | 10 luminal subtype C | 5 luminal subtype B | 32 luminal subtype A |

(i)

(ii)

(iii)

(iv)

(v)

1:8 1:6 1:4 1:2 1:1 2:1 4:1 6:1 8:1

Figure 10.19 Using expression microarrays to distinguish tumor subclasses. (A) The breast cancer gene expression patterns shown here were obtained by hybridizing 476 selected human cDNA probes (rows) against 85 labeled experimental samples (columns), comprising 78 carcinomas, 3 benign tumors, and 4 normal tissues. The sample columns and gene rows are grouped by *hierarchical clustering* so that samples with similar expression profiles are placed next to each other, as is more evident in the expanded view shown in (B). The tumor specimens could be divided into five (or six) subtypes based on differences in gene expression, as shown by the magnification of five regions (labeled from i to v). The major division is between luminal carcinomas (which are mostly estrogen receptor-positive; shown by the purple, deep blue, and light blue tags at top) and other types of breast carcinoma. The latter comprise basal-like (orange tag at top—they largely correspond to triple-negative breast cancers that lack estrogen and progesterone receptors and HER2, the product of the *ERBB2* oncogene), ERBB2+ (pink tag—they show *ERRB2* overexpression as a result of amplification), and cancers that appear very similar to normal breast epithelial cells (green tag). The magnified expression data show five clusters of genes, each containing genes with similar expression profiles: (i) *ERBB2* amplicon cluster; (ii) novel (previously unrecognized) cluster; (iii) basal epithelial cell-enriched cluster; (iv) normal breast-like cluster; (v) luminal epithelial gene cluster containing estrogen receptor. (Adapted from Sørlie T et al. [2001] *Proc Natl Acad Sci USA* 98:10869–10874; PMID 11553815. With permission from the National Academy of Sciences.)

from simple tumor classification based on histology and has led to the development of clinical assays that are now used to identify patients who have such a low risk of cancer recurrence that surveillance is employed instead of chemotherapy.

Subsequently, smaller gene sets have been used to provide gene expression signatures such as the 70-gene assay marketed by Agendia as MammaPrint®, which received US FDA (Federal Drug Agency) approval for prognostic prediction. Various other gene expression signatures have been developed that are intended to provide information on comparative risks and recurrence risks, for example.

Combinations of genomic, transcriptomic, and proteomic signatures may also be used for classifying tumors, as in the report by the Cancer Genome Atlas Research Network in 2013 on an integrated approach to endometrial carcinoma, which is listed under Further Reading.

Genome sequencing reveals extraordinary mutational diversity in tumors and insights into cancer evolution

Massively parallel DNA sequencing (sometimes called *next-generation sequencing* or second-generation sequencing) is transforming genetics: in contrast with standard Sanger dideoxy DNA sequencing, it offers a huge step up in DNA sequencing output (see Box 11.2 on page 448 for a brief summary). Because of the extraordinary complexity of cancer evolution it has been extensively applied to sequencing cancer genomes.

Initially, exome sequencing was used to analyze the DNA of cancer cells (the great majority of cancer genes make proteins, and many of the mutations occur within exons). Exome sequencing cannot readily detect copy number variation, and so cancer exome sequencing projects are supplemented by genomewide screens for copy number variation. More recently, sequencing of whole (euchromatic) cancer genomes has also been carried out to reveal all classes of change in somatic DNA in a tumor (when referenced against a corresponding normal tissue genome from the individual).

The first whole cancer genome—an acute myeloid leukemia genome—was reported in 2008. Since then, large numbers of whole cancer genome sequences have been determined; within a decade, by 2018, the sequences of many tens of thousands of such genomes can be expected.

The extraordinary volume of data pouring out of cancer genome sequencing is collated in different databases, notably the COSMIC database (see Table 10.9). Working out what the huge amount of sequence data means is inevitably a challenge, but already some valuable insights have been revealed.

Mutation number

How many mutations are there in a cancer? There are certainly more than we used to think. From multiple sequenced cancer genomes we now know that adult cancers often have between 1000 and 10,000 somatic substitutions across the genome. However, some types of cancer—medulloblastomas, testicular germ cell tumors, and acute leukemias, for example—have relatively few mutations; others, such as lung cancers and melanomas, have many more mutations (sometimes more than 100,000). If we focus on just the coding sequence (1.2% of the genome) and consider only the nonsynonymous mutations (which, by changing an amino acid, are more likely to have an effect on cell function than, say, synonymous mutations), the number of nonsynonymous mutations per tumor continues to shows a clear dependence on the type of tumor (**Figure 10.20**).

How can we explain the differences in mutation number? In part this is because different cancers can vary in the number of cell divisions that separate the fertilized egg and the cancer cell. And differences in mutation rate at the cell divisions from the fertilized egg cell to the cancer

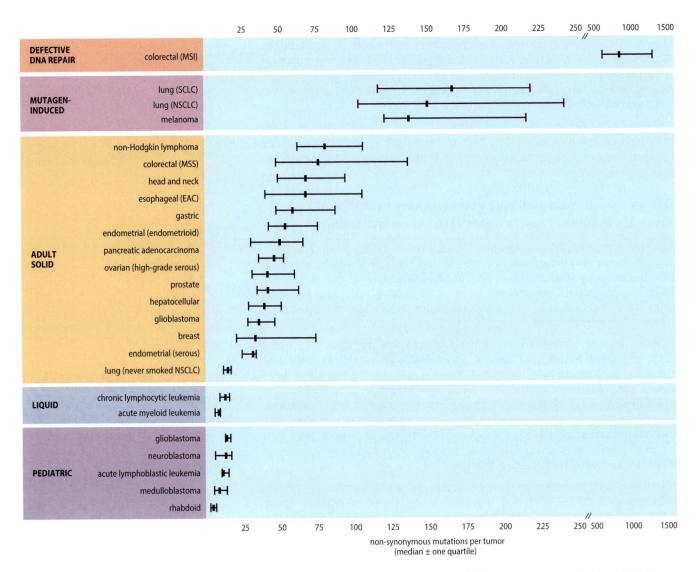

Figure 10.20 Variation in the number of somatic nonsynonymous mutations per tumor in representative human cancers. The median number of nonsynonymous mutations per tumor is estimated from genomewide sequencing in tumors. Horizontal bars indicate the 25% and 75% quartiles. MSI, microsatellite instability; SCLC, small-cell lung cancers; NSCLC, non-small-cell lung cancers; ESCC, esophageal squamous cell carcinomas; MSS, microsatellite stable; EAC, esophageal adenocarcinomas. (Data from Vogelstein B et al. [2013] *Science* 339:1546–1558; PMID 23539594.)

cell must be a factor. Tumors in children or young adults might have lower mutation prevalence simply because the cancer cell has been through comparatively few mitoses. The high mutation prevalence in lung cancers and melanomas most probably reflects an exceptionally high exposure or vulnerability to specific mutagens (tobacco carcinogens and UV radiation, respectively). The highest mutation frequencies are, perhaps inevitably, found in cancers that have incurred a mutation in their ability to repair replication errors (because of a defect in mismatch repair; see Figure 10.20).

Mutational processes and cancer evolution

Cancer genome and exome sequencing has permitted comprehensive studies of the mutational processes involved in the evolution of a cancer. In 2012 a series of papers provided the first comprehensive dissection of breast cancer. To the cancer surgeon, one of the most striking things about breast cancer is how it progresses differently in each patient, and how each patient responds differently to therapy. This is where molecular genetics might make a difference by helping identify subclasses of tumors with distinct properties that allow different treatment options to be applied, depending on the tumor subtype. The breast cancer studies revealed extraordinary mutational diversity, with multiple independent mutational signatures; they also indicated that in most such cancers more than one mutational process has been operative.

Specific mutational signatures found in some cancers simply reflect excessive exposure to specific environmental mutagens that preferentially cause particular types of mutation (for example, UV radiation causes preferential C:G→T:A transitions in melanoma, and tobacco carcinogens cause preferential C:G→A:T transversions in lung cancer). Splicing mutations are particularly common in some types of cancer, notably myelodysplastic syndrome (MDS) and chronic lymphocytic leukemia (CLL). This happens because in these cancers some genes that make components of the RNA splicing machinery (such as *U2AF1* in MDS, and *SF3B1* in both MDS and CLL) are frequently mutated.

But the comprehensive studies of breast cancer were the first to illustrate just how complex the mutational processes are (interested readers should consult the papers by Nik-Zainal et al. [2012] and Alexandrov et al. [2013] under Further Reading). One novel mutation process initially discovered from sequencing breast cancer genomes is a type of hypermutation called *kataegis* (from the Greek word for thunderstorm). If a breast cancer genome shows, say, 10,000 tumor-specific mutations, one might expect that the mutations would be mostly randomly distributed across the genome. In that case the average density of the mutations would be 10,000 per 3 Gb or roughly 1 mutation in every 300 kb of DNA. But sometimes highly clustered mutations of the same type are seen, such as C→T mutations (**Figure 10.21**). This type of hypermutation is thought to be due to excess activity of cellular APOBEC proteins, which naturally act as cytidine deaminases in processes such as antibody diversification (and RNA editing). By promoting excess activity by these enzymes, tumors find yet another way of generating multiple mutations that natural selection can work on to promote cancer development.

Intertumor and intratumor heterogeneity

Genome sequencing has provided the first full understanding of mutational differences between different tumors of the same type and mutational differences within tumors. An early indication of differences between colorectal tumors is shown in **Figure 10.22**. Here, the locations of mutations in two different tumors of the same type were found to be quite different. Most differences are due to passenger mutations. Some driver mutations may be found in common key cancer-susceptibility genes, such as the mutations in *APC* and *TP53* in Figure 10.22; others reside in different cancer-susceptibility genes.

The first comprehensive insights into intratumor heterogeneity came from a study of renal cancer: exome sequencing analyses were carried out on multiple biopsies taken from a primary renal carcinoma and

Figure 10.21 A rainfall plot showing an example of kataegis, a form of clustered hypermutation, in breast cancer. Cancer-specific mutations (a total of more than 10,000 from across the genome in this case) are ordered on the horizontal axis according to their position in the genome, starting from the first variant on the short arm of chromosome 1 (at the extreme left) to the last variant on the long arm of chromosome X (position number about 10,500) and are colored according to mutation type (see the color key at the right). The vertical axis shows the distance between each mutation and the one before it (the intermutation distance), plotted on a logarithmic scale. Most mutations in this genome have an intermutation distance of about 10^5 bp to about 10^6 bp, but the plot clearly shows a major region of hypermutation roughly centered on mutation position 4000 (corresponding to a 14 Mb region on the long arm of chromosome 6), where there is an extraordinary clustering of C→T mutations (red dots) that are spaced from their nearest neighbors by very short distances (often 100 bp or less). Within this region there are defined very short regions of intense C→T mutation clustering as shown in Figure 4 of the original article. (From Nik-Zainal S et al. [2012] *Cell* 149:979–993; PMID 22608084. With permission from Elsevier.)

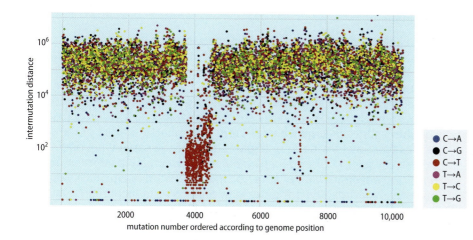

(A)

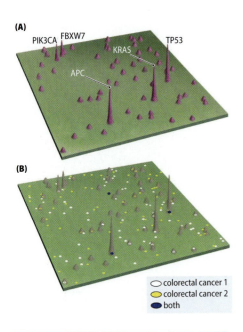

(B)

○ colorectal cancer 1
● colorectal cancer 2
● both

Figure 10.22 An early glimpse of the degree of intertumor mutational heterogeneity in colorectal cancer. (A) A pictorial display indicating the difference between genes that are commonly mutated in colorectal cancer (gene 'mountains' such as *APC*, *KRAS*, and *TP53*) and recurrent genes that are much less commonly mutated (gene 'hills'). (B) The very large difference between the mutational spectra of two colorectal cancers as revealed by exome sequencing. Apart from mutations in the *APC* and *TP53* genes, and in one other shared gene, the mutational spectra of the two cancers are distinct; the great majority of the differences are due to passenger mutations. (From Wood LD et al. [2007] *Science* 308:1108–1113; PMID 17932254. With permission from the AAAS.)

associated metastases from a single individual. Only about one-third of the 128 somatic mutations detected by exome sequencing in the different biopsies from this individual were present in all regions (the ubiquitous class shown in **Figure 10.23B**), and only one driver gene—*VHL*, the von Hippel–Lindau tumor suppressor gene—was mutated in all analyzed regions.

Another driver gene, *SETD2* (which encodes the histone H3K36 methyl-transferase), showed three distinct mutations associated with different regions (**Figure 10.23C**): the metastases had a missense mutation, region R4 exhibited a splice site mutation, and all other regions had a frameshift deletion (also present in R4). That is, selection pressure found three different ways of inactivating *SETD2* to produce similar tumor phenotypes. (A similar example of *convergent evolution* was apparent in the *KDM5C* gene, which encodes a histone H3K4 demethylase; see Figure 10.23C.)

In addition to the differences in mutation profiles between regions, even a single biopsy at R4 appeared to consist of two different clonal populations. The cells that seeded metastases seem to have diverged at an early stage from those that formed the primary tumor; the two groups had a differentiating series of mutations, and the cells that would form metastases lacked a mutation in a driver gene, *MTOR* (which encodes the mammalian target of rapamycin kinase).

Defining the landscape of driver mutations in cancer and the quest to establish a complete inventory of cancer-susceptibility genes

As described below, proteins made by known cancer genes have become targets for successful anti-cancer drug development. Identifying new cancer-susceptibility genes has therefore been a key goal of cancer genome studies. Until quite recently, most known cancer genes had been identified by three approaches: analyses of associated chromosome abnormalities (notably translocations) in which breakpoints could be identified by FISH, candidate gene studies (using information from experimental model organisms), and studies of copy number variation (by detecting oncogene amplification or by scanning for loss of heterozygosity). Linkage analyses had also been important in defining some genes that underlie inherited cancers, such as the *BRCA1* and *BRCA2* genes that are important susceptibility genes in breast and ovarian cancer.

To identify novel cancer-susceptibility genes, genomewide association studies were initially employed, but they have had limited success. Instead, genome or exome sequencing of multiple tumors has become the preferred approach: the sequences are referenced against normal somatic cells from the relevant individuals to identify tumor-specific mutations.

How can genomewide sets of tumor-specific mutations allow us to identify cancer-susceptibility genes? The driver mutations for any type of cancer might be expected to be confined to a comparatively small set of key cancer genes that are frequently mutated in that type of cancer (and which are presumably crucial to its evolution). Passenger mutations, by contrast, might be expected to be rather randomly distributed across the genome and to be somewhat different in unrelated tumors of the same cancer type. By looking at multiple tumors of the same type, one might expect to quickly discriminate between driver and passenger mutations. (However, it is not always so straightforward: a cluster of somatic mutations may also be attributable to an increased local mutation rate; in that case, passenger mutations may initially be confused with driver mutations.)

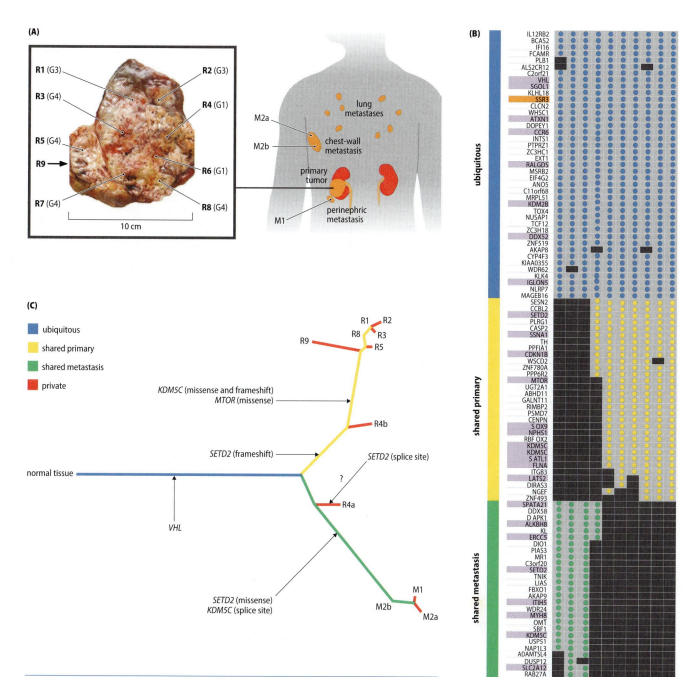

Figure 10.23 Intratumor genetic heterogeneity and branching evolution revealed by exome sequencing in a patient with a primary renal carcinoma and associated metastases. (A) Sites of core biopsies and regions harvested from nephrectomy of a primary renal carcinoma (regions R1 to R9; G indicates tumor grade) and from metastases M1 (perinephric), M2a, and M2b (chest wall). (B) Heatmap showing regional distribution of 133 point mutations (pale gray boxes with colored circle mean that mutation is present; dark gray boxes signify an absence of mutation) in seven primary tumor regions of the nephrectomy specimen, and in three metastases. (C) Phylogenetic relationships of the tumor regions, showing branching evolution. R4a and R4b are the subclones detected in R4. A question mark indicates that a detected *SETD2* splice-site mutation probably resides in R4a; R4b most probably shares a *SETD2* frameshift mutation found in other primary tumor regions. Branch lengths are proportional to the number of nonsynonymous mutations separating the branching points. Potential driver mutations were acquired by the indicated genes in the branch (arrows). (Adapted from Gerlinger M et al. [2012] *N Engl J Med* 366:883–892; PMID 22397650. With permission from the Massachusetts Medical Society.)

Cancer gene and driver mutation distribution

Genomewide sequencing approaches have been enormously successful, not just in identifying novel cancer-susceptibility genes but also in defining the distribution of the cancer-susceptibility genes in different cancers and the profile of associated driver mutations. In a study of 100 breast cancer tumors, for example, a total of 250 driver mutations were found; the number of driver mutations per tumor ranged up to a maximum of six, with an average of 2.5 (**Figure 10.24**).

In the above study, seven genes—*TP53*, *PIK3CA*, *ERBB2*, *MYC*, *FGFR1/ ZNF703*, *GATA3*, and *CCND1*—were found to be mutated in 10% or more of tumors, and collectively these genes were the source of almost 60% of the driver mutations detected (see Figure 10.24). In a parallel exome sequencing study of 510 breast cancers, also published in 2012, *TP53* and *PIK3CA* were also found to be the most frequently mutated genes.

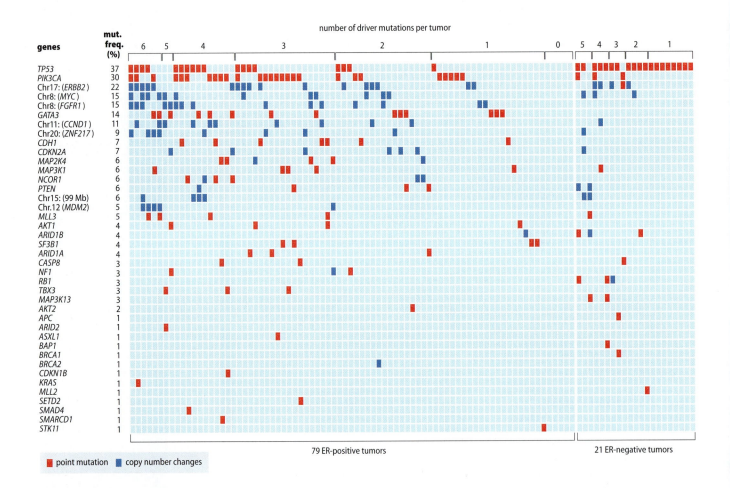

Figure 10.24 The landscape of driver mutations in a study of 100 primary breast cancers. Of the 100 cancers, 79 expressed estrogen receptor (ER-positive), and 21 were ER-negative. By referencing against control DNA samples from normal cells, somatic mutations were identified in 40 cancer-susceptibility genes (shown in the left-hand column). Point mutations were identified by whole-exome sequencing. Changes in copy number (shown in blue) include amplification of oncogenes and loss of alleles in tumor suppressors; they were identified by hybridization to whole-genome SNP (single nucleotide polymorphism) arrays (described at http://www.sanger.ac.uk/genetics/CGP/CopyNumberMapping/ Affy_SNP6.shtml). At least one of these genes or loci was mutated in all of the tumors, except in five ER-positive tumors. A maximum of six of the genes were mutated in any one tumor. The most significant (frequent) cancer genes were found to be *TP53* (mutated in 37% of all tumors, and in close to 90% of ER-negative tumors) and *PIK3CA* (mutated in 30% of the tumors). Mut. freq., mutation frequency. (Adapted from Stephens PJ et al. [2012] *Nature* 486:400–404; PMID 22722201. With permission from Macmillan Publishers Ltd.)

BREAST (100 TUMORS, WGS), PMID 22722201	COLORECTAL (224 MSI-NEGATIVE TUMORS), PMID 22810696	GLIOBLASTOMA (291 TUMORS), PMID 24120142	LUNG SCC (178 TUMORS), PMID 22960745	MELANOMA (135 TUMORS), PMID 22817889	OVARIAN (316 TUMORS), PMID 22720365
TP53 (37%)	APC (81%)	PTEN (31%)	TP53 (81%)	BRAF (63%)	TP53 (93%)
PIK3CA (30%)	TP53 (60%	TP53 (29%)	MLL2 (20%)	NRAS (26%)	CSMD3 (6%)
ERBB2 (22%)	KRAS (43%)	EGFR (26%)	PIK3CA (16%)	TP53 (19%)	FAT3 (6%)
MYC (15%)	TTN (31%)	NF1 (11%)	CDKN2A (15%)	CDKN2A (19%)	NF1 (4%)
FGFR1 (15%)	PIK3CA (18%)	PIK3CA (11%)	NFE2L2 (15%)	PTEN (11%)	BRCA1 (3%)
GATA3 (14%)	FBXW7 (11%)	PIK3R1 (11%)	KEAP1 (12%)	CCND1 (11%)	BRCA2 (3%)

Different types of tumor have different mutational spectra (Table 10.10), but in some cases, such as melanomas, a single gene may be an extremely important contributor to cancer development.

Novel cancer-susceptibility genes

The genomewide sequencing projects are delivering many novel cancer-susceptibility genes. However, the genes that are being discovered are ones that are infrequently mutated; as with other complex diseases, the major cancer-susceptibility genes have previously been identified. It may be that there will be a large number of low-frequency cancer-susceptibility genes. For example, the breast cancer genome sequencing study described in Figure 10.24 identified nine novel breast cancer-susceptibility genes, and a parallel exome sequencing study by the Cancer Genome Atlas Research Network of 510 breast cancers that was published at much the same time revealed a total of 10 novel breast cancer-susceptibility genes and yet only one of the novel breast cancer-susceptibility genes identified was common to the two studies: TBX3.

By March 2014 the Cancer Gene Census (at http://cancer.sanger.ac.uk/cancergenome/projects/census/) had listed a total of 522 identified human cancer genes. The quest is now very much on to identify all cancer-susceptibility genes. Until recently, the focus has very much been on looking at coding sequences. However, a study of regulatory regions in melanomas has emphasized the need to look at noncoding regions: whole-genome sequencing found that highly recurrent somatic mutations occur at two specific nucleotides in the promoter of the telomerase reverse transcriptase gene, TERT. Follow-up functional studies show that the effect of the mutations is to generate a binding site for the ETS transcription factor that up-regulates TERT expression. The TERT promoter mutations were found to occur in more than 70% of melanomas and about one in six of the other types of tumor examined in the study.

Novel cancer-susceptibility genes are going to be drawn from the genes that support the different biological capabilities of cancers, and many might not be the conventional oncogenes and tumor suppressor genes that we have become familiar with—the next subsection describes an important example.

Non-classical cancer genes linking metabolism to the epigenome

One of the surprises emerging from cancer genome sequencing has been the extent to which genes that work in metabolism are important in cancer. These genes can be non-classical oncogenes or tumor suppressor genes, and many of them have are been linked to epigenetic regulation.

Take, for example, the IDH1 and IDH2 genes that make respectively cytosolic and mitochondrial isocitrate dehydrogenase, enzymes that work in the tricarboxylic acid (Krebs) cycle to convert isocitrate to 2-oxoglutarate

Table 10.10 The six most frequently mutated genes identified in selected cancer genome/exome sequencing projects. Most studies used exome sequencing, but the breast cancer study involved whole-genome sequencing (WGS). Some genes are frequently mutated in many types of cancer cell, notably TP53 and to a smaller extent PIK3CA. Many other minor genes have especially important roles in individual cancers, such as BRAF in melanoma, and APC in colorectal cancer. MSI, microsatellite instability (apart from having a high frequency of APC mutations, the MSI-positive colorectal tumors had a rather different mutation spectrum). PMID, PubMed indentifier for relevant journal article.

(also known as α-ketoglutarate). One of these two genes is (heterozygously) mutated in 80–90% of adult grade II/III gliomas and secondary glioblastoma, in more than 50% of chondrosarcomas, in a significant proportion of acute myeloid leukemias, and in some other cancers. In terms of mutation types and distribution, the genes clearly fall in the oncogene camp (as previously noted for *IDH1* in Figure 10.8).

The predominant *IDH1/IDH2* cancer-associated mutations are specific missense mutations producing mutant enzymes that convert 2-oxoglutarate (produced by the normal allele) to 2-hydroxyglutarate. At high concentrations, 2-hydroxyglutarate inhibits multiple enzymes that depend on 2-oxoglutarate as a cofactor and work in epigenetic modification, including certain DNA demethylases, such as TET2, and various histone demethylases. That can cause reprogramming of the cell to make it less differentiated (**Figure 10.25**).

As well as oncogenes, tumor suppressor genes regulate the epigenetic–metabolic link in cancer cells. See Sebastian et al. (2012) under Further Reading for the example of the SIRT6 tumor suppressor, a histone deacetylase that normally suppresses aerobic glycolysis.

10.5 Genetic Inroads into Cancer Therapy

As described in Section 8.3, complex diseases are caused by a combination of genetic and environmental factors, and cancers are no different in this respect. Before we go on to look at therapeutic approaches directed at genetic control points in cancer, it is important to acknowledge the huge effect of environmental factors in cancer. In addition to well-established connections between UV radiation and melanoma, and between tobacco carcinogens and lung cancer, many other cancers are strongly determined by environmental factors. For example, rates of colon cancer

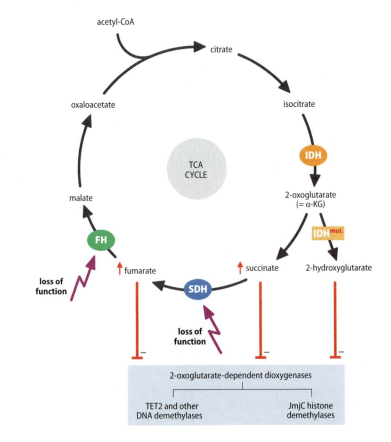

Figure 10.25 Mutation of certain genes encoding enzymes of the tricarboxylic acid (TCA) cycle can cause epigenetic modifications that contribute to cancer. Normal alleles of the *IDH1* and *IDH2* genes produce an isocitrate dehydrogenase enzyme that converts isocitrate to 2-oxoglutarate (also called α-ketoglutarate; α-KG). In certain cancers, certain missense mutations in *IDH1* (R132H, R132C) or in *IDH2* (R140Q, R172K) result in a mutant isocitrate dehydrogenase that can convert the 2-oxoglutarate made by a wild-type (normal) allele to 2-hydroxyglutarate. This abnormal oncometabolite changes the epigenetic profile of the cell, reversing differentiation to make the cell more like a stem cell. It does that by inhibiting multiple enzymes that depend on using 2-oxoglutarate as a cofactor, including some DNA demethylases such as TET2 and certain histone demethylases of the JumanjiC (JmjC) class. High levels of succinate and fumarate can also inhibit the 2-oxoglutarate-dependent enzymes, as when two loss-of-function alleles result in a genetic deficiency of fumarate dehydrogenase (FH) or succinate dehydrogenase (SDH), causing a buildup of substrate (red arrows).

vary as much as 20-fold between countries; the dramatic differences are due to environmental factors, specifically dietary components, rather than genetic susceptibility. That is evident when a population migrating from one country to another exhibits a colon cancer rate typical of the new country within one or two generations of settling there. Microbial infections, play their part too, and not just viral infections associated with cancers but also certain chronic bacterial infections – almost 1 in 11 of all cancer deaths arise because chronic infections of the stomach by *Helicobacter pylori* lead to gastric carcinomas.

How can the burgeoning knowledge of the underlying genetics of cancer have a clinical impact? As previous sections of this chapter testify, the revolution in cancer genomics has made clear the complexity of cancer evolution, and also the extraordinary degree of both intratumor and intertumor heterogeneity. This can pose difficulties in validating biomarkers for the oncogenic process: biopsies from the same tumor may show different genetic profiles (see Figure 10.23). And, because of intratumor heterogeneity, natural selection can be expected to propel the growth of drug-resistant clones.

Treatment or prevention?

Faced with these problems, should we simply accept that treating cancers is never going to be anything more than damage limitation, disease management rather than cure? Maybe. But genetics—and especially genomics—has shone a bright torch into the gloom that used to shroud the inner workings of many cancers. The result is a much more informed understanding of the fine, granular detail of the underlying mutation mechanisms, a greater appreciation of the molecular characteristics of cancers, and detailed insights into how cancers evolve. Once we have fully understood the molecular pathways of cancer and cancer evolution in fine detail, we may be in a much better position to devise novel treatments.

There are grounds for optimism in certain types of cancer therapy. In general, though, formidable challenges remain. This has prompted increased interest in a very different approach: prevention. Take the observation that most cancer deaths are due to carcinomas—notably breast, colorectal, lung, pancreatic, prostate, and ovarian—that have a long latency period (two decades or more). By the time they come to medical attention, the cells of these carcinomas may have accumulated mutations in hundreds of genes; there may be no single target for therapy.

But what if we could intervene during the long latency period? Imagine administering preventive drugs at an early stage, before the cancer becomes genetically complex, and long before the development of invasive and metastatic disease. That might seem a profitable approach. Because this type of cancer prevention depends on screening individuals to identify people at risk, we will consider it within the general context of genetic testing and screening in Chapter 11.

The efficacy of cancer therapy

Until quite recently, cancer therapy was very limited, relying heavily on three types of treatment. Where possible, tumors were surgically removed. Additional radiotherapy and cytotoxic chemotherapies, designed to kill actively dividing cells, have commonly been used; the problem here is that, in addition to cancer cells, actively dividing normal cells are killed (with significant adverse effects on the health of the patient). Despite their limitations, the long-established triad of surgical intervention, chemotherapy, and radiation therapy account for the majority of cancer treatment today.

Modern approaches to cancer therapy have had variable success. As detailed below, there have been some important successes using drugs designed to inhibit specific cancer targets. But cancer gene therapy, for example, has been generally disappointing. We considered the substantial recent successes in treating certain recessive disorders using gene therapy in Section 9.4. But, in essence, cancer gene therapy has led to minimal clinical improvement. In retrospect, the general failure of cancer gene therapy should not be surprising. The efficiency of gene transfection and of expression of the therapeutic gene in gene therapy trials has never been very high; even if the therapeutic gene were to be expressed at high levels, a substantial number of cancer cells might survive treatment so that tumor growth would be stunted for only a short period.

The more recent cancer gene therapy approaches are now mostly directed not at the tumor itself, but at healthy host tissue. Here, the strategy is sometimes to disrupt the tumor microenvironment (that supports tumor development), or to increase the resistance of healthy tissue to cytotoxic drugs. By far the most common way, however, is to transfer genes that are expected to provoke amplified immune responses against the tumor. Genetic modification of a patient's T cells using *ex vivo* gene therapy (see Section 9.3) may have particular therapeutic promise. One interesting development is to transfect a patient's T cells with a transgene that makes an artificial single-chain chimeric antigen receptor containing scFv variable domains from a tumor-recognizing antibody (see the review by Brenner et al. [2013] under Further Reading).

At various points in this book we have touched on the issue of **personalized medicine**, the idea that treatment should be individually tailored to the specific needs of individuals. Extensive intratumor and intertumor heterogeneity means that every tumor is different, and so cancer treatment has the potential to be the ultimate in personalized medicine. But in practice, and not least for economic reasons, **stratified medicine** is likely to be the way forward: tumors will be divided into major subtypes with more predictable responses to drugs.

We also need to consider rethinking the approaches that are taken. For example, in breast cancer most of the effort has been focused on limiting tumor growth, but a greater priority might be the need to prevent metastasis (metastasis is responsible for more than 90% of cancer-associated mortality). Currently, there are significant impediments to translational advances in this area, as summed up by Brabletz et al. (2013) under Further Reading.

In the sections below, we consider certain facets of how genetics and genomics are providing important inroads into cancer treatment, and some of the remaining challenges that need to be confronted in treating cancer.

The different biological capabilities of cancer cells afford many different potential therapeutic entry points

In our introduction to cancer we referred to the many different biological capabilities that are hallmarks of cancer cells. Each of these presents a focus for attacking cancer cells, so that quite different cancer treatments can be developed (**Figure 10.26**). If a biological property is very important for a cancer to develop, and if we could find some way of disabling that property, we might hope to stop the cancer in its tracks.

There is an enduring problem, however: the biological capabilities illustrated in Figure 10.26 are regulated by partially redundant signaling pathways. A therapeutic agent, designed to inhibit a key tumor pathway, may not completely shut down the relevant biological capability: an

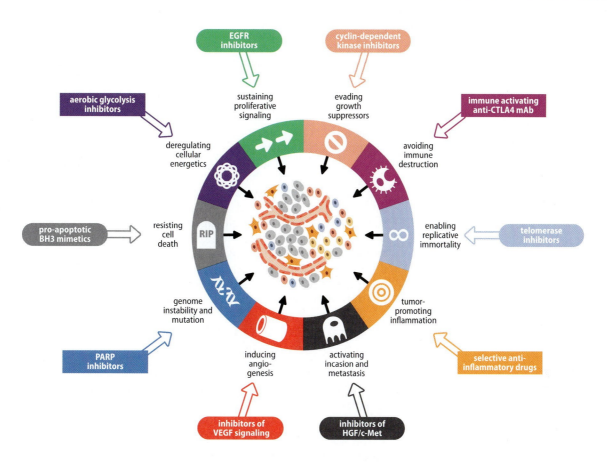

Figure 10.26 Therapeutic targeting of hallmarks of cancer. The 10 symbols in the inner ring with accompanying text descriptions denote 10 major biological capabilities proposed as characteristic properties (hallmarks) of cancers. The outer ring of text with colored backgrounds indicates relevant therapeutic options to address the indicated hallmarks (note that the drugs listed are simply illustrative—many different drugs have been used in clinical trials or have been approved for clinical use). EGFR, epidermal growth factor receptor; CTLA4, cytotoxic T lymphocyte-associated protein 4; HGF, hepatocyte growth factor; c-Met, receptor for hepatocyte growth factor (also called HGFR); VEGF, vascular endothelial growth factor, an inducer of angiogenesis; PARP, poly(ADP-ribose) polymerase, which detects single-strand breaks in DNA and signals their presence to the enzymatic machinery involved in repairing single-strand breaks. Pro-apoptotic BH3 mimetics are small molecule drugs that mimic the BH3 domain to antagonize BCL-2, a powerful inhibitor of the mitochondrial apoptosis pathway that is frequently strongly expressed by tumor cells. (From Hanahan D & Weinberg R [2011] *Cell* 144:646–674; PMID 21376230. With permission from Elsevier.)

alternative pathway can be activated to ensure tumor survival. For example, cancer development is strongly assisted by switching on telomerase; but in response to drugs that inhibit telomerase, it is common for cancer cells to find alternative ways of maintaining telomeres to ensure continued cell proliferation. A telomerase-independent alternative mechanism known as ALT (alternative lengthening of telomeres) is naturally used by the minority of human tumor cells that lack significant telomerase. It appears to involve sequence exchanges between different telomeres and is not an easy therapeutic target.

Control of angiogenesis seemed another promising line of attack. During the development of the vasculature in normal embryogenesis, new endothelial cells are formed and assembled into tubes (vasculogenesis); in addition, new blood vessels can sprout from existing ones (angiogenesis). Thereafter, the normal vasculature becomes largely quiescent. But during processes such as wound healing, angiogenesis can be turned on in adults, if only transiently. Like normal tissues, tumors need nutrients and oxygen and must get rid of metabolic waste and carbon dioxide. To fulfill these needs, angiogenesis is almost always activated and remains switched on to generate new tumor-associated blood vessels. In some preclinical models, potent inhibitors of angiogenesis are successful in suppressing this property; however, when confronted with angiogenesis inhibitors, tumors simply adapt by developing enhanced properties of tissue invasion and metastasis. By invading nearby tissues, cancer cells can gain access to alternative sources of preexisting tissue vasculature.

A further example of trying to find the Achilles heel of cancer is to exploit the propensity for cancer cells to disable DNA repair systems (so that genome instability is promoted). For example, homologous recombination-mediated repair of double-strand breaks is often disabled in tumor cells such as breast cancers (for example by inactivating BRCA1 or

BRCA2, components of double-stranded DNA repair; see Figure 10.15). In such cases, the therapeutic strategy is to disable a different, complementary DNA repair system. The loss of both DNA repair systems might be expected to lead to the death of cancer cells but not of normal cells in which the double-stranded DNA repair system remains functional. This is a type of *synthetic lethality*.

In the most common approach, drugs inhibit single-stranded DNA repair to counteract tumors in which homologous recombination-mediated repair of double-stranded DNA is defective. Small molecule drugs are designed to inhibit poly(ADP-ribose) polymerase (PARP), which senses any single-strand breaks and reports them to the enzymatic machinery that repairs single-strand breaks. When tumors deficient in double-stranded DNA repair are challenged with PARP inhibitors, they are initially unable to repair single-strand breaks, leading to collapse of the replication fork and to the presence of more serious double-strand breaks that might be expected to be lethal for the cell if not repaired. However, natural selection can foster the outgrowth of tumor cells that become resistant in different ways.

Targeted cancer therapy after genetic studies define a precise molecular focus for therapy

For many cancers, genetic analyses have identified an aberrant gene product that results from the mutation of a driver gene and which is highly characteristic of that cancer. In that case, conventional small molecule drugs or specific monoclonal antibodies might be directed to treating the cancer with considerable success.

The paradigm of targeted cancer therapy has been the treatment of chronic myeloid leukemia (CML) with imatinib. CML accounts for 20% of adult leukemia, and more than 90% of CML cases have the Philadelphia translocation chromosome (with breakpoints in the *ABL1* proto-oncogene and *BCR* genes; see Figure 10.7). The resulting hybrid gene makes the BCR–ABL1 fusion protein that is a constitutively active tyrosine kinase powered by ATP. The small molecule drug imatinib (marketed as Gleevec™) was developed to inhibit this tyrosine kinase by acting as a competitive inhibitor—it binds to the ATP binding site of the kinase. By blocking ATP binding, imatinib prevents the switch from the inactive enzyme conformation to the active conformation.

The specificity of imatinib is not perfect (it inhibits certain other kinases in addition to ABL1 kinase). It has nevertheless transformed the treatment of CML: it became the first-line therapy for CML almost immediately after its introduction and has had very significant success in prolonging survival times. Because resistance to imatinib can be conferred by point mutations within the ABL1 kinase domain, second-generation tyrosine kinase inhibitors (notably nilotinib and dasatinib) are used to provide alternative treatments.

The success of imatinib has driven a large number of other targeted cancer therapy trials (currently about 1200 different studies worldwide). The targets are typically the products of oncogenes; in particular, kinases are particularly promising targets for targeted cancer therapy because they are generally very susceptible to drugs. The *BRAF* oncogene, which encodes a serine kinase, is mutated in about 60% of malignant melanomas, and BRAF V600E, a particularly common mutant, seems to be responsive to therapeutic drugs, with some promising results in phase I trials (**Figure 10.27**).

Targeted cancer therapy can also be applied to over-expressed oncogenes; in this case, specific monoclonal antibodies can be raised to bind

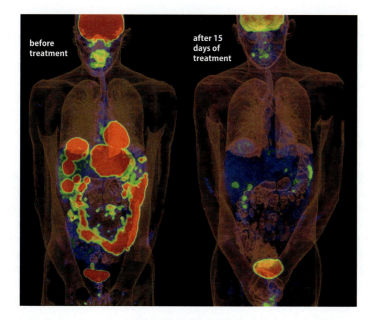

Figure 10.27 Targeted therapy for melanoma using a BRAF inhibitor. Images show three-dimensional representations of glucose metabolism in ^{18}F-fluorodeoxyglucose positron emission tomography scans obtained at baseline (before treatment) and 15 days after the initiation of treatment in a patient with melanoma carrying the V600E *BRAF* mutation. The patient was treated with the BRAF inhibitor PLX4032. Hypermetabolism of injected radioactive glucose is indicated by the red, green, and yellow signals; as well as being a characteristic feature of dividing cancer cells, it is a feature of normal brain and bladder metabolism or excretion. (Courtesy of Grant McArthur, Jason Callaghan, and Rod Hicks, Peter MacCallum Cancer Centre, Melbourne.)

to the amplified oncogene product. By blocking ligand–receptor interactions needed for cell growth and survival pathways, the antibodies can induce tumor cell death. For example, the *ERBB2* (*HER-2*) oncogene is amplified in many breast cancers, and targeted therapy often involves administration of the monoclonal antibody trastuzumab (Herceptin™).

The downside of targeted cancer therapies is that the resulting clinical responses have generally been transitory, and they are almost always followed by relapses when tumors develop resistance to the drugs that are being administered. In some cases, such as the treatment of chronic myeloid leukemia with imatinib, survival rates are nevertheless high (89% after one year); for others, such as targeting BRAF mutations in malignant melanoma with the drug PLX4032, the evolution of drug resistance is more rapid. We turn to a consideration of this problem in the final section of this chapter.

The molecular basis of tumor recurrence and the evolution of drug resistance in cancers

Cancer therapies usually do not kill all tumor cells—tumor recurrence is a major problem. Take, for example, glioblastoma multiforme, the most common primary malignant brain tumor. It has a median survival of about 1 year; the poor prognosis is due to therapeutic resistance and tumor recurrence after surgical removal. Although therapies that target specific molecular changes in tumors can be very successful (as described in the section above), the tumors do grow back, and quite quickly in some cases.

The basis of tumor recurrence

Why should tumors recur so quickly? In a genetically engineered mouse model of glioblastoma, a relatively quiescent subset of endogenous glioma cells, with properties resembling cancer stem cells, was recently found to be responsible for sustaining long-term tumor growth through the production of transient populations of highly proliferative cells. This suggested the possibility that cancer stem cells might be relatively resistant to therapy, thus surviving to repopulate a vastly shrunken tumor. If so, that would pose the problem of how to effectively target and kill populations of cancer stem cells about which we know little.

Another model implicates tumor heterogeneity, which has been forcefully demonstrated in various studies, such as that profiled in Figure 10.23. If a malignant tumor consists of genetically different populations, some cells might survive drug treatment and natural selection could foster the development of tumor subclones with mutations that render the therapeutic drug ineffective in some way. (There are therefore parallels with infectious diseases and the evolution of drug resistance in microbes.)

The evolution of drug resistance

The evolution of drug resistance in targeted cancer therapy can occur in different ways. Sometimes, mutations develop in the gene encoding the drug target itself. For example, in the treatment of chronic myeloid leukemia with imatinib, tumor subclones develop imatinib resistance by developing point mutations that alter the kinase domain of the BCR–ABL1 protein. The mutant kinase retains the catalytic activity required for tumor formation, but imatinib can no longer bind to it effectively to inhibit it. Drug resistance for many other kinase inhibitors works by a similar mechanism: often the mutations confer resistance by blocking interactions between drug and target through steric hindrance.

An alternative way of developing drug resistance occurs when the tumor mutates to amplify the drug target gene. Occasionally, for example, resistance to kinase inhibitors in CML is achieved when tumors succeed in amplifying the *BCR–ABL1* gene. Prostate cancers often acquire resistance to drug-mediated androgen deprivation by amplifying the androgen receptor gene.

Yet another option for a tumor to develop drug resistance is to find a way of bypassing the primary drug target (which remains unaltered, and continues to be inhibited by the drug). This can take the form of mutating a downstream effector in the same pathway to render cells insensitive to drug inhibition of a cell surface receptor, for example; or an alternative pathway is activated. For example, the monoclonal antibody trastuzumab is designed to treat breast cancer by binding to and interfering with the human epidermal growth factor receptor 2 (HER2), but tumors can bypass the effects of the drug by activating expression of the alternative receptors, including HER3.

Combinatorial drug therapies

Because of tumor heterogeneity, is targeted drug therapy always doomed to eventual failure? One potential approach is to develop therapies that rely on using combinations of drugs that might be targeted to different components, including the simultaneous targeting of upstream and downstream components in one pathway, or targeting parallel pathways (see the review by Al-Lazikani et al. [2012] under Further Reading). In the future, knowledge of the different genetic variants in individual malignant tumors might be expected to guide the choices of different drugs to be used.

Summary

- Cancers are diseases in which there is an unregulated increase in cell growth that leads to cells invading neighboring tissues and spreading to distant sites in the body.

- The genetic contribution to cancers predominantly occurs through somatic (post-zygotic) mutations. Germ-line mutations may result in inherited cancers, but even in these cases additional somatic mutations are required for cancers to form.

- Cancer development occurs only after a series of successive regulatory controls have gone wrong in cells, leading to increased cell proliferation or reduced apoptosis.

- Cancers are primarily diseases of later life, because it takes time for multiple cell controls to be disrupted.

- Tumors originate ultimately from a single cell but they are genetically heterogeneous. Descendants of the founder cell can acquire genetic mutations that afford a growth advantage; they form a dominant subclone that is then surpassed in growth by successive subclones (which acquire additional mutations conferring further growth advantages).

- In some types of cancer, undifferentiated cells are found with stem cell properties; they can self-replicate and also give rise to more differentiated cells within the tumor. Genetically different cancer stem cells may also arise by clonal evolution.

- Some cancers can arise through mutations in stem cells; others may arise through genetic and epigenetic changes in differentiated cells that cause the cells to become progressively less differentiated and progressively acquire other characteristics of cancer cells.

- Intratumor heterogeneity includes not just genetically different descendants from a single founding cell, but also non-tumor cells that are recruited to the tumor microenvironment, including some types of infiltrating immune cell.

- Cancer is a battle between Darwinian natural selection operating at the level of the individual (over generations) and at the level of the cell (within a single individual). Although cancer cells can successfully proliferate and form tumors within a person, they cannot leave progeny beyond the life of their host; tumorigenesis processes must start afresh in a new individual.

- Cancer cells usually contain thousands of somatic mutations. A small number, often from one to eight, are driver mutations that are crucially important in cancer development and are positively selected. The rest are chance (passenger) mutations resulting from genomic instability.

- Cancer genes can be grouped into two classes according to how they work in cells. In some cases mutation of a single allele is sufficient to make a major contribution to the development of cancer. For other cancer genes both alleles need to be inactivated to make a significant contribution to cancer.

- Oncogenes are dominantly acting cancer genes that result from an activating mutation in one allele of a cellular proto-oncogene. Classical proto-oncogenes typically work in growth signaling pathways to promote cell proliferation or inhibit apoptosis.

- Proto-oncogenes can be activated to become oncogenes by acquiring gain-of-function mutations; by being over-expressed as a result of gene amplification; or through activated expression resulting from a translocation (which repositions a transcriptionally silenced gene so that it comes under the control of transcription-activating regulatory elements).

- Classical tumor suppressors are recessively acting cancer genes in which the inactivation of both alleles promotes cell proliferation or inhibits apoptosis. Additional tumor suppressors work in other areas such as in genome maintenance.

- The two-hit hypothesis describes how cancer develops from two successive inactivating mutations in a tumor suppressor gene. It explains why dominantly inherited cancers are recessive at the cellular level (the first mutation occurs in the germ line and so there is a very high chance that the second allele is inactivated in at least one cell in the body to form a tumor). In sporadic cancers of the same type, both the first and second inactivating mutations occur in a somatic cell.

- Cancer cells become more plastic by shaking off normal controls on genome and epigenome stability.

- Genome instability ensures additional mutations for natural selection to work on to drive tumor formation. It is often manifested as chromosomal instability (resulting in aneuploidies, translocations, and so on) but can also be apparent at the DNA level as microsatellite instability (resulting from mutations in genes that work in mismatch DNA repair).

- Epigenetic dysregulation is important in both cancer initiation and cancer progression. It can be induced by genetic changes (notably mutation in genes that make epigenetic regulators) or by tissue inflammation causing altered cell signaling that results in altered chromatin states.

- Aberrant chromatin states produced by epigenetic dysregulation can allow cancer cells to become unspecialized (poorly differentiated) and can silence alleles of cancer-susceptibility genes. Additionally, DNA hypomethylation can result in widespread chromosome instability.

- Genomewide gene expression profiling of tumors can subdivide cancers of the same type, such as breast carcinomas, into different groups with different biological characteristics and different drug responses.

- Two tumors of the same type show very different mutational spectra—the great majority of passenger mutations are often distributed randomly across the genome; although some key cancer genes might be mutated in both tumors, other driver mutations may be located in different cancer-susceptibility genes.

- Tumors evolve, so cells in different regions of the same tumor can show regional mutational differences; metastatic cells typically share mutations that distinguish them from the primary tumor.

- Human cancer-susceptibility genes have been identified by analyzing associated chromosome breakpoints or associated changes in copy number (oncogene amplification, or loss of heterozygosity in the case of tumor suppressor genes); by studying candidate genes suggested by analyses of experimental organisms; and by exome or genome sequencing.

- In targeted cancer therapies, a drug or other treatment agent is directed at counteracting the effects of a specific genetic mutation that is known to be crucial for development of the cancer.

- Recurrence of tumors may be driven by cancer stem cells that are comparatively resistant to therapy.

- After initial success in shrinking tumors, cancer therapies often fail, causing a clinical relapse. Tumor cells evolve to become resistant to the drug as a result of natural selection (which promotes the growth of tumor cells that develop mutations to combat the effects of the drug).

- Tumors often develop drug resistance by changing the conformation of the drug target so that the drug is sterically hindered from binding to it; by amplifying the gene encoding the drug target; or by activating an alternative pathway that bypasses the effect on the drug target.

Questions

Help on answering these questions and a multiple-choice quiz can be found at www.garlandscience.com/ggm-students

1. There are more than 100 different cancers. What two key characteristics define these diseases?

2. The replication of DNA at the very ends of the telomeres of each of our chromosomes is problematic. Why is this, and what are the consequences?

3. One characteristic of cancer cells is that they can become immortal. How does this happen?

4. Intratumor heterogeneity involves various types of functional differences between the cells of a tumor. How do these differences arise?

5. DNA studies show that specific cancers are associated with distinctive mutational signatures. Give some examples.

6. Cancer whole-genome sequencing has enabled assays of the number of driver mutations in cancer genomes. Driver mutations can be assigned to specific gene loci according to whether distinctive types of mutation that might be expected to disturb gene expression are commonly identified in specific genes. However, the numbers of driver mutations identified can be quite small. In the breast cancer study shown in Figure 10.24, the number of driver mutations in 100 tumors, including both point mutations and changes in copy number, ranges from 0 to 6 per tumor. How do you interpret the small number of driver mutations identified?

7. The majority of clinical gene therapy trials are aimed at treating cancer, but unlike gene therapy for monogenic disorders, cancer gene therapy has been greatly disappointing. What factors make cancer gene therapy such a difficult prospect?

8. What is meant by targeted cancer therapies, and what advantages do they offer?

Further Reading

Cancer Biology

Weinberg, RA (2014) The Biology of Cancer, 2nd edn., Garland Science.

General Molecular Characteristics of Cancer

Hanahan D & Weinberg R (2011) Hallmarks of cancer: the next generation. *Cell* 144:646–674; PMID 21376230.

Shay JW & Wright WE (2011) Role of telomeres and telomerase in cancer. *Semin Cancer Biol* 21:349–353; PMID 22015685.

Vander Heiden MG, Cantley LC & Thompson CB (2009) Understanding the Warburg effect: the metabolic requirements of cell proliferation. *Science* 324:1029–1033; PMID 19460998.

Cancer Evolution, Cancer Stem Cells, and Intratumor Heterogeneity

Burrell RA, McGranahan N, Bartek J & Swanton C (2013) The causes and consequences of genetic heterogeneity in cancer evolution. *Nature* 501:338–345; PMID 24048066.

Greaves M (2007) Darwinian medicine: a case for cancer. *Nature Rev Cancer* 7:213–221; PMID 17301845. [Natural selection operating at the cell level in cancer.]

Greaves M & Maley CC (2012) Clonal evolution in cancer. *Nature* 481:306–313; PMID 22258609.

Hanahan D & Coussens LM (2012) Accessories to the crime: functions of cells recruited to the tumor microenvironment. *Cancer Cell* 21:309–322; PMID 22439926.

Landau DA, Carter SL, Stojanov P et al. (2013) Evolution and impact of subclonal mutations in chronic lymphocytic leukemia. *Cell* 152:714–726; PMID 23415222.

Leung CT & Brugge JS (2012) Outgrowth of single oncogene-expressing cells from suppressive epithelial environments. *Nature* 482:410–414; PMID 22318515.

Magee JA, Piskounova E & Morrison SJ (2012) Cancer stem cells: impact, heterogeneity and uncertainty. *Cancer Cell* 21:283–296; PMID 22439924.

Marusyk A, Almendro V & Polyak K (2012) Intra-tumour heterogeneity: a looking glass for cancer. *Nature Rev Cancer* 12:323–333; PMID 22513401.

Wu X, Northcott PA, Dubuc A et al. (2012) Clonal selection drives genetic divergence of medulloblastoma. *Nature* 482:529–533; PMID 22343890.

Oncogenes, Tumor Suppressor Genes, and Haploinsufficiency in Cancer

Berger AH & Knudson AG, Pandolfi PP (2011) A continuum model for tumor suppression. *Nature* 476:163–169; PMID 21833082.

Berger AH & Pandolfi PP (2011) Haploinsufficiency: a driving force in cancer. *J Pathol* 223:137–146; PMID 21125671.

Freed-Pastor WA & Prives C (2012) Mutant p53: one name, many proteins. *Genes Dev* 26:1268–1286; PMID 22713868.

Knudson AG (2001) Two genetic hits (more or less) to cancer. *Nature Rev Cancer* 1:157–162; PMID 11905807. [A historical perspective of the development of the two-hit tumor suppressor hypothesis.]

Roukos V, Burman B & Misteli T (2013) The cellular etiology of chromosome translocations. *Curr Opin Cell Biol* 25:357–364; PMID 23498663. [Gives some background on recurring chromosome translocations.]

Solimini NL, Xu Q, Mermel CH et al. (2012) Recurring hemizygous deletions in cancer may optimize proliferative potential. *Science* 337:104–109; PMID 22628553.

Storlazzi CT, Lonoce A, Guastadisegni MC et al. (2010) Gene amplification as double minutes or homogenously staining regions in solid tumors: origin and structure. *Genome Res* 20:1198–1208; PMID 20631050.

Vogelstein B & Kinzler KW (2004) Cancer genes and the pathways they control. *Nature Med* 10:789–799; PMID 15286780. [A historical perspective of oncogenes and tumor suppressor genes.]

RNA Genes In Cancer

Cheetham SW, Gruhl F, Mattick JS & Dinger ME (2013) Long noncoding RNAs and the genetics of cancer. *Br J Cancer* 108:2419–2425; PMID 23660942.

Esquela-Kerscher A & Slack FJ (2006) Oncomirs—microRNAs with a role in cancer. *Nature Rev Cancer* 6:259–269; PMID 16557279.

Farazi TA, Hoell JI, Morozov P & Tuschl T (2013) microRNAs in human cancer. *Adv Exp Med Biol* 774:1–20; PMID 23377965.

Genome Instability in Cancer

Forment JV, Kaidi A & Jackson SP (2012) Chromothripsis and cancer: causes and consequences of chromosome shattering. *Nature Rev Cancer* 12:663–670; PMID 22972457.

Lord CJ & Ashworth A (2012) The DNA damage response and cancer therapy. *Nature* 481:287–294; PMID 22258607.

Pena-Diaz J & Jiricny J (2012) Mammalian mismatch repair: error-free or error-prone? *Trends Biochem Sci* 37:206–214; PMID 22475811.

Roy R, Chun J & Powell SN (2012) BRCA1 and BRCA2: different roles in a common pathway of genome protection. *Nature Rev Cancer* 12:68–78; PMID 22193408.

Epigenetic and Metabolic Dysregulation in Cancer

Schwitalla S, Fingerle AA, Cammareri P et al. (2013) Intestinal tumorigenesis initiated by dedifferentiation and acquisition of stem cell-like properties. *Cell* 152: 25–38; PMID 23273993.

Sebastián C, Zwaans BM, Silberman DM et al. (2012) The histone deacetylase SIRT6 is a tumor suppressor that controls cancer metabolism. *Cell* 151:1185–1199; PMID 23217706.

Shen H & Laird PW (2013) Interplay between the cancer genome and epigenome. *Cell* 153:38–55; PMID 23540689.

Timp W & Feinberg AP (2013) Cancer as a dysregulated epigenome allowing cellular growth advantage at the expense of the host. *Nature Rev Cancer* 13:497–510; PMID 23760024.

You JS & Jones PA (2012) Cancer genetics and epigenetics: two sides of the same coin. *Cancer Cell* 22:9–20; PMID 22789535.

Cancer Genomics: Review Articles

Garraway LA & Lander ES (2013) Lessons from the cancer genome. *Cell* 153:17–37; PMID 23540688.

Stratton MR (2011) Exploring the genomes of cancer cells: progress and promise. *Science* 331:1553–1558; PMID 21436442.

Vogelstein B, Papadopoulos N, Velculescu VE et al. (2013) Cancer genome landscapes. *Science* 339:1546–1558; PMID 23539594.

Watson IR, Takahashi K, Futreal PA & Chin L (2013) Emerging patterns of somatic mutations in cancer. *Nature Rev Genet* 14:703–717; PMID 24022702.

Cancer Genomics: Original Articles

Alexandrov LB, Nik-Zainal S, Wedge DC et al. (2013) Signatures of mutational processes in human cancer. *Nature* 500:415–421; PMID 23945592. [Identifies more than 20 distinct mutational signatures from close to 5 million mutations in over 7000 cancers.]

Cancer Genome Atlas Research Network (2013) Integrated genomic characterization of endometrial carcinoma. *Nature* 497:67–73; PMID 23636398. [Genomic features of endometrial carcinomas permit a reclassification that may affect post-surgical treatment for women with aggressive tumors.]

Curtis C, Shah SP, Chin SF et al. (2012) The genomic and transcriptomic architecture of 2000 breast tumours reveals novel subgroups. *Nature* 486:346–352; PMID 22522925.

Gerlinger M, Rowan AJ, Horswell S et al. (2012) Intratumor heterogeneity and branched evolution revealed by multiregion sequencing. *N Engl J Med* 366:883–892; PMID 22397650.

Kandoth C, McLellan MD, Vandin F et al. (2013) Mutational landscape and significance across 12 major cancer types. *Nature* 502:333–339; PMID 24132290. [Analysis by the Cancer Genome Atlas (TCGA) research network of point mutations and small indels from 3281 tumors belonging to 12 major cancers.]

Nature Focus on TCGA Pan-Cancer Analysis. http://www.nature.com/tcga

Nik-Zainal S, Alexandrov LB, Wedge DC et al. (2012) Mutational processes molding the genomes of 21 breast cancers. *Cell* 149:979–993; PMID 22608084.

Stephens PJ, Tarpey PS, Davies H et al. (2012) The landscape of cancer genes and mutational processes in breast cancer. *Nature* 486:400–404; PMID 22722201.

Cancer Therapeutics

Al-Lazikani B, Banerji U & Workman P (2012) Combinatorial drug therapy for cancer in the post-genomic era. *Nature Biotechnol* 30:1–13; PMID 22781697.

Brabletz T, Lyden D, Steeg PS & Werb Z (2013) Roadblocks to translational advances on metastasis research. *Nature Med* 19:1104–1109; PMID 24013756.

Brenner MK, Gottschalk S, Leen AM & Vera JF (2013) Is cancer gene therapy an empty suit? *Lancet Oncol* 14: e447-456; PMID 24079872.

Chen J, Li Y, Yu TS et al. (2012) A restricted cell population propagates glioblastoma growth after chemotherapy. *Nature* 488:522–531; PMID 22854781.

Dancey JE, Bedard PL, Onetto N & Hudson TJ (2012) The genetic basis for cancer treatment decisions. *Cell* 148:409–420; PMID 22304912.

Dawson MA & Kouzarides T (2012) Cancer epigenetics: from mechanism to therapy. *Cell* 150:12–26; PMID 22770212.

Glickmann MS & Sawyers CL (2012) Converting cancer therapies into cures: lessons from infectious diseases. *Cell* 148:1089–1098; PMID 22424221.

Holohan C, Van Schaeybroeck S, Longley DB & Johnston PG (2013) Cancer drug resistance: an evolving paradigm. *Nature Rev Cancer* 13:714–726; PMID 24060863.

Jones SJ, Laskin J, Li YY et al. (2010) Evolution of an adenocarcinoma in response to selection by targeted kinase inhibitors. *Genome Biol* 11:R82; PMID 20696054.

McDermott U, Downing JR & Stratton MR (2011) Genomics and the continuum of cancer care. *N Engl J Med* 364:340–350; PMID 21268726.

Nature Medicine Focus on Targeted Cancer Therapies (2013) *Nature Med* 19:1380–1464 [A collection of various reviews in this area in the November 2013 issue.]

Genetic Testing from Genes to Genomes, and the Ethics of Genetic Testing and Therapy

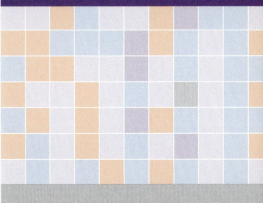

We end this book by considering an important aspect of how the ever-expanding knowledge of our genome, our genes, and genetics is being used to make a positive impact on the health of society: genetic testing. And we discuss ethical concerns and the impact on society of the genetic technologies that are being applied, or might be applied in the near future, toward both diagnosis and treatment.

In some of the previous chapters we looked at how genetics and genomics are illuminating our understanding of the molecular basis of disease, and how this knowledge has brought about significant—and sometimes profound—changes in how we diagnose and treat human disease. For many genetic disorders there remains no adequate treatment, however; genetic approaches to treatment are still mostly at the pioneering stage. (There may be a range of licensed genetically engineered therapeutic proteins, but despite the recent impressive progress made in using gene therapies, the first official license to be granted for a gene therapy as a fully accredited treatment was issued as recently as November 2012.) Nevertheless, one can expect that genetic technologies in treating disease will make a substantial contribution to medicine in the coming decades; there is an important need to assess ethical concerns and the impact of the new genetic technologies on society.

There has been a much longer history of successfully using genetic testing. Analysis of chromosome abnormalities in fetal cell samples first began in the late 1960s. The subsequent DNA cloning revolution allowed rapid developments in DNA-based diagnosis. Initially used in just a few medical settings (notably clinical genetics), DNA technologies were then democratized by PCR, an inexpensive DNA technology that was very easy to use. PCR-based testing became the standard way of identifying pathogens and so is a key tool used by microbiologists and virologists, but it is also used widely in clinical genetics services, and increasingly in other medical specialties such as hematology and oncology.

As the genetic basis of common diseases becomes known, and with increasingly inexpensive sequencing, DNA-based diagnosis is moving from genes to genomes and is reaching out to all the major divisions of medicine. As the revolution in genome and exome sequencing brings us ever closer to personalized medicine, direct-to-consumer genetic testing is becoming available (in which testing is offered by private companies rather than within the usual health care systems).

In Section 11.1 we give an overview of genetic testing before describing the technology of genetic testing for detecting chromosome abnormalities and larger-scale DNA changes (Section 11.2), and testing for point mutations and DNA methylation changes (Section 11.3). In Section 11.4 we describe how genetic services are organized and the practical applications of genetic testing. (Note that we have previously described applications in pharmacogenetic testing within the context of treatment for genetic disorders; readers interested in this application should consult Section 9.1.) Finally, in Section 11.5, we take a look at ethical considerations plus the impact on society of both genetic testing and the application of genetic technologies to treat disease.

11.1 An Overview of Genetic Testing

Genetic testing of individuals is carried out for different reasons. Checking identity or biological relationships can be conducted for forensic or legal purposes and in tracing ancestry. Genetic testing is also important in understanding the normal genetic variation in different human populations.

The genetic testing outlined in this chapter is primarily concerned with detecting the relatively small portion of human genetic variation that confers susceptibility to disease. There are different general strategies for carrying out the testing, and different levels and environments at which it is carried out.

Evaluating genetic tests

Genetic testing can be evaluated by an ACCE framework that was established by the Evaluation of Genomic Applications in Practice and Prevention initiative (http://www.egappreviews.org/) of the US Centers for Disease Control and Prevention. It gets its name from four aspects of how the test performs, as follows:

- **A**nalytical validity: how well does the test assay measure what it claims to measure?

- **C**linical validity: how well does the test predict the projected health outcome?

- **C**linical utility: how useful is the test result?

- **E**thical validity: how well does the test meet the expected ethical standards?

The analytical validity of the test is determined by two key performance indicators: the **sensitivity**, the proportion of all people with the condition who are correctly identified as such by the test assay; and the **specificity**, the proportion of all people who do not have the condition and who are correctly identified as such by the test assay. See **Table 11.1** for a worked example and for how related measures are defined.

Genetic testing: direct genotyping assays, mutation scanning, downstream assays, and indirect linkage analyses

In some types of genetic test, a disease-causing genetic variant, or a variant that confers susceptibility to a complex disease, can be assayed directly by genotyping that variant. We might know (or suspect) the identity of a disease-causing variant simply because the disease shows exceptional mutational homogeneity, as in sickle-cell disease or Huntington disease, for example.

For the vast majority of single-gene disorders, however, if we do not have prior knowledge of the molecular pathology in a family, we first need to identify the causative mutation in the family. After that, we can carry out a direct assay to determine whether an individual at risk carries the disease-causing variant or not. To identify the causative mutation in the

		CONDITION	
		Present	Absent
TEST	+ve	a (90)	c (30)
	−ve	b (10)	d (1870)

Sensitivity	a/a + b	(90/100 = 90%)
Specificity	d/c + d	(1870/1900 = 98.4%)
False positive rate	c/a + c	(30/120 = 25%)
Positive predictive value	a/a + c	(90/120 = 75%)
False negative rate	b/b + d	(10/1880 = 0.5%)
Negative predictive value	d/b + d	(1870/1880 = 99.5%)

Table 11.1 Parameters relating to the analytical validity of a test. Numbers in parentheses are specific values for illustrative purposes only, drawn from 100 people with a hypothetical condition and 1900 lacking the condition. The false positive rate is the proportion of people who test positive for the factor that is being assayed but who do not have the condition. The false negative rate is the proportion of people who test negative for the factor that is being assayed but who have the condition. The positive predictive value is the proportion of people testing positive who have the condition. The negative predictive value is the proportion of people testing negative who do not have the condition. Note that the sum of the false positive rate and the positive predictive value is always 100%, as is the sum of the false negative rate and the negative predictive value.

first place, some kind of **mutation scanning** is carried out. That usually means DNA sequencing to look for point mutations within genes known to cause the disorder, or scanning candidate genes for deletions and duplications. By comparison with reference sequences, candidate genetic variants are examined in an effort to identify a likely pathogenic variant. Sometimes, however, mutation scanning involves simultaneously genotyping a battery of many different specific mutations commonly found in affected individuals to see whether an individual tests positive for any of them.

Some genetic tests assay some property that is a consequence of genetic variation, rather than the genetic variation itself. The testing might seek evidence of abnormal RNA or protein expression products, or a characteristic disease-associated biomarker such as an abnormally elevated metabolite; sometimes a functional assay is used. At the end of the day, a test is sufficient if it lets us know whether or not the gene is faulty in the way that we expect.

For recessive disorders, a single functional assay might be sufficient to detect a loss of function and can be conveniently carried out in cultured cells in which the gene is expressed. The DNA repair disorder Fanconi anemia (PMID 20301575) provides a paradigm. DNA-based testing for this disorder is extraordinarily complicated because the disease can be caused by mutations in any one of 15 different genes, some of which have large numbers of exons. A more convenient laboratory test for this disorder involves a simple type of DNA repair assay. Cultured lymphocytes from an individual are treated with a DNA interstrand cross-linking agent such as diepoxybutane or mitomycin C and examined to identify chromosomal aberrations that result because of defective repair of the induced DNA cross-linking.

Indirect linkage analyses

Direct assays for specific genetic variants are now the norm in genetic testing. In some situations, however, an indirect assay is used to track the inheritance of a disease allele. Here, we need to be confident that we know the identity of the disease gene; that would be possible when the disorder is known to be caused by mutations in just the one gene.

Indirect assays may sometimes be used because of time pressure. A woman who is 15 weeks pregnant, and who has previously had a child with autosomal recessive polycystic kidney disease who died in the neonatal period, requests prenatal diagnosis. This disease is associated with inactivating mutations in the *PKHD1* gene (PMID 20301501). To carry out a direct assay, the causative mutations would need to be found; however, scanning the multiple exons of a complicated gene such as *PKHD1* (which has 66 coding exons) takes time. A quick alternative would be to genotype closely linked nonpathogenic polymorphic DNA markers that are known to map in the immediate vicinity, or even within the disease gene.

Figure 11.1 Gene tracking using a linked marker. In this X-linked recessive condition, the question marks signify initial uncertainty about the risk to the sisters in generation III of having an affected son. After testing for a polymorphic marker that is very closely linked to the disease locus, the genotypes obtained are shown in blue underneath the pedigree symbols (males have a single allele; females have two alleles, separated by a hyphen). Allele 3 of the marker locus appears to be on the same X chromosome as the disease allele (red asterisk), and the women in generations I and II appear to be carriers. The older sister in generation III has inherited allele 1 from her mother at the marker locus. She would be inferred to have inherited a normal (N) allele at the disease locus, and not to be at risk of having an affected son. However, the younger sister would be predicted to be a carrier with a 50% disease risk for any son she produces (and could be offered prenatal diagnosis using linkage for any male fetus, if she wished). There is a small chance of recombination between a single closely linked marker and the disease locus; however, if informative, tightly linked proximal and distal markers are used, accurate predictions can be expected.

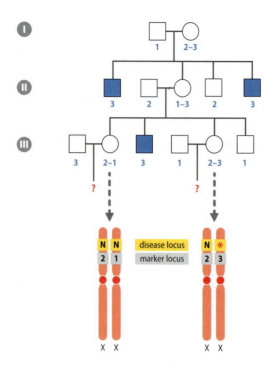

The indirect assays are based on genetic linkage and so require DNA samples from key family members to track the inheritance of the linked markers through two or more generations. By identifying which allele is segregating with the disease at multiple closely linked marker loci in the family, we can predict with high accuracy whether an individual will inherit the unidentified disease allele(s) or not (see **Figure 11.1** for a simple illustration using just a single marker locus).

Different levels at which genetic testing can be carried out

Genetic testing can be carried out at different molecular levels, and for different purposes. The testing often involves analyzing DNA or chromosomes; however, sometimes RNA, gene products, metabolites, and other biomarkers are assayed (and on rare occasions, a functional assay is used, as described above). The source material can be cells (usually blood, tumor, skin, embryonic, or fetal cells; see **Table 11.2**) or body fluids such as blood, urine, and even exhaled breath (increasingly used in assaying certain cancer biomarkers). In addition, testing is sometimes carried out on archived material from deceased persons to provide information that can be of clinical help to surviving family members.

Like other clinical tests, genetic tests may be conducted on individuals, but often they are carried out in different environments that involve couples, families, communities, and even populations. We describe the details in Section 11.4. Here, by way of introduction, we provide a quick snapshot of the different levels.

Genetic testing used to be dominated by testing for chromosome abnormalities and for rare highly penetrant mutations within genes underlying monogenic diseases. These services are predominantly directed at prospective parents or at families in which prior incidence of genetic disorders in family members or other known factors confer elevated risk of disease-susceptibility. Testing is often carried out at the prenatal level on cells taken at different stages of pregnancy (see Table 11.2), but in preimplantation diagnosis, tests are carried out for genetic abnormalities in the context of *in vitro* fertilization.

SOURCE OF CELLS OR DNA	TYPE OF TESTING OR SCREENING
Embryonic/fetal	
Single cell from a blastomere or a few cells from a blastocyst	preimplantation diagnosis
Fetal DNA in maternal blood	prenatal diagnosis, as early as 6 weeks (testing for paternal alleles). Fetal sexing (Table 11.4)
Chorionic villus	prenatal diagnosis at about 9–14 weeks
Amniotic fluid	prenatal diagnosis at about 15–20 weeks
Umbilical cord blood	prenatal diagnosis at about 18–24 weeks
Adult/postnatal	
Peripheral blood	screening for heterozygote carriers. Testing for defined heterozygous carrier genotype. Pre-symptomatic genotype screening or testing. Identity testing (DNA fingerprinting or profiling). Testing for chromosome abnormalities
Mouthwash/buccal scrape	
Biopsy of skin, muscle or other tissue	RNA-based testing
Tumor biopsy	cancer-associated genotypes or gene expression patterns
Guthrie card	neonatal screening
Archived material from deceased persons	
Pathological specimens	genotyping
Guthrie card	possible source of DNA from a dead individual (not all of the blood spots on the card might have been used in neonatal screening)

Table 11.2 Sources of material for genetic testing.

Pre-symptomatic testing is increasingly being carried out in adults, and sometimes in children. Here, the aim is to predict whether an asymptomatic person is at risk of a genetic disorder that develops later in life. If the test result is positive, there may be the opportunity to take some medication and/or alter lifestyle factors to reduce the disease risk; a negative test result offers psychological benefit. As the costs of personal genome sequencing continue to fall, whole-exome and then whole-genome sequencing can be expected to become routine, and this type of testing will become more prevalent. As we will see, various personal genomics companies have recently offered genetic testing for complex disorders directly to consumers, outside the normal health care environments.

Genetic testing is also carried out on apparently asymptomatic individuals in communities and populations to identify individuals carrying harmful mutations (**genetic screening**). The aim is usually to identify a high-risk subset of the population who can then be offered additional specific testing (such as follow-up prenatal diagnosis after identifying couples who are both carriers of a recessive condition). Unless the diseases are dominated by a few known types of genetic variant, we typically do not know what mutations may be present in the individuals. As a result, genetic screening quite often involves assays of gene products or biomarkers associated with the pathogenesis, such as altered metabolites.

11.2 The Technology of Genetic Testing for Chromosome Abnormalities and Large-Scale DNA Changes

In Section 7.4 we described two fundamental classes of chromosome abnormality: numerical abnormalities (in which abnormal chromosome segregation leads to aneuploidy (with fewer or more chromosomes copies than normal) and structural abnormalities (in which chromosome rearrangements cause large-scale deletions, duplications, inversions, or translocations).

Traditionally, chromosome abnormalities have been identified by standard cytogenetic karyotyping, which relies on chemically staining chromosome preparations to reveal chromosome banding patterns (Box 7.4). More recently, alternative DNA technologies have moved to the forefront, being widely used in first-pass screening for both chromosome abnormalities and other large-scale DNA changes (that are too small to be detected by standard cytogenetic analyses). Chromosome banding is still used, however, for certain types of screen as described below.

Interpretation of chromosome abnormalities and large-scale DNA changes is assisted by various electronic resources. The Database of Chromosomal Imbalance and Phenotype in Humans using Ensembl Resources (DECIPHER) at http://decipher.sanger.ac.uk/ collects clinical information about chromosomal microdeletions/duplications/insertions, translocations, and inversions; it displays this information on the human genome map. The International Collaboration for Clinical Genomics (ICCG) at http://www.iccg.org/ organizes data sharing.

Detecting aneuploidies with the use of quantitative fluorescence PCR

Until recently, standard karyotyping of fetal cells by using chromosome banding was the most commonly used method to screen for any evidence of aneuploidy (the common aneuploidies are trisomies 13, 18, and 21 and altered numbers of sex chromosomes). But that requires quite a long time to get a result (about 2 weeks; it takes time to grow sufficient numbers of fetal cells in culture from samples of amniotic fluid or chorionic villi). When ultrasound screening or serum screening identifies women at high risk of aneuploidies such as trisomy 21, it becomes important to rapidly confirm or exclude the aneuploidy diagnosis. As a result, the modern trend is to use faster DNA-based diagnoses as a front-line screening system.

Principles of quantitative fluorescence PCR

Quantitative fluorescence PCR (QF-PCR) on uncultured fetal cells is rapidly becoming a front-line screen for aneuploidy: it is fast, robust, highly accurate, and largely automated. Several pairs of fluorescently labeled primers are used in a *multiplex* PCR—the idea is to simultaneously amplify multiple polymorphic markers on the chromosomes most frequently involved in aneuploidies. For each marker, the amplification products will fall within a characteristic size range of different lengths; as required, two or more markers that have overlapping allele sizes can be distinguished by labeling them using fluorophores that fluoresce at different wavelengths.

Certain polymorphic short tandem repeat polymorphisms are usually selected, often based on tetranucleotide or pentanucleotide repeats to maximize the length difference between alleles. Fluorescently labeled products from the exponential phase of the PCR reaction (Figure 3.4) are separated according to size by electrophoresis through long and extremely thin tubes containing polyacrylamide (capillary electrophoresis). That happens in a commercial DNA analyzer of the type that is used in capillary DNA sequencing: a detector at a fixed position records the intensity of fluorescence signals as fragments migrate through the capillary tubes and past the detector (Box 3.3 on page 74 describes the principle of capillary electrophoresis).

Autosomal aneuploidies

To monitor the common autosomal aneuploidies, highly polymorphic short tandem repeat markers are used. An individual marker might

(A)

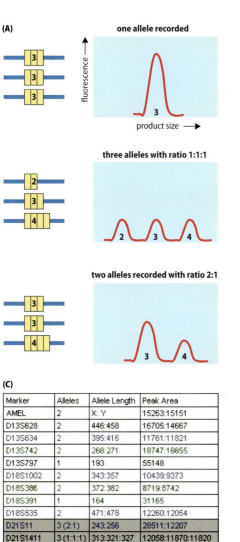

(B)

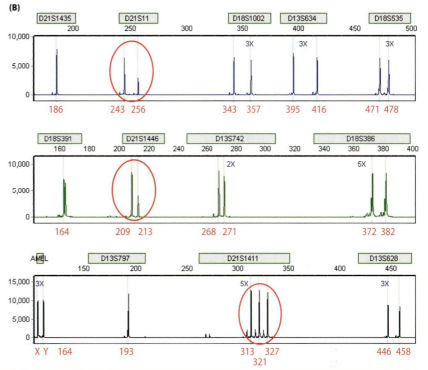

(C)

Marker	Alleles	Allele Length	Peak Area
AMEL	2	X: Y	15263:15151
D13S628	2	446:458	16705:14667
D13S634	2	395:416	11761:11821
D13S742	2	268:271	18747:16855
D13S797	1	193	55148
D18S1002	2	343:357	10439:9373
D18S386	2	372:382	8719:8742
D18S391	1	164	31165
D18S535	2	471:478	12260:12054
D21S11	3 (2:1)	243:256	28511:12207
D21S1411	3 (1:1:1)	313:321:327	12058:11870:11820
D21S1435	1	186	34476
D21S1446	3 (2:1)	209:213	37508:18331

Figure 11.2 Autosomal trisomy screening using QF-PCR. (A) Interpreting marker data (on the right) from an imagined locus with three common alleles that have two, three, or four tandem repeats, as illustrated on the left. The top trace is uninformative (just one length variant is recorded). The middle trace is fully informative: the presence of three alleles with different lengths strongly suggests trisomy. The bottom trace is suggestive of trisomy: two length variants are evident, but the fluorescence associated with allele 3 seems to be approximately twice that associated with allele 4 (the area under the peaks is normally used for quantitation). (B) A practical example. The output shows traces for three sets of markers (shown in blue at top, green in the middle, and black at the bottom) that collectively represent assays for microsatellite markers on chromosomes 13, 18, and 21 (the three autosomes associated with viable trisomies), plus control X and Y markers from the amelogenin genes (see Figure 11.3). The data highlighted by red ovals strongly suggest trisomy in this individual: three alleles of different sizes for *D21S1411*, and a 2:1 ratio for the two length variants for each of *D21S11* and *D21S1446*. The other chromosome 21 marker, *D21S1435*, is uninformative, showing only one length variant, which is presumably due to three alleles of identical lengths. (C) The calculation of peak areas and interpretation by SoftGenetics software (rows highlighted in gray are significant). (B, C, Data courtesy of Jerome Evans, NHS Northern Genetics Service, Newcastle upon Tyne, UK.)

not always be informative: in a trisomy, for example, the marker might show identical repeat numbers for all three chromosome copies, just by chance, resulting in an uninformative, single PCR product. The most informative situation occurs when the marker exhibits different numbers of repeats on the three chromosomes. But quite often only two length variants are recorded for a single marker; then quantification becomes important (**Figure 11.2A**). Because four or more different markers are used per chromosome, however, there is little difficulty with interpretation (two or more markers are often informative for each chromosome—see **Figure 11.2B, C** for a practical example).

Sex chromosome aneuploidies

The copy number of our sex chromosomes is more varied than that of autosomes, ranging from monosomy (45,X) to different types of trisomy, tetrasomy, and occasionally pentasomy. Identifying a monosomy using PCR might seem particularly challenging—how can 45,X be distinguished

(A)

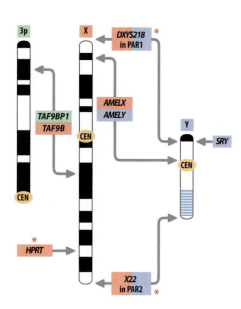

(B)

Marker	Alleles	Allele Length	Peak Area
D13S628	2	454.4:461.8	7680:7262
D13S634	2	391.4:406.3	7003:7181
D18S1002	2	341.8:354	6083:5638
D18S386	2	352.2:381.7	7291:5847
D21S1411	2	316.9:333.4	6558:6037
D21S1446	1	212.4	14090
DXYS218	1	239.5	7669
HPRT	1	282.0	6169
SRY	0		
TAF_9	3 (2:1)	3:X	40238:18508
X22	1	218	6680

Figure 11.3 Detecting sex chromosome aneuploidies using QF-PCR. (A) Marker sets. Primer pairs are designed to amplify X-specific markers (*HPRT*), Y-specific markers (*SRY*), markers in the pseudoautosomal regions PAR1 or PAR2 (shared by the X and Y), and highly homologous sequences on the X and Y chromosomes, such as the amelogenin genes *AMELX* and *AMELY* in which a single set of primers can amplify both sequences (which can be differentiated because of small length differences due to insertion or deletion). To gauge the ratio of X chromosomes to autosomes, primers are used to amplify equivalent segments of the *TAF9B* gene on Xq and a highly related pseudogene *TAF9BP1* on 3p. CEN, centromere. (B) Data from a practical example. The interpretation would be monosomy X (Turner syndrome) on the basis of the absence of the SRY marker and ratio of 2:1 for the length variants from *TAF9BP1* in chromosome 2 and *TAF9* on the X chromosome. (Data courtesy of Jerome Evans, NHS Northern Genetics Service, Newcastle upon Tyne, UK.)

from 46,XX? However, counting the sex chromosomes is possible by using primer sets that are specific for the X or Y chromosome plus primer sets that simultaneously amplify conserved sequences on both sex chromosomes or on both the X and an autosome (**Figure 11.3**).

Noninvasive fetal aneuploidy screening

Recently, screening for fetal aneuploidies has been made possible by high-throughput sequencing of fetal DNA in maternal plasma, an advanced form of noninvasive prenatal screening. We describe recent major advances in this area in Section 11.4.

Detecting large-scale DNA copy number changes with the use of microarray-based genomic copy number analysis

As described above, aneuploidies can be detected by quantitative fluorescent PCR. Because the change in copy number applies to whole chromosomes, and because there is only a small number of viable human aneuploidies, it is relatively easy to design a series of QF-PCR assays for this purpose.

If we wish to scan for any subchromosomal changes in DNA copy number across the genome, a scanning method is needed. Traditional chromosome banding techniques (karyotyping) have been used for this purpose, but the resolution is not high: deletions and duplications involving less than 5–10 Mb of DNA are frequently not detected (even when using extended prometaphase chromosome banding). Once the human genome had been sequenced, however, DNA-based methods were developed to allow high-resolution scanning. **Chromosomal microarray analysis** is the clinical application of microarray-based DNA hybridization assays to scan the DNA of each chromosome for changes in copy number (deletions or duplications) of DNA segments from tens of kilobases to tens of megabases. There are two principal techniques, as described in the first two subsections below.

Array comparative genome hybridization (aCGH)

In comparative genome hybridization, two genomic DNA populations, a test sample and a normal control, are labeled with different fluorophores, then mixed and hybridized to a panel of unlabeled DNA probes collectively representing the genome (**Figure 11.4A, C**). By comparing the hybridization patterns of component DNA sequences in the control and test samples, we can screen for changes in copy number over quite large regions of DNA.

Figure 11.4 Principles of array CGH. The object is to screen a test sample of genomic DNA for evidence of copy number variation of a large region of DNA (tens of kilobases to tens of megabases). (A) To do this, the test sample is labeled with a fluorophore (often the cyanine CY3) and then mixed with a control genomic DNA sample that has been labeled with a different fluorophore (often CY5). The labeled DNA mixture is denatured and allowed to hybridize to a panel of (usually) single-stranded oligonucleotides that collectively represent all regions of the genome and that have been fixed at specific grid positions on a microarray. (B) Each *feature* of the array (each individual position with its bound oligonucleotide; shown here as a gray oval) has many identical copies of just one oligonucleotide sequence and can therefore bind any complementary sequences in the mixture of labeled DNAs. If a region of DNA in the test sample has sustained a heterozygous deletion, the fluorescence emitted by the fluorophore in the test sample DNA might therefore be expected to be one-half that of the control sample's fluorophore; if there has been a duplication, the expected ratio is 3:2. (C) A small section of a microarray (from the practical example in Figure 11.5), showing some features where the fluorescence signal was skewed toward the red, indicating a deletion (green:red ratio = 1:2, shown by red arrows) or a duplication (green:red ratio = 3:2, shown by green arrow) of that sequence in the test sample.

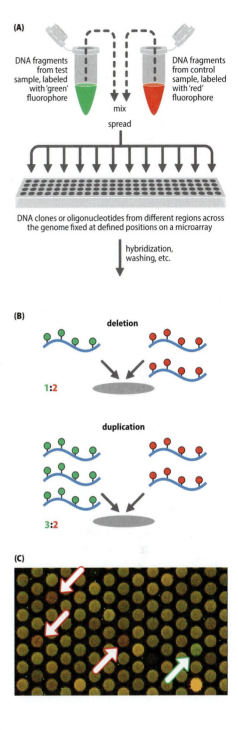

In the most widely used version of CGH, the probes are long oligonucleotides (often 55–65 nucleotides) fixed in a grid formation on a microarray (array CGH or aCGH). At each position in the microarray grid, the probe consists of very many identical copies of the oligonucleotide to drive the hybridization reaction by binding to complementary sequences in both types of labeled genomic DNA. The hybridization conditions are designed so that highly repetitive DNA sequences in the labeled populations are blocked from hybridizing to the probe oligonucleotides on the microarray.

When using a diploid test sample DNA and control DNA, any non-repetitive DNA sequence will be present in two copies. For sequences such as this, the ratio of the two fluorophores bound by probes should be approximately constant across the genome, but in chromosomal regions where the test sample has more than two copies (such as partial trisomy) or fewer than two copies (deletion), the ratio of the two fluorophores will change (see Figure 11.4B).

Because gains and losses of DNA sequences are readily detected by aCGH, it is used in detecting chromosomal imbalances and large-scale changes in cancer (using tumor samples referenced against a control of lymphocyte DNA from the same individual). It is also very commonly used in a pediatric setting to identify suspected gene imbalances in children with developmental disabilities or congenital anomalies—see **Figure 11.5** for an example.

SNP microarray hybridization

Single nucleotide polymorphism (SNP) microarrays comprise panels of oligonucleotides that are shorter in length than those used in aCGH (because they are designed to hybridize to individual alleles only). That is, for each SNP locus, different oligonucleotides will be present that are designed to hybridize to individual alleles at that locus.

In contrast with aCGH, SNP arrays do not directly compare a patient's test sample with a control sample. Instead, the assay compares the dosage of the individual being tested at any given locus with the equivalent values in a database of SNP array results from control individuals. Like aCGH, SNP arrays can detect gains and losses of sequences across the genome.

Deletions can be identified because of the absence of heterozygosity: the SNPs in the deleted area should show just a single allele. For duplications,

Figure 11.5 A pediatric example of using array CGH to detect large-scale copy number variation. (A) Array CGH analysis of DNA from a male infant with global developmental delay, dysmorphic features, broad thumbs, and heart abnormalities. The child's DNA and control DNA had been labeled with CY3 and CY5, respectively, and the mixed labeled DNA samples were hybridized to an array with 60,000 different long oligonucleotides (with a backbone of one oligonucleotide every 75 kb plus higher densities in gene regions). The small green circles connected by lines represent individual oligonucleotide probes that are located on a vertical scale marking their position on chromosome 16 from nucleotide number 1 at the top to 90,249,800 at the bottom (the gap in the central region marks heterochromatic regions at the centromere and 16q11.2). The CY3:CY5 fluorescence ratio is given on the horizontal axis using a logarithmic scale (base 2); the thin brown and green vertical lines at positions −0.30 and 0.30 represent the limits for normal copy number. Outliers beyond these limits indicate deletions (thick vertical red bar on the left) and duplications (thick vertical green bars on the right). (B) Interpretation of the array CGH data (×3 denotes three copies; ×1 denotes a single copy). ISCN nomenclature (see Box 7.4) is used: for example, del.16q23.3q24.1 means a deletion from 16q23.3 to 16q24.1, and dup.16p13.2p13.12 means a duplication of the region from 16p13.2 to 16p13.12. The presence of three regions showing gene imbalance is unusual; the phenotype might conceivably result from dosage-sensitive genes in each region. (Data courtesy of Simon Zwolinski, NHS Northern Genetics Service, Newcastle upon Tyne, UK.)

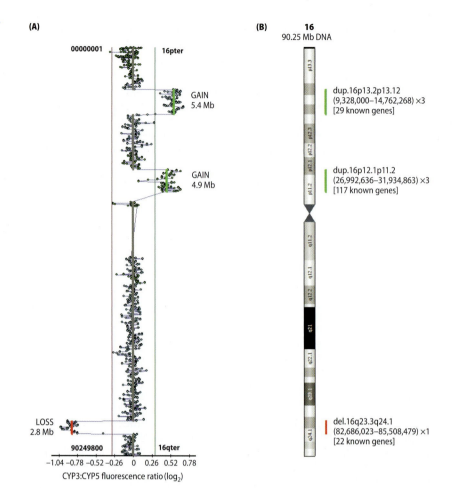

the ratios of alleles will vary: if we imagine a SNP locus as having two alleles, say A and B, then a normal heterozygote will be scored as AB (with equal representations of alleles A and B); however, in regions of partial trisomy, loci in which both alleles are evident might show skewed allele ratios and might appear as AAB (twice the signal for allele A compared with B) or ABB, instead of AB. In practice, inferring DNA gains and losses using SNP microarrays is less secure than in aCGH, and aCGH is the more widely used method for this purpose.

Some commercial microarrays use components of both chromosomal microarray methods. For example, the Affymetrix Genome-Wide Human SNP Array 6.0 has 946,000 long oligonucleotide probes that are dedicated to detecting copy number variation, and short oligonucleotides for typing of alleles at more than 906,600 SNP loci across the human genome.

Unclassified variants and incidental findings

The interpretation of large-scale DNA changes identified in CGH is assisted by reference to the DECIPHER database and ICCG resources described on page 436. In pediatric applications it is important to obtain follow-up parental DNA samples to establish whether any gain or loss is present in a healthy parent and is therefore likely to be a benign copy number variant. For *de novo* copy number variants the interpretation of the likely effect can sometimes be difficult to predict, simply because of lack of data; a small percentage of tests will therefore report unclassified variants.

Any genome scan can reveal incidental (secondary) findings—discoveries made as a result of genetic testing that are unrelated to the medical reason for ordering the test but which can have important health consequences. That would happen only occasionally in traditional (chromosome banding) karyotyping, but in aCGH incidental findings are more frequent because of its higher resolution. For example, when aCGH is used to scan the DNA of a young child to investigate a developmental or neurocognitive disorder, it might identify a variant associated with an adult-onset cancer syndrome. We consider problems with incidental findings in greater detail in Section 11.5 within the context of interpreting clinical whole-exome and whole-genome sequencing.

The need for conventional karyotyping and chromosome FISH (fluorescence *in situ* hybridization)

We described in Box 7.4 the methodology of human chromosome banding (in which metaphase and prometaphase chromosome preparations are treated with chemical stains to produce alternating dark and light bands). Because the resolution of chromosome banding is poor, traditional karyotyping has been overtaken by modern DNA methods, such as array CGH.

Conventional karyotyping using chromosome banding is nevertheless still clinically important. That is so because molecular genetic methods such as array CGH are not suited to detecting balanced chromosome rearrangements in which there is no net gain or loss of DNA. Inversions and balanced translocations would normally be invisible to these methods, but they can be detected by chromosome banding (**Figure 11.6**).

Chromosome FISH

The essence of chromosome FISH (fluorescence *in situ* hybridization) is to fix chromosome preparations on microscopic slides, treat the slides so as to denature the DNA, and hybridize fluorescently labeled probes of interest to the denatured DNA. The locations of the fluorescent signals are recorded against a background stain that binds to all DNA sequences—see **Figure 11.7A** for the principle.

Chromosome FISH is often used to confirm regions of chromosome duplication or deletion that have been suggested by other screening methods, such as array CGH. It can also be used to screen for the amplification of specific oncogenes that are associated with particular types of cancer, such as amplification of the *MYCN* gene in neuroblastoma (Figure 10.6A).

Another major application is in detecting translocations, notably acquired translocations that are common in cancer. Recurrent translocations are associated with certain types of cancer; often, the translocation involves breakages in specific genes, producing hybrid genes that are inappropriately expressed. For example, as illustrated schematically in Figure 10.7 on page 390, translocations involving the *BCR* gene and the *ABL1* oncogene are common in chronic myeloid leukemia.

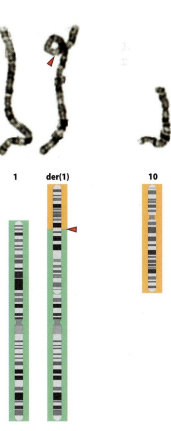

1   der(1)   10   der(10)

Figure 11.6 Use of traditional karyotyping to detect a carrier of a balanced translocation. This translocation—46,XX,t(1;10)(p36.22;q22.3)—could not have been identified by array CGH because it is balanced (with no net loss or gain of DNA). The derivative (der) chromosomes contain sequences from chromosomes 1 and 10 that are fused at the junction points shown by the red darts; they are named der(1) or der(10), according to which chromosome has provided the centromere. The translocation is important to detect because it predisposes the woman in whom it was identified to transmit unbalanced forms of the translocation by meiotic malsegregation. (Courtesy of Gareth Breese, NHS Northern Genetics Service, Newcastle upon Tyne, UK.)

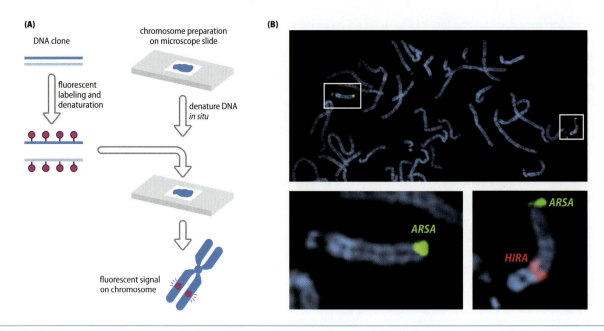

Figure 11.7 Chromosome FISH (fluorescence *in situ* hybridization).
(A) Principle of chromosome FISH. A labeled DNA clone of interest is hybridized to a (metaphase or interphase) chromosome preparation on a microscope slide that has been treated with a DNA-denaturing agent such as formamide. When metaphase chromosome preparations are used, as shown here, a double fluorescent signal is often seen, representing hybridization to target sequences on the sister chromatids. Figure 11.8 gives a practical example of interphase FISH. (B) Metaphase FISH detects a 22q11.2 deletion in a patient with suspected 22q11.2 deletion syndrome. The background blue color of the chromosomes is due to staining with the general DNA-binding stain, DAPI (4′,6-diamidino-2-phenylindole.) The green fluorescent signal comes from a control probe, the aryl sulfatase A gene, *ARSA*, which maps at 22q13.33, close to the telomere on the long arm of chromosome 22. The contents of the small white boxes at the top are expanded to give the lower panels. Both chromosomes 22 can be seen to test positive; in each case, the strong green signals are due to overlapping signals from the two sister chromatids. The red fluorescent signal derives from a test probe, the histone cell cycle regulator gene, *HIRA* (which maps to 22q11.2). The chromosome 22 on the right shows a double red signal (from the two sister chromatids), but the chromosome 22 on the left lacks any *HIRA* signal, confirming a deletion in the 22q11.2 region. (B, courtesy of Gareth Breese, NHS Northern Genetics Service, Newcastle upon Tyne, UK.)

Well-defined translocations, such as the t(9;22) translocation that produces the *BCR-ABL1* hybrid gene, can be screened by interphase FISH. In the absence of visible chromosomes, probes from the two genes that participate in a translocation are labeled with different fluorophores, so that one produces a red fluorescent signal, for example, and the other produces a green fluorescent signal. The translocation chromosomes can be identified because here the green and red fluorescent signals are superimposed (**Figure 11.8**).

DNA technologies for detecting pathogenic changes in copy number of specific DNA sequences

As long as the clinical phenotype does not immediately suggest a defined location for a pathogenic copy number variant, array CGH is the preferred scanning option, but sometimes, however, the phenotype does suggest the chromosomal location; it might suggest a particular syndrome associated with large-scale deletions or duplications of a specific chromosomal region. That region might be prone to some instability, and a resulting change in gene copy number contributes to disease.

To identify pathogenic copy number variants at specific disease loci, different methods can be used. For example, if a disorder is associated with large deletions, chromosome FISH may often be used (see Figure 11.7B). For comparatively small changes, such as expansions of unstable oligonucleotide repeats, Southern hybridization; or PCR assays can be used (**Box 11.1**). An additional convenient PCR-based method has become a

BOX 11.1 Detecting large-scale expansions: Southern blots and PCR assays.

As described in Section 7.3, various single-gene disorders arise through dynamic unstable expansion of an oligonucleotide repeat. They include modest expansions of polyglutamine-encoding CAG repeats and sometimes large expansions of noncoding oligonucleotide repeats, notably in myotonic dystrophy and fragile X-linked intellectual disability. In the case of very large expansions, diagnostic assays have traditionally been based on Southern blot hybridization, but increasingly PCR assays are used.

Southern blot hybridization assays

Large oligonucleotide expansions used to be assayed exclusively by Southern blot hybridization, a technique that is also used to analyze large-scale expansions of repeat arrays in facioscapulohumeral dystrophy. The method begins with the digestion of genomic DNA by a suitable restriction nuclease; the resulting DNA fragments are then separated according to size by electrophoresis on agarose gels. Subsequently, a nylon membrane is placed in contact with the gel to allow the transfer of separated DNA fragments from the gel to the nylon membrane, and a labeled probe from a DNA region of interest is allowed to bind to denatured DNA transferred to the nylon membrane (**Figure 1A**). The labeled probe detects restriction fragments from the region of interest, and changes in the DNA in that region may be reflected by changes in size of the detected DNA fragments.

Southern blot hybridization assays have often been used to detect abnormally long arrays of tandem repeats in myotonic dystrophy and facioscapulohumeral dystrophy. In type 2 myotonic dystrophy, the arrays can expand to as many as 5000 repeats (see Figure 1B for an example of a Southern blot assay).

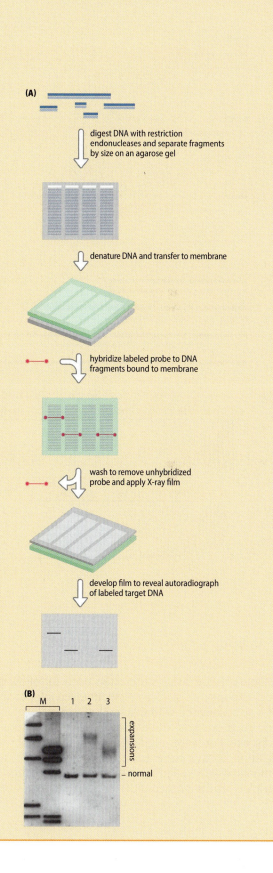

(A)

digest DNA with restriction endonucleases and separate fragments by size on an agarose gel

denature DNA and transfer to membrane

hybridize labeled probe to DNA fragments bound to membrane

wash to remove unhybridized probe and apply X-ray film

develop film to reveal autoradiograph of labeled target DNA

(B)   M   1   2   3

expansions

— normal

Figure 1 Using Southern blot hybridization to detect mutant alleles with expanded CCTG repeat arrays in type 2 myotonic dystrophy. (A) Principle of Southern blot hybridization. (B) Detection of mutant alleles in type 2 myotonic dystrophy. Genomic DNA samples were digested with the restriction nuclease *BgI*I, and the resulting fragments were separated according to size by agarose gel electrophoresis. Southern blots of the *BgI*I-digested genomic DNA samples were hybridized with a *CNBP* (formerly *ZNF9*) probe. M, size markers; 1, normal control; 2, 3, patients with one mutant *CNBP* allele that has expanded in size because of an increase in the number of tandem CCTG repeats in intron 1. (B, Courtesy of David Bourn and colleagues, NHS Northern Genetics Service, Newcastle upon Tyne, UK.)

BOX 11.1 (continued)

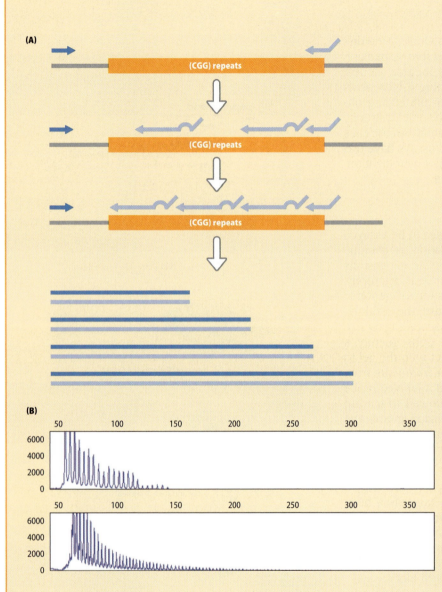

Figure 2 Triplet repeat-primed PCR (TP-PCR) assay of oligonucleotide repeat expansion. (A) Principle of the method, as developed to analyze large triplet repeat expansion in the *FMR1* gene in fragile X syndrome. Forward and reverse primers are shown by deep blue and pale blue arrows, respectively. The reverse primer hybridizes at the junction of the 3′ end of the CGG tract and downstream sequences, but it can also hybridize randomly across the CGG tract. After initial rounds of PCR, extended reverse primers can themselves serve as primers. This results in various PCR product sizes as shown at the bottom (this will give a 'stutter' pattern on electrophoresis). (B) A practical example of TP-PCR: electrophoretic profile of the CTG tandem repeat array in the 3′ UTR of the *DMPK* gene in type 1 myotonic dystrophy. The horizontal and vertical axes of the electropherograms measure the size in base pairs (at the top) and the amount of amplification products, respectively. The electropherograms show a stutter series of PCR products with a 3 bp periodicity, corresponding to incremental numbers of CTG. The traces are from a normal individual (top) and a patient with type 1 myotonic dystrophy (bottom). (A, Adapted from Hantash FM et al. [2010] *Genet Med* 12:162–173; PMID 20168238. With permission from Macmillan Publishers Ltd; B, data courtesy of David Bourn and colleagues, NHS Northern Genetics Service, Newcastle upon Tyne, UK.

Triplet repeat-primed PCR assays

Standard PCR assays are often used to follow modest expansions in oligonucleotide repeats, including those associated with CAG codon expansions in various neurodegenerative disorders such as Huntington disease. For larger expansions, Southern blot hybridization assays have been used, but increasingly a modified PCR reaction is the preferred method. The triplet repeat-primed PCR (TP-PCR assay) uses an external primer (which hybridizes to a sequence flanking the oligonucleotide repeat array) plus a primer that can hybridize to target sequences within the oligonucleotide repeat array as well as outside it. Because of the tandem repetition, the internal primer can hybridize to multiple possible binding sites within the array, producing a series of peaks of increasing size. Increased sizes are apparent in patients when referenced against controls (**Figure 2**).

Figure 11.8 Interphase FISH to detect recurrent t(9;22) translocations in chronic myeloid leukemia. Cases with chronic myeloid leukemia often show translocations with breakpoints in the *ABL1* gene (on 9q) and the *BCR* gene (on 22q; see Figure 10.7). Here, the *ABL1* and *BCR1* probes (selected from regions of these genes that are retained on the translocation chromosomes) give, respectively, red and green fluorescent signals. The white arrows show characteristic signals for the fusion genes on the translocation chromosomes; they are readily identified because of the very close positioning of the red and green signals, with sometimes overlapping signals that appear orange–yellow. By contrast, the red and green signals at bottom are well separated and represent the normal chromosome 9 and normal chromosome 22, respectively. (Note that RT (reverse transcriptase)-PCR is an alternative assay. RNA from blood lymphocytes is converted to cDNA by using a reverse transcriptase. The PCR assay requires an oligonucleotide primer from the *BCR* gene and one from the *ABL1* gene to allow amplification across the fusion gene only.) (Courtesy of Fiona Harding, Northern Genetics Service, Newcastle upon Tyne, UK.)

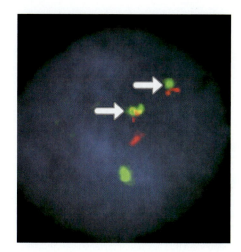

very popular way of detecting copy number changes, as described in the next subsection.

Multiplex ligation-dependent probe amplification (MLPA)

The MLPA method is a quick and versatile method that can detect copy number changes over a broad range of DNA lengths. It is most frequently used to scan for intragenic deletions and duplications by monitoring the copy number of individual exons within genes; it is therefore especially useful in disorders in which there is a high frequency of intragenic deletions (as in Duchenne muscular dystrophy, for example). It may also serve as a general method of scanning for copy number changes that complements the ability of DNA sequencing to scan for point mutations.

The MLPA method uses pairs of short single-stranded sequences (called probes) that are designed to bind to specific exons or other sequences (called target sequences) whose relative copy number we wish to determine. Each pair of probes is designed to hybridize collectively to a *continuous* target DNA sequence; that is, when they bind to the target DNA, the pair of probes align immediately next to each other. The gap between them can then be sealed using DNA ligase to give a single probe that is complementary to the target (**Figure 11.9**).

The 5′ end of one of the probe pair and the 3′ end of the other are designed to contain unique sequences not present in our genome. By designing oligonucleotide primers that will bind to regions in the unique end sequences only, the probe sequences are selectively amplified in a PCR reaction.

A key feature of MLPA is that the amount of amplified probe product is proportional to the number of bound copies of the probe, which in turn depends on the number of target sequences that the probe has bound to. With a heterozygous deletion, for example, there is one copy of the target sequence instead of two; the amount of bound (and therefore ligated) probe is one-half of the normal amount, and the amount of amplified product is proportionally reduced.

Often, MLPA is designed to be a multiplex reaction: multiple pairs of probes can be used to bind simultaneously to different target sequences. The left and right probes for each target sequence all have the same set of left and right unique end sequences, and so all ligated probes can be amplified by a common set of primers (that are specific for the unique end sequences). But the stuffer sequences (see Figure 11.9) are designed to

Figure 11.9 The principle of multiplex ligation probe amplification (MLPA). For each target sequence (such as an individual exon), a pair of probes is designed that will hybridize to adjacent sequences within the target and will carry unique end sequences not present in the genome. The aim is to use DNA ligase to seal the left and right probes to give a continuous sequence flanked by the unique end sequences and then to amplify the continuous sequence by using primers complementary in sequence to the unique end sequences. Probe pairs for multiple different target sequences (such as multiple exons within a gene) are simultaneously hybridized to their target sequences and ligated to form continuous sequences that are then simultaneously amplified in a multiplex reaction. The point of the stuffer fragment is simply to provide a spacer sequence whose length can be varied. This can ensure different sizes of the PCR products from a multiplex reaction (in which multiple probe sets are used simultaneously) so that the products can be readily separated by capillary gel electrophoresis.

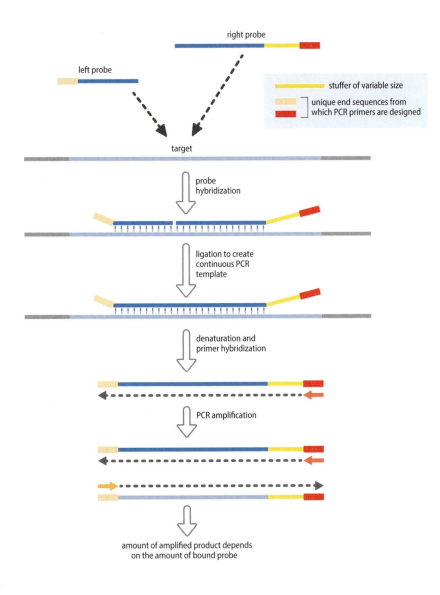

be of different lengths for different probes, enabling the amplified probes to be physically separated by capillary gel electrophoresis and quantified independently. See **Figure 11.10** for a practical example of how MLPA can be used to screen simultaneously for large numbers of different exons.

The biotechnology company MRC-Holland has a useful summary of MLPA technology that can be accessed from their home page at http://www.mlpa.com.

11.3 The Technology of Genetic Testing for Small-Scale DNA Changes

With the exception of some common chromosome aneuploidies (such as trisomy 21, causing Down syndrome), large-scale mutations are comparatively rare contributors to disease. The most common pathogenic DNA changes are small-scale DNA changes. They are dominated by point mutations, mostly single nucleotide changes. Another type of small-scale pathogenic DNA change involves aberrant cytosine methylations that induce inappropriate gene silencing or inappropriate gene expression. As in the previous section, diagnostic techniques can be applied to scan for

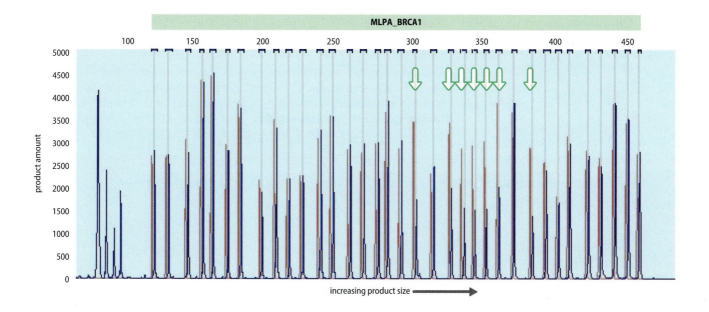

increasing product size ⟶

DNA changes whose precise identity is unknown, or in assaying defined specific DNA changes.

Scanning for undefined point mutations in single and multiple genes, and in whole exomes and genomes

Because it is readily applicable to any gene, DNA sequencing has traditionally been used for mutation scanning. In single-gene disorders in which the causative mutations are limited to one gene, or one of a few genes, Sanger (dideoxy) sequencing remains the foremost approach for mutation scanning. Here mutation scanning usually involves amplifying sequences from individual exons of a single gene, plus short regions of intron sequence adjacent to exons, and sequencing the amplified DNA; other mutation scanning methods are sometimes also used, as described below. Recently, massively parallel DNA sequencing has been used to extend mutation scanning to multiple genes or even the whole exome or whole genome, as described below.

Gene-specific microarray-based mutation scanning

Oligonucleotide microarrays (also known as gene chips) have been devised for scanning mutations in certain genes of high interest, such as for *TP53*, *BRCA1*, and *BRCA2*. Unlike in the chromosome microarrays described above, the oligonucleotides are designed to hybridize to just one gene in a test DNA sample. Nevertheless, there are often thousands of different oligonucleotides because the oligonucleotide sequences are chosen to comprise a series of overlapping sequences that collectively cover all the known functionally important parts of the gene of interest: the coding sequence, exon–intron boundaries, and any other known functionally important sequence. The method relies on a high hybridization stringency so that only perfect base matching is tolerated: if the test sample has a mutation, oligonucleotide probes spanning the mutation site can detect the abnormality.

Multiplex mutation scanning: multiple genes to whole exomes

Traditionally, mutation scanning meant that an individual gene of interest would be analyzed. More recently, mutation scanning has been extended in scope, to simultaneously cover multiple genes from any desired

Figure 11.10 Using MLPA to scan for constitutional copy number changes in the exons of the *BRCA1* gene. MLPA scan. The blue peaks at the left from 0 to 110 bp on the horizontal size scale are internal controls. The paired blue and red peaks in the size range 125–475 bp represent comparative MLPA results in a normal control sample (blue) and a test sample (red) for individual exons of the *BRCA1* gene in most cases (however, for some large exons two partly overlapping probes were used). The test sample came from an individual with breast cancer in whom previous DNA sequencing investigations were unable to identify changes in the exons of the *BRCA1* gene. The MLPA analysis shown here identified a deletion that encompassed seven consecutive exons (marked by vertical green-outlined arrows). In each case the blue peak is reduced by roughly one-half, as expected for a heterozygous deletion. Note that the order of the peaks is not the same as the order of the exons in the gene. (Data courtesy of Louise Stanley, NHS Northern Genetics Service, Newcastle upon Tyne, UK.)

regions of the genome. Increasingly, whole exomes, and sometimes whole genomes, are scanned using massively parallel DNA sequencing (**Box 11.2**).

Multiplex mutation scanning typically involves selectively purifying certain DNA sequences of interest (target sequences) that can be captured from an individual genomic DNA sample, and then sequenced. In **target enrichment sequencing** (also called *targeted sequencing*), the desired sequences are captured from a genomic DNA sample by a DNA

BOX 11.2 Massively parallel ('next-generation') DNA sequencing and whole-exome/ whole-genome sequencing.

In standard dideoxy sequencing, individual DNA sequences of interest must first be purified; they are then sequenced, one after another. The sequencing involves DNA synthesis reactions, producing a series of reaction products of different lengths that are then separated by gel electrophoresis (Figure 3.10, page 72). By contrast, massively parallel DNA sequencing (often called next-generation sequencing) is indiscriminate: all of the different DNA fragments in a complex starting DNA sample can be simultaneously sequenced without any need for gel electrophoresis. That allows a vastly greater sequencing output.

There are many different types of massively parallel DNA sequencing, but they can be separated into two broad categories: those in which the starting DNA sequences are first amplified by PCR, and those that involve single molecule sequencing (that is, sequencing of unamplified DNA molecules). We give details of run parameters for major commercially available technologies in Table 3.3 on page 75.

Massively parallel DNA sequencing often involves *sequencing-by-synthesis*. That is, the sequencing reaction is monitored as each consecutive nucleotide is inserted during DNA synthesis. For example, the Roche/454 GS-FLX sequencer uses reiterative pyrosequencing. Figure 11.16 shows how pyrosequencing is used to monitor the identity of an individual nucleotide as it is incorporated into a growing DNA chain; now imagine that step being repeated each time a nucleotide is incorporated until we establish the sequence over hundreds of nucleotides.

Figure 1 shows the workflow that is involved in massively parallel sequencing and gives a simplified illustration of another popular form of sequencing-by-synthesis used by the Illumina company. Alternative methods are used by some other companies, such as the sequencing-by-ligation method used in the ABI SOLiD system.

Whole-exome and whole-genome sequencing

Until recently, the sequencing of whole human genomes was rather demanding. Whole-exome sequencing offered a much more affordable alternative, and it has been increasingly used for diagnostic purposes.

Commercial exome capture typically uses DNA hybridization in solution to capture exome sequences from genomic DNA samples (see Figure 11.11). For example, the popular Roche NimbleGen exome capture kit uses a very large set of long oligonucleotides to retrieve mostly coding DNA sequences and some microRNA sequences from more than 20,000 human genes. (Because both microRNAs and intronic sequences that immediately flank exons are also included, the total size of the regions covered by the oligonucleotide probes is more than 60 Mb, or about 2% of the genome.)

As DNA sequencing costs drop even more, the alternative of whole-genome sequencing is becoming increasingly attractive as a diagnostic tool and will probably supplant whole-exome sequencing. It has the advantage that mutations in regulatory sequences that might lie in extragenic space or within central regions of introns will be recorded (but might be difficult to assess). The disadvantage is that the number of identified sequence variants is much larger than in exome sequencing. This means that much more extensive bioinformatics analysis is required, and there is a much greater prospect of incidentally finding additional pathogenic variants unrelated to the condition that prompted the initial clinical investigation. We consider ethical issues that stem from incidental findings in Section 11.5.

hybridization method and submitted for DNA sequencing. The capture method relies on the extraordinarily high affinity of streptavidin, a bacterial protein, for the vitamin biotin. A series of oligonucleotides are synthesized to represent all the desired target sequences and are each designed to have a biotin group covalently attached to one end. The biotinylated oligonucleotides are then mixed with magnetized beads coated with streptavidin; the strong biotin–streptavidin affinity means that the oligonucleotides bind strongly to the beads. The genomic DNA sample

BOX 11.2 (*continued*)

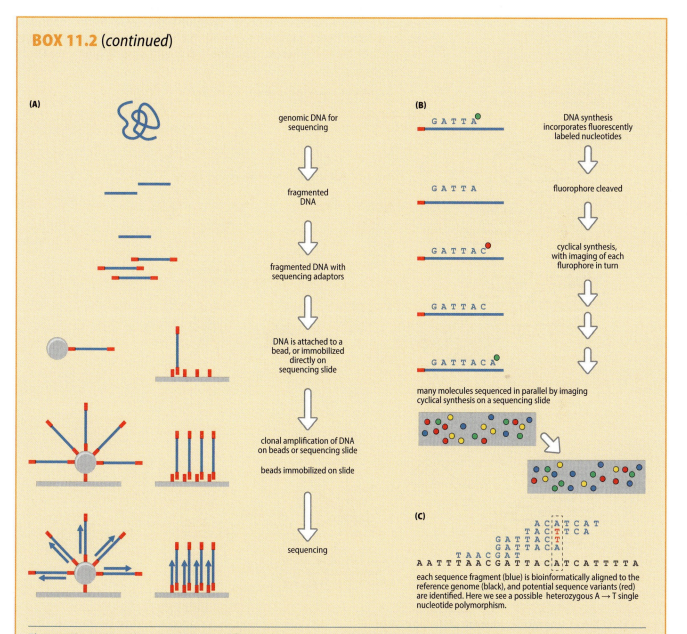

Figure 1 Next-generation sequencing workflow. (A) Genomic DNA is fragmented, and adaptor oligonucleotides are attached. The DNA is then attached either to a bead or directly to the sequencing slide. In either case, the DNA is clonally amplified in this location to provide a cluster of molecules with identical sequences. If beads are used, they are then immobilized on a sequencing slide. (B) The Illumina Genome Analyzer system of sequencing-by-synthesis. The sequence of each fragment is read by decoding the sequence of fluorophores imaged at each physical position on a sequencing slide. Advanced optics permit massively parallel sequencing. (C) Each DNA fragment yields one or two end sequences, depending on whether it is sequenced from one or both ends. These end sequences are computationally aligned with a reference sequence and mismatches are identified. (From Ware JS et al. [2012] *Heart* 98:276–281; PMID 22128206. With permission from the BMJ Publishing Group Ltd.)

is fragmented, denatured, and mixed with bead–oligonucleotide complexes. Target sequences in the genomic DNA samples will hybridize to complementary oligonucleotide sequences on the beads. Having fished out the target sequences from the genomic DNA sample, the beads can be removed with a magnet, and the target sequences can be eluted, amplified, and sequenced (**Figure 11.11**).

A popular application of target enrichment sequencing is to capture and sequence whole exomes from genomic DNA samples (see Box 11.2). It has also been used to capture smaller subsets of the genome and sequence exons (plus sequences spanning the exon–intron boundaries, and other regions known to harbor pathogenic mutations) from sets of genes of clinical interest. Illumina's TruSight One Sequencing Panel

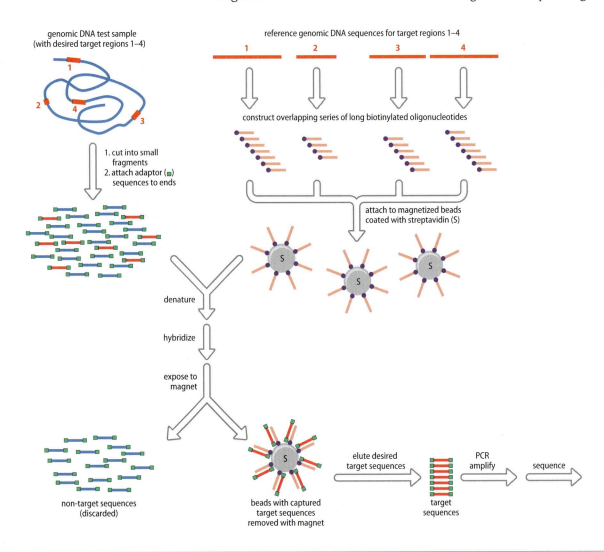

Figure 11.11 Target sequence enrichment: capturing desired target sequences from a genomic DNA sample for DNA sequencing. In this example, genomic regions 1–4 (shown in red at the top left) are the regions of interest in the genomic DNA samples to be analyzed, and they will be captured by DNA hybridization using complementary synthetic oligonucleotides. The reference DNA sequences for the four regions are examined; for each region, a series of overlapping long oligonucleotides (top right) are synthesized with a biotin group covalently attached at one end (purple circle). Streptavidin-coated magnetic beads are used to bind the resulting biotinylated oligonucleotide probes. The genomic DNA test sample is cleaved to generate small DNA fragments, and short adaptor sequences are attached to each end (the adaptors will be needed for the final steps of PCR amplification and DNA sequencing). The resulting fragments are denatured, and sequences from the desired genomic regions hybridize to the oligonucleotide probes on the magnetized beads. After hybridization, the beads (with bound complementary target sequences) are removed from solution with a magnet, and the target sequences are eluted (by gently disrupting hydrogen bonding to the bound oligonucleotides), amplified using primers specific for the adaptor sequences, and submitted for DNA sequencing.

REFERENCE SEQUENCE FOR GENE OR PRODUCT	EXAMPLE	INTERPRETATION
Genomic DNA (g.) Nucleotide numbering starts with 1 for the first nucleotide of the cited reference genomic sequence	g.1021C>T	C at position 1021 is replaced by T
	g.275_276insG	G inserted between nucleotides 275 and 276
cDNA (c.)—for coding and untranslated regions The first nucleotide of the initiator ATG codon is numbered +1. The base immediately preceding it is –1, and all positions in the 5′ UTR are preceded by a minus sign. Bases in the 3′ UTR are preceded by an asterisk	c.–107A>G	replacement of A by G at position number –107 in the 5′ untranslated region (107 nucleotides before the A of the start codon)
	c.872_875del	deletion of nucleotides 872–875 in the coding sequence
	c.*57C>G	replacement of C by G at nucleotide position 57 in the 3′ untranslated region
cDNA (c.)—for neighboring intronic sequence The first number is the nucleotide number of the cDNA sequence for a nucleotide at the start or end of an exon. The second number is the number of nucleotides required to reach the intron nucleotide(s) at which the variant occurs. An intervening plus sign means that the first number is the last nucleotide of an exon and we are counting forward from the 5′ end of an intron. An intervening minus sign means that the first number is the first nucleotide of an exon and we are counting backward from the 3′ end of the preceding intron	c.178+9A>G	replacement of A by G at the ninth nucleotide within the intron that follows nucleotide number 178 in the cDNA (the last nucleotide of the preceding exon)
	c.179–3C>T	replacement of C by T at third nucleotide preceding nucleotide 179 in the cDNA (the first nucleotide of the following exon)
Protein (p.) The three-letter code is used and an asterisk means a stop codon. Note: an older, but still quite widely used, nomenclature for nonsense mutations uses an X to indicate a stop, such as G542X	p.Asp107His (= p.D107H)	aspartate at amino acid position 107 is replaced by histidine
	p.Gly542* (= p.G542*)	the codon specifying glycine at amino acid position 542 is replaced by a stop codon

allows the simultaneous screening of close to 5000 genes associated with known clinical phenotypes. More specific sequence capture can also be conducted for more specific disease categories, allowing multiple gene testing. For example, Illumina's TruSight sequencing panels include individual panels focused at autism (101 genes), cancer (94 genes), cardiomyopathy (46 genes), and inherited disease with a focus on pediatric recessive disorders (552 genes). Multiplex gene scanning is beginning to be widely used in clinical diagnostics.

Table 11.3. Examples of nomenclature for describing DNA and amino acid variants. The nomenclature conventions are described in detail in the Human Genome Variation Society's Website (http://www.hgvs.org/mutnomen). The computer program Mutalyzer (http://www.lovd.nl/mutalyzer) will generate the correct name of any sequence variant that a user inputs.

Interpreting sequence variants and the problem of variants of uncertain clinical significance

Mutation scanning of a known disease-associated gene (or candidate gene) in a panel of patients or across whole individual exomes and genomes will reveal many different genetic variants (see **Table 11.3** for the recommended nomenclature for sequence variants). Many variants will not be related to pathogenesis; the challenge is to interpret the variants and, as required, test potentially promising ones to identify those involved in pathogenesis.

Interpreting sequence variants in a known or candidate disease locus can be straightforward. In genes associated with loss-of-function phenotypes, some mutation classes—large deletions, frameshifts, nonsense mutations, and changes to the canonical GT and AG motifs at splice sites—are highly likely to be pathogenic. Assessing nucleotide substitutions is much less easy, and they are particularly common. The problem is massively compounded, of course, in whole-exome and whole-genome sequencing.

For people with a dominant disorder but unaffected parents, a *de novo* mutation (not present in either parent) has a generally much higher likelihood of being pathogenic than an inherited mutation. If the disorder is familial, the mutation might be checked in other family members. If the mutation does not segregate with disease, it is highly unlikely to be implicated in the disease, assuming a high penetrance. But the reverse is not true: co-segregation with disease is not evidence that a variant is pathogenic (a nonpathogenic variant at a disease locus has a 50% chance of residing within the same allele as the true disease-causing mutation, and a 50% chance of co-segregating with disease).

Ethnically matched control DNA samples may be used (but results are often compared with data stored in databases—see below). A sequence variant found in healthy male and female controls would usually be eliminated from consideration in a highly penetrant early-onset dominant or X-linked condition, but could be pathogenic in an autosomal recessive or a low-penetrance dominant condition. In general, however, testing controls is very poor at ruling out any variant with a low (less than 0.01) frequency in the population (the virtual absence of any phenotyping in control cohorts means that they may include pre-symptomatic individuals with disease-associated variants).

Information on previously recorded variants can be obtained from different mutation databases, as described below. Further testing of candidate sequence variants usually involves bioinformatic analyses (and sometimes laboratory analyses, often in a research context), but classifying variants can be challenging, as described below.

Mutation interpretation and databases

Checking mutation databases is important to determine whether a variant has previously been reported and what information might be available. First steps include checking the exome variant server of the NHLBI Exome Sequencing Project at http://evs.gs.washington.edu/ and the relevant mutation database for the gene containing the variant, if listed within locus-specific mutation databases (for a list, see http://www.hgvs.org/dblist/glsdb.html).

More generally, information about previously recorded mutations and single nucleotide polymorphisms is available in databases such as the Human Gene Mutation Database (http://www.hgmd.org/) and dbSNP (http://www.ncbi.nlm.nih.gov/projects/SNP). If the variant has previously been reported and is common (with a frequency of more than 0.01), it is unlikely to be associated with a rare single-gene disorder (but many rare nonpathogenic variants will not be in the database). Attempts are being made to establish a universal clinical genomics database, as described below.

Evaluation of the predicted effects of point mutations may often be relatively straightforward for mutations that introduce a premature termination codon (Section 7.2). For amino acid substitutions, a nonconservative change (replacing one amino acid by another of a different chemical class) is much more likely to be pathogenic than a conservative amino acid change. Different types of computer program can be helpful here, such as those included within the commercial Alamut 2.3 software (see http://www.interactive-biosoftware.com) and also Web-available programs such as PolyPhen-2 (http://genetics.bwh.harvard.edu/pph2/) and SIFT (http://sift.jcvi.org/).

Evolutionary conservation is another important factor. A substitution is unlikely to be pathogenic if it changes the amino acid to one that is the normal (wild-type) amino acid at an equivalent position in a clearly orthologous protein from another species. Conversely, if the normal amino acid is very highly conserved across a range of species, a mutation producing a nonconservative amino acid change at this position becomes highly significant. **Figure 11.12** provides a recent example of the utility of this approach in a recent case in which whole-exome sequencing allowed a definitive diagnosis in a child with intractable inflammatory bowel disease (the precise molecular diagnosis led to more specific, and possibly life-saving, clinical management).

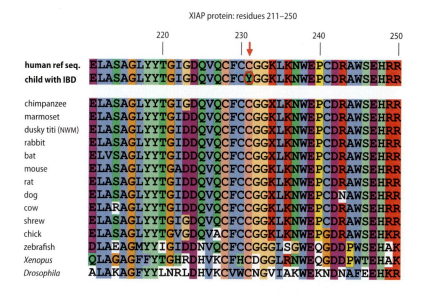

XIAP protein: residues 211–250

Figure 11.12 Evolutionary conservation analyses helped obtain a definitive diagnosis in a child with inflammatory bowel disease (IBD). In this example, 6799 nonsynonymous coding variants were identified after sequencing the exome of a child with intractable IBD born to unaffected parents. The disease severity and unique clinical presentation suggested a recessive disorder. A total of 66 genes were identified as containing potential compound heterozygous mutations that the PolyPhen program predicted could be possibly harmful, but they were excluded on the basis of a lack of novelty of the mutations and/or a lack of strong evolutionary conservation of the original amino acid. Eight out of a further 70 homozygous or hemizygous nonsynonymous variants were predicted to be damaging by PolyPhen, but in only two of them were the original amino acids strongly conserved during evolution. One was a variant in the *GSTM1* gene (but could be excluded because this gene has a high null-genotype frequency in the general population). The other was a novel variant in the X-linked inhibitor of apoptosis (*XIAP*) gene causing a p.Cys231Tyr substitution. This mutation appeared highly significant on the basis of its very strong evolutionary conservation as shown here (amino acids of similar function are given the same or similar colors, such as the acidic amino acids aspartate [D] and glutamate [E]). The mutant protein XIAP protein showed a loss of the normal function in apoptosis and *NOD2* signaling, confirming the mutation as the causative one. NWM, New world monkey. (Adapted from Worthey EA et al. [2011] *Genet Med* 13:255–262; PMID 21173700. With permission from Macmillan Publishers Ltd.)

Functional studies can be carried out to examine how a sequence variant might affect how a gene behaves, especially if the product is easily assayed in cultured cells (such as for many enzymes). Often, however, a gene-specific functional test is too impractical to be easily adopted by a standard diagnostic laboratory.

Candidate splice-site mutations (other than substitutions of invariant terminal GT and AG dinucleotides of introns) can be evaluated by *in silico* splice-site prediction, such as using programs accessed through the Alamut 2.3 software described above. Follow-up laboratory analyses are sometimes carried out to check how specific mutations affect splicing ability, such as by using RT (reverse transcriptase)-PCR on RNA isolated from blood or skin samples (when the gene is expressed in these cells). Much less is known about noncoding DNA variants that affect gene regulation—the review by Jarinova & Ekker (2012) under Further Reading provides a recent summary.

Variants of uncertain clinical significance

Interpreting the effect of noncoding DNA variants is challenging. However, except for those that occur close to splice junctions (which are generally easier to interpret), noncoding DNA variants make a modest contribution to highly penetrant disease phenotypes. But even within coding DNA the effects of many sequence variants—notably in-frame short deletions and insertions, missense variants, and some synonymous variants—are difficult to predict. Genetic test results that report variants of uncertain clinical significance pose a significant problem in risk assessment, counseling, and preventive care. When genetic testing is extended to genome-wide sequencing the scale of the problem is hugely extended. Each individual will exhibit several thousand variants at the exome level and a few million variants at the whole genome level: each one of us will have a large number of DNA variants of uncertain clinical significance.

For some genes, significant effort is being devoted to carrying out comprehensive functional assays to assess the effects of recorded variants. For example, the ENIGMA consortium (Evidence-based Network for the Interpretation of Germline Mutant Alleles) is focusing on sequence variation in the *BRCA1* and *BRCA2* gene regions. More than 1500 unique variants have been recorded in the *BRCA1* gene alone, and a recent report from the ENIGMA consortium describes efforts to assess some of the variants (see Further Reading).

Scanning for possible pathogenic changes in cytosine methylation patterns

Altered cytosine methylation patterns represent another type of small-scale DNA change that can contribute to pathogenesis. Scans for disease-associated changes in cytosine methylation are important in cancer studies (epigenetic changes are very common in tumorigenesis), and in testing for certain inherited disorders, notably imprinting disorders (according to the sex of the transmitting parent, alleles at imprinted gene loci are subject to epigenetic silencing and hypermethylation). Additionally, in some disorders, such as fragile X-linked intellectual disability, the pathogenesis involves the expansion of tandem repeats that then predisposes the amplified region to hypermethylation.

Methylated and unmethylated cytosines can be distinguished by making the DNA single stranded and treating it with sodium bisulfite (Na_2SO_3). Under controlled conditions, the unmethylated cytosines are deaminated to produce uracils, but 5-methylcytosines remain unchanged. After treatment with sodium bisulfite, the relevant region can be amplified by PCR, during which newly created uracils are read and propagated as thymines. New DNA strands are synthesized without incorporating methyl groups so that any retained methylated cytosines in the template DNA are propagated as unmethylated cytosines. That allows different ways of distinguishing the methylated cytosines from the original unmethylated cytosines.

Figure 11.13 shows how, after treatment with sodium bisulfite, samples can be amplified by PCR and sequenced to distinguish methylated cytosines from unmethylated cytosines. Methylation-specific PCR assays can also be devised by designing alternative PCR primers to have 3' nucleotides that are specific for one of the variable nucleotides after treatment with sodium bisulfite (that is, a U or T versus a C).

Figure 11.13 Distinguishing methylated cytosines from unmethylated cytosines with the use of sodium bisulfite. Sodium bisulfite converts unmethylated cytosines (left panel) to uracils; after DNA replication in a PCR reaction, they become thymines in newly synthesized DNA. Sodium bisulfite does not react with methylated cytosines, which remain unchanged (right panel). Newly synthesized DNA strands in a PCR reaction are not methylated, and so although the starting DNA is methylated in the right panel, the PCR product is unmethylated. DNA sequencing can identify all unmethylated cytosines because after treatment with sodium bisulfite each unmethylated C becomes a T (shown by boxes in sequencing panels at the bottom); if the cytosines are methylated, the sequence obtained is the same as in DNA that has not been treated with sodium bisulfite. DNA sequencing of PCR products is therefore one way of distinguishing between the two patterns. Alternative assays use methylation-specific PCR by designing primers with a nucleotide at the 3' end that corresponds to a variable site, pairing with either U/T (from an unmethylated cytosine that has been chemically converted by sodium bisulfite) or C (representing an originally methylated cytosine). Other assays take advantage of methylation-sensitive restriction enzymes. For example, the enzyme TaqI recognizes and cuts the unmethylated sequence TCGA, but not when the cytosine is methylated.

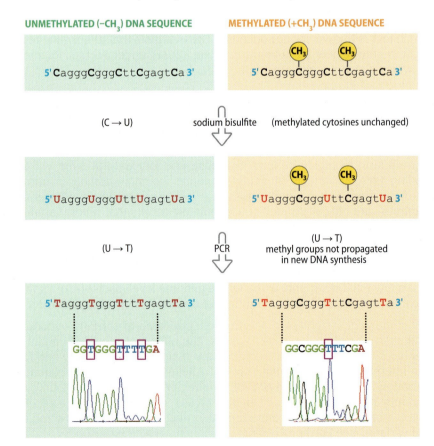

Other approaches use restriction enzymes that do not cut their normal target sites if they contain a methylated cytosine, such as *Hha*I (which cuts at GCGC sites, but not at GC^meGC). For example, MS-MLPA, a methylation-specific variant of the multiplex ligation-dependent probe amplification method described above, depends on methylation-sensitive restriction nucleases (see the MLPA technology component in the MRC-Holland Website at http://www.mlpa.com). It can differentiate cases of imprinting disorders such as Angelman syndrome or Prader–Willi syndrome that result from deletion from those that do not (as a result of uniparental disomy, for example).

Technologies for genotyping a specific point mutation or SNP

Instead of simply scanning for any potentially pathogenic mutations, many types of genetic test seek to identify a *specific* point mutation. DNA sequencing can be used to identify such variants, but it is often more convenient to use alternative detection methods, such as when large numbers of samples are involved.

Different methods can be found to discriminate between alleles that differ by just a single nucleotide. Most of them rely on forming duplexes that are perfectly matched or mismatched at a single nucleotide. That can allow allele-specific hybridization (as in the oligonucleotide microarray approach above), allele-specific PCR, and allele-specific ligation to a neighboring oligonucleotide, as described below. Some examples of popular mutation detection methods are listed below.

Amplification refractory mutation system (ARMS). If the 3′ end nucleotide of a PCR primer is not correctly base paired, PCR amplification is not possible when using a DNA polymerase such as *Taq*I polymerase that lacks a proofreading activity. That requirement allows a form of allele-specific PCR: normal and mutant alleles that differ by a single nucleotide are discriminated by a PCR assay using a common primer plus an allele-specific primer. The allele-specific primer has a terminal 3′ nucleotide that base pairs correctly with either the normal or the mutant nucleotide. That is, the primer is designed so that the 3′ end nucleotide is placed opposite the mutation site (**Figure 11.14**).

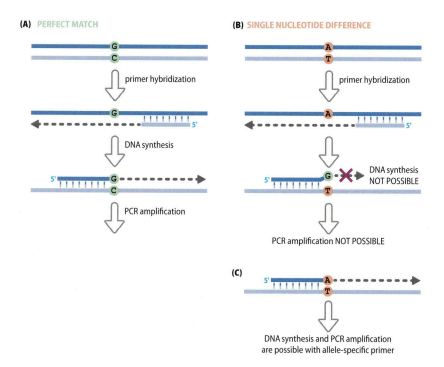

Figure 11.14 The principle underlying the amplification refractory mutation system (ARMS) for detecting a specific single nucleotide change. In a PCR assay, when using a DNA polymerase that lacks a proof-reading activity, the nucleotide at the 3′ end of each oligonucleotide primer needs to be correctly base paired to the template DNA to allow DNA synthesis. (A) The deep blue primer terminating in a G at its 3′ end will allow DNA synthesis (and consequent PCR amplification) because it is correctly base paired to the normal template DNA. (B) The template DNA has a mutation with a single difference at the nucleotide position that corresponds to the 3′ terminal nucleotide of the oligonucleotide primer. The lack of base pairing at the 3′ end means that DNA synthesis cannot be primed. (C) However, if an allele-specific primer is used that terminates in a A at its 3′ end, DNA synthesis can occur and PCR amplification is possible. Thus, the primer terminating in a G would be specific for normal alleles, and the primer terminating in an A would be specific for the mutant allele.

(A) **PERFECT MATCH**

primer hybridization

DNA synthesis

PCR amplification

(B) **SINGLE NUCLEOTIDE DIFFERENCE**

primer hybridization

DNA synthesis
NOT POSSIBLE

PCR amplification NOT POSSIBLE

(C)

DNA synthesis and PCR amplification are possible with allele-specific primer

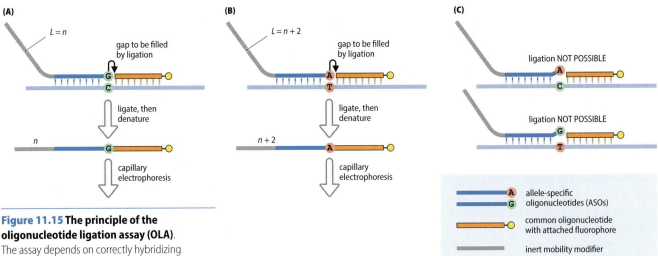

Figure 11.15 The principle of the oligonucleotide ligation assay (OLA).
The assay depends on correctly hybridizing two oligonucleotides to adjacent sequences on the same template DNA: an allele-specific oligonucleotide (ASO), whose specificity depends on the terminal nucleotide at its 3′ end, and a common oligonucleotide (which base pairs with different alleles). The ASO is attached at its 5′ end to an inert molecule whose size can vary—equivalent to a length (L) of n or $n + 2$ nucleotides in this example—and the common oligonucleotide has a fluorophore at its 3′ end. Ligation is possible only if nucleotides at the junction of the aligned oligonucleotides are correctly base paired; when that happens, the 3′-terminal hydroxyl group of the ASO becomes covalently linked to the 5′-terminal phosphate of the common oligonucleotide. The resulting larger molecule can be size-separated by capillary electrophoresis on a suitable DNA sequencing machine. By having different sizes for the inert molecule attached to the 5′ end of the ASO, different alleles can be detected simultaneously. (A) The template DNA (light blue strand) is normal and the ASO terminating in a G is perfectly base paired, allowing ligation. (B) The template DNA has a mutation (a T instead of a C), but an ASO terminating in an A can base pair with it perfectly. (C) Ligation is not possible when an ASO is not perfectly base paired at its 3′ end. Note that multiplexing allows multiple mutations at different sites to be simultaneously assayed by adjusting the size of the inert molecule attached to the ASOs.

Oligonucleotide ligation assay. Like the ARMS method, this method depends on binding by an allele-specific oligonucleotide (ASO) in which the specificity depends on correct base pairing at the terminal nucleotide at the 3′ end (which is intended to base pair with the mutant nucleotide or its normal equivalent). When the ASO and another oligonucleotide are simultaneously annealed to adjacent sequences on a DNA template, they can be ligated together only if the oligonucleotides perfectly match the template at the junction. That is, ligation is impossible if, at the junction, the end nucleotide of either oligonucleotide is incorrectly base paired (**Figure 11.15**).

Pyrosequencing. This is a DNA synthesis reaction. When DNA is synthesized from deoxynucleoside triphosphate (dNTP) precursors, the bond between the α and β phosphates of the dNTP is cleaved, so that a dNMP (containing the α phosphate) is incorporated into the growing DNA chain. That leaves a pyrophosphate (PPi) residue made up of the β and γ phosphates, which is normally discarded. In pyrosequencing, sequential enzyme reactions detect the pyrophosphate that is released when a nucleotide is successfully incorporated into a growing DNA chain (**Figure 11.16A**). Pyrosequencing can be used to discriminate between normal and mutant sequence variants because individual dNTPs needed for DNA synthesis are provided in separate reactions in a defined order (**Figure 11.16B**). It has been particularly helpful in detecting mutations in low-level mosaicism, and in heterogeneous tumor samples.

Multiplex genotyping of specific disease-associated variants as a form of mutation scanning

In most genetic disorders, the disease can be caused by any number of mutations, but for some types of genetic disorder there is limited or very limited mutational heterogeneity. Sickle-cell anemia is the outstanding example: its very specific phenotype is always due to a nucleotide substitution that replaces a valine residue at position 6 in the β-globin chain by glutamate. Unstable oligonucleotide repeat disorders such as Huntington disease usually show a very limited range of mutations.

For some other disorders, much of the pathogenesis is attributable to a comparatively small number of mutations. In cystic fibrosis, for example, the c.1521_1523delTTC mutation causing the p.Phe508 del variant is very common in populations of European origin. Genotyping for this variant

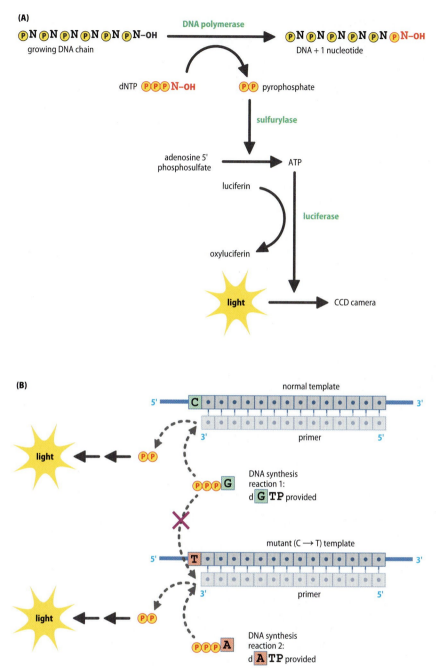

(A)

growing DNA chain → DNA polymerase → DNA + 1 nucleotide

dNTP → pyrophosphate

sulfurylase

adenosine 5' phosphosulfate → ATP

luciferin

luciferase

oxyluciferin

light → CCD camera

(B)

normal template

primer

light

DNA synthesis reaction 1: dGTP provided

mutant (C → T) template

primer

light

DNA synthesis reaction 2: dATP provided

Figure 11.16 The principle underlying pyrosequencing. (A) Insertion of a nucleotide into a growing DNA chain involves cleavage of the dNTP precursor and insertion of a dNMP residue. The remaining pyrophosphate is detected in pyrosequencing by a two-step enzymatic reaction. First, ATP sulfurylase quantitatively converts PP$_i$ to ATP in the presence of adenosine 5' phosphosulfate. Then the released ATP drives a reaction in which luciferase converts luciferin to oxyluciferin, a product that generates visible light in amounts that are proportional to the amount of ATP. Thus, each time a nucleotide is incorporated, a light signal is produced and recorded by a CCD (charge-coupled device) camera. (B) The dNTP precursors for DNA synthesis are provided *individually* and in a set order. If in the first reaction dGTP is the only nucleotide precursor, a G would be incorporated opposite the highlighted C in the normal template at the top, and a pyrophosphate residue would be released and trigger light production as shown in (A). But if a mutant DNA template were used (with a T replacing the C), no light would be produced in the first reaction (a G would not be inserted opposite the T residue). If, however, dATP is provided in a second reaction, an A would be inserted opposite the T in the mutant template and produce pyrophosphate and light, but there would be no base incorporation opposite the C in the normal template.

plus a limited number of other quite common variants provides a useful initial mutation scan.

Multiplex genotyping is often carried out with ARMS or oligonucleotide ligation assays described in the previous section. Various commercial kits are available, such as GEN-PROBE's Elucigene CF-EU2v1 kit for cystic fibrosis (ARMS testing for 50 disease-associated mutations). The Elucigene FH20 kit for familial hypercholesterolemia (ARMS testing for 20 common point mutations in three separate reactions) picks up more than 50% of mutations associated with familial hypercholesterolemia. However, as target enrichment sequencing develops and costs fall, it is likely to be increasingly applied to mutation scanning of complex individual genes.

11.4 Genetic Testing: Organization of Services and Practical Applications

Genetic services evolved in many countries in the early 1960s to translate new chromosome findings into clinical services. Initially, the services were staffed by cytogeneticists and clinicians; DNA testing was added in the mid-1980s. In some cases, a high prevalence of certain disorders in a population has strongly affected how services have developed. In Cyprus, for example, the exceptional prevalence of β-thalassemia has meant that much of the health budget has been focused on this disorder.

In economically developing countries where there is a high frequency of serious infectious diseases, such as malaria and HIV-AIDS, a service focused on genetic disorders has not been the priority. When such countries do introduce genetics into medical practice, they may develop different models tailored to their needs rather than adopt the genetic services model used in economically more advanced countries.

Future developments in genetic services will occur against the background of an exponential increase in understanding of the molecular basis of genetic disorders. About 10 years ago, at the turn of the new millennium, genetic testing was available for few disorders; patients were generally seen by clinical geneticists before having a test. Now, with an increasing number of genetic tests that are becoming more widely applicable, genetic testing will increasingly be devolved.

The devolution of genetic services is occurring in different ways. Within the health care sector, a recent policy of *mainstreaming genetics* envisages genetics being rapidly incorporated into mainstream medicine—diagnostic genetic testing will increasingly become the responsibility of the clinicians to whom patients are initially referred.

Clinicians in medical specialties with limited experience of genetic testing will need to develop the necessary knowledge to interpret and convey the results of the testing, and to recognize when referral to a genetic service is appropriate. Outside the health care sector, genetic testing is being marketed by private companies, but there are concerns about different aspects of direct-to-consumer genetic testing.

In this section we are primarily concerned with the environment in which genetic testing is offered, the practical organization that is required, the range of tests that are offered, and the object of the tests. Genetic testing may be offered to individuals and their close relatives in response to a clinical referral of affected individuals, or to individuals who are expected to be at risk of developing a disorder. In addition, genetic screening programs may be proactively directed at communities and populations to identify individuals at risk of transmitting or developing disease; they may then be offered more specific testing.

Genetic testing of affected individuals and their close relatives

Within health care systems, genetic tests are often directed at persons presenting with a clinical problem, or at close relatives of a person with a genetic disorder. For affected individuals the aim is to identify causative genetic variants, or subsequently to monitor a disorder or the response to treatment. Follow-up tests on close relatives can identify if they carry the mutant allele. Genetic tests have traditionally been arranged through clinical geneticists and genetic counselors, but some specialist services—hemophilia centers in the UK, for example—have been accustomed to cover all aspects of care for a patient group, including genetic testing and counseling of extended families.

Until quite recently, genetic testing was expensive, but technological advances have allowed the enrichment of target genes for sequencing (Figure 11.11). As DNA sequencing costs plummet, multiple genes can be studied in parallel, and the costs of testing are falling. As the extent and utility of genetic testing increases, the old model in which clinical geneticists arrange most genetic tests is becoming unsustainable—increasingly, the clinical team investigating and managing patients will request investigations directly.

The clinicians most skilled in a particular group of disorders will add genetic investigation to the diagnostic tests available to them. That will become increasingly important if genetic diagnosis changes therapeutic decisions. Take, for example, the effect of the discovery that PARP-1 (poly[ADP-ribose] polymerase) inhibitors are particularly effective in treating breast cancer in patients with *BRCA1* and *BRCA2* mutations. Homologous recombination (one of the two major DNA repair methods) is nonfunctional in cells with either of these mutations, but base-excision repair is unaffected. PARP-1 inhibition disables base-excision repair, and thus cells with *BRCA1* or *BRCA2* mutations are no longer able to repair DNA. As described below, genetic testing for cancer is undergoing a revolution.

When funding for a genetic test exists, it clearly makes sense for the clinician making the diagnosis to request the test directly. Concerns about mainstreaming genetics revolve around whether medical professionals in other medical disciplines will interpret complex reports correctly: will they be less conscious than clinical geneticists of the need to consider the extended family?

There is often little information to help clinicians interpret sometimes complex DNA or chromosome results and to guide them on what to do on receiving such a report. More genetics education for different medical specialties and communication with staff in clinical genetics services will be a priority.

Cascade testing

Cascade testing refers to testing of relatives after the identification of a genetic condition in a family. The relatives might be at risk of going on to develop the same single-gene disorder (predictive testing, discussed below). Unaffected relatives may also be at risk of transmitting a disorder if they carry a harmful allele (heterozygote carriers in recessive disorders, non-penetrance in dominant disorders) or a balanced translocation.

The family and the genetic counselor (see below) need to consider different issues. How important is it for the relatives to be made aware of the information on the basis of the severity of the condition and the level of risk of a relative developing the condition, or having a child with the condition? How might the information change things? How easy will it be for family members to pass information on to relatives?

Take the example of a child with multiple malformations and developmental delay who has inherited unbalanced chromosome translocation products from a parent with a balanced translocation. The translocation will be explained to the parents along with information about their future pregnancies, and also the possibility of other family members carrying the same balanced translocation. In addition to addressing questions from the couple about the risk to future pregnancies and about their child's future, health professionals need to consider which additional family members should be contacted who might have the same balanced translocation (and be at risk of producing children with unbalanced translocation products), and how to go about this. The same principles apply to cascade testing for carriers of autosomal or X-linked recessive disorders.

Traditional prenatal diagnosis uses fetal tissue samples recovered by an invasive procedure

Couples who have a family history of a serious genetic disorder usually want to know whether they are at risk themselves of having an affected child. If they are at risk, they might choose not to have children but to adopt instead. More frequently, they wish to have children.

Occasionally, such a couple will choose to have a child in a way that circumvents the genetic risk: by egg or sperm donation, or by preimplantation diagnosis so that only healthy embryos are selected, as described below. More commonly, the couple opt for natural conception and prenatal diagnosis in which the fetus is tested to see if it carries the harmful genetic abnormality, or not. Couples who request this type of diagnosis usually wish to terminate the pregnancy if the fetus is affected. Genetic counseling and risk assessment is important in prenatal diagnosis (**Box 11.3**).

Accurate genetic testing is possible for single-gene disorders in which the major genetic variant contributing to disease has been identified in an affected family member. Prenatal diagnosis is also often carried out in situations in which there is an increased risk of transmitting a chromosomal aneuploidy (advanced maternal age is an important risk factor). Or one parent might have been identified as a carrier of a balanced translocation, and there is a risk that a fetus with unbalanced translocation products might be viable but have severe problems.

Traditionally, prenatal diagnosis has involved collecting a sample of fetal tissue recovered by an invasive procedure. A sample may be taken from the chorion (the outermost extra-embryonic membrane), and fetal DNA can be isolated from the cells obtained (**Figure 11.17A**); there is a roughly 1% excess risk of miscarriage. The sample can be taken any time in the pregnancy from 11 weeks onward, but typically in the first trimester (to allow the possibility of early termination of pregnancy).

Amniocentesis is the other major alternative sampling method, and it also has a small risk of miscarriage. A sample of amniotic fluid is taken at, or close to, 16 weeks of gestation (**Figure 11.17B**); it provides fetal cells that are processed to give either chromosome preparations to check for chromosome abnormalities, or fetal DNA samples for analysis.

Preimplantation genetic testing often analyzes a single cell in the context of assisted reproduction (*in vitro* fertilization)

Preimplantation genetic testing is technically challenging: the procedure often involves analyzing an individual cell as a way of monitoring the genotype of the oocyte or of the early embryo. To infer the genotype of an oocyte, polar bodies are sometimes analyzed. More commonly, a single cell (blastomere) is removed from the very early embryo for testing (**Figure 11.18**). For technical reasons, some centers prefer to analyze a few cells taken from the outer trophectoderm at the later blastocyst stage (the trophectoderm will give rise to extra-embryonic membranes). In either case, the remaining embryo can be implanted successfully and is viable.

Standard assisted reproduction techniques are used to obtain embryos for testing: ovarian stimulation (to produce eggs that are then collected under sedation), addition of sperm, and assessment of the *in vitro* fertilization (IVF) and of the embryos produced. In the case of single blastomere

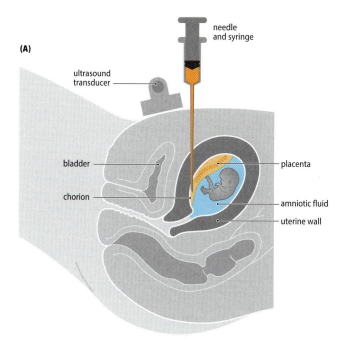

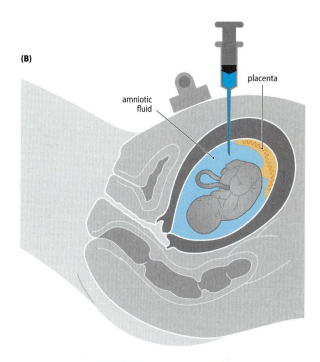

analyses, individual embryos are grown in culture to reach the 6–10-cell stage. At this stage a small hole is made in the zona pellucida and a single cell is removed through the hole for testing. Despite the loss of one cell for analysis, the embryo will go on to develop normally.

There are two broad categories of preimplantation genetic testing: diagnosis and screening. Diagnosis applies to couples who are at risk of transmitting a specific genetic abnormality: one or both parents have previously been shown to carry a harmful mutation or chromosome abnormality that the test is designed to identify. By contrast, screening is performed on couples who may have difficulty conceiving but have no

Figure 11.17 Invasive prenatal diagnosis using chorionic villus sampling or amniocentesis. (A) Chorionic villus sampling. As shown here, this is usually carried out by a transabdominal approach guided by ultrasound under local anesthetic. (B) Amniocentesis.

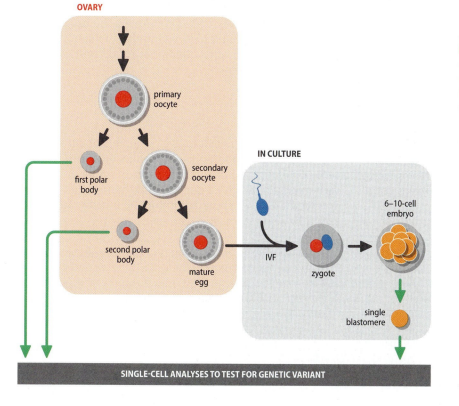

Figure 11.18 Preimplantation genetic diagnosis often involves analyzing single cells. The object is to prevent the transmission of a harmful genetic defect by using *in vitro* fertilization (IVF). Unlike for sperm cells, the meiotic divisions giving rise to an egg cell are asymmetric: the primary oocyte divides to give a secondary oocyte and a polar body, and the secondary oocyte divides to give the mature egg cell and a second polar body. The polar bodies are disposable and can be analyzed to infer whether the egg cell is carrying a specific harmful genetic variant or a chromosomal aneuploidy. If not, IVF proceeds with what appears to be a normal egg cell. More commonly, a single cell is sampled from the early embryo and tested for the presence of the harmful genetic variant. If the test result is negative, the remaining embryo is implanted in the uterus, and development can proceed normally. Because it can be challenging to obtain data from a single cell, some centers prefer to allow the embryo to develop further and remove a few cells from the blastocyst for testing.

BOX 11.3 Genetic consultations, genetic counseling, and risk assessment.

Genetic consultations

Clinical geneticists have traditionally made diagnoses in patients with genetic diseases, providing appropriate information to those affected and their families so that they can make life choices. A distinctive feature of clinical genetics has been the parallel focus on the family and on the person at the appointment. Many children are referred without an explanation for their problems; the clinical geneticist takes the family history and examines the child to establish a differential diagnosis and arrange appropriate tests.

Genetic consultations often start with a child or adult with a genetic disorder—but many consultations are initiated by a person's concern about other family members. Rather than focus on his or her own health issues, the person may bring up the subject of a family history of a medical problem such as cancer. Together, the patient and geneticist construct the pedigree, the basic tool in the genetics clinic. Diagnoses are then confirmed by using registries, or death certificates in the case of deceased individuals, or by requesting consent to access medical information of living relatives.

It is essential that the diagnoses are confirmed, whereupon the geneticist considers whether genetic tests are appropriate, and whom it is most appropriate to test. If there is a relative who has the disorder, it would be more appropriate to test the relative first to establish the causative mutation, which would then be the basis of a predictive test.

Genetic counseling and risk assessment

Parents at risk of having a child with a genetic disorder, and affected individuals and relatives of an affected family member, benefit from **genetic counseling**, the process by which they are informed of the consequences and nature of the disorder, the probability of developing or transmitting it, and the options open to them. Genetic counseling may be provided by doctors or by professionals trained as genetic counselors or by nurse specialists.

A non-directive approach to counseling is taken: the counselor should provide the necessary information that will help family members to make a decision rather than direct them toward the decision.

As well as offering general support, the counseling process has at its core the determination of recurrence risks. That may be relatively simple for single-gene disorders based on Mendelian principles, but as detailed in Section 5.3 there are often complications, such as lack of penetrance or variable expressivity.

The risk estimate may be determined by a Bayesian calculation in which a prior probability (such as the risk predicted from Mendelian principles alone, for a single-gene disorder) is modified by some other relevant information. For an X-linked recessive disorder, for example, the daughter of an obligate carrier would have a 50% risk of herself being a carrier. However, the carrier risk for a woman whose maternal grandmother is an obligate carrier but whose own mother's status is unknown (a 50% chance of being a carrier) can be modified by circumstance.

In **Figure 1**, individual I-2 is an obligate carrier of the X-linked recessive condition because she has two affected boys. III-3 is concerned that her mother, II-3, might be a carrier. Because we do not know her status, II-3 has a 50% chance of being a carrier; if she were a carrier, she would have a 50% chance of transmitting the mutant allele to III-3. That is, the probability that III-3 is a carrier, based on this information alone, would be 50% × 50% = 25%. On drawing the pedigree, however, III-3 is found to have four brothers, none of whom are affected—this additional conditional information alters the risk.

Bayesian analysis to account for conditional information

If II-3 were a carrier, it would be possible, but unusual, that she would have had four unaffected sons. The new conditional information suggests that she is more likely not to be a carrier, and the risk that III-3 would be a carrier should therefore be much reduced.

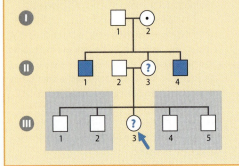

Figure 1 Genetic risk in a pedigree with a childhood-onset X-linked recessive condition. If III-3 were an only child, her risk of being a carrier would be 1 in 4 (her mother, II-3, is the daughter of obligate carrier I-2 and has a 1 in 2 chance of being a carrier; if so, III-3 also has a 1 in 2 risk of inheriting the mutant allele). If she were an only child, III-3 would therefore have a 1 in 4 risk of being a carrier. However, III-3 subsequently mentions that she has four grown-up brothers, none of whom are affected. That additional information suggests that the probability that II-3 is a carrier is much less than 0.5, and that means the chance that III-3 is a carrier is greatly reduced—but by how much?

BOX 11.3 (continued)

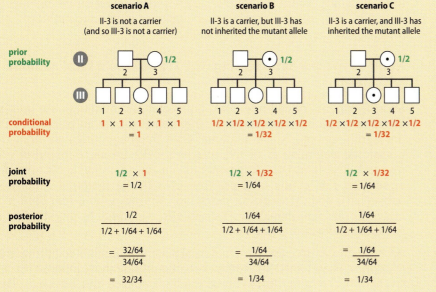

scenario A	scenario B	scenario C
II-3 is not a carrier (and so III-3 is not a carrier)	II-3 is a carrier, but III-3 has not inherited the mutant allele	II-3 is a carrier, and III-3 has inherited the mutant allele

prior probability

conditional probability

scenario A: $1 \times 1 \times 1 \times 1 \times 1 = 1$

scenario B: $1/2 \times 1/2 \times 1/2 \times 1/2 \times 1/2 = 1/32$

scenario C: $1/2 \times 1/2 \times 1/2 \times 1/2 \times 1/2 = 1/32$

joint probability

scenario A: $1/2 \times 1 = 1/2$

scenario B: $1/2 \times 1/32 = 1/64$

scenario C: $1/2 \times 1/32 = 1/64$

posterior probability

scenario A:
$$\frac{1/2}{1/2 + 1/64 + 1/64} = \frac{32/64}{34/64} = 32/34$$

scenario B:
$$\frac{1/64}{1/2 + 1/64 + 1/64} = \frac{1/64}{34/64} = 1/34$$

scenario C:
$$\frac{1/64}{1/2 + 1/64 + 1/64} = \frac{1/64}{34/64} = 1/34$$

Figure 2 According to Bayesian calculations, the risk that III-3 in Figure 1 is a carrier is only about 3%. Here, the prior probability is the standard risk due to Mendelian segregation, and the conditional probability is the *multiplicative product* of the individual probabilities that individuals in generation III have the status that is attributed to them. The probability that an individual male in generation III is unaffected is 1 in 2 if II-3 is a carrier (scenarios B and C), or 1 if II-3 is not a carrier (scenario A). The probability that III-3 is not a carrier is 1 in scenario A (because her mother is not a carrier), or 1 in 2 when her mother is a carrier (scenarios B and C). The joint probability is the product of the prior probability and conditional probability, and the posterior probability is the fraction of the total joint probabilities (for all scenarios) that is attributable to one scenario.

The question is: by how much? To answer that question and to give a new risk estimate based on all the information, Bayesian analysis is used. Four steps are involved, as listed below.

- Identify all the different scenarios that can explain the observations.

- For each scenario, calculate the prior probability and conditional probability.

- Multiply the prior probability by the conditional probability to obtain a joint probability for each scenario.

- Determine what fraction of the total joint probability is represented by each individual scenario

to get a posterior probability for each of the three scenarios.

If we discount fresh mutation, there are three possible scenarios in this case: (A) II-3 is not a carrier, and so III-3 is also not a carrier; (B) II-3 is a carrier, but III-3 is not a carrier (because she did not inherit the mutant allele); (C) II-3 is a carrier and III-3 is also a carrier (because she inherited the mutant allele). As detailed in **Figure 2**, Bayesian analysis suggests that scenario A is by far the most likely—the ratio of the probability for the three scenarios is 32:1:1 for A:B:C. Coming back to the original question, the probability that III-3 is a carrier is given by the posterior probability for scenario C, which is 1/34 (or close to 3%), substantially less than the prior probability of 25%.

known genetic abnormality. Here, the embryo is screened for the presence of any chromosomal aneuploidy. In both cases, the object is to implant normal embryos only, to avoid the birth of an affected child (in diagnostic cases) or to improve the pregnancy success rate (in screening cases).

Because the biopsy provides very limited starting material (just one or a few cells), preimplantation genetic diagnosis testing can be technically challenging and is not widely available. For preimplantation genetic diagnosis, prior identification of mutant alleles in one or both parents allows a test in which one or more relevant DNA regions in the DNA from the biopsy are PCR-amplified and sequenced. If there has been difficulty in identifying a parental mutation, indirect genetic linkage tests can be conducted using a well-established set of polymorphic markers that span the disease gene locus. Occasionally, the test seeks to identify the transmission of a chromosomal abnormality, and involves interphase FISH.

The process of achieving a pregnancy also becomes medicalized (with potential side effects associated with ovarian hyperstimulation). Additionally, the likelihood of a successful pregnancy outcome is quite low: it is only about 1 in 5 at the start of an IVF treatment cycle (sometimes no embryos are suitable for transfer, depending on the number of eggs fertilized, and the number and quality of unaffected embryos), but increases to 1 in 3 after embryo transfer.

Noninvasive prenatal testing (NIPT) and whole fetal genome screening

Short fragments of cell-free DNA, both fetal and maternal, are present in maternal blood. The fetal DNA fragments arise from placental cells undergoing apoptosis; the maternal DNA fragments originate from the occasional degradation of the mother's cells through apoptosis and necrosis. The fetal DNA fragments are in the minority, accounting for around 10–15% of the total cell-free DNA in the maternal circulation between 10 and 20 weeks of gestation. Analysis of cell-free DNA in a maternal plasma sample can therefore be sufficient to investigate the genetic composition of the fetus.

Because the cell-free DNA in a maternal plasma is dominated by maternal DNA, the easiest fetal DNA sequences to identify are those inherited exclusively from the father (they can be readily amplified and detected). That includes Y chromosome DNA sequences, and noninvasive fetal sexing is now routinely available with a sensitivity of about 90% and a specificity of 98%. Testing has also been possible for other exclusively paternal sequences in certain situations (see **Table 11.4** for some applications).

Technological breakthroughs

It is technically easy to test cell-free DNA in maternal plasma for the presence of exclusively paternal DNA sequences. More comprehensive testing has not been easy (fetal markers are not readily distinguished from maternal homologs, and there is a large background of circulating maternal DNA in maternal plasma). Very recent technological advances in noninvasive prenatal testing and screening have dramatically opened up this field, offering an exciting new window on fetal diagnosis and fetal screening.

A major breakthrough came from overcoming technical obstacles to what is a very simple principle: counting the parental haplotypes. For any very short genome region, three haplotypes exist in the freely circulating DNA in maternal plasma: the maternal haplotype that is transmitted to

GENETIC DISORDER	NONINVASIVE TESTING FOR PATERNAL DETERMINANTS
Serious X-linked recessive disorders	fetal sexing test: identification of a female fetus avoids the need for subsequent invasive prenatal diagnosis with its associated miscarriage risk, but not for a male fetus
Congenital adrenal hyperplasia	fetal sexing test: abnormal androgen production in affected female fetuses results in virilization of the external genitalia. After early identification of a female fetus, the fetal adrenals can be suppressed by the oral administration of dexamethasone to the mother. Treatment has to be started before 8 weeks of gestation to be effective. Invasive prenatal diagnosis is subsequently undertaken so that treatment can be stopped in all but the pregnancies of affected females
Hemolytic disease of the newborn	testing for paternal rhesus D blood group: rhesus D-negative women may be at increased risk of hemolytic disease of the newborn (because of a previous affected pregnancy or raised antibody titer). If a paternal rhesus D is identified, the pregnancy needs to be monitored closely because of the risk of fetal anemia

Table 11.4 Common applications of noninvasive testing for paternal determinants in genetic testing.

CONTRIBUTION MADE BY DNA FROM:	EXPECTED COUNT OF ALLELES AT MARKER LOCI ON:	
	Transmitted maternal haplotype, M_t	Untransmitted maternal haplotype, M_u
Mother ($M_t + M_u$)	$N(1 - \varepsilon)$	$N(1 - \varepsilon)$
Fetus ($M_t + P_t$)	$N\varepsilon$	0
Total	N	$N(1 - \varepsilon)$

Table 11.5 Distinguishing between the two maternal haplotypes by counting alleles at maternally heterozygous marker loci in DNA from maternal plasma. ε is the fraction of the DNA in maternal plasma that originates from fetal cells. N is the number of haplotypes with the allele of interest that have been analyzed.

the fetus (M_t), the maternal haplotype that is not transmitted to the fetus (M_u), and the paternally transmitted haplotype (P_t).

Because the DNA in maternal plasma will be a mix of DNA originating from degraded maternal cells plus a relatively small amount of fetal DNA, typing for individual DNA markers will show a small excess of alleles from M_t haplotypes over alleles from M_u haplotypes (**Table 11.5**). To detect such a small difference reliably, a very specific test would be needed; however, with massively parallel DNA sequencing it is comparatively easy to count millions (or even billions) of DNA molecules—permitting very specific testing.

From fetal aneuploidy screening to genome screening

NIPT can be used with targeted massively parallel DNA sequencing: certain genome regions of interest are captured from the genomic DNA (see Figure 11.11 for the principle) and sequenced.

One application of NIPT is fetal aneuploidy screening. Previously, the problem here had been distinguishing fetal autosomes from the maternal equivalents. When fetal DNA accounts for 10% of the DNA in maternal plasma, the amount of chromosome 21 DNA increases by just 5% if the fetus has trisomy 21. Because of massively parallel DNA sequencing, that small difference can be readily detected (trisomy 21 can be detected with a sensitivity of 99% and a specificity of 99% using this method). Current evidence suggests that NIPT is more cost-effective as a screening tool (to define a high-risk group that can then be offered confirmatory amniocentesis) than as a diagnostic procedure.

It has also been possible to obtain complete fetal genome sequences noninvasively (see under Further Reading). That opens up huge possibilities for the future, which we consider later.

Pre-symptomatic and predictive testing for single-gene disorders in asymptomatic individuals

An individual who is at risk of a childhood-onset or adult-onset genetic disorder may be tested before they have signs or symptoms of having the disorder, such as an asymptomatic male infant with a positive family history for Duchenne muscular dystrophy, or a young adult at risk of Huntington disease. The testing is often described as pre-symptomatic testing when a patient possessing the mutant allele(s) will develop symptoms, for example in Huntington disease, or predictive testing if they are at high risk of developing symptoms, for example in carriers of mutant *BRCA1* or *BRCA2* alleles.

Early diagnosis might allow medical benefits such as the possibility of preventive or therapeutic intervention, targeted screening tests (surveillance) to check for early disease (that might be more amenable to treatment), refinement of the prognosis, and clarification of the diagnosis.

For some diseases the clinical benefit of predictive testing is clear. Take, for example, familial hypercholesterolemia (OMIM 143890), an autosomal dominant disorder that is commonly due to pathogenic mutations in the *LDLR* (low-density lipoprotein receptor) gene. Affected individuals normally develop premature cardiovascular disease in the third decade, but early detection of a pathogenic *LDLR* mutation offers the possibility of prevention by lowering LDL-cholesterol through dietary changes and medication. That has led to recommendations of cascade testing either by measuring LDL-cholesterol or by testing for the familial *LDLR* mutation from the age of 10 years.

Imagine a more difficult situation, an autosomal dominant disorder in which preventive treatment has significant side effects, or in which screening tests for potential complications are burdensome or have associated risks. Certain forms of colorectal cancer are dominantly inherited, but early detection of a germ-line mutation that predisposes to these diseases can be followed up by annual colonoscopy surveillance. By identifying and surgically removing polyps before they get larger, the risk of developing late-stage cancer is much reduced. However, there is a small risk of perforation of the bowel during the required colonoscopy. You might wish to know whether you definitely carry the mutation before opting for the treatment or entering the screening program.

Note that the surveillance carried out for some familial cancer syndromes (**Box 11.4**) does not prevent cancer—its aim is to identify early cancers while they are amenable to therapy. To reduce their risk of developing cancer, women from families with the *BRCA1* or *BRCA2* mutation may opt for mastectomy and/or surgical removal of the ovaries together with the associated fallopian tubes; again, they would ask for predictive testing before making this decision.

Weighing the relative merits and disadvantages of having a predictive test is not easy. According to current guidelines, therefore, predictive tests for adult-onset disorders should not be undertaken in children unless a medical intervention is possible that is applicable to children and shows clear medical benefit (as in familial hypercholesterolemia).

Pre-symptomatic testing without medical intervention

There are no current interventions to delay the onset of most of the late-onset neurological disorders. Nevertheless, Huntington's disease, a devastating neurological disorder that often does not manifest itself until later stages in life, was one of the first conditions for which predictive testing was offered. After initial concerns that individuals testing positive for this disorder might commit suicide, predictive testing was introduced with caution; experience has shown that predictive testing precipitates a catastrophic event (suicide, suicide attempt, or psychiatric hospitalization) in around 1% of cases.

The uptake of testing in people at 50% risk of Huntington's disease is about 10–20%. Young adults who undergo testing generally do so to assist in making career and family choices. Another group opting for testing is those who have reached the age by which signs and symptoms would usually have presented; they wish to be tested so that children and grandchildren can be reassured that they are not at risk (assuming, of course, that the result is negative).

An overview of the different types and levels of genetic screening

The genetic testing described above is reactive: it is carried out in response to individuals seeking medical help or advice about the risk of developing or transmitting a genetic disorder. In genetic screening, the genetic tests are carried out in a more proactive way. Here, the focus is not probands and their relatives but communities and populations.

Genetic screening can be carried out at different levels, and with different aims; it is often directed at biochemical and physiological markers as well as DNA variants and aberrant chromosomes. In Section 8.3 we considered a type of longitudinal population screening, exemplified by the UK Biobank project, in which comprehensive testing is carried out on people at regular intervals over decades. However, that is a primarily research-led type of screening. In contrast, the three types of genetic screening listed below are primarily directed at providing clinical benefit to the subjects tested.

Pregnancy screening. The object is to identify whether or not the pregnancy is at a very high risk of leading to the birth of a child with a specific, serious genetic disorder. The motivation for the test is usually to prevent the birth of an affected child. (Sometimes, however, the test might be requested to allow psychological preparation and medical management planning for such a birth, while simultaneously offering psychological benefit, should the test indicate that the fetus is unaffected.) Although it can entail screening for a serious single-gene disorder in communities in which that disorder is prevalent, it often involves screening for aneuploidies. As we describe below, technological advances are beginning to allow comprehensive genetic profiles to be obtained for a fetus, and that might lead to new ways of treating disease *in utero*.

Newborn screening. This is carried out in many countries, but to variable extents. A major motivation has been to target early treatment in serious disorders for which early intervention can make a substantial difference and may lead to disease prevention. Genetic screening of newborns began with certain metabolic disorders, and this class of disorder is still a major focus.

Carrier screening. The object is to identify carriers of a mutant allele for a severe autosomal recessive disorder that has a high prevalence in the community or population. The ultimate aim is disease prevention: a couple who have both been identified as carriers by the screen can subsequently elect to have prenatal diagnosis and termination of pregnancy for affected fetuses.

Maternal screening for fetal abnormalities

Specific maternal screening programs have been undertaken in the first trimester to identify fetuses at high risk of common and serious single-gene disorders that are prevalent in their communities (such as sickle-cell disease and thalassemia). However, the focus for most prenatal screening is maternal screening for fetal aneuploidy, notably the commonest chromosomal abnormality, trisomy 21 (causing Down syndrome).

As described above, massively parallel DNA sequencing is beginning to be used to screen DNA in maternal plasma (which includes small amounts of DNA from fetal cells) for evidence of aneuploidies such as trisomy 21. That might be expected to displace the traditional 'combined' screening system, which is based on three parameters. One is the skin thickness at the back of the neck (nuchal translucency) measured by ultrasound scanning between 11 and 14 weeks of gestation; it is determined by the amount of fluid that collects here (which is often greater in babies with Down syndrome). A second factor is the mother's age (the risk increases 16-fold as the maternal age increases from 35 to 45 years). The third factor is based on altered levels of certain maternal serum proteins, such as an increased level of free β-HCG (human chorionic gonadotropin) and a decrease in PAPP-A (pregnancy-associated plasma protein A).

On the basis of combined screening, approximately 2% of women will have a greater than 1 in 150 risk (compared with an overall population risk

BOX 11.4 Lynch syndrome and familial (*BRCA1/BRCA2*) breast cancer: cancer risks and cancer screening.

Lynch syndrome (hereditary nonpolyposis cancer)

The diagnosis is determined on the basis of the pattern of cancers in a family and the age at diagnosis (at least one diagnosis under the age of 50 years), or on the finding of microsatellite instability in tumor tissue. Although colorectal cancer is the commonest cancer in the condition, there are several associated cancers (**Table 1**).

A typical screening protocol for affected and high-risk individuals would be colonoscopy every 18–24 months, starting at the age of 25 years, followed by additional endoscopy examination of the esophagus, stomach, and duodenum from 50 years of age. The efficacy of screening for other associated tumors is not proven. Women who have the condition, or who are at high risk, may opt for total hysterectomy and surgical removal of ovaries plus fallopian tubes after completion of their family.

Familial breast cancer due to *BRCA1* or *BRCA2* mutations

In the general UK population, the lifetime risks of breast cancer and ovarian cancer are 12% and 2%,

CANCER TYPE	LIFETIME (TO AGE 80 YEARS) RISK (%)	
	BRCA1	*BRCA2*
Unaffected carriers		
Breast cancer	60–90	30–85
Ovarian cancer[a]	30–60	10–30
Male breast cancer	0.1–1	5
Prostate cancer	8[b]	25
Other cancers	<5	<5
Affected women carriers (with unilateral breast cancer)		
Cancer in other breast	50% (overall 5 year risk = 10%	50% (overall 5 year risk = 5–10%)

Table 2 Cancer risks for carriers of harmful *BRCA1* or *BRCA2* mutations. [a]Majority of lifetime risk after 40 years of age. [b]Similar to population risk.

CANCER	GENERAL POPULATION RISK (%)	RISK IN LYNCH SYNDROME (%)
Colorectal	5.5	80
Endometrial	2.7	20–60
Gastric	<1	11–19
Ovarian	1.6	9–12
Hepatobiliary tract	<1	2–7
Urinary tract	<1	4–5
Small bowel	<1	1–4
Brain and central nervous system	<1	1–3

Table 1 Cancer risks in Lynch syndrome compared with the normal population.

respectively. A person aged 20–25 years with a pathogenic *BRCA1* or *BRCA2* mutation has a 70% risk of going on to develop cancer, notably breast or ovarian cancer (**Table 2**).

When a pedigree indicates a high likelihood of familial breast (breast or ovarian) cancer, a typical screening program would commence with annual mammography at 40 years of age in women at high risk. Mammography is less sensitive in women aged less than 40 years. MRI breast screening may be offered from 30 years of age. Bilateral mastectomy reduces the risk of developing breast cancer by 95%, but cancer can still occur in remaining breast tissue on the chest wall.

There is no evidence that ovarian screening is helpful. Women may consider the surgical removal of ovaries and fallopian tubes—it reduces the risk of ovarian cancer by 95% (with a small residual risk of peritoneal carcinoma) and also reduces the risk of breast cancer.

of about 1 in 670); they will be offered chorion biopsy for definitive aneuploidy testing. There will be an adverse outcome in 20% of these women (which includes trisomy 13 and trisomy 18 in addition to trisomy 21); that still means there is no chromosome abnormality in 80% of women who take up chorion biopsy—the development of a reliable test based on cell-free fetal DNA in the maternal serum will be a major advance. The combined screening detects 90% of all affected pregnancies.

First-trimester ultrasound is important in estimating the date of delivery and for nuchal measurement, and is essential for accurate estimates of gestational age needed for risk calculations based on the levels of the maternal serum proteins described above. Additional ultrasound is routinely offered in pregnancy at around 20 weeks of gestation to look for structural anomalies (a significant proportion of which are due to chromosomal or Mendelian disorders).

Newborn screening to allow early medical intervention

Newborn screening was pioneered in the late 1960s. Screening for phenylketonuria (PMID 20301677) used dried blood spots collected on a filter-paper card (the Guthrie card) at 5 days of age. Assays for congenital hypothyroidism (which has a number of causes, not all of which are genetic), were added shortly afterward. For both conditions the rationale was prevention of the mental handicap that would inevitably ensue in the absence of medical intervention (which involves dietary changes for phenylketonuria and hormone replacement for congenital hypothyroidism—see Table 11.6).

Inborn errors of metabolism have been a major focus of newborn screening for two reasons. First, they have been studied for decades, and there is a highly developed understanding of the molecular basis of disease, allowing useful early medical interventions in some cases.

The second advantage is that inborn errors of metabolism are typically amenable to easy-to-use screening systems that work at the gene-product or metabolite level and are applicable to easy-to-access patient samples such as blood or urine. A disease allele may have any one of a potentially very large number of different mutations; if the gene has many exons, the screening can be hard work. However, all that heterogeneity at the DNA level often has a rather uniform effect at the gene-product level: a single assay can often detect abnormalities in the product or characteristic changes in certain metabolites.

As a result, it is usual to use assays at the gene-product level, or assays for disease-associated metabolites (tandem mass spectrometry—which allows the parallel testing of multiple metabolites in blood and urine samples—can efficiently screen for a range of metabolic disorders at low cost).

GENETIC DISORDER	PREVALENCE	TYPE OF SCREENING	TREATMENT OF AFFECTED INDIVIDUALS
Congenital hypothyroidism	1 in 5000	assay of free thyroxine or thyroid-stimulating hormone in serum	hormone replacement
Cystic fibrosis	1 in 2500 in European populations	screen for immunoreactive trypsinogen, then confirm by scan for *CFTR* mutations	antibiotics, chest physiotherapy, pancreatic enzyme replacement for those with pancreatic insufficiency
Galactosemia	1 in 75,000	assay of levels of erythrocyte galactose-1-phosphate and galactose-1-phosphate uridyltransferases	change of diet to reduce intake of galactose
Phenylketonuria	~1 in 12,000	plasma amino acid analysis to show increased phenylalanine: tyrosine ratio	change of diet to reduce intake of phenylalanine (see Box 7.8)
Sickle-cell disease	~1 in 500 with African ancestry	hemoglobin separation by electrophoresis, IEF, or HPLC. DNA studies may be used to confirm genotype	hydroxyurea (increases HbF in red blood cells, reducing transfusion requirement and decreasing frequency and severity of vaso-occlusive events). Prophylactic penicillin

Table 11.6 Newborn screening programs for selected autosomal recessive disorders and congenital hypothyroidism. Data from guidelines proposed by the American College of Medical Genetics and Genomics, which make information on screening programs for individual disorders available through PubMed (PMID 21938795) and the NCBI bookshelf (http://www.ncbi.nlm.nih.gov/books/NBK55827/). *CFTR*, cystic fibrosis transmembrane conductance regulator gene; HbF, fetal hemoglobin; HPLC, high-performance liquid chromatography; IEF, isoelectric focusing.

Benefits versus disadvantages

More recently, other disorders have been added to screening lists, and the huge advances in massively parallel DNA sequencing have led to proposals to greatly increase the number of disorders that are screened for. As well as the benefits of screening, however, there are disadvantages, and getting the balance right is not easy.

In addition to the large costs of implementing national screening programs, any screening program will include false positives. Anxiety can be generated in families who receive a positive screen result but whose child is unaffected on second testing (and the more tests that are taken, the greater is the chance of receiving a false positive result). Accordingly, some countries have taken a quite conservative approach. In the UK, for example, national newborn screening is restricted to five conditions: phenylketonuria, congenital hypothyroidism, medium-chain acyl-CoA dehydrogenase deficiency (MCAD), sickle-cell disease, and cystic fibrosis.

By contrast, the American College of Medical Genetics and Genomics (ACMG) has recommended screening for 54 conditions (including hemoglobin abnormalities, various inborn errors of amino acid, fatty acid, or organic acid metabolism, biotinidase deficiency, congenital adrenal hyperplasia, galactosemia, and cystic fibrosis).

Early treatment might not be of clinical benefit in all of the conditions screened, but there can be other benefits. One benefit of the unified approach and implementation of the ACGM guidelines is a greater awareness of the disorders and a greater sharing of information, increasing knowledge of the natural history of these very rare disorders.

Another side benefit of earlier diagnosis is that parents will be informed about the condition and recurrence risks before they have further children. Some countries have piloted newborn screening for Duchenne muscular dystrophy, not because of therapeutic benefit but because if a child does not present until 4 years of age, couples may already have a second affected child at the time of diagnosis.

Carrier screening for disease prevention in serious autosomal recessive disorders

The identification of people who are asymptomatic carriers for a serious recessive disorder allows disease prevention. Couples who have both been identified as heterozygous carriers can elect to have prenatal diagnosis, with a view to terminating pregnancies in which a homozygous mutant genotype is identified. Carrier screening can be very effective in reducing the incidence of a harmful disorder in communities, and has been established in a variety of countries. The screening is variable, however, and there have been ethical concerns in some cases.

Screening for β-thalassemia

Carrier screening for β-thalassemia, an autosomal recessive disorder with diminished production of β-globin, is an illustrative example. Approximately 70,000 babies are born each year with this disorder, the incidence being highest in Mediterranean countries, India, Africa, Central America, the Middle East, and Southeast Asia. To treat the resulting anemia, affected individuals require repeated blood transfusions; however, that causes iron overload, which in turn leads to liver damage and cardiomyopathy. Iron chelation therapy is then used to increase iron excretion, prolonging life expectancy well into the fourth decade of life and usually beyond that.

Carrier screening can be undertaken using mean corpuscular volume and mean corpuscular hemoglobin levels in the standard full blood examination; various methods are used to confirm the diagnosis. In 1973, carrier testing was introduced across Greece and Cyprus after educational programs at schools, in the armed forces, maternity clinics, through the mass media, and, in Cyprus, through the Orthodox Church. Sardinia introduced screening a few years later.

Subsequently, many countries have developed screening programs. In Iran, many provinces of Turkey, the Gaza Strip, and Saudi Arabia the testing is mandatory for couples registering for marriage. In the Gaza Strip, couples have to sign a declaration that they are aware both are carriers if they continue with the marriage. These countries have opted for screening before marriage; in other countries, screening occurs in antenatal clinics—if the woman is found to be a carrier, testing is offered to her partner. Although this is intended to follow informed consent, screening evaluations have indicated that patient awareness and understanding of the program are very variable.

There has been a significant reduction in affected births in countries with screening programs, partly due to altered marriage plans but mainly due to the uptake of prenatal diagnosis and termination of pregnancy. For example, the incidence of β-thalassemia in Sardinia when screening was introduced in 1975 was 1 in 250 births; by 1995 it was 1 in 4000 births. In Cyprus the number of affected births in 1974 was 51, in 1979 it was 8, and there were no affected births between 2002 and 2007. Similar marked reductions have been reported after the introduction of antenatal screening programs in Taiwan and Guangdong China.

Community screening for Tay–Sachs disease

Carrier screening programs have also sometimes been directed to particular groups with a high incidence of a serious disorder. For example, the recessive disorder Tay–Sachs disease (PMID 20301397) is a progressive neurodegenerative disorder that is normally rare (with a carrier frequency of about 1 in 300 in Europe and America), but is especially common in Ashkenazi Jews (about 1 in 27 is a carrier). This inborn error of metabolism presents with progressive weakness and loss of motor skills at between 3 and 6 months, followed by seizures, blindness, spasticity, and death usually before 5 years of age. It is caused by failure to produce the enzyme hexosaminidase A, as a result of genetic mutation in the *HEXA* gene. As a consequence, a fatty substance, GM2 ganglioside, accumulates in brain cells and nerves, damaging and eventually destroying them.

Carrier testing based on assaying serum hexosaminidase A began in 1970, when it was recognized that carriers may be distinguished from non-carriers by this assay. Testing is available through health services in many countries; the Dor Yeshorim organization also offers genetic screening to Ashkenazi Jews worldwide through orthodox Jewish High Schools and in community testing sessions. When testing is undertaken in orthodox schools, the results may not be given directly but instead be available for couples considering marriage. This screening program has led to a significant reduction in the number of children born with Tay–Sachs disease in this community.

New genomic technologies are beginning to transform cancer diagnostics

Recent advances are transforming clinical applications of molecular genetics in identifying susceptibility to cancer and in prognosis. Genetic testing can identify different types of cancer markers and also aid in making decisions about the best treatments and drugs to use.

ROLE	DNA/GENE EXPRESSION BIOMARKER	CANCER TYPE (COMMENT)
Diagnostic	BCR-ABL1	chronic myeloid leukemia (see Figure 10.7)
	JAK2	myeloproliferative disease (mutations confirm diagnosis of clonal MPD)
	EWS-FLI1	Ewing sarcoma
Predictive	HER2	breast cancer (amplification predicts response to anti-HER2 antibodies)
	BRAF	melanoma (mutations predict response to specific BRAF inhibitors)
	KIT, PDGFRA	gastrointestinal stromal tumors (mutations predict response to c-KIT/PDGFRA inhibitors)
Prognostic	TP53	chronic lymphocytic leukemia (mutations are indicative of poor outcome)
	BRAF	metastatic colorectal cancer (mutations are indicative of poor outcome)
	MammaPrint (70-gene expression signature)	breast cancer (risk stratification)
	OncotypeDx (21-gene expression signature)	breast cancer (risk stratification)
Disease monitoring	BCR-ABL1	chronic myeloid leukemia (detection of minimal residual disease)
	PML-RARA	acute promyelocytic leukemia (detection of minimal residual disease)

Table 11.7 Examples of different roles for DNA biomarkers and gene expression biomarkers in cancer testing.

Different types of new cancer biomarkers are becoming available. In addition to DNA biomarkers, some gene expression signatures on microarrays have served as cancer biomarkers (see **Table 11.7** for examples). Prominent among the latter is Agendia's MammaPrint, which received approval from the United States Food and Drug Administration (FDA) in 2011. By following the expression patterns of 70 key genes, MammaPrint can classify early stage tumors into those with high or low risk (with 98.5% accuracy and with a technical reproducibility of 98.5%); the classification is helpful in directing different types of treatment, according to the perceived risk.

Multiplex testing using targeted genome sequencing

As described above in Figure 11.11, target enrichment sequencing allows DNA sequences from any genome region of interest to be selectively captured and sequenced. Multiplex testing for panels of cancer susceptibility genes recently became available for research purposes from companies such as Ambry Genetics and Illumina (whose TruSight Cancer test captures genomic sequences for more than 90 cancer-susceptibility genes).

Tests such as this can be expected to be incorporated into diagnostic settings in the near future. In the UK, the goal of the Mainstreaming Cancer Genetics Programme (http://mcgprogramme.com/), led by the Institute of Cancer Research in London, is to make genetic testing part of the routine care of cancer patients within the National Health Service. A cancer genetic testing service is being piloted, based initially on the Illumina Trusight Cancer panel (but likely to be replaced by genomewide sequencing in the future as sequencing costs fall). At present, cancer predisposition gene testing in the UK is carried out by geneticists; however, in the new model, cancer specialists will organize all aspects of the testing.

From 2013 to 2016 the Mainstreaming Cancer Genetics Programme plans to roll out genetic testing in 35 different cancers. Testing patients to identify key cancer-susceptibility variants will permit more personalized treatment (which may include more comprehensive surgery, different medicines, or extra monitoring). It will also improve the information available to relatives of cancer patients, a high-risk group, about their own cancer risks. Follow-up testing can provide the reassuring news that a relative is not at increased risk of cancer and does not need medical interventions. If, instead, a relative is found to also have an increased cancer risk, additional monitoring and screening can be carried out, and sometimes measures can be taken to prevent cancer from developing. For example, in ovarian cancer, an asymptomatic woman identified to be at increased risk might seek to have her ovaries removed by keyhole surgery after completing her family.

Noninvasive cancer testing

Another promising recent development is noninvasive cancer testing. Instead of taking a tumor biopsy (which can be difficult, according to the type of cancer), different approaches allow the analysis of freely circulating tumor DNA in plasma. (They were stimulated by the application of high-throughput DNA sequencing in noninvasive prenatal testing of the fetal genome, as described above.)

The freely circulating DNA originates from cells undergoing apoptosis or necrosis, which includes originally healthy cells as well as inflamed cells and diseased cells, such as cancer cells. This means that tumor-specific variants need to be detected against a background of circulating DNA from non-tumor cells in the same individual.

Massively parallel DNA sequencing is one of the newer methods that has been applied to the analysis of plasma DNA. Tumor-specific chromosome alterations can be detected with comparative ease (**Figure 11.19**). In addition, it has been possible to detect mutant alleles through quantification studies (as in studying acquired resistance to cancer therapy—see the paper by Murtaza et al. [2013] under Further Reading). Although promising, there are significant hurdles to overcome before this type of testing is suitable for diagnostic work.

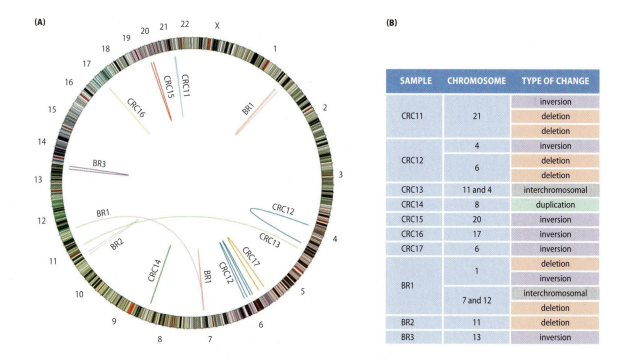

(B)

SAMPLE	CHROMOSOME	TYPE OF CHANGE
CRC11	21	inversion
		deletion
		deletion
CRC12	4	inversion
	6	deletion
		deletion
CRC13	11 and 4	interchromosomal
CRC14	8	duplication
CRC15	20	inversion
CRC16	17	inversion
CRC17	6	inversion
BR1	1	deletion
		inversion
	7 and 12	interchromosomal
		deletion
BR2	11	deletion
BR3	13	inversion

Figure 11.19 Detection of tumor-specific rearrangements by massively parallel DNA sequencing in plasma samples. (A) A Circos plot indicating rearrangements identified in tumor-cell DNA present within plasma samples from 10 cancer patients, 7 with colorectal cancer (CRC11 to CRC17), and 3 with breast cancer (BR1 to BR3). No rearrangements were identified in DNA obtained from plasma samples from 10 unaffected controls. (B) Observed DNA rearrangements. A table with details of the breakpoint coordinates is found in the original paper. (Adapted from Leary RJ et al. [2012] *Sci Transl Med* 4:162ra154; PMID 23197571. With permission from the AAAS.)

Genetic testing for complex disease and direct-to-consumer genetic testing

Up to this point we have considered genetic testing offered through health care services. That was the only option until recently, when commercial genetic testing began to be offered directly to consumers. Two stimuli in particular led to the growth of these services. The first was the recent rapid decrease in the cost of genetic testing. Then there was the increasing identification of genetic variants that conferred susceptibility to complex diseases that are common in populations.

In 2007 personal genomics testing began to be offered by various companies based in the United States. Consumers would reply to Internet adverts from the personal genomics company and the company would send out a kit to the consumer for collecting saliva. Cells from the returned saliva sample would then be processed to isolate and amplify genomic DNA fragments for testing, with the results fed back to the consumer within a few weeks. The services offered included: tests to establish carrier status for various disorders; testing for risk of developing various complex diseases; drugs response tests; and tests for comparative resistance to some infectious diseases.

Because direct-to-consumer genetic testing is a recent development, regulatory frameworks have been slow to develop in many countries. Advisory reports to governments and professional bodies stress the importance of accreditation of the testing laboratories by a relevant authority, the need for consumers to be informed of the benefits, risks, and limits of the tests, and the need to safeguard the confidentiality of genetic information (which could become a major issue if a company were to become insolvent). A principal concern has been the lack of direct interaction with professionally qualified clinicians and genetic counselors who could offer expert pre-test and post-test advice (an important consideration because interpretation of the multiplex tests may not be straightforward). Because the common genetic variants that are tested are generally of quite low penetrance, individual genotypes usually confer at most a very modest increase in disease risk, and the clinical utility of much of the testing is questionable.

In 2010 the FDA ruled that the personal genome services offered by the US companies were medical devices that would require review and approval before they could be offered on the market. Within the next four years almost all of the 17 companies contacted by the FDA had stopped selling personal genome tests; some have exited the business completely and some others have re-focused on different types of testing (paternity and ancestry testing), or matured into biotech companies. Those who still offer personal genome testing are required by the FDA to inform clients that the tests they offer need to be ordered by physicians.

11.5 Ethical Considerations and Societal Impact of Genetic Testing and Genetic Approaches to Treating Disease

We have previously described the advantages of genetic testing and of the application of genetic technologies in treating genetic disorders. Here we are mostly concerned with some of the relevant ethical issues, but we also touch upon the impact on society.

For monogenic disorders, close relatives and potential offspring of persons who carry relevant mutant alleles are at high risk of developing the disorder. Positive identification of harmful genetic variants in one person therefore raises the stakes for unaffected relatives who may subsequently be found to carry similarly harmful genetic variants. Prenatal diagnoses are quite frequently carried out for serious monogenic disorders and in cases where there is a high risk of transmitting a chromosomal abnormality. Where testing has additional implications for current and future family members—rather than simply the person being tested— exceptional ethical issues arise. Three important issues are consent for genetic testing, sharing of genetic information, and the confidentiality of the test results. Genetic testing for complex disorders is much less developed than for single-gene disorders, and there are issues here regarding the usefulness of the tests, the complexity in interpreting the results, and the provision of genetic advice to the tested individuals.

As genetic technologies develop, increasing inroads have been made into applying them to the treatment of genetic disorders. The (physical) effect of somatic gene therapy is limited to the individual who is treated, but the prospect of germ-line gene therapy carries with it direct implications for future generations. We also consider the controversy concerning a prospective treatment for, or (more accurately) prevention of, mitochondrial disorders that envisages artificial intervention to produce embryos in which the genetic material comes from three parents.

Consent issues in genetic testing

In economically advanced societies, health professionals are normally required to obtain the consent of individuals who are giving a sample or undergoing a medical intervention. The consent process is meant to ensure that a person understands the nature and purpose of giving a sample, or of undergoing the medical intervention—see **Table 11.8** for the types of genetic issue discussed.

The issue of who can give consent for genetic testing mostly revolves around age and mental competence. The person who is being tested may give consent if he or she is an adult and capable of understanding the implications of the test. For adults with mental incapacity (who are judged incapable of giving consent), a genetic test can be undertaken if it is believed to be in the best interests of the adult concerned, and a legal guardian typically provides the informed consent. For children below age 16 years (or some accepted age of maturity), parents (or those with parental responsibility) may give consent for genetic testing.

Informed understanding of the nature of genetic testing is an important part of the consenting process, but adequate consent may be difficult to achieve: the often complex genetic techniques used in testing are not easy to convey, and the predictive value of the testing can also be complicated in many cases.

Except in issues of sharing genetic information (which we cover in the next section), genetic testing of adults is normally straightforward. However, there can be ethical concerns when the testing in adults with mental incapacity might be mostly of value to the relatives. And for some types of genetic testing, the process of consent is currently denied in certain societies. For example, carrier testing for β-thalassemia is currently mandatory for couples registering for marriage in Iran, many provinces of Turkey, the Gaza Strip, and Saudi Arabia.

Consent issues in testing children

Genetic testing of children raises additional consent issues. Should children at risk be tested for adult-onset single-gene disorders? Or screened for carrier status for serious recessive disorders? The answer to both

GENETIC ISSUE
The use and sharing of information (pedigree, diagnosis, affected/carrier status, test results) with other family members for their benefit
The nature of the testing to be undertaken and its implications
The possible prolonged nature of the testing process
The possibility that testing may reveal unexpected results, depending on the particular analyses used
The storing of samples
That samples may be used for quality assurance, education, and training
That information may be shared with health professionals, including the primary care team

Table 11.8 Genetic issues to be discussed during the consent process: recommendations by the UK's Joint Committee on Medical Genetics.

questions is usually no; unless there is clear medical benefit, testing should be delayed until the child has the capacity to make the choice.

Two ethical principles are particularly relevant. First, there is the principle of beneficence that considers the benefit to the child: it urges clinicians to test for clinically relevant and *actionable* genetic variants. Second, there is the 'right to an open future' principle—disclosure of any genetic information from the above type of tests could infringe a child's future autonomy. This applies especially to conditions in which adults might sometimes choose not to be tested—testing during childhood would deny the child the right not to give consent that he or she could exercise as an adult.

Current professional guidelines in many countries therefore stipulate that children should not normally be tested in this way unless there is clear medical benefit in early testing. Testing for familial hypercholesterolemia is one such example: early detection of a pathogenic *LDLR* mutation offers the possibility of prevention by lowering LDL-cholesterol through dietary changes and medication, and testing is recommended from the age of 10 years.

Problems with sharing of genetic information and the limits on confidentiality

For monogenic disorders and other conditions in which there is a high risk to unaffected family members, genetic findings in one person may be very pertinent to the relatives at risk, but when can disclosing the findings to other family members be justified? What are the limits on confidentiality?

Genetic health professionals may be conflicted in how they should deal with families. There is the moral obligation to contact family members who have not sought advice, but individuals also have the right not to know. Medical professionals may also be in the rather uncomfortable position of meeting, and having access to information on, different family members who do not know about each other. They may also discover through genetic testing that the biological relationships between some family members are different from the assumed relationships (as a result of misattributed paternity, unsuspected adoption, or sperm donation for example). Although it is not standard practice to do so, verification of relationships between family members may be obtained by *identity testing*: genomic DNA samples from the individuals concerned are submitted for multiplex genotyping of various highly polymorphic simple tandem repeat loci, such as Promega's PowerPlex STR systems. (Note that informed consent would always be obtained before any analysis relating to paternity were carried out.)

Medical information can be disclosed when consent has been obtained to allow this. Because most family members wish their information to be available to help relatives, it is good practice to obtain and document consent for disclosure of medical information. Sometimes, however, consent may have been obtained some time previously; because the information that is discussed during the consenting process is continually changing as the technology advances, the scope of the former consent may no longer be clear.

Verifying past consent, or updating consent, may not be possible because contact may have been lost, or it may not be clinically appropriate because the family member seeking information may be concerned about his or her confidentiality being compromised. For example, a pregnant woman who wishes to undergo prenatal diagnosis may not want anyone to know about the pregnancy until the test results are available. The necessity to seek consent from another family member for release of the information would lead to a breach of confidentiality for the pregnant woman.

Sometimes a person refuses to give his or her consent to release information. However, the rule of confidentiality cannot be absolute: a breach is justified if the avoidance of serious harm by disclosure of information substantially outweighs the patient's claim to confidentiality. A breach of confidentiality would be justified, for example, if a person declined to inform relatives of a substantial genetic risk of which they might be unaware, or if it allowed information to be released that would enable specific, clinically important, genetic testing to be undertaken.

Ethical and societal issues in prenatal diagnosis and testing

Prenatal diagnosis for serious genetic disorders has long been available in many developed societies. As noninvasive prenatal screening technology develops and is standardized, the miscarriage risk of invasive procedures (chorion biopsy or amniocentesis) is soon likely to become irrelevant. After a positive diagnosis for a harmful genetic variant, terminating a pregnancy is accepted in many societies—but support is far from universal: it is important to acknowledge the significant numbers of people who are morally opposed to all abortion.

For people who feel that terminating a pregnancy can be justified for serious genetic disorders, another issue remains: where do we draw the line that divides serious disorders from non-serious disorders? Some couples might wish to contemplate termination for what many other people might consider mild disorders, such as congenital deafness. There is the additional question of termination of pregnancy to prevent a gene from being passed on rather than a condition, as in cases where a prospective father is an asymptomatic carrier of a *BRCA1* mutation.

As the extent of genetic testing proliferates—and costs plummet—might we see a time when noninvasive prenatal testing using whole-exome (or even whole-genome) screening becomes universal? How far would we be moving then toward an obsession with genetic perfection and eugenics? Could we ever contemplate going so far and releasing huge amounts of genetic information if we have an imperfect knowledge about what all the genetic variants means? (Variants of uncertain clinical significance that have arisen *de novo* are always going to generate anxiety.) We consider this issue later in the section below when we describe ethical issues brought up by genomewide sequencing.

Preimplantation genetic diagnosis (PGD)

Two relevant issues are sex selection and human leukocyte antigen (HLA) selection. Screening for Y-linked markers allows the sex to be identified. Because of cultural preferences for male children in some societies, this can result in biased sex selection when male embryos are preferentially selected for implantation. HLA selection occurs when an embryo is selected for implantation to produce a child that can provide HLA-matched transplant tissue to save the life of a sick brother or sister.

Some highly publicized cases of HLA selection involve a transplant of stem cells from cord blood to repopulate the bone marrow. This would do no harm to the baby—it would be different if it were proposed that the child should donate a kidney. It is understandable that parents might wish to do all that they can to help their affected child, but knowing that he or she was born primarily to act as a cell donor might harm the psychosocial development of the unaffected child.

Restrictions on genetic testing as a result of gene patenting

Over a period of decades the US Patent and Trademark Office (USPTO) has granted thousands of patents on cloned genes, on the basis of an

Figure 11.20 Demonstrating against human gene patenting. Outside the US Supreme Court, Lisa Schlager, a Vice President at FORCE (Facing Our Risk of Cancer Empowered—an advocacy organization for people affected by hereditary breast and ovarian cancer) addresses protesters opposed to gene patenting. (Courtesy of Lisa Schlager.)

argument that although the genes are natural they need to be 'isolated' from nature (by severing chemical bonds, and so on). Companies were then able to patent human genes, including medically significant genes, such as the *BRCA1* and *BRCA2* genes for which patents were filed by Myriad Genetics in 1997–1998.

The idea of patenting pieces of our common genetic heritage appeared offensive to many people, who feel that no company should be able to own or exploit our DNA (**Figure 11.20**). And it placed the companies in a privileged position: by aggressively enforcing the patents they can create exclusive rights to profit from genetic testing and therapeutic applications.

Challenges to the Myriad patents soon came from many quarters. Within Europe, oppositions and appeals to the *BRCA1/2* patents began in early 2001 after the *BRCA1* patent was granted. As a result of a long and vigorous campaign of opposition, the European *BRCA1/2* patents have been drastically reduced in scope. Within the USA, the patents were legally challenged by plaintiffs led by the Association for Molecular Pathology. After a lengthy set of legal hearings, a final and unanimous decision was reached in June 2013 when the US Supreme Court ruled that merely isolating genes found in nature does not make them patentable, and that Myriad Genetics had not created or altered any of the genetic information encoded in the *BRCA1* and *BRCA2* genes, invalidating the company's claims.

Although the Supreme Court's decision clearly marks a change in US law regarding gene patents, it also ruled that cDNAs could be patented, arguing that cDNA does not occur naturally. The petitioners had argued that a cDNA should be ineligible for a patent, because the sequence of a cDNA is dictated by nature, not by the lab technician. And it is undeniable that endogenous reverse transcriptases occasionally make natural cDNA copies of mRNAs. However, the Supreme Court was not persuaded by the arguments put forward by the petitioners.

More recently, the focus of patenting within the genetic testing area appears to have moved toward patenting biomarkers that are often based on DNA variants.

Genetic discrimination and ethical, societal, and practical issues raised by clinical genomewide sequencing

At the outset of the Human Genome Project there were fears that as our genetic profiles became known, *genetic discrimination* might become an issue in health insurance and employment. Those fears have generally not been realized. In some countries, protective legislation has been passed. In the USA, for example, a Genetic Information Nondiscrimination Act was passed by United States Congress in 2008. It is designed to protect the privacy of a citizen's genetic information and to prevent genetic discrimination involving decisions about employment and health insurance coverage that are made on the basis of an individual's genetic testing or screening results. In some other countries, a moratorium has been agreed with insurance companies.

Until recently, the genomes of very few people had been decoded, and so ethical issues associated with genome sequencing had been a relatively minor concern. That is set to change: the volume of genomewide sequencing (of whole human exomes or whole human genomes) is about to expand enormously. Exome sequencing has increasingly been used in a diagnostic setting, and as sequencing costs fall to $1000 or less per genome, whole-genome sequencing will increasingly be applied. And at this price level, genetic testing companies might actively market whole-genome sequencing tests. Might we soon live in a world where everyone—or at least those living in areas with ready access to the technology—will have their genome sequenced?

Alongside the effort to identify clinically important variants at the genome level are parallel efforts to develop databases that link information on the phenotype to the variants. The ClinVar database (at http://www.ncbi.nlm.nih.gov/clinvar/) offers a freely available archive of reports of relationships between medically important variants and phenotypes. In addition to being tightly coupled with genome variation databases such as dbSNP and dbVar, the ClinVar database is also based on the phenotype descriptions stored in the MedGen database (at http://www.ncbi.nlm.nih.gov/medgen).

Looking to the future, pilot projects such as the 100K Genome Project are investigating the possibility of integrating clinical genome sequencing into health care systems (Box 11.5). If fetal or newborn genome sequencing were ever to be contemplated on a large scale, related ethical questions would spring to prominence, including issues relating to incidental findings and the disclosure of genetic information, as described below.

Incidental findings

Here, incidental (or secondary) findings are defined as discoveries made as a result of genetic testing that are unrelated to the medical reason for ordering the test. Incidental findings can have important health consequences, and questions arise regarding the obligation to disclose incidental findings and the circumstances in which they might be disclosed.

Genetic testing is not so very different from other types of clinical test in reporting incidental findings, such as detecting an unsuspected aneurysm on an MRI scan, or high cholesterol on a biochemical screen. But the scale can be different. The increasing adoption of array CGH over traditional karyotyping has led to a marked increase in incidental findings. However, using genomewide sequencing for clinical purposes takes the frequency of incidental findings to a much higher level given that on average there might be around 400 variants that contribute to disease in each person.

Incidental variants (also known as the incidentalome) have been variously regarded as a threat to, or an opportunity for, genomic medicine. In the former, there is the worry that clinicians might be overwhelmed by unexpected genomic findings and that patients might be subjected to unnecessary follow-up tests, possibly provoking anxiety and certainly leading to increasing costs, with quite often little associated benefit. Alternatively, by actively searching them out, studying how they link to phenotypes, and returning the information on variants to patients, incidental variants have been imagined to be the future of genomic medicine. That rosy scenario is currently beset, however, by our inadequate ability to interpret many sequence variants, and by logistical problems.

How should we report the incidental variants picked up in genomewide sequencing? There is a broad consensus that analytically valid and medically actionable variants should be returned to patients. A report from The American College of Medical Genetics and Genomics (ACMG) published in 2013 by Green et al. (see Further Reading) identified 57 genes associated with severe diseases for which it recommended that mutations previously known or expected to be pathogenic should always be reported in clinical germ-line genomewide sequencing in all subjects, irrespective of age, but excluding fetal samples.

BOX 11.5 From clinical genomics to public health genomics.

Until quite recently, human genetic studies were limited by our meagre understanding of human genetic variation, which in turn was due to a bottleneck in acquiring the necessary data. Massively parallel DNA sequencing changed all that, paving the way for genome studies to supplant gene studies. In January 2014 the milestone of the $1000 genome (sequencing a human genome at a cost of $1000) was achieved at last. Faster and cheaper sequencing in the near future can be expected, and will lead to an explosion in clinical genome sequencing.

The potential of clinical genomics to make an impact is massive. Cancer genome sequencing is providing new therapeutic targets and new forms of testing to stratify tumors to enable more targeted screening and treatment approaches. Applications in identifying constitutional mutations in genetic disorders were first reported in the late 2000s. They have been followed by multiple successes in disease gene identification (notably by whole-exome sequencing), and even in making clinical diagnoses. In the latter case, the ability to sequence genomes, and to obtain a diagnosis within a day or two (or even much less in the near future), might be life-saving for newborns undergoing a medical crisis.

Once clinical genome sequencing has really taken hold, the guesswork can be taken out of drug doses (even now, prescriptions for drugs such as warfarin are usually routinely made on a one-size fits all approach; knowing a person's genome sequence will allow more optimal dosing according to the profile of variants in the key drug-metabolizing enzymes). And in the case of infectious disease, there is the prospect of rapid-response sequencing of the genomes of individual pathogen strains within hours of disease outbreaks, allowing rapid, innovative responses to epidemics across the globe.

Large-scale pilot projects have begun that are examining the possibility of integrating genomewide sequencing into national health care services, and the era of public health genomics has begun with the ambitious 100k Genome Project (http://www.genomicsengland.co.uk/100k-genome-project/). Commissioned by the UK Government's Department of Health, the project seeks to sequence the genomes of about 30,000 patients per year from 2014, with a target of obtaining 100,000 genome sequences by 2017. The initial program will focus on rare inherited diseases, cancer, and infectious pathogens, and the plan is to link the data to electronic record systems within the UK's National Health Service (NHS). In the USA, a 5-year $25 million National Institutes of Health program was launched in 2013 to explore the possible use of genome sequencing in newborn screening. That possibility raises multiple ethical issues, notably the autonomy of the individual.

Bottlenecks in data analysis and networking challenges

The massively parallel DNA sequencing revolution has meant that the bottleneck has moved from data acquisition to data analysis. We cannot expect physicians to read genome reports (when even highly trained geneticists struggle to understand them), and there is some heavy lifting to do to work out what all the variants mean (see Incidental Findings in the main text). And even if we were to surmount that obstacle quickly there are many other challenges.

One of the biggest challenges will be to find ways of electronically integrating the data into health care services so that personal genome information can be stored safely and readily accessed without compromising security. As the volume of clinical genome sequence data increases exponentially, the sheer amount of data threatens to overwhelm current infrastructure systems (partly prompting the visual metaphor in **Figure 1**). There is an urgent need for system upgrades and innovation to remedy current deficiencies as well as improved bioinformatic capabilities.

If millions of individuals were to be involved, the volume of genome data becomes very substantial. 3.2 billion nucleotides means a minimum of about 800 megabytes of computer storage per individual. Add in 'deep sequencing' (in which each nucleotide is recorded multiple times) and the considerable descriptive data generated *about* individual nucleotides and it is quite common to save about 100 gigabytes of data when sequencing a single human genome.

Storing large amounts of genome data might be expected to require *cloud computing* services using dedicated banks of computers connected through the internet to healthcare services. By October 2013, the largest cloud-based genomics effort—a collaboration between Baylor College of Medicine, DNAnexus, and Amazon Web Services—had stored sequences from more than 3700 whole human genomes and close to 10,800 whole exome sequences, requiring hundreds of terabytes of storage space. In the future, however, maybe it will be easier to store and relay information only on the comparatively small number of differences by which each individual genome sequence differs from a reference genome sequence.

BOX 11.5 (*continued*)

Efficient computer programs need to be designed to ensure smooth linkage to existing networks, and to facilitate easy web-based data access while maintaining data security and confidentiality. In the latter case, for example, a problem that emerged recently was the ease with which the confidentiality of male genome sequences could be undermined. The policy of disseminating genome sequence data had permitted the disclosure of non-recombining Y sequences as part of the genome sequence of males. That proved to be a weak link: the lack of recombination in non-recombining Y-specific sequences and the frequent adoption of paternal surnames means that the pattern of variation in these sequences is linked to surnames. When genetic genealogy databases were interrogated, specific patterns of variation in the non-recombining Y could be linked to specific surnames, and the anonymity of the associated male genomes was able to be breached (see PMID 23329047).

Figure 1 A metaphor for the rate-limiting step in clinical genomics. The forthcoming deluge of clinical genome sequencing data will require a huge upgrading of bioinformatic and electronic network capabilities to interpret the data and to provide necessary software to relay and access the information efficiently. If there is not a major overhaul of current systems, the flood of data might be expected to block up the system, just like having too many vehicles on the roads at any one time, as in this example of a traffic jam in Delhi in 2010.

What to do with the others is the subject of much debate. Some of the incidental variants might have social rather than health implications, such as when the sequence results show that assumed biological relationships are wrong. Nevertheless, some people advocate full disclosure of genome sequence data; others argue that the raw genome sequencing data should have filters on to allow targeted genome screening: for certain genes that we wish to exclude, the raw sequencing data would be left unprocessed but could be retained to be processed and analyzed in the future.

Neonatal genome sequencing

Some people contend that complete genome sequencing will almost certainly become part of neonatal screening in the very near future, perhaps as part of the existing heel-prick screening of newborns. Others are strongly opposed, arguing that analyzing a newborn's complete genome violates the ethical principle of respecting an individual's autonomy.

The screening would reveal all manner of variants that are of no importance to the child's immediate health, but also some that would be expected to have serious consequences in later life. Proactively seeking out all genetic variants and disclosing the genetic findings would mean that the child would be robbed of the right not to know about his or her genome profile or to delay disclosure of that information until a later stage in life. And if a child were to be burdened by the knowledge that he or she had a high risk of developing some severe condition later in life, there could be the prospect of psychosocial harm.

The ethics of genetic manipulation of the germ line to prevent disease or to enhance normal human traits

As we detailed in Section 9.4, gene therapy is now offering successful treatment for a variety of different disorders. In each case the approved therapy is a somatic gene therapy. That is, the cells of the patient chosen to be genetically modified are somatic cells. Germ-line gene therapy, by contrast, would involve making a genetic change to germ-line cells that can be transmitted down the generations.

A motivation for germ-line gene therapy is to prevent disease, but the direct consequences would not be limited to one person; instead, they would be extended to potentially many descendants. Germ-line gene therapy has been the subject of ethical controversy, and genetic manipulation of the germ line is prohibited by law in many countries.

If germ-line gene therapy were ever to be attempted, it would most probably be done by genetic manipulation of a preimplantation embryo. (But it might occur incidentally when a treatment aimed at somatic cells modified the patient's germ cells unexpectedly.) Technical difficulties currently make the general prospect of germ-line therapy unrealistic. Even if these problems can be solved, ethical concerns will remain.

A possible argument in favor of germ-line gene therapy might be the desire to eliminate the risk of an inherited disease to future generations. However, for recessive conditions, only a very few of the disease alleles are carried by affected people (the great majority are in healthy heterozygotes), and most serious dominant or X-linked diseases are largely maintained in the population by recurrent mutation.

There are multiple arguments against germ-line therapy. Even if the technology could be improved so that it was risk-free, and the initial treatment were done with informed consent, later generations would be given no choice in the matter. A particularly compelling argument against germ-line gene therapy that involves genetic modification of the nuclear genome is that it is simply not necessary. Candidate couples would most probably have dominant or recessive Mendelian disorders (recurrence risk 50% and 25%, respectively). Given a dish containing half a dozen IVF embryos from the couple, it would seem bizarre to select the affected ones and then subject them to an uncertain procedure of genetic manipulation, rather than simply select the 50% or 75% of unaffected ones for reimplantation.

Genetic enhancement and designer babies

Instead of attempting to prevent disease, genetic manipulation of the germ line might conceivably allow improvements in normal human traits (genetic enhancement). In the decades to come, as we come to understand the biology of our genome, we will begin to understand the fine genetic control over many of our traits. (Already, large genomewide association studies are beginning to identify genetic variants associated with different human traits and abilities, such as educational attainment—see PMID 23722424.)

Caught up with the notion of genetic enhancement is the prospect that people will use *in vitro* fertilization and preimplantation diagnosis to select embryos with certain desired qualities, and reject the rest even though they are normal. That scenario is already with us in the form of preimplantation sex selection and HLA selection.

Might the above cases lead inexorably to demands for 'designer babies' with multiple desired qualities? Not if the embryos are simply selected on the basis of multiple desired criteria (most IVF procedures produce only a handful of embryos, and usually two or three are implanted to maximize the chance of success, leaving too few to guarantee that one of them would have all the desirable genetic variants). It would be a different story if we had a good idea about which genes to enhance and if genetic manipulations on the germ line were to be so efficient that the technology became extremely safe. Then, genetic enhancement would be a difficult question. It does raise the question, however, of how we could ever ethically justify getting the technology of manipulating the human germ line to such an advanced stage (if, on the way, unsafe procedures could have adverse effects on future descendants).

Ethical considerations and societal sensitivity to three-person IVF treatment for mitochondrial DNA disorders

In single-gene disorders caused by a mutation in nuclear DNA, some children inherit mutant alleles from one or both parents, and some children inherit normal alleles. In mitochondrial DNA (mtDNA) disorders, by contrast, a woman with homoplasmic mutant mtDNA would be expected to transmit the mutation in all of her eggs. All her children might then be expected to inherit the faulty mitochondria.

Because mitochondria act as the batteries of a cell, mtDNA disorders tend to affect parts of the body that need the most energy (notably the brain, muscles, and heart). Such disorders can therefore be very serious, they are incurable, and there has been no way of preventing transmission of the faulty mtDNA from mother to children (but preimplantation genetic diagnosis might be able to lower the risk by selecting embryos with the highest percentage of normal healthy mtDNA).

In a proposed *in vitro* fertilization treatment for mitochondrial DNA disorders, an embryo would be produced with DNA from three people: nuclear DNA from each of the two parents who seek treatment, and mtDNA from a woman who donates an oocyte with healthy mtDNA (see Figure 9.25 for the treatment protocol). The treatment would be a first, allowing parents in this situation to have normal children by preventing transmission of the mother's faulty mtDNA. However, after the treatment was proposed and publicized in the UK, it attracted considerable controversy—newspaper headlines almost inevitably seized upon the aspect of producing a child with three biological parents.

People opposed to the mtDNA treatment offer various arguments: destruction of IVF embryos is involved (but only in one of the two possible approaches shown in Figure 9.25); an embryo is being created simply as a source of spare parts; and the treatment is a form of germ-line gene therapy in which genetic modifications are made to the embryo that might be transmitted to future generations.

The germ-line gene therapy argument is strongly contested by many scientists who argue that the procedure does not involve modifying DNA: whole mitochondria with intact genomes containing a total of just 37 genes are simply being provided by the donor oocyte to replace faulty mitochondria with mutated DNA. They argue that the method is comparatively safe and would offer only a small risk; although it would be important to perfect the mitochondrial transfer and ensure safety, the approach could be justified by the urgent clinical need to provide help for families who are at risk of severe mtDNA disorders.

The UK's Human Fertilisation and Embryology Authority (HFEA) released the results of a national consultation in March 2013, showing that there was majority support from the British public, and the UK government has lent its support. The matter is to be debated in Parliament at the end of 2014; if parliamentary and HFEA approval are obtained, the treatment could be implemented from 2015.

Summary

- The analytical validity of a test evaluates how well the assay measures what it claims to measure.

- A genetic test assay is said to have a high sensitivity if a high proportion of all people with the condition are correctly identified as such, and a high specificity if a high proportion of all people who do not have the condition are correctly identified as such.

- Genetic tests are usually used to confirm a provisional clinical diagnosis, or to predict the likelihood of developing or transmitting a genetic disorder.

- Most genetic tests are designed to detect abnormal chromosomes or pathogenic DNA variants. These tests may involve scanning for an undefined abnormal DNA variant (such as DNA sequencing to look for any abnormal point mutations), or testing for a specific known pathogenic variant.

- Unidentified disease alleles at a known disease locus can also be tracked by assaying markers closely linked to the disease locus. The test is performed on multiple family members to predict the likelihood that an asymptomatic family member or fetus has inherited the disease-associated allele.

- Some genetic tests are assays that detect consequences of genetic variation: altered RNA or protein expression products, altered gene function, or a characteristic disease biomarker such as an abnormally elevated metabolite. Assays such as these can be convenient because they often test for a common disease-associated characteristic without the need to establish precisely what genetic variant caused it.

- Genetic screening means carrying out proactive assays to identify individuals in communities and populations at increased risk of carrying abnormal chromosomes or harmful genetic variants. For example, in populations with an elevated incidence of a particular autosomal recessive disorder, genetic screening can be used to identify asymptomatic carriers at high risk of transmitting the disorder and to be able to offer prenatal diagnosis to prospective parents who are both identified as carriers.

- Quantitative fluorescence PCR provides a rapid way of scanning for aneuploidy in fetal DNA. It uses panels of fluorescently labeled microsatellite markers for the chromosomes most frequently involved in aneuploidies.

- In practice, chromosome microarray analysis means using microarrays with oligonucleotide probes from across the genome to scan for subchromosomal changes in copy number and/or single nucleotide polymorphisms.

- Comparative genome hybridization is a popular way of screening for changes in copy number of subchromosomal regions. It usually relies on labeling the test DNA sample and a normal control sample with different fluorophores, and hybridizing them simultaneously to defined oligonucleotides probes on a microarray.

- Multiplex ligation-dependent probe amplification is a rapid way of scanning for copy number changes in defined sequences of interest; it is often used to scan for deletions or duplications of individual exons.

- Target enrichment sequencing is used to scan any desired subset of the genome—multiple disease genes or the whole exome—for evidence of point mutations.

- Treatment of DNA with sodium bisulfite allows the differentiation of unmethylated cytosines (which are modified to give first uracils and then thymines after DNA replication) from methylated cytosines (which are unchanged). Certain restriction nucleases will cleave sites containing a CpG dinucleotide only if the cytosine is unmethylated.

- Traditionally, prenatal diagnosis has used invasive procedures to recover and analyze fetal cells from early pregnancy. In preimplantation diagnosis, genetic testing occurs on embryos produced by *in vitro* fertilization in the context of assisted reproduction.

- In noninvasive prenatal testing, samples of freely circulating DNA recovered from maternal plasma are analyzed. The plasma DNA is a mixture of fetal DNA and maternal DNA that originated from degraded cells. It can be analyzed by massively parallel DNA sequencing to infer fetal DNA variants and the fetal genome sequence.

- Cascade testing means the testing of relatives after identifying a person with a pathogenic mutation. The relatives will be at a higher risk (than the general population) of being carriers in the case of a recessive disorder or chromosome translocation, or of developing disease in the case of a childhood-onset or late-onset dominant disorder.

- Pre-symptomatic diagnosis can be carried out on asymptomatic individuals who are at risk of developing a genetic disorder later in life. If a person is identified as carrying the mutant allele, follow-up screening can be carried out, and in some cases treatment regimes can be followed to reduce disease risk.

- In direct-to-consumer genetic testing, commercial companies carry out the tests and feed back results without the involvement of health care professionals.

- Genetic testing for susceptibility to complex diseases can identify individuals at increased disease risk, but much of the current testing is of limited predictive value because the tested susceptibility factors are of low effect.

- Mainstreaming genetics envisages that genetic testing will be incorporated into mainstream medicine—diagnostic genetic testing will increasingly become the responsibility of the clinicians to whom patients are initially referred.

- Clinical genome sequencing is the ultimate scan for pathogenic variants, but in each person a significant number of variants will be identified whose clinical significance is uncertain. Ethical concerns include the question of how to deal with incidental findings in which pathogenic variants are identified that are unrelated to the disorder for which the genome sequencing test was ordered.

- Clinical genome sequencing is likely to be incorporated soon into the existing health care systems of many economically advanced societies, but significant bioinformatic and electronic networking challenges will need to be addressed. There are also ethical concerns about releasing data when we currently have imperfect knowledge of the clinical significance of many variants.

- Testing for a genetic disorder is unusual in that the results have potential implications for close relatives as well as for the person tested. The person tested may refuse to give consent to releasing the information, but a breach of confidentiality can sometimes be ethically justified if failure to disclose the information means that a relative is seriously disadvantaged in some way.

- Ethical principles dictate that consent for genetic testing should always be obtained from the participating individuals or the legal guardians of children or of adults with mental incapacity.

- The ethical principle of beneficence requires that genetic testing in children should be of benefit to the child: it should test for clinically relevant and actionable genetic variants only (that is, variants whose results might lead to some kind of beneficial clinical intervention).

- The principle of the 'right to an open future' respects an individual's autonomy. It requires that information on non-actionable genetic variants in a child that confer susceptibility to late-onset disorders should not be disclosed before the child becomes an adult and subsequently wishes to obtain that information.

- Because the current genetic technology is imperfect, genetic modification of the germ line has potentially harmful consequences for future descendants and is widely prohibited, unlike somatic gene therapy, which is designed to have consequences only for the person undergoing the therapy.

Questions

Help on answering these questions and a multiple-choice quiz can be found at www.garlandscience.com/ggm-students

1. As a genetic test, indirect linkage analyses have mostly been supplanted by direct mutation screening. When are they still sometimes useful? What are the possible drawbacks of using linkage analyses to predict the inheritance of disease alleles, and how can they be minimized?

2. The table below shows the peak areas for individual markers tested by quantitative fluorescence PCR to assess the possibility of a human trisomy. Numbers within the peak area row are quantitative estimates of the fluorescence recorded for individual peaks at the corresponding marker locus (numbers are separated by a colon if there are two or more peaks). Give an interpretation of the results.

MARKER	D13S268	D13S634	D13S797	D18S386	D18S391	D18S535	D21S11	D21S1411	D21S1446
PEAK AREA	12790: 12596	32165	15695: 13894	51670	12557: 28261	31052: 14941	7911: 8267	41294	11098: 10786

3. The amplification refractory mutation system (ARMS) is a type of genetic test. What is it used for, and what is the principle on which it is based?

4. Genetic testing for cytosine methylation is important in some diseases. What kind of diseases are tested, and what form does the testing take?

5. What is meant by genetic screening? Illustrate your answer by giving examples of genetic screening with different objectives.

6. How is noninvasive prenatal genetic testing carried out, and how does it compare with standard invasive prenatal genetic testing?

7. What is meant by variants of uncertain clinical significance and incidental findings, and why have they increasingly become a problem in genetic testing?

Further Reading

Genetic Testing Overviews and Resources

Genetic Testing Registry [An electronic resource established by the US National Institutes of Health to serve as a central repository of genetic tests]. Available at: https://www.ncbi.nlm.nih.gov/gtr/

Katsanis SH & Katsanis N (2013) Molecular genetic testing and the future of clinical genomics. *Nature Rev Genet* 14:415–426; PMID 23681062.

Korf BR & Rehm HL (2013) New approaches to molecular diagnosis. *J Am Med Assoc* 309:1511–1521; PMID 23571590.

Identifying Chromosome Abnormalities and Large-Scale DNA Changes

Mann K & Ogilvie CM (2012) QF-PCR: application, overview and review of the literature. *Prenatal Diagn* 32:309–314; PMID 22467160.

Schaffer LG (2013) Microarray-based cytogenetics. In The Principles of Clinical Cytogenetics, 3rd ed. (Gersen SL & Keagle MB eds), pp. 441–450. Springer.

Vermeesch JR et al. (2012) Guidelines for molecular karyotyping in constitutional genetic diagnosis. *Eur J Hum Genet* 15:1105–1114; PMID 17637806.

Wapner RJ et al. (2012) Chromosomal microarray versus karyotyping for prenatal diagnosis. *N Engl J Med* 367:2175–2184; PMID 23215555.

Willis AS et al. (2012) Multiplex ligation-dependent probe amplification (MLPA) and prenatal diagnosis. *Prenatal Diagn* 32:315–320; PMID 22467161.

Microarray-Based Mutation Scanning in Individual Target Genes

Stef MA et al. (2013) A DNA microarray for the detection of point mutations and copy number variation causing familial hypercholesterolemia in Europe. *J Mol Diagn* 15:362–372; PMID 23537714.

Wen W-H et al. (2000) Comparison of *TP53* mutations identified by oligonucleotide microarray and conventional DNA sequence analysis. *Cancer Res* 60:2716–2722; PMID 10825146.

Genomewide and Disease-Targeted Sequencing in Mutation Scanning

Choi M et al. (2009) Genetic diagnosis by whole exome capture and massively parallel DNA sequencing. *Proc Natl Acad Sci USA* 106:19096–19101; PMID 19861545.

Rehm HL (2013) Disease-targeted sequencing: a cornerstone in the clinic. *Nature Rev Genet* 14:295–299; PMID 23478348.

Saunders CJ et al. (2012) Rapid whole-genome sequencing for genetic disease diagnosis in neonatal intensive care units. *Sci Transl Med* 4:1–13; PMID 23035047.

Yang Y et al. (2013) Clinical whole exome sequencing for the diagnosis of mendelian disorders. *N Engl J Med* 369:1502–1511; PMID 24088041.

Interpreting and Classifying Sequence Variants

Duzkale H et al. (2013) A systematic approach to assessing the clinical significance of genetic variants. *Clin Genet* 84:453–463; PMID 24033266.

Houdayer C (2011) In silico prediction of splice-affecting nucleotide variants. *Methods Mol Biol* 760:269–281; PMID 21780003.

Jarinova O & Ekker M (2012) Regulatory variations in the era of next generation sequencing: implications for clinical molecular diagnostics. *Hum Mut* 33:1021–1030; PMID 22431194.

Millot GA et al. (2012) A guide for functional analysis of BRCA1 variants of uncertain significance. *Hum Mut* 33:1526–1533; PMID 22753008.

Raynal C et al. (2013) A classification model relative to splicing for variants of unknown clinical significance: application to the CFTR gene. *Hum Mut* 34:773–784; PMID 23381846.

Genotyping Point Mutations and DNA Methylation Profiling

Heyn H & Esteller M (2012) DNA methylation profiling in the clinic: applications and challenges. *Nature Rev Genet* 13:679–692; PMID 22945394.

Syvanen A-C (2001) Accessing genetic variation: genotyping single nucleotide polymorphisms. *Nature Rev Genet* 2:930–942; PMID 11733746.

von Kanel T & Huber AR (2013) DNA methylation analysis. *Swiss Med Wkly* 143:w13799; PMID 23740463.

Preimplantation Genetic Testing

Brezina PR et al. (2012) Preimplantation genetic testing. *Br Med J* 345:e5908; PMID 22990995.

Harper C & SenGupta SB (2011) Preimplantation genetic diagnosis: state of the ART 2011. *Hum Genet* 131:175–186; PMID 21748341.

Noninvasive Prenatal Testing

Benn P et al. (2013) Non-invasive prenatal testing for aneuploidy: current status and future prospects. *Ultrasound Obstet Gynecol* 42:15–33; PMID 23765643.

Fan HC et al. (2012) Non-invasive prenatal measurement of the fetal genome. *Nature* 487:320–324; PMID 22763444.

Hui L & Bianchi DW (2013) Recent advances in the prenatal interrogation of the human fetal genome. *Trends Genet* 29:84–91; PMID 23158400.

Kitzman JO et al. (2012) Noninvasive whole-genome sequencing of a human fetus. *Sci Transl Med* 4:137ra6; PMID 22674554.

Lo YMD & Chiu RWK (2012) Genomic analysis of fetal nucleic acids in maternal blood. *Annu Rev Genom Hum Genet* 13:285–306; PMID 22657389.

Genetic Counseling

Harper PS (2010) Practical Genetic Counselling, 7th ed. Hodder-Arnold.

Predictive Testing and Genetic Screening

Cairns SR et al. (2010) Guidelines for colorectal cancer screening and surveillance in moderate and high risk groups (update from 2002). *Gut* 59:666–690; PMID 20427401.

Caskey CT et al. (2014) Adult genetic risk screening. *Annu Rev Med* 65, 1-17; PMID 24188662.

Cousens NE et al. (2010) Carrier screening for beta-thalassaemia: a review of international practice. *Eur J Hum Genet* 18:1077–1083; PMID 20571509.

Hawkins AK et al. (2011) Lessons from predictive testing for Huntington disease: 25 years on. *J Med Genet* 48:649–650; PMID 21931167.

Umbarger MA et al. Next-generation carrier screening. *Genet Med* 16:132–140; PMID 23765052.

Watson MS et al. (2006) Newborn screening: toward a uniform screening panel and system—executive summary. *Pediatrics* 117:S296–S307; PMID 16735256.

Wilcken B (2011) Newborn screening: how are we travelling, and where should we be going? *J Inher Metab Dis* 34:569–574; PMID 21499716.

New Approaches in Cancer Diagnostics

Crowley E et al. (2013) Liquid biopsy: monitoring cancer-genetics in the blood. *Nature Rev Clin Oncol* 10:472–484; PMID 23836314.

Dawson SJ et al. (2013) Analysis of circulating tumor DNA to monitor metastatic breast cancer. *N Engl J Med* 368:1199–1209; PMID 23484797.

Gonzalez de Castro D et al. (2013) Personalized cancer medicine: molecular diagnostics, predictive biomarkers and drug resistance. *Clin Pharmacol Ther* 93:252–259; PMID 23361103.

Kilpivaara O & Aaltonen LA (2013) Diagnostic cancer genome sequencing and the contribution of germline variants. *Science* 339:1559–1562; PMID 23539595.

Leary RJ (2012) Detection of chromosomal alterations in the circulation of cancer patients with whole-genome sequencing. *Sci Transl Med* 4:162ra154; PMID 23197571.

Murtaza M et al. (2013) Non-invasive analysis of acquired resistance to cancer therapy by sequencing of plasma DNA. *Nature* 497:108–112; PMID 23563269.

Wang L & Wheeler DA (2014) Genome sequencing for cancer diagnosis and therapy. *Annu Rev Med* 65, 33-48; PMID 24274147.

Genetic Testing of Complex Diseases

Caulfield T & McGuire AL (2012) Direct-to-consumer genetic testing: perceptions, problems and policy responses. *Annu Rev Med* 63:23–33; PMID 21888511.

Janssens AC & van Duijn CM (2008) Genome-based prediction of common diseases: advances and prospects. *Hum Molec Genet* 17:R166–R173; PMID 18852206.

Ethical Issues in Genetic Testing

Almond B (2006) Genetic profiling of newborns: ethical and social issues. *Nature Rev Genet* 7:67–71; PMID 16369573.

de Jong A et al. (2011) Advances in prenatal screening: the ethical dimension. *Nature Rev Genet* 12:657–663; PMID 21850045.

Graff GD et al. (2013) Not quite a myriad of gene patents. *Nature Biotechnol* 31:404–410; PMID 23657391.

Nowland W (2002) Human genetics. A rational view of insurance and genetic discrimination. *Science* 297:195–196; PMID 12114609.

Ross LF et al. (2013) Technical report: ethical and policy issues in genetic testing and screening of children. *Genet Med* 15:234–245; PMID 23429433.

Royal College of Physicians, Royal College of Pathologists and British Society for Human Genetics (2011) Consent and Confidentiality in Genetic Practice: Guidance on Genetic Testing and Sharing Information. A Report of the Joint Committee on Medical Genetics, 2nd ed. Royal College of Physicians and Royal College of Pathologists. Also available electronically at http://www.bsgm.org.uk/media/678746/consent_and_confidentiality_2011.pdf

Clinical and Public Health Genomics: Challenges and Ethics

Biesecker LG (2013) Incidental variants are critical for genomics. *Am J Hum Genet* 92:648–651; PMID 23643378.

Dondorp WJ & de Wert GM (2013) The 'thousand-dollar genome': an ethical exploration. *Eur J Hum Genet* 21:S6–S26; PMID 23677179.

Green RC et al. (2013) ACMG recommendations for reporting of incidental findings in clinical exome and genome sequencing. *Genet Med* 15:565–574; PMID 23788249.

McEwen JE et al. (2013) Evolving approaches to the ethical management of genome data. *Trends Genet* 29:375–382; PMID 23453621.

GLOSSARY

3′ end
The end of a DNA or RNA strand that is linked to the rest of the chain only by carbon 5′ of the sugar, not carbon 3′ (Figure 1.2).

5′ end
The end of a DNA or RNA strand that is linked to the rest of the chain only by carbon 3′ of the sugar, not carbon 5′ (Figure 1.2).

adaptive immune responses (or **system**)
Specific immune responses that rely on recognition of foreign antigen by antibodies and T-cell receptors.

allele frequency
The frequency of an allele in a population; that is, the proportion of all alleles at a locus that are the allele in question (often inaccurately represented as gene frequency).

allele
Individual version of a gene or DNA sequence at a *locus* on a single chromosome; often also used loosely to describe the encoded protein variants.

allogeneic
Describing cell and organ transplantation (or the transplanted cells) in which the donor cells are genetically different from that of the recipient. Compare *autologous*.

amino acid
The fundamental repeating unit of a polypeptide; a building block for a protein (Figure 2.2 and Table 7.2).

amplification
1. An artificial increase in the copy number of a DNA sequence as a result of *cloning* or *PCR* (Section 3.1).
2. A cellular mechanism that is responsible for a rapid increase in the number of copies of a gene in certain situations, such as in some cancers (Figure 10.6).

aneuploidy
A chromosome constitution with one or more chromosomes extra or missing from a full (euploid) set (Section 7.4).

angiogenesis
Process whereby new blood vessels are formed by sprouting from existing vessels.

annealing
Process whereby two single-stranded nucleic acids form a stable double-stranded nucleic acid by *base pairing*. The reverse of denaturation.

anticipation
The tendency for the severity of a condition to increase in successive generations (Section 5.3). Commonly due to bias of ascertainment, but a genuine outcome in the case of some *dynamic mutations*.

antigen
A molecule that can induce an adaptive immune response or that can bind to an antibody or T-cell receptor.

antigen presentation
The process by which antigen is presented in combination with an *MHC* (*major histocompatibility complex*) protein on the surface of certain cells so that it can be recognized by receptors on lymphocytes (Section 4.5 and Box 8.3).

antisense RNA
An RNA transcript that has a *complementary sequence* to an mRNA (or some functional noncoding RNA). Naturally occurring antisense RNAs, made using the non-template strand of a gene, are important regulators of gene expression.

antisense (or **template**) **strand**
The DNA strand of a gene that, during transcription, is used as a template by RNA polymerase for the synthesis of the RNA transcript (Figure 2.1).

apoptosis
A natural way of getting rid of unwanted or diseased cells in which the cell is targeted for destruction by various stimuli. Rapid fragmentation of the cell follows, after which the resulting cell fragments are phagocytosed by neighboring cells.

association
A tendency of two *characters* (such as diseases or marker alleles) to occur together at nonrandom frequencies. Association is a simple statistical observation, not a genetic phenomenon, but can be caused by *linkage disequilibrium* (Section 8.2).

augmentation therapy
Therapy that is intended to supplement some deficiency, as opposed to the great majority of drug therapies that are designed to inhibit some disease process.

autoimmune disorders
Diseases that arise because the distinction between self and nonself fails so that the body mounts an abnormal immune response against one or more self molecules.

autologous
Describing cell and organ transplantation in which the donor cells originally came from the recipient and have been modified in some way before being transplanted back again.

autosome
Any chromosome other than the sex chromosomes, X and Y.

autozygosity
In an inbred person, homozygosity for alleles identical by descent.

balancing selection
Selection working simultaneously in opposite directions on the same variant; can result in heterozygotes for a harmful mutation having a higher biological *fitness* than normal homozygotes (Section 5.4).

base complementarity
The relationship between bases on opposite strands of a double-stranded nucleic acid: A always occurs opposite T (or U in RNA) and G always occurs opposite C in DNA (but in RNA, U can also sometimes base pair with G).

base pair/base pairing
The outcome/process of stable hydrogen bonding between two complementary bases, a purine and a pyrimidine (Figure 1.4). The bases may reside on opposing strands of a duplex nucleic acid (Figure 1.5), or on the same RNA strand (Figure 2.4A).

benign tumor
An abnormal cell growth that is confined to a specific site within a tissue and shows no evidence of invading adjacent tissue.

biomarker
Any characteristic that is objectively measured and evaluated as an indicator of normal biological processes, pathogenic processes, or pharmacologic responses to a therapeutic intervention.

biotin–streptavidin system
A tool for isolating labeled molecules. The bacterial protein streptavidin happens to bind biotin (vitamin B7) with exceptionally high affinity. Biotinylated molecules can be isolated by using streptavidin-coated magnetic beads (Figure 8.7).

blastocyst
An embryo at a very early stage of development when it consists of a hollow ball of cells with a fluid-filled internal compartment (Figure 2 in Box 9.2).

blastomere
One of the multiple cells formed when the fertilized egg undergoes cleavage divisions.

capping
A stage in RNA processing. A special nucleotide, 7-methylguanosine triphosphate, is joined by a 5′–5′ phosphodiester bond to the 5′ end of a primary transcript. Capping is important for the stability of the RNA.

cancer
1. One of a heterogeneous group of disorders whose common features are uncontrolled cell growth and cell spreading. 2. A tumor that has become *malignant*.

carrier
A person, usually asymptomatic, who carries a genetic variant that can cause disease after being transmitted to the next generation, or that can contribute to disease in later life.

case-control study
A study in which samples from affected individuals (cases) are analyzed and compared with equivalent samples from unaffected control individuals.

cDNA (complementary DNA)
DNA synthesized by the enzyme reverse transcriptase using RNA (often mRNA) as a template.

centromere
The primary constriction of a chromosome, separating the short arm from the long arm, and the point at which spindle fibers attach to pull chromatids apart during cell division.

CGH See *comparative genome hybridization*.

character (or **trait**)
An observable property of an individual, such as eye color or ABO blood group type.

chimera
An organism derived from more than one zygote.

chromatid
One of a pair of sister chromatids that form when a chromosome replicates and persist until the anaphase stage of mitosis (see Figure 1.11).

chromatin
The nucleoprotein material of a chromosome.

chromatin remodeling
Movement, dissociation, or reconstitution of nucleosomes in chromatin, as part of the systems controlling chromatin conformation.

chromosomal microarray analysis
Clinical application of *microarray hybridization*. The usual object is to scan a genomic DNA sample for changes in copy number (deletions or duplications) of large DNA segments.

chromosome
In eukaryotes, a nucleoprotein structure formed when a nuclear DNA molecule is complexed with various types of proteins and occasionally some RNAs. The complexing helps compact the immensely long DNA molecules.

cis-acting (of gene regulation)
Term used to describe any gene regulation in which a regulatory DNA or RNA sequence controls the expression of some other sequence that is present on the same nucleic acid molecule (Box 6.1).

clones/cloning
Identical copies (of a DNA sequence, a cell, or an organism)/process of making the same. In genetic research, this often means cells containing identical recombinant DNA molecules.

CNV/CNP See *copy number variation*.

coding DNA
A segment of DNA whose sequence is used directly to specify a polypeptide (via a mRNA).

co-dominant
Term used to describe a heterozygous state in which both alleles are fully expressed.

codon
A sequence of three nucleotides (strictly in mRNA, but

by extension, in genomic coding DNA) that specifies an amino acid or a translation stop signal.

coefficient of inbreeding
The proportion of loci at which a person is homozygous by virtue of the consanguinity of their parents (Section 5.2).

coefficient of relationship
Of two people, the proportion of loci at which they share alleles identical by descent (Box 5.2).

comparative genome hybridization (CGH)
Simultaneous hybridization of test and control nucleic acids (usually to a microarray of mapped DNA clones from across the genome = array CGH). The object is to detect chromosomal regions in the test sample that are amplified or deleted compared with the control sample (Section 11.2 and Figure 11.4).

complementary sequences (or **strands**)
Nucleic acid sequences (or strands) that can form a stable double-stranded nucleic acid by *base pairing*.

complementary DNA See *cDNA*.

compound heterozygote
A person with two different mutant alleles at a locus.

conformation
Of a complex molecule, the three-dimensional shape—the result of the combined effects of many weak noncovalent bonds.

consanguineous
Description of persons who are closely related because they have descended from a very recent common ancestor (often within the previous three or four generations), usually as a result of a marriage between cousins.

conservative substitution
A nucleotide substitution that changes a codon so that it makes a different, but chemically similar, amino acid.

conserved sequence
DNA or amino acid sequence that is identical or recognizably similar across a range of organisms, suggestive of an important function.

constitutional (of genetic variation, mutation, chromosome abnormality)
Present in the genetic material of the zygote, and therefore present in every nucleated cell of a person.

copy number variation (CNV)
Variation between individuals in the number of copies in their genomes of a specific, moderately long to large DNA sequence (from hundreds of base pairs to many megabases). The term CNV is also used to denote a rare copy number variant (frequency less than 1%); if the frequency is above 1%, copy number polymorphism (CNP) is often used (Sections 4.3 and 8.2).

CpG island
Short stretch of DNA, often less than 1 kb long, containing frequent unmethylated CpG dinucleotides. CpG islands tend to mark the 5′ ends of genes (Box 6.3).

cryptic splice site
A sequence in pre-mRNA with significant homology to a splice site. Cryptic splice sites may be used as splice sites when splicing is disturbed or after a base substitution mutation that increases the resemblance to a normal splice site (Figure 7.4).

cross-linking (in DNA)
Abnormal occurrence of covalent bonds directly linking two bases. The cross-linked bases may be on the same strand or on opposite strands (Figure 4.1). In proteins, the disulfide bond is a natural form of cross-linking (Figure 2.5).

crossover
An act of meiotic *recombination*, or the physical manifestation of that (as seen under the microscope) (Figures 1.14 and 1.15).

cytokines
Extracellular signaling proteins or peptides that act as local mediators in cell–cell communication.

dedifferentiation
Epigenetic reprogramming of a *differentiated cell* so that the cell becomes less specialized (Box 9.2).

denaturation
Dissociation of double-stranded nucleic acid to give single strands. Also destruction of the three-dimensional structure of a protein by heat or high pH.

derivative chromosome
A chromosome that has been structurally rearranged, for example by translocation, but retains a centromere (Figure 7.12).

differentiation (of a cell)
Natural process of epigenetic modification that causes a cell to become more specialized.

diploid
Having two copies of each type of chromosome; the normal constitution of most human somatic cells.

direct repeats
Two or more copies of a sequence that occur in the same 5′→3′ direction on a single DNA strand. Usually used to mean repeats that are separated on the DNA; repeats that are directly adjacent to one another are normally described as *tandem repeats*.

distal (of chromosome)
Comparatively distant from the centromere (Box 7.4).

DNA libraries
The result of cloning random DNA fragments or molecules to produce a collection of cells containing different recombinant DNAs (which must then be screened to find any desired sequence).

dominant
In human genetics, any trait that is expressed in a heterozygote.

dominant-negative effect
The situation in which a mutant protein interferes with the function of its normal counterpart in a heterozygous person (Figure 7.17).

dosage-sensitive gene
A chromosomal gene that, when present in one copy instead of the normal two copies (causing reduced expression), is associated with disease. Disease can also sometimes result from an increased number of copies (with consequent overexpression) (Box 7.6).

driver mutations
In cancer, mutations that are subject to positive selection during tumorigenesis because they assist development of the tumor, as opposed to passenger mutations that occur during tumorigenesis but that are not positively selected or causally implicated in cancer development.

duplex
A double-stranded nucleic acid.

dynamic mutation
An unstable expanded repeat that changes in size between parent and child (Section 7.3).

embryonic stem (ES) cell line
Embryonic stem cells that have continued to proliferate after subculturing for a period of 6 months or longer and that are judged to be *pluripotent* and genetically normal.

endonuclease
An enzyme that cuts DNA or RNA at an internal position in the chain.

enhancer
A set of clustered short sequence elements that stimulate the transcription of a gene and whose function is not critically dependent on their precise position or orientation (Section 6.1).

epigenetic
Heritable (from mother cell to daughter cell, or sometimes from parent to child), but not produced by a change in DNA sequence.

epigenetic marks (or **settings**)
Patterns of epigenetic modification, notably DNA methylation, histone modifications, and nucleosome spacing.

epigenome
The totality of epigenetic marks in a cell.

epimutation
A change in chromatin organization that causes a change in expression of one or more genes without any change to the DNA sequence (Figure 6.19). Epimutations can be induced by mutation at a distant gene locus that regulates chromatin modification, and by environmental factors (resulting in metabolic changes or inflammation, for example, that can lead to changes in chromatin modification). Certain chromosome abnormalities (*position effects*) can lead to a similar outcome.

episome
Any DNA sequence that can exist in an autonomous (self-replicating) extrachromosomal form in the cell.

epistasis
Literally 'standing above'. Gene *A* is epistatic to gene *B* if *A* functions upstream of *B* in a common pathway.

Loss of function of *A* will cause all the effects of loss of function of *B*, and maybe other effects as well.

epitope
The part of an immunogenic molecule to which an antibody responds.

euchromatin
The fraction of the nuclear genome that contains transcriptionally active DNA and that, unlike heterochromatin, adopts a relatively extended conformation.

exon
Originally, any segment of an RNA transcript that is retained during RNA *splicing*, but now used widely to mean the corresponding sequence in genomic DNA. Individual exons may contain coding sequences that are translated and/or noncoding sequences (Figure 2.1).

exome
The totality of exons in a genome.

exon shuffling
An evolutionary process in which exons from one gene are copied and inserted into a different gene (Figure 2.15).

exon skipping
Occasional failure to include an exon within an RNA transcript (Figure 6.5D).

exonuclease
An enzyme that digests a DNA or RNA strand from one end. It may be a 3' or 5' exonuclease.

FISH See *fluorescence in situ hybridization*.

fitness (*f*)
In population genetics, a measure of the success in transmitting genotypes to the next generation, relative to the most successful genotype. Also called biological or reproductive fitness. *f* always lies between 0 and 1.

fluorescence *in situ* hybridization (FISH)
Hybridization of a fluorescently labeled probe to the denatured DNA of chromosome preparations that have been immobilized on a solid surface (Figures 11.7 and 11.8), or to the RNA of cells that have been similarly immobilized.

fluorophore (or **fluorochrome**)
A fluorescent chemical group, used for labeling nucleic acids or proteins (Box 3.2).

founder effect
High frequency of a particular allele in a population because the population is derived from a small number of founders, one or more of whom carried that allele (Section 5.4).

fragile site
Location on a chromosome where the chromatin of metaphase chromosomes can appear condensed under certain culture conditions. Most examples do not cause disease, but see Box 7.2 for one that does.

frameshift
A change in the base sequence that removes or

adds nucleotides in *coding DNA* so as to change the translational reading frame (Box 2.1).

gain-of-function mutations
Mutations that cause the gene product to do something abnormal, rather than simply to lose function. Usually the gain is a change in the timing or level of expression (Sections 7.5 and 10.2).

gamete
Sperm or egg; a haploid cell formed when a germ cell precursor undergoes meiosis.

gene
1. A functional DNA that is used to make a valuable product. 2. A factor that controls a phenotype and segregates in pedigrees according to Mendel's laws.

gene conversion
A naturally occurring nonreciprocal genetic exchange in which a short sequence of one DNA strand is altered so as to become identical to the sequence of another DNA strand (Figure 7.8).

gene dosage
The copy number of a gene. Alteration of the normal number of gene copies causes reduced expression (too little gene product) or overexpression (too much gene product). For *dosage-sensitive genes*, the amount of gene product made is critically important (Box 7.6).

gene family
A set of related genes that arose by some type of gene duplication (Section 2.4).

gene frequency See *allele frequency*.

gene knockdown
Targeted inhibition of expression of a specific gene by various methods, for example using *siRNA* (Box 9.5).

gene knockout
The targeted inactivation of a predetermined gene within intact cells so as to artificially create a *null allele*.

gene pool
All the genes (in the whole genome or at a specified locus) in a particular population.

gene silencing
Gross reduction in gene expression that occurs naturally by altering epigenetic settings, and that can occur both naturally and artificially through *RNA interference* (Boxes 6.2 and 9.5).

gene targeting
Artificial genetic modification of a specific predetermined gene in intact cells, such as embryonic stem cells (Figure 2 in Box 9.3).

gene therapy
Treating disease by genetically modifying the cells of a patient. May involve adding a functional copy of a gene that has lost its function, inhibiting a gene showing a pathological gain of function, or, more generally, replacing a defective gene.

genetic background
The genotypes at all loci other than one under active investigation. Variations in genetic background (*modifier*

genes) are a major reason for imperfect genotype–phenotype correlations (Section 7.7).

genetic code
The relationship between a codon and the amino acid it specifies (Figure 7.2).

genetic counseling
The process in which one or more members of a family who have, or are at risk of developing or transmitting, an inherited disease are informed by health professionals of the consequences and nature of the disorder, the probability of developing or transmitting it, and the options open to them.

genetic drift
Random changes in gene frequencies over generations because of random fluctuations in the proportions of the alleles in the parental population that are transmitted to offspring. Only significant in small populations.

genetic redundancy
Partly or completely overlapping function of genes at more than one locus, so that *loss-of-function mutations* at one locus do not cause overall loss of function.

genome/genomics
The total set of different DNA molecules of an organelle, cell, or organism/study of the same. The human genome consists of 3×10^9 *base pairs* of DNA divided between 24 different chromosomal DNA molecules and one mitochondrial DNA molecule.

genome browser
A computer program that provides a graphical interface for interrogating genome databases (Box 2.3).

genome editing
Artificial manipulation of an intact cell that is designed to make a double-strand break at just one locus and subsequently to make a desired change to the base sequence at that locus. See Figure 9.23 for an example.

genome (or **gene**) **imprinting** See *imprinting*.

genomewide association study or **scan (GWAS)**
The standard approach to identifying factors governing susceptibility to complex disease (Figure 8.13).

genotype
The genetic constitution of an individual, either overall or at a specific locus.

germ line
The germ cells (gametes) and those cells that give rise to them; other cells of the body constitute the soma.

germline (or **gonadal**) **mosaic**
An individual who has a subset of germ-line cells carrying a mutation that is not found in other germ-line cells.

haploid
Term used to describe a cell (typically a gamete) that has only a single copy of each chromosome (for example the 23 chromosomes in a human sperm or egg).

haploinsufficiency
A locus shows haploinsufficiency if producing a normal

phenotype requires more gene product than the amount produced by a single functional allele (Box 7.6).

haplotype
A series of alleles found at linked loci on a single chromosome (Box 4.4 and Figure 8.2).

haplotype block
A region of DNA showing limited haplotype diversity (Box 8.4).

Hardy–Weinberg law (or **equilibrium**)
The simple relationship between allele frequencies and genotype frequencies that is found in a population under ideal conditions (Section 5.4).

hemizygous
Having only one copy of a gene or DNA sequence in diploid cells. Males are hemizygous for most genes on the sex chromosomes. Deletions occurring on one autosome produce hemizygosity in males and in females.

heritability
The proportion of the causation of a character that is due to genetic causes (Section 8.2).

heterochromatin
Chromatin that is highly condensed and shows little or no evidence of active gene expression. Facultative heterochromatin may reversibly decondense to form *euchromatin*, depending on the requirements of the cell. Constitutive heterochromatin, which remains condensed throughout the cell cycle, is found at centromeres plus some other regions of human chromosomes (see Figure 2.8).

heteroduplex
Double-stranded DNA in which there is some mismatch between the two strands.

heteroplasmy
Mosaicism, usually within a single cell, for mitochondrial DNA variants (Section 5.2).

heterozygous/heterozygote
Having two different alleles at a particular locus/an individual with this property.

heterozygote advantage
The situation when a person heterozygous for a mutation has a reproductive advantage over both homozygotes for this mutation and also normal homozygotes. Sometimes called overdominance. Heterozygote advantage is one reason why severe recessive diseases may remain common (Section 5.4).

homologs (homologous chromosomes)
The two copies of a chromosome in a diploid cell. Unlike sister chromatids, homologous chromosomes are not copies of each other: one was inherited from the father and the other from the mother.

homologs (genes)
Two or more genes whose sequences are significantly related because of a close evolutionary relationship. They include *orthologs*, equivalent genes in two or more species that evolved from a single gene present in a common evolutionary ancestor, and paralogs that evolved by gene duplication such as the two α-globin genes present in humans.

homoplasmy
Of a cell or organism, having all copies of the mitochondrial DNA identical, as opposed to *heteroplasmy*.

homozygous/homozygote
Having two identical alleles at a particular locus/a person with this property. For clinical purposes a person is often described as homozygous *AA* if they have two normally functioning alleles, or homozygous *aa* if they have two pathogenic alleles at a locus, regardless of whether the alleles are in fact completely identical at the DNA sequence level. See also *autozygosity*.

hybridization (of nucleic acids and oligonucleotides)
Process in which complementary single strands are allowed to base pair (*anneal*) to form duplexes.

hybridization stringency
The degree to which the conditions (temperature, salt concentration, and so on) during a hybridization assay permit sequences with some mismatches to hybridize. High stringency conditions allow perfect matches only (Figure 3.7).

imprinting (of certain mammalian genes)
An epigenetic phenomenon in which the expression of the gene is determined by its parental origin (Sections 6.2 and 6.3).

indels
Insertion/deletion variants, often involving a single nucleotide, but sometimes involving more nucleotides. (The definition is a little imprecise, but in practice it usually includes variants that differ by possessing or lacking a sequence of up to 50 nucleotides.)

induced pluripotent stem (iPS) cells
Somatic cells that have been treated with specific genes, gene products, or other agents to reprogram them to resemble pluripotent stem cells. They can then be induced to differentiate into desired cell types (Box 9.2).

innate immune system
System of nonspecific response to a pathogen using the natural defenses of the body, as opposed to the *adaptive immune system*.

inner cell mass (ICM)
A group of cells located internally within the blastocyst which will give rise to the embryo proper (Figure 2 in Box 9.2).

insulator
DNA element that acts as a barrier to the spread of chromatin changes or the influence of *cis*-acting elements.

interphase
All the time in the cell cycle when a cell is not dividing.

intron
Originally any segment of a transcript that is cut out and discarded during RNA *splicing*, but now widely used to mean the corresponding sequence in genomic DNA (Figure 2.1).

isochromosome
An abnormal symmetrical chromosome consisting of

two identical arms, usually either the short arm or the long arm of a normal chromosome.

isoform
Alternative form of a protein as a result of differential expression of the same gene or through the production of different but highly related proteins from two or more loci.

karyotype
A summary of the chromosome constitution of a cell or person, such as 46,XY, but widely used loosely to mean an image showing the chromosomes of a cell sorted in order and arranged in pairs.

ligand
Any molecule that binds specifically to a receptor or other molecule. (An example is the FASLG ligand that binds to the FAS receptor in Figure 10.12.)

ligase
DNA ligase is an enzyme that can seal single-strand *nicks* in double-stranded DNA or covalently join two oligonucleotides that are hybridized at adjacent positions on a DNA strand.

lineage (of cells)
In development, the ancestry and descendants of a cell, as traced backward or forward through successive cell divisions.

linkage disequilibrium
A statistical association between particular alleles at separate but linked loci, normally the result of a particular ancestral haplotype being common in the population studied. An important tool for high-resolution mapping (Section 8.2).

locus (plural: **loci**)
A unique chromosomal location defining the position of an individual gene or DNA sequence.

lod score (Z)
A measure of the likelihood of genetic linkage between loci. The log (base 10) of the odds that the loci are linked (with recombination fraction q) rather than unlinked. For Mendelian characters a lod score greater than +3 provides minimal evidence of linkage; one that is less than –2 is evidence against linkage (Box 8.1).

loss-of-function mutations
Mutations that cause a gene product to lose its function, partly or totally (Section 7.5).

loss of heterozygosity (LOH)
Homozygosity or hemizygosity in a tumor or other somatic cell when the constitutional genotype is heterozygous. Evidence of a somatic genetic change (Section 10.2 and Figure 10.10).

major histocompatibility complex (MHC)
A large gene cluster containing multiple genes including, notably, genes that function in antigen recognition by binding fragments of antigens and presenting them on the surface of T cells. The human version is known as the HLA complex (see Boxes 4.4 and 8.3).

malignant tumor
A tumor whose cells show evidence of spreading

(invading adjacent tissue, disseminating through the bloodstream and/or lymphatic system).

marker (molecular)
A chemical group or molecule that can be assayed in some way.

meiosis
The specialized reductive form of cell division used exclusively to produce gametes (Figures 1.14 and 1.15).

Mendelian
Description for a character whose pattern of inheritance suggests it is caused by variation at a single chromosomal locus.

mesenchyme
Connective tissues.

messenger RNA (mRNA)
A processed gene transcript that carries protein-coding information to cytoplasmic ribosomes.

metastasis
The process whereby cells from a primary malignant tumor are disseminated via the blood stream or lymphatic system to establish secondary tumors at distant sites in the body.

MHC See *major histocompatibility complex.*

microarray hybridization
A nucleic acid hybridization assay in which thousands to millions of different oligonucleotide (or DNA) probes are fixed at specific grid coordinates on a miniature solid surface and allowed to hybridize to complementary sequences within a solution containing a heterogeneous test sample population of labeled DNA or RNA molecules (Figure 3.9).

microbiome (or **microbiota**)
The aggregate of microorganisms that share our body space; most of them are found in the gastrointestinal tract.

microRNAs (miRNAs)
Short (21–22-nucleotide) RNA molecules encoded within normal genomes that have a major role in the regulation of gene expression (Figure 6.8).

microsatellite
Small array of *tandem repeats* of a very simple DNA sequence, usually 1–4 *base pairs*, for example $(CA)_n$. The total length of the array is usually less than 0.1 kb. A polymorphic microsatellite is alternatively known as a short tandem repeat polymorphism (Figure 4.7).

mismatch repair
A form of DNA repair in which very simple DNA replication errors (nucleotide substitutions and deletions/insertions of one or two nucleotides) are repaired (Figure 10.17).

missense mutations
Changes in a coding sequence that cause one amino acid in the gene product to be replaced by a different one (Section 7.2).

mitosis
The normal process of cell division, which usually

produces daughter cells genetically identical to the parent cell (Figure 1.13).

modifier (gene)
A gene whose expression can influence a phenotype resulting from a mutation at another locus (Section 7.7).

monozygotic
Originating from a single zygote, as in identical twins (other twins are dizygotic, having originated from different zygotes).

mosaic
An individual who has two or more genetically different cell lines derived from a single zygote. The difference may be point mutations, large-scale mutations, or chromosomal abnormalities (Box 5.3).

mRNA See *messenger RNA*.

mtDNA
Mitochondrial DNA (Figure 2.11).

multifactorial
A character that is determined by some unspecified combination of genetic and environmental factors.

mutagen
An agent that results in an increased mutation frequency.

mutation
1. A localized change in the base sequence of a DNA molecule.
2. The process that creates it.

mutation scanning
Testing for any non-defined change in the base sequence of a genome or a genome component (such as an exon, gene, or exome) in the hope of identifying abnormal mutations that correlate with disease. (As opposed to testing for specific mutations.)

natural selection
Process whereby the population frequencies of alleles change by causing a change in the biological *fitness* of the individuals who carry them. Many alleles cause reduced biological fitness (*purifying* or *negative selection*); a few alleles cause increased biological fitness of the individuals who carry them (*positive selection*). See also *balancing selection*.

ncRNA See *noncoding RNA*.

nick (in DNA)
Cleavage of a single phosphodiester bond on one DNA strand only.

non-allelic homologous recombination (NAHR)
Recombination between misaligned DNA repeats, either on the same chromosome, on sister chromatids or on homologous chromosomes. NAHR generates recurrent deletions, duplications, or inversions (Section 7.3).

noncoding RNA (ncRNA)
Mature RNA transcript that is not translated to make a polypeptide (Figure 2.7).

nondisjunction
Failure of chromosomes (sister chromatids in mitosis

or meiosis II; paired homologs in meiosis I) to separate (disjoin) at anaphase (Figure 7.14). The major cause of numerical chromosome abnormalities.

nonhomologous end joining
Form of repair of double-strand breaks in DNA that involves the fusion of broken ends without copying from a DNA template.

non-penetrance
The situation when somebody carrying an allele that normally causes a phenotype to be expressed does not show that phenotype, as a result of interaction with alleles of other genes (*modifier genes*) or with non-genetic factors (Figure 5.12).

nonsense mutation
A nucleotide substitution that changes a codon specifying an amino acid so that it becomes a premature termination codon (Section 7.2).

nonsense-mediated mRNA decay
A cellular mechanism that degrades mRNA molecules that contain a premature termination codon (more than 50 nucleotides upstream of the last splice junction) (Box 7.1).

nonsynonymous substitution (or **mutation**)
A change in the sequence of a codon that results in a different codon interpretation. Table 7.1 gives the different classes.

nucleosome
The basic structural unit of chromatin, comprising 146 *base pairs* of DNA wound around an octamer of histone molecules (Figures 1.8 and 6.11A).

nucleotide
The fundamental repeating unit of a nucleic acid, consisting of a sugar to which is covalently attached a base and a phosphate group (Figure 1.2).

null allele
Any mutant allele where the normal gene product is not made or is completely non-functional.

odds ratio
In *case-control* studies, the relative odds of a person with or without a factor under study being a case (Table 8.6).

OMIM
Online Mendelian Inheritance in Man database (Box 5.1). www.omim.org

oncogene
A gene that when activated in some way (often by a change that stimulates its expression) can help to transform a normal cell into a tumor cell. Originally the word was reserved for activated forms of the gene (while the normal unactivated cellular gene was called a proto-oncogene), but this distinction is now widely ignored.

open reading frame
A continuous sequence of *coding DNA*.

origin of replication
A site on a DNA molecule where replication can be initiated.

orthologs
Homologous genes present in different organisms having descended from a common ancestral gene.

PCR (polymerase chain reaction)
The standard technique used to amplify short DNA sequences (Figure 3.3).

penetrance
The frequency with which a genotype manifests itself in a given phenotype.

pedigree
A limited family tree; a more extensive family tree is a kindred.

personalized medicine
A model of health care in which medical decisions and practice are tailored to the individual patient. Knowledge of a person's genome, for example, can allow more informed decisions about the suitability of prescribing certain drugs, and knowledge of cancer mutations may allow suitably targeted therapies.

pharmacodynamics
The study of the response of a target organ or cell to a drug.

pharmacogenetics
The study of the influence of individual genes or alleles on the metabolism or function of drugs.

pharmacokinetics
The study of the absorption, activation, catabolism, and elimination of a drug.

phenocopy
A person or organism that has a phenotype normally caused by a certain genotype but does not have that genotype. Phenocopies may be the result of a different genetic variant, or of an environmental factor.

phenome
The totality of phenotypes of an individual organism.

phenotype
The observable characteristics of a cell or organism, including the result of any test that is not a direct test of the genotype.

phosphodiester bond
The link between adjacent nucleotides in DNA or RNA.

plasmid
A small circular DNA molecule that can replicate independently in a cell. Modified plasmids are widely used as cloning vectors (Section 3.1).

pleiotropy
The common situation in which variation in one gene affects several different aspects of the phenotype.

ploidy
The number of complete sets of chromosomes in a cell. Gametes are *haploid* and most normal cells are *diploid*, but some of our cells naturally have multiple chromosome sets (polyploidy) or none at all (nulliploidy).

pluripotent (of a mammalian stem cell)
Capable of giving rise to descendant cells that participate in the formation of all of the tissues of an embryo except the extraembryonic membranes.

PMID
PubMed identifier, a seven-digit or eight-digit number that, when typed into the query box at the NCBI PubMed database (http://www.ncbi.nlm.nih.gov/pubmed/), allows electronic access to a specific article in a biomedical journal.

point mutation
A mutation causing a small alteration in the DNA sequence at a locus, often changing just a single nucleotide.

polyadenylation/poly(A) tail
Addition of 200 or so adenosines to the 3′ end of an mRNA. The resulting poly(A) tail is important for stabilizing mRNA (Section 2.1).

polygenic
Description of a character determined by the combined action of a number of genetic loci. Polygenic theory (Box 8.2) assumes that there are very many loci, each with a small effect.

polymorphism
The existence of two or more variants (alleles, phenotypes, sequence variants, chromosomal structure variants) at significant frequencies in the population. Often also used more loosely to mean any sequence variant present at a frequency of more than 1% in a population.

polypeptide
A string of amino acids linked by peptide bonds. Proteins may contain one or more polypeptide chains.

position effect
Complete or partial silencing of a gene when the gene is moved to a different chromosomal location close to heterochromatin. Also sometimes used more generally to include the effect of a rearrangement causing physical separation of elements of a gene.

positive selection
Selection in favor of a particular genotype that confers increased biological *fitness* (Section 4.4).

potency
Of a cell, its potential for dividing into different cell types. Cells can be totipotent, *pluripotent*, multipotent, or committed to one fate.

premutation allele
Among diseases caused by *dynamic mutations*, a repeat expansion that is large enough to be unstable on transmission but not large enough to cause disease (Box 7.2).

primary structure
Of a polypeptide or nucleic acid, the linear sequence of amino acids or nucleotides in the molecule.

primary transcript
The RNA product of transcription of a gene by RNA polymerase, before splicing. The primary transcript of a gene contains all the exons and introns.

primer
A short oligonucleotide, often 16–25 bases long, which base pairs specifically to a target sequence to allow a polymerase to initiate the synthesis of a complementary strand.

primordial germ cells
Cells in the embryo and fetus that will ultimately give rise to germ-line cells.

probe
A known DNA or RNA fragment (or a collection of such fragments) used in a hybridization assay to identify closely related target sequences within a complex, poorly understood population of nucleic acid molecules (the test sample)—see Section 3.2.

prodrug
An inactive precursor to a therapeutic drug that is administered to a patient and activated within the body after natural conversion by a drug-metabolizing enzyme or other component (Section 9.2).

promoter
A combination of short sequence elements, usually just upstream of a gene, to which RNA polymerase binds so as to initiate transcription of the gene (Figure 6.1).

protective factor
A variant that reduces susceptibility to disease (Table 8.11).

proofreading
An enzymatic mechanism by which DNA replication errors are identified and corrected.

proteome/proteomics
All the different proteins in a cell or organism/study of the same.

proximal (of a chromosomal location)
Comparatively close to the centromere.

pseudoautosomal regions (or **sequences**) **(PAR)**
Regions with identical genes at the tip of the short arms of the X and Y chromosomes, and at the tip of the long arms of the X and Y chromosomes (Figure 5.7). Because of X–Y recombination, these genes move between the X and the Y (Figure 5.8), behaving as alleles that show an apparently autosomal mode of inheritance.

pseudogene
A DNA sequence that shows a high degree of sequence homology to a non-allelic functional gene but is itself nonfunctional or does not make a protein like its closely related homolog (but it may, however, make a functional non-coding RNA) (Box 2.4).

purifying (negative) selection
A form of natural selection in which harmful mutations that wreck or disturb the function of an important DNA sequence tend to be removed from the population.

purine
A double-ringed organic nitrogenous base that is a constituent of a nucleic acid, notably adenine (A) and guanine (G)—see Figure 1.3.

pyrimidine
A single-ringed organic nitrogenous base that is a constituent of a nucleic acid, notably cytosine (C), thymine (T), and uracil (U)—see Figure 1.3.

quantitative character
A character such as height, which everybody has but to differing degrees (in contrast with a dichotomous character such as polydactyly, which some people have and others do not).

quantitative PCR (qPCR)
PCR methods that allow accurate estimation of the amount of template present (Section 3.1). See also *real-time PCR*.

quantitative trait locus (QTL)
A locus that contributes to determining the phenotype of a continuous character.

reactive oxygen species (ROS)
Chemically reactive molecules or atoms containing oxygen, such as oxygen ions, oxygen radicals, and peroxides. Formed within cells as a natural by-product of normal oxygen metabolism, they have important roles in cell signaling and homeostasis but cause DNA damage (see Section 4.1).

reading frame
During translation, the way in which the continuous sequence of the mRNA is read as a series of triplet codons. There are three possible forward reading frames for any mRNA, and the correct reading frame is set by correct recognition of the AUG initiation codon (see Box 2.1).

real-time PCR
A form of quantitative PCR in which the accumulation of product is followed in real time, allowing accurate quantitation of the amount of template present (Section 3.1).

recessive
Referring to a character that is manifest in the homozygote but not in the heterozygous state.

recombinant
In linkage analysis, a gamete that contains a haplotype with a combination of alleles that is different from the combination that the parent had inherited (Figure 8.5).

recombinant DNA
An artificially constructed hybrid DNA containing covalently linked sequences (Figure 2 in Box 3.1).

recombination (or **crossover**)
Exchange of DNA sequences between paired homologous chromosomes at meiosis (Figures 1.14 and 1.15).

relative risk
In epidemiology, the relative risks of developing a condition in people with and without a *susceptibility factor* (Table 8.3).

replication fork
In DNA replication, the point along a DNA strand where the replication machinery is currently at work (Figure 1.6).

replication origin See *origin of replication*.

replication slippage
A mistake in replication of a short tandemly repeated DNA sequence that results in newly synthesized DNA strands with more or fewer copies of the tandem repeats than in the template DNA (Figure 4.8).

reprogramming (cellular, nuclear, or epigenetic)
Large-scale epigenetic changes to convert the pattern of gene expression in a cell to that typical of another cell type or cell state. Often occurs in cancers (Section 10.3) and can be artificially induced (Box 9.2).

restriction endonuclease
A bacterial enzyme that cuts double-stranded DNA at a short (normally 4, 6, or 8 *base pairs* long) recognition sequence (Box 3.1).

restriction fragment length polymorphism (RFLP)
A DNA polymorphism that creates or abolishes a recognition sequence for a restriction endonuclease. When DNA is digested with the relevant enzyme, the sizes of the fragments will differ, depending on the presence or absence of the restriction site (Figure 4.6).

restriction site
A site on a DNA molecule that is cleaved by a restriction endonuclease.

retrogene
A functional gene that appears to be derived from a reverse-transcribed RNA (Box 2.4).

retroposon (or **retrotransposon**)
A member of a family of mobile DNA elements that transpose by making an RNA that is copied into a cDNA which integrates elsewhere in the genome (Section 2.4).

retrovirus
An RNA virus with a reverse transcriptase function, enabling the RNA genome to be copied into cDNA before integration into the chromosomes of a host cell (Figure 9.19).

reverse transcriptase
An enzyme that makes a DNA copy of an RNA template; an RNA-dependent DNA polymerase (Table 1.1).

ribozyme
A natural or synthetic catalytic RNA molecule.

risk ratio
In family studies, the relative risk of disease in a relative of an affected person compared with that of a member of the general population (Section 8.2).

RNA gene
A gene that makes a functional noncoding RNA (Figure 2.7).

RNA interference (RNAi)
A cellular defense system activated by the presence of long double-stranded RNA sequences and designed to protect against viruses and excessive transposon activity within cells (see Box 6.2). Its discovery allowed specific *gene silencing* using siRNAs (Box 9.5).

RNA polymerase
An enzyme that can add ribonucleotides to the 3′ end of an RNA chain. Most RNA polymerases use a DNA template to make an RNA transcript.

RNA processing
The processes required to convert a primary transcript into a mature messenger RNA, notably capping, splicing, and polyadenylation.

RNA splicing See *splicing*.

secondary structure
The path of the backbone of a folded polypeptide or single-stranded nucleic acid, determined by weak interactions between residues in different parts of the sequence (Box 2.2).

segmental duplication
The existence of very highly related DNA sequence blocks on different chromosomes, or at more than one location within a chromosome.

segregation
1. The distribution of allelic sequences between daughter cells at meiosis. Allelic sequences are said to segregate, non-allelic sequences to assort.
2. In pedigree analysis, the probability of a child inheriting a phenotype from a parent.

selection See *natural selection*.

selective sweep
Process whereby *positive selection* for a favorable DNA variant causes a reduction in variation in the population at the immediately neighboring nucleotide sequences (Box 4.3).

sense strand
The DNA strand of a gene that is complementary in sequence to the template (antisense) strand and identical to the transcribed RNA sequence (except that DNA contains T where RNA has U). Quoted gene sequences always give the sense strand, in the 5′→3′ direction (Figure 2.1).

sensitivity (of a test)
The proportion of all true positives that the test is able to detect (Table 11.1).

sib
Brother or sister.

silencer
Combination of short DNA sequence elements that suppress the transcription of a gene.

silent mutation
Has the same meaning as *synonymous substitution*, which is the preferred term because sometimes this type of change can result in altered gene expression and disease (Figure 7.4B).

single nucleotide polymorphism/variant See *SNP/SNV*.

siRNA (small or **short interfering RNA)**
Double-stranded RNA molecules 21–22 nucleotides long that can dramatically shut down the expression of genes through RNA interference (Boxes 6.2 and 9.5).

sister chromatid
One of the two paired chromatids of a single chromosome that form after DNA replication and remain joined at the centromere until the anaphase stage of mitosis. Non-sister chromatids are present on different but homologous chromosomes (Figure 1.11).

SNP (single nucleotide polymorphism)
A nucleotide position in the genome where two or occasionally three alternative nucleotides are common in the population. May be pathogenic or neutral. The dbSNP database lists human SNPs but includes some rare pathogenic variants and some variants that involve two or more contiguous nucleotides.

SNV (single nucleotide variant)
A rare DNA variant (frequency less than 0.01) that can be seen to differ at a single nucleotide position from the consensus sequence in the population.

somatic cell
Any cell in the body that is not part of the germ line.

specificity (of a test for a condition)
A measure of the performance of a test that assesses the proportion of all people who do not have the condition who are correctly identified as such by the test assay. Specificity = (1 – false positive rate) (Table 11.1).

splice acceptor site
The site that defines the junction between the end of an intron in RNA and the start of the following exon. The junction sequence often conforms to the consensus sequence yyyyyyyyyyyyny**ag**R, where y is a pyrimidine, n is any nucleotide, and R is a purine that is the first nucleotide of the exon.

splice donor site
The site that defines the junction between the end of an exon in RNA and the start of the following intron. The junction sequence often conforms to the consensus sequence (C/A)AG**gu**ragu, where r is a purine and capital letters denote the end nucleotides of the exon.

splicing
The process whereby some precursor RNA transcripts are cleaved into sequences, some of which (exons) are retained and fused (spliced) to give the mature RNA whereas others are discarded (introns).

stem cell
A cell that can act as a precursor to differentiated cells but retains the capacity for self-renewal. Can be a tissue stem cell that gives rise to a limited number of cell types (Figure 9.20 and Box 10.2) or a *pluripotent* stem cell (Box 9.2).

stop (termination) codon
An in-frame codon that does not specify an amino acid but instead acts as a signal for the ribosome to dissociate from the mRNA and release the nascent polypeptide. See Figure 2.3 for the principle and Figure 7.2 for the different types of stop codon.

stratification
A population is stratified if it consists of several subpopulations that do not interbreed freely.

Stratification is a source of error in association studies and risk estimation.

stratified medicine
A model of health care in which different medical treatments are targeted to subsets of the same disease according to which disease-associated genetic variants a person possesses.

stringency (of hybridization)
The choice of conditions that will allow either imperfectly matched sequences or only perfectly matched sequences to hybridize (Figure 3.7).

stroma
Supportive tissue of an epithelial organ, tumor, and so on, consisting of connective tissues and blood vessels.

structural variation
Large-scale DNA variation that involves moving or changing the copy number of moderately long to very long DNA sequences, by one of various mechanisms: translocation, inversion, insertion, deletion, or duplication (Section 4.3).

susceptibility factor
A variant that provides increased risk of developing a specific disease.

synonymous substitution (or **silent mutation**)
A nucleotide substitution that changes the sequence of a codon without any change in the amino acid that it specifies, but sometimes causes altered splicing and disease (Figure 7.4B).

tandem repeats
Any pattern in which a sequence of one or more nucleotides in DNA is repeated and the repetitions are directly adjacent to each other. See Figure 2.12A for an example.

target enrichment sequencing (or **targeted sequencing**)
The process in which a defined subset of a genome (containing target sequences of interest) is captured by a DNA hybridization assay and then submitted for DNA sequencing (Figure 11.11).

telomere
Specialized structure that stabilizes the ends of linear chromosomes. (See Figure 1.10 for telomeric DNA structure.)

termination codon See *stop codon*.

terminal differentiation
The state of a cell that has ceased dividing and has become irreversibly committed to some specialized function.

tissue
A set of contiguous functionally related cells.

trait See *character*.

trans-acting
The term used to describe any gene regulation in which the expression of some sequence on a DNA or RNA molecule is regulated by a different molecule or molecular assembly (in practice, a different RNA or a

protein that is usually expressed from a remote gene and needs to diffuse to its site of action) (Box 6.1).

transcription factor
DNA-binding protein that promotes the transcription of genes. Some are ubiquitous, promoting transcription in all cells, but many are tissue-specific.

transcription unit
A segment of DNA that is used to make a primary RNA transcript (see Figure 2.1). May occasionally span multiple genes, as in the transcription of mitochondrial DNA (Figure 2.11) and in the transcription of adjacent 28S, 5.8S, and 18S rRNA genes.

transcriptome/transcriptomics
All the different RNA transcripts in a cell or tissue/the study of the same.

transdifferentiation
Epigenetic reprogramming of the nucleus of a cell, causing it to change from one cell type to another, such as from a skin cell to a neuron.

transduction
1. Relaying a signal from a cell surface receptor to a target within a cell.
2. Using recombinant viruses to introduce foreign DNA into a cell.

transfection
Direct introduction of an exogenous DNA molecule into a cell without using a vector.

transformation (of a cell)
1. Uptake by a competent microbial cell of naked high-molecular-weight DNA from the environment.
2. Alteration of the growth properties of a normal eukaryotic cell as a step toward evolving into a tumor cell.

transgene
An exogenous gene that has been transfected into cells of an animal or plant. It may be present in some tissues (as in human gene therapies) or in all tissues (as in germ-line engineering, for example in the mouse—see Box 9.3). Introduced transgenes may integrate into host cell chromosomes or replicate extrachromosomally and be transiently expressed.

transgenic animal
An animal in which artificially introduced foreign DNA (a transgene) becomes stably incorporated into the germ line (Box 9.3).

transit amplifying cells
The immediate progeny by which stem cells give rise to differentiated cells. Transit amplifying cells go through many cycles of division, but they eventually differentiate (Boxes 9.2 and 10.2).

translocation
Transfer of chromosomal regions between nonhomologous chromosomes (Figure 7.12).

transposon/transposon repeat
A mobile genetic element/a member of a repetitive DNA family containing some members that are able to transpose but also many inactivated copies of transposons (Figure 2.14).

trophoblast
Outer layer of polarized cells in the blastocyst that will go on to form the chorion, the embryonic component of the placenta (Figure 2 of Box 9.2).

tropism
The specificity of a virus for a particular cell type, determined in part by the interaction of viral surface structures with receptors present on the surface of the cell.

tumor suppressor gene
A gene that is commonly inactivated in tumors (by an inactivating mutation, by deletion as a result of abnormal chromosome segregation/recombination, or by epigenetic silencing). Classic tumor suppressor genes normally work to inhibit or control cell division.

uniparental diploidy
A 46,XX diploid conceptus in which both genomes derive from the same parent. Such conceptuses never develop normally (Figure 6.17).

uniparental disomy
A cell or organism in which both copies of one particular chromosome pair are derived from one parent. Depending on the chromosome involved, this may or may not cause disease (Figure 6.21).

unrelated
Ultimately everybody is related; the word is used in this book to mean people who do not have an identified common ancestor in the last four or so generations.

untranslated region (5′ UTR, 3′ UTR)
Regions at the 5′ end of mRNA before the AUG translation start codon, or at the 3′ end after the stop codon (Figures 2.1 and 2.3).

variant (in relation to DNA)
A sequence that is different from the majority sequence but exists at a low frequency (<0.01; that is, less than 1%) in the population.

vector
A nucleic acid that is able to replicate and maintain itself within a host cell and that can be used to confer similar properties on any sequence covalently linked to it.

X-chromosome inactivation (or X-inactivation)
The *epigenetic* inactivation of all except one of the X chromosomes in the cells of humans and other mammals that have more than one X (Figure 6.18).

zinc finger nucleases
Synthetic enzymes that combine an endonuclease module with a sequence-specific targeting module, so as to cleave DNA at a selected sequence (Figure 9.23).

zygote
The fertilized egg cell.

INDEX

Note: The index covers the main text but not the summaries, questions or glossary.

Prefixes have been ignored for filing unless integral to a topic (so 'β–sheets' at 'beta' but 'β–thalassemia' at 'thalassemia'). The same apples to numeric prefixes but unavoidable numerics have been sorted as though spelled out (so '5-methlycytosine' will be at 'methyl' but '7SL RNA' at 'seven').

Page numbers ending with 'B', 'F' or 'T' indicate that the listed topic is dealt with on that page *only* in a box, figure or table. On pages where there is also coverage in the text, that distinction is not made.